T0180491

Lecture Notes in Computer Science   12347

More information about this series at http://www.springer.com/series/7412

Andrea Vedaldi · Horst Bischof ·
Thomas Brox · Jan-Michael Frahm (Eds.)

Computer Vision – ECCV 2020

16th European Conference
Glasgow, UK, August 23–28, 2020
Proceedings, Part II

Springer

Editors
Andrea Vedaldi 🆔
University of Oxford
Oxford, UK

Horst Bischof 🆔
Graz University of Technology
Graz, Austria

Thomas Brox 🆔
University of Freiburg
Freiburg im Breisgau, Germany

Jan-Michael Frahm
University of North Carolina at Chapel Hill
Chapel Hill, NC, USA

ISSN 0302-9743   ISSN 1611-3349 (electronic)
Lecture Notes in Computer Science
ISBN 978-3-030-58535-8   ISBN 978-3-030-58536-5 (eBook)
https://doi.org/10.1007/978-3-030-58536-5

LNCS Sublibrary: SL6 – Image Processing, Computer Vision, Pattern Recognition, and Graphics

This Springer imprint is published by the registered company Springer Nature Switzerland AG
The registered company address is: Gewerbestrasse 11, 6330 Cham, Switzerland

Foreword

Hosting the European Conference on Computer Vision (ECCV 2020) was certainly an exciting journey. From the 2016 plan to hold it at the Edinburgh International Conference Centre (hosting 1,800 delegates) to the 2018 plan to hold it at Glasgow's Scottish Exhibition Centre (up to 6,000 delegates), we finally ended with moving online because of the COVID-19 outbreak. While possibly having fewer delegates than expected because of the online format, ECCV 2020 still had over 3,100 registered participants.

Although online, the conference delivered most of the activities expected at a face-to-face conference: peer-reviewed papers, industrial exhibitors, demonstrations, and messaging between delegates. In addition to the main technical sessions, the conference included a strong program of satellite events with 16 tutorials and 44 workshops.

Furthermore, the online conference format enabled new conference features. Every paper had an associated teaser video and a longer full presentation video. Along with the papers and slides from the videos, all these materials were available the week before the conference. This allowed delegates to become familiar with the paper content and be ready for the live interaction with the authors during the conference week. The live event consisted of brief presentations by the oral and spotlight authors and industrial sponsors. Question and answer sessions for all papers were timed to occur twice so delegates from around the world had convenient access to the authors.

As with ECCV 2018, authors' draft versions of the papers appeared online with open access, now on both the Computer Vision Foundation (CVF) and the European Computer Vision Association (ECVA) websites. An archival publication arrangement was put in place with the cooperation of Springer. SpringerLink hosts the final version of the papers with further improvements, such as activating reference links and supplementary materials. These two approaches benefit all potential readers: a version available freely for all researchers, and an authoritative and citable version with additional benefits for SpringerLink subscribers. We thank Alfred Hofmann and Aliaksandr Birukou from Springer for helping to negotiate this agreement, which we expect will continue for future versions of ECCV.

August 2020

Vittorio Ferrari
Bob Fisher
Cordelia Schmid
Emanuele Trucco

Preface

Welcome to the proceedings of the European Conference on Computer Vision (ECCV 2020). This is a unique edition of ECCV in many ways. Due to the COVID-19 pandemic, this is the first time the conference was held online, in a virtual format. This was also the first time the conference relied exclusively on the Open Review platform to manage the review process. Despite these challenges ECCV is thriving. The conference received 5,150 valid paper submissions, of which 1,360 were accepted for publication (27%) and, of those, 160 were presented as spotlights (3%) and 104 as orals (2%). This amounts to more than twice the number of submissions to ECCV 2018 (2,439). Furthermore, CVPR, the largest conference on computer vision, received 5,850 submissions this year, meaning that ECCV is now 87% the size of CVPR in terms of submissions. By comparison, in 2018 the size of ECCV was only 73% of CVPR.

The review model was similar to previous editions of ECCV; in particular, it was double blind in the sense that the authors did not know the name of the reviewers and vice versa. Furthermore, each conference submission was held confidentially, and was only publicly revealed if and once accepted for publication. Each paper received at least three reviews, totalling more than 15,000 reviews. Handling the review process at this scale was a significant challenge. In order to ensure that each submission received as fair and high-quality reviews as possible, we recruited 2,830 reviewers (a 130% increase with reference to 2018) and 207 area chairs (a 60% increase). The area chairs were selected based on their technical expertise and reputation, largely among people that served as area chair in previous top computer vision and machine learning conferences (ECCV, ICCV, CVPR, NeurIPS, etc.). Reviewers were similarly invited from previous conferences. We also encouraged experienced area chairs to suggest additional chairs and reviewers in the initial phase of recruiting.

Despite doubling the number of submissions, the reviewer load was slightly reduced from 2018, from a maximum of 8 papers down to 7 (with some reviewers offering to handle 6 papers plus an emergency review). The area chair load increased slightly, from 18 papers on average to 22 papers on average.

Conflicts of interest between authors, area chairs, and reviewers were handled largely automatically by the Open Review platform via their curated list of user profiles. Many authors submitting to ECCV already had a profile in Open Review. We set a paper registration deadline one week before the paper submission deadline in order to encourage all missing authors to register and create their Open Review profiles well on time (in practice, we allowed authors to create/change papers arbitrarily until the submission deadline). Except for minor issues with users creating duplicate profiles, this allowed us to easily and quickly identify institutional conflicts, and avoid them, while matching papers to area chairs and reviewers.

Papers were matched to area chairs based on: an affinity score computed by the Open Review platform, which is based on paper titles and abstracts, and an affinity score computed by the Toronto Paper Matching System (TPMS), which is based on the paper's full text, the area chair bids for individual papers, load balancing, and conflict avoidance. Open Review provides the program chairs a convenient web interface to experiment with different configurations of the matching algorithm. The chosen configuration resulted in about 50% of the assigned papers to be highly ranked by the area chair bids, and 50% to be ranked in the middle, with very few low bids assigned.

Assignments to reviewers were similar, with two differences. First, there was a maximum of 7 papers assigned to each reviewer. Second, area chairs recommended up to seven reviewers per paper, providing another highly-weighed term to the affinity scores used for matching.

The assignment of papers to area chairs was smooth. However, it was more difficult to find suitable reviewers for all papers. Having a ratio of 5.6 papers per reviewer with a maximum load of 7 (due to emergency reviewer commitment), which did not allow for much wiggle room in order to also satisfy conflict and expertise constraints. We received some complaints from reviewers who did not feel qualified to review specific papers and we reassigned them wherever possible. However, the large scale of the conference, the many constraints, and the fact that a large fraction of such complaints arrived very late in the review process made this process very difficult and not all complaints could be addressed.

Reviewers had six weeks to complete their assignments. Possibly due to COVID-19 or the fact that the NeurIPS deadline was moved closer to the review deadline, a record 30% of the reviews were still missing after the deadline. By comparison, ECCV 2018 experienced only 10% missing reviews at this stage of the process. In the subsequent week, area chairs chased the missing reviews intensely, found replacement reviewers in their own team, and managed to reach 10% missing reviews. Eventually, we could provide almost all reviews (more than 99.9%) with a delay of only a couple of days on the initial schedule by a significant use of emergency reviews. If this trend is confirmed, it might be a major challenge to run a smooth review process in future editions of ECCV. The community must reconsider prioritization of the time spent on paper writing (the number of submissions increased a lot despite COVID-19) and time spent on paper reviewing (the number of reviews delivered in time decreased a lot presumably due to COVID-19 or NeurIPS deadline). With this imbalance the peer-review system that ensures the quality of our top conferences may break soon.

Reviewers submitted their reviews independently. In the reviews, they had the opportunity to ask questions to the authors to be addressed in the rebuttal. However, reviewers were told not to request any significant new experiment. Using the Open Review interface, authors could provide an answer to each individual review, but were also allowed to cross-reference reviews and responses in their answers. Rather than PDF files, we allowed the use of formatted text for the rebuttal. The rebuttal and initial reviews were then made visible to all reviewers and the primary area chair for a given paper. The area chair encouraged and moderated the reviewer discussion. During the discussions, reviewers were invited to reach a consensus and possibly adjust their ratings as a result of the discussion and of the evidence in the rebuttal.

After the discussion period ended, most reviewers entered a final rating and recommendation, although in many cases this did not differ from their initial recommendation. Based on the updated reviews and discussion, the primary area chair then made a preliminary decision to accept or reject the paper and wrote a justification for it (meta-review). Except for cases where the outcome of this process was absolutely clear (as indicated by the three reviewers and primary area chairs all recommending clear rejection), the decision was then examined and potentially challenged by a secondary area chair. This led to further discussion and overturning a small number of preliminary decisions. Needless to say, there was no in-person area chair meeting, which would have been impossible due to COVID-19.

Area chairs were invited to observe the consensus of the reviewers whenever possible and use extreme caution in overturning a clear consensus to accept or reject a paper. If an area chair still decided to do so, she/he was asked to clearly justify it in the meta-review and to explicitly obtain the agreement of the secondary area chair. In practice, very few papers were rejected after being confidently accepted by the reviewers.

This was the first time Open Review was used as the main platform to run ECCV. In 2018, the program chairs used CMT3 for the user-facing interface and Open Review internally, for matching and conflict resolution. Since it is clearly preferable to only use a single platform, this year we switched to using Open Review in full. The experience was largely positive. The platform is highly-configurable, scalable, and open source. Being written in Python, it is easy to write scripts to extract data programmatically. The paper matching and conflict resolution algorithms and interfaces are top-notch, also due to the excellent author profiles in the platform. Naturally, there were a few kinks along the way due to the fact that the ECCV Open Review configuration was created from scratch for this event and it differs in substantial ways from many other Open Review conferences. However, the Open Review development and support team did a fantastic job in helping us to get the configuration right and to address issues in a timely manner as they unavoidably occurred. We cannot thank them enough for the tremendous effort they put into this project.

Finally, we would like to thank everyone involved in making ECCV 2020 possible in these very strange and difficult times. This starts with our authors, followed by the area chairs and reviewers, who ran the review process at an unprecedented scale. The whole Open Review team (and in particular Melisa Bok, Mohit Unyal, Carlos Mondragon Chapa, and Celeste Martinez Gomez) worked incredibly hard for the entire duration of the process. We would also like to thank René Vidal for contributing to the adoption of Open Review. Our thanks also go to Laurent Charling for TPMS and to the program chairs of ICML, ICLR, and NeurIPS for cross checking double submissions. We thank the website chair, Giovanni Farinella, and the CPI team (in particular Ashley Cook, Miriam Verdon, Nicola McGrane, and Sharon Kerr) for promptly adding material to the website as needed in the various phases of the process. Finally, we thank the publication chairs, Albert Ali Salah, Hamdi Dibeklioglu, Metehan Doyran, Henry Howard-Jenkins, Victor Prisacariu, Siyu Tang, and Gul Varol, who managed to compile these substantial proceedings in an exceedingly compressed schedule. We express our thanks to the ECVA team, in particular Kristina Scherbaum for allowing open access of the proceedings. We thank Alfred Hofmann from Springer who again serve as the publisher. Finally, we thank the other chairs of ECCV 2020, including in particular the general chairs for very useful feedback with the handling of the program.

August 2020

Andrea Vedaldi
Horst Bischof
Thomas Brox
Jan-Michael Frahm

Organization

General Chairs

Vittorio Ferrari   Google Research, Switzerland
Bob Fisher   University of Edinburgh, UK
Cordelia Schmid   Google and Inria, France
Emanuele Trucco   University of Dundee, UK

Program Chairs

Andrea Vedaldi   University of Oxford, UK
Horst Bischof   Graz University of Technology, Austria
Thomas Brox   University of Freiburg, Germany
Jan-Michael Frahm   University of North Carolina, USA

Industrial Liaison Chairs

Jim Ashe   University of Edinburgh, UK
Helmut Grabner   Zurich University of Applied Sciences, Switzerland
Diane Larlus   NAVER LABS Europe, France
Cristian Novotny   University of Edinburgh, UK

Local Arrangement Chairs

Yvan Petillot   Heriot-Watt University, UK
Paul Siebert   University of Glasgow, UK

Academic Demonstration Chair

Thomas Mensink   Google Research and University of Amsterdam,
The Netherlands

Poster Chair

Stephen Mckenna   University of Dundee, UK

Technology Chair

Gerardo Aragon Camarasa   University of Glasgow, UK

Tutorial Chairs

Carlo Colombo	University of Florence, Italy
Sotirios Tsaftaris	University of Edinburgh, UK

Publication Chairs

Albert Ali Salah	Utrecht University, The Netherlands
Hamdi Dibeklioglu	Bilkent University, Turkey
Metehan Doyran	Utrecht University, The Netherlands
Henry Howard-Jenkins	University of Oxford, UK
Victor Adrian Prisacariu	University of Oxford, UK
Siyu Tang	ETH Zurich, Switzerland
Gul Varol	University of Oxford, UK

Website Chair

Giovanni Maria Farinella	University of Catania, Italy

Workshops Chairs

Adrien Bartoli	University of Clermont Auvergne, France
Andrea Fusiello	University of Udine, Italy

Area Chairs

Lourdes Agapito	University College London, UK
Zeynep Akata	University of Tübingen, Germany
Karteek Alahari	Inria, France
Antonis Argyros	University of Crete, Greece
Hossein Azizpour	KTH Royal Institute of Technology, Sweden
Joao P. Barreto	Universidade de Coimbra, Portugal
Alexander C. Berg	University of North Carolina at Chapel Hill, USA
Matthew B. Blaschko	KU Leuven, Belgium
Lubomir D. Bourdev	WaveOne, Inc., USA
Edmond Boyer	Inria, France
Yuri Boykov	University of Waterloo, Canada
Gabriel Brostow	University College London, UK
Michael S. Brown	National University of Singapore, Singapore
Jianfei Cai	Monash University, Australia
Barbara Caputo	Politecnico di Torino, Italy
Ayan Chakrabarti	Washington University, St. Louis, USA
Tat-Jen Cham	Nanyang Technological University, Singapore
Manmohan Chandraker	University of California, San Diego, USA
Rama Chellappa	Johns Hopkins University, USA
Liang-Chieh Chen	Google, USA

Haibin Ling	Stony Brooks, State University of New York, USA
Jiaying Liu	Peking University, China
Ming-Yu Liu	NVIDIA, USA
Si Liu	Beihang University, China
Xiaoming Liu	Michigan State University, USA
Huchuan Lu	Dalian University of Technology, China
Simon Lucey	Carnegie Mellon University, USA
Jiebo Luo	University of Rochester, USA
Julien Mairal	Inria, France
Michael Maire	University of Chicago, USA
Subhransu Maji	University of Massachusetts, Amherst, USA
Yasushi Makihara	Osaka University, Japan
Jiri Matas	Czech Technical University in Prague, Czech Republic
Yasuyuki Matsushita	Osaka University, Japan
Philippos Mordohai	Stevens Institute of Technology, USA
Vittorio Murino	University of Verona, Italy
Naila Murray	NAVER LABS Europe, France
Hajime Nagahara	Osaka University, Japan
P. J. Narayanan	International Institute of Information Technology (IIIT), Hyderabad, India
Nassir Navab	Technical University of Munich, Germany
Natalia Neverova	Facebook AI Research, France
Matthias Niessner	Technical University of Munich, Germany
Jean-Marc Odobez	Idiap Research Institute and Swiss Federal Institute of Technology Lausanne, Switzerland
Francesca Odone	Universita di Genova, Italy
Takeshi Oishi	The University of Tokyo, Tokyo Institute of Technology, Japan
Vicente Ordonez	University of Virginia, USA
Manohar Paluri	Facebook AI Research, USA
Maja Pantic	Imperial College London, UK
In Kyu Park	Inha University, South Korea
Ioannis Patras	Queen Mary University of London, UK
Patrick Perez	Valeo, France
Bryan A. Plummer	Boston University, USA
Thomas Pock	Graz University of Technology, Austria
Marc Pollefeys	ETH Zurich and Microsoft MR & AI Zurich Lab, Switzerland
Jean Ponce	Inria, France
Gerard Pons-Moll	MPII, Saarland Informatics Campus, Germany
Jordi Pont-Tuset	Google, Switzerland
James Matthew Rehg	Georgia Institute of Technology, USA
Ian Reid	University of Adelaide, Australia
Olaf Ronneberger	DeepMind London, UK
Stefan Roth	TU Darmstadt, Germany
Bryan Russell	Adobe Research, USA

Kwang Moo Yi   University of Victoria, Canada
Zhaozheng Yin   Stony Brook, State University of New York, USA
Chang D. Yoo   Korea Advanced Institute of Science and Technology,
                                South Korea
Shaodi You   University of Amsterdam, The Netherlands
Jingyi Yu   ShanghaiTech University, China
Stella Yu   University of California, Berkeley, and ICSI, USA
Stefanos Zafeiriou   Imperial College London, UK
Hongbin Zha   Peking University, China
Tianzhu Zhang   University of Science and Technology of China, China
Liang Zheng   Australian National University, Australia
Todd E. Zickler   Harvard University, USA
Andrew Zisserman   University of Oxford, UK

Technical Program Committee

Sathyanarayanan	Samuel Albanie	Pablo Arbelaez
N. Aakur	Shadi Albarqouni	Shervin Ardeshir
Wael Abd Almgaeed	Cenek Albl	Sercan O. Arik
Abdelrahman	Hassan Abu Alhaija	Anil Armagan
Abdelhamed	Daniel Aliaga	Anurag Arnab
Abdullah Abuolaim	Mohammad	Chetan Arora
Supreeth Achar	S. Aliakbarian	Federica Arrigoni
Hanno Ackermann	Rahaf Aljundi	Mathieu Aubry
Ehsan Adeli	Thiemo Alldieck	Shai Avidan
Triantafyllos Afouras	Jon Almazan	Angelica I. Aviles-Rivero
Sameer Agarwal	Jose M. Alvarez	Yannis Avrithis
Aishwarya Agrawal	Senjian An	Ismail Ben Ayed
Harsh Agrawal	Saket Anand	Shekoofeh Azizi
Pulkit Agrawal	Codruta Ancuti	Ioan Andrei Bârsan
Antonio Agudo	Cosmin Ancuti	Artem Babenko
Eirikur Agustsson	Peter Anderson	Deepak Babu Sam
Karim Ahmed	Juan Andrade-Cetto	Seung-Hwan Baek
Byeongjoo Ahn	Alexander Andreopoulos	Seungryul Baek
Unaiza Ahsan	Misha Andriluka	Andrew D. Bagdanov
Thalaiyasingam Ajanthan	Dragomir Anguelov	Shai Bagon
Kenan E. Ak	Rushil Anirudh	Yuval Bahat
Emre Akbas	Michel Antunes	Junjie Bai
Naveed Akhtar	Oisin Mac Aodha	Song Bai
Derya Akkaynak	Srikar Appalaraju	Xiang Bai
Yagiz Aksoy	Relja Arandjelovic	Yalong Bai
Ziad Al-Halah	Nikita Araslanov	Yancheng Bai
Xavier Alameda-Pineda	Andre Araujo	Peter Bajcsy
Jean-Baptiste Alayrac	Helder Araujo	Slawomir Bak

Mahsa Baktashmotlagh
Kavita Bala
Yogesh Balaji
Guha Balakrishnan
V. N. Balasubramanian
Federico Baldassarre
Vassileios Balntas
Shurjo Banerjee
Aayush Bansal
Ankan Bansal
Jianmin Bao
Linchao Bao
Wenbo Bao
Yingze Bao
Akash Bapat
Md Jawadul Hasan Bappy
Fabien Baradel
Lorenzo Baraldi
Daniel Barath
Adrian Barbu
Kobus Barnard
Nick Barnes
Francisco Barranco
Jonathan T. Barron
Arslan Basharat
Chaim Baskin
Anil S. Baslamisli
Jorge Batista
Kayhan Batmanghelich
Konstantinos Batsos
David Bau
Luis Baumela
Christoph Baur
Eduardo
  Bayro-Corrochano
Paul Beardsley
Jan Bednavr'ik
Oscar Beijbom
Philippe Bekaert
Esube Bekele
Vasileios Belagiannis
Ohad Ben-Shahar
Abhijit Bendale
Róger Bermúdez-Chacón
Maxim Berman
Jesus Bermudez-cameo

Florian Bernard
Stefano Berretti
Marcelo Bertalmio
Gedas Bertasius
Cigdem Beyan
Lucas Beyer
Vijayakumar Bhagavatula
Arjun Nitin Bhagoji
Apratim Bhattacharyya
Binod Bhattarai
Sai Bi
Jia-Wang Bian
Simone Bianco
Adel Bibi
Tolga Birdal
Tom Bishop
Soma Biswas
Mårten Björkman
Volker Blanz
Vishnu Boddeti
Navaneeth Bodla
Simion-Vlad Bogolin
Xavier Boix
Piotr Bojanowski
Timo Bolkart
Guido Borghi
Larbi Boubchir
Guillaume Bourmaud
Adrien Bousseau
Thierry Bouwmans
Richard Bowden
Hakan Boyraz
Mathieu Brédif
Samarth Brahmbhatt
Steve Branson
Nikolas Brasch
Biagio Brattoli
Ernesto Brau
Toby P. Breckon
Francois Bremond
Jesus Briales
Sofia Broomé
Marcus A. Brubaker
Luc Brun
Silvia Bucci
Shyamal Buch

Pradeep Buddharaju
Uta Buechler
Mai Bui
Tu Bui
Adrian Bulat
Giedrius T. Burachas
Elena Burceanu
Xavier P. Burgos-Artizzu
Kaylee Burns
Andrei Bursuc
Benjamin Busam
Wonmin Byeon
Zoya Bylinskii
Sergi Caelles
Jianrui Cai
Minjie Cai
Yujun Cai
Zhaowei Cai
Zhipeng Cai
Juan C. Caicedo
Simone Calderara
Necati Cihan Camgoz
Dylan Campbell
Octavia Camps
Jiale Cao
Kaidi Cao
Liangliang Cao
Xiangyong Cao
Xiaochun Cao
Yang Cao
Yu Cao
Yue Cao
Zhangjie Cao
Luca Carlone
Mathilde Caron
Dan Casas
Thomas J. Cashman
Umberto Castellani
Lluis Castrejon
Jacopo Cavazza
Fabio Cermelli
Hakan Cevikalp
Menglei Chai
Ishani Chakraborty
Rudrasis Chakraborty
Antoni B. Chan

Rozenn Dahyot
Bo Dai
Dengxin Dai
Hang Dai
Longquan Dai
Shuyang Dai
Xiyang Dai
Yuchao Dai
Adrian V. Dalca
Dima Damen
Bharath B. Damodaran
Kristin Dana
Martin Danelljan
Zheng Dang
Zachary Alan Daniels
Donald G. Dansereau
Abhishek Das
Samyak Datta
Achal Dave
Titas De
Rodrigo de Bem
Teo de Campos
Raoul de Charette
Shalini De Mello
Joseph DeGol
Herve Delingette
Haowen Deng
Jiankang Deng
Weijian Deng
Zhiwei Deng
Joachim Denzler
Konstantinos G. Derpanis
Aditya Deshpande
Frederic Devernay
Somdip Dey
Arturo Deza
Abhinav Dhall
Helisa Dhamo
Vikas Dhiman
Fillipe Dias Moreira
  de Souza
Ali Diba
Ferran Diego
Guiguang Ding
Henghui Ding
Jian Ding

Mingyu Ding
Xinghao Ding
Zhengming Ding
Robert DiPietro
Cosimo Distante
Ajay Divakaran
Mandar Dixit
Abdelaziz Djelouah
Thanh-Toan Do
Jose Dolz
Bo Dong
Chao Dong
Jiangxin Dong
Weiming Dong
Weisheng Dong
Xingping Dong
Xuanyi Dong
Yinpeng Dong
Gianfranco Doretto
Hazel Doughty
Hassen Drira
Bertram Drost
Dawei Du
Ye Duan
Yueqi Duan
Abhimanyu Dubey
Anastasia Dubrovina
Stefan Duffner
Chi Nhan Duong
Thibaut Durand
Zoran Duric
Iulia Duta
Debidatta Dwibedi
Benjamin Eckart
Marc Eder
Marzieh Edraki
Alexei A. Efros
Kiana Ehsani
Hazm Kemal Ekenel
James H. Elder
Mohamed Elgharib
Shireen Elhabian
Ehsan Elhamifar
Mohamed Elhoseiny
Ian Endres
N. Benjamin Erichson

Jan Ernst
Sergio Escalera
Francisco Escolano
Victor Escorcia
Carlos Esteves
Francisco J. Estrada
Bin Fan
Chenyou Fan
Deng-Ping Fan
Haoqi Fan
Hehe Fan
Heng Fan
Kai Fan
Lijie Fan
Linxi Fan
Quanfu Fan
Shaojing Fan
Xiaochuan Fan
Xin Fan
Yuchen Fan
Sean Fanello
Hao-Shu Fang
Haoyang Fang
Kuan Fang
Yi Fang
Yuming Fang
Azade Farshad
Alireza Fathi
Raanan Fattal
Joao Fayad
Xiaohan Fei
Christoph Feichtenhofer
Michael Felsberg
Chen Feng
Jiashi Feng
Junyi Feng
Mengyang Feng
Qianli Feng
Zhenhua Feng
Michele Fenzi
Andras Ferencz
Martin Fergie
Basura Fernando
Ethan Fetaya
Michael Firman
John W. Fisher

Matthew Fisher
Boris Flach
Corneliu Florea
Wolfgang Foerstner
David Fofi
Gian Luca Foresti
Per-Erik Forssen
David Fouhey
Katerina Fragkiadaki
Victor Fragoso
Jean-Sébastien Franco
Ohad Fried
Iuri Frosio
Cheng-Yang Fu
Huazhu Fu
Jianlong Fu
Jingjing Fu
Xueyang Fu
Yanwei Fu
Ying Fu
Yun Fu
Olac Fuentes
Kent Fujiwara
Takuya Funatomi
Christopher Funk
Thomas Funkhouser
Antonino Furnari
Ryo Furukawa
Erik Gärtner
Raghudeep Gadde
Matheus Gadelha
Vandit Gajjar
Trevor Gale
Juergen Gall
Mathias Gallardo
Guillermo Gallego
Orazio Gallo
Chuang Gan
Zhe Gan
Madan Ravi Ganesh
Aditya Ganeshan
Siddha Ganju
Bin-Bin Gao
Changxin Gao
Feng Gao
Hongchang Gao

Jin Gao
Jiyang Gao
Junbin Gao
Katelyn Gao
Lin Gao
Mingfei Gao
Ruiqi Gao
Ruohan Gao
Shenghua Gao
Yuan Gao
Yue Gao
Noa Garcia
Alberto Garcia-Garcia
Guillermo
    Garcia-Hernando
Jacob R. Gardner
Animesh Garg
Kshitiz Garg
Rahul Garg
Ravi Garg
Philip N. Garner
Kirill Gavrilyuk
Paul Gay
Shiming Ge
Weifeng Ge
Baris Gecer
Xin Geng
Kyle Genova
Stamatios Georgoulis
Bernard Ghanem
Michael Gharbi
Kamran Ghasedi
Golnaz Ghiasi
Arnab Ghosh
Partha Ghosh
Silvio Giancola
Andrew Gilbert
Rohit Girdhar
Xavier Giro-i-Nieto
Thomas Gittings
Ioannis Gkioulekas
Clement Godard
Vaibhava Goel
Bastian Goldluecke
Lluis Gomez
Nuno Gonçalves

Dong Gong
Ke Gong
Mingming Gong
Abel Gonzalez-Garcia
Ariel Gordon
Daniel Gordon
Paulo Gotardo
Venu Madhav Govindu
Ankit Goyal
Priya Goyal
Raghav Goyal
Benjamin Graham
Douglas Gray
Brent A. Griffin
Etienne Grossmann
David Gu
Jiayuan Gu
Jiuxiang Gu
Lin Gu
Qiao Gu
Shuhang Gu
Jose J. Guerrero
Paul Guerrero
Jie Gui
Jean-Yves Guillemaut
Riza Alp Guler
Erhan Gundogdu
Fatma Guney
Guodong Guo
Kaiwen Guo
Qi Guo
Sheng Guo
Shi Guo
Tiantong Guo
Xiaojie Guo
Yijie Guo
Yiluan Guo
Yuanfang Guo
Yulan Guo
Agrim Gupta
Ankush Gupta
Mohit Gupta
Saurabh Gupta
Tanmay Gupta
Danna Gurari
Abner Guzman-Rivera

JunYoung Gwak
Michael Gygli
Jung-Woo Ha
Simon Hadfield
Isma Hadji
Bjoern Haefner
Taeyoung Hahn
Levente Hajder
Peter Hall
Emanuela Haller
Stefan Haller
Bumsub Ham
Abdullah Hamdi
Dongyoon Han
Hu Han
Jungong Han
Junwei Han
Kai Han
Tian Han
Xiaoguang Han
Xintong Han
Yahong Han
Ankur Handa
Zekun Hao
Albert Haque
Tatsuya Harada
Mehrtash Harandi
Adam W. Harley
Mahmudul Hasan
Atsushi Hashimoto
Ali Hatamizadeh
Munawar Hayat
Dongliang He
Jingrui He
Junfeng He
Kaiming He
Kun He
Lei He
Pan He
Ran He
Shengfeng He
Tong He
Weipeng He
Xuming He
Yang He
Yihui He

Zhihai He
Chinmay Hegde
Janne Heikkila
Mattias P. Heinrich
Stéphane Herbin
Alexander Hermans
Luis Herranz
John R. Hershey
Aaron Hertzmann
Roei Herzig
Anders Heyden
Steven Hickson
Otmar Hilliges
Tomas Hodan
Judy Hoffman
Michael Hofmann
Yannick Hold-Geoffroy
Namdar Homayounfar
Sina Honari
Richang Hong
Seunghoon Hong
Xiaopeng Hong
Yi Hong
Hidekata Hontani
Anthony Hoogs
Yedid Hoshen
Mir Rayat Imtiaz Hossain
Junhui Hou
Le Hou
Lu Hou
Tingbo Hou
Wei-Lin Hsiao
Cheng-Chun Hsu
Gee-Sern Jison Hsu
Kuang-jui Hsu
Changbo Hu
Di Hu
Guosheng Hu
Han Hu
Hao Hu
Hexiang Hu
Hou-Ning Hu
Jie Hu
Junlin Hu
Nan Hu
Ping Hu

Ronghang Hu
Xiaowei Hu
Yinlin Hu
Yuan-Ting Hu
Zhe Hu
Binh-Son Hua
Yang Hua
Bingyao Huang
Di Huang
Dong Huang
Fay Huang
Haibin Huang
Haozhi Huang
Heng Huang
Huaibo Huang
Jia-Bin Huang
Jing Huang
Jingwei Huang
Kaizhu Huang
Lei Huang
Qiangui Huang
Qiaoying Huang
Qingqiu Huang
Qixing Huang
Shaoli Huang
Sheng Huang
Siyuan Huang
Weilin Huang
Wenbing Huang
Xiangru Huang
Xun Huang
Yan Huang
Yifei Huang
Yue Huang
Zhiwu Huang
Zilong Huang
Minyoung Huh
Zhuo Hui
Matthias B. Hullin
Martin Humenberger
Wei-Chih Hung
Zhouyuan Huo
Junhwa Hur
Noureldien Hussein
Jyh-Jing Hwang
Seong Jae Hwang

Sung Ju Hwang
Ichiro Ide
Ivo Ihrke
Daiki Ikami
Satoshi Ikehata
Nazli Ikizler-Cinbis
Sunghoon Im
Yani Ioannou
Radu Tudor Ionescu
Umar Iqbal
Go Irie
Ahmet Iscen
Md Amirul Islam
Vamsi Ithapu
Nathan Jacobs
Arpit Jain
Himalaya Jain
Suyog Jain
Stuart James
Won-Dong Jang
Yunseok Jang
Ronnachai Jaroensri
Dinesh Jayaraman
Sadeep Jayasumana
Suren Jayasuriya
Herve Jegou
Simon Jenni
Hae-Gon Jeon
Yunho Jeon
Koteswar R. Jerripothula
Hueihan Jhuang
I-hong Jhuo
Dinghuang Ji
Hui Ji
Jingwei Ji
Pan Ji
Yanli Ji
Baoxiong Jia
Kui Jia
Xu Jia
Chiyu Max Jiang
Haiyong Jiang
Hao Jiang
Huaizu Jiang
Huajie Jiang
Ke Jiang

Lai Jiang
Li Jiang
Lu Jiang
Ming Jiang
Peng Jiang
Shuqiang Jiang
Wei Jiang
Xudong Jiang
Zhuolin Jiang
Jianbo Jiao
Zequn Jie
Dakai Jin
Kyong Hwan Jin
Lianwen Jin
SouYoung Jin
Xiaojie Jin
Xin Jin
Nebojsa Jojic
Alexis Joly
Michael Jeffrey Jones
Hanbyul Joo
Jungseock Joo
Kyungdon Joo
Ajjen Joshi
Shantanu H. Joshi
Da-Cheng Juan
Marco Körner
Kevin Köser
Asim Kadav
Christine Kaeser-Chen
Kushal Kafle
Dagmar Kainmueller
Ioannis A. Kakadiaris
Zdenek Kalal
Nima Kalantari
Yannis Kalantidis
Mahdi M. Kalayeh
Anmol Kalia
Sinan Kalkan
Vicky Kalogeiton
Ashwin Kalyan
Joni-kristian Kamarainen
Gerda Kamberova
Chandra Kambhamettu
Martin Kampel
Meina Kan

Christopher Kanan
Kenichi Kanatani
Angjoo Kanazawa
Atsushi Kanehira
Takuhiro Kaneko
Asako Kanezaki
Bingyi Kang
Di Kang
Sunghun Kang
Zhao Kang
Vadim Kantorov
Abhishek Kar
Amlan Kar
Theofanis Karaletsos
Leonid Karlinsky
Kevin Karsch
Angelos Katharopoulos
Isinsu Katircioglu
Hiroharu Kato
Zoltan Kato
Dotan Kaufman
Jan Kautz
Rei Kawakami
Qiuhong Ke
Wadim Kehl
Petr Kellnhofer
Aniruddha Kembhavi
Cem Keskin
Margret Keuper
Daniel Keysers
Ashkan Khakzar
Fahad Khan
Naeemullah Khan
Salman Khan
Siddhesh Khandelwal
Rawal Khirodkar
Anna Khoreva
Tejas Khot
Parmeshwar Khurd
Hadi Kiapour
Joe Kileel
Chanho Kim
Dahun Kim
Edward Kim
Eunwoo Kim
Han-ul Kim

Hansung Kim
Heewon Kim
Hyo Jin Kim
Hyunwoo J. Kim
Jinkyu Kim
Jiwon Kim
Jongmin Kim
Junsik Kim
Junyeong Kim
Min H. Kim
Namil Kim
Pyojin Kim
Seon Joo Kim
Seong Tae Kim
Seungryong Kim
Sungwoong Kim
Tae Hyun Kim
Vladimir Kim
Won Hwa Kim
Yonghyun Kim
Benjamin Kimia
Akisato Kimura
Pieter-Jan Kindermans
Zsolt Kira
Itaru Kitahara
Hedvig Kjellstrom
Jan Knopp
Takumi Kobayashi
Erich Kobler
Parker Koch
Reinhard Koch
Elyor Kodirov
Amir Kolaman
Nicholas Kolkin
Dimitrios Kollias
Stefanos Kollias
Soheil Kolouri
Adams Wai-Kin Kong
Naejin Kong
Shu Kong
Tao Kong
Yu Kong
Yoshinori Konishi
Daniil Kononenko
Theodora Kontogianni
Simon Korman

Adam Kortylewski
Jana Kosecka
Jean Kossaifi
Satwik Kottur
Rigas Kouskouridas
Adriana Kovashka
Rama Kovvuri
Adarsh Kowdle
Jedrzej Kozerawski
Mateusz Kozinski
Philipp Kraehenbuehl
Gregory Kramida
Josip Krapac
Dmitry Kravchenko
Ranjay Krishna
Pavel Krsek
Alexander Krull
Jakob Kruse
Hiroyuki Kubo
Hilde Kuehne
Jason Kuen
Andreas Kuhn
Arjan Kuijper
Zuzana Kukelova
Ajay Kumar
Amit Kumar
Avinash Kumar
Suryansh Kumar
Vijay Kumar
Kaustav Kundu
Weicheng Kuo
Nojun Kwak
Suha Kwak
Junseok Kwon
Nikolaos Kyriazis
Zorah Lähner
Ankit Laddha
Florent Lafarge
Jean Lahoud
Kevin Lai
Shang-Hong Lai
Wei-Sheng Lai
Yu-Kun Lai
Iro Laina
Antony Lam
John Wheatley Lambert

Xiangyuan lan
Xu Lan
Charis Lanaras
Georg Langs
Oswald Lanz
Dong Lao
Yizhen Lao
Agata Lapedriza
Gustav Larsson
Viktor Larsson
Katrin Lasinger
Christoph Lassner
Longin Jan Latecki
Stéphane Lathuilière
Rynson Lau
Hei Law
Justin Lazarow
Svetlana Lazebnik
Hieu Le
Huu Le
Ngan Hoang Le
Trung-Nghia Le
Vuong Le
Colin Lea
Erik Learned-Miller
Chen-Yu Lee
Gim Hee Lee
Hsin-Ying Lee
Hyungtae Lee
Jae-Han Lee
Jimmy Addison Lee
Joonseok Lee
Kibok Lee
Kuang-Huei Lee
Kwonjoon Lee
Minsik Lee
Sang-chul Lee
Seungkyu Lee
Soochan Lee
Stefan Lee
Taehee Lee
Andreas Lehrmann
Jie Lei
Peng Lei
Matthew Joseph Leotta
Wee Kheng Leow

Gil Levi
Evgeny Levinkov
Aviad Levis
Jose Lezama
Ang Li
Bin Li
Bing Li
Boyi Li
Changsheng Li
Chao Li
Chen Li
Cheng Li
Chenglong Li
Chi Li
Chun-Guang Li
Chun-Liang Li
Chunyuan Li
Dong Li
Guanbin Li
Hao Li
Haoxiang Li
Hongsheng Li
Hongyang Li
Houqiang Li
Huibin Li
Jia Li
Jianan Li
Jianguo Li
Junnan Li
Junxuan Li
Kai Li
Ke Li
Kejie Li
Kunpeng Li
Lerenhan Li
Li Erran Li
Mengtian Li
Mu Li
Peihua Li
Peiyi Li
Ping Li
Qi Li
Qing Li
Ruiyu Li
Ruoteng Li
Shaozi Li

Sheng Li
Shiwei Li
Shuang Li
Siyang Li
Stan Z. Li
Tianye Li
Wei Li
Weixin Li
Wen Li
Wenbo Li
Xiaomeng Li
Xin Li
Xiu Li
Xuelong Li
Xueting Li
Yan Li
Yandong Li
Yanghao Li
Yehao Li
Yi Li
Yijun Li
Yikang LI
Yining Li
Yongjie Li
Yu Li
Yu-Jhe Li
Yunpeng Li
Yunsheng Li
Yunzhu Li
Zhe Li
Zhen Li
Zhengqi Li
Zhenyang Li
Zhuwen Li
Dongze Lian
Xiaochen Lian
Zhouhui Lian
Chen Liang
Jie Liang
Ming Liang
Paul Pu Liang
Pengpeng Liang
Shu Liang
Wei Liang
Jing Liao
Minghui Liao

Renjie Liao
Shengcai Liao
Shuai Liao
Yiyi Liao
Ser-Nam Lim
Chen-Hsuan Lin
Chung-Ching Lin
Dahua Lin
Ji Lin
Kevin Lin
Tianwei Lin
Tsung-Yi Lin
Tsung-Yu Lin
Wei-An Lin
Weiyao Lin
Yen-Chen Lin
Yuewei Lin
David B. Lindell
Drew Linsley
Krzysztof Lis
Roee Litman
Jim Little
An-An Liu
Bo Liu
Buyu Liu
Chao Liu
Chen Liu
Cheng-lin Liu
Chenxi Liu
Dong Liu
Feng Liu
Guilin Liu
Haomiao Liu
Heshan Liu
Hong Liu
Ji Liu
Jingen Liu
Jun Liu
Lanlan Liu
Li Liu
Liu Liu
Mengyuan Liu
Miaomiao Liu
Nian Liu
Ping Liu
Risheng Liu

Sheng Liu
Shu Liu
Shuaicheng Liu
Sifei Liu
Siqi Liu
Siying Liu
Songtao Liu
Ting Liu
Tongliang Liu
Tyng-Luh Liu
Wanquan Liu
Wei Liu
Weiyang Liu
Weizhe Liu
Wenyu Liu
Wu Liu
Xialei Liu
Xianglong Liu
Xiaodong Liu
Xiaofeng Liu
Xihui Liu
Xingyu Liu
Xinwang Liu
Xuanqing Liu
Xuebo Liu
Yang Liu
Yaojie Liu
Yebin Liu
Yen-Cheng Liu
Yiming Liu
Yu Liu
Yu-Shen Liu
Yufan Liu
Yun Liu
Zheng Liu
Zhijian Liu
Zhuang Liu
Zichuan Liu
Ziwei Liu
Zongyi Liu
Stephan Liwicki
Liliana Lo Presti
Chengjiang Long
Fuchen Long
Mingsheng Long
Xiang Long

Yang Long
Charles T. Loop
Antonio Lopez
Roberto J. Lopez-Sastre
Javier Lorenzo-Navarro
Manolis Lourakis
Boyu Lu
Canyi Lu
Feng Lu
Guoyu Lu
Hongtao Lu
Jiajun Lu
Jiasen Lu
Jiwen Lu
Kaiyue Lu
Le Lu
Shao-Ping Lu
Shijian Lu
Xiankai Lu
Xin Lu
Yao Lu
Yiping Lu
Yongxi Lu
Yongyi Lu
Zhiwu Lu
Fujun Luan
Benjamin E. Lundell
Hao Luo
Jian-Hao Luo
Ruotian Luo
Weixin Luo
Wenhan Luo
Wenjie Luo
Yan Luo
Zelun Luo
Zixin Luo
Khoa Luu
Zhaoyang Lv
Pengyuan Lyu
Thomas Möllenhoff
Matthias Müller
Bingpeng Ma
Chih-Yao Ma
Chongyang Ma
Huimin Ma
Jiayi Ma

K. T. Ma
Ke Ma
Lin Ma
Liqian Ma
Shugao Ma
Wei-Chiu Ma
Xiaojian Ma
Xingjun Ma
Zhanyu Ma
Zheng Ma
Radek Jakob Mackowiak
Ludovic Magerand
Shweta Mahajan
Siddharth Mahendran
Long Mai
Ameesh Makadia
Oscar Mendez Maldonado
Mateusz Malinowski
Yury Malkov
Arun Mallya
Dipu Manandhar
Massimiliano Mancini
Fabian Manhardt
Kevis-kokitsi Maninis
Varun Manjunatha
Junhua Mao
Xudong Mao
Alina Marcu
Edgar Margffoy-Tuay
Dmitrii Marin
Manuel J. Marin-Jimenez
Kenneth Marino
Niki Martinel
Julieta Martinez
Jonathan Masci
Tomohiro Mashita
Iacopo Masi
David Masip
Daniela Massiceti
Stefan Mathe
Yusuke Matsui
Tetsu Matsukawa
Iain A. Matthews
Kevin James Matzen
Bruce Allen Maxwell
Stephen Maybank

Helmut Mayer
Amir Mazaheri
David McAllester
Steven McDonagh
Stephen J. Mckenna
Roey Mechrez
Prakhar Mehrotra
Christopher Mei
Xue Mei
Paulo R. S. Mendonca
Lili Meng
Zibo Meng
Thomas Mensink
Bjoern Menze
Michele Merler
Kourosh Meshgi
Pascal Mettes
Christopher Metzler
Liang Mi
Qiguang Miao
Xin Miao
Tomer Michaeli
Frank Michel
Antoine Miech
Krystian Mikolajczyk
Peyman Milanfar
Ben Mildenhall
Gregor Miller
Fausto Milletari
Dongbo Min
Kyle Min
Pedro Miraldo
Dmytro Mishkin
Anand Mishra
Ashish Mishra
Ishan Misra
Niluthpol C. Mithun
Kaushik Mitra
Niloy Mitra
Anton Mitrokhin
Ikuhisa Mitsugami
Anurag Mittal
Kaichun Mo
Zhipeng Mo
Davide Modolo
Michael Moeller

Pritish Mohapatra
Pavlo Molchanov
Davide Moltisanti
Pascal Monasse
Mathew Monfort
Aron Monszpart
Sean Moran
Vlad I. Morariu
Francesc Moreno-Noguer
Pietro Morerio
Stylianos Moschoglou
Yael Moses
Roozbeh Mottaghi
Pierre Moulon
Arsalan Mousavian
Yadong Mu
Yasuhiro Mukaigawa
Lopamudra Mukherjee
Yusuke Mukuta
Ravi Teja Mullapudi
Mario Enrique Munich
Zachary Murez
Ana C. Murillo
J. Krishna Murthy
Damien Muselet
Armin Mustafa
Siva Karthik Mustikovela
Carlo Dal Mutto
Moin Nabi
Varun K. Nagaraja
Tushar Nagarajan
Arsha Nagrani
Seungjun Nah
Nikhil Naik
Yoshikatsu Nakajima
Yuta Nakashima
Atsushi Nakazawa
Seonghyeon Nam
Vinay P. Namboodiri
Medhini Narasimhan
Srinivasa Narasimhan
Sanath Narayan
Erickson Rangel
  Nascimento
Jacinto Nascimento
Tayyab Naseer

Lakshmanan Nataraj
Neda Nategh
Nelson Isao Nauata
Fernando Navarro
Shah Nawaz
Lukas Neumann
Ram Nevatia
Alejandro Newell
Shawn Newsam
Joe Yue-Hei Ng
Trung Thanh Ngo
Duc Thanh Nguyen
Lam M. Nguyen
Phuc Xuan Nguyen
Thuong Nguyen Canh
Mihalis Nicolaou
Andrei Liviu Nicolicioiu
Xuecheng Nie
Michael Niemeyer
Simon Niklaus
Christophoros Nikou
David Nilsson
Jifeng Ning
Yuval Nirkin
Li Niu
Yuzhen Niu
Zhenxing Niu
Shohei Nobuhara
Nicoletta Noceti
Hyeonwoo Noh
Junhyug Noh
Mehdi Noroozi
Sotiris Nousias
Valsamis Ntouskos
Matthew O'Toole
Peter Ochs
Ferda Ofli
Seong Joon Oh
Seoung Wug Oh
Iason Oikonomidis
Utkarsh Ojha
Takahiro Okabe
Takayuki Okatani
Fumio Okura
Aude Oliva
Kyle Olszewski

Björn Ommer
Mohamed Omran
Elisabeta Oneata
Michael Opitz
Jose Oramas
Tribhuvanesh Orekondy
Shaul Oron
Sergio Orts-Escolano
Ivan Oseledets
Aljosa Osep
Magnus Oskarsson
Anton Osokin
Martin R. Oswald
Wanli Ouyang
Andrew Owens
Mete Ozay
Mustafa Ozuysal
Eduardo Pérez-Pellitero
Gautam Pai
Dipan Kumar Pal
P. H. Pamplona Savarese
Jinshan Pan
Junting Pan
Xingang Pan
Yingwei Pan
Yannis Panagakis
Rameswar Panda
Guan Pang
Jiahao Pang
Jiangmiao Pang
Tianyu Pang
Sharath Pankanti
Nicolas Papadakis
Dim Papadopoulos
George Papandreou
Toufiq Parag
Shaifali Parashar
Sarah Parisot
Eunhyeok Park
Hyun Soo Park
Jaesik Park
Min-Gyu Park
Taesung Park
Alvaro Parra
C. Alejandro Parraga
Despoina Paschalidou

Nikolaos Passalis
Vishal Patel
Viorica Patraucean
Badri Narayana Patro
Danda Pani Paudel
Sujoy Paul
Georgios Pavlakos
Ioannis Pavlidis
Vladimir Pavlovic
Nick Pears
Kim Steenstrup Pedersen
Selen Pehlivan
Shmuel Peleg
Chao Peng
Houwen Peng
Wen-Hsiao Peng
Xi Peng
Xiaojiang Peng
Xingchao Peng
Yuxin Peng
Federico Perazzi
Juan Camilo Perez
Vishwanath Peri
Federico Pernici
Luca Del Pero
Florent Perronnin
Stavros Petridis
Henning Petzka
Patrick Peursum
Michael Pfeiffer
Hanspeter Pfister
Roman Pflugfelder
Minh Tri Pham
Yongri Piao
David Picard
Tomasz Pieciak
A. J. Piergiovanni
Andrea Pilzer
Pedro O. Pinheiro
Silvia Laura Pintea
Lerrel Pinto
Axel Pinz
Robinson Piramuthu
Fiora Pirri
Leonid Pishchulin
Francesco Pittaluga

Daniel Pizarro
Tobias Plötz
Mirco Planamente
Matteo Poggi
Moacir A. Ponti
Parita Pooj
Fatih Porikli
Horst Possegger
Omid Poursaeed
Ameya Prabhu
Viraj Uday Prabhu
Dilip Prasad
Brian L. Price
True Price
Maria Priisalu
Veronique Prinet
Victor Adrian Prisacariu
Jan Prokaj
Sergey Prokudin
Nicolas Pugeault
Xavier Puig
Albert Pumarola
Pulak Purkait
Senthil Purushwalkam
Charles R. Qi
Hang Qi
Haozhi Qi
Lu Qi
Mengshi Qi
Siyuan Qi
Xiaojuan Qi
Yuankai Qi
Shengju Qian
Xuelin Qian
Siyuan Qiao
Yu Qiao
Jie Qin
Qiang Qiu
Weichao Qiu
Zhaofan Qiu
Kha Gia Quach
Yuhui Quan
Yvain Queau
Julian Quiroga
Faisal Qureshi
Mahdi Rad

Filip Radenovic
Petia Radeva
Venkatesh
  B. Radhakrishnan
Ilija Radosavovic
Noha Radwan
Rahul Raguram
Tanzila Rahman
Amit Raj
Ajit Rajwade
Kandan Ramakrishnan
Santhosh
  K. Ramakrishnan
Srikumar Ramalingam
Ravi Ramamoorthi
Vasili Ramanishka
Ramprasaath R. Selvaraju
Francois Rameau
Visvanathan Ramesh
Santu Rana
Rene Ranftl
Anand Rangarajan
Anurag Ranjan
Viresh Ranjan
Yongming Rao
Carolina Raposo
Vivek Rathod
Sathya N. Ravi
Avinash Ravichandran
Tammy Riklin Raviv
Daniel Rebain
Sylvestre-Alvise Rebuffi
N. Dinesh Reddy
Timo Rehfeld
Paolo Remagnino
Konstantinos Rematas
Edoardo Remelli
Dongwei Ren
Haibing Ren
Jian Ren
Jimmy Ren
Mengye Ren
Weihong Ren
Wenqi Ren
Zhile Ren
Zhongzheng Ren

Zhou Ren
Vijay Rengarajan
Md A. Reza
Farzaneh Rezaeianaran
Hamed R. Tavakoli
Nicholas Rhinehart
Helge Rhodin
Elisa Ricci
Alexander Richard
Eitan Richardson
Elad Richardson
Christian Richardt
Stephan Richter
Gernot Riegler
Daniel Ritchie
Tobias Ritschel
Samuel Rivera
Yong Man Ro
Richard Roberts
Joseph Robinson
Ignacio Rocco
Mrigank Rochan
Emanuele Rodolà
Mikel D. Rodriguez
Giorgio Roffo
Grégory Rogez
Gemma Roig
Javier Romero
Xuejian Rong
Yu Rong
Amir Rosenfeld
Bodo Rosenhahn
Guy Rosman
Arun Ross
Paolo Rota
Peter M. Roth
Anastasios Roussos
Anirban Roy
Sebastien Roy
Aruni RoyChowdhury
Artem Rozantsev
Ognjen Rudovic
Daniel Rueckert
Adria Ruiz
Javier Ruiz-del-solar
Christian Rupprecht

Chris Russell
Dan Ruta
Jongbin Ryu
Ömer Sümer
Alexandre Sablayrolles
Faraz Saeedan
Ryusuke Sagawa
Christos Sagonas
Tonmoy Saikia
Hideo Saito
Kuniaki Saito
Shunsuke Saito
Shunta Saito
Ken Sakurada
Joaquin Salas
Fatemeh Sadat Saleh
Mahdi Saleh
Pouya Samangouei
Leo Sampaio
  Ferraz Ribeiro
Artsiom Olegovich
  Sanakoyeu
Enrique Sanchez
Patsorn Sangkloy
Anush Sankaran
Aswin Sankaranarayanan
Swami Sankaranarayanan
Rodrigo Santa Cruz
Amartya Sanyal
Archana Sapkota
Nikolaos Sarafianos
Jun Sato
Shin'ichi Satoh
Hosnieh Sattar
Arman Savran
Manolis Savva
Alexander Sax
Hanno Scharr
Simone Schaub-Meyer
Konrad Schindler
Dmitrij Schlesinger
Uwe Schmidt
Dirk Schnieders
Björn Schuller
Samuel Schulter
Idan Schwartz

William Robson Schwartz
Alex Schwing
Sinisa Segvic
Lorenzo Seidenari
Pradeep Sen
Ozan Sener
Soumyadip Sengupta
Arda Senocak
Mojtaba Seyedhosseini
Shishir Shah
Shital Shah
Sohil Atul Shah
Tamar Rott Shaham
Huasong Shan
Qi Shan
Shiguang Shan
Jing Shao
Roman Shapovalov
Gaurav Sharma
Vivek Sharma
Viktoriia Sharmanska
Dongyu She
Sumit Shekhar
Evan Shelhamer
Chengyao Shen
Chunhua Shen
Falong Shen
Jie Shen
Li Shen
Liyue Shen
Shuhan Shen
Tianwei Shen
Wei Shen
William B. Shen
Yantao Shen
Ying Shen
Yiru Shen
Yujun Shen
Yuming Shen
Zhiqiang Shen
Ziyi Shen
Lu Sheng
Yu Sheng
Rakshith Shetty
Baoguang Shi
Guangming Shi

Hailin Shi
Miaojing Shi
Yemin Shi
Zhenmei Shi
Zhiyuan Shi
Kevin Jonathan Shih
Shiliang Shiliang
Hyunjung Shim
Atsushi Shimada
Nobutaka Shimada
Daeyun Shin
Young Min Shin
Koichi Shinoda
Konstantin Shmelkov
Michael Zheng Shou
Abhinav Shrivastava
Tianmin Shu
Zhixin Shu
Hong-Han Shuai
Pushkar Shukla
Christian Siagian
Mennatullah M. Siam
Kaleem Siddiqi
Karan Sikka
Jae-Young Sim
Christian Simon
Martin Simonovsky
Dheeraj Singaraju
Bharat Singh
Gurkirt Singh
Krishna Kumar Singh
Maneesh Kumar Singh
Richa Singh
Saurabh Singh
Suriya Singh
Vikas Singh
Sudipta N. Sinha
Vincent Sitzmann
Josef Sivic
Gregory Slabaugh
Miroslava Slavcheva
Ron Slossberg
Brandon Smith
Kevin Smith
Vladimir Smutny
Noah Snavely

Roger
    D. Soberanis-Mukul
Kihyuk Sohn
Francesco Solera
Eric Sommerlade
Sanghyun Son
Byung Cheol Song
Chunfeng Song
Dongjin Song
Jiaming Song
Jie Song
Jifei Song
Jingkuan Song
Mingli Song
Shiyu Song
Shuran Song
Xiao Song
Yafei Song
Yale Song
Yang Song
Yi-Zhe Song
Yibing Song
Humberto Sossa
Cesar de Souza
Adrian Spurr
Srinath Sridhar
Suraj Srinivas
Pratul P. Srinivasan
Anuj Srivastava
Tania Stathaki
Christopher Stauffer
Simon Stent
Rainer Stiefelhagen
Pierre Stock
Julian Straub
Jonathan C. Stroud
Joerg Stueckler
Jan Stuehmer
David Stutz
Chi Su
Hang Su
Jong-Chyi Su
Shuochen Su
Yu-Chuan Su
Ramanathan Subramanian
Yusuke Sugano

Subeesh Vasu
Mayank Vatsa
David Vazquez
Javier Vazquez-Corral
Ashok Veeraraghavan
Erik Velasco-Salido
Raviteja Vemulapalli
Jonathan Ventura
Manisha Verma
Roberto Vezzani
Ruben Villegas
Minh Vo
MinhDuc Vo
Nam Vo
Michele Volpi
Riccardo Volpi
Carl Vondrick
Konstantinos Vougioukas
Tuan-Hung Vu
Sven Wachsmuth
Neal Wadhwa
Catherine Wah
Jacob C. Walker
Thomas S. A. Wallis
Chengde Wan
Jun Wan
Liang Wan
Renjie Wan
Baoyuan Wang
Boyu Wang
Cheng Wang
Chu Wang
Chuan Wang
Chunyu Wang
Dequan Wang
Di Wang
Dilin Wang
Dong Wang
Fang Wang
Guanzhi Wang
Guoyin Wang
Hanzi Wang
Hao Wang
He Wang
Heng Wang
Hongcheng Wang

Hongxing Wang
Hua Wang
Jian Wang
Jingbo Wang
Jinglu Wang
Jingya Wang
Jinjun Wang
Jinqiao Wang
Jue Wang
Ke Wang
Keze Wang
Le Wang
Lei Wang
Lezi Wang
Li Wang
Liang Wang
Lijun Wang
Limin Wang
Linwei Wang
Lizhi Wang
Mengjiao Wang
Mingzhe Wang
Minsi Wang
Naiyan Wang
Nannan Wang
Ning Wang
Oliver Wang
Pei Wang
Peng Wang
Pichao Wang
Qi Wang
Qian Wang
Qiaosong Wang
Qifei Wang
Qilong Wang
Qing Wang
Qingzhong Wang
Quan Wang
Rui Wang
Ruiping Wang
Ruixing Wang
Shangfei Wang
Shenlong Wang
Shiyao Wang
Shuhui Wang
Song Wang

Tao Wang
Tianlu Wang
Tiantian Wang
Ting-chun Wang
Tingwu Wang
Wei Wang
Weiyue Wang
Wenguan Wang
Wenlin Wang
Wenqi Wang
Xiang Wang
Xiaobo Wang
Xiaofang Wang
Xiaoling Wang
Xiaolong Wang
Xiaosong Wang
Xiaoyu Wang
Xin Eric Wang
Xinchao Wang
Xinggang Wang
Xintao Wang
Yali Wang
Yan Wang
Yang Wang
Yangang Wang
Yaxing Wang
Yi Wang
Yida Wang
Yilin Wang
Yiming Wang
Yisen Wang
Yongtao Wang
Yu-Xiong Wang
Yue Wang
Yujiang Wang
Yunbo Wang
Yunhe Wang
Zengmao Wang
Zhangyang Wang
Zhaowen Wang
Zhe Wang
Zhecan Wang
Zheng Wang
Zhixiang Wang
Zilei Wang
Jianqiao Wangni

Anne S. Wannenwetsch
Jan Dirk Wegner
Scott Wehrwein
Donglai Wei
Kaixuan Wei
Longhui Wei
Pengxu Wei
Ping Wei
Qi Wei
Shih-En Wei
Xing Wei
Yunchao Wei
Zijun Wei
Jerod Weinman
Michael Weinmann
Philippe Weinzaepfel
Yair Weiss
Bihan Wen
Longyin Wen
Wei Wen
Junwu Weng
Tsui-Wei Weng
Xinshuo Weng
Eric Wengrowski
Tomas Werner
Gordon Wetzstein
Tobias Weyand
Patrick Wieschollek
Maggie Wigness
Erik Wijmans
Richard Wildes
Olivia Wiles
Chris Williams
Williem Williem
Kyle Wilson
Calden Wloka
Nicolai Wojke
Christian Wolf
Yongkang Wong
Sanghyun Woo
Scott Workman
Baoyuan Wu
Bichen Wu
Chao-Yuan Wu
Huikai Wu
Jiajun Wu

Jialin Wu
Jiaxiang Wu
Jiqing Wu
Jonathan Wu
Lifang Wu
Qi Wu
Qiang Wu
Ruizheng Wu
Shangzhe Wu
Shun-Cheng Wu
Tianfu Wu
Wayne Wu
Wenxuan Wu
Xiao Wu
Xiaohe Wu
Xinxiao Wu
Yang Wu
Yi Wu
Yiming Wu
Ying Nian Wu
Yue Wu
Zheng Wu
Zhenyu Wu
Zhirong Wu
Zuxuan Wu
Stefanie Wuhrer
Jonas Wulff
Changqun Xia
Fangting Xia
Fei Xia
Gui-Song Xia
Lu Xia
Xide Xia
Yin Xia
Yingce Xia
Yongqin Xian
Lei Xiang
Shiming Xiang
Bin Xiao
Fanyi Xiao
Guobao Xiao
Huaxin Xiao
Taihong Xiao
Tete Xiao
Tong Xiao
Wang Xiao

Yang Xiao
Cihang Xie
Guosen Xie
Jianwen Xie
Lingxi Xie
Sirui Xie
Weidi Xie
Wenxuan Xie
Xiaohua Xie
Fuyong Xing
Jun Xing
Junliang Xing
Bo Xiong
Peixi Xiong
Yu Xiong
Yuanjun Xiong
Zhiwei Xiong
Chang Xu
Chenliang Xu
Dan Xu
Danfei Xu
Hang Xu
Hongteng Xu
Huijuan Xu
Jingwei Xu
Jun Xu
Kai Xu
Mengmeng Xu
Mingze Xu
Qianqian Xu
Ran Xu
Weijian Xu
Xiangyu Xu
Xiaogang Xu
Xing Xu
Xun Xu
Yanyu Xu
Yichao Xu
Yong Xu
Yongchao Xu
Yuanlu Xu
Zenglin Xu
Zheng Xu
Chuhui Xue
Jia Xue
Nan Xue

Tianfan Xue
Xiangyang Xue
Abhay Yadav
Yasushi Yagi
I. Zeki Yalniz
Kota Yamaguchi
Toshihiko Yamasaki
Takayoshi Yamashita
Junchi Yan
Ke Yan
Qingan Yan
Sijie Yan
Xinchen Yan
Yan Yan
Yichao Yan
Zhicheng Yan
Keiji Yanai
Bin Yang
Ceyuan Yang
Dawei Yang
Dong Yang
Fan Yang
Guandao Yang
Guorun Yang
Haichuan Yang
Hao Yang
Jianwei Yang
Jiaolong Yang
Jie Yang
Jing Yang
Kaiyu Yang
Linjie Yang
Meng Yang
Michael Ying Yang
Nan Yang
Shuai Yang
Shuo Yang
Tianyu Yang
Tien-Ju Yang
Tsun-Yi Yang
Wei Yang
Wenhan Yang
Xiao Yang
Xiaodong Yang
Xin Yang
Yan Yang

Yanchao Yang
Yee Hong Yang
Yezhou Yang
Zhenheng Yang
Anbang Yao
Angela Yao
Cong Yao
Jian Yao
Li Yao
Ting Yao
Yao Yao
Zhewei Yao
Chengxi Ye
Jianbo Ye
Keren Ye
Linwei Ye
Mang Ye
Mao Ye
Qi Ye
Qixiang Ye
Mei-Chen Yeh
Raymond Yeh
Yu-Ying Yeh
Sai-Kit Yeung
Serena Yeung
Kwang Moo Yi
Li Yi
Renjiao Yi
Alper Yilmaz
Junho Yim
Lijun Yin
Weidong Yin
Xi Yin
Zhichao Yin
Tatsuya Yokota
Ryo Yonetani
Donggeun Yoo
Jae Shin Yoon
Ju Hong Yoon
Sung-eui Yoon
Laurent Younes
Changqian Yu
Fisher Yu
Gang Yu
Jiahui Yu
Kaicheng Yu

Ke Yu
Lequan Yu
Ning Yu
Qian Yu
Ronald Yu
Ruichi Yu
Shoou-I Yu
Tao Yu
Tianshu Yu
Xiang Yu
Xin Yu
Xiyu Yu
Youngjae Yu
Yu Yu
Zhiding Yu
Chunfeng Yuan
Ganzhao Yuan
Jinwei Yuan
Lu Yuan
Quan Yuan
Shanxin Yuan
Tongtong Yuan
Wenjia Yuan
Ye Yuan
Yuan Yuan
Yuhui Yuan
Huanjing Yue
Xiangyu Yue
Ersin Yumer
Sergey Zagoruyko
Egor Zakharov
Amir Zamir
Andrei Zanfir
Mihai Zanfir
Pablo Zegers
Bernhard Zeisl
John S. Zelek
Niclas Zeller
Huayi Zeng
Jiabei Zeng
Wenjun Zeng
Yu Zeng
Xiaohua Zhai
Fangneng Zhan
Huangying Zhan
Kun Zhan

Xiaohang Zhan
Baochang Zhang
Bowen Zhang
Cecilia Zhang
Changqing Zhang
Chao Zhang
Chengquan Zhang
Chi Zhang
Chongyang Zhang
Dingwen Zhang
Dong Zhang
Feihu Zhang
Hang Zhang
Hanwang Zhang
Hao Zhang
He Zhang
Hongguang Zhang
Hua Zhang
Ji Zhang
Jianguo Zhang
Jianming Zhang
Jiawei Zhang
Jie Zhang
Jing Zhang
Juyong Zhang
Kai Zhang
Kaipeng Zhang
Ke Zhang
Le Zhang
Lei Zhang
Li Zhang
Lihe Zhang
Linguang Zhang
Lu Zhang
Mi Zhang
Mingda Zhang
Peng Zhang
Pingping Zhang
Qian Zhang
Qilin Zhang
Quanshi Zhang
Richard Zhang
Rui Zhang
Runze Zhang
Shengping Zhang
Shifeng Zhang

Shuai Zhang
Songyang Zhang
Tao Zhang
Ting Zhang
Tong Zhang
Wayne Zhang
Wei Zhang
Weizhong Zhang
Wenwei Zhang
Xiangyu Zhang
Xiaolin Zhang
Xiaopeng Zhang
Xiaoqin Zhang
Xiuming Zhang
Ya Zhang
Yang Zhang
Yimin Zhang
Yinda Zhang
Ying Zhang
Yongfei Zhang
Yu Zhang
Yulun Zhang
Yunhua Zhang
Yuting Zhang
Zhanpeng Zhang
Zhao Zhang
Zhaoxiang Zhang
Zhen Zhang
Zheng Zhang
Zhifei Zhang
Zhijin Zhang
Zhishuai Zhang
Ziming Zhang
Bo Zhao
Chen Zhao
Fang Zhao
Haiyu Zhao
Han Zhao
Hang Zhao
Hengshuang Zhao
Jian Zhao
Kai Zhao
Liang Zhao
Long Zhao
Qian Zhao
Qibin Zhao

Qijun Zhao
Rui Zhao
Shenglin Zhao
Sicheng Zhao
Tianyi Zhao
Wenda Zhao
Xiangyun Zhao
Xin Zhao
Yang Zhao
Yue Zhao
Zhichen Zhao
Zijing Zhao
Xiantong Zhen
Chuanxia Zheng
Feng Zheng
Haiyong Zheng
Jia Zheng
Kang Zheng
Shuai Kyle Zheng
Wei-Shi Zheng
Yinqiang Zheng
Zerong Zheng
Zhedong Zheng
Zilong Zheng
Bineng Zhong
Fangwei Zhong
Guangyu Zhong
Yiran Zhong
Yujie Zhong
Zhun Zhong
Chunluan Zhou
Huiyu Zhou
Jiahuan Zhou
Jun Zhou
Lei Zhou
Luowei Zhou
Luping Zhou
Mo Zhou
Ning Zhou
Pan Zhou
Peng Zhou
Qianyi Zhou
S. Kevin Zhou
Sanping Zhou
Wengang Zhou
Xingyi Zhou

Yanzhao Zhou	Wei Zhu	Christian Zimmermann
Yi Zhou	Xiangyu Zhu	Karel Zimmermann
Yin Zhou	Xinge Zhu	Larry Zitnick
Yipin Zhou	Xizhou Zhu	Mohammadreza
Yuyin Zhou	Yanjun Zhu	Zolfaghari
Zihan Zhou	Yi Zhu	Maria Zontak
Alex Zihao Zhu	Yixin Zhu	Daniel Zoran
Chenchen Zhu	Yizhe Zhu	Changqing Zou
Feng Zhu	Yousong Zhu	Chuhang Zou
Guangming Zhu	Zhe Zhu	Danping Zou
Ji Zhu	Zhen Zhu	Qi Zou
Jun-Yan Zhu	Zheng Zhu	Yang Zou
Lei Zhu	Zhenyao Zhu	Yuliang Zou
Linchao Zhu	Zhihui Zhu	Georgios Zoumpourlis
Rui Zhu	Zhuotun Zhu	Wangmeng Zuo
Shizhan Zhu	Bingbing Zhuang	Xinxin Zuo
Tyler Lixuan Zhu	Wei Zhuo	

Additional Reviewers

Victoria Fernandez	Jonathan P. Crall	Jaedong Hwang
Abrevaya	Kenan Dai	Andrey Ignatov
Maya Aghaei	Lucas Deecke	Muhammad
Allam Allam	Karan Desai	Abdullah Jamal
Christine	Prithviraj Dhar	Saumya Jetley
Allen-Blanchette	Jing Dong	Meiguang Jin
Nicolas Aziere	Wei Dong	Jeff Johnson
Assia Benbihi	Turan Kaan Elgin	Minsoo Kang
Neha Bhargava	Francis Engelmann	Saeed Khorram
Bharat Lal Bhatnagar	Erik Englesson	Mohammad Rami Koujan
Joanna Bitton	Fartash Faghri	Nilesh Kulkarni
Judy Borowski	Zicong Fan	Sudhakar Kumawat
Amine Bourki	Yang Fu	Abdelhak Lemkhenter
Romain Brégier	Risheek Garrepalli	Alexander Levine
Tali Brayer	Yifan Ge	Jiachen Li
Sebastian Bujwid	Marco Godi	Jing Li
Andrea Burns	Helmut Grabner	Jun Li
Yun-Hao Cao	Shuxuan Guo	Yi Li
Yuning Chai	Jianfeng He	Liang Liao
Xiaojun Chang	Zhezhi He	Ruochen Liao
Bo Chen	Samitha Herath	Tzu-Heng Lin
Shuo Chen	Chih-Hui Ho	Phillip Lippe
Zhixiang Chen	Yicong Hong	Bao-di Liu
Junsuk Choe	Vincent Tao Hu	Bo Liu
Hung-Kuo Chu	Julio Hurtado	Fangchen Liu

Hanxiao Liu
Hongyu Liu
Huidong Liu
Miao Liu
Xinxin Liu
Yongfei Liu
Yu-Lun Liu
Amir Livne
Tiange Luo
Wei Ma
Xiaoxuan Ma
Ioannis Marras
Georg Martius
Effrosyni Mavroudi
Tim Meinhardt
Givi Meishvili
Meng Meng
Zihang Meng
Zhongqi Miao
Gyeongsik Moon
Khoi Nguyen
Yung-Kyun Noh
Antonio Norelli
Jaeyoo Park
Alexander Pashevich
Mandela Patrick
Mary Phuong
Bingqiao Qian
Yu Qiao
Zhen Qiao
Sai Saketh Rambhatla
Aniket Roy
Amelie Royer
Parikshit Vishwas
    Sakurikar
Mark Sandler
Mert Bülent Sarıyıldız
Tanner Schmidt
Anshul B. Shah

Ketul Shah
Rajvi Shah
Hengcan Shi
Xiangxi Shi
Yujiao Shi
William A. P. Smith
Guoxian Song
Robin Strudel
Abby Stylianou
Xinwei Sun
Reuben Tan
Qingyi Tao
Kedar S. Tatwawadi
Anh Tuan Tran
Son Dinh Tran
Eleni Triantafillou
Aristeidis Tsitiridis
Md Zasim Uddin
Andrea Vedaldi
Evangelos Ververas
Vidit Vidit
Paul Voigtlaender
Bo Wan
Huanyu Wang
Huiyu Wang
Junqiu Wang
Pengxiao Wang
Tai Wang
Xinyao Wang
Tomoki Watanabe
Mark Weber
Xi Wei
Botong Wu
James Wu
Jiamin Wu
Rujie Wu
Yu Wu
Rongchang Xie
Wei Xiong

Yunyang Xiong
An Xu
Chi Xu
Yinghao Xu
Fei Xue
Tingyun Yan
Zike Yan
Chao Yang
Heran Yang
Ren Yang
Wenfei Yang
Xu Yang
Rajeev Yasarla
Shaokai Ye
Yufei Ye
Kun Yi
Haichao Yu
Hanchao Yu
Ruixuan Yu
Liangzhe Yuan
Chen-Lin Zhang
Fandong Zhang
Tianyi Zhang
Yang Zhang
Yiyi Zhang
Yongshun Zhang
Yu Zhang
Zhiwei Zhang
Jiaojiao Zhao
Yipu Zhao
Xingjian Zhen
Haizhong Zheng
Tiancheng Zhi
Chengju Zhou
Hao Zhou
Hao Zhu
Alexander Zimin

Contents – Part II

Diffraction Line Imaging

Mark Sheinin$^{(\boxtimes)}$, Dinesh N. Reddy, Matthew O'Toole,
and Srinivasa G. Narasimhan

Carnegie Mellon University, Pittsburgh, PA 15213, USA
marksheinin@gmail.com

Abstract. We present a novel computational imaging principle that combines diffractive optics with line (1D) sensing. When light passes through a diffraction grating, it disperses as a function of wavelength. We exploit this principle to recover 2D and even 3D positions from only line images. We derive a detailed image formation model and a learning-based algorithm for 2D position estimation. We show several extensions of our system to improve the accuracy of the 2D positioning and expand the effective field of view. We demonstrate our approach in two applications: (a) fast passive imaging of sparse light sources like street lamps, headlights at night and LED-based motion capture, and (b) structured light 3D scanning with line illumination *and* line sensing. Line imaging has several advantages over 2D sensors: high frame rate, high dynamic range, high fill-factor with additional on-chip computation, low cost beyond the visible spectrum, and high energy efficiency when used with line illumination. Thus, our system is able to achieve high-speed and high-accuracy 2D positioning of light sources and 3D scanning of scenes.

Keywords: Line sensor · Diffraction grating · 3D sensing · Motion capture · Computational imaging

1 Introduction

Artificial light sources are widely used in computer vision. Whether observed directly (Fig. 1[a–b]) or indirectly (Fig. 1[c]), artificial lights act as strong features to track [4,29], reconstruct [21,36], and interpret the scene and its objects. In this work, we rely on a key observation: these light sources occupy the image domain sparsely. But positioning sparse light sources with 2D sensors wastes pixel resources and limits the image acquisition rate.[1] Specifically, fast operation requires short exposures which leave most of the captured 2D image pixels completely dark (see Fig. 1[c]). Thus, most of the system's bandwidth is wasted.

---

[1] Event-based cameras known as dynamic vision sensors [8,14] output changes in intensity on a per-pixel basis, but the prototype sensors have limited spatial resolution.

Electronic supplementary material The online version of this chapter (https://doi.org/10.1007/978-3-030-58536-5_1) contains supplementary material, which is available to authorized users.

© Springer Nature Switzerland AG 2020
A. Vedaldi et al. (Eds.): ECCV 2020, LNCS 12347, pp. 1–16, 2020.
https://doi.org/10.1007/978-3-030-58536-5_1

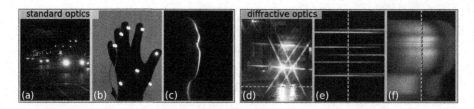

Fig. 1. When light sources or bright projections of light **(a-c)** are viewed through a transmissive diffraction grating, the incoming light is dispersed as a function of wavelength, creating streaks of light on the camera's sensor plane **(d-f)**. If a 1D color sensor (dotted green) or very few rows in a 2D sensor, intersects with these streaks, the measured colors can be used to efficiently determine the 2D spatial positions for each light source (or projected-line reflection) at high frame rates. (Color figure online)

Instead of using the full 2D sensor, we take a novel approach for saving bandwidth by imaging the scene using 1D (line) sensors.

Light passing through a diffraction grating is dispersed as a function of wavelength. When imaged by a 2D camera, the dispersed light manifests as colorful streaks. The micro-structure of the grating influences the shapes and number of streaks. And the brightness and appearance of the streaks depend on the 2D spatial locations of the sources in the image. Diffraction gratings are used to create artistic effects [16] (as in Fig. 1[d]), as well as in many imaging and scientific applications including spectroscopy [18,31], multi-spectral sensing [13,26], and rainbow particle velocimetry [35]. Unlike prior works, our method uses diffraction to encode the spatial position of scene light sources.

In this work, we introduce a novel class of imaging systems that use one or more diffraction gratings in conjunction with line (1D) sensors. We call this "Diffraction Line Imaging". We demonstrate how line sensors can yield 2D and even 3D information. Line sensors offer several advantages over 2D sensors: high frame rates, high dynamic range, and additional on-chip computation near the sensing area. Hence, this imaging approach results in fast 2D positioning and 3D reconstruction in many applications including night imaging, LED-based motion capture, and industrial product scanning.

We derive a detailed image formation model that maps a 3D point source to a 2D location on a virtual image plane. We then develop a learning-based algorithm to estimate the 2D locations of a sparse set of light sources or a line illumination projected onto a 3D scene. We numerically evaluate the uncertainty of the 2D positioning and the achieved field of view. To improve positioning accuracy significantly, the imaging system is extended to include multiple diffraction gratings and/or an additional cylindrical lens. Finally, we extend the approach to multiple line sensors (or multiple regions of interest in a 2D sensor) to increase the imaging system's field of view. Our approach can also be thought of as a variant of compressive sensing [2,3,32,34] with direct decoding of 2D positions, in place of computationally intensive and noise-prone decoding algorithms.

Our imaging systems are demonstrated in the following two applications.

Passive Imaging of Light Sources: Night scenes include a variety of light sources such as street lamps, vehicle headlamps, tail lights, turn signals, and bicycle lights (Fig. 1[a]). The flicker of sources can determine the AC phase of electrical circuits and even analyze power grids [27,28]. The glows around street lamps reveal the weather condition (fog, haze, rain) and visibility [19]. Finally, motion capture systems often attach light sources (LEDs) to estimate the object's motion in 3D (Fig. 1[b]) [22,24]. Our experiments show that we are able to estimate 2D light source positions from line images at high frame rates (up to 2220 fps in our camera). In contrast to previous works that use line sensors [22,24] or use spatio-temporal illumination to light a subject [15,25], our approach is based purely on passive observation of light sources without requiring special modulation or synchronization.

Structured Light Line Scanning: Structured light 3D scanning often projects or sweeps an illumination line across a scene [6]. A 2D camera observes the intersection of the line with the 3D scene. The 3D scene is then reconstructed by intersecting the pixel ray with the plane of illumination. We show how to accurately reconstruct the scene using a line light source and diffraction line imaging. Our image acquisition is mainly limited by the signal-to-noise ratio (SNR) (*i.e.*, exposure time) and not by the bandwidth. Interestingly, our system is the first instance of a structured light system with a 1D source and a 1D camera, since prior methods used either a 2D projector and/or a 2D camera [33]. Further, line illumination and imaging is significantly more energy efficient than 2D illumination and imaging [23]. Hence, bright light sources enable scan rates up to tens of thousands of lines per second, making this approach very useful in industrial/logistics scanning applications [5,11]. **See supplementary material for videos of results.**

2   Background on Diffraction Gratings

Our approach to light source positioning exploits *diffraction*: the wavelength-dependent optical phenomenon that causes light to bend (*i.e.*, diffract) when passing through and around obstacles and slits. A *diffraction grating* is an optical element that produces diffraction patterns, useful in scientific applications such as spectroscopy [18,31], holography [17], and hyperspectral imaging [13,26].

Figure 2(Left) shows a grating consisting of a periodic structure repeated every d microns. In a *transmissive* grating [17], incident light diffracts according to:

$$d[\sin(\theta_m) - \sin(\theta_i)] = m\lambda, \tag{1}$$

where θ_m is the angle of the mth diffraction order, θ_i is the angle of incident light, and λ is the wavelength in microns.[2] In Eq. (1), $m = 0$ corresponds to the zeroth-order component which passes unbent through the grating, while each

---

[2] Eq. (1) assumes a collimated incident beam and an identical index of refraction for the medium on both sides of the grating.

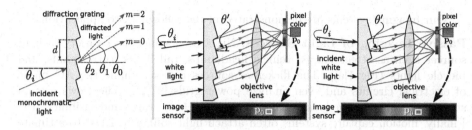

Fig. 2. Diffraction-based positioning. **Left:** A monochromatic collimated beam passes through a diffraction grating. The grating diffracts the beam yielding new beam directions according to Eq. (1). **Middle:** Per diffraction order m, each wavelength is diffracted in a different spatial direction. When imaged by a camera, this results in a horizontal rainbow pattern on the image plane. Here, pixel $\mathbf{p_0}$ measures the energy of some wavelength in the green range. **Right:** Shifting the incident angle θ_i results in a spatial shift of the spectral pattern on the image plane. Each θ_i maps a unique wavelength to $\mathbf{p_0}$. Thus, the color value at $\mathbf{p_0}$ provides information about θ_i. (Color figure online)

wavelength disperses into a unique angle for $m \neq 0$, producing the characteristic rainbow color streaks (see Fig. 1).

For a *fixed* θ_i, a camera imaging the exiting light maps each wavelength onto unique sensor position. This mapping is the basis for spectroscopy, whose purpose is the spectral analysis of the incident light. In contrast, we propose to *invert* this process, using color to efficiently and precisely recover the incident direction of light θ_i for unknown light sources.

Now, consider light exiting the diffraction grating at a fixed angle θ' belonging to the first diffraction order (*i.e.*, $m = 1$). Suppose that on the camera image plane, light exiting at θ' is focused at some camera pixel $\mathbf{p_0}$, as illustrated in Fig. 2(Middle). Then, from Eq. (1), the color (wavelength) measured at $\mathbf{p_0}$ is:

$$\lambda(\theta_i) = d[\sin(\theta') - \sin(\theta_i)], \qquad (2)$$

and depends on the incident light direction θ_i. Thus, for a *fixed* θ', the measured color at $\mathbf{p_0}$ indicates the direction of incident light.

3   Diffraction-Based 2D Positioning

The diffraction equation (Eq. (2)) has two key properties that enable computing the direction of scene light sources. First, the function $\lambda(\theta_i)$ is an injective (*i.e.*, one-to-one) function across the domain $\theta_i \in [-pi/2, \theta']$. The inverse of $\lambda(\theta_i)$ is therefore well-defined and given by

$$\theta_i(\lambda) = \arcsin\left(\sin(\theta') - \frac{\lambda}{d}\right). \qquad (3)$$

Second, Eq. (2) and its inverse do not depend on the intensity of light. This property makes our imaging system effectively invariant to the incident light intensity

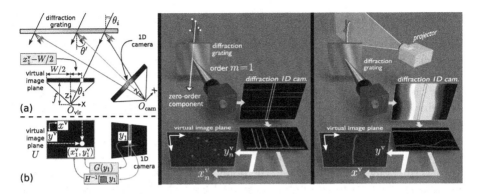

Fig. 3. Positioning system schematic. The first diffraction order is imaged by a RGB line scan camera, yielding 2D light source positions based on the measured color. **Left**: System schematic with (**a**) Top-view and (**b**) Front-view. **Middle**: Sparse light-sources yield intensity spikes on the line scan camera (green rectangle). Spike location along the vertical line scan camera encodes the vertical coordinate in the virtual image plane, while the spike color encodes the horizontal coordinate. **Right**: Line projection 3D scanning. The projected line yields a piece-wise continuous signal in the 1D camera. (Color figure online)

and, more specifically, its spectral characteristics (see derivation in Sect. 3.1). Note that this assumes no under- and over-saturated pixel measurements.

A basic schematic of our optical system is shown in Fig. 3(Left). The system consists of a diffraction grating and a color line scan camera. The line scan camera is positioned vertically, containing pixels along the Y-direction (in the camera's reference frame) denoted by the single coordinate y. We use the terms 'line scan camera' and '1D camera' interchangeably. For ease of explanation, we sometimes refer to the line scan camera's image plane as the full 2D image, had the camera been equipped with a standard 2D sensor.

The proposed system can then track both 2D spatial coordinates. For each point source, the vertical source coordinate is trivially measured by the rainbow's streak position along the 1D vertical sensor (*i.e.*, y_1 in Fig. 3[Left]b). Computing the source's horizontal coordinate amounts to computing its θ_i angle, which involves three simple steps: (1) measure the response in RGB space, (2) compute the dominant wavelength λ that produces the RGB signal, and (3) evaluate the inverse function $\theta_i(\lambda)$. Note that this procedure requires the light sources to be sufficiently broadband, such that the RGB measurement is non-zero for the given incident angle θ_i and the corresponding wavelength $\lambda(\theta_i)$.

As mentioned in the introduction, we tackle two types of imaging regimes. (a) *Sparse light sources*: scenes having a sparse set of point sources distributed across the image plane (Fig. 3[Middle]), and (b) *Structured light line scanning*: scenes illuminated by a vertically projected line, which yields a vertical curve on the image plane (Fig. 3[Right]). The projected line (*i.e.*, plane in space) in (b) can either be swept across a static object by a projector, or can have a fixed direction while the object moves through the plane (*e.g.*, by placing the object on a conveyor belt or turn table).

3.1   Image Formation Model

We derive the model which connects the projection of light from 3D positions in space onto a 2D virtual camera image plane. The line scan camera, modeled as a pinhole, is set to image the first-order diffracted light from the scene. Let O_{cam} denote the line scan camera's position and orientation. As shown in Fig. 3(Left), we define a virtual image plane U with pixel coordinates $\mathbf{p}^{\mathrm{v}} = (x^{\mathrm{v}}, y^{\mathrm{v}})$, that belong to a virtual 2D camera positioned at O_{vir}.

For simplicity, we begin by describing the model for a single point source indexed by $n = 1$, having homogeneous world coordinates $\mathbf{w}_1 \equiv [X, Y, Z, 1]^T$. On the line scan camera's image plane, the point source creates a holographic image in the shape of a *horizontal* rainbow line (see Figs. 1, 4). The camera's line sensor is orthogonal to the rainbow line, intersecting it over a few pixels centered at y_1. The geometric relationship between y_1 and virtual image coordinate y_1^{v} is:

$$y_1^{\mathrm{v}} \equiv G(y_1). \tag{4}$$

Neglecting distortions that arise from light entering at highly oblique angles [10], the rainbow line's y_1 coordinate is given by standard projective geometry:

$$\gamma \begin{pmatrix} y_1 \\ 1 \end{pmatrix} = \begin{pmatrix} 0\ 1\ 0 \\ 0\ 0\ 1 \end{pmatrix} \mathbf{P} \mathbf{w}_1, \tag{5}$$

where $\mathbf{P}$ is the camera's projection matrix and γ is an arbitrary scale factor [9]. In turn, since the Y-axes of both the virtual and line sensors are identical, G can be approximated by an affine transformation.

As seen in Fig. 3(a), the angle of the incident light with respect to the grating can be expressed as:

$$\theta_i(x_1^{\mathrm{v}}) = \arctan([x_1^{\mathrm{v}} - W/2]/f), \tag{6}$$

where f the virtual camera's focal length in pixel units and W is the virtual image plane width. Combining this with Eq. (3) yields:

$$\lambda(x_1^{\mathrm{v}}) = d\left(\sin(\theta') - \sin[\arctan([x_1^{\mathrm{v}} - W/2]/f)]\right). \tag{7}$$

The RGB intensity measured by the camera is modeled as:

$$\mathbf{I}_\sigma(y_1) = T \mathbf{c}_\sigma[\lambda(x_1^{\mathrm{v}})]\ s[\lambda(x_1^{\mathrm{v}})], \tag{8}$$

where T is the exposure time, $\mathbf{c}_\sigma(\lambda)$ is the camera's spectral response function [grayscale/Joule] for every color channel $\sigma \in \{R, G, B\}$, and $s(\lambda)$ represents the source's spectral radiant flux [Joule/sec] falling on y_1. In Eq. (8) we assumed that each pixel y integrates a very narrow band of wavelengths due to its the very narrow field of view.

Normalizing the 3x1 vector $\mathbf{I}_\sigma(y_1)$ by its L^2 norm removes the dependence on exposure time and source spectral flux:

$$\bar{\mathbf{I}}_\sigma(y_1) = \frac{\mathbf{I}_\sigma(y_1)}{||\mathbf{I}_\sigma(y_1)||_2} = \frac{\mathbf{c}_\sigma[\lambda(\theta', x_1^{\mathrm{v}})]}{||\mathbf{c}_\sigma[\lambda(\theta', x_1^{\mathrm{v}})]||_2} \equiv H_\sigma(x_1^{\mathrm{v}}). \tag{9}$$

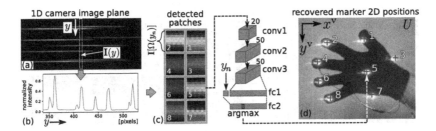

Fig. 4. Recovering sparse point sources. (a) Rainbow streaks incident on the line scan camera image plane. (b) The vertical rainbow locations along the line scan camera y_n are computed by detecting peaks in the averaged grayscale intensity along y. (c) For each detected peak y_n, a small 8×9 patch is extracted within sensor's narrow region of interest (ROI). (d) Each patch and its corresponding y_n is processed by a CNN to obtain x_n^{v}. Coordinate y_n^{v} is computed directly from y_n using Eq. (4). The 2D glove image from the helper camera is shown for reference only.

Finally, x_1^{v} is given by inverting Eq. (9):

$$x_1^{\mathrm{v}} = H^{-1}[\bar{\mathbf{I}}_\sigma(y_1)]. \tag{10}$$

The model in Eqs. (6–10) assumes that only a single source is predominately projected onto every line-camera pixel. Recovering x_1^{v} is described next.

3.2 Learning to Recover Horizontal Coordinates

Computing H^{-1} in practice requires accounting for various factors—sources not truly at infinity, sources having a small (non-point) projected surface area, sensor-diffraction grating geometry, the line scan camera's radiometric response, sensor saturation, and more. And, due to camera lens distortions, sensor-diffraction grating misalignment, and deviations to Eq. (1) due to incident light in oblique angles, H^{-1} generally depends on y_1 as well.[3] These factors make computing H^{-1} directly a laborious task. Instead, we adopt a learning-based approach.

We train a neural network to approximate H^{-1}, denoted by $\hat{H}^{-1}$. The network receives sensor color measurements in RGB space along with y, termed RGBy, and outputs the x^{v} coordinates. We tailored two different network architectures for our two recovery regimes: sparse points and vertical line projection.

Recovering Sparse Point Sources. For sparse sources, we train a convolutional neural network (CNN). The network, denoted by H_{point}^{-1}, maps RGBy values from a small patch into a predefined discrete set of possible x^{v} coordinates (see Fig. 4). Our network consists of three convolutional layers (channel widths 20, 50, 50 respectively) followed by two fully connected layers; see Fig. 4. Consider a scene with N point sources, indexed by $n = 1, 2, \ldots, N$. Let $\mathbf{I}(y)$

---

[3] Equation (1) strictly holds when the grating groves are perpendicular to the incidence plane. The incidence plane is the plane containing the beam direction and the normal to diffraction grating plane.

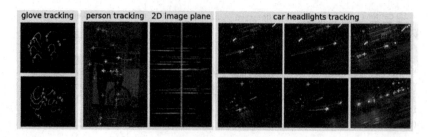

Fig. 5. Detection and tracking of light sources on a fast moving glove fitted with 8 LEDs, a suit fitted with 18 LEDs and headlamps of vehicles at multiple road intersections. The low-exposure background image from the helper camera is only shown to aid visualization. See supplementary material for videos and tracker details. (Color figure online)

denote the $8 \times Q$ color image from our line scan camera. Here, 8 denotes the number of image columns, since line scan cameras can be a few pixels wide in the horizontal direction as well (*e.g.*, due to RGB Bayer color filter).

The first step is to identify the rainbow line positions from peaks in $\mathbf{I}(y)$. Then, for every detected rainbow with coordinate y_n, we provide the CNN with a normalized small 8×9 image patch $\Omega(y_n)$ vertically centered at y_n, and concatenate the coordinate y_n with the input to the first fully connected layer.[4] The network outputs a $W \times 1$ vector $\mathbf{s}_n$ with scores for every possible horizontal virtual image location x^{v}. Then, x_n^{v} are recovered as:

$$x_n^{\mathrm{v}} = \arg\max(S[\mathbf{s}_n]), \tag{11}$$

where S is a softmax function.

The training dataset consists of ground truth 2D coordinates captured with a 2D helper camera, line scan camera coordinates y_m, and the corresponding image patches $\bar{\mathbf{I}}[\Omega(y_m)]$:

$$\left\{x_m^{\mathrm{v,GT}}\right\}_{m=1}^{M} \longleftrightarrow \left\{\bar{\mathbf{I}}[\Omega(y_m)], y_m\right\}_{m=1}^{M}, \tag{12}$$

where M is the number of training examples. The training loss is given by:

$$L = \frac{1}{M} \sum_{m=1}^{M} \mathrm{BCE}(S[\mathbf{s}_m], D[x_m^{\mathrm{v,GT}}, \sigma]), \tag{13}$$

where $BCE(\cdot, \cdot)$ is the Binary Cross Entropy function and function $D(\cdot)$ generates a Gaussian probability distribution with mean $x_m^{\mathrm{v,GT}}$ and standard deviation σ. Intuitively, the Binary Cross Entropy function drives the output distribution $S[\mathbf{s}_m]$ to match a narrow Gaussian centered at $x_m^{\mathrm{v,GT}}$. Using a Gaussian instead of a Delta function for $D(\cdot)$ provides a tolerance for small deviation due to image noise. See additional results in Fig. 5.

---

[4] Normalizing a 8×9 consists of dividing all RGB pixel values by a scalar. For example, for L^∞, $\bar{\mathbf{I}}[\Omega(y_m)] = \mathbf{I}[\Omega(y_m)]/\max_{\mathrm{RGB}}$, where $\max_{\mathrm{RGB}}$ is the maximum value across all patch pixels and color channels.

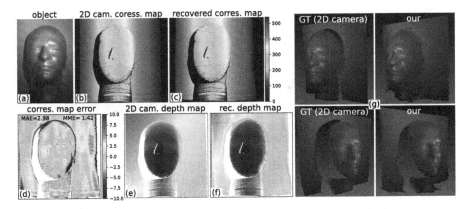

Fig. 6. Structured light with a 1D sensor and a 1D illumination. An object **(a)** is scanned using line projection. The recovered lines using a 2D camera (estimated ground truth) **(b)** and our method **(c)** are visualized with a correspondence map, in which each color indicates the projected column index. **(d)** Correspondence map error. Depth maps recovered using a 2D camera **(e)** and our method **(f)**. Displayed range 50.6 mm to 71.1 mm. **(g)** Recovered dense point clouds. We believe that this is the first instance of structured light scanning with both a 1D sensor and 1D camera. (Color figure online)

Recovering a Vertically Projected Line. As detailed in Sect. 7, 3D reconstruction relies on an aggregate of many continuous measurements through interpolation. Thus, here the network implements regression. Let the 3×1 RGB vector $\mathbf{u}_y$ denote the mean color of patch $\Omega(y)$. Namely, $\mathbf{u}_y$ is computed by averaging $\Omega(y)$ over the spatial domain.

The network H_{line}^{-1} is fed with a normalized $\mathbf{u}_y$ along with y, and outputs a continuous scalar approximation for x^{v}:

$$x^{\text{v}} = \begin{cases} H_{\text{line}}^{-1}\left(\frac{\mathbf{u}_y}{\|\mathbf{u}_y\|_2}, y\right), & \text{if } \mathbf{u}_y \in \mathcal{A} \\ \text{none}, & \text{otherwise}, \end{cases} \tag{14}$$

where $\mathcal{A}$ is a subspace of 'valid' RGB input values such that

$$\mathcal{A} = \{\mathbf{u}_y : \|\mathbf{u}_y\|_{\inf} > t, \ \|\mathbf{u}_y\|_0 > 1\},$$

where t is a predefined intensity threshold. Subspace $\mathcal{A}$ excludes low intensity (thus noisy) measurements as well as wavelengths that map to only a single color channel and thus can't be distinguished after normalization. The latter occurs at the edges of the rainbow streak (*i.e.*, deep reds and blues).

The network consists of four fully connected layers of size 300, and is trained using the Huber loss [12]. The training set is $\left\{x_m^{\text{v,GT}}\right\}_{m=1}^{M} \longleftrightarrow \{\mathbf{u}_m, y_m\}_{m=1}^{M}$, and was obtained by scanning a vertical line on a white wall. See Sect. 6 for more calibration details and Fig. 6 for example result.

4   Expanding the FOV Using Multiple ROIs

Our original prototype's readout was set to a single ROI consisting of a few columns (rows in a 90° rotated camera) at the center of the camera's image

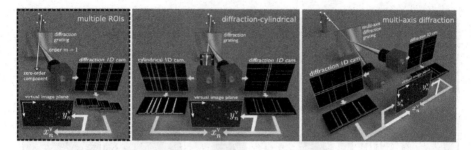

Fig. 7. Left: Using a 2D camera with multiple ROIs (three here) extends the system's FOV by 'catching' streaks outside the central ROI's domain. **Middle**: The additional left camera images the zeroth-order diffracted light through a cylindrical lens which focuses the sources onto horizontal white streaks. The intersection of the white streaks with the left line sensor yields additional horizontal point coordinate measurements which improve source positioning. **Right**: A multi-axis diffraction grating additionally yields horizontal rainbow streaks. Capturing these streaks using an additional line sensor mitigates the vertically-overlapping sources ambiguity.

plane. However, reading multiple ROIs from different spatial locations on the sensor, as shown in Fig. 7(Left), increases the horizontal FOV without sacrificing acquisition speed too much. This is because multiple ROIs can catch rainbow streaks whose visible spectrum does not intersect with the center column. Multiple ROIs also reduce recovery uncertainty since they may increase the signal per point, when the point's rainbow streak has sufficient signal over several ROIs.

Let $\mathbf{I}_\sigma^r$ denote the camera image from ROIs indexed $r = 1, 2, .., R$. For sparse points, we concatenate the R patches $\mathbf{I}_\sigma^r[\Omega(y_n)]$ from y_n to form an extended $8 \times (9R)$ color patch and feed it to the network along with y_n. For vertical line projection, we similarly concatenate the RGB measurements $\mathbf{u}_y^r$ and feed the resulting $(3R+1) \times 1$ vector to the network. As in Eq. (14), only 'valid' y's are considered, where now a measurement at y is valid if any one of the R terms $\mathbf{u}_y^r$ is valid. To preserve the network's invariance to object albedo, we augment each individual $\mathbf{u}_y^r$ during training (before concatenation) by a random scale factor, followed by adding simulated sensor noise. Figure 6 shows a result where $R = 5$. See supplementary for FOV analysis.

5 Reducing Sparse Point Uncertainty

The system described so far has two limitations: (1) uncertainty in the horizontal coordinates is typically higher than the vertical one, and (2) ambiguities in position estimates for certain source alignments (vertically overlapping sources). Here we describe two hardware variations that mitigate these limitations, for the sparse point case: (1) using two line sensors with a diffraction grating and a cylindrical lens, respectively (Sect. 5.1), and (2) using two line sensors with a double-axis diffraction grating (Sect. 5.2).

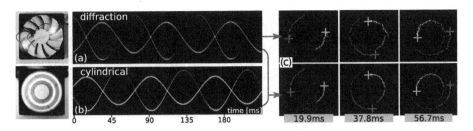

Fig. 8. High speed operation. A fan is mounted with two LEDs at different radii and is rotated at 780 RPM. A diffraction-cylindrical system (Sect. 5.1) images the fan at 2220 FPS. **(a)** Spectral space-time diagram for the diffraction camera where columns show the 1D signal (along the rows) at different times. **(b)** Space-time diagram for the additional cylindrical lens camera. **(c)** Recovered source positions and trajectories at three times within the first revolution. Top row shows results using only the diffraction signal. Bottom row shows the improvement by incorporating cylindrical measurements.

5.1 Horizontal Cylindrical Lens

To improve the accuracy of x^v, we place an additional *horizontal* line scan camera to image the zero-order component through a cylindrical lens (see Fig. 7[Middle]). The cylindrical lens focuses scene points onto vertical lines which intersect the horizontal line sensor. These white streaks yield precise measurement for coordinates x^v. However, unlike the rainbow streaks, the white streaks provide no information about y^v. Nevertheless, this data can improve 2D positioning.

Merging the recovered coordinates from the cylindrical lens sensor is fully detailed in the supplementary material. Here we describe the basic idea. Let $\tilde{x}_k^v$ denote the recovered cylindrical lens coordinates, indexed by $k = 1, 2, .., K$. For every recovered point (x_n^v, y_n^v), we compute the distance of its x_n^v coordinates to all $\tilde{x}_k^v$. If for any k, $\tilde{x}_k^v$ is very 'close' to x_n^v (*e.g.* below four pixels), we assume that measurement k originated from point n, and thus replace $x_n^v \leftarrow \tilde{x}_k^v$. Otherwise, we discard (x_n^v, y_n^v) as an inaccurate measurement. See Fig. 8 for example result.

5.2 Double-Axis Diffraction

The transmissive grating of Sect. 2 has spatial variation in only one axis, thus it diffracts light along said axis. A *double-axis* grating has spatial variation in both axes (*i.e.*, a star filter) and thus diffracts light both horizontally and vertically. Using a double-axis grating allows for a direct generalization of Sect. 3.

We replace the single-axis grating in Fig. 7(Left) with a double-axis grating and add an additional line sensor camera to image the *vertical* first-order diffraction component as well (see Fig. 7[Right]). The vertical diffraction component creates a vertical rainbow streak on the second line sensor camera (which now requires no tilting). Each line sensor now provides a separate measurement for (x^v, y^v). Merging these pair of measurements follows the same logic as in Sect. 5.1. Namely, if two points fall within the predefined distance on the virtual image plane, they are assumed to originate from the same scene point and are

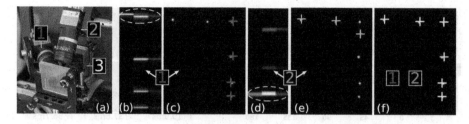

Fig. 9. A double-axis diffraction system with 3-ROIs **(a)** yields signals from both vertical (cam 1) **(b)**, and horizontal (cam 2) **(d)** diffraction. Recovered points from the vertical **(c)** and horizontal **(e)** diffraction superimposed on a 2D helper camera (ground truth, cam 3). In **(c)**, the three top points share the same y and thus not fully recovered. Similarly, in **(e)** the right vertical points share the same x and thus are not fully recovered. **(f)** Merging both measurements yields correct positions for all points.

merged by taking y^v from the vertical sensor, while taking x^v from the horizontal one. See the supplementary for details. See example result in Fig. 9.

6   Hardware and Calibration

Our prototype is shown in Fig. 9(a). The setup consists of three IDS UI-3070CP-C-HQ Rev.2 color cameras, which support up to 64 ROIs. Camera 1 and 2 capture the vertical and horizontal diffraction streaks, respectively. Camera 3 is the helper camera, used for gathering ground-truth training data and visualizations. Camera 3 is also used in Sect. 5.1 in conjunction with a cylindrical lens. At least two rows are required per ROIs to yield color measurements (due to Bayer filter). Camera 1 is rotated by 90° to obtain vertical 1D measurements. Cameras 1 and 2 were stripped of their built-in IR filter and mounted with Fujinon 1.5 MP 9 mm lenses. Camera 3 was mounted with a 8 mm M12 lens using an adapter. We used Thorlabs 50 mm 1200 grooves/mm transmission gratings (GT50-12). The double-axis diffraction system had two stacked gratings, where one is rotated 90° with respect to the other. Otherwise, a single grating was used. Our motion-capture suit and glove prototypes are fitted with 5 mm 12V white diffuse LEDs. A second glove was fitted with 20 mm 3.3V white LEDs. For 3D scanning, we used an InFocus IN124STa Short Throw DLP Projector.

Calibration: Calibration is done using camera 3, whose image plane serves as the virtual plane in Fig. 3. For sparse points, we simultaneously recorded about 40K samples of a single moving source using all cameras. For sources, we used three types of LEDs (two clear one diffuse). This procedure yields the training set of Eq. (12), and is used to compute H^{-1} and G. For 3D scanning, we sweep and image a vertical line on a white wall. For each projected line, we compute its line equation $y^v = cx^v + d$ and use it to determine the ground truth disparity $x^{v,GT}(y) = [G(y) - d]/c$. Together, all lines yield about 700K RGBy/position samples. Helper-camera and projector intrinsic and extrinsic calibration is done using a checkerboard [9].

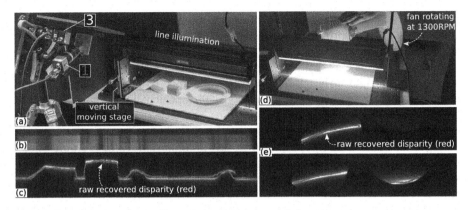

Fig. 10. Fast line-illumination scanning. **(a)** The experimental setup scans objects moving through the illuminated area, *e.g.*, for in-line inspection of objects on a conveyor belt. The bright source enables very fast scan speeds, up to 1743 scan lines per second. **(b)** The measured signal of scene (a) from the three ROIs. **(c)** The raw recovered disparity from (a), superimposed on the ground truth helper camera. **(d)** Fan rotating at 1300RPM. **(e)** Recovered disparity in two frames where the helper camera and our system temporally coincide. As shown in the supplementary video, the helper camera is 30x slower than our system and thus it is unable to capture the fan's motion.

7 Experimental Evaluation Details

Before using a neural network for H^{-1}, we tried various hand-crafted color-based methods (*e.g.*, mapping LAB or RGB values to position using optimization). The networks-based approach outperformed these methods. For sparse points, the mean absolute error (MAE) over the test set was 1.71 pixels with a 1.54 standard deviation. For line projection, MAE was 2.27 pixels with a 2.32 standard deviation. See supplementary for evaluation and network training details.

Point source detection and tracking is shown in Figs. 4, 5, and 8. Glove and person keypoints are accurately tracked for long durations even in the case of overlap in the diffraction readings. Figure 5(Right) shows tracking of headlights at multiple intersections. Observe that we are able to detect and track multiple light sources in the wild using a CNN trained on only three LEDs.

Figure 6 show 3D scanning using the setup shown in Fig. 3(Right). We used an off-the-shelf projector, and a 2D camera configured with five ROIs, each yielding a 8×2056 measurement. For each projected line, the algorithm yields up to 2056 continuous measurements, tracing that line in the virtual image. After imaging all lines, the measurements from all lines are used to interpolate the final correspondence map – a virtual camera image where each pixel is identified with one projected line index (or none). Then, 3D reconstruction follows from simple triangulation; see supplementary for more details.

In Fig. 6, the projector has limited contrast. Namely, when projecting a white vertical line, a significant amount of light was leaking to the 'dark' pixels outside the line. To compensate, we used longer exposures and averaged multiple frames per line. In Fig. 6, we averaged 55 frames of 50ms exposure each. Additionally, we captured a black projector image prior to scanning and subtract it from all subsequently measurements.

Inter-reflections may degrade reconstruction quality by yielding a mixture of signal from multiple surface points on the same line. To effectively reduce their effect, we use high frequency illumination [20]. Specifically, we split each projected line into three high-frequency patterns, and extract the direct component as in [20]. High-frequency patterns are used in Fig. 6.

Figure 10 shows our experimental fast line illumination scanner. Applications for this setup include scanning products on rapidly moving conveyor-belts, or scanning objects rotated by a turntable. The camera was configured to readout three 8×2056 ROIs at 1743 FPS with an exposure of $300\,\mu s$. The system could compute disparity for fast moving objects (e.g., fan rotating at 1300 RPM) captured under regular lighting conditions (with room and sunlight ambient light present). See supplementary for additional details.

8 Concluding Remarks

Diffraction line imaging is a novel computational imaging principle based on light diffraction and 1D sensing. Using the principle, we showed proof of concepts for applications like motion capture of uncontrolled and unmodulated LEDs, tracking car headlights, and 3D scanning using both 1D sensing and 1D illumination. Using line sensors significantly decreases bandwidth, which leads to speed. Speed is crucial for motion capture since it greatly eases tracking. In 3D scanning, speed is vital when scanning moving objects (e.g., industrial conveyor belt).

Our prototype mimicked a line sensor using multiple rows from a conventional 2D sensor, resulting in fast readout rates of up to 2220 FPS. Faster line sensors [7] could reach up to 45,000 FPS (x20 faster than our prototype) and improve light efficiency with large pixels up to 4×32 microns in size (x10 larger). Conversely, using 2D sensors with multiple ROIs gives smooth control over the speed vs. quality trade-off, namely more ROIs reduce speed but increase accuracy.

As with any sensor though, our system's performance depends on the available SNR. Using hyper-spectral sensors may improve position decoding by raising the discrimination between signals from adjacent wavelengths. For 3D scanning, bright broadband sources, such as supercontinuum/swept-frequency lasers can additionally increase SNR many folds [1,30]. Learning-based or dictionary-based approaches may improve reconstruction quality by extracting multiple vertically overlapping points, which yield a linear color mixture on the 1D line scan sensor. Finally, we believe that our approach is an important step towards achieving a simple high-speed and low-cost solution to light source positioning with potential applications from vision to robotics.

Acknowledgments. We thank A. Sankaranarayanan and V. Saragadam for help with building the hardware prototype and S. Panev and F. Moreno for neural network-related advice. We were supported in parts by NSF Grants IIS-1900821 and CCF-1730147 and DARPA REVEAL Contract HR0011-16-C-0025.

References

1. Alfano, R.R.: The Supercontinuum Laser Source. Springer, Heidelberg (1989). https://doi.org/10.1007/b106776
2. Antipa, N.: Diffusercam: lensless single-exposure 3D imaging. Optica **5**(1), 1–9 (2018)
3. Antipa, N., Oare, P., Bostan, E., Ng, R., Waller, L.: Video from stills: lensless imaging with rolling shutter. In: Proceedings of IEEE ICCP, pp. 1–8 (2019)
4. Chen, Y.L., Wu, B.F., Huang, H.Y., Fan, C.J.: A real-time vision system for night-time vehicle detection and traffic surveillance. IEEE Trans. Ind. Elect. **58**(5), 2030–2044 (2010)
5. Loadscan: load management solutions (2020). https://www.loadscan.com/
6. Curless, B., Levoy, M.: Better optical triangulation through spacetime analysis. In: Proceedings of IEEE ICCV, pp. 987–994 (1995)
7. Dlis2k: ultra configurable digital output (2020). http://dynamax-imaging.com/products/line-scan-product/dlis2k-2/
8. Gallego, G., et al.: Event-based vision: a survey (2019). arXiv preprint arXiv:1904.08405
9. Hartley, R., Zisserman, A.: Multiple View Geometry in Computer Vision. Cambridge University Press, Cambridge (2003)
10. Harvey, J.E., Vernold, C.L.: Description of diffraction grating behavior in direction cosine space. Appl. Opt. **37**(34), 8158–8159 (1998)
11. Hossain, F., PK, M.K., Yousuf, M.A.: Hardware design and implementation of adaptive canny edge detection algorithm. Int. J. Comput. Appl. **124**(9), 31–38 (2015)
12. Huber, P.J.: Robust estimation of a location parameter. In: Kotz, S., Johnson, N.L. (eds.) Breakthroughs in statistics. Springer Series in Statistics (Perspectives in Statistics), pp. 492–518. Springer, Heidelberg (1992). https://doi.org/10.1007/978-1-4612-4380-9_35
13. Jeon, D.S., et al.: Compact snapshot hyperspectral imaging with diffracted rotation. ACM TOG **38**(4), 117 (2019)
14. Kim, H., Leutenegger, S., Davison, A.J.: Real-time 3D reconstruction and 6-DoF tracking with an event camera. In: Leibe, B., Matas, J., Sebe, N., Welling, M. (eds.) ECCV 2016. LNCS, vol. 9910, pp. 349–364. Springer, Cham (2016). https://doi.org/10.1007/978-3-319-46466-4_21
15. Kim, J., Han, G., Lim, H., Izadi, S., Ghosh, A.: Thirdlight: Low-cost and high-speed 3D interaction using photosensor markers. In: Proceedingd of CVMP, p. 4. ACM (2017)
16. Liu, D., Geng, H., Liu, T., Klette, R.: Star-effect simulation for photography. Comput. Graph. **61**, 19–28 (2016)
17. Loewen, E.G., Popov, E.: Diffraction Gratings and Applications. CRC Press, Boca Raton (2018)
18. Nagaoka, H., Mishima, T.: A combination of a concave grating with a Lummer-Gehrcke plate or an echelon grating for examining fine structure of spectral lines. Astrophys. J. **57**, 92 (1923)
19. Narasimhan, S.G., Nayar, S.K.: Shedding light on the weather. In: Proceedings of IEEE CVPR, pp. 665–672 (2003)
20. Nayar, S.K., Krishnan, G., Grossberg, M.D., Raskar, R.: Fast separation of direct and global components of a scene using high frequency illumination. In: ACM SIGGRAPH, pp. 935–944 (2006)

21. Nelson, P., Churchill, W., Posner, I., Newman, P.: From dusk till dawn: localisation at night using artificial light sources. In: Proceedings of IEEE ICRA (2015)
22. Optotrak certus (2020). https://www.ndigital.com/msci/products/optotrak-certus/
23. O'Toole, M., Achar, S., Narasimhan, S.G., Kutulakos, K.N.: Homogeneous codes for energy-efficient illumination and imaging. ACM TOG **34**(4), 1–13 (2015)
24. Phase space inc. (2020). http://www.phasespace.com/
25. Raskar, R., et al.: Prakash: lighting aware motion capture using photosensing markers and multiplexed illuminators. ACM TOG **26**(3), 36 (2007)
26. Saragadam, V., Sankaranarayanan, A.C.: KRISM: Krylov subspace-based optical computing of hyperspectral images. ACM TOG **38**(5), 1–14 (2019)
27. Sheinin, M., Schechner, Y.Y., Kutulakos, K.N.: Computational imaging on the electric grid. In: Proceedings of IEEE CVPR, pp. 2363–2372 (2017)
28. Sheinin, M., Schechner, Y.Y., Kutulakos, K.N.: Rolling shutter imaging on the electric grid. In: Proceedings of IEEE ICCP, pp. 1–12 (2018)
29. Tamburo, R., et al.: Programmable automotive headlights. In: Fleet, D., Pajdla, T., Schiele, B., Tuytelaars, T. (eds.) ECCV 2014. LNCS, vol. 8692, pp. 750–765. Springer, Cham (2014). https://doi.org/10.1007/978-3-319-10593-2_49
30. Vasilyev, A.: The optoelectronic swept-frequency laser and its applications in ranging, three-dimensional imaging, and coherent beam combining of chirped-seed amplifiers. Ph.D. thesis, Caltech (2013)
31. Vogt, S.S., et al.: HIRES: the high-resolution echelle spectrometer on the Keck 10-m Telescope. In: Instrumentation in Astronomy VIII, vol. 2198, pp. 362–375. International Society for Optics and Photonics (1994)
32. Wang, J., Gupta, M., Sankaranarayanan, A.C.: Lisens-a scalable architecture for video compressive sensing. In: Proceedings of IEEE ICCP, pp. 1–9. IEEE (2015)
33. Wang, J., Sankaranarayanan, A.C., Gupta, M., Narasimhan, S.G.: Dual structured light 3D using a 1D sensor. In: Leibe, B., Matas, J., Sebe, N., Welling, M. (eds.) ECCV 2016. LNCS, vol. 9910, pp. 383–398. Springer, Cham (2016). https://doi.org/10.1007/978-3-319-46466-4_23
34. Weinberg, G., Katz, O.: 100,000 frames-per-second compressive imaging with a conventional rolling-shutter camera by random point-spread-function engineering. arXiv preprint arXiv:2004.09614 (2020)
35. Xiong, J., et al.: Rainbow particle imaging velocimetry for dense 3D fluid velocity imaging. ACM TOG **36**(4), 36 (2017)
36. Zhi, T., Pires, B.R., Hebert, M., Narasimhan, S.G.: Deep material-aware cross-spectral stereo matching. In: Proceedings of IEEE CVPR, pp. 1916–1925 (2018)

Transforming and Projecting Images into Class-Conditional Generative Networks

Minyoung Huh[1,2]($\boxtimes$), Richard Zhang[2], Jun-Yan Zhu[2], Sylvain Paris[2], and Aaron Hertzmann[2]

[1] MIT CSAIL, Cambridge, USA
minhuh@mit.edu
[2] Adobe Research, San Francisco, USA

Abstract. We present a method for projecting an input image into the space of a class-conditional generative neural network. We propose a method that optimizes for transformation to counteract the model biases in generative neural networks. Specifically, we demonstrate that one can solve for image translation, scale, and global color transformation, during the projection optimization to address the object-center bias and color bias of a Generative Adversarial Network. This projection process poses a difficult optimization problem, and purely gradient-based optimizations fail to find good solutions. We describe a hybrid optimization strategy that finds good projections by estimating transformations and class parameters. We show the effectiveness of our method on real images and further demonstrate how the corresponding projections lead to better editability of these images. The project page and the code is available at https://minyoungg.github.io/GAN-Transform-and-Project/.

1 Introduction

Deep generative models, particularly Generative Adversarial Networks (GANs) [24], can create a diverse set of realistic images, with a number of controls for transforming the output, e.g., [7,30,33,48]. However, most of these methods apply only to synthetic images that are generated by GANs in the first place. In many real-world cases, a user would like to edit their own image. One approach is to train a network for each separate image transformation. However, this would require a combinatorial explosion of training time and model parameters.

Instead, a user could "project" their image to the manifold of images produced by the GAN, by searching for an appropriate latent code [60]. Then, any transformations available within the GAN could be applied to the user's image. This could allow a powerful range of editing operations within a relatively compact representation. However, projection is a challenging problem. Previous methods have

---

M. Huh—Work started during an internship at Adobe Research.

Electronic supplementary material The online version of this chapter (https://doi.org/10.1007/978-3-030-58536-5_2) contains supplementary material, which is available to authorized users.

© Springer Nature Switzerland AG 2020
A. Vedaldi et al. (Eds.): ECCV 2020, LNCS 12347, pp. 17–34, 2020.
https://doi.org/10.1007/978-3-030-58536-5_2

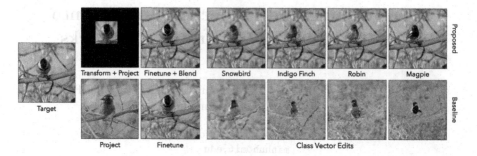

Fig. 1. Given a pre-trained BigGAN [9] and a target image (left), our method uses gradient-free BasinCMA to transform the image and find a latent vector to closely reconstruct the image. Our method (top) can better fit the input image, compared to the baseline (bottom), which does not model image transformation and uses gradient-based ADAM optimization. Finding an accurate solution to the inversion problem allows us to further fine-tune the model weights to match the target image without losing downstream editing capabilities. For example, our method allows for changing the class of the object (top row), compared to the baseline (bottom).

focused on class-specific models, for example, for objects [60], faces [10,46], or specific scenes such as bedrooms and churches [6,8]. With the challenges in both optimization and generative model's limited capacity, we wish to find a generic method that can fit *real* images from diverse categories into the same generative model.

This paper proposes the first method for projecting images into class-conditional models. In particular, we focus on BigGAN [9]. We address the main problems with these tasks, mainly, the challenges of optimization, object alignment, and class label estimation:

- To help avoid local minima during the optimization process, we systematically study choices of both gradient-based and gradient-free optimizers and show Covariance Matrix Adaptation (CMA) [26] to be more effective than standalone gradient-based optimizers, such as L-BFGS [41] and Adam [34].
- To better fit a real image into the latent space, we account for the model's center bias by simultaneously estimating both spatial image transformation (translation, scale, and color) and latent variable. Such a transformation can then be inverted back to the input image frame. Our simultaneous transformation and projection method largely expands the scope and diversity of the images that a GAN can reconstruct.
- Finally, we show that estimating and jointly optimizing the continuous embedding of the class variable leads to better projections. This ultimately leads to more expressive editing by harnessing the representation of the class-conditional generative model.

We evaluate our method against various baselines on projecting real images from ImageNet. We quantitatively and qualitatively demonstrate that it is crucial to simultaneously estimate the correct transformation during the projection step. Furthermore, we show that CMA, a non-parametric gradient-free optimization technique, significantly improves the robustness of the optimization and

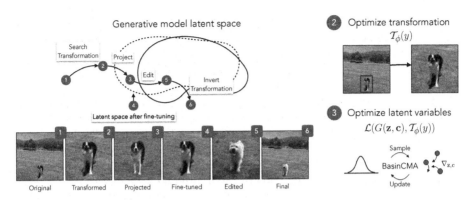

Fig. 2. Overview: Our method first searches for a transformation to apply to the input target image. We then solve for the latent vector that closely resembles the object in the target image, using our proposed optimization method, also referred to as "projection". The generative model can then be further fine-tuned to reconstruct the missing details that the original model could not generate. Finally, we can edit the image by altering the latent code or the class vector (e.g., changing the border collie to a west highland white terrier), and invert and blend the edited image back into the original image.

leads to better solutions. As shown in Fig. 1, our method allows us to fine-tune our model to recover the missing details without losing the editing capabilities of the generative model (Fig. 2).

2   Related Work

Image Editing with Generative Models.   Image editing tools allow a user to manipulate a photograph according to their goal while producing realistic visual content. Seminal work is often built on low-level visual properties, such as patch-based texture synthesis [5,18,19,28], gradient-domain image blending [47], and image matting with locally affine color model [38]. Different from previous hand-crafted low-level methods, several recent works [10,60] proposed to build editing tools based on a deep generative model, with the hope that a generative model can capture high-level information about the image manifold.

Many prior works have investigated using trained generative models as a tool to edit images [6,7,10,60]. The same image prior from deep generative models has also been used in face editing, image inpainting, colorization, and deblurring prior [3,25,46,50,56]. Unlike these works that focuses on single-class and fixated image, our method presents a new ways of embedding an image into a class-conditional generative model, which allows the same GAN to be applied to many more "in-the-wild" scenarios.

Inverting Networks.   Our work is closely related to methods for inverting pre-trained networks. Earlier work proposes to invert CNN classifiers and intermediate features for visualizing recognition networks [16,42–44]. More recently, researchers adopted the above methods to invert generative models. The common techniques include: (1) Optimization-based methods: they find the latent vector that can closely reconstruct the input image using gradient-based method (e.g.., ADAM, LBFGS) [2,10,11,40,49,56,60] or MCMC [52], (2) Encoder-based methods: they learn an encoder to directly predict the latent vector given a real image [10,13,14,17,46,60], (3) Hybrid methods [6,8,60]: they use the encoder to initialize the latent vector and then solve the optimization problem. Although the optimized latent vector roughly approximates the real input image, many important visual details are missing in the reconstruction [6]. To address the issue, GANPaint [6] generates residual features to adapt to the individual image. Image2StyleGAN [1] optimizes StyleGAN's intermediate representation rather than the input latent vector. Unfortunately, the above techniques still cannot handle images in many scenarios due to the limited model capacity [8], the lack of generalization ability [1], and their single-class assumption. As noted by prior work [1], the reconstruction quality severely degrades under simple image transformation, and translation has been found to cause most of the damage. Compared to prior work, we consider two new aspects in the reconstruction pipeline: image transformation and class vector. Together, these two aspects significantly expand the diversity of the images that we can reconstruct and edit.

3   Image Projection Methods

We aim to project an image into a class-conditional generative model (e.g., BigGAN [9]) for the purposes of downstream editing. We first introduce the basic objective function that we slowly build upon. Next, since BigGAN is an object-centric model for most classes, we infer an object mask from the input image and focus on fitting the pixels inside the mask.

Furthermore, to better fit our desired image into the generative model, we propose to optimize for various image transformation (scale, translation, and color) to be applied to the target image. Lastly, we explain how we optimize the aforementioned objective loss function.

3.1   Basic Loss Function

Class-Conditional Generative Model. A class-conditional generative network can synthesize an image $\hat{\mathbf{y}} \in \mathbb{R}^{H \times W \times 3}$, given a latent code $\mathbf{z} \in \mathbb{R}^{Z}$ that models intra-class variations and a one-hot class-conditioning vector $\tilde{\mathbf{c}} \in \Delta^{C}$ to choose over C classes. We focus on the 256×256 BigGAN model [9] specifically, where $Z = 128$ and $C = 1,000$ ImageNet classes.

The BigGAN architecture first maps the one-hot $\tilde{\mathbf{c}}$ into a continuous vector $\mathbf{c} \in \mathbb{R}^{128}$ with a linear layer $\mathbf{W} \in \mathbb{R}^{128 \times 1000}$, before injecting into the main network G_θ, with learned parameters θ.

$$\hat{\mathbf{y}} = G_\theta(\mathbf{z}, \mathbf{c}) = G_\theta(\mathbf{z}, \mathbf{W}\tilde{\mathbf{c}}). \tag{1}$$

Here, a choice must be made whether to optimize over the discrete $\tilde{\mathbf{c}}$ or continuous $\mathbf{c}$. As optimizing a discrete class vector is non-trivial, we optimize over the continuous embedding.

Optimization Setup. Given a target image $\mathbf{y}$, we would like to find a $\mathbf{z}^*$ and $\mathbf{c}^*$ that generates the image.

$$\mathbf{z}^*, \mathbf{c}^* = \arg\min_{\mathbf{z},\mathbf{c}} \mathcal{L}(G_\theta(\mathbf{z},\mathbf{c}), \mathbf{y}) \qquad \text{s.t. } C(\mathbf{z}) \leq C_{\max}. \qquad (2)$$

During training, the latent code is sampled from a multivariate Gaussian $\mathbf{z} \sim \mathcal{N}(\mathbf{0}, \mathbf{I})$. Interestingly, recent methods [9, 35] find that restricting the distribution at *test time* produces higher-quality samples. We follow this and constrain our search space to match the sampling distribution from Brock et al. [9]. Specifically, we use $C(\mathbf{z}) = ||\mathbf{z}||_\infty$ and $C_{\max} = 2$. During optimization, elements of $\mathbf{z}$ that fall outside the threshold are clamped to $+2$, if positive, or -2, if negative. Allowing larger values of $\mathbf{z}$ produces better fits but compromises editing ability.

Loss Function. The loss function $\mathcal{L}$ attempts to capture how close the approximate solution is to the target. A loss function that perfectly corresponds to human perceptual similarity is a longstanding open research problem [54], and evaluating the difference solely on a per-pixel basis leads to blurry results [58]. Distances in the feature space of a pre-trained CNN correspond more closely with human perception [15, 21, 31, 59]. We use the LPIPS metric [59], which calibrates a pre-trained model using human perceptual judgments. Here, we define our basic loss function, which combines per-pixel ℓ_1 and LPIPS.

$$\mathcal{L}_{\text{basic}}(\mathbf{y}, \hat{\mathbf{y}}) = \frac{1}{HW}||\hat{\mathbf{y}} - \mathbf{y}||_1 + \beta\mathcal{L}_{\text{LPIPS}}(\hat{\mathbf{y}}, \mathbf{y}). \qquad (3)$$

In preliminary experiments, we tried various loss combinations and found $\beta = 10$ to work well. We now expand upon this loss function by leveraging object mask information.

3.2   Object Localization

Real images are often more complex than the ones generated by BigGAN. For example, objects may be off-centered and partially occluded, or multiple objects appear in an image. Moreover, it is possible that the object in the image can be approximated by GANs but not the background.

Accordingly, we focus on fitting a single foreground object in an image and develop a loss funciton to emphasize foreground pixels. We automatically produce a foreground rectangular mask $\mathbf{m} \in [0, 1]^{H \times W \times 1}$ using the bounding box of an object detector [27]. Here, we opt for bounding boxes for simplicity, but one could consider using segmentation mask, saliency maps, user-provided masks, etc. The foreground and background values within mask $\mathbf{m}$ are set to 1 and 0.3, respectively. We adjust the objective function to spatially weigh the loss:

$$\mathcal{L}_{\text{mask}}(\mathbf{y}, \hat{\mathbf{y}}, \mathbf{m}) = \frac{1}{M}||\mathbf{m} \odot (\hat{\mathbf{y}} - \mathbf{y})||_1 + \beta\mathcal{L}_{\text{mLPIPS}}(\hat{\mathbf{y}}, \mathbf{y}, \mathbf{m}), \qquad (4)$$

where normalization parameter $M = ||\mathbf{m}||_1$ and $\odot$ represents element-wise multiplication across the spatial dimensions. Given a mask of all foreground (all ones),

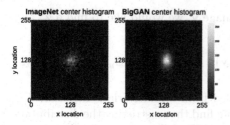
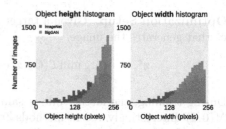

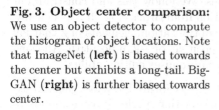

Fig. 3. Object center comparison: We use an object detector to compute the histogram of object locations. Note that ImageNet (**left**) is biased towards the center but exhibits a long-tail. Big-GAN (**right**) is further biased towards center.

Fig. 4. Object size comparison: We use an object detector to compute the distribution of object widths (**left**) and heights (**right**). Note that ImageNet (black) has a long-tail, whereas the BigGAN (blue) accentuates the mode.(Color figure online)

the objective function is equivalent to Eq. 3. We calculate the masked version of the perceptual loss $\mathcal{L}_{\mathrm{mLPIPS}}(\hat{\mathbf{y}}, \mathbf{y}, \mathbf{m})$ by bilinearly downsampling the mask at the resolution of the intermediate spatial feature maps within the perceptual loss. The details are described in Appendix B. With the provided mask, we now explore how one can optimize for image transformation to better fit the object in the image.

3.3   Transformation Model and Loss

Generative models may exhibit biases for two reasons: (a) inherited biases from the training distribution and (b) bias introduced by mode collapse [23], where the generative model only captures a portion of the distribution. We mitigate two types of biases, *spatial* and *color* during image reconstruction process.

Studying Spatial Biases. To study spatial bias, we first use a pre-trained object detector, MaskRCNN [27], over 10,000 real and generated images to compute the statistics of object locations. We show the statistics regarding the center locations and object sizes in Figs. 3 and 4, respectively.

Figure 3 (left) demonstrates that ImageNet images exhibit clear center bias over the location of objects, albeit with a long tail. While the BigGAN learns to mimic this distribution, it further accentuates the bias [8,30], largely forgoing the long tail to generate high-quality samples in the middle of the image. In Fig. 4, we see similar trends with object height and width. Abdal et al. [1] noted that the quality of image reconstruction degrades given a simple translation in the target image. Motivated by this, we propose to incorporate spatial alignment in the inversion process.

Searching over Spatial Alignments. We propose to transform the generated image using $\mathcal{T}_\psi^{\mathrm{spatial}}(\cdot)$, which shifts and scales the image using parameters $\psi = [s_x, s_y, t_x, t_y]$. The parameters ψ are used to generate a sampling grid which in turn is used by a grid-sampler to construct a new transformed image [29]. The corresponding inverse parameters are $\psi^{-1} = \left[\frac{1}{s_x}, \frac{1}{s_y}, -\frac{t_x}{s_x}, -\frac{t_y}{s_y}\right]$.

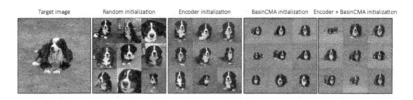

Fig. 5. Initialization from various methods: We show samples drawn from different methods, before the final gradient descent optimization. In "random initialization", seeds are drawn from the normal distribution; the results show higher variation. For the "encoder initialization", we use a trained encoder network to predict the latent vector and apply a minor perturbation. Our method uses CMA to find a good starting distribution. For "Encoder+BasinCMA", we initialize CMA with the output of the encoder. The results are more consistent and better reconstruct the target image.

Transforming the generated image allows for more flexibility in the optimization. For example, if G can perfectly generate the target image, but at different scales or at off-centered locations, this framework allows it to do so.

Searching over Color Transformations. Furthermore, we show that the same framework allows us to search over color transformations $\mathcal{T}_\gamma^{\mathrm{color}}(\cdot)$. We experimented with various color transformations such as hue, brightness, gamma, saturation, contrast, and found brightness and contrast to work the best. Specifically, we optimize for brightness, which is parameterized by scalar γ with inverse value $\gamma^{-1} = -\gamma$. If the generator can perfectly generate the target image, but slightly darker or brighter, this allows a learned brightness transformation to compensate for the difference (Fig. 5).

Final Objective. Let transformation function $\mathcal{T}_\phi = \mathcal{T}_\psi^{\mathrm{spatial}} \circ \mathcal{T}_\gamma^{\mathrm{color}}$ be a composition of spatial and color transformation functions, where transformation parameters ϕ is a concatenation of spatial and color parameters ψ, γ, respectively. The inverse function is $\mathcal{T}_{\phi^{-1}}$. Our final optimization objective function, with consideration for (a) the foreground object and (b) spatial and color biases, is to minimize the following loss:

$$\underset{\mathbf{z},\mathbf{c},\phi}{\arg\min} \ \mathcal{L}_{\mathrm{mask}}(\mathcal{T}_{\phi^{-1}}(G_\theta(\mathbf{z},\mathbf{c})),\mathbf{y},\mathbf{m}) \quad \text{s.t. } C(\mathbf{z}) \le C_{\max} \tag{5}$$

Our optimization algorithm, described next, has a mix of gradient-free and gradient-based updates. Alternatively, instead of inverse transforming the *generated* image, we can transform the *target* and mask images during gradient-based updates and compute the following loss: $\mathcal{L}_{\mathrm{mask}}(G_\theta(\mathbf{z},\mathbf{c}),\mathcal{T}_\phi(\mathbf{y}),\mathcal{T}_\phi(\mathbf{m}))$. We will discuss when to use each variant in the next section.

3.4   Optimization Algorithms

Unfortunately, the objective function is highly non-convex. Gradient-based optimization, as used in previous inversion methods, frequently fall into poor local minima. Bau et al. [8] note that recent large-scale GAN models [32,33] are significantly harder to invert due to a large number of layers, compared to earlier

---

Algorithm 1. Transformation-aware projection algorithm

> **Input:** Image $\mathbf{y}$, initial class vector $\mathbf{c}_0$, mask $\mathbf{m}$
> **Output:** Transformation parameter ϕ^*, latent variable $\mathbf{z}^*$, class vector $\mathbf{c}^*$

1: # Optimize for transformation ϕ
2: Initialize $(\mu_\phi, \Sigma_\phi) \leftarrow (\phi_0, 0.1 \cdot \mathbf{I})$ ▷ ϕ_0 precomputed in Section 3.3
3: **for** n iterations **do**
4:     $\phi_{1:N} \sim \text{SampleCMA}(\mu_\phi, \Sigma_\phi)$ ▷ Draw N samples of ϕ
5:     $\mathbf{z}_{1:N} \sim \mathcal{N}(\mathbf{0}, \mathbf{I})$, reset $\mathbf{c}_{1:N} \leftarrow \mathbf{c}_0$ ▷ Reinitialize $\mathbf{z}$ and $\mathbf{c}$
6:     **for** m iterations **do**
7:        **for** $i \leftarrow 1$ **to** N **do**   ▷ This loop is batched
8:           $g_i \leftarrow \mathcal{L}_{\text{mask}}(G_\theta(\mathbf{z}_i, \mathbf{c}_i), \mathcal{T}_{\phi_i}(\mathbf{y}), \mathcal{T}_{\phi_i}(\mathbf{m}))$
9:           $(\mathbf{z}_i, \mathbf{c}_i) \leftarrow (\mathbf{z}_i, \mathbf{c}_i) - \eta \cdot \nabla_{\mathbf{z}, \mathbf{c}} \, g_i$ ▷ Update each sample $\mathbf{z}, \mathbf{c}$
10:     $g_{1:N}^{inv} \leftarrow \mathcal{L}_{\text{mask}}(\mathcal{T}_{\phi_{1:N}}^{-1}(G_\theta(\mathbf{z}_{1:N}, \mathbf{c}_{1:N})), \mathbf{y}, \mathbf{m})$ ▷ Recompute loss with inverse
11:     $\mu_\phi, \phi_{1:N} \leftarrow \text{UpdateCMA}(\phi_{1:N}, g_{1:N}^{inv}, \mu_\phi, \Sigma_\phi)$ ▷ Section 3.4
12: Set $\phi^* \leftarrow \mu_\phi$
13: # Optimize for latent variables $\mathbf{z}, \mathbf{c}$
14: Initialize $(\mu_\mathbf{z}, \Sigma_\mathbf{z}) \leftarrow (\mathbf{0}, \mathbf{I})$
15: **for** p iterations **do**
16:     $\mathbf{z}_{1:M} \sim \text{SampleCMA}(\mu_\mathbf{z}, \Sigma_\mathbf{z})$, reset $\mathbf{c}_{1:M} \leftarrow \mathbf{c}_0$ ▷ Draw M samples of $\mathbf{z}$
17:     **for** q iterations **do**
18:        **for** $i \leftarrow 1$ **to** M **do**   ▷ This loop is batched
19:           $g_i \leftarrow \mathcal{L}_{\text{mask}}(G_\theta(\mathbf{z}_i, \mathbf{c}_i), \mathcal{T}_{\phi^*}(\mathbf{y}), \mathcal{T}_{\phi^*}(\mathbf{m}))$
20:           $(\mathbf{z}_i, \mathbf{c}_i) \leftarrow (\mathbf{z}_i, \mathbf{c}_i) - \nabla_{\mathbf{z}, \mathbf{c}} \, g_i$
21:     $g_{1:N}^{inv} \leftarrow \mathcal{L}_{\text{mask}}(\mathcal{T}_{\phi_{1:N}}^{-1}(G_\theta(\mathbf{z}_{1:N}, \mathbf{c}_{1:N})), \mathbf{y}, \mathbf{m})$ ▷ Recompute loss with inverse
22:     $\mu_\mathbf{z}, \Sigma_\mathbf{z} \leftarrow \text{UpdateCMA}_\mathbf{z}(\mathbf{z}_{1:M}, g_{1:M}^{inv}, \mu_\mathbf{z}, \Sigma_\mathbf{z})$ ▷ Section 3.4
23: Set $\mathbf{z}^*, \mathbf{c}^* \leftarrow \arg\min_{\mathbf{z}, \mathbf{c}}(g_{1:M})$ ▷ Choose the best $\mathbf{z}, \mathbf{c}$

---

models [48]. Thus, formulating an optimizer that reliably finds good solutions is a significant challenge. We evaluate our method against various baselines and ablations in Sect. 4. Given the input image $\mathbf{y}$ and foreground rectangular mask $\mathbf{m}$ (which is automatically computed), we present the following algorithm.

Class and Transform Initialization. We first predict the class of the image with a pre-trained ResNeXt101 classifier [55] and multiply it by $\mathbf{W}$ to obtain our initial class vector $\mathbf{c}_0$.

Next, we initialize the spatial transformation vector $\psi_0 = [s_{x_0}, s_{y_0}, t_{y_0}, t_{x_0}]$ such that the foreground object is well-aligned with the statistics of the Big-GAN model. As visualized in Figs. 3 and 4, $(\bar{h}, \bar{w}) = (137, 127)$ is the center of BigGAN-generated objects and $(\bar{y}, \bar{x}) = (213, 210)$ is the mode of object sizes. We define $(h_\mathbf{m}, w_\mathbf{m})$ to be the height and width and $(y_\mathbf{m}, x_\mathbf{m})$ to be center of the masked region. We initialize scale factors as $s_{y_0} = s_{x_0} = \max\left(\frac{h_\mathbf{m}}{\bar{h}}, \frac{w_\mathbf{m}}{\bar{w}}\right)$ and translations as $(t_{y_0}, t_{x_0}) = \left(\frac{\bar{y} - y_\mathbf{m}}{2}, \frac{\bar{x} - x_\mathbf{m}}{2}\right)$. Finally, initial brightness transformation parameter is initialized as $\gamma_0 = 1$.

Choice of Optimizer. We find the choice of optimizer critical and that BasinCMA [8] provides better results than previously used optimizers for the GAN inversion problem. Previous work [1,60] has exclusively used gradient-based optimization, such as LBFGS [41] and ADAM [34]. However, such methods are prone to obtaining poor results due to local minima, requiring the use of multiple random initial seeds. Covariance Matrix Adaptation (CMA) [26], a *gradient-free* optimizer, finds better solutions than gradient-based methods. CMA maintains a Gaussian distribution in parameter space $\mathbf{z} \sim \mathcal{N}(\mu, \mathbf{\Sigma})$. At each iteration, N samples are drawn, and the Gaussian is updated using the loss. The details of this update are described in Hansen and Ostermeier [26]. A weakness of CMA is that when it nears a solution, it is slow to refine results, as it does not use gradients. To address this, we use a variant, BasinCMA [53], that alternates between CMA updates and ADAM optimization, where CMA distribution is updated after taking M gradient steps.

Next, we describe the optimization procedure between the transformation parameters ϕ and latent variables $\mathbf{z}, \mathbf{c}$.

Choice of Loss Function. In Eq. 5, we described two variants of our optimization objective. Ideally, we would like to optimize the former variant $\mathcal{L}_{\mathrm{mask}}(\mathcal{T}_{\phi^{-1}}(G_\theta(\mathbf{z}, \mathbf{c})), \mathbf{y}, \mathbf{m})$ such that the target image $\mathbf{y}$ is consistent through-out optimization; and we do so for all CMA updates. However for gradient optimization, we found that back-propagating through a grid-sampler to hurt performance, especially for small objects. A potential reason is that when shrinking a generated image, the grid-sampling operation sparsely samples the image. Without low-pass filtering, this produces a noisy and aliased result [22,45]. Therefore, for gradient-based optimization, we optimize the latter version $\mathcal{L}_{\mathrm{mask}}(G_\theta(\mathbf{z}, \mathbf{c}), \mathcal{T}_\phi(\mathbf{y}), \mathcal{T}_\phi(\mathbf{m}))$.

Two-Stage Approach. Historically, searching over spatial transformations with reconstruction loss as guidance has proven to be a difficult task in computer vision [4]. We find this to be the case in our application as well, and that joint optimization over the transformation ϕ, and variables $\mathbf{z}, \mathbf{c}$ is unstable. We use a two-stage approach, as shown in Algorithm 1, where we first search for ϕ^* and use ϕ^* to optimize for $\mathbf{z}^*$ and $\mathbf{c}^*$. In both stages, a gradient-free CMA outer loop maintains a distribution over the variable of interest in that stage. In the inner loop, ADAM is used to quickly find the local optimum over latent variables $\mathbf{z}, \mathbf{c}$.

To optimize for the transformation parameter, we initialize CMA distribution for ϕ. The mean μ_ϕ is initialized with pre-computed statistics ϕ_0, and Σ_ϕ is set to $0.1 \cdot \mathbf{I}$ (Algorithm 1, line 2). A set of transformations $\phi_{1:N}$ is drawn from CMA, and latent variables $\mathbf{z}_{1:N}$ are randomly initialized (Algorithm 1, line 4–5). To evaluate the sampled transformation, we take gradient updates w.r.t. $\mathbf{z}_{1:N}, \mathbf{c}_{1:N}$ for $m = 30$ iterations (Algorithm 1, line 6–9). This inner loop can be interpreted as quickly assessing the viability of a given spatial transform. The final samples of $\mathbf{z}_{1:N}, \mathbf{c}_{1:N}, \phi_{1:N}$ are used to compute the loss for the CMA update (Algorithm 1, line 10–11). This procedure is repeated for $n = 30$ iterations, and the final transformation ϕ^* is set to the mean of the current estimate of CMA (Algorithm 1, line 12).

After solving for the transformation ϕ^*, a similar procedure is used to optimize for $\mathbf{z}$. We initialize CMA distribution for $\mathbf{z}$ with $\mu_\mathbf{z} = \mathbf{0}$ and $\Sigma_\mathbf{z} = \mathbf{I}$ (Algorithm 1, line 14). M samples of $\mathbf{z}_{1:M}$ are drawn from the CMA distribution and $\mathbf{c}_{1:M}$ is set to the initial predicted class vector (Algorithm 1, line 16). The drawn samples are evaluated by taking $q = 30$ gradient updates w.r.t $\mathbf{z}_{1:M}$ and $\mathbf{c}_{1:M}$ (Algorithm 1, line 17–20). The optimized samples are used to compute the loss for the CMA update (Algorithm 1, line 21–22). This procedure is repeated for $p = 30$ iterations. On the final iteration, we take 300 gradient updates instead to obtain the final solution $\mathbf{z}, \mathbf{c}$ (Algorithm 1, line 23).

Target	Mask	ADAM	L-BFGS	CMA	BasinCMA	ADAM + Transform	BasinCMA + Transform

Fig. 6. ImageNet comparisons: Comparison across various methods on inverting ImageNet images without fine-tuning. A rectangular mask centered around the object of interest is provided for all methods using MaskRCNN [27]. The losses are weighted by the mask. BasinCMA+Transform is our full method.

3.5   Fine-Tuning

So far, we have located an approximate match within a generative model. We hypothesize that if a high-quality match is found, fine-tuning to fit the image will preserve the editability of the generative model. On the contrary, if a poor match is found, the fine-tuning will corrupt the network and result in low-quality images after editing. Next, we describe this fine-tuning process.

To synthesize the missing details that the generator could not produce, we wish to fine-tune our model after solving for the latent vector $\mathbf{z}$, the class vector $\mathbf{c}$, and transformation parameters ϕ. Unlike previous work [6], which proposed to produce the residual features using a small, auxiliary network, we update the weights of the original GAN directly. This allows us to perform edits that spatially deform the image. After obtaining the values for $\phi, \mathbf{z}, \mathbf{c}$ in our projection step, we fine-tune the weights of the generative model. During fine-tuning, the full objective function is:

$$\underset{\mathbf{z},\mathbf{c},\phi,\theta}{\arg\min} \ \ \mathcal{L}_{\text{mask}}(\mathcal{T}_{\phi^{-1}}(G_\theta(\mathbf{z},\mathbf{c})),\mathbf{y},\mathbf{m}) \ + \ \lambda\|\theta - \theta_0\|_2 \quad \text{s.t.} \ \ C(\mathbf{z}) \leq C_{\max} \quad (6)$$

We put an ℓ_2-regularization on the weights, such that the fine-tuned weights do not deviate too much from the original weights θ_0. In doing so, we can prevent overfitting and preserve the generative model's ability to edit the final image. We use $\lambda = 10^3$ for our results with fine-tuning.

4   Results

We demonstrate results on images from ImageNet [12], compare against baselines and ablations, examine cases that BigGAN cannot generate, and show failure cases. We further demonstrate the validity of our method on out-of-distribution data such as COCO and conduct perceptual studies on the edited images.

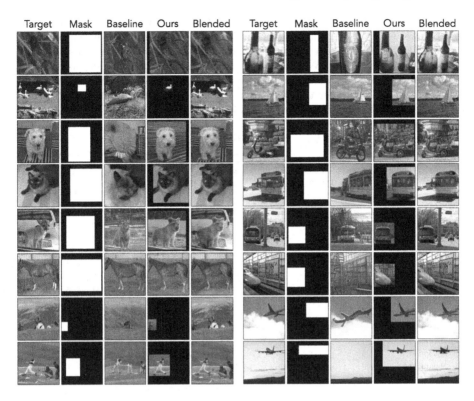

| Target | Mask | Baseline | Ours | Blended | Target | Mask | Baseline | Ours | Blended |

Fig. 7. ImageNet results: Results using our final method without fine-tuning. The final method uses BasinCMA as well as spatial and color transformation. Our generated results are inverted back for visualization. We also provide the ADAM baseline along with the blended result using Poisson blending [47].

The ImageNet dataset consists of 1.3 million images with 1,000 classes. We construct a test set by using PASCAL [20] classes as super-classes. There are a total of 229 classes from ImageNet that map to 16 out of 20 classes in PASCAL. We select 10 images at random from each super-class to construct a dataset of 160 images. We run off-the-shelf Mask-RCNN [27] and take the highest activating class to generate the detection boxes. We use the same bounding box for all baselines, and the optimization hyper-parameters are tuned on a separate set of ImageNet images.

Experimental Details. We use a learning rate of 0.05 for z and 0.0001 for c. We use AlexNet-LPIPS [37,59] as our perceptual loss for all our methods. We did observe an improvement using VGG-LPIPS [51,59] but found it to be 1.5 times slower. In our experiments, we use a total of 18 seeds for each method. After we project and edit the object, we blend the newly edited object with the original background using Poisson blending [47].

For all of our baselines, we optimize both the latent vector z and class embedding c. We use the same mask m, and the same loss function throughout all of

Table 1. ImageNet: We compare various methods for inverting images from ImageNet (lower is better). The last row is our full method. The model is optimized using $L1$ and AlexNet-LPIPS perceptual loss. The mask and ground-truth class vector is provided for each method. We show the error using different metrics: per-pixel and perceptual [59]. We show the average and the best score among 18 random seeds. Methods that optimized for transformation are inverted to the original location and the loss is computed on the masked region for a fair comparison. All the results here are not fine-tuned.

Method				Average of 18 seeds				Best of 18 seeds			
				Per-pixel		LPIPS		Per-pixel		LPIPS	
Optimizer	Spatial Transform	Color Transform	Encoder	L1	L2	Alex	VGG	L1	L2	Alex	VGG
ADAM				0.98	0.62	0.41	0.58	0.83	0.47	0.33	0.51
L-BFGS				1.04	0.68	0.45	0.61	0.85	0.49	0.35	0.53
CMA				0.96	0.61	0.39	0.55	0.91	0.54	0.37	0.54
None			✓	1.61	1.39	0.62	0.68	1.35	1.00	0.55	0.64
ADAM			✓	0.96	0.60	0.39	0.56	0.82	0.46	0.32	0.51
ADAM		✓		0.98	0.62	0.42	0.58	0.83	0.47	0.33	0.51
ADAM	✓			0.90	0.54	0.44	0.57	0.76	0.41	0.36	0.50
ADAM	✓	✓	✓	0.88	0.52	0.42	0.55	0.76	0.40	0.36	0.49
CMA+ADAM				0.93	0.57	0.37	0.55	0.83	0.47	0.32	0.51
BasinCMA				0.82	0.48	0.29	0.51	0.78	0.43	0.26	0.49
BasinCMA			✓	0.82	0.47	0.29	0.50	0.78	0.43	0.26	0.49
BasinCMA		✓		0.81	0.46	**0.29**	0.50	0.77	0.42	**0.25**	0.49
BasinCMA	✓			0.72	0.38	0.33	0.48	0.69	0.35	0.31	**0.46**
BasinCMA	✓	✓	✓	**0.71**	**0.37**	0.32	**0.47**	**0.68**	**0.34**	0.31	**0.46**

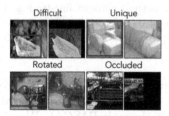

Fig. 8. Failure cases: Our method fails to invert images that are not well represented by BigGAN. The mask is overlayed on the target image in blue.(Color figure online)

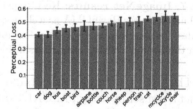

Fig. 9. Projection error by class: The average VGG-perceptual loss with standard error. The ImageNet images are sampled from the PASCAL superclass.

our experiments. The optimization details of our method and the baselines are in the Appendix A.

Experiments. We show qualitative comparisons of various optimization methods for ImageNet images in Fig. 6. We show results of our final method with blending in Fig. 7. We then quantify these results by comparing against each method using various metrics in Table 1. For all methods, we do *not* fine-tune our results and we only compute the loss inside the mask for a fair comparison. For methods optimized with transformation, the projected images are inverted

Table 2. Class search: Given a fixed optimization method (ADAM), we compare different methods for initializing the class vector (lower is better). Baselines are: initialized from $\mathcal{N}(\mathbf{0}, \mathbf{I})$, a random class, and the ground truth class.

Class search	Best of 18 seeds			
	Per-pixel		LPIPS	
	L1	L2	Alex	VGG
Random Gaussian	1.26	0.88	0.69	0.86
Random Class	0.88	0.51	0.40	0.59
Predicted	0.84	0.47	0.33	0.52
Ground Truth	0.83	0.47	0.33	0.51

Table 3. Out-of-distribution: We compare different methods on the COCO-dataset (lower is better). BigGAN was not trained on COCO images. The class labels are predicted using ResNext-101 and the masks are predicted using MaskRCNN.

Method	Best of 18 seeds			
	Per-pixel		LPIPS	
	L1	L2	Alex	VGG
ADAM	0.96	0.57	0.32	0.56
ADAM + Transform	0.81	0.45	0.39	0.52
BasinCMA	0.93	0.18	**0.81**	0.53
BasinCMA + Transform	**0.78**	**0.42**	0.36	**0.49**

back before computing the loss. We further evaluate on COCO dataset [39] in Table 3, and observed our findings to hold true on out-of-distribution dataset. The success of hybrid optimization over purely gradient-based optimization techniques may indicate that the generative model latent space is locally smooth but not globally.

Without transforming the object, we observed that the optimization often fails to find an approximate solution, specifically when the objects are off-centered or contain multiple objects. We observed that optimizing over color transformation does not lead to drastic improvements. Possibly because BigGAN can closely match the color gamut statistics of ImageNet images. Nonetheless, we found that optimizing for color transformation can slightly improve visual aesthetics. Out of the experimented color transformations, optimizing for brightness gave us the best result, and we use this for color transformation throughout our experiments. We further experimented with composing multiple color transformations but did not observe additional improvements.

We found that using CMA/BasinCMA is robust to initialization and is a better optimization technique regardless of whether the transform was applied. Note that we did not observe any benefits of optimizing the class vectors $\mathbf{c}$ with CMA compared to gradient-based methods, perhaps because the embedding between the continuous class vectors is not necessarily meaningful. Qualitatively, we often found the class embeddings to be meaningful when it is either in the close vicinity of original class embeddings or between the interpolation of 2 similar classes and not more. As a result, we use gradient descent to search within the local neighborhood of the initial class embedding space.

We also provide ablation study on how the number of CMA and ADAM updates for BasinCMA affects performance, and how other gradient-free optimizers compare against CMA in Appendix D. We further provide additional qualitative results for our final method in Appendix C.

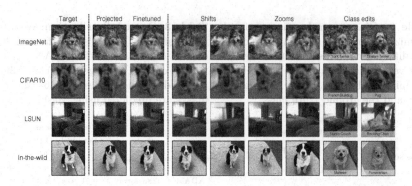

Fig. 10. Fine-tuned edits: Inversion results on various datasets. We use BasinCMA and transformation to optimize for the latent variables. After obtaining the projections, we fine-tune the model weights and perform edits in the latent and class vector space.

Class Initialization.   In downstream editing application, the user may not know the exact ImageNet class the image belongs to. In Table 2, we compare different strategies for initializing the class vector. Here the classifier makes an incorrect prediction 20% of the time. We found that using the predicted class of an ImageNet classifier performs almost as well as the ground truth class. Since we optimize the class vector, we can potentially recover from a wrong initial guess if the predicted class is sufficiently close to the ground-truth.

Failure Cases.   Figure 8 shows some typical failure cases. We observed that our method fails to embed images that are not well modeled by BigGAN – outlier modes that may have been dropped. For example, we failed to project images that are unique, complicated, rotated, or heavily occluded. More sophisticated transformations such as rotations and perspective transformation could address many of these failure cases and are left for future work.

Which Classes Does BigGAN Struggle to Generate?   Given our method, we analyze which classes BigGAN, or our method has difficulty generating. In Fig. 9, we plot the mean and the standard error for each class. The plot is from the output of the method optimized with ADAM + CMA + Transform. We observed a general tendency for the model to struggle in generating objects with delicate structures or with large inter-class variance.

Image Edits.   A good approximate solution allows us to fine-tune the generative model and recover the details easily. Good approximations require less fine-tuning and therefore preserve the original generative model editing capabilities. In Fig. 10, we embed images from various datasets including CIFAR [36], LSUN [57], and images in-the-wild. We then fine-tune and edit the results by changing the latent vector or class vector. Prior works [30,48] have found that certain latent vectors can consistently control the appearance change of GANs-generated images such as shifting an image horizontally or zooming an image in and out. We used the "shift" and "zoom" vectors [30] to modify our images.

Additionally, we also varied the class vector to a similar class and observed the editability to stay consistent. Even for images like CIFAR, our method was able to find good solutions that allowed us to edit the image. In cases like LSUN, where there is no corresponding class for the scene, we observed that the edits ended up being meaningless.

5 Discussion

Projecting an image into the "space" of a generative model is a crucial step for editing applications. We have systematically explored methods for this projection. We show that using a gradient-free optimizer, CMA, produces higher quality matches. We account for biases in the generative model by enabling spatial and color transformations in the search, and the combination of these techniques finds a closer match and better serves downstream editing pipelines. Future work includes exploring more transformations, such as local geometric changes and global appearance changes, as well as modeling generation of multiple objects or foreground/background.

Acknowledgements. We thank David Bau, Phillip Isola, Lucy Chai, and Erik Härkönen for discussions, and David Bau for encoder training code.

References

1. Abdal, R., Qin, Y., Wonka, P.: Image2StyleGAN: how to embed images into the StyleGAN latent space? In: International Conference on Computer Vision (2019)
2. Ankit, R., Li, Y., Bresler, Y.: GAN-based projector for faster recovery with convergence guarantees in linear inverse problems. In: International Conference on Computer Vision (2019)
3. Asim, M., Shamshad, F., Ahmed, A.: Blind image deconvolution using deep generative priors. In: British Machine Vision Conference (2018)
4. Baker, S., Matthews, I.: Lucas-kanade 20 years on: a unifying framework. Int. J. Comput. Vis. **56**(3), 221–255 (2004). https://doi.org/10.1023/B:VISI.0000011205.11775.fd
5. Barnes, C., Shechtman, E., Finkelstein, A., Goldman, D.B.: PatchMatch: a randomized correspondence algorithm for structural image editing. ACM Trans. Graph. (TOG) **28**, 24 (2009)
6. Bau, D., et al.: Semantic photo manipulation with a generative image prior. ACM Trans. Graph. (TOG) 38 (2019)
7. Bau, D., et al.: GAN dissection: visualizing and understanding generative adversarial networks. In: International Conference on Learning Representations (2019)
8. Bau, D., et al.: Seeing what a GAN cannot generate. In: International Conference on Computer Vision (2019)
9. Brock, A., Donahue, J., Simonyan, K.: Large scale GAN training for high fidelity natural image synthesis. In: International Conference on Learning Representations (2019)
10. Brock, A., Lim, T., Ritchie, J.M., Weston, N.: Neural photo editing with introspective adversarial networks. In: International Conference on Learning Representations (2017)

11. Creswell, A., Bharath, A.A.: Inverting the generator of a generative adversarial network. IEEE Trans. Neural Netw. Learn. Syst. **30**(7), 1967–1974 (2018)
12. Deng, J., Dong, W., Socher, R., Li, L.J., Li, K., Fei-Fei, L.: ImageNet: a large-scale hierarchical image database. In: IEEE Conference on Computer Vision and Pattern Recognition (2009)
13. Donahue, J., Krähenbühl, P., Darrell, T.: Adversarial feature learning. In: International Conference on Learning Representations (2017)
14. Donahue, J., Simonyan, K.: Large scale adversarial representation learning. In: Advances in Neural Information Processing Systems (2019)
15. Dosovitskiy, A., Brox, T.: Generating images with perceptual similarity metrics based on deep networks. In: Advances in Neural Information Processing Systems (2016)
16. Dosovitskiy, A., Brox, T.: Inverting visual representations with convolutional networks. In: IEEE Conference on Computer Vision and Pattern Recognition (2016)
17. Dumoulin, V., et al.: Adversarially learned inference. In: International Conference on Learning Representations (2017)
18. Efros, A.A., Freeman, W.T.: Image quilting for texture synthesis and transfer. In: ACM SIGGRAPH (2001)
19. Efros, A.A., Leung, T.K.: Texture synthesis by non-parametric sampling. In: International Conference on Computer Vision (1999)
20. Everingham, M., Eslami, S.M.A., Van Gool, L., Williams, C.K.I., Winn, J., Zisserman, A.: The PASCAL visual object classes challenge: a retrospective. Int. J. Comput. Vis. **111**(1), 98–136 (2015). https://doi.org/10.1007/s11263-014-0733-5
21. Gatys, L.A., Ecker, A.S., Bethge, M.: Image style transfer using convolutional neural networks. In: IEEE Conference on Computer Vision and Pattern Recognition (2016)
22. Gonzalez, R.C., Woods, R.E.: Digital Image Processing, 2nd edn. Pearson, London (1992)
23. Goodfellow, I.: NIPS 2016 tutorial: Generative adversarial networks. arXiv preprint arXiv:1701.00160 (2016)
24. Goodfellow, I., et al.: Generative adversarial nets. In: Advances in Neural Information Processing Systems (2014)
25. Gu, J., Shen, Y., Zhou, B.: Image processing using multi-code GAN prior. In: IEEE Conference on Computer Vision and Pattern Recognition (2020)
26. Hansen, N., Ostermeier, A.: Completely derandomized self-adaptation in evolution strategies. Evol. Comput. **9**, 159–195 (2001)
27. He, K., Gkioxari, G., Dollár, P., Girshick, R.: Mask R-CNN. In: International Conference on Computer Vision (2017)
28. Hertzmann, A., Jacobs, C.E., Oliver, N., Curless, B., Salesin, D.H.: Image analogies. In: ACM SIGGRAPH (2001)
29. Jaderberg, M., Simonyan, K., Zisserman, A., Kavukcuoglu, K.: Spatial transformer networks. In: Advances in Neural Information Processing Systems (2015)
30. Jahanian, A., Chai, L., Isola, P.: On the"steerability" of generative adversarial networks. In: International Conference on Learning Representations (2020)
31. Johnson, J., Alahi, A., Fei-Fei, L.: Perceptual losses for real-time style transfer and super-resolution. In: Leibe, B., Matas, J., Sebe, N., Welling, M. (eds.) European Conference on Computer Vision, vol. 9906, pp. 694–711. Springer, Heidelberg (2016). https://doi.org/10.1007/978-3-319-46475-6_43
32. Karras, T., Aila, T., Laine, S., Lehtinen, J.: Progressive growing of GANs for improved quality, stability, and variation. In: International Conference on Learning Representations (2018)

33. Karras, T., Laine, S., Aila, T.: A style-based generator architecture for generative adversarial networks. In: IEEE Conference on Computer Vision and Pattern Recognition (2019)
34. Kingma, D.P., Ba, J.: Adam: a method for stochastic optimization. In: International Conference on Learning Representations (2015)
35. Kingma, D.P., Dhariwal, P.: Glow: generative flow with invertible 1x1 convolutions. In: Advances in Neural Information Processing Systems (2018)
36. Krizhevsky, A.: Learning Multiple Layers of Features from Tiny Images. Master's Thesis, University of Toronto (2009)
37. Krizhevsky, A., Sutskever, I., Hinton, G.E.: ImageNet classification with deep convolutional neural networks. In: Advances in Neural Information Processing Systems (2012)
38. Levin, A., Lischinski, D., Weiss, Y.: A closed-form solution to natural image matting. IEEE Trans. Pattern Anal. Mach. Intell. **30**(2), 228–242 (2007)
39. Lin, T., et al.: Microsoft COCO: common objects in context. In: Fleet, D., Pajdla, T., Schiele, B., Tuytelaars, T. (eds.) European Conference on Computer Vision, vol. 8693, pp. 740–755. Springer, Heidelberg (2014). https://doi.org/10.1007/978-3-319-10602-1_48
40. Lipton, Z.C., Tripathi, S.: Precise recovery of latent vectors from generative adversarial networks. ICLR Workshop (2017)
41. Liu, D.C., Nocedal, J.: On the limited memory BFGS method for large scale optimization. Math. Program. **45**(1–3), 503–528 (1989). https://doi.org/10.1007/BF01589116
42. Mahendran, A., Vedaldi, A.: Understanding deep image representations by inverting them. In: IEEE Conference on Computer Vision and Pattern Recognition (2015)
43. Olah, C., Mordvintsev, A., Schubert, L.: Feature visualization. Distill **2**(11), e7 (2017)
44. Olah, C., et al.: The building blocks of interpretability. Distill **3**(3), e10 (2018)
45. Oppenheim, A.V., Schafer, R.W., Buck, J.R.: Discrete-Time Signal Processing, 2nd edn. Pearson, London (1999)
46. Perarnau, G., Van De Weijer, J., Raducanu, B., Álvarez, J.M.: Invertible conditional GANs for image editing. In: NIPS 2016 Workshop on Adversarial Training (2016)
47. Pérez, P., Gangnet, M., Blake, A.: Poisson image editing. ACM Trans. Graph. (TOG) **22**(3), 313–318 (2003)
48. Radford, A., Metz, L., Chintala, S.: Unsupervised representation learning with deep convolutional generative adversarial networks. In: International Conference on Learning Representations (2016)
49. Shah, V., Hegde, C.: Solving linear inverse problems using GAN priors: an algorithm with provable guarantees. In: ICASSP (2018)
50. Shen, Y., Gu, J., Tang, X., Zhou, B.: Interpreting the latent space of GANs for semantic face editing. In: IEEE Conference on Computer Vision and Pattern Recognition (2020)
51. Simonyan, K., Zisserman, A.: Very deep convolutional networks for large-scale image recognition. In: International Conference on Learning Representations (2015)
52. Tiantian, F., Schwing, A.: Co-generation with GANs using AIS based HMC. In: Advances in Neural Information Processing Systems (2019)
53. Wampler, K., Popović, Z.: Optimal gait and form for animal locomotion. ACM Trans. Graph. (TOG) **28**, 1–8 (2009)

54. Wang, Z., Bovik, A.C., Sheikh, H.R., Simoncelli, E.P., et al.: Image quality assessment: from error visibility to structural similarity. IEEE Trans. Image Process. **13**(4), 600–612 (2004)
55. Xie, S., Girshick, R., Dollár, P., Tu, Z., He, K.: Aggregated residual transformations for deep neural networks. arXiv preprint arXiv:1611.05431 (2016)
56. Yeh, R.A., Chen, C., Yian Lim, T., Schwing, A.G., Hasegawa-Johnson, M., Do, M.N.: Semantic image inpainting with deep generative models. In: IEEE Conference on Computer Vision and Pattern Recognition (2017)
57. Yu, F., Zhang, Y., Song, S., Seff, A., Xiao, J.: LSUN: construction of a large-scale image dataset using deep learning with humans in the loop. arXiv preprint arXiv:1506.03365 (2015)
58. Zhang, R., Isola, P., Efros, A.A.: Colorful image colorization. In: Leibe, B., Matas, J., Sebe, N., Welling, M. (eds.) European Conference on Computer Vision, vol. 9907, pp. 649–666 Springer, Heidelberg (2016). https://doi.org/10.1007/978-3-319-46487-9_40
59. Zhang, R., Isola, P., Efros, A.A., Shechtman, E., Wang, O.: The unreasonable effectiveness of deep networks as a perceptual metric. In: IEEE Conference on Computer Vision and Pattern Recognition (2018)
60. Zhu, J.Y., Krähenbühl, P., Shechtman, E., Efros, A.A.: Generative visual manipulation on the natural image manifold. In: Leibe, B., Matas, J., Sebe, N., Welling, M. (eds.) European Conference on Computer Vision, vol. 9909, pp. 597–613. Springer, Heidelberg (2016). https://doi.org/10.1007/978-3-319-46454-1_36

Suppress and Balance: A Simple Gated Network for Salient Object Detection

Xiaoqi Zhao[1], Youwei Pang[1], Lihe Zhang[1(✉)], Huchuan Lu[1,2], and Lei Zhang[3,4]

[1] Dalian University of Technology, Dalian, China
{zxq,lartpang}@mail.dlut.edu.cn,{zhanglihe,lhchuan}@dlut.edu.cn
[2] Peng Cheng Laboratory, Shenzhen, China
[3] Department of Computing, The Hong Kong Polytechnic University,
Hong Kong, China
cslzhang@comppolyu.edu.hk
[4] DAMO Academy, Alibaba Group, Hangzhou, China

Abstract. Most salient object detection approaches use U-Net or feature pyramid networks (FPN) as their basic structures. These methods ignore two key problems when the encoder exchanges information with the decoder: one is the lack of interference control between them, the other is without considering the disparity of the contributions of different encoder blocks. In this work, we propose a simple gated network (GateNet) to solve both issues at once. With the help of multilevel gate units, the valuable context information from the encoder can be optimally transmitted to the decoder. We design a novel gated dual branch structure to build the cooperation among different levels of features and improve the discriminability of the whole network. Through the dual branch design, more details of the saliency map can be further restored. In addition, we adopt the atrous spatial pyramid pooling based on the proposed "Fold" operation (Fold-ASPP) to accurately localize salient objects of various scales. Extensive experiments on five challenging datasets demonstrate that the proposed model performs favorably against most state-of-the-art methods under different evaluation metrics.

Keywords: Salient object detection · Gated network · Dual branch · Fold-ASPP

1 Introduction

Salient object detection aims to identify the visually distinctive regions or objects in a scene and then accurately segment them. In many computer vision applications, it is used as a pre-processing step, such as scene classification [22], visual tracking [19], person re-identification [24], light field image segmentation [30] and image captioning [8], etc.

X. Zhao and Y. Pang—These authors contributed equally to this work.

Electronic supplementary material The online version of this chapter (https://doi.org/10.1007/978-3-030-58536-5_3) contains supplementary material, which is available to authorized users.

© Springer Nature Switzerland AG 2020
A. Vedaldi et al. (Eds.): ECCV 2020, LNCS 12347, pp. 35–51, 2020.
https://doi.org/10.1007/978-3-030-58536-5_3

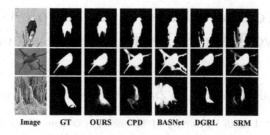

Fig. 1. Visual comparison of different CNN based methods.

With the development of deep learning, salient object detection has gradually evolved from the traditional method based on manual design features to the deep learning method. In recent years, U-shape based structures [16,23] have received the most attention due to their ability to utilize multilevel information to reconstruct high-resolution feature maps. Therefore, most state-of-the-art saliency detection networks [10,17,20,21,31,34,42,44,45] adopt U-shape as the encoder-decoder architecture. And many methods aim at combining multilevel features in either the encoder [21,31,34,36,42,44] or the decoder [10,17,36,45]. For each convolutional block, they separately formulate the relationships of internal features for forward update. It is well known that the high-quality saliency maps predicted in the decoder rely heavily on the effective features provided by the encoder. Nevertheless, the aforementioned methods directly use an all-pass skip-layer structure to concatenate the features of the encoder to the decoder, and the effectiveness of feature aggregation at different levels is not quantified. These restrictions not only introduce misleading context information into the decoder but also result in that the really useful features can not be adequately utilized. In cognitive science, Yang *et al.* [40] show that inhibitory neurons play an important role in how the human brain chooses to process the most important information from all the information presented to us. And inhibitory neurons ensure that humans respond appropriately to external stimuli by inhibiting other neurons and balancing excitatory neurons that stimulate neuronal activity. Inspired by this work, we think that it is necessary to set up an information screening unit between each pair of encoder and decoder blocks in saliency detection. It can help distinguish the most intense features of salient regions and suppress background interference, as shown in Fig. 1, in which these images have easily-confused backgrounds or low-contrast objects.

Moreover, due to the limited receptive field, a single-scale convolutional kernel is difficult to capture context information of size-varying objects. This motivates some efforts [6,42] to investigate multiscale feature extraction. These methods directly equip an atrous spatial pyramid pooling module [3] (ASPP) in their networks. However, when using a convolution with a large dilation rate, the information under the kernel seriously lacks correlation due to inserting too many zeros. This may be detrimental to the discrimination of subtle image structures.

In this paper, we propose a simple gated network (GateNet) for salient object detection. Based on the feature pyramid network (FPN), we construct multilevel gate units to combine the features from the decoder and the encoder. We use convolution operation and nonlinear functions to calculate the correlations among features and assign gate values to different blocks. In this process, a partnership is established between different blocks by using weight distribution and the decoder can obtain more efficient information from the encoder and pay more attention to the salient regions. Since the top-layer features of the encoder network contain rich contextual information, we construct a folded atrous spatial pyramid pooling (Fold-ASPP) module to gather multiscale high-level saliency cues. With the "Fold" operation, the atrous convolution is implemented on a group of local neighborhoods rather than a group of isolated sampling points, which can help generate more stable features and more adequately depict finer structure. In addition, we design a parallel branch by concatenating the output of the FPN branch and the features of the gated encoder, so that the residual information complementary to the FPN branch is supplemented to generate the final saliency map.

Our main contributions can be summarized as follows.

– We propose a simple gated network to adaptively control the amount of information that flows into the decoder from each encoder block. With multilevel gate units, the network can balance the contribution of each encoder block to the decoder block and suppress the features of non-salient regions.
– We design a Fold-ASPP module to capture richer context information and localize salient objects of various sizes. By the "Fold" operation, we can obtain more effective feature representation.
– We build a dual branch architecture. They form a residual structure, complement each other through the gated processing and generate better results.

We compare the proposed model with seventeen state-of-the-art methods on five challenging datasets. The results show that our method performs much better than other competitors. And, it achieves a real-time speed of 30 fps.

2   Related Work

2.1   Salient Object Detection

Early saliency detection methods are based on low-level features and some heuristics prior knowledge, such as color contrast [1], background prior [39] and center prior [12]. Most of them using hand-crafted features, and more details about the traditional methods are discussed in [32].

With the breakthrough of deep learning in the field of computer vision, a large number of convolutional neural networks-based salient object detection methods have been proposed and their performance had been improved gradually. Especially, fully convolutional networks (FCN), which avoid the problems caused by the fully-connected layer, become the mainstream for dense prediction tasks. Wang et al. [28] use weight sharing methods to iteratively refine features and promote mutual fusion between features. Hou et al. [10] achieve efficient feature expression by continuously blending features from deep layers into shallow layers. However, the single-scale feature cannot roundly characterize various objects as well as image contexts. How to get multiscale features and integrate context information is an important problem in saliency detection.

2.2 Multiscale Feature Extraction

Recently, the atrous spatial pyramid pooling module (ASPP) [3] is widely applied in many tasks and networks. The atrous convolution can enlarge the receptive field to obtain large-scale features and does not increase the computational cost. Therefore, it is often used in saliency detection networks. Zhang *et al.* [42] insert several ASPP modules into the encoder blocks of different levels, while Deng *et al.* [6] install it on the highest-level encoder block. Nevertheless, the repeated stride and pooling operations already make the top-layer features lose much fine information. With the increase of atrous rate, the correlation of sampling points further degrades, which leads to difficulties in capturing the changes of image details (e.g., lathy background regions between adjacent objects or spindly parts of objects). In this work, we propose a folded ASPP to alleviate these issues and achieve a *local-in-local* effect.

2.3 Gated Mechanisms

The gated mechanism plays an important role in controlling the flow of information and is widely used in the long short term memory (LSTM). In [2], the gate unit combines two consecutive feature maps of different resolutions from the encoder to generate rich contextual information. Zhang *et al.* [42] adopt gate function to control the message passing when combining feature maps at all levels of the encoder. Due to the ability to filter information, the gated mechanism can also be seen as a special kind of attention mechanism. Some saliency methods [4,34,45] employ attention networks. Zhang *et al.* [45] apply both spatial and channel attention to each layer of the decoder. Wang *et al.* [34] exploit the pyramid attention module to enhance saliency representations for each layer in the encoder and enlarge the receptive field. The above methods all unilaterally consider the information interaction between different levels either in the encoder or in the decoder. We integrate the features from the encoder and the decoder to formulate gate function, which plays the role of block-wise attention and model the overall distribution of all blocks in the network from the global perspective. While previous methods actually utilize the block-specific feature to compute dense attention weights for the corresponding block. Moreover, in order to take advantage of rich contextual information in the encoder, these methods directly feed the encoder features into the decoder and do not consider their mutual interference. Our proposed gate unit can naturally balance their contributions, thereby suppressing the response of the encoder to non-salient regions. Experimental results in Fig. 4 and Fig. 9 intuitively demonstrate the effect of multilevel gate units on the above two aspects, respectively.

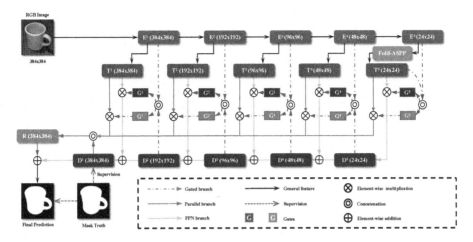

Fig. 2. The overall architecture of the gated network. It consists of the VGG-16 encoder ($\mathbf{E}^1$–$\mathbf{E}^5$), five transition layers ($\mathbf{T}^1$–$\mathbf{T}^5$), five gate units ($\mathbf{G}^1$–$\mathbf{G}^5$), five decoder blocks ($\mathbf{D}^1$–$\mathbf{D}^5$) and the Fold-ASPP module. We employ twice supervision in this network. Once acts at the end of the FPN branch D^1. The other is used to guide the fusion of the two branches.

3   Proposed Method

The gated network architecture is shown in Fig. 2, in which encoder blocks, transition layers, decoder blocks and gate units are respectively denoted as $\mathbf{E}^i$, $\mathbf{T}^i$, $\mathbf{D}^i$ and $\mathbf{G}^i$ ($i \in \{1, 2, 3, 4, 5\}$ indexes different levels). And their output feature maps are denoted as E^i, T^i, D^i and G^i, respectively. The final prediction is obtained by combining the FPN branch and the parallel branch. In this section, we first describe the overall architecture, then detail the gated dual branch structure and the folded atrous spatial pyramid pooling module.

3.1   Network Overview

Encoder Network. In our model, the encoder is based on a common pretrained backbone network, *e.g.*, the VGG [25], ResNet [9] or ResNeXt [37]. We take the VGG-16 network as an example, which contains thirteen Conv layers, five max-pooling layers and two fully connected layers. In order to fit saliency detection task, similar to most previous approaches [10,42,44,45], we cast away all the fully-connected layers of the VGG-16 and remove the last pooling layer to retain details of last convolutional layer.

Decoder Network. The decoder comprises three main components. i) The FPN branch, which continually fuses different level features from $T^1 \sim T^5$ by element-wise addition. ii) The parallel branch, which combines the saliency map of the FPN branch with the feature maps of transition layers by cross-channel concatenation. At the same time, multilevel gate units ($\mathbf{G}^1 \sim \mathbf{G}^5$) are inserted

Fig. 3. Detailed illustration of the gate unit. E^i, D^{i+1} indicates feature maps of the current encoder block and those of the previous decoder block, respectively. ⓢ is sigmoid function.

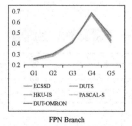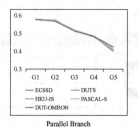

FPN Branch   Parallel Branch

Fig. 4. The distributions of the gate weights on five datasets. We calculate the average gate values for each level of the FPN branch and the parallel branch across all images in every dataset. For the FPN branch, the low-level gate values are significantly smaller than the high-level ones. For the parallel branch, the gate values gradually decrease with the promotion of levels.

between the transition layer and the decoder layer. iii) The Fold-ASPP module, which improves the original atrous spatial pyramid pooling (ASPP) by using a "Fold" operation. It can take advantage of semantic features learned from E^5 to provide multiscale information to the decoder.

3.2 Gated Dual Branch

The gate unit can control the message passing between scale-matching encoder and decoder blocks. By combining the feature maps of the previous decoder block, the gate value also characterizes the contribution that the current block of the encoder can provide. Figure 3 shows the internal structure of the proposed gate unit. In particular, encoder feature E^i and decoder feature D^{i+1} are integrated to obtain feature F^i, and then it is fed into two branches, which includes a series of convolution, activation and pooling operations, to compute a pair of gate values G^i. The entire gated process can be formulated as,

$$
G^i = \begin{cases} P(S(Conv(Cat(E^i, D^{i+1})))) & \text{if } i = 1,2,3,4 \\ P(S(Conv(Cat(E^i, T^i)))) & \text{if } i = 5 \end{cases} \tag{1}
$$

where $Cat(\cdot)$ is the concatenation operation among channel axis, $Conv(\cdot)$ refers to the convolution layer, $S(\cdot)$ is the element-wise sigmoid function, and $P(\cdot)$ is the global average pooling. The output channel of $Conv(\cdot)$ is 2. The resulted gate vector G^i has two different elements which correspond to two gate values in Fig. 3.

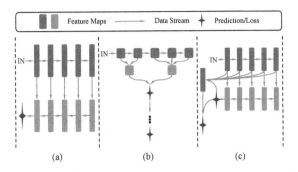

Fig. 5. Illustration of different decoder architectures. (a) Progressive structure, (b) Parallel structure and (c) Our dual branch structure.

Given the gate values, they are applied to the FPN branch and the parallel branch for weighting the transition-layer features $T^1 \sim T^5$, which are generated by exploiting 3×3 convolution to reduce the dimension of $E^1 \sim E^4$ and the Fold-ASPP to finely process E^5 (Please see Fig. 2 for details). Through multilevel gate units, we can suppress and balance the information flowing from different encoder blocks to the decoder.

In Fig. 4, we statistically demonstrate the curves of gate value with a convolutional level as the horizontal axis. It can be seen that the high-level encoder features contribute more contextual guidance to the decoder than the low-level encoder features in the FPN branch. This trend is just the opposite in the parallel branch. It is because the FPN branch is responsible to predict the main body of the salient object by progressively combining multilevel features, which needs more high-level semantic knowledge. While the parallel branch, as a residual structure, aims to fill in the details, which are mainly contained in the low-level features. In addition, some visual examples are shown in Fig. 9 demonstrate that multilevel gate units can significantly suppress the interference from each encoder block and enhance the contrast between salient and non-salient regions. Since the proposed gate unit is simple yet effective, a raw FPN network with multilevel gate units can be viewed as a new baseline for saliency detection task.

Most existing models either use progressive decoder [31,34,42,45] or parallel decoder [6,46], as shown in Fig. 5. The progressive structure begins with the top layer and gradually utilizes the output of the higher layer as prior knowledge to fuse the encoder features. This mechanism is not conducive to the recovery of details because the high-level features lack fine information. While the parallel structure easily results in inaccurate localization of objects since the low-level features without semantic information directly interfere with the capture of global structure cues. In this work, we mix the two structures to build a dual branch decoder to overcome the above restrictions. We briefly describe the FPN branch. Taking D^i as an example, we firstly apply bilinear interpolation to upsample the higher-level feature D^{i+1} to the same size as T^i. Next, to decrease the number of parameters, T^i is reduced to 32 channels and fed into gate unit G^i. Lastly, the gated feature is fused with the upsampled feature of D^{i+1} by element-wise addition and convolutional layers. This process can be formulated as follows:

$$D^i = \begin{cases} Conv(G_1^i \cdot T^i + Up(D^{i+1})) & \text{if } i = 1, 2, 3, 4 \\ Conv(G_1^i \cdot T^i) & \text{if } i = 5, \end{cases} \quad (2)$$

where D^1 is a single-channel feature map with the same size as the input image.

In the parallel branch, we firstly upsample $T^1 \sim T^5$ to the same size of D^1. Next, the multilevel gate units are followed to weight the corresponding transition-layer features. Lastly, we combine D^1 and the gated features by cross-channel concatenation. The whole process is written as follows:

$$F_{Cat} = Cat(D^1, Up(G_2^1 \cdot T^1), Up(G_2^2 \cdot T^2),$$
$$Up(G_2^3 \cdot T^3), Up(G_2^4 \cdot T^4), Up(G_2^5 \cdot T^5)). \quad (3)$$

The final saliency map S^F is generated by integrating the predictions of the two branches with a residual connection as shown in Fig. 5(c),

$$S^F = S(Conv(F_{Cat}) + D^1)), \quad (4)$$

where $S(\cdot)$ is the element-wise sigmoid function.

3.3   Folded Atrous Spatial Pyramid Pooling

In order to obtain robust segmentation results by integrating multiscale information, atrous spatial pyramid pooling (ASPP) is proposed in Deeplab [3]. And some works [6,42] also show its effectiveness in saliency detection. The ASPP uses multiple parallel atrous convolutional layers with different dilation rates. The sparsity of atrous convolution kernel, especially when using a large dilation rate, results in that the association relationships among sampling points are too weak to extract stable features. In this paper, we apply a simple "Fold" operation to effectively relieve this issue. We visualize the folded convolution structure in Fig. 6, which not only further enlarges the receptive field but also extends each valid sampling position from an isolate point to a 2×2 connected region.

Let $\mathbf{X}$ represent feature maps with the size of $N \times N \times C$ (C is the channel number). We slide a 2×2 window on $\mathbf{X}$ in stride 2 and then conduct atrous convolution with kernel size $K \times K$ in different dilation rates. Figure 6 shows the computational process when $K = 3$ and dilation rate is 2. Firstly, we collect $2 \times 2 \times C$ feature points in each window from $\mathbf{X}$ and then it is stacked by channel direction, we call this operation "Fold", which is shown in Fig. 6①. After the fold operation, we can get new feature maps with the size of $N/2 \times N/2 \times 4C$. A point on the new feature maps corresponds to a 2×2 area on the original feature maps. Secondly, we adopt an atrous convolution with a kernel size of 3×3 and dilation rate is 2. Followed by the reverse process of "Fold" which is called "Unfold" operation, the final feature maps are obtained. By using the folded atrous convolution, in the process of information transfer across

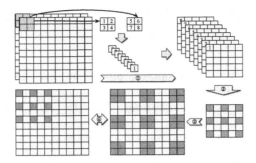

Fig. 6. Illustration of the folded convolution. We use ①, ② and ③ to respectively indicate "Fold" operation, atrous convolution and "Unfold" operation. ④ shows the comparison between atrous convolution (Left) and the folded atrous convolution (Right).

convolution layers, more contexts are merged and the certain local correlation is also preserved, which provides the fault-tolerance capability for subsequent operations.

As shown in Fig. 2, the Fold-ASPP is only equipped on the top of the encoder network, which consists of three folded convolutional layers with dilation rates [2, 4, 6] to fit the size of feature maps. Just as group convolution [37] is a trade-off between depthwise convolution [5,11] and vanilla convolution in the channel dimension, the proposed folded convolution is a trade-off between atrous convolution and vanilla convolution in the spatial dimension.

3.4   Supervision

As shown in Fig. 2, we use the cross-entropy loss for both the intermediate prediction from the FPN branch and the final prediction from the dual branch. In the dual branch decoder, since the FPN branch gradually combines all-level gated encoding and decoding features, it has very powerful prediction ability. We expect that it can predict salient objects as accurately as possible under the supervision of ground truth. While the parallel branch only combines the gated encoding features, which is helpful to remedy the ignored details with the design of residual structure. Moreover, the supervision on D^1 can drive gate units to learn the weight of the contribution of each encoder block to the final prediction. We use the cross-entropy loss. The total loss L could be written as:

$$L = l_{s1} + l_{sf}, \tag{5}$$

where l_{s1} and l_{sf} are respectively used to regularize the output of the FPN branch and the final prediction. The cross-entropy loss could be computed as:

$$l = Y log P + (1 - Y) log (1 - P), \tag{6}$$

where P and Y denote the predicted map and ground-truth, respectively.

4 Experiments

4.1 Experimental Setup

Dataset. We evaluate the proposed model on five benchmark datasets. *ECSSD* [38] contains $1,000$ semantically meaningful and complex images with pixel-accurate ground truth annotations. *HKU-IS* [13] has $4,447$ challenging images with multiple disconnected salient objects, overlapping the image boundary. *PASCAL-S* [15] contains 850 images selected from the PASCAL VOC 2009 segmentation dataset. *DUT-OMRON* [39] includes $5,168$ challenging images, each of which usually has complicated background and one or more foreground objects. *DUTS* [27] is the largest salient object detection dataset, which contains $10,553$ training and $5,019$ test images. These images contain very complex scenarios with high-diversity contents.

Evaluation Metrics. For quantitative evaluation, we adopt four widely-used metrics: precision-recall (PR) curve, F-measure score, mean absolute error (MAE) and S-measure score. *Precision-Recall curve:* The pairs of precision and recall are calculated by comparing the binary saliency maps with the ground truth to plot the PR curve, where the threshold for binarizing slides from 0 to 255. The closer the PR curve is to the upper right corner, the better the performance is. *F-measure*: It is an overall performance measurement that synthetically considers both precision and recall:

$$F_\beta = \frac{\left(1 + \beta^2\right) \cdot \text{precision} \cdot \text{recall}}{\beta^2 \cdot \text{precision} + \text{recall}}, \tag{7}$$

where β^2 is set to 0.3 as suggested in [1] to emphasize the precision. In this paper, we report the maximum F-measure score across the binary maps of different thresholds. *Mean Absolute Error*: As the supplement of the PR curve and F-measure, it computes the average absolute difference between the saliency map and the ground truth pixel by pixel. *S-measure*: It is more sensitive to foreground structural information than the F-measure. It considers the region-aware structural similarity S_r and the object-aware structural similarity S_o:

$$S_m = \alpha * S_o + (1 - \alpha) * S_r, \tag{8}$$

where α is set to 0.5 [7].

Implementation Details. We follow most state-of-the-art saliency detection methods $[21,26,31,33,34,36,41,42,45]$ to use the DUTS-TR as the training dataset which contains $10,553$ images. Our model is implemented based on the Pytorch repository and the hyper-parameters are set as follows: We train the GateNet on a PC with GTX 1080 Ti GPU for 40 epochs with mini-batch size 4. For the optimizer, we adopt the stochastic gradient descent (SGD). The momentum, weight decay, and learning rate are set as 0.9, 0.0005 and 0.001, respectively. The "poly" policy [18] with the power of 0.9 is used to adjust the learning rate. We adopt some data augmentation techniques to avoid overfitting and make the learned model more robust, which include random horizontally flipping, random rotation, random brightness, saturation and contrast changing. In order to preserve the integrity of the image semantic information, we only resize the image to 384×384 instead of using a random crop. The source code will be publicly available at https://github.com/Xiaoqi-Zhao-DLUT/GateNet-RGB-Saliency.

4.2   Performance Comparison with State-of-the-Art

We compare the proposed algorithm with seventeen state-of-the-art saliency detection methods, including the DCL [14], DSS [10], Amulet [44], SRM [29], DGRL [31], RAS [4], PAGRN [45], BMPM [42], R3Net [6], HRS [41], MLMS [35], PAGE [34], ICNet [33], CPD [36], BANet [26], BASNet [21] and Capsal [43]. For fair comparisons, all the saliency map of these methods are directly provided by their respective authors or computed by their released codes. To further show the effectiveness of our GateNet, we test its performance in both **RGBD SOD** and **Video Object Segmentation** tasks and include the results in supplementary materials.

Quantitative Evaluation. Table 1 shows the experimental comparison results in terms of the F-measure, S-measure and MAE scores, from which we can see that the GateNet can consistently outperform other approaches across all five datasets and different metrics. In particular, the GateNet achieves significant performance improvement in terms of the F-measure compared to the second best method BANet [26] on the challenging DUTS-test (0.870 vs 0.852 and 0.888 vs 0.872) and PASCAL-S (0.882 vs 0.866 and 0.883 vs 0.877) datasets. This clearly demonstrates its superior performance in complex scenes. Moreover, some methods [6,10,14,29] apply the post-processing techniques to refine their saliency maps. Our GateNet still performs better than them without any post-processing. We evaluate different algorithms using the standard PR curves in Fig. 7. It can be seen that our PR curves are significantly higher than those of other methods on five datasets.

Qualitative Evaluation. Fig. 1 and Fig. 8 illustrate some visual comparisons. In Fig. 1, other methods are severely disturbed by branches and weeds while ours can precisely identify the whole objects. And the GateNet can significantly suppress the background with similar shapes to salient objects (see the 1st row in Fig. 8). Since the Fold-ASPP can obtain more stable structural features, it can help to accurately locate objects and separate adjacent objects well, but some competitors make adjacent objects stick together (see the 3th and 4th rows in Fig. 8). Besides, the proposed parallel branch can restore more details, therefore, the boundary information is retained well.

4.3   Ablation Studies

We detail the contribution of each component to the overall network.

Effectiveness of Backbones. Table 1 demonstrates that the performance of the gated network can be significantly improved by using better backbones such as ResNet-50, ResNet-101 or ResNeXt-101.

Table 1. Quantitative comparisons. **Blue** indicates the best performance under each backbone setting, while red indicates the best performance among all settings. The subscript in the first column regards the publication year. "†", "S" and "X" mean using the post-processing, ResNet-101 and ResNeXt-101 backbone, respectively. "—" represents that the results are not available. ↑ and ↓ indicate that the larger and smaller scores are better, respectively.

Method	DUTS-test			DUT-OMRON			PASCAL-S			HKU-IS			ECSSD		
	F_β↑	S_m↑	MAE↓	F_β↑	S_m↑	MAE↓	F_β↑	S_m↑	MAE↓	F_β↑	S_m↑	MAE↓	F_β↑	S_m↑	MAE↓
						VGG-16 backbone									
$DCL^\dagger_{16}$	0.782	0.796	0.088	0.757	0.770	0.080	0.829	0.793	0.109	0.907	0.877	0.048	0.901	0.868	0.068
$DSS^\dagger_{17}$	—	—	—	0.781	0.789	0.063	0.840	0.792	0.098	0.916	0.878	0.040	0.921	0.882	0.052
$Amulet_{17}$	0.778	0.804	0.085	0.743	0.780	0.098	0.839	0.819	0.099	0.899	0.886	0.050	0.915	0.894	0.059
$BMPM_{18}$	0.852	0.860	0.049	0.774	0.808	0.064	0.862	0.842	0.076	0.921	0.906	0.039	0.928	0.911	0.045
RAS_{18}	0.831	0.838	0.059	0.786	0.813	0.062	0.836	0.793	0.106	0.913	0.887	0.045	0.921	0.893	0.056
$PAGRN_{18}$	0.854	0.837	0.056	0.771	0.774	0.071	0.855	0.814	0.095	0.919	0.889	0.048	0.927	0.889	0.061
HRS_{19}	0.843	0.828	0.051	0.762	0.771	0.066	0.850	0.798	0.092	0.913	0.882	0.042	0.920	0.883	0.054
$MLMS_{19}$	0.852	0.861	0.049	0.774	0.808	0.064	0.864	0.844	0.075	0.921	0.906	0.039	0.928	0.911	0.045
$PAGE_{19}$	0.838	0.853	0.052	0.792	**0.824**	0.062	0.858	0.837	0.079	0.920	0.904	0.036	0.931	0.912	0.042
$BANet_{19}$	0.852	0.860	0.046	0.793	0.822	**0.061**	0.866	0.838	0.079	0.919	0.901	0.037	0.935	0.913	**0.041**
GateNet	**0.870**	**0.869**	**0.045**	**0.794**	0.820	**0.061**	**0.882**	**0.855**	**0.070**	**0.928**	**0.909**	**0.035**	**0.941**	**0.917**	**0.041**
						ResNet-50 backbone									
$SRM^\dagger_{17}$	0.826	0.835	0.059	0.769	0.797	0.069	0.848	0.830	0.087	0.906	0.886	0.046	0.917	0.895	0.054
$DGRL_{18}$	0.828	0.841	0.050	0.774	0.805	0.062	0.856	0.836	0.073	0.911	0.895	0.036	0.922	0.903	0.041
CPD_{19}	0.865	0.868	0.043	0.797	0.824	0.056	0.870	0.844	0.074	0.925	0.906	0.034	0.939	0.918	0.037
$ICNet_{19}$	0.855	0.864	0.048	0.813	**0.837**	0.061	0.865	0.849	0.072	0.925	0.908	0.037	0.938	0.918	0.041
$BASNet_{19}$	0.860	0.864	0.048	0.805	0.835	0.057	0.860	0.834	0.079	0.930	0.907	**0.033**	0.943	0.916	0.037
$BANet_{19}$	0.872	0.878	**0.040**	0.803	0.832	0.059	0.877	0.851	0.072	0.930	0.913	**0.033**	0.944	**0.924**	**0.035**
GateNet	**0.888**	**0.884**	**0.040**	**0.818**	**0.837**	**0.055**	**0.883**	**0.857**	**0.069**	**0.933**	**0.915**	**0.033**	**0.945**	0.920	0.040
						ResNet/ResNeXt-101 backbone									
$R3Net^\dagger_{18}{}^X$	0.819	0.827	0.063	0.795	0.816	0.063	0.844	0.802	0.095	0.915	0.895	0.035	0.934	0.910	0.040
$Capsal^S_{19}$	0.819	0.818	0.063	0.639	0.673	0.101	0.869	0.837	0.074	0.883	0.851	0.058	0.863	0.826	0.077
$GateNet^S$	**0.893**	**0.889**	**0.038**	0.821	0.844	0.054	0.883	0.862	0.067	**0.937**	0.920	0.031	**0.951**	**0.930**	0.035
$GateNet^X$	0.898	0.895	0.035	0.829	0.848	0.051	0.888	0.865	0.065	0.943	0.925	0.029	0.952	**0.929**	0.035

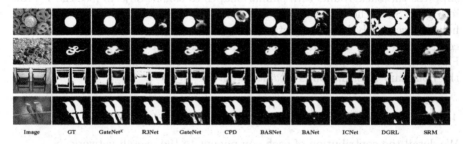

Fig. 7. Precision (vertical axis) recall (horizontal axis) curves on six popular rgb-salient object datasets.

Fig. 8. Visual comparison between our results and state-of-the-art methods.

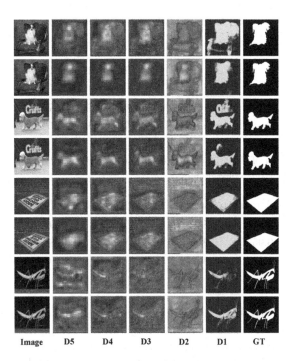

| Image | D5 | D4 | D3 | D2 | D1 | GT |

Fig. 9. Visual comparison of feature maps for showing the effect of the multilevel gate units. D5 ∼ D1 represent the feature maps of each decoder block from high level to low level. Odd rows and even rows are the results of the FPN baseline without or with multilevel gate units, respectively.

Effectiveness of Components. We quantitatively show the benefit of each component in Table 2. We take the results of the VGG-16 backbone with the FPN branch as the baseline. Firstly, the multilevel gate units are added to the baseline network. The performance is significantly improved with the gain of 2.94%, 2.17% and 11.67% in terms of the F-measure, S-measure and MAE, respectively. To show the effect of the gate units more intuitively, we visualize the features of different levels in Fig. 9. It can be observed that even if the dog has a very low contrast with the chair or the billboard (see the $1^{st} \sim 4^{th}$ rows), through using multilevel gate units, the high contrast between the object region and the background is always maintained at each layer while the detail information is continually regained, thereby making salient objects be effectively distinguished. Besides, the gate units can avoid excessive suppression for the slender parts of objects (see the $5^{th} \sim 8^{th}$ rows). The corners of the poster, the limbs and even tentacles of the mantis are retained well. Secondly, based on the gated baseline network, we design a series of experimental options to verify the effectiveness of the folded convolution and Fold-ASPP. Table 3 illustrates the results in detail. We adopt the atrous convolution with dilation rates of $[2, 4, 6]$ and the same dilation rates are also applied to the folded convolution. It can be observed that the folded convolution

Table 2. Ablation analysis on the DUTS dataset.

	F_β	S_m	MAE
Baseline (FPN)	0.816	0.829	0.060
+ Gate Units	0.840	0.847	0.053
+ Fold-ASPP	0.866	0.863	0.047
+ Parallel Branch	0.870	0.869	0.045

Table 3. Evaluation of the folded convolution and Fold-ASPP. (x) stands for different sampling rates of atrous convolution.

	Atrous(2)	Atrous(4)	Atrous(6)	Fold(2)	Fold(4)	Fold(6)	ASPP	Fold-ASPP
F_β	0.840	0.845	0.848	0.853	0.856	0.860	0.856	**0.866**
MAE	0.055	0.053	0.051	0.051	0.050	0.048	0.051	**0.047**
S_m	0.847	0.849	0.851	0.856	0.858	0.859	0.860	**0.863**

consistently yields significant performance improvement at each dilation rate than the corresponding atrous convolution in terms of all three metrics. And the single-layer Fold(6) already performs better than the ASPP of aggregating three atrous convolution layers. The Fold-ASPP also naturally outperforms the ASPP with the gain of 1.17% and 8.0% in terms of the F-measure and MAE, respectively. Finally, we add the parallel branch to further restore the details of objects. In this process, the gate units, Fold-ASPP and parallel branch complement each other without repulsion.

5 Conclusions

In this paper, we propose a novel gated network architecture for saliency detection. We first adopt multilevel gate units to balance the contribution of each encoder block and suppress the activation of the features of non-salient regions, which can provide useful context information for the decoder while minimizing interference. The gate unit is simple yet effective, therefore, a gated FPN network can be used as a new baseline for dense prediction tasks. Next, we use the Fold-ASPP to gather multiscale semantic information for the decoder. By the folded operation, the atrous convolution achieves a local-in-local effect, which not only expands the receptive field but also retains the correlation among local sampling points. Finally, to further supplement the details, we combine all encoder features in parallel and construct a residual structure. Experimental results on five benchmark datasets demonstrate that the proposed model outperforms seventeen state-of-the-art methods under different evaluation metrics.

Acknowledgements. This work was supported in part by the National Natural Science Foundation of China #61876202, #61725202, #61751212 and #61829102, the Dalian Science and Technology Innovation Foundation #2019J12GX039, and the Fundamental Research Funds for the Central Universities # DUT20ZD212.

References

1. Achanta, R., Hemami, S., Estrada, F., Süsstrunk, S.: Frequency-tuned salient region detection. In: Proceedings of IEEE Conference on Computer Vision and Pattern Recognition, pp. 1597–1604 (2009)
2. Amirul Islam, M., Rochan, M., Bruce, N.D., Wang, Y.: Gated feedback refinement network for dense image labeling. In: Proceedings of IEEE Conference on Computer Vision and Pattern Recognition, pp. 3751–3759 (2017)
3. Chen, L.C., Papandreou, G., Kokkinos, I., Murphy, K., Yuille, A.L.: DeepLab: Semantic image segmentation with deep convolutional nets, atrous convolution, and fully connected CRFs. IEEE Trans. Pattern Anal. Mach. Intell. **40**(4), 834–848 (2017)
4. Chen, S., Tan, X., Wang, B., Hu, X.: Reverse attention for salient object detection. In: Proceedings of European Conference on Computer Vision, pp. 234–250 (2018)
5. Chollet, F.: Xception: deep learning with depthwise separable convolutions. In: Proceedings of IEEE Conference on Computer Vision and Pattern Recognition, pp. 1251–1258 (2017)
6. Deng, Z., et al.: R3Net: recurrent residual refinement network for saliency detection. In: Proceedings of International Joint Conference on Artificial Intelligence, pp. 684–690 (2018)
7. Fan, D.P., Cheng, M.M., Liu, Y., Li, T., Borji, A.: Structure-measure: a new way to evaluate foreground maps. In: Proceedings of IEEE International Conference on Computer Vision, pp. 4548–4557 (2017)
8. Fang, H., et al.: From captions to visual concepts and back. In: Proceedings of IEEE Conference on Computer Vision and Pattern Recognition, pp. 1473–1482 (2015)
9. He, K., Zhang, X., Ren, S., Sun, J.: Deep residual learning for image recognition. In: Proceedings of IEEE Conference on Computer Vision and Pattern Recognition, pp. 770–778 (2016)
10. Hou, Q., Cheng, M.M., Hu, X., Borji, A., Tu, Z., Torr, P.H.: Deeply supervised salient object detection with short connections. In: Proceedings of IEEE Conference on Computer Vision and Pattern Recognition, pp. 3203–3212 (2017)
11. Howard, A.G., et al.: MobileNets: efficient convolutional neural networks for mobile vision applications. arXiv preprint arXiv:1704.04861 (2017)
12. Jiang, Z., Davis, L.S.: Submodular salient region detection. In: Proceedings of IEEE Conference on Computer Vision and Pattern Recognition, pp. 2043–2050 (2013)
13. Li, G., Yu, Y.: Visual saliency based on multiscale deep features. In: Proceedings of IEEE Conference on Computer Vision and Pattern Recognition, pp. 5455–5463 (2015)
14. Li, G., Yu, Y.: Deep contrast learning for salient object detection. In: Proceedings of IEEE Conference on Computer Vision and Pattern Recognition, pp. 478–487 (2016)
15. Li, Y., Hou, X., Koch, C., Rehg, J.M., Yuille, A.L.: The secrets of salient object segmentation. In: Proceedings of IEEE Conference on Computer Vision and Pattern Recognition, pp. 280–287 (2014)

16. Lin, T.Y., Dollár, P., Girshick, R., He, K., Hariharan, B., Belongie, S.: Feature pyramid networks for object detection. In: Proceedings of IEEE Conference on Computer Vision and Pattern Recognition, pp. 2117–2125 (2017)

17. Liu, N., Han, J.: DHSNet: deep hierarchical saliency network for salient object detection. In: Proceedings of IEEE Conference on Computer Vision and Pattern Recognition, pp. 678–686 (2016)

18. Liu, W., Rabinovich, A., Berg, A.C.: ParseNet: looking wider to see better. arXiv preprint arXiv:1506.04579 (2015)

19. Mahadevan, V., Vasconcelos, N.: Saliency-based discriminant tracking. In: Proceedings of IEEE Conference on Computer Vision and Pattern Recognition (2009)

20. Pang, Y., Zhao, X., Zhang, L., Lu, H.: Multi-scale interactive network for salient object detection. In: Proceedings of IEEE Conference on Computer Vision and Pattern Recognition, pp. 9413–9422 (2020)

21. Qin, X., Zhang, Z., Huang, C., Gao, C., Dehghan, M., Jagersand, M.: BASNet: boundary-aware salient object detection. In: Proceedings of IEEE Conference on Computer Vision and Pattern Recognition, pp. 7479–7489 (2019)

22. Ren, Z., Gao, S., Chia, L.T., Tsang, I.W.H.: Region-based saliency detection and its application in object recognition. IEEE Trans. Circuits Syst. Video Technol. **24**(5), 769–779 (2013)

23. Ronneberger, O., Fischer, P., Brox, T.: U-Net: convolutional networks for biomedical image segmentation. In: Navab, N., Hornegger, J., Wells, W.M., Frangi, A.F. (eds.) MICCAI 2015. LNCS, vol. 9351, pp. 234–241. Springer, Cham (2015). https://doi.org/10.1007/978-3-319-24574-4_28

24. Rui, Z., Ouyang, W., Wang, X.: Unsupervised salience learning for person re-identification. In: Proceedings of IEEE Conference on Computer Vision and Pattern Recognition (2013)

25. Simonyan, K., Zisserman, A.: Very deep convolutional networks for large-scale image recognition. arXiv preprint arXiv:1409.1556 (2014)

26. Su, J., Li, J., Zhang, Y., Xia, C., Tian, Y.: Selectivity or invariance: boundary-aware salient object detection. In: Proceedings of IEEE International Conference on Computer Vision, pp. 3799–3808 (2019)

27. Wang, L., et al.: Learning to detect salient objects with image-level supervision. In: Proceedings of IEEE Conference on Computer Vision and Pattern Recognition, pp. 136–145 (2017)

28. Wang, L., Wang, L., Lu, H., Zhang, P., Ruan, X.: Saliency detection with recurrent fully convolutional networks. In: Leibe, B., Matas, J., Sebe, N., Welling, M. (eds.) ECCV 2016. LNCS, vol. 9908, pp. 825–841. Springer, Cham (2016). https://doi.org/10.1007/978-3-319-46493-0_50

29. Wang, T., Borji, A., Zhang, L., Zhang, P., Lu, H.: A stagewise refinement model for detecting salient objects in images. In: Proceedings of IEEE International Conference on Computer Vision, pp. 4019–4028 (2017)

30. Wang, T., Piao, Y., Li, X., Zhang, L., Lu, H.: Deep learning for light field saliency detection. In: Proceedings of IEEE International Conference on Computer Vision, pp. 8838–8848 (2019)

31. Wang, T., et al.: Detect globally, refine locally: a novel approach to saliency detection. In: Proceedings of IEEE Conference on Computer Vision and Pattern Recognition, pp. 3127–3135 (2018)

32. Wang, W., Lai, Q., Fu, H., Shen, J., Ling, H.: Salient object detection in the deep learning era: an in-depth survey. arXiv preprint arXiv:1904.09146 (2019)

33. Wang, W., Shen, J., Cheng, M.M., Shao, L.: An iterative and cooperative top-down and bottom-up inference network for salient object detection. In: Proceedings of IEEE Conference on Computer Vision and Pattern Recognition, pp. 5968–5977 (2019)
34. Wang, W., Zhao, S., Shen, J., Hoi, S.C., Borji, A.: Salient object detection with pyramid attention and salient edges. In: Proceedings of IEEE Conference on Computer Vision and Pattern Recognition, pp. 1448–1457 (2019)
35. Wu, R., Feng, M., Guan, W., Wang, D., Lu, H., Ding, E.: A mutual learning method for salient object detection with intertwined multi-supervision. In: Proceedings of IEEE Conference on Computer Vision and Pattern Recognition, pp. 8150–8159 (2019)
36. Wu, Z., Su, L., Huang, Q.: Cascaded partial decoder for fast and accurate salient object detection. In: Proceedings of IEEE Conference on Computer Vision and Pattern Recognition, pp. 3907–3916 (2019)
37. Xie, S., Girshick, R., Dollár, P., Tu, Z., He, K.: Aggregated residual transformations for deep neural networks. In: Proceedings of IEEE Conference on Computer Vision and Pattern Recognition, pp. 1492–1500 (2017)
38. Yan, Q., Xu, L., Shi, J., Jia, J.: Hierarchical saliency detection. In: Proceedings of IEEE Conference on Computer Vision and Pattern Recognition, pp. 1155–1162 (2013)
39. Yang, C., Zhang, L., Lu, H., Ruan, X., Yang, M.H.: Saliency detection via graph-based manifold ranking. In: Proceedings of IEEE Conference on Computer Vision and Pattern Recognition, pp. 3166–3173 (2013)
40. Yang, G.R., Murray, J.D., Wang, X.J.: A dendritic disinhibitory circuit mechanism for pathway-specific gating. Nat. Commun. **7**, 12815 (2016)
41. Zeng, Y., Zhang, P., Zhang, J., Lin, Z., Lu, H.: Towards high-resolution salient object detection. In: Proceedings of IEEE International Conference on Computer Vision, pp. 7234–7243 (2019)
42. Zhang, L., Dai, J., Lu, H., He, Y., Wang, G.: A bi-directional message passing model for salient object detection. In: Proceedings of IEEE Conference on Computer Vision and Pattern Recognition, pp. 1741–1750 (2018)
43. Zhang, L., Zhang, J., Lin, Z., Lu, H., He, Y.: CapSal: leveraging captioning to boost semantics for salient object detection. In: Proceedings of IEEE Conference on Computer Vision and Pattern Recognition, pp. 6024–6033 (2019)
44. Zhang, P., Wang, D., Lu, H., Wang, H., Ruan, X.: Amulet: aggregating multi-level convolutional features for salient object detection. In: Proceedings of IEEE International Conference on Computer Vision, pp. 202–211 (2017)
45. Zhang, X., Wang, T., Qi, J., Lu, H., Wang, G.: Progressive attention guided recurrent network for salient object detection. In: Proceedings of IEEE Conference on Computer Vision and Pattern Recognition, pp. 714–722 (2018)
46. Zhao, T., Wu, X.: Pyramid feature attention network for saliency detection. In: Proceedings of IEEE Conference on Computer Vision and Pattern Recognition, pp. 3085–3094 (2019)

Visual Memorability for Robotic Interestingness via Unsupervised Online Learning

Chen Wang[iD], Wenshan Wang[iD], Yuheng Qiu[iD], Yafei Hu[iD],
and Sebastian Scherer[✉][iD]

Carnegie Mellon University, Pittsburgh, PA 15213, USA
chenwang@dr.com, {wenshanw,yuhengq,yafeih,basti}@andrew.cmu.edu
https://github.com/wang-chen/interestingness

Abstract. In this paper, we explore the problem of interesting scene prediction for mobile robots. This area is currently underexplored but is crucial for many practical applications such as autonomous exploration and decision making. Inspired by industrial demands, we first propose a novel translation-invariant visual memory for recalling and identifying interesting scenes, then design a three-stage architecture of long-term, short-term, and online learning. This enables our system to learn human-like experience, environmental knowledge, and online adaption, respectively. Our approach achieves much higher accuracy than the state-of-the-art algorithms on challenging robotic interestingness datasets.

Keywords: Unsupervised · Online · Memorability · Interestingness

1 Introduction

Interesting scene prediction is crucial for autonomous exploration [28], which is one of the most fundamental capabilities of mobile robots. It has a significant impact on decision making and robot cooperation. For example, the finding of a door shown in Fig. 1(f) may affect the future planing, the hole on the wall in Fig. 1(h) may attract more attentions. However, prior algorithms often have difficulty when they are deployed to unknown environments, as the robots not only have to find interesting scenes, but also have to lose the interests on repetitive scenes, i.e., interesting scenes may become uninteresting during robot exploration after repeatedly observing similar scenes or following moving objects. For example in Fig. 6, we expect to have high interests on the truck when it appears but loss the interests when it exists for a long time. Nevertheless, the recent approaches of interestingness detection [17,21], as well as saliency detection [41], anomaly detection [27,42], novelty detection [2], and meaningfulness detection [18] algorithms cannot achieve this online updates scheme.

Electronic supplementary material The online version of this chapter (https://doi.org/10.1007/978-3-030-58536-5_4) contains supplementary material, which is available to authorized users.

© Springer Nature Switzerland AG 2020
A. Vedaldi et al. (Eds.): ECCV 2020, LNCS 12347, pp. 52–68, 2020.
https://doi.org/10.1007/978-3-030-58536-5_4

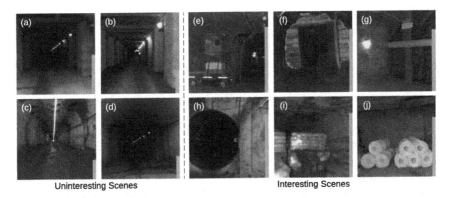

Fig. 1. In this paper, we aim to predict robotic interesting scenes, which are crucial for decision making and autonomous cooperation. To enable the behavior of online losing interests on repetitive scenes for exploration of mobile robots, we propose to establish an online update scheme for interesting scene prediction. This figure shows several examples of both uninteresting and interesting scenes in SubT data [1] taken by autonomous robots. The height of green strip located at the right of each image indicates the interestingness level predicted by our unsupervised online learning algorithm when it sees the scene for the first time. (Color figure online)

To this end, we propose to establish an **online learning** scheme to search for interesting scenes for robot exploration tasks. On the other hand, existing algorithms are heavily dependent on back-propagation algorithm [32] for learning, which is very computational expensive. To solve this problem, we introduce a novel *translation-invariant 4-D visual memory* to identify and recall visually interesting scenes. Human beings have a great capacity to direct visual attention and judge the interestingness of a scene [6]. For mobile robots, we find the following properties are necessary to establish a sense of visual interestingness.

Unsupervised: As shown in Fig. 1, the interesting scenes in robot operating environments are often unique and unknown, thus the labels are normally difficult to obtain, but prior research mainly focuses on supervised methods [3,6] and suffers in prior unseen environments. We hypothesize that a sense of interestingness can be established in an unsupervised manner.

Task-Dependent: In many practical applications, we might only know uninteresting scenes before a mission is started. In the example of tunnel exploration task in Fig. 1, the deployment will be more efficient and easier if the robots can be taught what is uninteresting within several minutes. In this sense, we argue that the visual interestingness prediction system should be able to learn from negative samples quickly without accessing to data from unsupervised learning, thus an incremental learning method is necessary. Note that we expect the model is capable of learning from negative samples, but it is not necessary for all tasks.

To achieve the above properties, we propose a three-stage architecture:

Long-Term Learning: In this stage, we expect a model to be trained off-line on a large amount of data in an unsupervised manner as human beings acquire common knowledge from experience. We also expect the training time on single machine to be no more than the order of days.

Short-Term Learning: For task-dependent knowledge, the model then should be able to learn from hundreds of uninteresting images in minutes. This can be done before a mission started and beneficial to quick robot deployment.

Online Learning: During mission execution the system should express the top interests in real-time and the detected interests should be lost online when similar scenes appear frequently, regardless if they exist in the uninteresting images or not. Another important aspect for online learning is no data leakage, *i.e.*, each frame is proceed without using information from its subsequent frames. This is in contrast to prior works [17,21] and datasets [8], where interesting frames are selected after an entire sequence is processed [15]. Since robots need to respond in real-time, we require that our algorithms are able to adapt quickly. To measure such capability of online response, we propose a new evaluation metric.

In summary, our contributions are:

- We introduce an extremely simplified three-stage architecture for robotic interesting scene prediction, which is crucial for practical applications. Concretely, we leverage long-term learning to acquire human-link experience, short-term learning for quick robot deployment and task-related knowledge, and online learning for environment adaption and real-time response.
- To accelerate the short-term and online learning, we propose a novel 4-D visual memory to replace back-propagation. Concretely, we introduce cross-correlation similarity for translational invariance, which is crucial for perceiving video stream, we also introduce tangent operator for safe writing, which is crucial for incremental learning from negative samples.
- To measure the online performance, we propose a strict evaluation metric, *i.e.*, the area under the curve of online precision (AUC-OP) to jointly consider precision, recall rate, and online performance.
- It is demonstrated that our approach achieves much higher overall performance than the state-of-the-art algorithms.

2   Related Work

A learning system that encodes the three-stage architecture for interesting scene prediction has not been achieved, thus the formulation as well as performance evaluation will be quite different from prior approaches. Some works on interestingness prediction have different objectives [3], *e.g.*, Shen *et al.* aimed to predict human interestingness on social media [34]. In this section we will mainly review the related techniques, as some methods used in saliency, anomaly, and novelty detection are also useful for our work.

The definition of interestingness is subjective, thus the annotation has to be averaged over different participants. To mimic the human judgment, prior works have paid great attentions to investigate the relationship between human visual interestingness and image features [3]. They are typically inspired by psycholog-ical cues and heavily leverage human annotation for training, which results in a large family of supervised learning methods. For instance, Dhar *et al.* designed three hand-crafted rules, including attributes of composition, content, and sky-illumination to approximate both aesthetics and interestingness of images [10]. Jiang *et al.* extended image interestingness to video and evaluated hand-crafted visual features for predicting interestingness on the YouTube and Flickr datasets [21]. Fu *et al.* formulated interestingness as a problem of unified learning to rank, which is able to jointly identify human annotation outliers [11,12].

Deep neural networks played more and more significant roles in recent works on interestingness prediction. For example, Gygli *et al.* introduced VGG features [35] and leveraged a support vector regression model to predict the interesting-ness of animated GIFs [17]. Chaabouni *et al.* constructed a customized CNN model to identify salient and non-salient windows for video interestingness pre-diction [5]. Inspired by a human annotation procedure of pairwise comparison, Wang *et al.* combined two deep ranking networks [39] to obtain better per-formance, and this method ranked first in the 2017 interestingness prediction competition [9]. Shen *et al.* combined both CNN and LSTM [19] for feature learning to predict video interestingness [34] for media contents.

However, the aforementioned methods are highly dependent on human anno-tation for training, which is labor expensive and not suitable for interestingness search [6]. Some efforts for unsupervised learning of interestigness have been made in [20], where interesting events of videos are detected using the density ratio estimation algorithm with the HOG feature [7]. However, in practice the approach cannot adapt well to changing distributions.

In the long-term stage, we introduce an autoencoder [23] for unsupervised learning, which has been widely used for feature extraction in many applica-tions. For example, Hasan *et al.* showed that an autoencoder is able to learn regular dynamics and identify irregularity in long-duration videos [18]. Zhang *et al.* introduced dropout into the autoencoder for pixel-wise saliency detection in images [41]. Zhao *et al.* proposed a spatio-temporal autoencoder to extract both spatial and temporal features for anomaly detection [43].

In order to learn online, we introduce a novel visual memory module into the convolutional neural networks. Visual memory has been widely investigated in neuroscience [30]. While in computer vision, memory aided neural networks received limited attentions and used for several different tasks. For example, Graves et al. proposed a differentiable neural Turning machines (NTM) [16], which coupled external memory with recurrent neural networks (RNN). Santoro et al. extended NTM and designed a module to efficiently access the memory [33]. Gong et al. introduced memory module into an auto-encoder to remember normal events for anomaly detection [13]. Kim et al. introduced the memory network into GANs to remember previously generated samples to alleviate the forgetting problem [22]. However, the memories in the above works are defined as flattened vectors, thus the spatial structural information cannot be retained. In this paper, we propose a translation-invariant memory module and introduce online learning to solve the problem of robotic interestingness prediction.

3 Visual Memory

To retain the structural information of visual inputs, the visual memory $\mathbf{M}$ is defined as a 4-D tensor, $i.e.$, $\mathbf{M} \in \mathbb{R}^{n \times c \times h \times w}$, where n is the number of memory cubes and c, h, and w are the channel, height, and width of each cube, respectively. Intuitively, memory writing is to encode visual inputs into the memory, while reading is to recall one's memory regarding the visual inputs.

3.1 Memory Writing

We desire that the visual memory is able to balance new visual inputs and old knowledge. To this end, we denote visual inputs at time t as $\mathbf{x}(t) \in \mathbb{R}^{c \times h \times w}$ and define the writing protocol for the i_{th} memory cube $\mathbf{M}_i$ at time t as

$$\mathbf{M}_i(t) = (1 - \mathbf{w}_i) \cdot \mathbf{M}_i(t-1) + \mathbf{w}_i \cdot \mathbf{x}(t), \tag{1}$$

where $\mathbf{w}_i$ is the i_{th} element of a weight vector $\mathbf{w} \in \mathbb{R}^n$,

$$\mathbf{w} = \sigma(\gamma_w \cdot \tan(\frac{\pi}{2} \cdot D(\mathbf{x}(t), \mathbf{M}(t-1)))), \tag{2}$$

where $\sigma(\cdot)$ is the softmax function and $D(\mathbf{x}, \mathbf{M})$ is a cosine similarity vector, in which the i_{th} element $D_i(\mathbf{x}, \mathbf{M})$ is

$$D_i(\mathbf{x}, \mathbf{M}) = \frac{\sum(\mathbf{x} \odot \mathbf{M}_i)}{\|\mathbf{x}\|_{\mathbf{F}} \cdot \|\mathbf{M}_i\|_{\mathbf{F}}}, \tag{3}$$

where $\odot$, $\sum$, and $\| \cdot \|_{\mathbf{F}}$ are element-wise product, elements summation, and Frobenius norm, respectively. The writing protocol in (1) is a moving average, whose learning speed can be controlled via the writing rate γ_w ($\gamma_w > 0$), so that the training samples can be learned with an expected speed.

It is worth noting that, to promote the sparsity of memory writing, we introduce a tangent operator in (2) to map the range of cosine similarity $[-1, 1]$ in (3) to $[-\inf, \inf]$, thus memory writing can be focused on fewer but more relevant cubes via the softmax function. This leads to easier incremental learning and efficient space usage, which will be further explained in Sect. 4.2 and Sect. 6.1.

3.2 Memory Reading

Recall that convolutional features (visual inputs) are invariant to small input translations due to the concatenation of pooling layers to convolutional layers [14]. To obtain invariance to large translations, we need other techniques such as data augmentation, which is very computationally heavy. To solve this problem, we introduce translation in memory reading, leveraging that the structural information of visual inputs are retained in memory writing. Denote 2-D circular translation along the width and height directions with (x, y) elements of the i_{th} memory cube at time t as $\mathbf{M}_i^{(x,y)}(t)$, memory reading $\mathbf{f}(t) \in \mathbb{R}^{c \times h \times w}$ is

$$\mathbf{f}(t) = \sum_{i=1}^{n} \mathbf{r}_i \cdot \mathbf{M}_i^{(x,y)}(t), \tag{4}$$

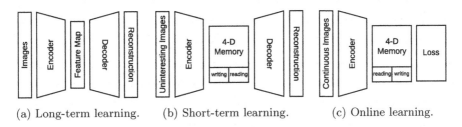

(a) Long-term learning.   (b) Short-term learning.   (c) Online learning.

Fig. 2. The proposed three learning stages. (a) In long-term learning, the parameters in both encoder and decoder are trainable. (b) In short-term learning, the parameters in the encoder and decoder are frozen; the memory writing is performed before reading. (c) In online learning, the parameters in the encoder are frozen; the memory reading is performed before writing.

where $\mathbf{r}_i$ is the i_{th} element of reading weight vector $\mathbf{r} \in \mathbb{R}^n$,

$$\mathbf{r} = \sigma(\gamma_r \cdot \tan(\frac{\pi}{2} \cdot S(\mathbf{x}(t), \mathbf{M}(t)))), \tag{5}$$

where $\gamma_r > 0$ is the reading rate. The i_{th} element of $S(\mathbf{x}, \mathbf{M})$ is the maximum cosine similarity of $\mathbf{x}$ with $\mathbf{M}_i^{(a,b)}$, where $a = 0 : h - 1$ and $b = 0 : w - 1$ imply all translations. Intuitively, to find the maximum cosine similarity, we need to repeatedly compute (3) for translated memory cube $h \times w$ times, resulting in a high computational complexity. To solve this problem, we leverage the fast Fourier transform (FFT) to compute the cross-correlation [36]. Recall that 2-D cross-correlation is the inner-products between the first signal and circular translations of the second signal [38], we can compute $S_i(\mathbf{x}, \mathbf{M}_i)$ as

$$S_i(\mathbf{x}, \mathbf{M}) = \frac{\max \mathcal{F}^{-1}(\sum^c \hat{\mathbf{x}}^* \odot \hat{\mathbf{M}}_i)}{\|\mathbf{x}\|_{\mathbf{F}} \cdot \|\mathbf{M}_i\|_{\mathbf{F}}}, \tag{6}$$

where $\hat{\cdot}$ is the 2-D FFT, $\cdot^*$ is the complex conjugate, and $\sum^c$ is element-wise summation along channel dimension. The translation (x, y) in (4) for the i_{th} memory cube is corresponding to the location of the maximum response, *i.e.*,

$$(x, y) = \arg\max_{(a,b)}(\sum^C \hat{\mathbf{x}}^* \odot \hat{\mathbf{M}}_i)[a, b]. \tag{7}$$

In this way, the computational complexity for each memory cube can be reduced from $\mathcal{O}(ch^2w^2)$ to $\mathcal{O}(chw \log hw)$. Another advantage of translation-invariance in memory reading is that memory usage becomes more efficient, since scene translation is common in video stream for many robotic applications, *e.g.*, robot exploration and object search, which will be further explained in Sect. 6.3.

4   Learning

4.1   Long-Term Learning

Inspired by the fact that human has a massive memory storage capacity [4], we use an autoencoder in Fig. 2a for long-term learning for the following reasons.

Unsupervised Knowledge: A reconstruction model can be trained in an unsupervised way, hence we can collect massive number of images from the internet or in real-time during execution to train the model without much efforts. This agrees with our objective of long-term learning that is to remember as many scenes as possible. In this stage, we still leverage back-propagation for the training, so that the large amount of knowledge will be 'stored' in the trainable parameters, which will be frozen afterwards. In this sense, the learned knowledge can be treated as unforgettable human-like experience.

Detailed and Semantic: To precisely reconstruct images by smaller feature maps, the output of the bottleneck layer has to contain both detailed and semantic information. This is crucial for visual interestingness, since both texture and object-level information may attract one's interests. Feature maps are invariant to small translations due to CNN, we leverage the invariance of visual memory to large translations in short-term and online learning. We construct the encoder following VGG [35] and concatenate 5 deconvolutional blocks [26] for decoder.

4.2   Short-Term Learning

As aforementioned, we normally only know the uninteresting scenes before a robotic mission is started. For known interesting objects, we prefer to use supervised object detectors. Therefore, we expect that our unsupervised model can be trained *incrementally* with negative labeled samples within *several minutes*. This is beneficial for learning environmental knowledge and quick robot deployment.

To this end, we propose the short-term learning architecture in Fig. 2b. The memory module is inserted into the trained reconstruction model, in which all parameters are frozen. For each sample, the output of the encoder is first written into the memory, then memory reading is taken as inputs of the decoder. Intuitively, the images cannot be reconstructed well initially, as feature maps are not fully learned by the memory, and memory reading will be different from the encoding outputs. In this sense, we can inspect the reconstruction error to know whether the memory has learned to encode the training samples or not.

The memory leaning is much faster than back-propagation and has several advantages. Recall that the gradient descent algorithms cannot be directly applied to neural networks for incremental learning, since all trainable parameters are changed during training, leading the model to be biased towards the augmented data (new negative labeled data), and forgetting the previously learned knowledge. Although we can train the model on the entire data, which takes the learned parameters from long-term learning as an initialization, it is too computationally expensive and cannot meet the requirements for short-term learning. Nevertheless, memory learning is able to solve this problem inherently. One of the reasons is that the tangent operator in (2) promotes writing sparsity, thus less memory cubes are affected, resulting in safer and faster incremental learning.

4.3   Online Learning

Online learning is one of the most important capabilities for a real-time visual interestingness prediction system, as human feelings always keep changing according to one's environments and experiences. Moreover, people tend to lose interests when repeatedly observing the same objects or exploring the same scenes, which is very common in a video stream from a mobile robot. Therefore, we aim to establish such an online learning capability for real-time robotic systems, instead of selecting interesting frames after processing an entire sequence [6]. We design a few control variables that can be simply adjusted for different applications, e.g., a hyper-parameter to control the rate of losing interests will be useful for objects search. To this end, we propose an architecture for online learning in Fig. 2c, in which only the frozen encoder and memory are involved.

In this stage, memory reading is performed before writing and the inputs are continuous image sequences (a video stream), which is different from short-term learning. If unobserved scenes or objects appear suddenly, memory reading confidence will be lower than before, which can be treated as a new interest. Moreover, since the new scenes or objects are then written into the memory, their reading confidence level will become higher in the following images. Therefore, the model will learn to lose interests on repetitive scenes once the scene is remembered by the memory. In this sense, a visual interestingness is negative correlated with the memory reading confidence. In experiments, we adopt averaged cosine similarity over feature channel to approximate the reading confidence.

During online learning, a large translation often happens during robot exploration, hence an invariance to large translations introduced in (4) is able to further reduce memory consumption and improve the system robustness.

5   Experiments

Evaluation Metric. Prior research typically only focused on the precision or recall rate and is not able to capture the online response of interestingness. Therefore, we propose a new metric, i.e., area under curve of online precision (AUC-OP) to evaluate one frame without using the information from its subsequent frames (no data leakage). This metric is stricter and jointly consider online response, precision, and recall rate. Intuitively, if K frames of a sequence are labeled as interesting in the ground truth, an algorithm is perfect if the set of its top K interesting frames are the same with the ground truth.

Consider a sequence $I_{[1:N]}$, we take an interestingness prediction $p(I_t)$ as a true positive (interesting) if and only if $p(I_t)$ ranks in the top $K_{t,n}$ among a subsequence $p(I_{t-n+1}), p(I_{t-n+2}) \cdots, p(I_t)$, where $K_{t,n}$ is the number of interesting frames in the ground truth. Note that the subsequence $I_{[t-n+1:t]}$ only contains frames before I_t, as data leakage is not allowed in the online performance. Therefore, we may calculate an online precision score for length n subsequences as $s(n) = \sum \text{TP}/(\sum \text{TP} + \sum \text{FP})$, where TP and FP denote the number of true positives and false positives, respectively. Since all true positives rank in the top $K_{t,n}$, this means that no false negative is allowed. Recall that a recall rate can be calculated as $r = \sum \text{TP}/(\sum \text{TP} + \sum \text{FN})$, which means that the proposed online precision score $s(n)$ requires a 100% recall rate. For a better comparison, we often accept true positive predictions as ranking in the top $\delta \cdot K_{t,n}$, where $\delta \geq 1$. Therefore, the overall performance of that jointly considers online performance,

Table 1. The SubT dataset. "Normal" and "Difficult" means that the percentage of frames labeled as interesting by at least 1 subjects or 2 subjects, respectively.

Video	I	II	III	IV	V	VI	VII	Overall
Length (min)	53.1	55.7	79.4	80.0	59.0	57.5	83.0	467.7
Normal (%)	11.11	15.07	9.37	17.51	24.52	22.77	11.04	15.14
Difficult (%)	2.76	4.49	3.02	4.29	4.07	3.30	3.21	3.58

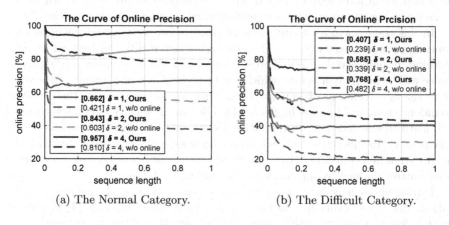

(a) The Normal Category.   (b) The Difficult Category.

Fig. 3. The performance on SubT with and without (w/o) online learning.

precision, and recall rate is the AUC of online precision $s(\frac{n}{N}, \delta)$ where $\frac{n}{N} \in (0, 1]$, which considers all subsequence length as $n = [1 : N]$. In practice, we often allow some false negatives and $\delta = 2$ is recommended for most of exploration task.

Dataset. To test the online performance on robotic systems for visual interesting scene prediction, we choose two datasets recorded by fully autonomous robots, *i.e.*, the SubT dataset [1] for unmanned ground vehicles (UGV) and the Drone Filming dataset [40] for unmanned aerial vehicles (UAV).

The SubT dataset is based on the DARPA Subterranean Challenge (SubT) Tunnel Circuit. In this challenge, the competitors are expected to build robotic systems to autonomously search and explore the subterranean environments. The environments pose significant challenges, including a lack of lighting, lack of GPS and wireless communication, dripping water, thick smoke, and cluttered or irregularly shaped environments. Each of the tunnels has a cumulative linear distance of 4–8 km. The dataset listed in Table 1 contains seven long videos (1 h) recorded by two fully autonomous UGV from Team Explorer[1]. Each sequence is evaluated by at least 3 persons. It can be seen that the SubT dataset is very challenging, as human annotation varies a lot, *i.e.*, only 15% and 3.6% of the frames are labeled as interesting by at least 1 (normal category) and 2 subjects

---

[1] Team Explorer won the first place at the DARPA SubT Tunnel Circuit.

Table 2. The comparison with the state-of-the-art method on AUC-OP.

(a) The SubT Normal Category.

Methods	$\delta = 1$	$\delta = 2$	$\delta = 3$
baseline [25]	0.622	0.798	0.904
ours	**0.662**	**0.842**	**0.923**

(b) The SubT Difficult Category.

Methods	$\delta = 1$	$\delta = 2$	$\delta = 4$
baseline [25]	0.352	0.544	0.700
ours	**0.407**	**0.585**	**0.768**

(difficult category), respectively. Some of the interesting scenes predicted by our algorithms are presented in Fig. 1, in which we can see that our method predicts many interesting scenes correctly.

The Drone Filming dataset [40] is recorded by quadcopters during autonomous aerial filming. It also contains challenging environments, *e.g.*, intensive light changes, severe vibrations, and motion blur, *etc.* . Different from other sources such as surveillance camera, robotic visual systems pose extra challenges due to fast background changes, limited computational resources, and unique and even dangerous operating environments in which human beings cannot get access to.

Implementation. In all experiments in this section, a memory capacity of 1000 and mean square error (MSE) loss are adopted. The memory reading and writing rate are set as $\gamma_r = \gamma_w = 5$. Our algorithm is implemented using the PyTorch library [29] and conducted on a single Nvidia GPU of GeForce GTX 1080Ti.

Efficiency. During long-term learning, we perform unsupervised training with the coco dataset [24]. It takes about 3 days running on single GPU. For short-term learning, our model takes about 10 min for learning 912 uninteresting images in the SubT dataset, which is feasible for deployment purpose of most practical applications. For online learning, it runs about 72.01 ms per frame on single GPU, which is feasible for real-time[2] robotic interestingness prediction.

Performance. Online learning is able to remove many repetitive scenes thus it is able to reduce the number of false positives dramatically. The curve of online precision of our model for the normal category and difficult category are presented in Fig. 3a and b, respectively, where the overall AUC-OP is shown in the associated square brackets. It can be seen that our model achieves an average of 20% higher overall performance than the model without online learning, which verifies the importance of the proposed online learning.

Comparison. To the best of our knowledge, robotic visual interestingness prediction is currently underexplored, and existing methods in saliency or anomaly

---

[2] Real-time means processing images as fast as human brain, *i.e.*, 100 ms/frame [31].

Table 3. The effects of the proposed modules on SubT (AUC-OP).

Methods	Normal Category			Difficult Category		
	$\delta = 1$	$\delta = 2$	$\delta = 4$	$\delta = 1$	$\delta = 2$	$\delta = 4$
w/o sparsity	0.437	0.633	0.846	0.260	0.373	0.523
w/o invariance	0.330	0.510	0.752	0.212	0.268	0.379
w/o short-term	0.508	0.711	0.913	0.329	0.450	0.621
Ours	**0.662**	**0.842**	**0.957**	**0.407**	**0.585**	**0.768**

detection have poor performance in this scenario. In this section, we select the state-of-the-art method, frame prediction in [25] as the baseline, which has very good generalization ability. Basically, it introduces temporal constraint into the video prediction task to detect anomaly. The overall performance of the AUC-OP of our method is presented in Table 2a and b, respectively. It can be seen that our method achieve an average of 4.0%, 4.4%, 1.9% and 5.5%, 4.1%, 6.8% higher overall accuracy in the two categories for $\delta = 1$, 2, 4, respectively, which verifies its effectiveness. We next present analysis to show the effects of the proposed writing sparsity, translational invariance, and short-term learning.

Effect of Writing Sparsity. To show its effectiveness, we replace our proposed writing protocol with the one used in [16], which is denoted as 'without (w/o) sparsity' in the first row of Table 3. It can be seen that our model achieves about 20–30% higher overall accuracy, which verifies the effectiveness of our method.

Effect of Translational Invariance. Without the large translational invariance, the performance will drop a lot, as translational movement is very common in robotic applications. As shown in the second row of Table 3, our model achieves about 20–30% higher accuracy than the one w/o translational invariance.

Effect of Short-term Learning. Short-term learning plays an important role for quick robot deployment. The performance can be largely improved if some uninteresting scenes are known before a mission. It can be seen in the fourth row of Table 3 that our model achieves about 10–20% higher accuracy than the one without short-term learning (w/o short-term).

6   Ablation Study

In this section, we further test the proposed algorithm and aim to provide intuitive explanations for the influences of the proposed writing protocol, memory capacity, translational invariance, and capability of losing interests. Following the ablation principle, all configurations are the same unless otherwise stated.

6.1   Writing Protocol

It has been pointed out that the memory learning is highly dependent on the writing vector in (1), in which a tangent operator is introduced for writing

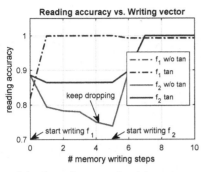

(a) The influence of writing vector.

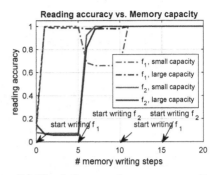

(b) The influence of memory capacity.

Fig. 4. The memory recall accuracy. (a) Writing vector with a tangent operator enables sparsity, thus less memory cubes are affected in learning. (b) Larger memory capacity leads to more computation but easy incremental learning.

sparsity. This section explores this effect and compare with the writing vector in (8) used in [16]. Note that memory defined in [16] is vectors, thus it is not invariant to large translation. Following the ablation principle, we use the same 4-D memory structure and the reading protocol proposed in this paper.

$$\mathbf{w} = \text{softmax}(\gamma \cdot D(\mathbf{x}, \mathbf{M})), \tag{8}$$

where γ is a parameter. To show the writing performance, we write two random 3-D tensors into the memory, $i.e.$, $\mathbf{f}_1$ and $\mathbf{f}_2$, and compare their reading accuracy in terms of cosine similarity in (9).

$$S^c(\mathbf{r}, \mathbf{f}) = \frac{\sum(\mathbf{r} \odot \mathbf{f})}{\|\mathbf{r}\|_{\mathbf{F}} \cdot \|\mathbf{f}\|_{\mathbf{F}}}, \tag{9}$$

where $\mathbf{r}$ and $\mathbf{f}$ are the memory reading and writing tensors, respectively. In experiments, we set $\gamma_w = \gamma = 5$ and write both $\mathbf{f}_1$ and $\mathbf{f}_2$ 5 times continuously and show their reading accuracy in terms of number of writing in Fig. 4a. It can be seen that both memories are able to remember the random tensors after repeatedly writing. However, when writing a vector without a tangent operator, the accuracy of $\mathbf{f}_2$ keeps dropping even when $\mathbf{f}_1$ is learned, $i.e.$, $S^c(\mathbf{r}_1, \mathbf{f}_1) \approx 1$. This is because all memory cubes are affected due to the non-sparse writing vector in (8). This will be a severe issue when a robot keeps learning the same thing (observing the same scene), since the learned knowledge may be forgotten due to the non-sparse writing. Nevertheless, our proposed writing vector with the tangent operator is able to map the weight of $\mathbf{f}_1$ to infinite when $\mathbf{f}_1$ is learned, resulting in safer writing as only a few memory cubes are affected. This verifies the effectiveness of the proposed writing vector. We notice that sparsity is also mentioned in [13,27,42], while it is designed for different objectives using different strategies. For instance, [13] introduced a simple threshold and an entropy loss to promote sparsity for reducing reconstruction accuracy to detect anomaly.

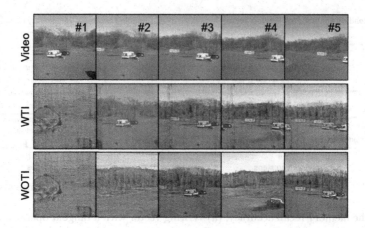

Fig. 5. Memory reading with translational invariance (WTI) recall translated scenes better than without translation-invariance (WOTI).

6.2 Memory Capacity

This section explores the effects of memory capacity, *i.e.*, the number of memory cubes c, which is an important hyper-parameter for incremental learning. To this end, we write two same random 3-D tensors $\mathbf{f}_1$ and $\mathbf{f}_2$ five times sequentially into two different memories in terms of the memory capacity c. Their reading accuracy for the performance comparison is shown in Fig. 4b.

As can be seen, both memories are able to learn random samples, while the accuracy of $\mathbf{f}_1$ drops a lot for smaller capacity when start to write $\mathbf{f}_2$, although it is remembered later when $\mathbf{f}_1$ is written again. We observe similar phenomenon when the number of samples is around the same or larger than the memory capacity. This means that a memory that has a small capacity quickly forgets old knowledge when learning new knowledge. We can also leverage this property for model design, since uninteresting objects can become interesting in some cases. This means that for larger capacity, reading accuracy is less affected by new knowledge, resulting in safer and easier incremental learning.

6.3 Translational Invariance

Although CNN features are invariant to small translations [37], they still fail to recall memory when large translations occur. To solve this problem, we introduce translational invariance by cross-correlation in (6). We next test it on the Drone Filming dataset [40] in Fig. 5. In this sequence an ambulance appears suddenly in the 1st frame and disappears in the 5th frame. We construct two memory modules to learn this video based on the online learning strategy presented in Sect. 4.3. The first module adopts the cross-correlation similarity presented in Sect. 3.2 for memory reading (denote as WTI), while another one adopts the cosine similarity (WOTI). It can be seen that both modules cannot recall the

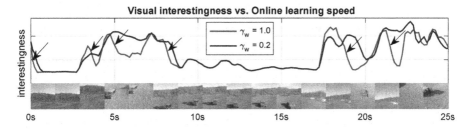

Fig. 6. The visual interestingness with different writing rates for drone video footage [40]. As indicated by the arrows, a larger writing rate results in a faster loss of interest for new objects during online learning.

memory for the 1st frame, since the ambulance is not seen before. However, the module WTI is able to recall the memory precisely in the subsequent frames, while the module WOTI quickly fails, although its reading is still meaningful, *e.g.*, the 2nd and 4th frames have correct patterns for sky, trees, and ground. It can be seen that the recalled memory for the 3rd frame from module WTI is roughly a translated replica of the 2nd frame of the video (this also occurs at the 4th and 5th frame), which means that the module WTI correctly takes the 2nd frame as the most similar scene to the 3rd frame. This phenomenon verifies the translational invariance of our proposed reading protocol.

Note that there is a small translation for the 2nd frame from WTI. This is because the invariance to small translations of CNN features, *i.e.*, the features look the same for visual memory, although they appear with a small translational difference. Therefore, our proposed cross-correlation similarity together with the CNN features contribute complete invariance of translation to memory recall.

6.4   Losing Interest

To test the capability of losing interest of the algorithm, we perform a qualitative test on the Drone Filming dataset [40]. The objects tracked in the videos, *e.g.*, cars or bikes, are relatively stable, while the background keeps changing due to the movement of the objects. This makes it suitable for testing the capability of online learning. One of the video clips is shown in Fig. 6, where two different online learning speeds are adopted, *i.e.*, $\gamma_w = 1.0$ and $\gamma_w = 0.2$. It can be seen that the interestingness level of both settings become high when new objects or scenes appear, *i.e.*, both settings are able to detect novel objects. However, the interestingness level with a larger writing rate always drops faster, meaning it is quicker to lose interest of the similar scenes. This verifies our objective that a simple hyper-parameter can be adjusted for different missions.

7   Conclusion

In this paper, we developed an unsupervised online learning algorithm for visual robotic interestingness prediction. We first proposed a novel translation-invariant

4-D visual memory, which can be trained without back-propagation. To better fit for practical applications, we designed a three-stage learning architecture, *i.e.*, long-term, short-term, and online learning. Concretely, the long-term learning stage is responsible for human-like life-time knowledge accumulation and trained on unlabeled data via back-propagation. The short-term learning is responsible for learning environmental knowledge and trained via visual memory for quick robot deployment. The online learning is responsible for environment adaption and leverage the visual memory to identify the interesting scenes. The experiments show that, implemented on a single machine, our approach is able to learn online and find interesting scenes efficiently in real-world robotic tasks.

Acknowledgements. This work was sponsored by ONR grant #N0014-19-1-2266. The human subject survey was approved under #2019_00000522.

References

1. http://theairlab.org/dataset/interestingness
2. Abati, D., Porrello, A., Calderara, S., Cucchiara, R.: Latent space autoregression for novelty detection. In: Proceedings of the IEEE Conference on Computer Vision and Pattern Recognition, pp. 481–490 (2019)
3. Amengual, X., Bosch, A., de la Rosa, J.L.: Review of methods to predict social image interestingness and memorability. In: Azzopardi, G., Petkov, N. (eds.) CAIP 2015. LNCS, vol. 9256, pp. 64–76. Springer, Cham (2015). https://doi.org/10.1007/978-3-319-23192-1_6
4. Brady, T.F., Konkle, T., Alvarez, G.A., Oliva, A.: Visual long-term memory has a massive storage capacity for object details. Proc. Natl. Acad. Sci. **105**(38), 14325–14329 (2008)
5. Chaabouni, S., Benois-Pineau, J., Zemmari, A., Ben Amar, C.: Deep saliency: prediction of interestingness in video with CNN. In: Benois-Pineau, J., Le Callet, P. (eds.) Visual Content Indexing and Retrieval with Psycho-Visual Models. MSA, pp. 43–74. Springer, Cham (2017). https://doi.org/10.1007/978-3-319-57687-9_3
6. Constantin, M.G., Redi, M., Zen, G., Ionescu, B.: Computational understanding of visual interestingness beyond semantics: literature survey and analysis of covariates. ACM Comput. Surv. (CSUR) **52**(2), 25 (2019)
7. Dalal, N., Triggs, B.: Histograms of oriented gradients for human detection. In: 2005 IEEE Computer Society Conference on Computer Vision and Pattern Recognition (CVPR 2005), vol. 1, pp. 886–893. IEEE (2005)
8. Demarty, C.-H., et al.: Predicting interestingness of visual content. In: Benois-Pineau, J., Le Callet, P. (eds.) Visual Content Indexing and Retrieval with Psycho-Visual Models. MSA, pp. 233–265. Springer, Cham (2017). https://doi.org/10.1007/978-3-319-57687-9_10
9. Demarty, C.H., Sjöberg, M., Ionescu, B., Do, T.T., Gygli, M., Duong, N.: Mediaeval 2017 predicting media interestingness task (2017)
10. Dhar, S., Ordonez, V., Berg, T.L.: High level describable attributes for predicting aesthetics and interestingness. In: CVPR 2011, pp. 1657–1664. IEEE (2011)
11. Fu, Y., Hospedales, T.M., Xiang, T., Gong, S., Yao, Y.: Interestingness prediction by robust learning to rank. In: Fleet, D., Pajdla, T., Schiele, B., Tuytelaars, T. (eds.) ECCV 2014. LNCS, vol. 8690, pp. 488–503. Springer, Cham (2014). https://doi.org/10.1007/978-3-319-10605-2_32

12. Fu, Y., et al.: Robust subjective visual property prediction from crowdsourced pairwise labels. IEEE Trans. Pattern Anal. Mach. Intell. **38**(3), 563–577 (2015)
13. Gong, D., et al.: Memorizing normality to detect anomaly: memory-augmented deep autoencoder for unsupervised anomaly detection. In: Proceedings of the IEEE International Conference on Computer Vision, pp. 1705–1714 (2019)
14. Goodfellow, I., Bengio, Y., Courville, A.: Deep Learning. MIT Press, New York (2016)
15. Grabner, H., Nater, F., Druey, M., Van Gool, L.: Visual interestingness in image sequences. In: Proceedings of the 21st ACM International Conference on Multimedia, pp. 1017–1026. ACM (2013)
16. Graves, A., Wayne, G., Danihelka, I.: Neural turing machines. arXiv preprint arXiv:1410.5401 (2014)
17. Gygli, M., Soleymani, M.: Analyzing and predicting gif interestingness. In: Proceedings of the 24th ACM International Conference on Multimedia, pp. 122–126. ACM (2016)
18. Hasan, M., Choi, J., Neumann, J., Roy-Chowdhury, A.K., Davis, L.S.: Learning temporal regularity in video sequences. In: Proceedings of the IEEE Conference on Computer Vision and Pattern Recognition, pp. 733–742 (2016)
19. Hochreiter, S., Schmidhuber, J.: Long short-term memory. Neural Comput. **9**(8), 1735–1780 (1997)
20. Ito, Y., Kitani, K.M., Bagnell, J.A., Hebert, M.: Detecting interesting events using unsupervised density ratio estimation. In: Fusiello, A., Murino, V., Cucchiara, R. (eds.) ECCV 2012. LNCS, vol. 7585, pp. 151–161. Springer, Heidelberg (2012). https://doi.org/10.1007/978-3-642-33885-4_16
21. Jiang, Y.G., Wang, Y., Feng, R., Xue, X., Zheng, Y., Yang, H.: Understanding and predicting interestingness of videos. In: Twenty-Seventh AAAI Conference on Artificial Intelligence (2013)
22. Kim, Y., Kim, M., Kim, G.: Memorization precedes generation: learning unsupervised GANs with memory networks. In: The International Conference on Learning Representations (ICLR) (2018)
23. Kramer, M.A.: Nonlinear principal component analysis using autoassociative neural networks. AIChE J. **37**(2), 233–243 (1991)
24. Lin, T.-Y., et al.: Microsoft COCO: common objects in context. In: Fleet, D., Pajdla, T., Schiele, B., Tuytelaars, T. (eds.) ECCV 2014. LNCS, vol. 8693, pp. 740–755. Springer, Cham (2014). https://doi.org/10.1007/978-3-319-10602-1_48
25. Liu, W., Luo, W., Lian, D., Gao, S.: Future frame prediction for anomaly detection-a new baseline. In: Proceedings of the IEEE Conference on Computer Vision and Pattern Recognition, pp. 6536–6545 (2018)
26. Long, J., Shelhamer, E., Darrell, T.: Fully convolutional networks for semantic segmentation. In: Proceedings of the IEEE Conference on Computer Vision and Pattern Recognition, pp. 3431–3440 (2015)
27. Luo, W., Liu, W., Gao, S.: A revisit of sparse coding based anomaly detection in stacked RNN framework. In: Proceedings of the IEEE International Conference on Computer Vision, pp. 341–349 (2017)
28. Oßwald, S., Bennewitz, M., Burgard, W., Stachniss, C.: Speeding-up robot exploration by exploiting background information. IEEE Robot. Autom. Lett. **1**(2), 716–723 (2016)
29. Paszke, A., et al.: Automatic differentiation in PyTorch (2017)
30. Phillips, W.: On the distinction between sensory storage and short-term visual memory. Percept. Psychophys. **16**(2), 283–290 (1974)

31. Potter, M.C., Levy, E.I.: Recognition memory for a rapid sequence of pictures. J. Exp. Psychol. **81**(1), 10 (1969)
32. Rumelhart, D.E., Hinton, G.E., Williams, R.J., et al.: Learning representations by back-propagating errors. Cognit. Model. **5**(3), 1 (1988)
33. Santoro, A., Bartunov, S., Botvinick, M., Wierstra, D., Lillicrap, T.: Meta-learning with memory-augmented neural networks. In: International Conference on Machine Learning, pp. 1842–1850 (2016)
34. Shen, Y., Demarty, C.H., Duong, N.Q.: Deep learning for multimodal-based video interestingness prediction. In: 2017 IEEE International Conference on Multimedia and Expo (ICME), pp. 1003–1008. IEEE (2017)
35. Simonyan, K., Zisserman, A.: Very deep convolutional networks for large-scale image recognition. In: International Conference on Learning Research (2015)
36. Wang, C.: Kernel learning for visual perception. Ph.D. thesis, Nanyang Technological University (2019)
37. Wang, C., Yang, J., Xie, L., Yuan, J.: Kervolutional neural networks. In: Proceedings of the IEEE Conference on Computer Vision and Pattern Recognition, pp. 31–40 (2019)
38. Wang, C., Zhang, L., Xie, L., Yuan, J.: Kernel cross-correlator. In: Thirty-Second AAAI Conference on Artificial Intelligence (2018)
39. Wang, S., Chen, S., Zhao, J., Jin, Q.: Video interestingness prediction based on ranking model. In: Proceedings of the Joint Workshop of the 4th Workshop on Affective Social Multimedia Computing and first Multi-Modal Affective Computing of Large-Scale Multimedia Data, pp. 55–61. ACM (2018)
40. Wang, W., Ahuja, A., Zhang, Y., Bonatti, R., Scherer, S.: Improved generalization of heading direction estimation for aerial filming using semi-supervised regression. In: 2019 International Conference on Robotics and Automation (ICRA), pp. 5901–5907. IEEE (2019)
41. Zhang, P., Wang, D., Lu, H., Wang, H., Yin, B.: Learning uncertain convolutional features for accurate saliency detection. In: Proceedings of the IEEE International Conference on Computer Vision, pp. 212–221 (2017)
42. Zhao, B., Fei-Fei, L., Xing, E.P.: Online detection of unusual events in videos via dynamic sparse coding. In: CVPR 2011, pp. 3313–3320. IEEE (2011)
43. Zhao, Y., Deng, B., Shen, C., Liu, Y., Lu, H., Hua, X.S.: Spatio-temporal autoencoder for video anomaly detection. In: Proceedings of the 25th ACM International Conference on Multimedia, pp. 1933–1941 (2017)

Post-training Piecewise Linear Quantization for Deep Neural Networks

Jun Fang[1]([⊠]), Ali Shafiee[1], Hamzah Abdel-Aziz[1], David Thorsley[1],
Georgios Georgiadis[2], and Joseph H. Hassoun[1]

[1] Samsung Semiconductor, Inc., San Jose, USA
{jun.fang,ali.shafiee,hamzah.a,d.thorsley,j.hassoun}@samsung.com
[2] Microsoft, Redmond, USA
georgios.georgiadis@microsoft.com

Abstract. Quantization plays an important role in the energy-efficient deployment of deep neural networks on resource-limited devices. Post-training quantization is highly desirable since it does not require retraining or access to the full training dataset. The well-established uniform scheme for post-training quantization achieves satisfactory results by converting neural networks from full-precision to 8-bit fixed-point integers. However, it suffers from significant performance degradation when quantizing to lower bit-widths. In this paper, we propose a piecewise linear quantization (PWLQ) scheme (Code will be made available at https://github.com/jun-fang/PWLQ) to enable accurate approximation for tensor values that have bell-shaped distributions with long tails. Our approach breaks the entire quantization range into non-overlapping regions for each tensor, with each region being assigned an equal number of quantization levels. Optimal breakpoints that divide the entire range are found by minimizing the quantization error. Compared to state-of-the-art post-training quantization methods, experimental results show that our proposed method achieves superior performance on image classification, semantic segmentation, and object detection with minor overhead.

Keywords: Deep neural networks · Post-training quantization · Piecewise linear quantization

1 Introduction

In recent years, deep neural networks (DNNs) have achieved state-of-the-art results in a variety of learning tasks including image classification [19, 23, 24, 29,

---

G. Georgiadis—Work performed while at Samsung Semiconductor, Inc.

Electronic supplementary material The online version of this chapter (https://doi.org/10.1007/978-3-030-58536-5_5) contains supplementary material, which is available to authorized users.

A. Vedaldi et al. (Eds.): ECCV 2020, LNCS 12347, pp. 69–86, 2020.
https://doi.org/10.1007/978-3-030-58536-5_5

53–55], segmentation [5,18,49] and detection [36,47,48]. Scaling up DNNs by one or all of the dimensions [55] of network depth [19], width [59] or image resolution [30] attains better accuracy, at a cost of higher computational complexity and increased memory requirements, which makes the deployment of these networks on embedded devices with limited resources impractical.

One feasible way to deploy DNNs on embedded systems is quantization of full-precision (32-bit floating-point, FP32) weights and activations to lower precision (such as 8-bit fixed-point, INT8) integers [25]. By decreasing the bit-width, the number of discrete values is reduced, while the quantization error, which generally correlates with model performance degradation increases. To minimize the quantization error and maintain the performance of a full-precision model, many recent studies [4,6,12,25,27,40,60,63] rely on training either from scratch ("quantization-aware" training) or by fine-tuning a pre-trained FP32 model.

However, post-training quantization is highly desirable since it does not require retraining or access to the full training dataset. It saves time-consuming fine-tuning effort, protects data privacy, and allows for easy and fast deployment of DNN applications. Among various post-training quantization schemes proposed in the literature [7,28,62], uniform quantization is the most popular approach to quantize weights and activations since it discretizes the domain of values to evenly-spaced low-precision integers which can be efficiently implemented on commodity hardware's integer-arithmetic units.

Recent work [28,31,42] shows that post-training quantization based on a uniform scheme with INT8 is sufficient to preserve near original FP32 pre-trained model performance for a wide variety of DNNs. However, ubiquitous usage of DNNs in resource-constrained settings requires even lower bit-width to achieve higher energy efficiency and smaller models. In lower bit-width scenarios, such as 4-bit, post-training uniform quantization causes significant accuracy drop [28, 62]. This is mainly because the distributions of weights and activations of pre-trained DNNs are bell-shaped such as Gaussian or Laplacian [17,35]. That is, most of the weights are clustered around zero while few of them are spread in a long tail. As a result, when operating at low bit-widths, uniform quantization assigns too few quantization levels to small magnitudes and too many to large ones, which leads to significant accuracy degradation [28,62].

To mitigate this issue, various quantization schemes [3,4,26,34,41,43] are designed to take advantage of the fact that weights and activations of pre-trained DNNs typically have bell-shaped distributions with long tails. Here, we present a new number representation via a piecewise linear approximation to be suited for these phenomena. It breaks the entire quantization range into *non-overlapping regions* where each region is assigned an equal number of quantization levels. Although our method works with an arbitrary number of regions, we suggest limiting them to two to simplify the complexity of the proposed approach and the hardware overhead. The *optimal breakpoints* that divide the entire range can be found by minimizing the quantization error. Compared to uniform quantization, our piecewise linear quantization (PWLQ) provides a richer representation that reduces the quantization error. This indicates its potential to reduce the gap between floating-point and low-bit precision models. It is also more hardware-friendly when compared to other non-linear approaches such as logarithm-based and clustering-based approaches [3,41,56], since in our method, the computation can still be carried out without the need of any transforms or look-up tables.

The main contributions of our work are as follows:

- We propose a piecewise linear quantization (PWLQ) scheme for efficient deployment of pre-trained DNNs without retraining or access to the full training dataset. We also investigate its impact on hardware implementation.
- We present a solution to find the optimal breakpoints and demonstrate that our method achieves a lower quantization error than the uniform scheme.
- We provide a comprehensive evaluation on image classification, semantic segmentation, and object detection benchmarks and show that our proposed method achieves state-of-the-art results.

2   Related Work

There is a wide variety of approaches in the literature that facilitate the efficient deployment of DNNs. The first group of techniques relies on designing network architectures that depend on more efficient building blocks. Notable examples include depth/point-wise layers [22,52] as well as group convolutions [38,61]. These methods require domain knowledge, training from scratch and full access to the task datasets. The second group of approaches optimizes network architectures in a typical task-agnostic fashion and may or may not require (re)training. Weight pruning [17,20,32,37], activation compression [9,10,14], knowledge distillation [21,45] and quantization [8,25,41,46,64,66] fall under this category.

In particular, quantization of activations and weights [6,15,16,35,57,60,62] leads to model compression and acceleration as well as to overall savings in power consumption. Model parameters can be stored in a fewer number of bits while the computation can be executed on integer-arithmetic units rather than on power-hungry floating-point ones [25]. There has been extensive research on quantization with and without (re)training. In the rest of this section, we focus on post-training quantization that directly converts full-precision pre-trained models to their low-precision counterparts.

Recent works [28,31,42] have demonstrated that 8-bit quantized models have been able to accomplish negligible accuracy loss for a variety of networks. To improve accuracy, per-channel (or channel-wise) quantization is introduced in [28,31] to address variations of the range of weight values across channels. Weight equalization/factorization is applied by [39,42] to rescale the difference of weight ranges between different layers. In addition, bias shifts in the mean and variance of quantized values are observed and counteracting methods are suggested by [2,13]. A comprehensive evaluation of clipping techniques is presented by [62] along with an outlier channel splitting method to improve quantization performance. Moreover, adaptive processes of assigning different bit-width for each layer are proposed in [35,65] to optimize the overall bit allocation.

There are also a few attempts to tackle 4-bit post-training quantization by combining multiple techniques. In [2], a combination of analytical clipping, bit allocation, and bias correction is used, while [7] minimizes the mean squared quantization error by representing one tensor with one or multiple 4-bit tensors as well as by optimizing the scaling factors.

Most of the aforementioned works utilize a linear or uniform quantization scheme. However, linear quantization cannot capture the bell-shaped distribution of weights and activations, which results in sub-optimal solutions. To overcome this deficiency, [3] proposes a quantile-based method to improve accuracy but their method works efficiently only on highly customized hardware; [26] employs two different scale factors on overlapping regions to reduce computation bits over fixed-point implementations. However, its scale factors restricted to powers of two and heuristic options limit the accuracy performance. Instead, we propose a piecewise linear approach that improves over the selection of optimal breakpoints that leads to state-of-the-art quantized model results. Our method can be implemented efficiently with minimal modification to commodity hardware.

3   Quantization Schemes

In this section, we review a uniform quantization scheme and discuss its limitations. We then present PWLQ, our piecewise linear quantization scheme and show that it has a stronger representational power (a smaller quantization error) compared to the uniform scheme.

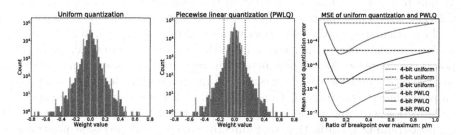

Fig. 1. Quantization of *conv*4 layer weights in a pre-trained Inception-v3. Left: uniform quantization. Middle: piecewise linear quantization (PWLQ) with one breakpoint, dotted line indicates the breakpoint. Right: Mean squared quantization error (MSE) for various bit-widths ($b = 4, 6, 8$). MSE of PWLQ is convex *w.r.t.* the breakpoint p, the b-bit PWLQ can achieve a smaller quantization error than the b-bit uniform scheme

3.1   Uniform Quantization

Uniform quantization (the left of Fig. 1) linearly maps full-precision real numbers r into low-precision integer representations. From [7,25], the approximated version $\hat{r}$ from uniform quantization scheme at b-bit can be defined as:

$$\hat{r} = \text{uni}(r; b, r_l, r_u, z) = s \times r_q + z,$$

$$r_q = \left\lceil \frac{\text{clamp}(r; r_l, r_u) - z}{s} \right\rfloor_{\mathbb{Z}_b},$$

$$\text{clamp}(r; r_l, r_u) = \min(\max(r, r_u), r_l), \tag{1}$$

$$s = \frac{\Delta}{N-1}, \quad \Delta = r_u - r_l, \quad N = 2^b,$$

where $[r_l, r_u]$ is the quantization range, s is the scaling factor, z is the offset, N is the number of quantization levels, r_q is the quantized integer computed by a rounding function $\lceil \cdot \rfloor_{\mathbb{Z}_b}$ followed by saturation to the integer domain $\mathbb{Z}_b$. We set the offset $z = 0$ for symmetric signed distributions combined with $\mathbb{Z}_b = \{-2^{b-1}, ..., 2^{b-1} - 1\}$ and $z = r_l$ for asymmetric unsigned distributions (e.g., ReLU-based activations) with $\mathbb{Z}_b = \{0, ..., 2^b - 1\}$. Since the scheme (1) introduces a quantization error defined as $\varepsilon_{uni} = \hat{r} - r$, the expected quantization error squared is given by:

$$\mathbb{E}(\varepsilon_{uni}^2; b, r_l, r_u) = \frac{s^2}{12} = C(b)\Delta^2, \tag{2}$$

with $C(b) = \frac{1}{12(2^b - 1)^2}$ under uniform distributions [58].

From the above definition, uniform quantization divides the range evenly despite the distribution of r. Empirically, the distributions of weights and activations of pre-trained DNNs are similar to bell-shaped Gaussian or Laplacian [17,35]. Therefore, uniform quantization is not always able to achieve small enough approximation error to maintain model accuracy, especially in low-bit cases.

3.2   Piecewise Linear Quantization (PWLQ)

To improve model accuracy for quantized models, we need to approximate the original model as accurately as possible by minimizing the quantization error. We follow this natural criterion to investigate the quantization performance, even though no direct relationship can easily be established between the quantization error and the final model accuracy [7].

Inspired from [26,43] that takes advantage of bell-shaped distributions, our approach based on piecewise linear quantization is designed to minimize the quantization error. It breaks the quantization range into two non-overlapping regions: the dense, central region and the sparse, high-magnitude region. An equal number of quantization levels $N = 2^b$ is assigned to these two regions. We chose to use two regions with one breakpoint to maintain simplicity in the inference algorithm (Sect. 5.1) and the hardware implementation (Sect. 4). Multiple-region cases are discussed in Sect. 5.1.

Therefore, we only consider one breakpoint p to divide the bounded quantization range[1] $[-m, m]$ $(m > 0)$ into two symmetric regions: the center region $R_1 = [-p, p]$ and the tail region $R_2 = [-m, -p) \cup (p, m]$. Each region consists of a negative piece and a positive piece. Within each of the four pieces, $(b - 1)$-bit

---

[1] Here we consider symmetric quantization range $[-m, m]$ $(m > 0)$ for simplicity, it is extendable to asymmetric ranges $[m_1, m_2]$ for any real numbers $m_1 < m_2$.

$(b \geq 2)$ uniform quantization (1) is applied such that including the sign every value in the quantization range is being represented into b-bit. We define the b-bit piecewise linear quantization (denoted by PWLQ) scheme as:

$$\mathrm{pw}(r;b,m,p) = \begin{cases} \mathrm{sign}(r) \times \mathrm{uni}(|r|;b-1,0,p,0), r \in R_1 \\ \mathrm{sign}(r) \times \mathrm{uni}(|r|;b-1,p,m,p), r \in R_2 \end{cases}, \qquad (3)$$

where the sign of full-precision real number r is denoted by $\mathrm{sign}(r)$. The associated quantization error is defined as $\varepsilon_{pw} = \mathrm{pw}(r;b,m,p) - r$.

Figure 1 shows the comparison between uniform quantization and PWLQ on the empirical distribution of the *conv4* layer weights in a pre-trained Inception-v3 model [54]. We emphasize that b-bit PWLQ represents FP32 values into b-bit integers to support b-bit multiply-accumulate operations, even though in total, it has the same number of quantization levels as $(b+1)$-bit uniform quantization. The implications of this are further discussed in Sect. 4.

3.3   Error Analysis

To study the quantization error for PWLQ, we suppose full-precision real number r has a symmetric probability density function (PDF) $f(r)$ on $[-m, m]$ with a cumulative distribution function (CDF) $F(r)$ satisfying $f(r) = f(-r)$, $F(-m) = 0, F(m) = 1$, and $F(r) = 1 - F(-r)$. Then, we calculate the expected quantization error squared of PWLQ from (2) based on the error of each piece:

$$\begin{aligned} \mathbb{E}(\varepsilon_{pw}^2; b, m, p) &= C(b-1)\Big\{(m-p)^2\big[F(-p) + 1 - F(p)\big] + p^2\big[F(p) - F(-p)\big]\Big\} \\ &= C(b-1)\Big\{(m-p)^2 + m(2p-m)\big[2F(p) - 1\big]\Big\}. \end{aligned}$$
$$(4)$$

The performance of a quantized model with PWLQ critically depends on the value of the breakpoint p. If $p = \frac{m}{2}$, then the PWLQ is essentially equivalent to uniform quantization, because the four pieces have equal quantization ranges and bit-widths. If $p < \frac{m}{2}$, the center region has a smaller range and greater precision than the tail region, as shown in the middle of Fig. 1. Conversely, if $p > \frac{m}{2}$, the tail region has greater precision than the center region. To reduce the overall quantization error for bell-shaped distributions found in DNNs, we increase the precision in the center region and decrease it in the tail region. Thus, we limit the breakpoint to the range $0 < p < \frac{m}{2}$. Accordingly, the optimal breakpoint p^* can be estimated by minimizing the expected squared quantization error:

$$p^* = \arg\min_{p \in (0, \frac{m}{2})} \mathbb{E}(\varepsilon_{pw}^2; b, m, p). \qquad (5)$$

Since bell-shaped distributions tend to zero as r becomes large, we consider a smooth $f(r)$ is decreasing when r is positive, i.e., $f'(r) < 0, \forall r > 0$. Then we prove that the optimization problem (5) is convex with respect to the breakpoint $p \in (0, \frac{m}{2})$. Therefore one unique p^* exists to minimize the quantization error (4), as demonstrated by the following Lemma 1.

Lemma 1. *If* $f(-r) = f(r)$, $f'(r) < 0$ *for all* $r > 0$, *then* $\mathbb{E}(\varepsilon_{pw}^2; b, m, p)$ *is a convex function of the breakpoint* $p \in (0, \frac{m}{2})$.

Proof. Taking the first and second derivatives of (4) yields:

$$
\begin{aligned}
\frac{\partial \mathbb{E}(\varepsilon_{pw}^2; b, m, p)}{\partial p} &= 2C(b-1)\Big[p - 2m + 2mF(p) + m(2p - m)f(p)\Big], \\
\frac{\partial^2 \mathbb{E}(\varepsilon_{pw}^2; b, m, p)}{\partial p^2} &= 2C(b-1)\Big[1 + 4mf(p) + m(2p - m)f'(p)\Big].
\end{aligned}
\tag{6}
$$

Since $f'(p) < 0$ and $p < \frac{m}{2}$, $m(2p - m)f'(p) > 0$, then $\frac{\partial^2 \mathbb{E}(\varepsilon_{pw}^2; b, m, p)}{\partial p^2} > 0$. Therefore, $\mathbb{E}(\varepsilon_{pw}^2; b, m, p)$ is convex w.r.t. p, and thus a unique p^* exists.

In practice, we can find the optimal breakpoint by solving (5) by assuming an underlying Gaussian or Laplacian distribution using gradient descent [50]. Once the optimal breakpoint p^* is found, both Lemma 2 and the numerical simulation in the right of Fig. 1 show that PWLQ achieves a smaller quantization error than uniform quantization, which indicates its stronger representational power.

Lemma 2. $\mathbb{E}(\varepsilon_{pw}^2; b, m, p^*) < \frac{C(b-1)}{16C(b)}\mathbb{E}(\varepsilon_{uni}^2; b, -m, m)$ *for* $b \geq 2$.

Proof. The b-bit uniform quantization error on $[-m, m]$ is calculated from (2):

$$
\mathbb{E}(\varepsilon_{uni}^2; b, -m, m) = C(b)(2m)^2 = 4C(b)m^2.
\tag{7}
$$

For b-bit PWLQ, we solve the convex problem (5) by letting the first derivative equal to zero in (6), and determine that the optimal breakpoint p^* satisfies:

$$
2mF(p^*) = 2m - p^* + m(m - 2p^*)f(p^*).
\tag{8}
$$

By substituting (8) in (4) and simplifying, we obtain:

$$
\mathbb{E}(\varepsilon_{pw}^2; b, m, p^*) = C(b-1)\Big[-(p^*)^2 + mp^* - m(m - 2p^*)^2 f(p^*)\Big].
\tag{9}
$$

Subtract the above from $\frac{C(b-1)}{16C(b)}$ of (7), we complete the proof:

$$
\begin{aligned}
&\mathbb{E}(\varepsilon_{pw}^2; b, m, p^*) - \frac{C(b-1)}{16C(b)}\mathbb{E}(\varepsilon_{uni}^2; b, -m, m) \\
&= \mathbb{E}(\varepsilon_{pw}^2; b, m, p^*) - C(b-1)(\tfrac{1}{4}m^2) \\
&\leq C(b-1)\Big[-(p^* - \tfrac{m}{2})^2 - m(m - 2p^*)^2 f(p^*)\Big] < 0.
\end{aligned}
\tag{10}
$$

Note that $C(b) = \frac{1}{12(2^b-1)^2}$ given from (2), for $b \geq 2$, $\frac{C(b-1)}{16C(b)} \leq \frac{9}{16}$. Therefore, b-bit PWLQ achieves a smaller quantization error, which is at most $\frac{9}{16}$ of b-bit uniform scheme. This improvement in performance requires only an extra bit for storage and no extra multiplication, as we discuss in the next section.

4   Hardware Impact

In this section, we discuss the hardware requirements for efficient deployment of DNNs quantized with PWLQ. In convolutional and fully-connected layers, every output can be computed using an inner product between vector X and vector W, which correspond to the input activation and weight (sub)tensors respectively.

From scheme (1), the approximated versions of uniform quantization are $\hat{X} = s_x X_q + z_x I$ and $\hat{W} = s_w W_q$ (assuming symmetric quantization for weights), where X_q and W_q are quantized integer vectors from X and W, I is an identity vector, s_x, s_w and z_x are associated constant-valued scaling factors and offset, respectively. The output of this uniform quantization is:

$$\langle \hat{X}, \hat{W} \rangle = \langle s_x X_q + z_x I, s_w W_q \rangle = C_0 \langle X_q, W_q \rangle + C_1, \qquad (11)$$

where $\langle \cdot, \cdot \rangle$ is defined as vector inner product, $C_0 = s_x s_w$ and $C_1 = z_x s_w \langle W_q, I \rangle$ denote floating-point constant terms that can be pre-computed offline.

Equation (11) implies that a uniformly quantized DNN requires two steps: (i) an integer-arithmetic (INT) inner product, and (ii) followed by a floating-point (FP) affine map. The expensive $O(|W|)$ (the size of vector W) FP operations $\langle \hat{X}, \hat{W} \rangle$ are then accelerated via INT operations $\langle X_q, W_q \rangle$, plus $O(1)$ FP re-scaling and adding operands using C_0 and C_1.

As we showed in Sect. 3.2 when applying PWLQ on weights with one break-point, the algorithm breaks the ranges into non-overlapping regions (R_1 and R_2), which requires separate computational paths (P_1 and P_2) as each region has a different scaling factor. We set offsets $z_{w_1} = 0, z_{w_2} = p$ and denote scaling factors by s_{w_1}, s_{w_2} in R_1, R_2, respectively. We also define by $\langle \cdot, \cdot \rangle_{R_i}$ the associated partial vector inner product, and W_{q_i} the associated quantized integer vector of W in region R_i for $i = 1, 2$. Then P_1 is computed using the following equation:

$$P_1 = \langle s_x X_q + z_x I, s_{w_1} W_{q_1} \rangle_{R_1} = C_2 \langle X_q, W_{q_1} \rangle_{R_1} + C_3. \qquad (12)$$

P_2 has additional terms as it has a non-zero offset p:

$$P_2 = \langle s_x X_q + z_x I, s_{w_2} W_{q_2} + p I \rangle_{R_2} = C_4 \langle X_q, W_{q_2} \rangle_{R_2} + C_5 \langle X_q, I \rangle_{R_2} + C_6, \quad (13)$$

where C_2, C_3, C_4, C_5, and C_6 are constant terms, which can be pre-computed similar to C_0 and C_1 in (11).

As indicated by (12) and (13) for PWLQ compared to uniform quantization (11), the extra term $\langle X_q, I \rangle_{R_2}$ is needed due to the non-zero offset p, which sums up the activations corresponding to weights in R_2. Since most of the weights[2] are in R_1, these extra computations in R_2 rarely happen. In addition, FP re-scaling and adding are needed in each region, which also increases the overall FP operation overhead.

In short, an efficient hardware implementation of PWLQ requires: (i) one multiplier for products in both of $\langle X_q, W_{q_1} \rangle_{R_1}$ and $\langle X_q, W_{q_2} \rangle_{R_2}$, (ii) three accumulators: one of each for sum of products in P_1 and P_2, and another one for

---

[2] Around 90% of the weights are locating in the center region R_1 in our experiments.

activations in P_2, and (iii) at most one extra bit for storage[3] per weight value to indicate the region. Note that this extra bit does not increase the multiply-accumulate (MAC) computation and it is only used to determine the appropriate accumulator, which can be done in hardware at negligible cost on the MAC unit.

Based on the above explanation, it is clear that more breakpoints require more accumulators and more storage bits per weight tensor. Also, applying PWLQ on both weights and activations requires accumulators for each combination of activation regions and weight regions, which translates to more hardware overhead. As a result, more than one breakpoint on the weight tensor or applying PWLQ on both weights and activations might not be feasible, from a hardware implementation perspective. We describe more details of the hardware implementation and its impact on energy and latency in the supplementary material.

5   Experiments

We evaluate the robustness of our proposed PWLQ for post-training quantization on popular networks of several computer vision benchmarks: ImageNet classification [51], semantic segmentation and object detection on the Pascal VOC challenge [11]. We perform all experiments in Pytorch 1.2.0 [44]. Unless stated otherwise, we always apply batch normalization folding [25] and then quantize all folded network weights *per-channel*.

5.1   Ablation Study on ImageNet

In this section, we conduct experiments on the ImageNet classification challenge [51] and investigate the effectiveness of our proposed PWLQ method. We evaluate the top-1 accuracy performance on the validation dataset for three popular network architectures: Inception-v3 [54], ResNet-50 [19] and MobileNet-v2 [52]. We use torchvision[4] 0.4.0 and its pre-trained models for our experiments.

Activation Quantization. Throughout this paper, we use a *top-k* median method[5] with $k = 10$ to calibrate the activation range boundaries $[r_l, r_u]$. After sampling from 512 random training images [62], we sort the activations into an array X_{sort} at every layer. We then compute the median of the *top-k* smallest and largest values in X_{sort}, i.e., $r_l = \text{median}(X_{sort}[: k])$ and $r_u = \text{median}(X_{sort}[-k :])$. During inference, unless stated otherwise we apply 8-bit uniform quantization *per-layer* after clipping with these ranges. We report additional experiments applying PWLQ on activations in the supplementary material.

---

[3] This extra storage cost can be further compressed by exploiting the non-uniform distribution of values [1,43].

[4] https://pytorch.org/docs/stable/torchvision.

[5] We test the *top-k* median and percentile-based approaches [33] in the supplementary material and use the top-10 median for better robustness of low-bit quantization.

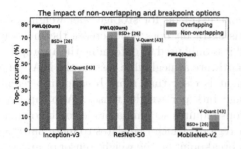 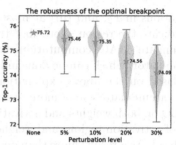

Fig. 2. Left: the impact of non-overlapping and breakpoint options on the top-1 accuracy for 4-bit post-training quantization models. Right: the robustness of the optimal breakpoint found by solving (5) with some perturbation levels from 5% to 30% for 4-bit Inception-v3. The star and the associated number indicate the median accuracy, the bold bar displays the accuracy range between the 25th and 75th percentiles

Optimal Breakpoint Selection. In order to apply PWLQ, we require the *optimal breakpoints* to divide the quantization ranges into *non-overlapping regions*. As stated in Sect. 3.3, we assume weights and activations satisfy Gaussian or Laplacian distributions, then we find the optimal breakpoints by solving (5).

For the case of one optimal breakpoint p^*, we can iteratively find it by gradient descent since the optimization problem (5) is convex; or using a simple and fast approximation of $p^*/m = \ln(0.8614m + 0.6079)$ for normalized Gaussian. Experimental results show that the approximation obtains almost the same accuracy compared to gradient descent, while also being considerably faster. Therefore, unless stated otherwise we use this approximated version of the optimal breakpoint for the rest of this paper. We report results with other assumptions such as Laplacian distributions in the supplementary material.

Other works treat the data distributions differently: BiScaled-DNN [26] proposes a ratio heuristic to divide the data into two overlapping regions; and V-Quant [43] introduces a value-aware method to split them into two non-overlapping regions, e.g., 2% (98%) of large (small) values located in the tail (center) region, respectively. Our implementation results in Fig. 2 (left) show that PWLQ with non-overlapping regions achieves a superior performance on low-bit quantization compared to BiScaled-DNN improved version[6] (denoted by BSD+) and V-Quant, especially with a large margin on 4-bit MobileNet-v2. Non-overlapping approach shortens the quantization ranges (Δ in (2)) for the tail regions by $1.25\times$ to $2\times$. Therefore, both our choices of *non-overlapping regions* and *optimal breakpoints* have a significant impact on reducing the quantization error and improving the performance of low-bit quantized models.

In Fig. 2 (right), we explore the robustness of the optimal breakpoint found by minimizing (5) for 4-bit Inception-v3. We randomly add perturbation levels from 5% to 30% on each optimal breakpoint p^* per-channel per-layer, e.g., the

---

[6] We improved the original BiScaled-DNN [26] by applying affine-based uniform scheme (1) on each region and per-channel quantization.

new breakpoint $\widehat{p}^* = 0.95p^*$ or $1.05p^*$ for 5% of perturbation. We run 100 random samples for each perturbation level to generate the results. Overall, model performance decreases as perturbation level increases, which indicates that our selection of the optimal breakpoint is crucial for accurate post-training quantization. Note that when 5% of perturbation is added to the optimal breakpoints, more than half of the experiments produce a lower accuracy, and can be as low as 74.05%, which is a 1.67% drop from the zero-perturbation baseline.

Multiple Breakpoints. In this section, we discuss the trade-off of multiple breakpoints on model accuracy and hardware overhead. Theoretically, as the number of breakpoints on weights increases, the associated hardware cost linearly rises. Meanwhile, the number of non-overlapping regions and the associated total number of quantization levels grows, indicating a stronger representational power. Numerically, the extension of finding the optimal multi-breakpoints is straightforward by calculating the same quantization error (4), and solving the same optimization problem (5) with gradient descent in an enlarged search space. Table 1 shows the accuracy performance up to three breakpoints. In general, using more breakpoints consistently improves model accuracy under the growing support of customized hardware. We suggest using one breakpoint to maintain the simplicity of the inference algorithm and its hardware implementation. Thus we only report PWLQ with one breakpoint for the rest of this paper.

Table 1. Top-1 accuracy (%) and requirement of hardware accumulators for PWLQ with multiple breakpoints on weights

Number of breakpoints	Hardware accumulators	Inception-v3 (77.49)			ResNet-50 (76.13)			MobileNet-v2 (71.88)		
		5-bit	4-bit	3-bit	5-bit	4-bit	3-bit	5-bit	4-bit	3-bit
One	**Three**	77.28	75.72	61.76	75.62	74.28	67.30	69.05	54.34	16.77
Two	Five	77.31	76.73	71.40	75.94	75.24	73.27	70.01	65.74	36.44
Three	Seven	**77.46**	**77.00**	**74.07**	**76.06**	**75.77**	**73.84**	**70.43**	**67.71**	**55.17**

PWLQ and Uniform Quantization. In Sect. 3.3, we analytically and numerically demonstrate that our method, PWLQ, obtains a smaller quantization error than uniform quantization. We compare these two schemes in Table 2. In this table, weights are quantized per-channel with the same computational bit-width $b = 4, 6, 8$; activations are uniformly quantized per-layer into 8-bit. Generally, PWLQ achieves higher accuracy than uniform quantization except for one minor case of 8-bit Inception-v3. When the bit-width is large enough ($b = 8$), the quantization error is small and both uniform quantization and PWLQ provide good accuracy. However, when the bit-width is decreased to 4, PWLQ obtains a notably higher accuracy, i.e., PWLQ attains 75.72% but uniform quantization only attains 44.28% for 4-bit Inception-v3. These results show that PWLQ is

Table 2. Comparison results of top-1 accuracy (%) for uniform and PWLQ schemes on weights. b+BC: b-bit with bias correction for bit-width $b = 4, 6, 8$. Each bold value indicates the best result from different methods for specified bit-width and network

Network	Weight Bit-width	8-bit	8+BC	6-bit	6+BC	4-bit	4+BC
Inception-v3 (77.49)	Uniform	77.53	77.52	76.87	77.24	44.28	62.46
	PWLQ (Ours)	77.52	**77.53**	77.42	**77.48**	75.72	**76.45**
ResNet-50 (76.13)	Uniform	76.10	**76.14**	75.61	75.92	65.48	72.45
	PWLQ (Ours)	76.10	76.10	76.03	**76.08**	74.28	**75.62**
MobileNet-v2 (71.88)	Uniform	71.35	71.58	67.76	70.81	11.37	41.80
	PWLQ (Ours)	71.59	**71.73**	70.82	**71.58**	54.34	**69.22**

a more powerful representation scheme in terms of both quantization error and model accuracy, making it a viable alternative for uniform quantization in low bit-width cases. Moreover, PWLQ applies uniform quantization on each piece, hence it features a simple computational scheme and can benefit from any tricks that improve uniform quantization performance such as bias correction.

Bias Correction. An inherent bias in the mean and variance of the tensor values was observed after the quantization process and the benefits of correcting this bias term have been demonstrated in [2,13,42]. This bias can be compensated by folding certain correction terms into the scale and the offset [2]. We adopt this idea into our PWLQ method and show the results in Table 2 (columns with "+BC"). Applying bias correction further improves the performance of low-bit quantized models. It allows 6-bit post-training quantization with piecewise linear scheme for all three networks to achieve near full-precision accuracy within a drop of 0.30%; 4-bit MobileNet-v2, also without retraining, achieves an accuracy of 69.22%. In general, a combination of low-bit PWLQ and bias correction on weights achieves minimal loss of full-precision model performance.

5.2   Comparison to Existing Approaches

In this section, we compare our PWLQ method with other existing approaches, by quoting the reported performance scores from the original literature.

An inclusive evaluation of clipping techniques along with outlier channel splitting (OCS) was presented in [62]. To fairly compare with these methods, we adopt the same setup of applying per-layer quantization on weights and without quantizing the first layer. In Table 3, we show that our PWLQ (no bias correction) outperforms the best results of clipping method combined with OCS. Besides, OCS needs to change the network architecture, in contrast to PWLQ.

Table 3. Comparison results of per-layer PWLQ and best clipping with OCS [62] on top-1 accuracy (%) loss. W/A indicate the bit-width on weights/activations. The accuracy difference values are measured from the full-precision (32/32) result

Network	W/A	32/32	8/8	7/8	6/8	5/8	4/8
Inception-v3	OCS + Best Clip	75.9	−0.6 (75.3)	−1.2 (74.7)	−3.4 (72.5)	−13.0 (62.9)	−71.1 (4.8)
	PWLQ (Ours)	77.5	**+0.1 (77.6)**	**−0.1 (77.4)**	**−0.3 (77.2)**	**−2.0 (75.5)**	**−12.8 (64.7)**
ResNet-50	OCS + Best Clip	76.1	−0.4 (75.7)	−0.5 (75.6)	−0.9 (75.2)	−2.7 (73.4)	−6.8 (69.3)
	PWLQ (Ours)	76.1	**−0.0 (76.1)**	**−0.1 (76.0)**	**−0.2 (75.9)**	**−0.7 (75.5)**	**−2.4 (73.7)**

In Table 4, we provide a comprehensive comparison result of PWLQ to other existing methods. Here we apply per-layer quantization on activations and per-channel PWLQ on weights with bias correction. Except for the 4/4 case where we apply 4-bit PWLQ on activations, we always apply 8-bit uniform quantization on activations for the rest of the 8/8 and 4/8 cases. Under the same bit-width of computational cost among all the methods, our PWLQ combined with bias correction achieves the state-of-the-art results on all cases and it outperforms all other methods with a large margin on 4/8 and 4/4 cases. We emphasize that our PWLQ method is simple and efficient. It achieves the desired accuracy at the small cost of a few more accumulations per MAC unit and a minor overhead of storage. More importantly, it is orthogonal and applicable to other methods.

Table 4. Comparison of our PWLQ and other methods on top-1 accuracy (%) loss. PWLQ: weights are piecewise linearly quantized per-channel with bias correction, activations are quantized per-layer

Network	W/A	PWLQ (Ours)	QWP [28]	ACIQ [2]	LBQ [7]	SSBD [39]	QRD [31]	UNIQ [3]	DFQ [42]
Inception-v3 (Top1%)	32/32	77.49	78.00	77.20	76.23	77.90	77.97	-	-
	8/8	**+0.04 (77.53)**	0.00 (78.00)	-	-	-0.03 (77.87)	-0.09 (77.88)	-	-
	4/8	**-1.04 (76.45)**	-7.00 (71.00)	-9.00 (68.20)	-1.44 (74.79)	-	-	-	-
	4/4	**-2.58 (74.91)**	-	-10.80 (66.40)	-4.62 (71.61)	-	-	-	-
ResNet-50 (Top1%)	32/32	76.13	75.20	76.10	76.01	75.20	-	76.02	-
	8/8	**-0.03 (76.10)**	-0.10 (75.10)	-	-	-0.25 (74.95)	-	-	-
	4/8	**-0.51 (75.62)**	-21.20 (54.00)	-0.80 (75.30)	-1.03 (74.98)	-	-	-2.56 (73.37)	-
	4/4	**-1.28 (74.85)**	-	-2.30 (73.80)	-3.41 (72.60)	-	-	-	-
MobileNet-v2 (Top1%)	32/32	71.88	71.90	-	-	71.80	71.23	-	71.72
	8/8	**-0.15 (71.73)**	-2.10 (69.80)	-	-	-0.61 (71.19)	-1.68 (69.55)	-	-0.53 (71.19)
	4/8	**-2.68 (69.22)**	-71.80 (0.10)	-	-	-	-	-	-

5.3   Other Applications

To show the robustness and applicability of our proposed approach, we extend the PWLQ idea to other computer vision tasks including semantic segmentation on DeepLab-v3+ [5] and object detection on SSD [36].

Semantic Segmentation. In this section, we apply PWLQ on DeepLab-v3+ with a backbone of MobileNet-v2. The performance is evaluated using mean intersection over union (mIoU) on the Pascal VOC segmentation challenge [11].

Table 5. Uniform quantization and PWLQ on DeepLab-v3+. Weights are quantized per-channel with bias correction, activations are uniformly quantized per-layer

Network	W/A	32/32	8/8	6/8	4/8
DeepLab-v3+ (mIoU%)	Uniform	70.81	−0.65 (70.16)	−1.54 (69.27)	−20.76 (50.05)
	PWLQ (Ours)	70.81	**−0.12 (70.69)**	**−0.42 (70.39)**	**−3.15 (67.66)**
	DFQ [42]	72.94	−0.61 (72.33)	–	–

In our experiments, we utilize the implementation of public Pytorch repository[7] to evaluate the performance. After folding batch normalization of the pre-trained model into the weights, we found that several layers of weight ranges become very large (e.g., $[−54.4, 64.4]$). Considering the fact that quantization range [27], especially in the early layers [7], has a profound impact on the performance of quantized models, we fix the configuration of some early layers in the backbone. More precisely, we apply 8-bit PWLQ on three depth-wise convolution layers with large ranges in all configurations shown in Table 5. Note that the MAC operations of these three layers are negligible in practice since they only contribute 0.2% of the entire network computation, but it is remarkably beneficial to the performance of low-bit quantized models.

As noticed in classification, low-bit uniform quantization causes significant accuracy drop from the full-precision models. In Table 5, applying PWLQ combined with bias correction, the 6-bit model on weights even outperforms 8-bit DFQ [42], which attains 0.42% degradation of the pre-trained model. Moreover, the 4-bit PWLQ significantly improves the mIoU by 17.61% from the 4-bit uniform quantized model, indicating the potential of low-bit post-training quantization via piecewise linear approximation for the semantic segmentation task.

Object Detection. We also test our PWLQ for object detection task. The experiments are performed on the public Pytorch implementation[8] of SSD-Lite version [36] with a backbone of MobileNet-v2. The performance is evaluated with mean average precision (mAP) on the Pascal VOC detection challenge [11].

Table 6. Uniform quantization and PWLQ of SSD-Lite version. Weights are quantized per-channel with bias correction, activations are uniformly quantized per-layer

Network	W/A	32/32	8/8	6/8	4/8
SSD-Lite (mAP%)	Uniform	68.70	−0.20 (68.50)	−0.43 (68.37)	−3.91 (64.79)
	PWLQ (Ours)	68.70	**−0.19 (68.51)**	**−0.28 (68.42)**	**−0.38 (68.32)**
	DFQ [42]	68.47	−0.56 (67.91)	–	–

Table 6 compares the results of the mAP score of quantized models using the uniform and PWLQ schemes. Similar to image classification and semantic segmentation tasks, even with bias correction and per-channel quantization

---

[7] https://github.com/jfzhang95/pytorch-deeplab-xception.
[8] https://github.com/qfgaohao/pytorch-ssd.

enhancements, 4-bit uniform scheme causes 3.91% performance drop from the full-precision model, while 4-bit PWLQ with these two enhancements is able to remove this notable gap down to 0.38%.

6   Conclusion

In this work, we present a piecewise linear quantization scheme for accurate post-training quantization of deep neural networks. It breaks the bell-shaped distributed values into non-overlapping regions per tensor where each region is assigned an equal number of quantization levels. We further analyze the resulting quantization error as well as the hardware requirements. We show that our approach achieves state-of-the-art low-bit post-training quantization performance on image classification, semantic segmentation, and object detection tasks under the same computational cost. It indicates its potential for efficient and rapid deployment of computer vision applications on resource-limited devices.

References

1. Bakunas-Milanowski, D., Rego, V., Sang, J., Chansu, Y.: Efficient algorithms for stream compaction on GPUs. Int. J. Netw. Comput. **7**(2), 208–226 (2017)
2. Banner, R., Nahshan, Y., Hoffer, E., Soudry, D.: Post training 4-bit quantization of convolution networks for rapid-deployment. CoRR, abs/1810.05723 1, 2 (2018)
3. Baskin, C., et al.: UNIQ: uniform noise injection for non-uniform quantization of neural networks. arXiv preprint arXiv:1804.10969 (2018)
4. Cai, Z., He, X., Sun, J., Vasconcelos, N.: Deep learning with low precision by half-wave gaussian quantization. In: Proceedings of the IEEE Conference on Computer Vision and Pattern Recognition, pp. 5918–5926 (2017)
5. Chen, L.-C., Zhu, Y., Papandreou, G., Schroff, F., Adam, H.: Encoder-decoder with atrous separable convolution for semantic image segmentation. In: Ferrari, V., Hebert, M., Sminchisescu, C., Weiss, Y. (eds.) ECCV 2018. LNCS, vol. 11211, pp. 833–851. Springer, Cham (2018). https://doi.org/10.1007/978-3-030-01234-2_49
6. Choi, J., Wang, Z., Venkataramani, S., Chuang, P.I.J., Srinivasan, V., Gopalakrishnan, K.: PACT: parameterized clipping activation for quantized neural networks. arXiv preprint arXiv:1805.06085 (2018)
7. Choukroun, Y., Kravchik, E., Kisilev, P.: Low-bit quantization of neural networks for efficient inference. arXiv preprint arXiv:1902.06822 (2019)
8. Courbariaux, M., Bengio, Y., David, J.P.: BinaryConnect: training deep neural networks with binary weights during propagations. In: Advances in Neural Information Processing Systems, pp. 3123–3131 (2015)
9. Dhillon, G.S., et al.: Stochastic activation pruning for robust adversarial defense. arXiv preprint arXiv:1803.01442 (2018)
10. Dong, X., Huang, J., Yang, Y., Yan, S.: More is less: a more complicated network with less inference complexity. In: Proceedings of the IEEE Conference on Computer Vision and Pattern Recognition, pp. 5840–5848 (2017)
11. Everingham, M., Van Gool, L., Williams, C.K., Winn, J., Zisserman, A.: The pascal visual object classes (VOC) challenge. Int. J. Comput. Vis. **88**(2), 303–338 (2010)

12. Faraone, J., Fraser, N., Blott, M., Leong, P.H.: SYQ: learning symmetric quantization for efficient deep neural networks. In: Proceedings of the IEEE Conference on Computer Vision and Pattern Recognition, pp. 4300–4309 (2018)

13. Finkelstein, A., Almog, U., Grobman, M.: Fighting quantization bias with bias. arXiv preprint arXiv:1906.03193 (2019)

14. Georgiadis, G.: Accelerating convolutional neural networks via activation map compression. In: Proceedings of the IEEE Conference on Computer Vision and Pattern Recognition, pp. 7085–7095 (2019)

15. Gong, Y., Liu, L., Yang, M., Bourdev, L.: Compressing deep convolutional networks using vector quantization. arXiv preprint arXiv:1412.6115 (2014)

16. Gupta, S., Agrawal, A., Gopalakrishnan, K., Narayanan, P.: Deep learning with limited numerical precision. In: International Conference on Machine Learning, pp. 1737–1746 (2015)

17. Han, S., Mao, H., Dally, W.J.: Deep compression: compressing deep neural networks with pruning, trained quantization and Huffman coding. arXiv preprint arXiv:1510.00149 (2015)

18. He, K., Gkioxari, G., Dollár, P., Girshick, R.: Mask R-CNN. In: Proceedings of the IEEE International Conference on Computer Vision, pp. 2961–2969 (2017)

19. He, K., Zhang, X., Ren, S., Sun, J.: Deep residual learning for image recognition. In: Proceedings of the IEEE Conference on Computer Vision and Pattern Recognition, pp. 770–778 (2016)

20. He, Y., Zhang, X., Sun, J.: Channel pruning for accelerating very deep neural networks. In: Proceedings of the IEEE International Conference on Computer Vision, pp. 1389–1397 (2017)

21. Hinton, G., Vinyals, O., Dean, J.: Distilling the knowledge in a neural network. arXiv preprint arXiv:1503.02531 (2015)

22. Howard, A.G., et al.: MobileNets: efficient convolutional neural networks for mobile vision applications. arXiv preprint arXiv:1704.04861 (2017)

23. Hu, J., Shen, L., Sun, G.: Squeeze-and-excitation networks. In: Proceedings of the IEEE Conference on Computer Vision and Pattern Recognition, pp. 7132–7141 (2018)

24. Huang, G., Liu, Z., Van Der Maaten, L., Weinberger, K.Q.: Densely connected convolutional networks. In: Proceedings of the IEEE Conference on Computer Vision and Pattern Recognition, pp. 4700–4708 (2017)

25. Jacob, B., et al.: Quantization and training of neural networks for efficient integer-arithmetic-only inference. In: Proceedings of the IEEE Conference on Computer Vision and Pattern Recognition, pp. 2704–2713 (2018)

26. Jain, S., Venkataramani, S., Srinivasan, V., Choi, J., Gopalakrishnan, K., Chang, L.: BiScaled-DNN: quantizing long-tailed datastructures with two scale factors for deep neural networks. In: 2019 56th ACM/IEEE Design Automation Conference (DAC), pp. 1–6. IEEE (2019)

27. Jung, S., et al.: Learning to quantize deep networks by optimizing quantization intervals with task loss. In: Proceedings of the IEEE Conference on Computer Vision and Pattern Recognition, pp. 4350–4359 (2019)

28. Krishnamoorthi, R.: Quantizing deep convolutional networks for efficient inference: a whitepaper. arXiv preprint arXiv:1806.08342 (2018)

29. Krizhevsky, A., Sutskever, I., Hinton, G.E.: ImageNet classification with deep convolutional neural networks. In: Advances in Neural Information Processing Systems, pp. 1097–1105 (2012)

30. Lai, W.S., Huang, J.B., Ahuja, N., Yang, M.H.: Fast and accurate image super-resolution with deep Laplacian pyramid networks. IEEE Trans. Pattern Anal. Mach. Intell. **41**, 2599–2613 (2018)
31. Lee, J.H., Ha, S., Choi, S., Lee, W.J., Lee, S.: Quantization for rapid deployment of deep neural networks. arXiv preprint arXiv:1810.05488 (2018)
32. Li, H., Kadav, A., Durdanovic, I., Samet, H., Graf, H.P.: Pruning filters for efficient convnets. arXiv preprint arXiv:1608.08710 (2016)
33. Li, R., Wang, Y., Liang, F., Qin, H., Yan, J., Fan, R.: Fully quantized network for object detection. In: Proceedings of the IEEE Conference on Computer Vision and Pattern Recognition, pp. 2810–2819 (2019)
34. Li, Y., Dong, X., Wang, W.: Additive powers-of-two quantization: an efficient non-uniform discretization for neural networks. In: International Conference on Learning Representations (2020). https://openreview.net/forum?id=BkgXT24tDS
35. Lin, D., Talathi, S., Annapureddy, S.: Fixed point quantization of deep convolutional networks. In: International Conference on Machine Learning, pp. 2849–2858 (2016)
36. Liu, W., et al.: SSD: single shot MultiBox detector. In: Leibe, B., Matas, J., Sebe, N., Welling, M. (eds.) ECCV 2016. LNCS, vol. 9905, pp. 21–37. Springer, Cham (2016). https://doi.org/10.1007/978-3-319-46448-0_2
37. Luo, J.H., Wu, J., Lin, W.: ThiNet: a filter level pruning method for deep neural network compression. In: Proceedings of the IEEE International Conference on Computer Vision, pp. 5058–5066 (2017)
38. Ma, N., Zhang, X., Zheng, H.-T., Sun, J.: ShuffleNet V2: practical guidelines for efficient CNN architecture design. In: Ferrari, V., Hebert, M., Sminchisescu, C., Weiss, Y. (eds.) Computer Vision – ECCV 2018. LNCS, vol. 11218, pp. 122–138. Springer, Cham (2018). https://doi.org/10.1007/978-3-030-01264-9_8
39. Meller, E., Finkelstein, A., Almog, U., Grobman, M.: Same, same but different-recovering neural network quantization error through weight factorization. arXiv preprint arXiv:1902.01917 (2019)
40. Micikevicius, P., et al.: Mixed precision training. arXiv preprint arXiv:1710.03740 (2017)
41. Miyashita, D., Lee, E.H., Murmann, B.: Convolutional neural networks using logarithmic data representation. arXiv preprint arXiv:1603.01025 (2016)
42. Nagel, M., van Baalen, M., Blankevoort, T., Welling, M.: Data-free quantization through weight equalization and bias correction. arXiv preprint arXiv:1906.04721 (2019)
43. Park, E., Yoo, S., Vajda, P.: Value-aware quantization for training and inference of neural networks. In: Ferrari, V., Hebert, M., Sminchisescu, C., Weiss, Y. (eds.) ECCV 2018. LNCS, vol. 11208, pp. 608–624. Springer, Cham (2018). https://doi.org/10.1007/978-3-030-01225-0_36
44. Paszke, A., et al.: Automatic differentiation in PyTorch. In: 31st Conference on Neural Information Processing Systems (2017)
45. Polino, A., Pascanu, R., Alistarh, D.: Model compression via distillation and quantization. arXiv preprint arXiv:1802.05668 (2018)
46. Rastegari, M., Ordonez, V., Redmon, J., Farhadi, A.: XNOR-Net: ImageNet classification using binary convolutional neural networks. In: Leibe, B., Matas, J., Sebe, N., Welling, M. (eds.) ECCV 2016. LNCS, vol. 9908, pp. 525–542. Springer, Cham (2016). https://doi.org/10.1007/978-3-319-46493-0_32
47. Redmon, J., Farhadi, A.: Yolo9000: better, faster, stronger. In: Proceedings of the IEEE Conference on Computer Vision and Pattern Recognition, pp. 7263–7271 (2017)

48. Ren, S., He, K., Girshick, R., Sun, J.: Faster R-CNN: towards real-time object detection with region proposal networks. In: Advances in Neural Information Processing Systems, pp. 91–99 (2015)
49. Ronneberger, O., Fischer, P., Brox, T.: U-Net: convolutional networks for biomedical image segmentation. In: Navab, N., Hornegger, J., Wells, W.M., Frangi, A.F. (eds.) MICCAI 2015. LNCS, vol. 9351, pp. 234–241. Springer, Cham (2015). https://doi.org/10.1007/978-3-319-24574-4_28
50. Rumelhart, D.E., Hinton, G.E., Williams, R.J.: Learning representations by back-propagating errors. Nature 323(6088), 533–536 (1986)
51. Russakovsky, O., Bernstein, M., et al.: ImageNet large scale visual recognition challenge. Int. J. Comput. Vis. 115(3), 211–252 (2015)
52. Sandler, M., Howard, A., Zhu, M., Zhmoginov, A., Chen, L.C.: MobileNetV2: inverted residuals and linear bottlenecks. In: Proceedings of the IEEE Conference on Computer Vision and Pattern Recognition, pp. 4510–4520 (2018)
53. Simonyan, K., Zisserman, A.: Very deep convolutional networks for large-scale image recognition. arXiv preprint arXiv:1409.1556 (2014)
54. Szegedy, C., Vanhoucke, V., Ioffe, S., Shlens, J., Wojna, Z.: Rethinking the inception architecture for computer vision. In: Proceedings of the IEEE Conference on Computer Vision and Pattern Recognition, pp. 2818–2826 (2016)
55. Tan, M., Le, Q.V.: EfficientNet: rethinking model scaling for convolutional neural networks. arXiv preprint arXiv:1905.11946 (2019)
56. Ullrich, K., Meeds, E., Welling, M.: Soft weight-sharing for neural network compression. arXiv preprint arXiv:1702.04008 (2017)
57. Wu, J., Leng, C., Wang, Y., Hu, Q., Cheng, J.: Quantized convolutional neural networks for mobile devices. In: Proceedings of the IEEE Conference on Computer Vision and Pattern Recognition, pp. 4820–4828 (2016)
58. You, Y.: Audio Coding: Theory and Applications. Springer, New York (2010). https://doi.org/10.1007/978-1-4419-1754-6
59. Zagoruyko, S., Komodakis, N.: Wide residual networks. arXiv preprint arXiv:1605.07146 (2016)
60. Zhang, D., Yang, J., Ye, D., Hua, G.: LQ-Nets: learned quantization for highly accurate and compact deep neural networks. In: Ferrari, V., Hebert, M., Sminchisescu, C., Weiss, Y. (eds.) ECCV 2018. LNCS, vol. 11212, pp. 373–390. Springer, Cham (2018). https://doi.org/10.1007/978-3-030-01237-3_23
61. Zhang, X., Zhou, X., Lin, M., Sun, J.: ShuffleNet: an extremely efficient convolutional neural network for mobile devices. In: Proceedings of the IEEE Conference on Computer Vision and Pattern Recognition, pp. 6848–6856 (2018)
62. Zhao, R., Hu, Y., Dotzel, J., De Sa, C., Zhang, Z.: Improving neural network quantization without retraining using outlier channel splitting. In: International Conference on Machine Learning, pp. 7543–7552 (2019)
63. Zhou, A., Yao, A., Guo, Y., Xu, L., Chen, Y.: Incremental network quantization: towards lossless CNNs with low-precision weights. arXiv preprint arXiv:1702.03044 (2017)
64. Zhou, S., Wu, Y., Ni, Z., Zhou, X., Wen, H., Zou, Y.: DoReFa-Net: training low bitwidth convolutional neural networks with low bitwidth gradients. arXiv preprint arXiv:1606.06160 (2016)
65. Zhou, Y., Moosavi-Dezfooli, S.M., Cheung, N.M., Frossard, P.: Adaptive quantization for deep neural network. In: Thirty-Second AAAI Conference on Artificial Intelligence (2018)
66. Zhu, C., Han, S., Mao, H., Dally, W.J.: Trained ternary quantization. arXiv preprint arXiv:1612.01064 (2016)

Joint Disentangling and Adaptation for Cross-Domain Person Re-Identification

Yang Zou[1]($\boxtimes$), Xiaodong Yang[2], Zhiding Yu[2], B.V.K. Vijaya Kumar[1], and Jan Kautz[2]

[1] Carnegie Mellon University, Pittsburgh, USA
yzou2@andrew.cmu.edu
[2] NVIDIA, Santa Clara, USA

Abstract. Although a significant progress has been witnessed in supervised person re-identification (re-id), it remains challenging to generalize re-id models to new domains due to the huge domain gaps. Recently, there has been a growing interest in using unsupervised domain adaptation to address this scalability issue. Existing methods typically conduct adaptation on the representation space that contains both id-related and id-unrelated factors, thus inevitably undermining the adaptation efficacy of id-related features. In this paper, we seek to improve adaptation by purifying the representation space to be adapted. To this end, we propose a joint learning framework that disentangles id-related/unrelated features and enforces adaptation to work on the id-related feature space exclusively. Our model involves a disentangling module that encodes cross-domain images into a shared appearance space and two separate structure spaces, and an adaptation module that performs adversarial alignment and self-training on the shared appearance space. The two modules are co-designed to be mutually beneficial. Extensive experiments demonstrate that the proposed joint learning framework outperforms the state-of-the-art methods by clear margins.

Keywords: Person re-id · Feature disentangling · Domain adaptation

1 Introduction

Person re-identification (re-id) is a task of retrieving the images that contain the person of interest across non-overlapping cameras given a query image. It has been receiving lots of attention as a popular benchmark for metric-learning and found wide real applications such as smart cities [35,47,48,55]. Current state-of-the-art re-id methods predominantly hinge on deep convolutional neural networks (CNNs) and have considerably boosted re-id performance in the

---

Y. Zou—Work done during an internship at NVIDIA Research.

Electronic supplementary material The online version of this chapter (https://doi.org/10.1007/978-3-030-58536-5_6) contains supplementary material, which is available to authorized users.

© Springer Nature Switzerland AG 2020
A. Vedaldi et al. (Eds.): ECCV 2020, LNCS 12347, pp. 87–104, 2020.
https://doi.org/10.1007/978-3-030-58536-5_6

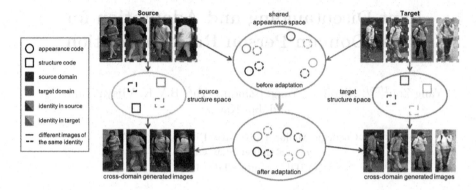

Fig. 1. An overview of the proposed joint disentangling and adaptation framework. The disentangling module encodes images of two domains into a shared appearance space (id-related) and a separate source/target structure space (id-unrelated) via cross-domain image generation. Our adaptation module is exclusively conducted on the id-related feature space, encouraging the intra-class similarity and inter-class difference of the disentangled appearance features.

supervised learning scenario [46,50,58,59]. However, this idealistic closed-world setting postulates that training and testing data has to be drawn from the same camera network or the same domain, which rarely holds in real-world deployments. As a result, these re-id models usually encounter a dramatic performance degradation when deployed to new domains, mainly due to the great domain gaps between training and testing data, such as the changes of season, background, viewpoint, illumination, camera, etc. This largely restricts the applicability of such domain-specific re-id models, in particular, relabeling a large identity corpus for every new domain is prohibitively costly.

To solve this problem, recent years have seen growing interests in person re-id under cross-domain settings. One popular solution to reduce the domain gap is unsupervised domain adaptation (UDA), which utilizes both labeled data in the source domain and unlabeled data in the target domain to improve the model performance in the target domain [18,68]. A fundamental design principle is to align feature distributions across domains to reduce the gap between source and target. A well-performing source model is expected to achieve similar performance in the target domain if the cross-domain gap is closed.

Compared to the conventional problems of UDA, such as image classification and semantic segmentation, person re-id is a more challenging open-set problem as two different domains contain disjoint or completely different identity class spaces. Recent methods mostly bridge the domain gap through adaptation at input-level and or feature-level. For input-level, the generative adversarial networks (GANs) are often utilized to transfer the holistic or factor-wise image style from source to target [6,31]. Adaptation at feature-level often employs self-training or distribution distance minimization to enforce similar cross-domain distributions [30,51]. Zhong et al. [64] combine the complementary benefits of both input-level and feature-level to further improve adaptation capability.

However, a common issue behind these methods is that such adaptations typically operate on the feature space, which encodes both id-related and id-unrelated factors. Therefore, the adaptation of id-related features is inevitably interfered with and impaired by id-unrelated features, restricting the performance gain from UDA. Since cross-domain person re-id is coupled with both disentangling and adaptation problems, and existing methods mostly treat the two problems separately, it is important to come up with a principled framework that solves both issues together. Although disentangling has been studied for supervised person re-id in [8,59], it remains an open question how to integrate with adaptation, and it is under-presented in unsupervised cross-domain re-id as a result of the large domain gap and lack of target supervision.

In light of the above observation, we propose a joint learning framework that disentangles id-related/unrelated factors so that adaptation can be more effectively performed on the id-related space to prevent id-unrelated interference. Our work is partly inspired by DG-Net [59], a recent supervised person re-id approach that performs within-domain image disentangling and leverages such disentanglement to augment training data towards better model training. We argue that successful cross-domain disentangling can create a desirable foundation for more targeted and effective domain adaptation. We thus propose a cross-domain and cycle-consistent image generation with three latent spaces modeled by corresponding encoders to decompose source and target images. The latent spaces incorporate a **shared appearance** space that captures id-related features (i.e., appearance and other semantics), a **source structure** space and a **target structure** space that contain id-unrelated features (i.e., pose, position, viewpoint, background and other variations). We refer to the encoded features in the three spaces as codes. Our adaptation module is exclusively conducted in the shared appearance space, as illustrated in Fig. 1.

This design forms a joint framework that creates mutually beneficial cooperation between the disentangling and adaptation modules: (1) disentanglement leads to better adaptation as we can make the latter focus on id-related features and mitigate the interference of id-unrelated features, and (2) adaptation in turn improves disentangling as the shared appearance encoder gets enhanced during adaptation. We refer the proposed cross-domain joint disentangling and adaptation learning framework as **DG-Net++**.

Our main contributions of this paper are summarized as follows. First, we propose a joint learning framework for unsupervised cross-domain person re-id to disentangle id-related/unrelated factors so that adaptation can be more effectively performed on the id-related space. Second, we introduce a cross-domain cycle-consistency paradigm to realize the desired disentanglement. Third, our disentangling and adaptation are co-designed to let the two modules mutually promote each other. Fourth, our approach achieves superior results on six benchmark pairs, largely pushing person re-id systems toward real-world deployment. Our code and model are available at https://github.com/NVlabs/DG-Net-PP.

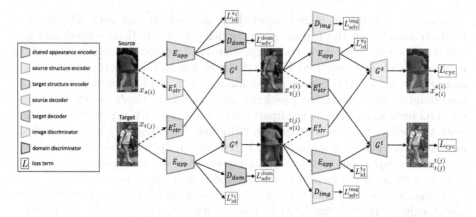

Fig. 2. A schematic overview of the cross-domain cycle-consistency image generation. Our disentangling and adaptation modules are connected by the shared appearance encoder. The two domains also share the image and domain discriminators, but have their own structure encoders and decoders. A dashed line indicates that the input image to the source/target structure encoder is converted to gray-scale.

2   Related Work

Disentangling. This task explores explanatory and independent factors among features in a representation. A generic framework combining deep convolutional auto-encoder with adversarial training is proposed in [34] to disentangle hidden factors within a set of labeled observations. InfoGAN [3] and β-VAE [17] are introduced to learn interpretable factorized features in an unsupervised manner. A two-step disentanglement method [14] is used to extract label relevant information for image classification. In [21,27], images are decomposed to content and style information to serve image-to-image translation.

Unsupervised Domain Adaptation. UDA has been gaining increasing attention in image classification, object detection, and semantic segmentation. Based on the typical closed-set assumption that label classes are shared across domains, UDA methods can be roughly categorized as input-level and or feature-level adaptation. At input-level, models are usually adapted by training with style translated images [7,21,27]. Adaptation at feature-level often minimizes certain distance or divergence between source and target feature distributions, such as correlation [44], maximum mean discrepancy (MMD) [33], sliced Wasserstein discrepancy [26], and lifelong learning [2]. Moreover, domain adversarial [19,49] and self-training [1,12,67,68] have also shown to be powerful feature-level alignment methods. CyCADA [18] adapts at both input-level and feature-level with the purpose of incorporating the effects of both.

Person Re-id. A large family of person re-id focuses on supervised learning. They usually approach re-id as deep metric learning problems [10,16], exploit pedestrian attributes as extra supervisions via multitask learning [42,51], utilize part-based matching or ensembling to reduce intra-class variations [41,53,56], make use of human pose and parsing to facilitate local feature learning [24,43,60], or resort to generative models to augment training data [13,37,59]. Although these methods have achieved tremendous progress in supervised setting, their performances degrade significantly on new domains.

Similar to the traditional problems of UDA, feature-level adaptation is widely used to seek source-target distribution alignment. In [28,30], feature adaptation is enforced by minimizing MMD between feature distributions in two domains. The self-training based methods also present promising results in [40]. Another line is at input-level using GANs to transfer source images into target styles. An adaptive transfer method is developed in [31] to decompose a holistic style to a set of imaging factors. Li et al. [28] propose to learn domain-invariant representation through pose-guided image translation. Chen et al. [4] present an instance-guided context rendering to enable supervised learning in target domain by transferring source person identities into target contexts.

Although DG-Net++ inherits (and extends) the appearance and structure spaces of DG-Net [59], there exist significant new designs in DG-Net++ to allow it to work for a very different problem. (1) DG-Net++ aims to address unsupervised cross-domain re-id, while DG-Net is developed under the fully supervised setting. (2) DG-Net++ is built upon a new cross-domain cycle-consistency scheme to disentangle id-related/unrelated factors without any target supervision. In comparison, DG-Net employs a within-domain disentanglement through latent code reconstruction with access to the ground truth identity. (3) DG-Net++ seamlessly integrates disentangling with adaptation in a unified manner to enable the two modules to mutually benefit each other, which is not considered in DG-Net. (4) DG-Net++ substantially outperforms DG-Net for unsupervised cross-domain re-id on six benchmark pairs.

3   Method

As illustrated in Fig. 2, DG-Net++ combines the disentangling and adaptation modules via the shared appearance encoder. We propose the cross-domain cycle-consistency generation to facilitate disentangling id-related (appearance) and id-unrelated (structure) factors. Our adaptation module involves adversarial alignment and self-training, which are co-designed with the disentangling module to target at id-related features and adapt more effectively.

3.1   Disentangling Module

Formulation. We denote real images and labels in source domain as $X_s = \{x_{s(i)}\}_{i=1}^{N_s}$ and $Y_s = \{y_{s(i)}\}_{i=1}^{N_s}$, where s indicates source domain, N_s is the number of source images, $y_{s(i)} \in [1, K_s]$ and K_s is the number of source identities. Similarly, $X_t = \{x_{t(i)}\}_{i=1}^{N_t}$ denotes N_t real images in target domain t. Given a source image $x_{s(i)}$ and a target image $x_{t(j)}$, a new cross-domain synthesized image can be generated by swapping the appearance or structure codes between the two images. As shown in Fig. 2, the disentangling module consists of a shared appearance encoder $E_{app} : x \rightarrow \nu$, a source structure encoder $E_{str}^s : x_{s(i)} \rightarrow \tau_{s(i)}$, a target structure encoder $E_{str}^t : x_{t(j)} \rightarrow \tau_{t(j)}$, a source decoder $G^s : (\nu_{t(j)}, \tau_{s(i)}) \rightarrow x_{s(i)}^{t(j)}$, a target decoder $G^t : (\nu_{s(i)}, \tau_{t(j)}) \rightarrow x_{t(j)}^{s(i)}$, an image discriminator D_{img} to distinguish between real and synthesized images, and a domain discriminator D_{dom} to distinguish between source and target domains. Note: for synthesized images, we use superscript to indicate the real image providing appearance code and subscript to denote the one giving structure code; for real images, they only have subscript as domain and image index. Our adaptation and re-id are conducted using the appearance codes.

Cross-Domain Generation. We introduce cross-domain cycle-consistency image generation to enforce disentangling between appearance and structure factors. Given a pair of source and target images, we first swap their appearance or structure codes to synthesize new images. Since there exists no ground-truth supervision for the synthetic images, we take advantage of cycle-consistency self-supervision to reconstruct the two real images by swapping the appearance or structure codes extracted from the synthetic images. As demonstrated in Fig. 2, given a source image $x_{s(i)}$ and a target image $x_{t(j)}$, the synthesized images $x_{t(j)}^{s(i)} = G^t(\nu_{s(i)}, \tau_{t(j)})$ and $x_{s(i)}^{t(j)} = G^s(\nu_{t(j)}, \tau_{s(i)})$ are required to respectively preserve the corresponding appearance and structure codes from $x_{s(i)}$ and $x_{t(j)}$ to be able to reconstruct the two original real images:

$$L_{\text{cyc}} = \mathbb{E}\left[\left\|x_{s(i)} - G^s(E_{app}(x_{t(j)}^{s(i)}), E_{str}^s(x_{s(i)}^{t(j)}))\right\|_1\right] + \\ \mathbb{E}\left[\left\|x_{t(j)} - G^t(E_{app}(x_{s(i)}^{t(j)}), E_{str}^t(x_{t(j)}^{s(i)}))\right\|_1\right]. \tag{1}$$

With the identity labels available in source domain, we then explicitly enforce the shared appearance encoder to capture the id-related information by using the identification loss:

$$L_{\text{id}}^{s_1} = \mathbb{E}[-\log(p(y_{s(i)}|x_{s(i)}))]. \tag{2}$$

where $p(y_{s(i)}|x_{s(i)})$ is the predicted probability that $x_{s(i)}$ belongs to the ground-truth label $y_{s(i)}$. We also apply the identification loss on the synthetic image that retains the appearance code from source image to keep identity consistency:

$$L_{\text{id}}^{s_2} = \mathbb{E}[-\log(p(y_{s(i)}|x_{t(j)}^{s(i)}))]. \tag{3}$$

where $p(y_{s(i)}|x_{t(j)}^{s(i)})$ is the predicted probability of $x_{t(j)}^{s(i)}$ belonging to the ground-truth label $y_{s(i)}$ of $x_{s(i)}$. In addition, we employ adversarial loss to match the distributions between the synthesized images and the real data:

$$L_{\text{adv}}^{\text{img}} = \mathbb{E}\left[\log D_{img}(x_{s(i)}) + \log(1 - D_{img}(x_{t(j)}^{s(i)}))\right] + \\ \mathbb{E}\left[\log D_{img}(x_{t(j)}) + \log(1 - D_{img}(x_{s(i)}^{t(j)}))\right]. \tag{4}$$

Note that the image discriminator D_{img} is shared across domains to force the synthesized images to be realistic regardless of domains. This can indirectly drive the shared appearance encoder to learn domain-invariant features. Apart from the cross-domain generation, our disentangling module is also flexible to incorporate the within-domain generation as [59], which can be used to further stabilize and regulate the within-domain disentanglement.

3.2   Adaptation Module

Adversarial Alignment. Although the weights of appearance encoder are shared between source and target domains, the appearance representations across domains are still not ensured to have similar distributions. To encourage the alignment of appearance features in two domains, we introduce a domain discriminator D_{dom}, which aims to distinguish the domain membership of the encoded appearance codes $\nu_{s(i)}$ and $\nu_{t(j)}$. During adversarial training, the shared appearance encoder learns to produce appearance features of which domain membership cannot be differentiated by D_{dom}, such that the distance between cross-domain appearance feature distributions can be reduced. We express this domain appearance adversarial alignment loss as:

$$
\begin{aligned}
L_{\text{adv}}^{\text{dom}} = {} & \mathbb{E}\left[\log D_{dom}(\nu_{s(i)}) + \log(1 - D_{dom}(\nu_{t(j)}))\right] + \\
& \mathbb{E}\left[\log D_{dom}(\nu_{t(j)}) + \log(1 - D_{dom}(\nu_{s(i)}))\right].
\end{aligned}
\tag{5}
$$

Self-training. In addition to the global feature alignment imposed by the above domain adversarial loss, we incorporate self-training in the adaptation module. Essentially, self-training with identification loss is an entropy minimization process that gradually reduces intra-class variations. It implicitly closes the cross-domain feature distribution distance in the shared appearance space, and meanwhile encourages discriminative appearance feature learning.

We iteratively generate a set of pseudo-labels $\hat{Y}_t = \{\hat{y}_{t(j)}\}$ based on the reliable identity predictions in target domain, and refine the network using the pseudo-labeled target images. Note the numbers of pseudo-identities and labeled target images may change during self-training. In practice, the pseudo-labels are produced by clustering the target features that are extracted by the shared appearance encoder E_{app}. We assign the same pseudo-label to the samples within the same cluster. We adopt an affinity based clustering method DBSCAN [9] that has shown promising results in re-id. We utilize the K-reciprocal encoding [63] to compute pairwise distances, and update pseudo-labels every two epochs. With the pseudo-labels obtained by self-training in target domain, we apply the identification loss on the shared appearance encoder:

$$
L_{\text{id}}^{\text{t}_1} = \mathbb{E}[-\log(p(\hat{y}_{t(j)}|x_{t(j)}))].
\tag{6}
$$

where $p(\hat{y}_{t(j)}|x_{t(j)})$ is the predicted probability that $x_{t(j)}$ belongs to the pseudo-label $\hat{y}_{t(j)}$. We furthermore enforce the identification loss with pseudo-label on the synthetic image that reserves the appearance code from target image to keep pseudo-identity consistency:

$$L_{id}^{t_2} = \mathbb{E}[-\log(p(\hat{y}_{t(j)}|x_{s(i)}^{t(j)}))]. \tag{7}$$

where $p(\hat{y}_{t(j)}|x_{s(i)}^{t(j)})$ is the predicted probability of $x_{s(i)}^{t(j)}$ belonging to the pseudo-label $\hat{y}_{t(j)}$ of $x_{t(j)}$. Overall, adaptation with self-training encourages the shared appearance encoder to learn both domain-invariant and discriminative features that can generalize and facilitate re-id in target domain.

3.3 Discussion

Our disentangling and adaptation are co-designed to let the two modules positively interact with each other. On the one hand, **disentangling promotes adaptation**. Based on the cross-domain cycle-consistency image generation, our disentangling module learns detached appearance and structure factors with explicit and explainable meanings, paving the way for adaptation to exclude id-unrelated noises and specifically operate on id-related features. With the help of sharing appearance encoder, the discrepancy between cross-domain feature distributions can be reduced. Also the adversarial loss for generating realistic images across domains encourages feature alignment through the shared image discriminator. On the other hand, **adaptation facilitates disentangling**. In addition to globally close the distribution gap, the adversarial alignment by the shared domain discriminator helps to find the common appearance embedding that can assist disentangling appearance and structure features. Besides implicitly aligning cross-domain features, the self-training with the identification loss supports disentangling since it forces the appearance features of different identities to stay apart while reduces the intra-class variation of the same identity. Therefore, through the adversarial loss and identification loss via self-training, the appearance encoder is enhanced in the adaptation process, and a better appearance encoder generates better synthetic images, eventually leading to the improvement of the disentangling module.

3.4 Optimization

We jointly train the shared appearance encoder, image discriminator, domain discriminator, as well as source and target structure encoders, and source and target decoders to optimize the total objective, which is a weighted sum of the following loss terms:

$$L_{\text{total}}(E_{app}, D_{img}, D_{dom}, E_{str}^s, E_{str}^t, G^s, G^t) = \\ \lambda_{\text{cyc}}L_{\text{cyc}} + L_{id}^{s_1} + L_{id}^{t_1} + \lambda_{id}L_{id}^{s_2} + \lambda_{id}L_{id}^{t_2} + L_{\text{adv}}^{\text{img}} + L_{\text{adv}}^{\text{dom}}. \tag{8}$$

where λ_{cyc} and λ_{id} are the weights to control the importance of cross-domain cycle-consistent self-supervision loss and identification loss on synthesized images. Following the common practice in image-to-image translations [21,27,66], we set a large weight $\lambda_{\text{cyc}} = 2$ for L_{cyc}. As the quality of cross-domain synthesized images is not great at the early stage of training, the two losses L_{id}^{s2} and L_{id}^{t2} on such images would make training unstable, so we use a relatively small weight $\lambda_{id} = 0.5$. We fix the weights during the entire training process in all experiments. We first warm up E_{app}, E_{str}^s, G^s and D_{img} with the disentangling module in source domain for 100K iterations, then bring in the adversarial alignment to train the whole network for another 50K before self-training. In the process of self-training, all components are co-trained, and the pseudo-labels are updated every two epochs. We follow the alternative updating policy in training GANs to alternatively train E_{app}, E_{str}^s, E_{str}^t, G^s, G^t, and D_{img}, D_{dom}.

4   Experiments

We evaluate the proposed framework DG-Net++ following the standard experimental protocols on six domain pairs formed by three benchmark datasets: Market-1501 [57], DukeMTMC-reID [39] and MSMT17 [52]. We report comparisons to the state-of-the-art methods and provide in-depth analysis. A variety of ablation studies are performed to understand the contributions of each individual component in our approach. The qualitative results of cross-domain image generation are also presented. Extensive evaluations reveal that our approach consistently produces realistic cross-domain images, and more importantly, outperforms the competing algorithms by clear margins over all benchmarks.

4.1   Implementation Details

We implement our framework in PyTorch. In the following descriptions, we use channel $\times$ height $\times$ width to denote the size of feature maps. (**1**) E_{app} is modified from ResNet50 [15] and pre-trained on ImageNet [5]. Its global average pooling layer and fully-connected layer are replaced with a max pooling layer that outputs the appearance code ν in $2048 \times 4 \times 1$, which is in the end mapped to a 1024-dim vector to perform re-id. (**2**) E_{str}^s and E_{str}^t share the same architecture with four convolutional layers followed by four residual blocks [15], and output the source/target structure code τ in $128 \times 64 \times 32$. (**3**) G^s and G^t use the same decoding scheme to process the source/target code τ through four residual blocks and four convolutional layers. And each residual block includes two adaptive instance normalization layers [20] to absorb the appearance code ν as scale and bias parameters. (**4**) D_{img} follows the popular multi-scale Patch-GAN [23] at three different input scales: 64×32, 128×64, and 256×128. (**5**) D_{dom} is a multi-layer perceptron containing four fully-connected layers to map the appearance code τ to a domain membership. (**6**) For training, input images are resized to 256×128. We use SGD to train E_{app}, and Adam [25] to optimize E_{str}^s, E_{str}^t, G^s, G^t, D_{img}, D_{dom}. (**7**) For generating pseudo-labels with DBSCAN in self-training, we set the neighbor maximum distance to 0.45 and the minimum number of points required to form a dense region to 7. (**8**) At test time, our re-id model only involves E_{app}, which has a comparable network capacity to most re-id models using ResNet50 as a backbone. We use the 1024-dim vector output by E_{app} as the final image representation.

4.2   Quantitative Results

Comparison with the State-of-the-Art. We extensively evaluate DG-Net++ on six cross-domain pairs among three benchmark datasets with a variety of competing algorithms. Table 1 shows the comparative results on the six cross-domain pairs. In particular, compared to the second best methods, we achieve the state-of-the-art results with considerable margins of 10.4%, 3.4%, 8.9%, 8.8%, 24.5%, 5.0% mAP and 5.9%, 2.1%, 16.8%, 16.6%, 14.6%, 3.2% Rank@1 on Market $\rightarrow$ Duke, Duke $\rightarrow$ Market, Market $\rightarrow$ MSMT, Duke $\rightarrow$ MSMT, MSMT $\rightarrow$ Duke, MSMT $\rightarrow$ Market, respectively. Moreover, DG-Net++ is found to even outperform or approach some recent supervised re-id methods [22,32,45,61,62] that have access to the full labels of the target domain.

These superior performances collectively and clearly show the advantages of the joint disentangling and adaptation design, which enables more effective adaptation in the disentangled id-related feature space and presents strong cross-domain adaptation capability. Additionally, we emphasize that the disentangling module in DG-Net++ is orthogonal and applicable to other adaptation methods without considering feature disentangling. Overall, our proposed cross-domain disentangling provides a better foundation to allow for more effective cross-domain re-id adaptation. Other adaptation methods, such as some recent approaches [11,12,65], can be readily applied to the disentangled id-related feature space, and their performances may even be boosted further.

Ablation Study. We perform a variety of ablation experiments primarily on the two cross-domain pairs: Market $\rightarrow$ Duke and Duke $\rightarrow$ Market to evaluate the contribution of each individual component in DG-Net++. As shown in Table 2, our baseline is an ImageNet pre-trained ResNet50 that is trained on the source domain and directly transferred to the target domain. By just using the proposed disentangling module, our approach can boost the baseline performance by 4.9%, 11.8% mAP and 7.1%, 10.4% Rank@1 respectively on the two cross-domain pairs. Note this improvement is achieved without using any adaptations. This suggests that by only removing the id-unrelated features through disentangling, the cross-domain discrepancy has already been reduced since the id-unrelated noises largely contribute to the domain gap. Based on the disentangled id-related features, either adversarial alignment or self-training consistently provides clear performance gains. By combining both, our full model obtains the best performances that are substantially improved over the baseline results.

Next we study the gains of disentangling to adaptation in DG-Net++. As shown in Fig. 3(a), compared with the space entangled with both id-related and id-unrelated factors, in the disentangled id-related space, adversarial alignment can be conducted more effectively with 8.6% and 6.4% mAP improvements on Market $\rightarrow$ Duke and Duke $\rightarrow$ Market, respectively. A similar observation can also be found for self-training. In comparison to self-training only, disentangling

Table 1. Comparison with the state-of-the-art unsupervised cross-domain re-id methods on the six cross-domain benchmark pairs.

Methods	Market-1501 → DukeMTMC-reID				DukeMTMC-reID → Market-1501			
	Rank@1	Rank@5	Rank@10	mAP	Rank@1	Rank@5	Rank@10	mAP
SPGAN [6]	41.1	56.6	63.0	22.3	51.5	70.1	76.8	22.8
AIDL [51]	44.3	59.6	65.0	23.0	58.2	74.8	81.1	26.5
MMFA [30]	45.3	59.8	66.3	24.7	56.7	75.0	81.8	27.4
HHL [64]	46.9	61.0	66.7	27.2	62.2	78.8	84.0	31.4
CAL [36]	55.4	–	–	36.7	64.3	–	–	34.5
ARN [29]	60.2	73.9	79.5	33.4	70.3	80.4	86.3	39.4
ECN [65]	63.3	75.8	80.4	40.4	75.1	87.6	91.6	43.0
PDA [28]	63.2	77.0	82.5	45.1	75.2	86.3	90.2	47.6
CR-GAN [4]	68.9	80.2	84.7	48.6	77.7	89.7	92.7	54.0
IPL [40]	68.4	80.1	83.5	49.0	75.8	89.5	93.2	53.7
SSG [11]	73.0	80.6	83.2	53.4	80.0	90.0	92.4	58.3
DG-Net++	**78.9**	**87.8**	**90.4**	**63.8**	**82.1**	**90.2**	**92.7**	**61.7**
Methods	Market-1501 → MSMT17				DukeMTMC-reID → MSMT17			
	Rank@1	Rank@5	Rank@10	mAP	Rank@1	Rank@5	Rank@10	mAP
PTGAN [52]	10.2	–	24.4	2.9	11.8	–	27.4	3.3
ENC [65]	25.3	36.3	42.1	8.5	30.2	41.5	46.8	10.2
SSG [11]	31.6	–	49.6	13.2	32.2	–	51.2	13.3
DG-Net++	**48.4**	**60.9**	**66.1**	**22.1**	**48.8**	**60.9**	**65.9**	**22.1**
Methods	MSMT17 → Market-1501				MSMT17 → DukeMTMC-reID			
	Rank@1	Rank@5	Rank@10	mAP	Rank@1	Rank@5	Rank@10	mAP
PAUL [54]	68.5	–	–	40.1	72.0	–	–	53.2
DG-Net++	**83.1**	**91.5**	**94.3**	**64.6**	**75.2**	**73.6**	**86.9**	**58.2**

Table 2. Ablation study on two cross-domain pairs: Market → Duke and Duke → Market. We use "D" to denote disentangling, "A" to adversarial alignment, and "ST" to self-training.

Methods	Market-1501 → DukeMTMC-reID				DukeMTMC-reID → Market-1501			
	Rank@1	Rank@5	Rank@10	mAP	Rank@1	Rank@5	Rank@10	mAP
Baseline	37.4	52.4	58.4	19.3	39.7	57.9	64.3	15.0
+A+ST	71.4	81.8	85.7	57.5	75.7	86.4	90.1	57.1
+D	44.5	60.6	66.7	24.2	50.1	68.0	73.9	26.8
+D+A	53.2	68.7	73.8	36.3	52.2	70.7	77.0	28.6
+D+ST	74.2	82.8	86.5	58.4	78.0	87.1	90.3	56.5
+D+A+ST	**78.9**	**87.8**	**90.4**	**63.8**	**82.1**	**90.2**	**92.7**	**61.7**

largely boosts the performance by 4.0% and 5.7% mAP on the two cross-domain pairs. This strongly indicates the advantages of disentangling to enable more effective adaptation in the separated id-related space.

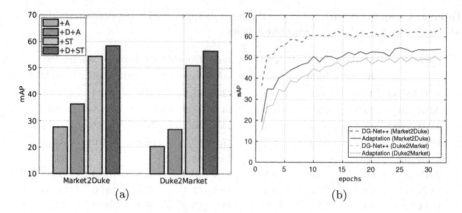

Fig. 3. (a) Improvements of disentangling to adaptation in DG-Net++. "A": adversarial alignment, "ST": self-training, and "D": disentangling. (b) Comparison of the training processes between our full model and the adaptation (self-training) alone model on the two cross-domain pairs.

To better understand the learning behavior of DG-Net++, we plot the training curves on the two cross-domain pairs in Fig. 3(b). Our full model consistently outperforms the self-training alone model by large margins during the training process thanks to the merits that the adaptation can be more effectively performed on the disentangled id-related space in our full model. In addition, as shown in the figure, the training curves are overall stable with slight fluctuations after 13 epochs, and we argue that such a stable learning behavior is quite desirable for model selection in the unsupervised cross-domain scenario where the target supervision is not available.

Comparison with DG-Net. To validate the superiority of DG-Net++ over DG-Net for unsupervised cross-domain adaptation, we conduct further ablation study on Market → Duke. (1) Based on DG-Net trained in source domain, we perform self-training with the trained model, i.e., the appearance encoder. It achieves 54.6% mAP, 9.2% inferior to 63.8% mAP of DG-Net++. This shows the necessity of joint disentangling and adaptation for cross-domain re-id. (2) We perform a semi-supervised training for DG-Net on two domains, where self-training is introduced to supervise the appearance encoder in target domain. It achieves 52.9% mAP, 10.9% inferior to DG-Net++. Note this result is even worse than self-training with only the appearance encoder (54.6%). This suggests that an inappropriate design of disentangling (the within-domain disentangling of DG-Net) can harm adaptation. In summary, DG-Net is designed to work on a single domain, while the proposed disentangling of DG-Net++ is vital for a joint disentangling and adaptation in cross-domain.

Sensitivity Analysis. We also study how sensitive the re-id performance is to the two important hyper-parameters in Eq. 8: one is λ_{cyc}, the weight to control the importance of L_{cyc}; the other is λ_{id} to weight the identification losses $L_{\text{id}}^{\text{s2}}$ and $L_{\text{id}}^{\text{t2}}$ on the synthesized images of source and target domains. This analysis

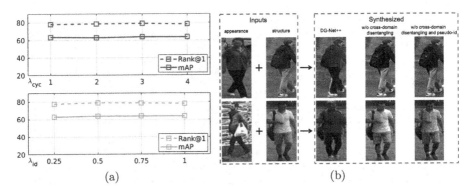

(a)   (b)

Fig. 4. (a) Analysis of the influence of hyper-parameters λ_{cyc} and λ_{id} on Market $\rightarrow$ Duke. (b) Comparison of the synthesized images by our full model, removing cross-domain disentangling, and further removing pseudo-identity supervision. We use source appearance and target structure in the first row, and target appearance and source structure in the second row.

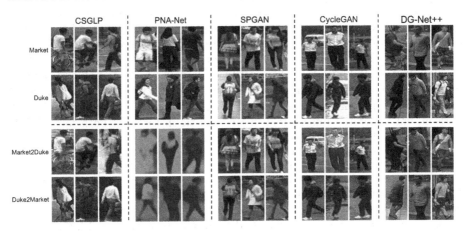

Fig. 5. Comparison of the generated images across two cross-domains between Market and Duke of different methods including CycleGAN [66], SPGAN [6], PNA-Net [28], CSGLP [38], and our approach DG-Net++. Please attention to both foreground and background of the synthetic images.

is conducted on Market $\rightarrow$ Duke. Figure 4(a) demonstrates that the re-id performances are overall stable and there are only slight variations when λ_{cyc} varies from 1 to 4 and λ_{id} from 0.25 to 1. Thus, our model is not sensitive to the two hyper-parameters, and we set $\lambda_{cyc} = 2$ and $\lambda_{id} = 0.5$ in all experiments.

4.3   Qualitative Results

Comparison with the State-of-the-Art. We also compare the image generation results between DG-Net++ and other representative image translation

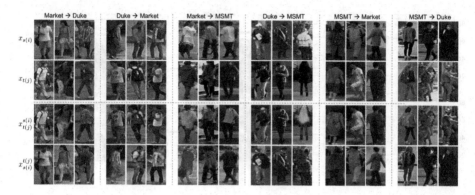

Fig. 6. Examples of our synthesized images on six cross-domain benchmark pairs. We show source images in the first row, target images in the second row, synthetic images with source appearance and target structure in the third row, and synthetic images with target appearance and source structure in the fourth row.

based methods for unsupervised cross-domain person re-id, including CycleGAN [66], SPGAN [52], PNA-Net [28] and CSGLP [38]. As shown in Fig. 5, Cycle-GAN and SPGAN virtually translate the illumination only. CSGLP can switch the illumination and background between two domains, but is not able to change foreground or person appearance. PDA-Net synthesizes various images by manipulating human poses, but the generated images are prone to be blurry. In comparison, our generated images look more realistic in terms of both foreground and background. This also verifies the effectiveness of the proposed framework to decompose id-related and id-unrelated factors, and therefore facilitating more effective cross-domain adaptation.

Cross-Domain Synthesized Images. Here we show more qualitative results of cross-domain generated images in Fig. 6, which shows the examples on six cross-domain pairs. Compared to the within-domain image generation [13,37,59], the cross-domain image synthesis is more challenging due to huge domain gap and lack of identity supervision in target domain. DG-Net++ is able to generate realistic images over different domain pairs, which present very diverse clothing styles, seasons, poses, viewpoints, backgrounds, illuminations, etc. This indicates that our approach is not just geared to solve a particular type of domain gap but is generalizable across different domains. The last column of this figure shows a failure case where the source and target appearances are not well retained in the synthetic images. We conjecture that this difficulty is caused by the occluded bottom right person in the target image as his appearance confuses the appearance feature extraction.

Ablation Study. We then qualitatively compare our full model DG-Net++ to its two variants without cross-domain disentangling and pseudo-identity supervision. As shown in Fig. 4(b), removing cross-domain disentangling or further pseudo-id, the synthetic images are unsatisfying as the models fail to translate the accurate clothing color or style. This again clearly shows the merits of our unified disentangling and adaptation for cross-domain image generation.

5 Conclusion

In this paper, we have proposed a joint learning framework that disentangles id-related/unrelated factors and performs adaptation exclusively on the id-related feature space. This design leads to more effective adaptation as the id-unrelated noises are segregated from the adaptation process. Our cross-domain cycle-consistent image generation as well as adversarial alignment and self-training are co-designed such that the disentangling and adaptation modules can mutually promote each other during joint training. Experimental results on the six benchmarks find that our approach consistently brings substantial performance gains. We hope the proposed approach would inspire more work of integrating disentangling and adaptation for unsupervised cross-domain person re-id.

References

1. Chen, B., et al.: Angular visual hardness. In: ICML (2020)
2. Chen, W., Yu, Z., Wang, Z., Anandkumar, A.: Automated synthetic-to-real generalization. In: ICML (2020)
3. Chen, X., Duan, Y., Houthooft, R., Schulman, J., Sutskever, I., Abbeel, P.: Info-GAN: interpretable representation learning by information maximizing generative adversarial nets. In: NeurIPS (2016)
4. Chen, Y., Zhu, X., Gong, S.: Instance-guided context rendering for cross-domain person re-identification. In: ICCV (2019)
5. Deng, J., Dong, W., Socher, R., Li, L.J., Li, K., Fei-Fei, L.: ImageNet: a large-scale hierarchical image database. In: CVPR (2009)
6. Deng, W., Zheng, L., Ye, Q., Kang, G., Yang, Y., Jiao, J.: Image-image domain adaptation with preserved self-similarity and domain-dissimilarity for person re-identification. In: CVPR (2018)
7. Dundar, A., Liu, M.Y., Yu, Z., Wang, T.C., Zedlewski, J., Kautz, J.: Domain stylization: a fast covariance matching framework towards domain adaptation. In: TPAMI (2020)
8. Eom, C., Ham, B.: Learning disentangled representation for robust person re-identification. In: NeurIPS (2019)
9. Ester, M., Kriegel, H.P., Sander, J., Xu, X.: A density-based algorithm for discovering clusters in large spatial databases with noise. In: KDD (1996)
10. Fan, L., Li, T., Fang, R., Hristov, R., Yuan, Y., Katabi, D.: Learning longterm representations for person re-identification using radio signals. In: CVPR (2020)
11. Fu, Y., Wei, Y., Wang, G., Zhou, Y., Shi, H., Huang, T.S.: Self-similarity grouping: a simple unsupervised cross domain adaptation approach for person re-identification. In: ICCV (2019)
12. Ge, Y., Chen, D., Li, H.: Mutual mean-teaching: pseudo label refinery for unsupervised domain adaptation on person re-identification. In: ICLR (2020)
13. Ge, Y., Li, Z., Zhao, H., Yin, G., Yi, S., Wang, X., et al.: FD-GAN: pose-guided feature distilling GAN for robust person re-identification. In: NeurIPS (2018)

14. Hadad, N., Wolf, L., Shahar, M.: A two-step disentanglement method. In: CVPR (2018)
15. He, K., Zhang, X., Ren, S., Sun, J.: Deep residual learning for image recognition. In: CVPR (2016)
16. Hermans, A., Beyer, L., Leibe, B.: In defense of the triplet loss for person re-identification. arXiv arXiv:1703.07737 (2017)
17. Higgins, I., et al.: β-VAE: learning basic visual concepts with a constrained variational framework. In: ICLR (2017)
18. Hoffman, J., et al.: CyCADA: cycle-consistent adversarial domain adaptation. In: ICML (2018)
19. Hong, W., Wang, Z., Yang, M., Yuan, J.: Conditional generative adversarial network for structured domain adaptation. In: CVPR (2018)
20. Huang, X., Belongie, S.: Arbitrary style transfer in real-time with adaptive instance normalization. In: ICCV (2017)
21. Huang, X., Liu, M.-Y., Belongie, S., Kautz, J.: Multimodal unsupervised image-to-image translation. In: Ferrari, V., Hebert, M., Sminchisescu, C., Weiss, Y. (eds.) ECCV 2018. LNCS, vol. 11207, pp. 179–196. Springer, Cham (2018). https://doi.org/10.1007/978-3-030-01219-9_11
22. Huang, Y., Xu, J., Wu, Q., Zheng, Z., Zhang, Z., Zhang, J.: Multi-pseudo regularized label for generated data in person re-identification. TIP **28**, 1391–1403 (2019)
23. Isola, P., Zhu, J.Y., Zhou, T., Efros, A.: Image-to-image translation with conditional adversarial networks. In: CVPR (2017)
24. Kalayeh, M., Basaran, E., Muhittin Gokmen, M.K., Shah, M.: Human semantic parsing for person re-identification. In: CVPR (2018)
25. Kingma, D., Ba, J.: Adam: a method for stochastic optimization. In: ICLR (2015)
26. Lee, C.Y., Batra, T., Baig, M.H., Ulbricht, D.: Sliced Wasserstein discrepancy for unsupervised domain adaptation. In: CVPR (2019)
27. Lee, H.Y., Tseng, H.Y., Huang, J.B., Singh, M., Yang, M.H.: Diverse image-to-image translation via disentangled representations. In: Ferrari, V., Hebert, M., Sminchisescu, C., Weiss, Y. (eds.) ECCV 2018. LNCS, vol. 11205. Springer, Cham (2018)
28. Li, Y.J., Lin, C.S., Lin, Y.B., Wang, Y.C.: Cross-dataset person re-identification via unsupervised pose disentanglement and adaptation. In: ICCV (2019)
29. Li, Y.J., Yang, F.E., Liu, Y.C., Yeh, Y.Y., Du, X., Wang, Y.C.: Adaptation and re-identification network: an unsupervised deep transfer learning approach to person re-identification. In: CVPR Workshop (2018)
30. Lin, S., Li, H., Li, C.T., Kot, A.C.: Multi-task mid-level feature alignment network for unsupervised cross-dataset person re-identification. In: BMVC (2018)
31. Liu, J., Zha, Z.J., Chen, D., Hong, R., Wang, M.: Adaptive transfer network for cross-domain person re-identification. In: CVPR (2019)
32. Liu, J., Ni, B., Yan, Y., Zhou, P., Cheng, S., Hu, J.: Pose transferrable person re-identification. In: CVPR (2018)
33. Long, M., Cao, Y., Wang, J., Jordan, M.I.: Learning transferable features with deep adaptation networks. In: ICML (2015)
34. Mathieu, M.F., Zhao, J.J., Zhao, J., Ramesh, A., Sprechmann, P., LeCun, Y.: Disentangling factors of variation in deep representation using adversarial training. In: NeurIPS (2016)
35. Naphade, M., et al.: The 4th AI city challenge. In: CVPR Workshop (2020)
36. Qi, L., Wang, L., Huo, J., Zhou, L., Shi, Y., Gao, Y.: A novel unsupervised camera-aware domain adaptation framework for person re-identification. In: ICCV (2019)

37. Qian, X., et al.: Pose-normalized image generation for person re-identification. In: Ferrari, V., Hebert, M., Sminchisescu, C., Weiss, Y. (eds.) ECCV 2018. LNCS, vol. 11213. Springer, Cham (2018). https://doi.org/10.1007/978-3-030-01240-3_40
38. Ren, C.X., Liang, B.H., Lei, Z.: Domain adaptive person re-identification via camera style generation and label propagation. arXiv arXiv:1905.05382 (2019)
39. Ristani, E., Solera, F., Zou, R., Cucchiara, R., Tomasi, C.: Performance measures and a data set for multi-target, multi-camera tracking. In: ECCV Workshop (2016)
40. Song, L., et al.: Unsupervised domain adaptive re-identification: Theory and practice. arXiv arXiv:1807.11334 (2018)
41. Su, C., Li, J., Zhang, S., Xing, J., Gao, W., Tian, Q.: Pose-driven deep convolutional model for person re-identification. In: ICCV (2017)
42. Su, C., Zhang, S., Xing, J., Gao, W., Tian, Q.: Deep attributes driven multi-camera person re-identification. In: Leibe, B., Matas, J., Sebe, N., Welling, M. (eds.) ECCV 2016. LNCS, vol. 9906. Springer, Cham (2016). https://doi.org/10.1007/978-3-319-46475-6_30
43. Suh, Y., Wang, J., Tang, S., Mei, T., Lee, K.M.: Part-aligned bilinear representations for person re-identification. In: Ferrari, V., Hebert, M., Sminchisescu, C., Weiss, Y. (eds.) ECCV 2018. LNCS, vol. 11218, pp. 418–437. Springer, Cham (2018). https://doi.org/10.1007/978-3-030-01264-9_25
44. Sun, B., Saenko, K.: Deep CORAL: correlation alignment for deep domain adaptation. In: Hua, G., Jégou, H. (eds.) ECCV 2016. LNCS, vol. 9915, pp. 443–450. Springer, Cham (2016). https://doi.org/10.1007/978-3-319-49409-8_35
45. Sun, Y., Zheng, L., Deng, W., Wang, S.: SVDNet for pedestrian retrieval. In: ICCV (2017)
46. Sun, Y., Zheng, L., Yang, Y., Tian, Q., Wang, S.: Beyond part models: person retrieval with refined part pooling (and a strong convolutional baseline). In: Ferrari, V., Hebert, M., Sminchisescu, C., Weiss, Y. (eds.) ECCV 2018. LNCS, vol. 11208. Springer, Cham (2018). https://doi.org/10.1007/978-3-030-01225-0_30
47. Tang, Z., et al.: PAMTRI: Pose-aware multi-task learning for vehicle re-identification using randomized synthetic data. In: ICCV (2019)
48. Tang, Z., et al.: CityFlow: a city-scale benchmark for multi-target multi-camera vehicle tracking and re-identification. In: CVPR (2019)
49. Tzeng, E., Hoffman, J., Saenko, K., Darrell, T.: Adversarial discriminative domain adaptation. In: CVPR (2017)
50. Wang, C., Zhang, Q., Huang, C., Liu, W., Wang, X.: Mancs: a multi-task attentional network with curriculum sampling for person re-identification. In: Ferrari, V., Hebert, M., Sminchisescu, C., Weiss, Y. (eds.) ECCV 2018. LNCS, vol. 11208. Springer, Cham (2018). https://doi.org/10.1007/978-3-030-01225-0_23
51. Wang, J., Zhu, X., Gong, S., Li, W.: Transferable joint attribute-identity deep learning for unsupervised person re-identification. In: CVPR (2018)
52. Wei, L., Zhang, S., Gao, W., Tian, Q.: Person transfer GAN to bridge domain gap for person re-identification. In: CVPR (2018)
53. Wei, L., Zhang, S., Yao, H., Gao, W., Tian, Q.: GLAD: global-local-alignment descriptor for pedestrian retrieval. In: ACM Multimedia (2017)
54. Yang, Q., Yu, H.X., Wu, A., Zheng, W.S.: Patch-based discriminative feature learning for unsupervised person re-identification. In: CVPR (2019)
55. Yao, Y., Zheng, L., Yang, X., Naphade, M., Gedeon, T.: Simulating content consistent vehicle datasets with attribute descent. In: ECCV (2020, to appear)
56. Zhao, H., et al.: Spindle Net: person re-identification with human body region guided feature decomposition and fusion. In: CVPR (2017)

57. Zheng, L., Shen, L., Tian, L., Wang, S., Wang, J., Tian, Q.: Scalable person re-identification: a benchmark. In: ICCV (2015)
58. Zheng, M., Karanam, S., Wu, Z., Radke, R.: Re-identification with consistent attentive Siamese networks. In: CVPR (2019)
59. Zheng, Z., Yang, X., Yu, Z., Zheng, L., Yang, Y., Kautz, J.: Joint discriminative and generative learning for person re-identification. In: CVPR (2019)
60. Zheng, Z., Yang, Y.: Person re-identification in the 3D space. arXiv arXiv:2006.04569 (2020)
61. Zheng, Z., Zheng, L., Yang, Y.: Unlabeled samples generated by GAN improve the person re-identification baseline in vitro. In: ICCV (2017)
62. Zheng, Z., Zheng, L., Yang, Y.: Pedestrian alignment network for large-scale person re-identification. In: TCSVT (2018)
63. Zhong, Z., Zheng, L., Cao, D., Li, S.: Re-ranking person re-identification with k-reciprocal encoding. In: CVPR (2017)
64. Zhong, Z., Zheng, L., Li, S., Yang, Y.: Generalizing a person retrieval model hetero- and homogeneously. In: Ferrari, V., Hebert, M., Sminchisescu, C., Weiss, Y. (eds.) ECCV 2018. LNCS, vol. 11217, pp. 176–192. Springer, Cham (2018). https://doi.org/10.1007/978-3-030-01261-8_11
65. Zhong, Z., Zheng, L., Luo, Z., Li, S., Yang, Y.: Invariance matters: exemplar memory for domain adaptive person re-identification. In: CVPR (2019)
66. Zhu, J.Y., Park, T., Isola, P., Efros, A.A.: Unpaired image-to-image translation using cycle-consistent adversarial networks. In: ICCV (2017)
67. Zou, Y., Yu, Z., Vijaya Kumar, B.V.K., Wang, J.: Unsupervised domain adaptation for semantic segmentation via class-balanced self-training. In: Ferrari, V., Hebert, M., Sminchisescu, C., Weiss, Y. (eds.) ECCV 2018. LNCS, vol. 11207, pp. 297–313. Springer, Cham (2018). https://doi.org/10.1007/978-3-030-01219-9_18
68. Zou, Y., Yu, Z., Liu, X., Kumar, B.V., Wang, J.: Confidence regularized self-training. In: ICCV (2019)

In-Home Daily-Life Captioning Using Radio Signals

Lijie Fan, Tianhong Li[⊠], Yuan Yuan, and Dina Katabi

MIT CSAIL, Cambridge, USA
`tianhong@mit.edu`

Abstract. This paper aims to caption daily life – i.e., to create a textual description of people's activities and interactions with objects in their homes. Addressing this problem requires novel methods beyond traditional video captioning, as most people would have privacy concerns about deploying cameras throughout their homes. We introduce RF-Diary, a new model for captioning daily life by analyzing the privacy-preserving radio signal in the home with the home's floormap. RF-Diary can further observe and caption people's life through walls and occlusions and in dark settings. In designing RF-Diary, we exploit the ability of radio signals to capture people's 3D dynamics, and use the floormap to help the model learn people's interactions with objects. We also use a multi-modal feature alignment training scheme that leverages existing video-based captioning datasets to improve the performance of our radio-based captioning model. Extensive experimental results demonstrate that RF-Diary generates accurate captions under visible conditions. It also sustains its good performance in dark or occluded settings, where video-based captioning approaches fail to generate meaningful captions.(For more information, please visit our project webpage: http://rf-diary.csail.mit.edu)

1 Introduction

Captioning is an important task in computer vision and natural language processing; it typically generates language descriptions of visual inputs such as images or videos [3,10,17,23,25,27,29,33,35–38]. This paper focuses on *in-home daily-life captioning*, that is, creating a system that observes people at home, and automatically generates a transcript of their everyday life. Such a system would help older people to age-in-place. Older people may have memory problems and some of them suffer from Alzheimer's. They may forget whether they took their medications, brushed their teeth, slept enough, woke up at night, ate

---

L. Fan, T. Li—Equal contribution.

Electronic supplementary material The online version of this chapter (https:// doi.org/10.1007/978-3-030-58536-5_7) contains supplementary material, which is available to authorized users.

ⓒ Springer Nature Switzerland AG 2020
A. Vedaldi et al. (Eds.): ECCV 2020, LNCS 12347, pp. 105–123, 2020.
https://doi.org/10.1007/978-3-030-58536-5_7

RGB Video   **Floormap Illustration**

<u>Video:</u> N/A.
RF+Floormap: A person is sleeping in the bed. Then he wears a coat. Then he walks to the table and plays laptop.
<u>GT:</u> A young man wakes up from his bed and puts on his shoes and clothes. He stands up from the bed and then sits down at the table and starts typing on a laptop.

<u>Video:</u> A person is standing next to the sink washing dishes.
RF+Floormap: A person is cooking food on the stove.
<u>GT:</u> A person is cooking with a black pan and spatula on the stove.

Fig. 1. Event descriptions generated from videos and RF+Floormap. The description generated from video shows its vulnerability to poor lighting and confusing images, while RF-Diary is robust to both. The visualization of floormap shown here is for illustration. The representation used by our model is person-centric and is described in detail in Subsect. 4.2.

their meals, etc. Daily life captioning enables a family caregiver, e.g., a daughter or son, to receive text updates about their parent's daily life, allowing them to care for mom or dad even if they live away, and providing them peace of mind about the wellness and safety of their elderly parents. More generally, daily-life captioning can help people track and analyze their habits and routine at home, which can empower them to change bad habits and improve their life-style.

But how do we caption people's daily life? One option would be to deploy cameras at home, and run existing video-captioning models on the recorded videos. However, most people would have privacy concerns about deploying cameras at home, particularly in the bedroom and bathroom. Also, a single camera usually has a limited field of view; thus, users would need to deploy multiple cameras covering different rooms, which would introduce a significant overhead. Moreover, cameras do not work well in dark settings and occlusions, which are common scenarios at home.

To address these limitations, we propose to use radio frequency (RF) signals for daily-life captioning. RF signals are more privacy-preserving than cameras since they are difficult to interpret by humans. Signals from a single RF device can traverse walls and occlusions and cover most of the home. Also, RF signals work in both bright and dark settings without performance degradation. Furthermore, the literature has shown that one can analyze the radio signals that bounce off people's bodies to capture people's movements [1,2], and track their 3D skeletons [43].

However, using RF signals also introduces new challenges, as described below:

- **Missing objects information:** RF signals do not have enough information to differentiate objects, since many objects are partially or fully transparent to radio signals. Their wavelength is on the order of a few centimeters, whereas the wavelength of visible light is hundreds nanometer [4]. Thus, it is also hard to capture the exact shape of objects using RF signals.
- **Limited training data:** Currently, there is no training dataset that contains RF signals from people's homes with the corresponding captions. Training a captioning system typically requires tens of thousands of labeled examples. However, collecting a new large captioning dataset with RF in people's homes would be a daunting task.

In this paper, we develop RF-Diary, an RF-based in-home daily-life captioning model that addresses both challenges. To capture objects information, besides RF signals, RF-Diary also takes as input the home floormap marked with the size and location of static objects like bed, sofa, TV, fridge, etc. Floormaps provide information about the surrounding environment, thus enabling the model to infer human interactions with objects. Moreover, floormaps are easy to measure with a cheap laser-meter in less than 10 min (Subsect. 4.2). Once measured, the floormap remains unchanged for potentially years, and can be used for all future daily-life captioning from that home.

RF-Diary proposes an effective representation to integrate the information in the floormap with that in RF signals. It encodes the floormap from the perspective of the person in the scene. It first extracts the 3D human skeletons from RF signals as in [43] and then at each time step, it shifts the reference system of floormap to the location of the extracted skeleton, and encodes the location and orientation of each object with respect to the person in the scene. This representation allows the model, at each time step, to focus on various objects depending on their proximity to the person.

To deal with the limited availability of training data, we propose a multimodal feature alignment training scheme to leverage existing video-captioning datasets for training RF-Diary. To transfer visual knowledge of event captioning to our model, we align the features generated from RF-Diary to the same space of features extracted from a video-captioning model trained on existing large video-captioning datasets. Once the features are aligned, we use a language model to generate text descriptions.

Figure 1 shows the performance of our model in two scenarios. In the first scenario, a person wakes up from bed, puts on his shoes and clothes, and goes to his desk to work on his laptop. RF-Diary generates a correct description of the events, while video-captioning fails due to poor lighting conditions. The second scenario shows a person cooking on the stove. Video-captioning confuses the person's activity as washing dishes because, in the camera view, the person looks as if he were near a sink full of dishes. In contrast, RF-Diary generates a correct caption because it can extract 3D information from RF signals, and hence can tell that the person is near the stove not the sink.

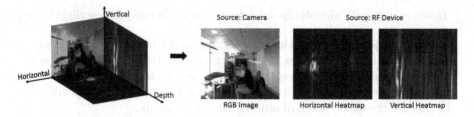

Fig. 2. RF heatmaps and an RGB image recorded at the same time. (Color figure online)

To evaluate RF-Diary, we collect a captioning dataset of RF signals and floormaps, as well as the synchronized RGB videos. Our experimental results demonstrate that: 1) RF-Diary can obtain comparable results to video-captioning in visible scenarios. Specifically, on our test set, RF-Diary achieves 41.5 average BLEU and 26.7 CIDEr, while RGB-based video-captioning achieves 41.1 average BLEU and 27.0 CIDEr. 2) RF-Diary continues to work effectively in dark and occluded conditions, where video-captioning methods fail. 3) Finally, our ablation study shows that the integration of the floormaps into the model and the multi-modal feature alignment both contribute significantly to improving performance.

Finally, we summarize our contributions as follows:

- We are the first to caption people's daily-life at home, in the presence of bad lighting and occlusions.
- We also introduce new modalities: the combination of RF and floormap, as well as new representations for both modalities that better tie them together.
- We further introduce a multi-modal feature alignment training strategy for knowledge transfer from a video-captioning model to RF-Diary.
- We evaluate our RF-based model and compare its performance to past work on video-captioning. Our results provide new insights into the strengths and weaknesses of these two types of inputs.

2   Related Work

(a) RGB-Based Video Captioning. Early works on video-captioning are direct extensions of image captioning. They pool features from individual frames across time, and apply image captioning models to video-captioning [34]. Such models cannot capture temporal dependencies in videos. Recent approaches, e.g., sequence-to-sequence video-to-text (S2VT), address this limitation by adopting recurrent neural networks (RNNs) [33]. In particular, S2VT customizes LSTM for video-captioning and generates natural language descriptions using an encoder-decoder architecture. Follow-up papers improve this model by introducing an attention mechanism [10,22,37], leveraging hierarchical architectures [3,12,13,17,29,38], or proposing new ways to improve feature extraction from

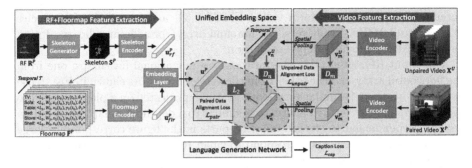

Fig. 3. Model architecture. It contains four parts: RF+Floormap feature extraction (the left yellow box), video feature extraction (the right blue box), unified embedding space for feature alignment (the center grey box), and language generation network (the bottom green box). RF-Diary extracts features from RF signals and floormaps and combines them into a unified human-environment feature map. The features are then taken by the language generation network to generate captions. RF-Diary also extracts features from both paired videos (synchronized videos with RF+Floormap) and unpaired videos (an existing video captioning dataset), and gets the video representation. These features are used to distill knowledge from existing video dataset to RF-Diary. During training, RF-Diary uses the caption loss and the feature alignment loss to train the network. During testing, RF-Diary takes only the RF+Floormap without videos as input and generates captions. (Color figure online)

video inputs, such as C3D features [37] or trajectories [36]. There have also been attempts to use reinforcement learning to generate descriptions from videos [25,27,35], in which they use the REINFORCE algorithm to optimize captioning scores.

(b) Human Behavior Analysis with Wireless Signals. Recently, there has been a significant interest in analyzing the radio signals that bounce off people's bodies to understand human movements and behavior. Past papers have used radio signals to track a person's location [1], extract 3D skeletons of nearby people [43], or do action classification [19]. To the best of our knowledge, we are the first to generate natural language descriptions of continuous and complex in-home activities using RF signals. Moreover, we introduce a new combined modality based on RF+Floormap and a novel representation that highlights the interaction between these two modalities, as well as a multi-modal feature alignment training scheme to allow RF-based captioning to learn from existing video captioning datasets.

3   RF Signal Preliminary

In this work, we use a radio commonly used in prior works on RF-based human sensing [7,11,14–16,20,26,31,39,40,42–44]. The radio has two antenna arrays organized vertically and horizontally, each equipped with 12 antennas. The antennas transmit a waveform called FMCW [30] and sweep the frequencies from 5.4 to 7.2 GHz. Intuitively, the antenna arrays provide angular resolution and the FMCW provides depth resolution.

Our input RF signal takes the form of two-dimensional heatmaps, one from the horizontal array and the other from the vertical array. As shown in Fig. 2, the horizontal heatmap is similar to a depth heatmap projected on a plane parallel to the ground, and the vertical heatmap is similar to a depth heatmap projected on a plane perpendicular to the ground. Red parts in the figure correspond to large RF power, while blue parts correspond to small RF power. The radio generates 30 pairs of heatmaps per second.

RF signals are different from vision data. They contain much less information than RGB images. This is because the wave-length of RF signals is few centimeters making it hard to capture objects' shape using RF signals; they may even totally miss small objects such as a pen or cellphone. However, the FMCW radio enables us to get a relatively high resolution on depth (~8 cm), making it much easier to locate a person. We harness this property to better associate RF signals and floormaps in the same coordinate system.

4   RF-Diary

RF-Diary generates event captions using RF signals and floormaps. As shown in Fig. 3, our model first performs feature extraction from RF signals and floormaps, then combine them into a unified feature (the left yellow box). The combined feature is taken by a language generation network to generate captions (the bottom green box). Below, we describe the model in detail. We also provide implementation details in Appendix A.

4.1   RF Signal Encoding

RF signals have properties totally different from visible light, which are usually not interpretable by human. Therefore, it can be hard to directly generate captions from RF signals. However, recent works demonstrate that it is possible to generate accurate human skeletons from radio signals [43], and that the human skeleton is a succinct yet informative representation for human behavior analysis [9,18]. In this work, we first generate 3D human skeletons from RF signals, then extract the feature representations of the 3D skeletons.

Thus, the first stage in RF-Diary is a skeleton generator network, which has an architecture similar to the one in [43], with 90-frame RF heatmaps (3 s) as input; we refer to these 90 frames as an RF segment. The skeleton generator first extracts information from the RF segment with a feature extraction network (12-layer ResNet). This is followed by a region proposal network (6-layer ResNet) to generate region proposals for potential human bounding boxes and a pose estimation network (2-layer ResNet) to generate the final 3D skeleton estimations based on the feature maps and the proposals. Note that these are dynamic skeletons similar to skeletons extracted from video segment.

After we obtain the 3D skeletons from RF signals, we extract the feature representation through a skeleton encoder from each skeleton segment **S**. The skeleton encoder is a Hierarchical Co-occurrence Network (HCN) [18], which is a CNN-based network architecture for skeleton-based action recognition. We use the features from the last convolutional layer of HCN, denoted as $\mathbf{u}_{rf}$, as the encoded features for RF signals.

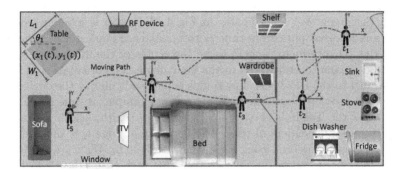

Fig. 4. Illustration of floormap representation. Noted that this figure is not the input to our model but only a visualization. Red dotted line in the apartment denotes the moving path of a person from time t_1 to t_5. Green axes X-Y centered at the person illustrate our person-centric coordinate system, where the origin of the coordinate system is changed through time along with people's location. Under this person-centric coordinate system, at t^{th} time step, we describe each object using a 5-element tuple: (length L, width W, center coordinates $x(t), y(t)$, and rotation θ), as exemplified using the **Table** in the figure. (Color figure online)

4.2   Floormap Encoding

Many objects are transparent or semi-transparent to RF signals and act in a manner similar to air [2]. Even objects that are not transparent to RF signals, they can be partially invisible; this is because they can deflect the incident RF signal away from the emitting radio, preventing the radio from sensing them [41,43]. Thus, to allow the model to understand interactions with objects, we must provide additional information about the surrounding environment. But we do not need to have every aspect of the home environment since most of the background information, e.g. the color or the texture of furniture, is irrelevant to captioning daily life. Therefore, we represent the in-home environment using the locations and sizes of objects – the floormap. The floormap is easily measured with a laser meter. Once measured, the floormap tends to remain valid for a long time. In our model, we only consider static objects relevant to people's in-home activities, e,g., bed, sofa, stove, or TV. To demonstrate the ease of collecting such measurements, we have asked 5 volunteers to measure the floormap for each of our test environments. The average time to measure one environment is about 7 mins.

Let M be the number of objects, N be the maximum number of instances of an object, then the floormap can be represented by a tensor $f \in \mathbb{R}^{M \times N \times O}$, where O denotes the dimension for the location and size of each object, which is typically 5, i.e., length L, width W, the coordinate of the center $x(t), y(t)$, and its rotation θ.

Since people are more likely to interact with objects close to them, we set the origin point of the floormap reference system to the location of the 3D skeleton extracted from RF. Specifically, we use a person-centric coordinate system as shown in the green X-Y coordinates in Fig. 4. A person is moving around at home in a red-dotted path from time t_1 to t_5. At each time step t_i, we set the origin of the 3D coordinate system to be the center of the 3D human skeleton and the X-Y plane to be parallel to the floor plane. The orientation of the X-axis is parallel to the RF device and Y-axis is perpendicular to the device. Each object is then projected onto the X-Y plane. For example, at time t_i, the center coordinates of the **Table** is $(x_1(t_i), y_1(t_i))$, while its width, length and rotation (L_1, W_1, θ_1) are independent of time. The floormap at time t_i is thus generated by describing each object k using a 5-element tuple: $(L_k, W_k, x_k(t_i), y_k(t_i), \theta_k)$, as shown in the left yellow box in Fig. 3. In this way, each object's location is encoded w.r.t. the person's location, allowing our model to pay different attention to objects depending on their proximity to the person at that time instance.

To extract features of the floormaps $\mathbf{F}$, we use a floormap encoder which is a two-layer fully-connected network which generates the encoded features for floormaps $\mathbf{u}_{flr}$.

Using the encoded RF signal features $\mathbf{u}_{rf}$ and floormap features $\mathbf{u}_{flr}$, we generate a unified human-environment feature:

$$\mathbf{u} = \psi(\mathbf{u}_{rf} \oplus \mathbf{u}_{flr}),$$

where $\oplus$ denotes the concatenation of vectors, and ψ denotes an encoder to map the concatenated features to a joint embedding space. Here we use another two-layer fully-connected sub-network for ψ.

4.3 Caption Generation

To generate language descriptions based on the extracted features, we use an attention-based sequence-to-sequence LSTM model similar to the one in [33,35]. During the encoding stage, given the unified human-environment feature $\mathbf{u} = \{u_t\}_1^T$ with time dimension T, the encoder LSTM operates on its time dimension to generate hidden states $\{h_t\}_1^T$ and outputs $\{o_t\}_1^T$. During the decoding stage, the decoder LSTM uses h_T as an initialization for hidden states and take inputs of previous ground-truth words with an attention module related to $\{u_t\}_1^T$, to output language sequence with m words $\{w_1, w_2, \ldots, w_m\}$. The event captioning loss $\mathcal{L}_{cap}(\mathbf{u})$ is then given by a negative-log-likelihood loss between the generated and the ground truth caption similar to [33,35].

5 Multi-modal Feature Alignment Training

Training RF-Diary requires a large labeled RF captioning dataset. Collecting such a dataset would be a daunting task. To make use of the existing large video-captioning dataset (e.g., Charades), we use a multi-model feature alignment strategy in the training process to transfer knowledge from video-captioning to RF-Diary. However, RGB videos from Charades and RF signals have totally different distributions both in terms of semantics and modality. To mitigate the gap between them and make the knowledge distillation possible, we collect a small dataset where we record synchronized videos and RF from people performing activities at home. We also collect the floormaps and provide the corresponding natural language descriptions (for dataset details see Sect. 6(a)). The videos in the small dataset are called paired videos, since they are in the same semantic space as their corresponding RF signal, while the videos in large existing datasets are unpaired videos with the RF signals. Both the paired and unpaired videos are in the same modality. Since the paired videos share the same semantics with the RF data, and the same modality with the unpaired videos, they can work as a bridge between video-captioning and RF-Diary, and distill knowledge from the video data to RF data.

Our multi-modal feature alignment training scheme operates by aligning the features from RF+Floormaps and those from RGB videos. During training, our model extracts features from not only RF+Floormaps, but also paired and unpaired RGB videos, as shown in Fig. 3 (the right blue box). Besides the captioning losses, we add additional paired-alignment loss between paired videos and RF+Floormaps and unpaired-alignment loss between paired and unpaired videos. This ensures features from the two modalities are mapped to a unified embedding space (the center grey box). Below, we describe the video encoder and the feature alignment method in detail.

5.1 Video Encoding

We use the I3D model [5] pre-trained on Kinetics dataset to extract the video features. For each 64-frame video segment, we extract the Mixed_5c features from I3D model, denoted as v_m. We then apply a spatial average pooling on top of the Mixed_5c feature and get the spatial pooled video-segment feature v_n. For a video containing T non-overlapping video-segment, its Mixed_5c features and the spatial-pooled features are denoted as $\mathbf{v}_m = \{v_m(t)\}_1^T$ and $\mathbf{v}_n = \{v_n(t)\}_1^T$. Therefore, the extracted features of paired videos $\mathbf{X}^P$ and unpaired videos $\mathbf{X}^U$ are denoted as $\mathbf{v}_m^P$, $\mathbf{v}_n^P$ and $\mathbf{v}_m^U$, $\mathbf{v}_n^U$. We use the spatial-pooled features to generate captions through the language generation model. The corresponding captioning loss is $\mathcal{L}_{cap}(\mathbf{v}_n^P)$ and $\mathcal{L}_{cap}(\mathbf{v}_n^U)$.

5.2 Alignment of Paired Data

Since the synchronized video and RF+Floormap correspond to the exact same event, we use L_2 loss to align the features from paired video $\mathbf{v}_n^P$ in Subsect. 5.1

and RF+Floormap $\mathbf{u}^P$ in Subsect. 4.2 (we denote a P here to indicate the paired data) to be consistent with each other in a unified embedding space, i.e., the paired data alignment loss $\mathcal{L}_{pair}(\mathbf{u}^P, \mathbf{v}_n^P) = \left\| \mathbf{u}^P - \mathbf{v}_n^P \right\|_2$.

Table 1. Statistics of our RCD dataset.

#environments	#clips	avg len	#action types	#object types	#sentences	#words	vocab	len (h)
10	1,035	31.3 s	157	38	3,610	77,762	6,910	8.99

5.3 Alignment of Unpaired Data

Existing large video-captioning datasets have neither synchronized RF signal nor the corresponding floormaps, so we cannot use the L_2-norm for alignment. Since we collect a small paired dataset, we can first train a video-captioning model on both paired and unpaired datasets, and then use the paired dataset to transfer knowledge to RF-Diary. However, the problem is that since the paired feature alignment is only applied on the paired dataset, the video-captioning model may overfit to the paired dataset and cause inconsistency between the distribution of features from paired and unpaired datasets. To solve this problem, we align the paired and unpaired datasets by making the two feature distributions similar. We achieve this goal by applying discriminators on different layers of video features that enforces the video encoder to generate indistinguishable features given $\mathbf{X}^P$ and $\mathbf{X}^U$. Specifically, we use two different layers of video features, i.e., $\mathbf{v}_m$ and $\mathbf{v}_n$ in Subsect. 5.1, to calculate the discriminator losses $\mathcal{L}_{unpair}(\mathbf{v}_m^P, \mathbf{v}_m^U)$ and $\mathcal{L}_{unpair}(\mathbf{v}_n^P, \mathbf{v}_n^U)$. Since features from the paired videos are also aligned with the RF+Floormap features, this strategy effectively aligns the feature distribution of the unpaired video with the feature distribution of RF+Floormaps. The total loss of training process is shown as below:

$$\mathcal{L} = \mathcal{L}_{cap}(\mathbf{u}^P) + \mathcal{L}_{cap}(\mathbf{v}_n^P) + \mathcal{L}_{cap}(\mathbf{v}_n^U)$$
$$+ \mathcal{L}_{pair}(\mathbf{u}^P, \mathbf{v}_n^P)$$
$$+ \mathcal{L}_{unpair}(\mathbf{v}_n^P, \mathbf{v}_n^U) + \mathcal{L}_{unpair}(\mathbf{v}_m^P, \mathbf{v}_m^U).$$

6 Experiments

(a) Datasets: We collect a new dataset named RF Captioning Dataset (RCD). It provides synchronized RF signals, RGB videos, floormaps, and human-labeled captions to describe each event. We generate floormaps using a commercial laser meter. The floormaps are marked with the following objects: cabinet, table, bed, wardrobe, shelf, drawer, stove, fridge, sink, sofa, television, door, window, air conditioner, bathtub, dishwasher, oven, bedside table. We use a radio device to collect RF signals, and a multi-camera system to collect multi-view videos, as the volunteers perform the activities. The synchronized radio signals and videos have a maximum synchronization error of 10 ms. The multi-camera system has 12 viewpoints to allow for accurate captioning even in cases where some viewpoints are occluded or the volunteers walk from one room to another room.

To generate captions, we follow the method used to create the Charades dataset [28] – i.e., we first generate instructions similar to those used in Charades, ask the volunteers to perform activities according to the instructions, and record the synchronized RF signals and multi-view RGB videos. We then provide each set of multi-view RGB videos to Amazon Mechanical Turk (AMT) workers and ask them to label 2–4 sentences as the ground-truth language descriptions.

We summarize our dataset statistics in Table 1. In total, we collect 1,035 clips in 10 different in-door environments, including bedroom, kitchen, living room, lounge, office, etc. Each clip on average spans 31.3 s. The RCD dataset exhibits two types of diversity. *1. Diversity of actions and objects:* Our RCD dataset contains 157 different actions and 38 different objects to interact with. The actions and objects are the same as the Charades dataset to ensure a similar action diversity. The same action is performed at different locations by different people, and different actions are performed at the same location. For example, all of the following actions are performed in the bathroom next to the sink: brushing teeth, washing hands, dressing, brushing hair, opening/closing a cabinet, putting something on a shelf, taking something off a shelf, washing something, etc. *2. Diversity of environments:* Environments in our dataset differ in their floormap, position of furniture, and the viewpoint of the RF device. Further, each environment and all actions performed in that environment are included either in testing or training, but not both.

(b) Train-Test Protocol: To evaluate RF-Diary under visible scenarios, we do a 10-fold cross-validation on our RCD Dataset. Each time 9 environments are used for training, and the other 1 environment is used for testing. We report the average performance of 10 experiments. To show the performance of RF-Diary under invisible scenarios, e.g., with occlusions or poor lighting conditions, we randomly choose 3 environments (with 175 clips) and collect corresponding clips under invisible conditions. Specifically, in these 3 environments, we ask the volunteers to perform the same series of activities twice under the visible and invisible conditions (with the light on and off, or with an occlusion in front of the RF device and cameras), respectively. Later we provide the same ground truth language descriptions for the clips under invisible conditions as the corresponding ones under visible conditions. During testing, clips under invisible scenarios in these 3 environments are used for testing, and clips in the other 7 environments are used to train the model.

During training, we only use RGB videos from 3 cameras with good views instead of all 12 views in the multi-modal feature alignment between the video-captioning model and RF-Diary model. Using multi-view videos will provide more training samples and help the feature space to be oblivious to the viewpoint. When testing the video-captioning model, we use the video from the master camera as it covers most of the scenes. The master camera is positioned atop of the RF device for a fair comparison.

We leverage the Charades caption dataset [28,35] as the unpaired dataset to train the video-captioning model. This dataset provides captions for different in-door human activities. It contains 6,963 training videos, 500 validation videos, and 1,760 test videos. Each video clip is annotated with 2–5 language descriptions by AMT workers.

(c) **Evaluation Metrics:**   We adopt 4 caption evaluation scores widely used in video-captioning: BLEU [24], METEOR [8], ROUGE-L [21] and CIDEr [32]. BLEU-n analyzes the co-occurrences of n words between the predicted and ground truth sentences. METEOR compares exact token matches, stemmed tokens, paraphrase matches, as well as semantically similar matches using Word-Net synonyms. ROUGE-L measures the longest common subsequence of two sentences. CIDEr measures consensus in captions by performing a Term Frequency Inverse Document Frequency (TF-IDF) weighting for each n words. According to [28], CIDEr offers the highest resolution and most similarity with human judgment on the captioning task for in-home events. We compute these scores using the standard evaluation code provided by the Microsoft COCO Evaluation Server [6]. All results are obtained as an average of 5 trials. We denote B@n, M, R, C short for BLEU-n, METEOR, ROUGE-L, CIDEr.

6.1   Quantitative Results

We compare RF-Diary with state-of-the-art video-captioning baselines [10,17,33, 37]. The video-based models are trained on RGB data from both the Charades and RCD training sets, and tested on the RGB data of the RCD test set. RF-Diary is trained on RF and floormap data from the RCD training sets. It also uses the RGB data from Charades and RCD training sets in multi-modal feature alignment training. It is then tested on the RF and floormap data of the RCD test set.

The results on the left side of Table 2 show that RF-Diary achieves comparable performance to state-of-the-art video captioning baselines in visible scenarios. The right side of Table 2 shows that RF-Diary also generates accurate language descriptions when the environment is dark or with occlusion, where video-captioning fails completely. The little reduction in RF-Diary's performance from the visible scenario is likely due to that occlusions attenuate RF signals and therefore introduce more noise in the RF heatmaps.

Table 2. Quantitative results for RF-Diary and video-based captioning models. All models are trained on Charades and RCD training set, and tested on the RCD test set. The left side of the Table shows the results under visible scenarios, and the right side of the Table shows the results under scenarios with occlusions or without light.

Methods	Visible scenario							Dark/occlusion ccenario						
	B@1	B@2	B@3	B@4	M	R	C	B@1	B@2	B@3	B@4	M	R	C
S2VT [33]	57.3	40.4	27.2	19.3	19.8	27.3	18.9	-	-	-	-	-	-	–
SA [37]	56.8	39.2	26.7	19.0	18.1	25.9	22.1	-	-	-	-	-	-	–
MAAM [10]	57.8	41.9	28.2	19.3	20.7	27.1	21.2	-	-	-	-	-	-	–
HTM [17]	61.3	44.6	32.2	22.1	21.3	28.3	26.5	-	-	-	-	-	-	–
HRL [35]	**62.5**	45.3	32.9	**23.8**	**21.7**	28.5	**27.0**	-	-	-	-	-	-	–
RF-Diary	62.3	**45.9**	**33.9**	23.5	21.1	**28.9**	26.7	**61.5**	**45.5**	**33.1**	**22.6**	**21.1**	**28.3**	**25.9**

6.2   Ablation Study

We conduct several ablation studies to demonstrate the necessity of each component in RF-Diary. All experiments here are evaluated on the visible test set of RCD.

3D Skeleton vs. Locations: One may wonder whether simply knowing the location of the person is enough to generate a good caption. This could happen if the RCD dataset has low diversity, i.e., each action is done in a specific location. This is however not the case in the RCD dataset, where each action is done in multiple locations, and each location may have different actions. To test this point empirically, we compare our model which extracts 3D skeletons from RF signals with a model that extracts only people locations from RF. We also compare with a model that extracts 2D skeletons with no locations (in this case the floormap's coordinate system is centered on the RF device).

Table 3 shows that replacing *3D skeletons* with *locations* or *2D skeletons* yields poor performance. This is because *locations* do not contain enough information of the actions performed by the person, and *2D skeletons* do not contain information of the person's position with respect to the objects on the floormaps. These results show that: 1) our dataset is diverse and hence locations are not enough to generate correct captioning, and 2) our choice of representation, i.e., *3D skeletons*, which combines information about both the people's locations and poses provides the right abstraction to learn meaningful features for proper captioning.

Person-Centric Floormap Representation: In this work, we use a person-centric coordinate representation for the floormap and its objects, as described in Subsect. 4.2. What if we simply use the image of the floormap with the objects, and mark the map with the person's location at each time instance? We compare this *image-based floormap* representation to our person-centric representation in Table 4. We use ResNet-18 pre-trained on ImageNet to extract features from the floormap image. The result shows that the image representation of floormap can achieve better performance than not having the floormap, but still worse

Table 3. Comparison between using different human representations.

Method	B@1	B@2	B@3	B@4	M	R	C
Locations	52.0	37.4	24.3	17.2	15.7	22.1	19.1
2D skeletons	56.5	39.8	26.9	18.8	18.0	24.1	22.3
3D skeletons	**62.3**	**45.9**	**33.9**	**23.5**	**21.1**	**28.9**	**26.7**

than our person-centric representation. This is because it is much harder for the network to interpret and extract features from an image representation, since the information is far less explicit than our person-centric coordinate-based representation.

Table 4. Performance of RF-Diary with or without using floormap, with different floormap representations, and with gaussian noise.

Method	B@1	B@2	B@3	B@4	M	R	C
w/o floormap	56.3	40.8	27.7	18.5	18.1	24.0	22.1
image-based floormap	60.1	43.9	31.5	21.6	20.1	26.7	24.4
person-centric floormap	**62.3**	**45.9**	**33.9**	**23.5**	**21.1**	**28.9**	**26.7**
person-centric floormap+noise	61.6	45.8	33.7	23.4	21.0	28.7	26.5

Measurement Errors: We analyze the influence of floormap measurement errors on our model's performance. We add a random Gaussian noise with a 20 cm standard deviation on location, 10 cm on size and 30° on object rotation. The results in the last row of Table 4 show that the noise has very little effect on performance. This demonstrates that our model is robust to measurement errors.

Feature Alignment: Our feature alignment framework consists of two parts: the L_2-norm between paired dataset, and the discriminator between unpaired datasets. Table 5 quantifies the contribution of each of these alignment mechanisms to RF-Diary's performance. The results demonstrate that our multi-modal feature alignment training scheme helps RF-Diary utilize the knowledge of the video-captioning model learned from the large video-captioning dataset to generate accurate descriptions, while training only on a rather small RCD dataset. We show a visualization of the features before and after alignment in the Appendix.

 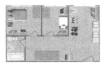

Video: A person walks into the room cleans the table. He sits down on chair.

RF+Floormap: A person walks to the table. Then he tidies the table and sits down at the table.

GT: A person enters the room and starts cleaning up the stuffs on the table. He then sits at the table when done.

(a)

Video: A person is texting on his phone. He then starts calling someone.

RF+Floormap: A person is working on his phone. Then he talks on his phone.

GT: A person plays with his phone. He then makes a phone call.

(b)

 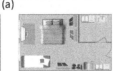

Video: N/A.

RF+Floormap: A person is brushing his teeth in the bathroom. Then he brushes hair.

GT: A person is in the bathroom. He takes out his tooth brush and cup, and then brushes his teeth. Then he starts brushing his hair.

(c)

Video: N/A.

RF+Floormap: A person removes his clothes and puts it in wardrobe. Then he sits at a table and starts typing.

GT: A person walks into the room. He takes off his coat and hangs it in the wardrobe. Then he walks to the table, sits down and works on his laptop.

(d)

 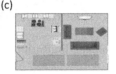

Video: A woman opens a fridge. She takes some juice and drink it and walks away.

RF+Floormap: A person opens the fridge, takes out something and then drinks water. Then he walks to the sink.

GT: In the kitchen, a woman takes some juice from fridge and drink it. She then closes the fridge door and goes to the sink.

(e)

Video: A person sits at a table. He opens snacks from the table and eats.

RF+Floormap: A person sits at a table. He then starts eating something.

GT: A person sits down at a dinning table. Then he takes a chocolate bar from the table and eats it.

Fig. 5. Examples from our RCD test set. Green words indicate actions. Blue words indicate objects included in floormap. Brown words indicate small objects not covered by floormap. Red words indicate the misprediction of small objects from RF-Diary. The first row shows RF-Diary can generate accurate captions compared to the video-based captioning model under visible scenarios. The second row shows that RF-Diary can still generate accurate captions when the video-based model does not work because of poor lighting conditions or occlusions. The third row shows the limitation of RF-Diary that it may miss object color and detailed descriptions of small objects. (Color figure online)

6.3   Qualitative Result

In Fig. 5, we show six examples from the RCD test set. The first row under each image is the caption generated by state of the art video-based captioning model [35], the second row is the caption generated by RF-Diary, and the third row is the ground truth caption labeled by a human.

The result shows that RF-Diary can generate accurate descriptions of the person's activities (green) and interaction with the surrounding environment (blue), and continue to work well even in the presence of occlusions (Fig. 5(c)), and poor lighting (Fig. 5(d)). Video-based captioning is limited by bad lighting,

Table 5. Performance of RF-Diary network on RCD with or without L_2 loss and discriminator. Note that without adding the L_2 loss, RF-Diary will not be affected by the video-captioning model. So if without the L_2 loss, then adding the discriminator loss on video-captioning model or not will not affect the RF-Diary's performance.

Method	B@1	B@2	B@3	B@4	M	R	C
w/o L_2	52.5	38.0	25.7	18.5	16.6	23.1	20.3
w/o discrim	59.4	44.1	31.4	21.0	19.8	26.3	24.6
RF-Diary	**62.3**	**45.9**	**33.9**	**23.5**	**21.1**	**28.9**	**26.7**

occlusions and the camera's field of view. So if the person exits the field of view, video captioning can miss some of the events (Fig. 5(e)).

Besides poor lighting conditions, occlusions and field of view, video-captioning is also faced with privacy problems. For example, in Fig. 5(b), the person just took a bath and is not well-dressed. The video will record this content which is quite privacy-invasive. However, RF signal can protect privacy since it is not interpretable by a human, and it does not contain detailed information because of the relatively low resolution.

We also observe that RF-Diary has certain limitations. Since RF signals cannot capture details of objects such as color, texture, and shape, the model can mispredict those features. It can also mistake small objects. For example, in Fig. 5(e), the person is actually drinking orange juice, but RF-Diary predicts he is drinking water. Similarly, in Fig. 5(f), our model reports that the person is eating but cannot tell that he is eating a chocolate bar. The model also cannot distinguish the person's gender, so it always predicts "he" as shown in Fig. 5(e).

6.4   Additional Notes on Privacy

In comparison to images, RF signal is privacy-preserving because it is difficult to interpret by humans. However, one may also argue that since RF signals can track people though walls, they could create privacy concerns. This issue can be addressed through a challenge-response authentication protocol that prevent people from maliciously using RF signals to see areas that they are not authorized to access. More specifically, previous work [1] demonstrates that RF signals can sense human trajectories and locate them in space. Thus, whenever the user sets up the system to monitor an area, the system first challenges the user to execute certain moves (e.g., take two steps to the right, or move one meter forward), to ensure that the monitored person is the user. The system also asks the user to walk around the area to be monitored, and only monitors that area. Hence, the system would not monitor an area which the user does not have access to.

7   Conclusion

In this paper, we introduce RF-Diary, a system that enables in-home daily-life captioning using RF signals and floormaps. We also introduce the combination of RF signal and floormap as new complementary input modalities, and propose a feature alignment training scheme to transfer the knowledge from large video-captioning dataset to RF-Diary. Extensive experimental results demonstrate that RF-Diary can generate accurate descriptions of in-home events even when the environment is under poor lighting conditions or has occlusions. We believe this work paves the way for many new applications in health monitoring and smart homes.

References

1. Adib, F., Kabelac, Z., Katabi, D., Miller, R.C.: 3D tracking via body radio reflections. In: 11th USENIX Symposium on Networked Systems Design and Implementation, NSDI 2014, pp. 317–329 (2014)
2. Adib, F., Katabi, D.: See through walls with WiFi!, vol. 43. ACM (2013)
3. Baraldi, L., Grana, C., Cucchiara, R.: Hierarchical boundary-aware neural encoder for video captioning. In: Proceedings of the IEEE Conference on Computer Vision and Pattern Recognition, pp. 1657–1666 (2017)
4. Barbrow, L.: International lighting vocabulary. J. SMPTE **73**(4), 331–332 (1964)
5. Carreira, J., Zisserman, A.: Quo vadis, action recognition? A new model and the kinetics dataset. In: Proceedings of the IEEE Conference on Computer Vision and Pattern Recognition, pp. 6299–6308 (2017)
6. Chen, X., et al.: Microsoft COCO captions: Data collection and evaluation server. arXiv preprint arXiv:1504.00325 (2015)
7. Chetty, K., Chen, Q., Ritchie, M., Woodbridge, K.: A low-cost through-the-wall FMCW radar for stand-off operation and activity detection. In: Radar Sensor Technology XXI, vol. 10188, p. 1018808. International Society for Optics and Photonics (2017)
8. Denkowski, M., Lavie, A.: Meteor universal: language specific translation evaluation for any target language. In: Proceedings of the 9th Workshop on Statistical Machine Translation, pp. 376–380 (2014)
9. Du, Y., Fu, Y., Wang, L.: Skeleton based action recognition with convolutional neural network. In: 2015 3rd IAPR Asian Conference on Pattern Recognition (ACPR), pp. 579–583. IEEE (2015)
10. Fakoor, R., Mohamed, A., Mitchell, M., Kang, S.B., Kohli, P.: Memory-augmented attention modelling for videos. arXiv preprint arXiv:1611.02261 (2016)
11. Fan, L., Li, T., Fang, R., Hristov, R., Yuan, Y., Katabi, D.: Learning longterm representations for person re-identification using radio signals. In: Proceedings of the IEEE/CVF Conference on Computer Vision and Pattern Recognition, pp. 10699–10709 (2020)
12. Gan, C., Gan, Z., He, X., Gao, J., Deng, L.: StyleNet: generating attractive visual captions with styles. In: CVPR, pp. 3137–3146 (2017)
13. Gan, Z., et al.: Semantic compositional networks for visual captioning. In: CVPR, pp. 5630–5639 (2017)
14. Hsu, C.Y., Ahuja, A., Yue, S., Hristov, R., Kabelac, Z., Katabi, D.: Zero-effort in-home sleep and insomnia monitoring using radio signals. In: Proceedings of the ACM on Interactive, Mobile, Wearable and Ubiquitous Technologies, vol. 1, no. 3, pp. 1–18 (2017)

15. Hsu, C.Y., Hristov, R., Lee, G.H., Zhao, M., Katabi, D.: Enabling identification and behavioral sensing in homes using radio reflections. In: Proceedings of the 2019 CHI Conference on Human Factors in Computing Systems, p. 548. ACM (2019)
16. Hsu, C.Y., Liu, Y., Kabelac, Z., Hristov, R., Katabi, D., Liu, C.: Extracting gait velocity and stride length from surrounding radio signals. In: Proceedings of the 2017 CHI Conference on Human Factors in Computing Systems, pp. 2116–2126 (2017)
17. Hu, Y., Chen, Z., Zha, Z.J., Wu, F.: Hierarchical global-local temporal modeling for video captioning. In: Proceedings of the 27th ACM International Conference on Multimedia, pp. 774–783 (2019)
18. Li, C., Zhong, Q., Xie, D., Pu, S.: Co-occurrence feature learning from skeleton data for action recognition and detection with hierarchical aggregation. In: Proceedings of the 27th International Joint Conference on Artificial Intelligence, pp. 786–792 (2018)
19. Li, T., Fan, L., Zhao, M., Liu, Y., Katabi, D.: Making the invisible visible: action recognition through walls and occlusions. In: Proceedings of the IEEE International Conference on Computer Vision, pp. 872–881 (2019)
20. Lien, J.: Soli: ubiquitous gesture sensing with millimeter wave radar. ACM Trans. Graph. (TOG) **35**(4), 142 (2016)
21. Lin, C.Y.: ROUGE: a package for automatic evaluation of summaries. In: Text Summarization Branches Out (2004)
22. Long, X., Gan, C., de Melo, G.: Video captioning with multi-faceted attention. Trans. Assoc. Comput. Linguist. **6**, 173–184 (2018)
23. Pan, P., Xu, Z., Yang, Y., Wu, F., Zhuang, Y.: Hierarchical recurrent neural encoder for video representation with application to captioning. In: Proceedings of the IEEE Conference on Computer Vision and Pattern Recognition, pp. 1029–1038 (2016)
24. Papineni, K., Roukos, S., Ward, T., Zhu, W.J.: BLEU: a method for automatic evaluation of machine translation. In: Proceedings of the 40th Annual Meeting on Association for Computational Linguistics, pp. 311–318. Association for Computational Linguistics (2002)
25. Pasunuru, R., Bansal, M.: Reinforced video captioning with entailment rewards. In: Proceedings of the 2017 Conference on Empirical Methods in Natural Language Processing, pp. 979–985 (2017)
26. Peng, Z., Muñoz-Ferreras, J.M., Gómez-García, R., Li, C.: FMCW radar fall detection based on ISAR processing utilizing the properties of RCS, range, and Doppler. In: 2016 IEEE MTT-S International Microwave Symposium (IMS), pp. 1–3. IEEE (2016)
27. Ranzato, M., Chopra, S., Auli, M., Zaremba, W.: Sequence level training with recurrent neural networks. arXiv preprint arXiv:1511.06732 (2015)
28. Sigurdsson, G.A., Varol, G., Wang, X., Farhadi, A., Laptev, I., Gupta, A.: Hollywood in homes: crowdsourcing data collection for activity understanding. In: Leibe, B., Matas, J., Sebe, N., Welling, M. (eds.) ECCV 2016. LNCS, vol. 9905, pp. 510–526. Springer, Cham (2016). https://doi.org/10.1007/978-3-319-46448-0_31
29. Song, J., Guo, Z., Gao, L., Liu, W., Zhang, D., Shen, H.T.: Hierarchical LSTM with adjusted temporal attention for video captioning. arXiv preprint arXiv:1706.01231 (2017)
30. Stove, A.G.: Linear FMCW radar techniques. In: IEE Proceedings F (Radar and Signal Processing), vol. 139, pp. 343–350. IET (1992)
31. Tian, Y., Lee, G.H., He, H., Hsu, C.Y., Katabi, D.: RF-based fall monitoring using convolutional neural networks. In: Proceedings of the ACM on Interactive, Mobile, Wearable and Ubiquitous Technologies, vol. 2, no. 3, p. 137 (2018)

32. Vedantam, R., Lawrence Zitnick, C., Parikh, D.: CIDEr: consensus-based image description evaluation. In: Proceedings of the IEEE Conference on Computer Vision and Pattern Recognition, pp. 4566–4575 (2015)
33. Venugopalan, S., Rohrbach, M., Donahue, J., Mooney, R., Darrell, T., Saenko, K.: Sequence to sequence-video to text. In: Proceedings of the IEEE International Conference on Computer Vision, pp. 4534–4542 (2015)
34. Venugopalan, S., Xu, H., Donahue, J., Rohrbach, M., Mooney, R., Saenko, K.: Translating videos to natural language using deep recurrent neural networks. In: Proceedings of the 2015 Conference of the North American Chapter of the Association for Computational Linguistics: Human Language Technologies, pp. 1494–1504 (2015)
35. Wang, X., Chen, W., Wu, J., Wang, Y.F., Yang Wang, W.: Video captioning via hierarchical reinforcement learning. In: Proceedings of the IEEE Conference on Computer Vision and Pattern Recognition, pp. 4213–4222 (2018)
36. Wu, X., Li, G., Cao, Q., Ji, Q., Lin, L.: Interpretable video captioning via trajectory structured localization. In: Proceedings of the IEEE conference on Computer Vision and Pattern Recognition, pp. 6829–6837 (2018)
37. Yao, L., et al.: Describing videos by exploiting temporal structure. In: Proceedings of the IEEE International Conference on Computer Vision, pp. 4507–4515 (2015)
38. Yu, H., Wang, J., Huang, Z., Yang, Y., Xu, W.: Video paragraph captioning using hierarchical recurrent neural networks. In: Proceedings of the IEEE Conference on Computer Vision and Pattern Recognition, pp. 4584–4593 (2016)
39. Zhang, Z., Tian, Z., Zhou, M.: Latern: dynamic continuous hand gesture recognition using FMCW radar sensor. IEEE Sens. J. **18**(8), 3278–3289 (2018)
40. Zhao, M., Adib, F., Katabi, D.: Emotion recognition using wireless signals. In: Proceedings of the 22nd Annual International Conference on Mobile Computing and Networking, pp. 95–108. ACM (2016)
41. Zhao, M., et al.: Through-wall human pose estimation using radio signals. In: Proceedings of the IEEE Conference on Computer Vision and Pattern Recognition, pp. 7356–7365 (2018)
42. Zhao, M., et al.: Through-wall human mesh recovery using radio signals. In: Proceedings of the IEEE International Conference on Computer Vision, pp. 10113–10122 (2019)
43. Zhao, M., et al.: RF-based 3D skeletons. In: Proceedings of the 2018 Conference of the ACM Special Interest Group on Data Communication, pp. 267–281. ACM (2018)
44. Zhao, M., Yue, S., Katabi, D., Jaakkola, T.S., Bianchi, M.T.: Learning sleep stages from radio signals: a conditional adversarial architecture. In: Proceedings of the 34th International Conference on Machine Learning, vol. 70, pp. 4100–4109. JMLR. org (2017)

Self-challenging Improves Cross-Domain Generalization

Zeyi Huang, Haohan Wang, Eric P. Xing, and Dong Huang$^{(\boxtimes)}$

School of Computer Science, Carnegie Mellon University, Pittsburgh, USA
zeyih@andrew.cmu.edu, {haohanw,epxing}@cs.cmu.edu, donghuang@cmu.edu

Abstract. Convolutional Neural Networks (CNN) conduct image classification by activating dominant features that correlated with labels. When the training and testing data are under similar distributions, their dominant features are similar, leading to decent test performance. The performance is nonetheless unmet when tested with different distributions, leading to the challenges in cross-domain image classification. We introduce a simple training heuristic, Representation Self-Challenging (RSC), that significantly improves the generalization of CNN to the out-of-domain data. RSC iteratively challenges (discards) the dominant features activated on the training data, and forces the network to activate remaining features that correlate with labels. This process appears to activate feature representations applicable to out-of-domain data without prior knowledge of the new domain and without learning extra network parameters. We present the theoretical properties and conditions of RSC for improving cross-domain generalization. The experiments endorse the simple, effective, and architecture-agnostic nature of our RSC method.

Keywords: Cross-domain generalization · Robustness

1 Introduction

Imagine teaching a child to visually differentiate "dog" from "cat": when presented with a collection of illustrations from her picture books, she may immediately answer that "cats tend to have chubby faces" and end the learning. However, if we continue to ask for more differences, she may start to notice other features like ears or body-size. We conjecture this follow-up challenge question plays a significant role in helping human reach the remarkable generalization ability. Most people should be able to differentiate "cat" from "dog" visually even when the images are presented in irregular qualities. After all, we did not stop learning after we picked up the first clue when we were children, even the first clue was good enough to help us recognize all the images in our textbook.

Z. Huang, H. Wang—Equal contribution; codes are available at https://github.com/DeLightCMU/RSC

Electronic supplementary material The online version of this chapter (https://doi.org/10.1007/978-3-030-58536-5_8) contains supplementary material, which is available to authorized users.

© Springer Nature Switzerland AG 2020
A. Vedaldi et al. (Eds.): ECCV 2020, LNCS 12347, pp. 124–140, 2020.
https://doi.org/10.1007/978-3-030-58536-5_8

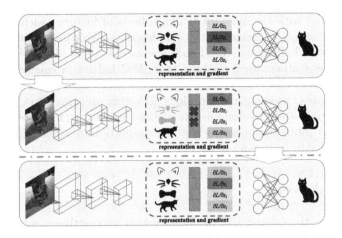

Fig. 1. The essence of our Representation Self-Challenging (RSC) training method: top two panels: the algorithm mutes the feature representations associated with the highest gradient, such that the network is forced to predict the labels through other features; bottom panel: after training, the model is expected to leverage more features for prediction in comparison to models trained conventionally.

Nowadays, deep neural networks have exhibited remarkable empirical results over various computer vision tasks, yet these impressive performances seem unmet when the models are tested with the samples in irregular qualities [32] (*i.e.*, out-of-domain data, samples collected from the distributions that are similar to, but different from the distributions of the training samples). To account for this discrepancy, technologies have been invented under the domain adaptation regime [2,3], where the goal is to train a model invariant to the distributional differences between the source domain (*i.e.*, the distribution of the training samples) and the target domain (*i.e.*, the distribution of the testing samples) [5,33].

As the influence of machine learning increases, the industry starts to demand the models that can be applied to the domains that are not seen during the training phase. Domain generalization [18], as an extension of domain adaptation, has been studied as a response. The central goal is to train a model that can align the signals from multiple source domains.

Further, Wang et al. extend the problem to ask how to train a model that generalizes to an arbitrary domain with only the training samples, but not the corresponding domain information, as these domain information may not be available in the real world [31]. Our paper builds upon this set-up and aims to offer a solution that allows the model to be robustly trained without domain information and to empirically perform well on unseen domains.

In this paper, we introduce a simple training heuristic that improves cross-domain generalization. This approach discards the representations associated with the higher gradients at each epoch, and forces the model to predict with remaining information. Intuitively, in a image classification problem, our heuristic works like a "self-challenging" mechanism as it prevents the fully-connected layers to predict with the most predictive subsets of features, such as the most frequent color, edges, or shapes in the training data. We name our method Representation Self Challenging (RSC) and illustrate its main idea in Fig. 1.

We present mathematical analysis that RSC induces a smaller generalization bound. We further demonstrate the empirical strength of our method with domain-agnostic cross-domain evaluations, following previous setup [31]. We also conduct ablation study to examine the alignment between its empirical performance and our intuitive understanding. The inspections also shed light upon the choices of its extra hyperparameter.

2   Related Work

We summarize the related DG works from two perspectives: learning domain invariant features and augmenting source domain data. Further, as RSC can be broadly viewed as a generic training heuristic for CNN, we also briefly discuss the general-purpose regularizations that appear similar to our method.

DG Through Learning Domain Invariant Features: These methods typically minimize the discrepancy between source domains assuming that the resulting features will be domain-invariant and generalize well for unseen target distributions. Along this track, Muandet et al. employed Maximum Mean Discrepancy (MMD) [18]. Ghifary et al. proposed a multi-domain reconstruction auto-encoder [10]. Li et al. applied MMD constraints to an autoencoder via adversarial training [15].

Recently, meta-learning based techniques start to be used to solve DG problems. Li et al. alternates domain-specific feature extractors and classifiers across domains via episodic training, but without using inner gradient descent update [14]. Balaji et al. proposed MetaReg that learns a regularization function (e.g., weighted ℓ_1 loss) particularly for the network's classification layer, while excluding the feature extractor [1].

Further, recent DG works forgo the requirement of source domains partitions and directly learn the cross-domain generalizable representations through a mixed collection of training data. Wang et al. extracted robust feature representation by projecting out superficial patterns like color and texture [31]. Wang et al. penalized model's tendency in predicting with local features in order to extract robust globe representation [30]. RSC follows this more recent path and directly activates more features in all source domain data for DG without knowledge of the partition of source domains.

DG Through Augmenting Source Domain: These methods augment the source domain to a wider span of the training data space, enlarging the possibility of covering the span of the data in the target domain. For example, An auxiliary domain classifier has been introduced to augment the data by perturbing input data based on the domain classification signal [23]. Volpi et al. developed an adversarial approach, in which samples are perturbed according to fictitious target distributions within a certain Wasserstein distance from the source [29].

A recent method with state-of-the art performance is JiGen [4], which leverages self-supervised signals by solving jigsaw puzzles.

Key Difference: These approaches usually introduce a model-specific DG model and rely on prior knowledge of the target domain, for instance, the target spatial permutation is assumed by JiGen [4]. In contrast, RSC is a model-agnostic training algorithm that aims to improve the cross-domain robustness of any given model. More importantly, RSC does not utilize any knowledge of partitions of domains, either source domain or target domain, which is the general scenario in real world application.

Generic Model Regularization: CNNs are powerful models and tend to overfit on source domain datasets. From this perspective, model regularization, *e.g.*, weight decay [19], early stopping, and shake-shake regularization [8], could also improve the DG performance. Dropout [25] mutes features by randomly zeroing each hidden unit of the neural network during the training phase. In this way, the network benefit from the assembling effect of small subnetworks to achieve a good regularization effect. Cutout [6] and HaS [24] randomly drop patches of input images. SpatialDropout [26] randomly drops channels of a feature map. DropBlock [9] drops contiguous regions from feature maps instead of random units. DropPath [11] zeroes out an entire layer in training, not just a particular unit. MaxDrop [20] selectively drops features of high activations across the feature map or across the channels. Adversarial Dropout [21] dropouts for maximizing the divergence between the training supervision and the outputs from the network. [12] leverages Adversarial Dropout [21] to learn discriminative features by enforcing the cluster assumption.

Key Difference: RSC differs from above methods in that RSC locates and mutes most predictive parts of feature maps by gradients instead of randomness, activation or prediction divergence maximization. This selective process plays an important role in improving the convergence, as we will briefly argue later.

3   Method

Notations: $(\mathbf{x}, \mathbf{y})$ denotes a sample-label pair from the data collection $(\mathbf{X}, \mathbf{Y})$ with n samples, and $\mathbf{z}$ (or $\mathbf{Z}$) denotes the feature representation of $(\mathbf{x}, \mathbf{y})$ learned by a neural network. $f(\cdot; \theta)$ denotes the CNN model, whose parameters are denoted as θ. $h(\cdot; \theta^{\text{top}})$ denotes the task component of $f(\cdot; \theta)$; $h(\cdot; \theta^{\text{top}})$ takes $\mathbf{z}$ as input and outputs the logits prior to a softmax function; θ^{top} denotes the parameters of $h(\cdot; \theta^{\text{top}})$. $l(\cdot, \cdot)$ denotes a generic loss function. RSC requires one extra scalar hyperparameter: the percentage of the representations to be discarded, denoted as p. Further, we use $\hat{\ }$ to denote the estimated quantities, use $\bar{\ }$ to denote the quantities after the representations are discarded, and use t in the subscript to index the iteration. For example, $\hat{\theta}_t$ means the estimated parameter at iteration t.

3.1   Self-challenging Algorithm

As a generic deep learning training method, RSC solves the same standard loss function as the ones used by many other neural networks, *i.e.*,

$$\widehat{\theta} = \arg\min_{\theta} \sum_{\langle \mathbf{x}, \mathbf{y} \rangle \sim \langle \mathbf{X}, \mathbf{Y} \rangle} l(f(\mathbf{x}; \theta), \mathbf{y}),$$

but RSC solves it in a different manner.

At each iteration, RSC inspects the gradient, identifies and then mutes the most predictive subset of the representation $\mathbf{z}$ (by setting the corresponding values to zero), and finally updates the entire model.

This simple heuristic has three steps (for simplicity, we drop the indices of samples and assume the batch size is 1 in the following equations):

1. Locate: RSC first calculates the gradient of upper layers with respect to the representation as follows:

$$\mathbf{g_z} = \partial(h(\mathbf{z}; \widehat{\theta}_t^{\text{top}}) \odot \mathbf{y}) / \partial \mathbf{z}, \tag{1}$$

   where $\odot$ denotes an element-wise product. Then RSC computes the $(100-p)^{\text{th}}$ percentile, denoted as q_p. Then it constructs a masking vector $\mathbf{m}$ in the same dimension of $\mathbf{g}$ as follows. For the ith element:

$$\mathbf{m}(i) = \begin{cases} 0, & \text{if } \mathbf{g_z}(i) \geq q_p \\ 1, & \text{otherwise} \end{cases} \tag{2}$$

   In other words, RSC creates a masking vector $\mathbf{m}$, whose element is set to 0 if the corresponding element in $\mathbf{g}$ is one of the top p percentage elements in $\mathbf{g}$, and set to 1 otherwise.

2. Mute: For every representation $\mathbf{z}$, RSC masks out the bits associated with larger gradients by:

$$\tilde{\mathbf{z}} = \mathbf{z} \odot \mathbf{m} \tag{3}$$

3. Update: RSC computes the softmax with perturbed representation with

$$\tilde{\mathbf{s}} = \text{softmax}(h(\tilde{\mathbf{z}}; \widehat{\theta}_t^{\text{top}})), \tag{4}$$

   and then use the gradient

$$\tilde{\mathbf{g}}_\theta = \partial l(\tilde{\mathbf{s}}, \mathbf{y}) / \partial \widehat{\theta}_t \tag{5}$$

to update the entire model for $\widehat{\theta}_{t+1}$ with optimizers such as SGD or ADAM.

We summarize the procedure of RSC in Algorithm 1. No that operations of RSC comprise of only few simple operations such as pooling, threshold and element-wise product. Besides the weights of the original network, no extra parameter needs to be learned.

Algorithm 1: RSC Update Algorithm

Input: data set $\langle \mathbf{X}, \mathbf{Y} \rangle$, percentage of representations to discard p, other configurations such as learning rate η, maximum number of epoches T, *etc*;

Output: Classifier $f(\cdot; \widehat{\theta})$;

random initialize the model $\widehat{\theta}_0$;

while $t \leq T$ **do**

    for *every sample (or batch)* $\mathbf{x}, \mathbf{y}$ **do**

        calculate $\mathbf{z}$ through forward pass;

        calculate $\mathbf{g_z}$ with Eq. 1;

        calculate q_p and $\mathbf{m}$ as in Eq. 2;

        generate $\tilde{\mathbf{z}}$ with Eq. 3;

        calculate gradient $\tilde{\mathbf{g}}_\theta$ with Eq. 4 and Eq. 5;

        update $\widehat{\theta}_{t+1}$ as a function of $\widehat{\theta}_t$ and $\tilde{\mathbf{g}}_\theta$

    end

end

3.2 Theoretical Evidence

To expand the theoretical discussion smoothly, we will refer to the "dog" vs. "cat" classification example repeatedly as we progress. The basic set-up, as we introduced in the beginning of this paper, is the scenario of a child trying to learn the concepts of "dog" vs. "cat" from illustrations in her book: while the hypothesis "cats tend to have chubby faces" is good enough to classify all the animals in her picture book, other hypotheses mapping ears or body-size to labels are also predictive.

On the other hand, if she wants to differentiate all the "dogs" from "cats" in the real world, she will have to rely on a complicated combination of the features mentioned about. Our main motivation of this paper is as follows: this complicated combination of these features is already illustrated in her picture book, but she does not have to learn the true concept to do well in her finite collection of animal pictures.

This disparity is officially known as "covariate shift" in domain adaptation literature: the conditional distribution (*i.e.*, the semantic of a cat) is the same across every domain, but the model may learn something else (*i.e.*, chubby faces) due to the variation of marginal distributions.

With this connection built, we now proceed to the theoretical discussion, where we will constantly refer back to this "dog" vs. "cat" example.

Background. As the large scale deep learning models, such as AlexNet or ResNet, are notoriously hard to be analyzed statistically, we only consider a simplified problem to argue for the theoretical strength of our method: we only concern with the upper layer $h(\cdot; \theta^{\text{top}})$ and illustrate that our algorithm helps improve the generalization of $h(\cdot; \theta^{\text{top}})$ when $\mathbf{Z}$ is fixed. Therefore, we can directly treat $\mathbf{Z}$ as the data (features). Also, for convenience, we overload θ to denote θ^{top} within the theoretical evidence section.

We expand our notation set for the theoretical analysis. As we study the domain-agnostic cross-domain setting, we no longer work with *i.i.d* data. Therefore, we use $\mathcal{Z}$ and $\mathcal{Y}$ to denote the collection of distributions of features and labels respectively. Let Θ be a hypothesis class, where each hypothesis $\theta \in \Theta$ maps $\mathcal{Z}$ to $\mathcal{Y}$. We use a set $\mathcal{D}$ (or $\mathcal{S}$) to index $\mathcal{Z}$, $\mathcal{Y}$ and θ. Therefore, $\theta^*(\mathcal{D})$ denotes the hypothesis with minimum error in the distributions specified with $\mathcal{D}$, but with no guarantees on the other distributions.

e.g., $\theta^*(\mathcal{D})$ can be "cats have chubby faces" when $\mathcal{D}$ specifies the distribution to be picture book.

Further, θ^* denotes the classifier with minimum error on every distribution considered. If the hypothesis space is large enough, θ^* should perform no worse than $\theta^*(\mathcal{D})$ on distributions specified by $\mathcal{D}$ for any $\mathcal{D}$.

e.g., θ^* is the true concept of "cat", and it should predict no worse than "cats have chubby faces" even when the distribution is picture book.

We use $\widehat{\theta}$ to denote any ERM and use $\widehat{\theta}_{\mathrm{RSC}}$ to denote the ERM estimated by the RSC method. Finally, following conventions, we consider $l(\cdot, \cdot)$ as the zero-one loss and use a shorthand notation $L(\theta; \mathcal{D}) = \mathbb{E}_{\langle \mathbf{z}, \mathbf{y} \rangle \sim \langle \mathcal{Z}(\mathcal{D}), \mathcal{Y}(\mathcal{D}) \rangle} l(h(\mathbf{z}; \theta), \mathbf{y})$ for convenience, and we only consider the finite hypothesis class case within the scope of this paper, which leads to the first formal result:

Corollary 1. *If*

$$|e(\mathbf{z}(\mathcal{S}); \theta_{\mathrm{RSC}}^\star) - e(\tilde{\mathbf{z}}(\mathcal{S}); \theta_{\mathrm{RSC}}^\star)| \leq \xi(p), \tag{6}$$

where $e(\cdot; \cdot)$ is a function defined as

$$e(\mathbf{z}; \theta^\star) := \mathbb{E}_{\langle \mathbf{z}, \mathbf{y} \rangle \sim \mathcal{S}} l(f(\mathbf{z}; \theta^\star); \mathbf{y})$$

and $\xi(p)$ is a small number and a function of RSC's hyperparameter p; $\tilde{\mathbf{z}}$ is the perturbed version of $\mathbf{z}$ generated by RSC, it is also a function of p, but we drop the notation for simplicity. If Assumptions A1, A2, and A3 (See Appendix) hold, we have, with probability at least $1 - \delta$

$$L(\widehat{\theta}_{\mathrm{RSC}}(\mathcal{S}); \mathcal{S}) - L(\theta_{\mathrm{RSC}}^\star(\mathcal{S}); \mathcal{D})$$

$$\leq (2\xi(p) + 1)\sqrt{\frac{2(\log(2|\Theta_{\mathrm{RSC}}|) + \log(2/\delta))}{n}}$$

As the result shows, whether RSC will succeed depends on the magnitude of $\xi(p)$. The smaller $\xi(p)$ is, the tighter the bound is, the better the generalization bound is. Interestingly, if $\xi(p) = 0$, our result degenerates to the classical generalization bound of *i.i.d* data.

While it seems the success of our method will depend on the choice of Θ to meet Condition 6, we will show RSC is applicable in general by presenting it forces the empirical counterpart $\widehat{\xi}(p)$ to be small. $\widehat{\xi}(p)$ is defined as

$$\widehat{\xi}(p) := |h(\widehat{\theta}_{\mathrm{RSC}}, \mathbf{z}) - h(\widehat{\theta}_{\mathrm{RSC}}, \tilde{\mathbf{z}})|,$$

where the function $h(\cdot, \cdot)$ is defined as

$$h(\widehat{\theta}_{\text{RSC}}, \mathbf{z}) = \sum_{(\mathbf{z}, \mathbf{y}) \sim \mathcal{S}} l(f(\mathbf{z}; \widehat{\theta}_{\text{RSC}}); \mathbf{y}). \tag{7}$$

We will show $\widehat{\xi}(p)$ decreases at every iteration with more assumptions:

A4: Discarding the most predictive features will increase the loss at current iteration.

A5: The learning rate η is sufficiently small (η^2 or higher order terms are negligible).

Formally,

Corollary 2. *If Assumption **A4** holds, we can simply denote*

$$h(\widehat{\theta}_{\text{RSC}}(t), \tilde{\mathbf{z}}_t) = \gamma_t(p) h(\widehat{\theta}_{\text{RSC}}(t), \mathbf{z}_t),$$

*where $h(\cdot, \cdot)$ is defined in Eq. 7. $\gamma_t(p)$ is an arbitrary number greater than 1, also a function of RSC's hyperparameter p. Also, if Assumption **A5** holds, we have:*

$$\Gamma(\widehat{\theta}_{\text{RSC}}(t+1)) = \Gamma(\widehat{\theta}_{\text{RSC}}(t)) - (1 - \frac{1}{\gamma_t(p)}) \|\tilde{\mathbf{g}}\|_2^2 \eta$$

where

$$\Gamma(\widehat{\theta}_{\text{RSC}}(t)) := |h(\widehat{\theta}_{\text{RSC}}(t), \mathbf{z}_t) - h(\widehat{\theta}_{\text{RSC}}(t), \tilde{\mathbf{z}}_t)|$$

t denotes the iteration, $\mathbf{z}_t$ (or $\tilde{\mathbf{z}}_t$) denotes the features (or perturbed features) at iteration t, and $\tilde{\mathbf{g}} = \partial h(\widehat{\theta}_{\text{RSC}}(t), \tilde{\mathbf{z}}_t) / \partial \widehat{\theta}_{\text{RSC}}(t)$

Notice that $\widehat{\xi}(p) = \Gamma(\widehat{\theta}_{\text{RSC}})$, where $\widehat{\theta}_{\text{RSC}}$ is $\widehat{\theta}_{\text{RSC}}(t)$ at the last iteration t. We can show that $\widehat{\xi}(p)$ is a small number because $\Gamma(\widehat{\theta}_{\text{RSC}}(t))$ gets smaller at every iteration. This discussion is also verified empirically, as shown in Fig. 2.

The decreasing speed of $\Gamma(\widehat{\theta}_{\text{RSC}}(t))$ depends on the scalar $\gamma_t(p)$: the greater $\gamma_t(p)$ is, the faster $\Gamma(\widehat{\theta}_{\text{RSC}}(t))$ descends. Further, intuitively, the scale of $\gamma_t(p)$ is highly related to the mechanism of RSC and its hyperparameter p. For example, RSC discards the most predictive representations, which intuitively guarantees the increment of the empirical loss (Assumption **A4**).

Finally, the choice of p governs the increment of the empirical loss: if p is

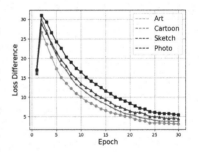

Fig. 2. $\Gamma(\widehat{\theta}_{\text{RSC}}(t))$, i.e., "Loss Difference", plotted for the PACS experiment (details of the experiment setup will be discussed later). Except for the first epoch, $\Gamma(\widehat{\theta}_{\text{RSC}}(t))$ decreases consistently along the training process.

small, the perturbation will barely affect the model, thus the increment will be small; while if p is large, the perturbation can alter the model's response dramatically, leading to significant ascend of the loss. However, we cannot blindly choose the largest possible p because if p is too large, the model may not be able to learn anything predictive at each iteration.

In summary, we offer the intuitive guidance of the choice of hyperparameter p: for the same model and setting,

- the smaller p is, the smaller the training error will be;
- the bigger p is, the smaller the (cross-domain) generalization error (*i.e.*, difference between testing error and training error) will be.

Therefore, the success of our method depends on the choice of p as a balance of the above two goals.

3.3   Engineering Specification and Extensions

For simplicity, we detail the RSC implementation on a ResNet backbone + FC classification network. RSC is applied to the training phase, and operates on the last convolution feature tensor of ResNet. Denote the feature tensor of an input sample as $\mathbf{Z}$ and its gradient tensor of as $\mathbf{G}$. $\mathbf{G}$ is computed by back propagating the classification score with respect to the ground truth category. Both of them are of size $[7 \times 7 \times 512]$.

Spatial-Wise RSC: In the training phase, global average pooling is applied along the channel dimension to the gradient tensor $\mathbf{G}$ to produce a weighting matrix w_i of size $[7 \times 7]$. Using this matrix, we select top p percentage of the $7 \times 7 = 49$ cells, and mute its corresponding features in $\mathbf{Z}$. Each of the 49 cells correspond to a $[1 \times 1 \times 512]$ feature vector in $\mathbf{Z}$. After that, the new feature tensor $\mathbf{Z_{new}}$ is forwarded to the new network output. Finally, the network is updated through back-propagation. We refer this setup as spatial-wise RSC, which is the default RSC for the rest of this paper.

Channel-Wise RSC: RSC can also be implemented by dropping features of the channels with high-gradients. The rational behind the channel-wise RSC lies in the convolutional nature of DNNs. The feature tensor of size $[7 \times 7 \times 512]$ can be considered a decomposed version of input image, where instead of the RGB colors, there are 512 different characteristics of the each pixels. The C characteristics of each pixel contains different statistics of training data from that of the spatial feature statistics.

For channel-wise RSC, global average pooling is applied along the spatial dimension of $\mathbf{G}$, and produce a weighting vector of size $[1 \times 512]$. Using this vector, we select top p percentage of its 512 cells, and mute its corresponding features in $\mathbf{Z}$. Here, each of the 512 cells correspond to a $[7 \times 7]$ feature matrix in $\mathbf{Z}$. After that, the new feature tensor $\mathbf{Z_{new}}$ is forwarded to the new network output. Finally, the network is updated through back-propagation.

Batch Percentage: Some dropout methods like curriculum dropout [17] do not apply dropout at the beginning of training, which improves CNNs by learning basic discriminative clues from unchanged feature maps. Inspired by these methods, we randomly apply RSC to some samples in each batch, leaving the other unchanged. This introduces one extra hyperparameter, namely Batch Percentage: the percentage of samples to apply RSC in each batch. We also apply RSC to top percentage of batch samples based on cross-entropy loss. This setup is slightly better than randomness.

Detailed ablation study on above extensions will be conducted in the experiment section below.

4   Experiments

4.1   Datasets

We consider the following four data collections as the battleground to evaluate RSC against previous methods.

- **PACS** [13]: seven classes over four domains (Artpaint, Cartoon, Sketches, and Photo). The experimental protocol is to train a model on three domains and test on the remaining domain.
- **VLCS** [27]: five classes over four domains. The domains are defined by four image origins, *i.e.*, images were taken from the PASCAL VOC 2007, LabelMe, Caltech and Sun datasets.
- **Office-Home** [28]: 65 object categories over 4 domains (Art, Clipart, Product, and Real-World).
- **ImageNet-Sketch** [30]: 1000 classes with two domains. The protocol is to train on standard ImageNet [22] training set and test on ImageNet-Sketch.

4.2   Ablation Study

We conducted five ablation studies on possible configurations for RSC on the PACS dataset [13]. All results were produced based on the ResNet18 baseline in [4] and were averaged over five runs.

(1) Feature Dropping Strategies (Table 1). We compared the two attention mechanisms to select the most discriminative spatial features. The "Top-Activation" [20] selects the features with highest norms, whereas the "Top-Gradient" (default in RSC) selects the features with high gradients. The comparison shows that "Top-Gradient" is better than "Top-Activation", while both are better than the random strategy. Without specific note, we will use "Top-Gradient" as default in the following ablation study.

(2) Feature Dropping Percentage (choice of p) (Table. 2): We ran RSC at different dropping percentages to mute spatial feature maps. The highest average accuracy was reached at $p = 33.3\%$. While the best choice of p is data-specific, our results align well with the theoretical discussion: the optimal p should be neither too large nor too small.

Table 1. Ablation study of Spatial-wise RSC on Feature Dropping Strategies. Feature Dropping Percentage 50.0% and Batch Percentage 50.0%.

Feature Drop Strategies	Backbone	Artpaint	Cartoon	Sketch	Photo	Avg ↑
Baseline [4]	ResNet18	78.96	73.93	70.59	**96.28**	79.94
Random	ResNet18	79.32	75.27	74.06	95.54	81.05
Top-Activation	ResNet18	80.31	76.05	76.13	**95.72**	82.03
Top-Gradient	ResNet18	**81.23**	**77.23**	**77.56**	95.61	**82.91**

Table 2. Ablation study of Spatial-wise RSC on Feature Dropping Percentage. We used "Top-Gradient" and fixed the Batch Percentage (50.0%) here.

Feature Dropping Percentage	Backbone	Artpaint	Cartoon	Sketch	Photo	Avg ↑
66.7%	ResNet18	80.11	76.35	76.24	95.16	81.97
50.0%	ResNet18	81.23	77.23	77.56	95.61	82.91
33.3%	ResNet18	**82.87**	**78.23**	**78.89**	95.82	**83.95**
25.0%	ResNet18	81.63	78.06	78.12	96.06	83.46
20.0%	ResNet18	81.22	77.43	77.83	96.25	83.18
13.7%	ResNet18	80.71	77.18	77.12	**96.36**	82.84

Table 3. Ablation study of Spatial-wise RSC on Batch Percentage. We used "Top-Gradient" and fixed Feature Dropping Percentage (33.3%).

Batch Percentage	Backbone	Artpaint	Cartoon	Sketch	Photo	Avg ↑
50.0%	ResNet18	**82.87**	78.23	78.89	95.82	83.95
33.3%	ResNet18	82.32	**78.75**	**79.56**	96.05	**84.17**
25.0%	ResNet18	81.85	78.32	78.75	**96.21**	83.78

Table 4. Ablation study of Spatial-wise RSC verse Spatial+Channel RSC. We used the best strategy and parameter by Table 3: "Top-Gradient", Feature Dropping Percentage(33.3%) and Batch Percentage(33.3%).

Method	Backbone	Artpaint	Cartoon	Sketch	Photo	Avg ↑
Spatial	ResNet18	82.32	78.75	79.56	**96.05**	84.17
Spatial+Channel	ResNet18	**83.43**	**80.31**	**80.85**	95.99	**85.15**

(3) Batch Percentage (Table 3): RSC has the option to be only randomly applied to a subset of samples in each batch. Table 3 shows that the performance is relatively constant. Nevertheless we still choose 33.3% as the best option on the PACS dataset.

(4) Spatial-wise plus Channel-wise RSC (Table 4): In "Spatial+Channel", both spatial-wise and channel-wise RSC were applied on a sample at 50% probability, respectively. (Better options of these probabilities could be explored.) Its

Table 5. Ablation study of Dropout methods. "S" and "C" represent spatial-wise and channel-wise respectively. For fair comparison, results of above methods are report at their best setting and hyperparameters. RSC used the hyperparameters selected in above ablation studies: "Top-Gradient", Feature Dropping Percentage (33.3%) and Batch Percentage (33.3%).

Method	Backbone	Artpaint	Cartoon	Sketch	Photo	Avg ↑
Baseline [4]	ResNet18	78.96	73.93	70.59	**96.28**	79.94
Cutout[6]	ResNet18	79.63	75.35	71.56	95.87	80.60
DropBlock[9]	ResNet18	80.25	77.54	76.42	95.64	82.46
AdversarialDropout[21]	ResNet18	82.35	78.23	75.86	96.12	83.07
Random(S+C)	ResNet18	79.55	75.56	74.39	95.36	81.22
Top-Activation(S+C)	ResNet18	81.03	77.86	76.65	96.11	82.91
RSC: Top-Gradient(S+C)	ResNet18	**83.43**	**80.31**	**80.85**	95.99	**85.15**

Table 6. DG results on PACS[13] (Best in bold).

PACS	Backbone	Artpaint	Cartoon	Sketch	Photo	Avg ↑
Baseline[4]	AlexNet	66.68	69.41	60.02	89.98	71.52
Hex[31]	AlexNet	66.80	69.70	56.20	87.90	70.20
PAR[30]	AlexNet	66.30	66.30	64.10	89.60	72.08
MetaReg[1]	AlexNet	69.82	70.35	59.26	**91.07**	72.62
Epi-FCR[14]	AlexNet	64.70	72.30	65.00	86.10	72.00
JiGen[4]	AlexNet	67.63	71.71	65.18	89.00	73.38
MASF[7]	AlexNet	70.35	72.46	**67.33**	90.68	75.21
RSC (ours)	AlexNet	**71.62**	**75.11**	66.62	90.88	**76.05**
Baseline[4]	ResNet18	78.96	73.93	70.59	**96.28**	79.94
MASF[7]	ResNet18	80.29	77.17	71.69	94.99	81.03
Epi-FCR[14]	ResNet18	82.10	77.00	73.00	93.90	81.50
JiGen[4]	ResNet18	79.42	75.25	71.35	96.03	80.51
MetaReg[1]	ResNet18	**83.70**	77.20	70.30	95.50	81.70
RSC (ours)	ResNet18	83.43	**80.31**	**80.85**	95.99	**85.15**
Baseline[4]	ResNet50	86.20	78.70	70.63	97.66	83.29
MASF[7]	ResNet50	82.89	80.49	72.29	95.01	82.67
MetaReg[1]	ResNet50	87.20	79.20	70.30	97.60	83.60
RSC (ours)	ResNet50	**87.89**	**82.16**	**83.35**	**97.92**	**87.83**

improvement over Spatial-wise RSC indicates that it further activated features beneficial to target domains.

(5) Comparison with different dropout methods (Table 5): Dropout has inspired a number of regularization methods for CNNs. The main differences between those methods lie in applying stochastic or non-stochastic dropout

Table 7. DG results on VLCS [27] (Best in bold).

VLCS	Backbone	Caltech	Labelme	Pascal	Sun	Avg ↑
Baseline[4]	AlexNet	96.25	59.72	70.58	64.51	72.76
Epi-FCR[14]	AlexNet	94.10	64.30	67.10	65.90	72.90
JiGen[4]	AlexNet	96.93	60.90	70.62	64.30	73.19
MASF[7]	AlexNet	94.78	**64.90**	69.14	67.64	74.11
RSC (ours)	AlexNet	**97.61**	61.86	**73.93**	**68.32**	**75.43**

mechanism at input data, convolutional or fully connected layers. Results shows that our gradient-based RSC is better. We believe that gradient is an efficient and straightforward way to encode the sensitivity of output prediction. To the best of our knowledge, we compare with the most related works and illustrate the impact of gradients. (a) Cutout [6]. Cutout conducts random dropout on input images, which shows limited improvement over the baseline. (b) Drop-Block [9]. DropBlock tends to dropout discriminative activated parts spatially. It is better than random dropout but inferior to non-stochastic dropout methods in Table 5 such as AdversarialDropout, Top-Activation and our RSC. (c) AdversarialDropout [12,21]. AdversarialDropout is based on divergence maximization, while RSC is based on top gradients in generating dropout masks. Results show evidence that the RSC is more effective than AdversarialDropout. (d) Random and Top-Activation dropout strategies at their best hyperparameter settings (Table 9).

Table 8. DG results on Office-Home [28] (Best in bold).

Office-Home	Backbone	Art	Clipart	Product	Real	Avg ↑
Baseline[4]	ResNet18	52.15	45.86	70.86	73.15	60.51
JiGen[4]	ResNet18	53.04	47.51	71.47	72.79	61.20
RSC (ours)	ResNet18	**58.42**	**47.90**	**71.63**	**74.54**	**63.12**

Table 9. DG results on ImageNet-Sketch [30].

ImageNet-Sketch	Backbone	Top-1 Acc ↑	Top-5 Acc ↑
Baseline[31]	AlexNet	12.04	24.80
Hex[31]	AlexNet	14.69	28.98
PAR [30]	AlexNet	15.01	29.57
RSC (ours)	AlexNet	**16.12**	**30.78**

4.3   Cross-Domain Evaluation

Through the following experiments, we used "Top-Gradient" as feature dropping strategy, 33.3% as Feature Dropping Percentages, 33.3% as Batch Percentage, and Spatial+Channel RSC. All results were averaged over five runs. In our RSC implementation, we used the SGD solver, 30 epochs, and batch size 128. The learning rate starts with 0.004 for ResNet and 0.001 for AlexNet, learning rate decayed by 0.1 after 24 epochs. For PACS experiment, we used the same data augmentation protocol of randomly cropping the images to retain between 80% to 100%, randomly applied horizontal flipping and randomly (10% probability) convert the RGB image to greyscale, following [4].

In Table 6, 7, and 8, we compare RSC with the latest domain generalization work, such as Hex [31], PAR [30], JiGen [4] and MetaReg [1]. All these work only report results on different small networks and datasets. For fair comparison, we compared RSC to their reported performances with their most common choices of DNNs (*i.e.*, AlexNet, ResNet18, and ResNet50) and datasets. RSC consistently outperforms other competing methods.

The empirical performance gain of RSC can be better appreciated if we have a closer look at the PACS experiment in Table 6. The improvement of RSC from the latest baselines [4] are significant and consistent: 4.5 on AlexNet, 5.2 on ResNet18, and 4.5 on ResNet50. It is noticeable that, with both ResNet18 and ResNet50, RSC boosts the performance significantly for sketch domain, which is the only colorless domain. The model may have to understand the semantics of the object to perform well on the sketch domain. On the other hand, RSC performs only marginally better than competing methods in photo domain, which is probably because that photo domain is the simplest one and every method has already achieved high accuracy on it.

5   Discussion

Standard ImageNet Benchmark: With the impressive performance observed in the cross-domain evaluation, we further explore to evaluate the benefit of RSC with other benchmark data and higher network capacity.

Table 10. Generalization results on ImageNet. Baseline was produced with official Pytorch implementation and their ImageNet models.

ImageNet	Backbone	Top-1 Acc ↑	Top-5 Acc ↑	#Param. ↓
Baseline	ResNet50	76.13	92.86	25.6M
RSC (ours)	ResNet50	77.18	93.53	25.6M
Baseline	ResNet101	77.37	93.55	44.5M
RSC (ours)	ResNet101	78.23	94.16	44.5M
Baseline	ResNet152	78.31	94.05	60.2M
RSC (ours)	ResNet152	78.89	94.43	60.2M

We conducted image classification experiments on the Imagenet database[22]. We chose three backbones with the same architectural design while with clear hierarchies in model capacities: ResNet50, ResNet101, and ResNet152. All models were finetuned for 80 epochs with learning rate decayed by 0.1 every 20 epochs. The initial learning rate for ResNet was 0.01. All models follow extra the same training prototype in default Pytorch ImageNet implementation, using original batch size of 256, standard data augmentation and 224 × 224 as input size.

The results in Table 10 shows that RSC exhibits the ability reduce the performance gap between networks of same family but different sizes (*i.e.*, ResNet50 with RSC approaches the results of baseline ResNet101, and ResNet101 with RSC approaches the results of baseline ResNet151). The practical implication is that, RSC could induce faster performance saturation than increasing model sizes. Therefore one could scale down the size of networks to be deployed at comparable performance.

6 Conclusion

We introduced a simple training heuristic method that can be directly applied to almost any CNN architecture with no extra model architecture, and almost no increment of computing efforts. We name our method Representation Self-challenging (RSC). RSC iteratively forces a CNN to activate features that are less dominant in the training domain, but still correlated with labels. Theoretical and empirical analysis of RSC validate that it is a fundamental and effective way of expanding feature distribution of the training domain. RSC produced the state-of-the-art improvement over baseline CNNs under the standard DG settings of small networks and small datasets. Moreover, our work went beyond the standard DG settings, to illustrate effectiveness of RSC on more prevalent problem scales, *e.g.*, the ImageNet database and network sizes up-to ResNet152.

Acknowledgement. This work was partially supported by the Intelligence Advanced Research Projects Activity (IARPA) via Department of Interior/ Interior Business Center (DOI/IBC) contract number D17PC00340. In addition, Haohan Wang is supported by NIH R01GM114311, NIH P30DA035778, and NSF IIS1617583.

References

1. Balaji, Y., Sankaranarayanan, S., Chellappa, R.: MetaReg: towards domain generalization using meta-regularization. In: Advances in Neural Information Processing Systems, pp. 998–1008 (2018)
2. Ben-David, S., Blitzer, J., Crammer, K., Kulesza, A., Pereira, F., Vaughan, J.W.: A theory of learning from different domains. Mach. Learn. **79**(1), 151–175 (2010)
3. Bridle, J.S., Cox, S.J.: RecNorm: simultaneous normalisation and classification applied to speech recognition. In: Advances in Neural Information Processing Systems. pp. 234–240 (1991)

4. Carlucci, F.M., D'Innocente, A., Bucci, S., Caputo, B., Tommasi, T.: Domain generalization by solving jigsaw puzzles. In: Proceedings of the IEEE Conference on Computer Vision and Pattern Recognition, pp. 2229–2238 (2019)
5. Csurka, G.: Domain adaptation for visual applications: A comprehensive survey. arXiv preprint arXiv:1702.05374 (2017)
6. DeVries, T., Taylor, G.W.: Improved regularization of convolutional neural networks with cutout. arXiv preprint arXiv:1708.04552 (2017)
7. Dou, Q., Castro, D.C., Kamnitsas, K., Glocker, B.: Domain generalization via model-agnostic learning of semantic features. arXiv preprint arXiv:1910.13580 (2019)
8. Gastaldi, X.: Shake-shake regularization. arXiv preprint arXiv:1705.07485 (2017)
9. Ghiasi, G., Lin, T.Y., Le, Q.V.: DropBlock: a regularization method for convolutional networks. In: Advances in Neural Information Processing Systems, pp. 10727–10737 (2018)
10. Ghifary, M., Bastiaan Kleijn, W., Zhang, M., Balduzzi, D.: Domain generalization for object recognition with multi-task autoencoders. In: Proceedings of the IEEE International Conference on Computer Vision, pp. 2551–2559 (2015)
11. Larsson, G., Maire, M., Shakhnarovich, G.: FractalNet: Ultra-deep neural networks without residuals. arXiv preprint arXiv:1605.07648 (2016)
12. Lee, S., Kim, D., Kim, N., Jeong, S.G.: Drop to adapt: learning discriminative features for unsupervised domain adaptation. In: Proceedings of the IEEE International Conference on Computer Vision, pp. 91–100 (2019)
13. Li, D., Yang, Y., Song, Y.Z., Hospedales, T.M.: Deeper, broader and artier domain generalization. In: Proceedings of the IEEE International Conference on Computer Vision, pp. 5542–5550 (2017)
14. Li, D., Zhang, J., Yang, Y., Liu, C., Song, Y.Z., Hospedales, T.M.: Episodic training for domain generalization. arXiv preprint arXiv:1902.00113 (2019)
15. Li, H., Jialin Pan, S., Wang, S., Kot, A.C.: Domain generalization with adversarial feature learning. In: Proceedings of the IEEE Conference on Computer Vision and Pattern Recognition, pp. 5400–5409 (2018)
16. Mitchell, T.M., et al.: Machine Learning, vol. 45, no. 37, pp. 870–877. McGraw Hill, Burr Ridge, IL (1997)
17. Morerio, P., Cavazza, J., Volpi, R., Vidal, R., Murino, V.: Curriculum dropout. In: Proceedings of the IEEE International Conference on Computer Vision, pp. 3544–3552 (2017)
18. Muandet, K., Balduzzi, D., Schölkopf, B.: Domain generalization via invariant feature representation. In: International Conference on Machine Learning, pp. 10–18 (2013)
19. Nowlan, S.J., Hinton, G.E.: Simplifying neural networks by soft weight-sharing. Neural Comput. 4(4), 473–493 (1992)
20. Park, S., Kwak, N.: Analysis on the dropout effect in convolutional neural networks. In: Lai, S.-H., Lepetit, V., Nishino, K., Sato, Y. (eds.) ACCV 2016. LNCS, vol. 10112, pp. 189–204. Springer, Cham (2017). https://doi.org/10.1007/978-3-319-54184-6_12
21. Park, S., Park, J., Shin, S.J., Moon, I.C.: Adversarial dropout for supervised and semi-supervised learning. In: 32nd AAAI Conference on Artificial Intelligence (2018)
22. Russakovsky, O.: ImageNet large scale visual recognition challenge. Int. J. Comput. Vis. (IJCV) 115(3), 211–252 (2015). https://doi.org/10.1007/s11263-015-0816-y

23. Shankar, S., Piratla, V., Chakrabarti, S., Chaudhuri, S., Jyothi, P., Sarawagi, S.: Generalizing across domains via cross-gradient training. arXiv preprint arXiv:1804.10745 (2018)
24. Singh, K.K., Lee, Y.J.: Hide-and-seek: forcing a network to be meticulous for weakly-supervised object and action localization. In: 2017 IEEE International Conference on Computer Vision (ICCV), pp. 3544–3553. IEEE (2017)
25. Srivastava, N., Hinton, G., Krizhevsky, A., Sutskever, I., Salakhutdinov, R.: Dropout: a simple way to prevent neural networks from overfitting. J. Mach. Learn. Res. **15**(1), 1929–1958 (2014)
26. Tompson, J., Goroshin, R., Jain, A., LeCun, Y., Bregler, C.: Efficient object localization using convolutional networks. In: Proceedings of the IEEE Conference on Computer Vision and Pattern Recognition, pp. 648–656 (2015)
27. Torralba, A., Efros, A.A., et al.: Unbiased look at dataset bias. In: CVPR, vol. 1, p. 7. Citeseer (2011)
28. Venkateswara, H., Eusebio, J., Chakraborty, S., Panchanathan, S.: Deep hashing network for unsupervised domain adaptation. In: Proceedings of the IEEE Conference on Computer Vision and Pattern Recognition, pp. 5018–5027 (2017)
29. Volpi, R., Namkoong, H., Sener, O., Duchi, J.C., Murino, V., Savarese, S.: Generalizing to unseen domains via adversarial data augmentation. In: Advances in Neural Information Processing Systems, pp. 5334–5344 (2018)
30. Wang, H., Ge, S., Xing, E.P., Lipton, Z.C.: Learning robust global representations by penalizing local predictive power. In: Advances in Neural Information Processing Systems, NeurIPS 2019 (2019)
31. Wang, H., He, Z., Lipton, Z.C., Xing, E.P.: Learning robust representations by projecting superficial statistics out. In: International Conference on Learning Representations (2019)
32. Wang, H., Wu, X., Huang, Z., Xing, E.P.: High-frequency component helps explain the generalization of convolutional neural networks. In: Proceedings of the IEEE/CVF Conference on Computer Vision and Pattern Recognition, pp. 8684–8694 (2020)
33. Wang, M., Deng, W.: Deep visual domain adaptation: a survey. Neurocomputing **312**, 135–153 (2018)

A Competence-Aware Curriculum for Visual Concepts Learning via Question Answering

Qing Li$^{(\boxtimes)}$, Siyuan Huang , Yining Hong , and Song-Chun Zhu

UCLA Center for Vision, Cognition, Learning, and Autonomy (VCLA),
Los Angeles, USA
{liqing,huangsiyuan,yininghong}@ucla.edu, sczhu@stat.ucla.edu

Abstract. Humans can progressively learn visual concepts from easy to hard questions. To mimic this efficient learning ability, we propose a competence-aware curriculum for visual concept learning in a question-answering manner. Specifically, we design a neural-symbolic concept learner for learning the visual concepts and a multi-dimensional Item Response Theory (mIRT) model for guiding the learning process with an adaptive curriculum. The mIRT effectively estimates the concept difficulty and the model competence at each learning step from accumulated model responses. The estimated concept difficulty and model competence are further utilized to select the most profitable training samples. Experimental results on CLEVR show that with a competence-aware curriculum, the proposed method achieves state-of-the-art performances with superior data efficiency and convergence speed. Specifically, the proposed model only uses **40% of training data** and converges **three times faster** compared with other state-of-the-art methods.

Keywords: Visual question answering · Visual concept learning · Curriculum learning · Model competence

1 Introduction

Humans excel at learning visual concepts and their compositions in a question-answering manner [10,16,18], which requires a joint understanding of vision and language. The essence of such learning skill is the superior capability to connect linguistic symbols (words/phrases) in question-answer pairs with visual cues (appearance/geometry) in images. Imagine a person without prior knowledge of colors is presented with two contrastive examples in Figure 1-I. The left images are the same except for color, and the right question-answer pairs differ only in the descriptions about color. By assuming that the differences in the question-answer pairs capture the differences in appearances, he can learn the concept of

---

Electronic supplementary material The online version of this chapter (https://doi.org/10.1007/978-3-030-58536-5_9) contains supplementary material, which is available to authorized users.

© Springer Nature Switzerland AG 2020
A. Vedaldi et al. (Eds.): ECCV 2020, LNCS 12347, pp. 141–157, 2020.
https://doi.org/10.1007/978-3-030-58536-5_9

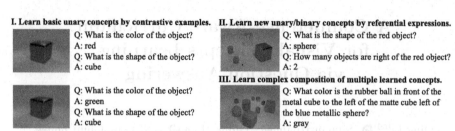

I. Learn basic unary concepts by contrastive examples.

Q: What is the color of the object?
A: red
Q: What is the shape of the object?
A: cube

Q: What is the color of the object?
A: green
Q: What is the shape of the object?
A: cube

II. Learn new unary/binary concepts by referential expressions.

Q: What is the shape of the red object?
A: sphere
Q: How many objects are right of the red object?
A: 2

III. Learn complex composition of multiple learned concepts.

Q: What color is the rubber ball in front of the metal cube to the left of the matte cube left of the blue metallic sphere?
A: gray

Fig. 1. The incremental learning of visual concepts in a question-answering manner. Three difficulty levels can be categorized into I) unary concepts from simple questions, II) binary (relational) concepts based on the learned concepts, and III) compositions of visual concepts from comprehensive questions. (Color figure online)

color and the appearance of specific colors (*i.e.*, red and green). Besides learning the basic unary concepts from contrastive examples, compositional relations from complex questions consisting of multiple concepts can be further learned, as shown in Fig. 1-II and -III.

Another crucial characteristic of the human learning process is to start *small* and learn *incrementally*. More specifically, the human learning process is well-organized with a curriculum that introduces concepts progressively and facilitates the learning of new abstract knowledge by exploiting learned concepts. A good curriculum serves as an experienced teacher. By ranking and selecting examples according to the learning state, it can guide the training process of the learner (student) and significantly increase the learning speed. This idea is originally examined in animal training as *shaping* [32,46,51] and then applied to machine learning as *curriculum learning* [7,13,20,21,44].

Inspired by the efficient curriculum, Mao *et al.* [41] proposes a neural-symbolic approach to learn visual concepts with a *fixed* curriculum. Their approach learns from image-question-answer triplets and does not require annotation on images or programs generated from questions. The model is trained with a manually-designed curriculum that includes four stages: (1) learning unary visual concepts; (2) learning relational concepts; (3) learning more complex questions with visual perception fixed; (4) joint fine-tuning all modules. They select questions for each stage by the depths of the latent programs. Their curriculum heavily relies on the manually-designed heuristic that measures the question difficulty and discretizes the curriculum. Such heuristic suffers from three limitations. First, it ignores the variance of difficulties for questions with the same program depths, where different concepts might have various difficulties. Second, the manually-designed curriculum relies on strong human prior knowledge for the difficulties, while such prior may conflict with the inherent difficulty distribution of the training examples. Last but most importantly, it neglects the progress of the learner that evolves along with the training process. More specifically, the order of training samples in the curriculum is nonadjustable based on the model state. This scheme is in stark contrast to the way that humans learn – by *actively* selecting learning samples based on our current learning state, instead of *passively* accepting specific training samples. A desirable learning system should be capable of automatically adjusting the curriculum during the learning process without requiring any prior knowledge, which makes the learning procedure more efficient with less data redundancy and faster convergence speed.

To address these issues and mimic human ability in adaptive learning, we propose a **competence-aware** curriculum for visual concept learning via question answering, where competence represents the capability of the model to recognize each concept. The proposed approach utilizes multi-dimensional Item Response Theory (mIRT) to estimate the **concept difficulty** and **model competence** at each learning step from accumulated model responses. Item Response Theory (IRT) [5,6] is a widely adopted method in psychometrics that estimates the human ability and the item difficulty from human responses on various items. We extend the IRT to a mIRT that matches the compositional nature of visual reasoning, and apply variational inference to get a Bayesian estimation for the parameters in mIRT. Based on the estimations of concept difficulty and model competence, we further define a continuous adaptive curriculum (instead of a discretized fixed regime) that selects the most profitable training samples according to the current learning state. More specifically, the learner can filter out samples with either too naive or too challenging questions. These questions bring either negligible or sharp gradients to the learner, which makes it slower and harder to converge.

With the proposed competence-aware curriculum, the learner can address the aforementioned limitations brought by a fixed curriculum with the following advantages:

1. The concept difficulty and the model competence at each learning step can be inferred effectively from accumulated model responses. It enables the model to distinguish difficulties among various concepts and be aware of its own capability for recognizing these concepts.
2. The question difficulty can be calculated with the estimated concept difficulty and model competence without requiring any heuristics.
3. The adaptive curriculum significantly contributes to the improvement of learning efficiency by relieving the data redundancy and accelerating the convergence, as well as the improvement of the final performance.

We explore the proposed method on the CLEVR dataset [29], an artificial universe where visual concepts are clearly defined and less correlated. We opt for this synthetic environment because there is little prior work on curriculum learning for visual concepts and there lacks a clear definition of visual concepts in real-world setting. CLEVR allows us to perform controllable diagnoses of the proposed mIRT model in building an adaptive curriculum. Section 5 further discusses the potentials and challenges of generalizing our method to other domains such as real-world images and natural language processing.

Experimental results show that the visual concept learner with the proposed competence-aware curriculum converges three times faster and consumes only 40% of the training data while achieving similar or even higher accuracy compared with other state-of-the-art models. We also evaluate individual modules in the proposed method and demonstrate their efficacy in Sect. 4.

2   Related Work

2.1   Neural-Symbolic Visual Question Answering

Visual question answering (VQA) [17,29,39] is a popular task for gauging the capability of visual reasoning systems. Some recent studies [2,3,24,30,57] focus on learning the neural module networks (NMNs) on the CLEVR dataset. NMNs translate questions into programs, which are further executed over image features to predict answers. The program generator is typically trained on human annotations. Several recent works target on reducing the supervision or increasing the generalization ability to new tasks in NMNs. For example, Johnson *et al.* [30] replaces the hand-designed syntactic parsers by a learned program generator. Neural-Symbolic VQA [58] explores an object-based visual representation and uses a symbolic executor for inferring the answer. Neural-symbolic concept learner [41] uses a symbolic reasoning process and manually-defined curriculum to bridge the learning of visual concepts, words, and the parsing of questions without explicit annotations. In this paper, we build our model on the neural-symbolic concept learner [41] and learn an adaptive curriculum to select the most profitable training samples.

Learning-by-asking (LBA) [42] proposes an interactive learning framework that allows the model to actively query an oracle and discover an easy-to-hard curriculum. LBA uses the expected accuracy improvement over candidate answers as an informativeness measure to pick questions. However, it is costly to compute the expected accuracy improvement for sampled questions since it requires to process all the questions and images through a VQA model. Moreover, the expected accuracy improvement cannot help to learn which specific component of the question contributes to the performance, especially while learning from the answers with little information such as "yes/no". In contrast, we select questions by explicitly modeling the difficulty of visual concepts, combined with model competence to infer the difficulty of each question.

2.2   Curriculum Learning and Machine Teaching

The competence-aware curriculum in our work is related to *curriculum learning* [7,20,21,44,47,50,52,54] and *machine teaching* [11,15,38,40,56,59,60]. *Curriculum learning* is firstly proposed by Bengio *et al.* [7] and demonstrates that a dataset order from easy instances to hard ones benefits learning process. The measures of hardness in curriculum learning approaches are usually determined by hand-designed heuristics [41,50,52,54]. Graves *et al.* [20] explore learning signals based on the increase rates in prediction accuracy and network complexity to adjust data distributions along with training. Self-paced learning [27,28,33,50] quantifies the sample hardness by the training loss and formulates curriculum learning as an optimization problem by jointly modeling the sample selection and the learning objective. These hand-designed heuristics are usually task-specific without any generalization ability to other domains.

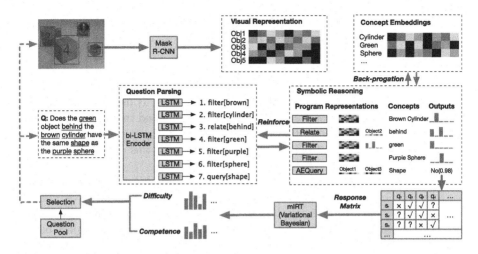

Fig. 2. The overview of the proposed approach. We use neural symbolic reasoning as a bridge to jointly learn concept embeddings and question parsing. The model responses in the training process are accumulated to estimate concept difficulty and model competence at each learning step with mIRT. The estimations help to select appropriate training samples for the current model. In the response matrix, '✓' or '✗' denotes that the snapshot predicts a correct or wrong answer, and '?' means the snapshot has no response to this question.

Machine teaching [38,59,60] introduces a teacher model that receives feedback from the student model and guides the learning of the student model accordingly. Zhu *et al.* [59,60] assume that the teacher knows the ground-truth model (*i.e.*, the Oracle) beforehand and constructs a minimal training set for the student model. The recent works *learning to teach* [15,56] break this strong assumption of the existence of the oracle model and endow the teacher with the capability of learning to teach via a reinforcement learning framework.

Our work explores curriculum learning in visual reasoning, which is highly compositional and more complex than tasks studied before. Different from previous works, our method requires neither hand-designed heuristics nor an extra teacher model. We combine the idea of *competence* with curriculum learning and propose a novel mIRT model that estimates the concept difficulty and model competence from accumulated model responses.

3 Methodology

In this section, we will discuss the proposed competence-aware curriculum for visual concept learning, as also shown in Fig. 2. We first describe a neural-symbolic approach to learn visual concepts from image-question-answer triplets. Next, we introduce the background of IRT model and discuss how we derive a mIRT model for estimating concept difficulty and model competence. Finally, we present how to select training samples based on the estimated concept difficulty and model competence to make the training process more efficient.

3.1 Neural-Symbolic Concept Learner

We briefly describe the neural-symbolic concept learner. It uses a symbolic reasoning process to bridge the learning of visual concepts and the semantic parsing of textual questions without any intermediate annotations except for the final answers. We refer readers to [41,58] for more details on this model.

Scene Parsing. A scene parsing module develops an object-based representation for each image. Concretely, we adopt a pre-trained Mask R-CNN [22] to generate object proposals from the image. The detected bounding boxes with the original image are sent to a ResNet-34 [23] to extract the object-based features.

Concept Embeddings. By assuming each visual attribute (*e.g.*, shape) contains a set of visual concepts (*e.g.*, cylinder), the extracted visual features are embedded into concept spaces by learnable neural operators of the attributes.

Question Parsing. The question parsing module translates a question in natural language into an executable program in a domain-specific language designed for VQA. The question parser generates the latent program from a question in a sequence-to-sequence manner. A bi-directional LSTM is used to encode the input question into a fixed-length representation. The decoder is an attention-based LSTM, which produces the operations in the program step-by-step. Some operations take concepts as their parameters, such as *Filter[Cube]* and *Relate[Left]*. These concepts are selected from the concepts appearing in the question by the attention mechanism.

Symbolic Reasoning. Given the latent program, the symbolic executor runs the operations in the program with the object-based image representation to derive an answer for the input question. The execution is fully differentiable with respect to the concept embeddings since the intermediate results are represented in a probabilistic manner. Specifically, we keep an attention mask on all object proposals, with each element in the mask denoting the probability that the corresponding object contains certain concepts. The attention mask is fed into the next operation, and the execution continues. The final operation predicts an answer to the question. We refer the readers to the supplementary materials for more details and examples of the symbolic execution process.

Joint Optimizing. We formulate the problem of jointly learning the question parser and the concept embeddings without the annotated programs. Suppose we have a training sample consisting of image I, question Q, and answer A, and we do not observe the latent program l. The goal of training the whole system is to maximize the following conditional probability:

$$p(A|I, Q) = \mathbb{E}_{l \sim p(l|Q)} [p(A|l, I)], \tag{1}$$

where $p(l|Q)$ is parametrized by the question parser with the parameters θ_l and $p(A|l, I)$ is parametrized by the concept embeddings θ_e (there are no learnable parameters in the symbolic reasoning module). Considering the expectation over the program space in Eq. 1 is intractable, we approximate the expectation with Monte Carlo sampling. Specifically, we first sample a program $\hat{l}$ from the question parser $p(l|Q; \theta_l)$ and then apply $\hat{l}$ to obtain a probability distribution over possible answers $p(A|\hat{l}, I; \theta_e)$.

Recalling the program execution is fully differentiable w.r.t. the concept embeddings, we learn the concept embeddings by directly maximizing $\log p(A|\hat{l}, I; \theta_e)$ using gradient descent and the gradient $\nabla_{\theta_e} \log p(A|\hat{l}, I; \theta_e)$ can be calculated through back-propagation. Since the hard selection of $\hat{l}$ through Monte Carlo sampling is non-differentiable, the gradients of the question parser cannot be computed by back-propagation. Instead we optimize the question parser using the REINFORCE algorithm [55]. The gradient of the reward function J over the parameters of the policy is:

$$\nabla J(\theta_l) = \mathbb{E}_{l \sim p(l|Q; \theta_l)} [\nabla \log p(l|Q; \theta_l) \cdot r], \tag{2}$$

where r denotes the reward. Defining the reward as the log-probability of the correct answer and again, we rewrite the intractable expectation with one Monte Carlo sample $\hat{l}$:

$$\nabla J(\theta_l) = \nabla \log p\left(\hat{l}|Q; \theta_l\right) \cdot [\log p(A|\hat{l}, I; \theta_e) - b], \tag{3}$$

where b is the exponential moving average of $\log p(A|\hat{l}, I; \theta_e)$, serving as a simple baseline to reduce the variance of gradients. Therefore, the update to the question parser at each learning step is simply the gradient of the log-probability of choosing the program, multiplied by the probability of the correct answer using that program.

3.2 Background of Item Response Theory (IRT)

Item response theory (IRT) [5,6] was initially created in the fields of educational measurement and psychometrics. It has been widely used to measure the latent abilities of subjects (e.g., human beings, robots or AI models) based on their responses to items (e.g., test questions) with different levels of difficulty. The core idea of IRT is that the probability of a correct response to an item can be modeled by a mathematical function of both individual ability and item characteristics. More formally, if we let i be an individual and j be an item, then the probability that the individual i answers the item j correctly can be modeled by a logistic model as:

$$p_{ij} = c_j + \frac{1 - c_j}{1 + e^{-a_j(\theta_i - b_j)}}, \tag{4}$$

where θ_i is the latent ability of the individual i and a_j, b_j, c_j are the characteristics of the item j. The item parameters can be interpreted as changing the shape of the standard logistic function: a_j (the discrimination parameter) controls the slope of the curve; b_j (the difficulty parameter) is the ability level, it is the point on θ_i where the probability of a correct response is the average of c_j (min) and 1 (max), also where the slope is maximized; c_j (the guessing parameter) is the asymptotic minimum of this function, which accounts for the effects of guessing on the probability of a correct response for a multi-choice item. Equation 4 is often referred to as the three-parameter logistic (3PL) model since it has three parameters describing the characteristics of items. We refer the readers to [5,6,14] for more background and details on IRT.

3.3   Multi-dimensional IRT Using Model Responses

Traditional IRT is proposed to model the human responses to several hundred items. However, datasets used in machine learning, especially deep neural networks, often consist of hundreds of thousands of samples or even more. It is costly to collect human responses for large datasets, and more importantly, human responses are not distinguishable enough to estimate the sample difficulties since samples in machine learning datasets are usually straightforward for humans. Lalor *et al.* [34,35] empirically shows on two NLP tasks that IRT models can be fit using machine responses by comparing item parameters learned from the human responses and the responses from an artificial crowd of thousands of machine learning models.

Similarly, we propose to fit IRT models with accumulated model responses (*i.e.*, the predictions of model snapshots) from the training process. Considering the compositional nature of visual reasoning, we propose a multi-dimensional IRT (mIRT) model to estimate the concept difficulty and model competence (corresponding to the subject ability in original IRT), from which the question difficulty can be further calculated.

Formally, we have C concepts, M model snapshots saved from all time steps, and N questions. Let $\Theta = \{\theta_{ic}\}_{i=1..M}^{c=1...C}$, where θ_{ic} is the i-th snapshot's competence on the c-th concept, and $B = \{b_c\}^{c=1...C}$, where b_c is the difficulty of the c-th concept, $Q = \{q_{jc}\}_{j=1...N}^{c=1...C}$, where q_{jc} is the number of the c-th concept in the j-th question and g_j is the probability of guessing the correct answer to the j-th question, $Z = \{z_{ij}\}_{i=1...M}^{j=1...N}$, where $z_{ij} \in \{0,1\}$ be the response of the i-th snapshot to the j-th question (1 if the model answers the question correctly and 0 otherwise). The probability that the snapshot i can correctly recognize the concept c is formulated by a logistic function:

$$p_{ic}(\theta_{ic}, b_c) = \frac{1}{1 + e^{-(\theta_{ic} - b_c)}}. \tag{5}$$

Then the probability that the snapshot i answers the question j correctly is calculated as:

$$p(z_{ij} = 1 | \theta_i, B) = g_j + (1 - g_j) \prod_{c=1}^{C} p_{ic}^{q_{jc}}. \tag{6}$$

The probability that the snapshot i answers the question j incorrectly is:

$$p(z_{ij} = 0|\theta_i, B) = 1 - p(z_{ij} = 1|\theta_i, B).\tag{7}$$

The total data likelihood is:

$$p(\mathcal{Z}|\Theta, B) = \prod_{i=1}^{M}\prod_{j=1}^{N} p(z_{ij}|\theta_i, B).\tag{8}$$

This formulation is also referred to as conjunctive multi-dimensional IRT [48,49].

3.4   Variational Bayesian Inference for mIRT

The goal of fitting an IRT model on observed responses is to estimate the latent subject abilities and item parameters. In traditional IRT, the item parameters are usually estimated by Marginal Maximum Likelihood (MML) via an Expectation-Maximization (EM) algorithm [9], where the subject ability parameters are randomly sampled from a normal distribution and marginalized out. Once the item parameters are estimated, the subject abilities are scored by maximum a posterior (MAP) estimation based on their responses to items. However, the EM algorithm is not computational efficient on large datasets. One feasible way for scaling up is to perform variational Bayesian inference on IRT [35,43]. The posterior probability of the parameters in mIRT can be written as:

$$p(\Theta, B|\mathcal{Z}) = \frac{p(\mathcal{Z}|\Theta, B)p(\Theta)p(B)}{\int_{\Theta, B} p(\Theta, B, \mathcal{Z})},\tag{9}$$

where $p(\Theta), p(B)$ are the priors distribution of Θ and B. The integral over the parameter space in Eq. 9 is intractable. Therefore, we approximate it by a factorized variational distribution on top of an independence assumption of Θ and B:

$$q(\Theta, B) = \prod_{i=1,c=1}^{M,C} \pi_{ic}^{\theta}(\theta_{ic}) \prod_{c=1}^{C} \pi_c^b(b_c),\tag{10}$$

where π_{ic}^{θ} and π_c^b denote Gaussian distributions for model competences and concept difficulties, respectively. We adopt the Kullback-Leibler divergence (KL-divergence) to measure the distance of p from q, which is defined as:

$$D_{\mathrm{KL}}(q\|p) := \mathbb{E}_{q(\Theta, B)} \log \frac{q(\Theta, B)}{p(\Theta, B|\mathcal{Z})},\tag{11}$$

where $p(\Theta, B|\mathcal{Z})$ is still intractable. We can further decompose the KL-divergence as:

$$D_{\mathrm{KL}}(q\|p) = \mathbb{E}_{q(\Theta, B)}\left[\log \frac{q(\Theta, B)}{p(\Theta, B, \mathcal{Z})} + \log p(\mathcal{Z})\right].\tag{12}$$

In other words, we also have:

$$\log p(\mathcal{Z}) = D_{\mathrm{KL}}(q\|p) - \mathbb{E}_{q(\Theta,B)} \log \frac{q(\Theta,B)}{p(\Theta,B,\mathcal{Z})} \tag{13}$$

$$= D_{\mathrm{KL}}(q\|p) + \mathcal{L}(q). \tag{14}$$

As the log evidence $\log p(\mathcal{Z})$ is fixed with respect to q, maximizing the final term $\mathcal{L}(q)$ minimizes the KL divergence of q from p. And since $q(\Theta,B)$ is a parametric distribution we can sample from, we can use Monte Carlo sampling to estimate this quantity. Since the KL-divergence is non-negative, $\mathcal{L}(q)$ is an evidence lower bound (ELBO) of $\log p(\mathcal{Z})$. By maximizing the ELBO with an Adam optimizer [31] in Pyro [8], we can estimate the parameters in mIRT.

3.5   Training Samples Selection Strategy

The proposed model can estimate the question difficulty for the current model competence without looking at the ground-truth images and answers. It facilitates the active selection for future training samples. More specifically, we can easily calculate the probability that the model answers a given question correctly from Eq. 5 and Eq. 6 (without guessing) using estimated Θ and b. This probability serves as an indicator of the question difficulty for the learner in each stage. The higher the probability, the easier the question. To select appropriate training samples, we rank the questions and filter out the hardest questions by setting a probability lower bound (LB) and the easiest questions by a probability upper bound (UB). Algorithm 1 summarizes the overall training process. We will discuss the influence of LB and UB on the learning process in Sect. 4.5.

Algorithm 1. Competence-aware Curriculum Learning

   Initialization: the training set $\mathcal{D} = \{(I_j, Q_j, A_j)\}_{j=1}^{N}$, concept difficulty $B^{(0)}$, model competence $\Theta^{(0)}$, concept learner $\phi^{(0)}$, accumulated responses $\mathcal{Z} = \{\}$
   for $t = 1$ to T **do**
      $\Theta^{(t)}, B^{(t)} = \arg\min_{\Theta,B} \mathcal{L}(q; \Theta^{(t-1)}, B^{(t-1)}, \mathcal{Z})$
      $\mathcal{D}^{(t)} = \{(I, Q, A) : \mathrm{LB} \leq p(Q; \Theta^{(t)}, B^{(t)}) \leq \mathrm{UB}\}$
      $\phi^{(t)}, \mathcal{Z}^{(t)} = \mathrm{Train}(\phi^{(t-1)}, \mathcal{D}^{(t)})$
      $\mathcal{Z} = \mathcal{Z} \cup \mathcal{Z}^{(t)}$
   end for

---

4   Experiments

4.1   Experimental Setup

Dataset. We evaluate the proposed method on the CLEVR dataset [29], which consists of a training set of 70 k images and ~700 k questions, and a validation set of 15 k images and ~150 k questions. The proposed model selects questions

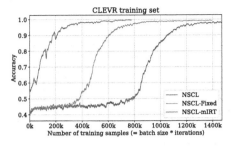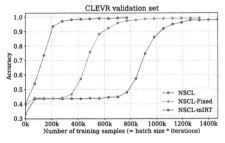

Fig. 3. The learning curves of different model variants on the CLEVR dataset.

from the training set during learning, and we evaluate our model on the entire validation set.

Models. To analyze the performance of the proposed approach, We conduct experiments by comparing with several model variants:

- **FiLM-LBA:** the best model from [42].
- **NSCL:** the neural-symbolic concept learner [41] without using any curriculum. Questions are randomly sampled from the training set.
- **NSCL-Fixed:** NSCL following a manually-designed discretized curriculum.
- **NSCL-mIRT:** NSCL following a continuous curriculum built by the proposed mIRT estimator.

Please refer to the supplementary materials for detailed model settings and learning techniques during training.

4.2 Training Process and Model Performance

Figure 3 shows the accuracies of the model variants at different timesteps on the training set (left) and validation set (right). Notably, the proposed NSCL-mIRT converges almost 2 times faster than NSCL-Fixed and 3 times faster than NSCL (*i.e.*, 400k v.s. 800k v.s. 1200k). Although NSCL-mIRT spends extra time to estimate the parameters of the mIRT model, such time cost is negligible compared to other time spent in training (less than 1%). From Table 1, we can see that NSCL-mIRT consistently outperforms FiLM-LBA at various iterations, which demonstrates the preeminence of mIRT in building an adaptive curriculum.

Besides, NSCL-mIRT consumes less than 300k unique questions for training when it converges. It indicates that NSCL-mIRT saves about 60% of the training data, which largely eases the data redundancy problems. It provides a promising direction for designing a data-efficient curriculum and helping current data-hungry deep learning models save time and money cost during data annotation and model training.

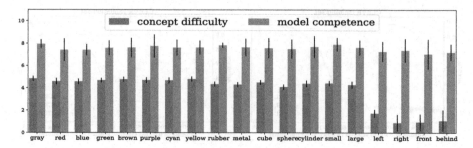

Fig. 4. The estimated concept difficulty and model competence at the final iteration.

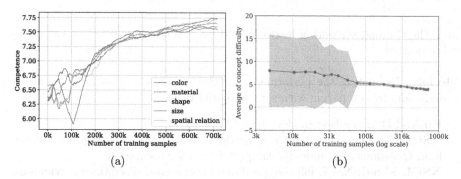

Fig. 5. (a) The estimated model competence at various iterations for different attributes. The value for each attribute type is averaged from the visual concept it contains. (b) The estimated concept difficulty at various iterations. The shaded area represents the variance of the estimations.

Moreover, NSCL-mIRT obtains even higher accuracy than NSCL and NSCL-Fixed. This indicates that the adaptive curriculum built by the multi-dimensional IRT model not only remarkably increases the speed of convergence and reduces the data consumption during the training process, but also leads to better performance, which also verifies the hypothesis made by Bengio *et al.* [7].

4.3 Multi-dimensional IRT

The estimated concept difficulty and model competence after converging is shown in Fig. 4 for studying the performance of the mIRT model. Several critical observations are: (1) The spatial relations (*i.e.*, left/right/front/behind) are the easiest concepts. It satisfies our intuition since the model only needs to exploit the object positions to determine their spatial relations without dealing with appearance. The spatial relations are learned during the late stages since they appear more frequently in complex questions to connect multiple concepts. (2) Colors are the most difficult concepts. The model needs to capture the subtle differences in the appearance of objects to distinguish eight different colors. (3) The model competence scores surpass the concept difficulty scores for all the

Table 1. The VQA accuracy of different models on the CLEVR validation set at various iterations. NSCL and NSCL-Fixed continue to improve with longer training steps, which is not shown for space limit.

Models	70k	140k	280k	420k	630k	700k
FiLM-LBA [42]	51.2	**76.2**	92.9	94.8	95.2	97.3
NSCL	43.3	43.4	43.3	43.4	44.5	44.7
NSCL-Fixed	44.1	43.9	44.0	57.2	92.4	95.9
NSCL-mIRT	**53.9**	73.4	**97.1**	**98.5**	**98.9**	**99.3**

Table 2. The accuracy of the visual attributes of different models. Please refer to the supplementary materials for detailed performance on each visual concept (*i.e.*, "gray" and "red" in color attribute).

Model	Overall	Color	Material	Shape	Size
IEP [29]	90.6	91.0	90.0	89.9	90.6
MAC [25]	95.9	98.0	91.4	94.4	94.2
NSCL-Fixed [41]	98.7	99.0	98.7	98.1	99.1
NSCL-mIRT	**99.5**	**99.5**	**99.7**	**99.4**	**99.6**

Table 3. Comparisons of the VQA accuracy on the CLEVR validation set with other models.

Model	Overall	Count	Cmp Num.	Exist	Query Attr.	Cmp Attr.
Human	92.6	86.7	86.4	96.6	95.0	96.0
IEP [29]	96.9	92.7	98.7	97.1	98.1	98.9
FiLM [45]	97.6	94.5	93.8	99.2	99.2	99.0
MAC [25]	98.9	97.2	99.4	99.5	99.3	99.5
NSCL [41]	98.9	98.2	99.0	98.8	99.3	99.1
NS-VQA [58]	**99.8**	**99.7**	**99.9**	**99.9**	**99.8**	**99.8**
NSCL-mIRT	99.5	98.9	99.0	99.7	99.7	99.6

Table 4. The VQA accuracy on CLEVR validation set with different LBs and UBs in the question selection strategy. Both LB and UB are in log scale.

(LB,UB)	70 k	140 k	210 k	280 k	560 k	770 k
(−10, 0)	44.39	52.01	63.04	73.5	97.93	99.01
(−5, 0)	53.75	69.55	82.44	95.31	98.92	99.27
(−3, 0)	51.38	55.97	58.33	65.11	69.57	70.01
(−5, −0.5)	42.06	52.67	80.46	95.54	98.41	99.06
(−5, −0.75)	**53.91**	**73.42**	**93.6**	**97.07**	99.04	**99.50**
(−5, −1)	44.57	63.65	82.95	94.38	**99.15**	99.48

concepts. This result corresponds to the nearly perfect accuracy (>99%) on all questions and concepts.

Figure 5(a) shows the estimation of the model competence for each attribute type at various iterations. We can observe that model competence consistently increases throughout the training. Figure 5(b) shows the estimations of the concept difficulty at different learning steps. As the training progresses, the estimations become more stable with smaller variance since more model responses are accumulated.

4.4   Concept Learner

We apply the count-based concept evaluation metric proposed in [41] to measure the performance of the concept learner, which evaluates the visual concepts on synthetic questions with a single concept such as "How many *red* objects are there?" Table 2 presents the results by comparing with several state-of-the-art methods, which includes methods based on neural module network with programs (IEP [29]) and neural attentions without programs (MAC [24]). Our model achieves nearly perfect performance across visual concepts and outperforms all other approaches. This means the model can learn visual concepts better with an adaptive curriculum. Our model can also be applied to the VQA. Table 3 summarizes the VQA accuracy on the CLEVR validation split. Our approach achieves comparable performance with state-of-the-art methods.

4.5   Question Selection Strategy

The question selection strategy is controlled by two hyper-parameters: the lower bound (LB) and upper bound (UB). We conduct experiments by learning with different LBs and UBs, and Table 4 shows the VQA accuracy at various iterations. It reveals that the proper lower bound can effectively filter out too hard questions and accelerate the learning at the early stage of the training, as shown in the first three rows. Similarly, a proper upper bound helps to filter out too easy questions at the late stage of the training when the model has learned most concepts. Please refer to the supplementary material for the visualization of selected questions at various iterations.

5   Conclusions and Discussions

We propose a competence-aware curriculum for visual concepts learning via question answering. We design a multi-dimensional IRT model to estimate concept difficulty and model competence at each training step from the accumulated model responses generated by different model snapshots. The estimated concept difficulty and model competence are further used to build an adaptive curriculum for the visual concept learner. Experiments on the CLEVR dataset show that the concept learner with the proposed competence-aware curriculum converges three times faster and consumes only 40% of the training data while achieving similar or even higher accuracy compared with other state-of-the-art models.

In the future, our work can be potentially applied to *real-world images* like GQA [26] and VQA-v2 [19] datasets, by explicitly modeling the relationship among visual concepts. However, there are still unsolved challenges for real-world images. Specifically, compared with synthetic images in CLEVR, real-world images have a much larger vocabulary of visual concepts. For example, as shown in [1], there are over 2,000 visual concepts in MSCOCO images. Usually, these concepts are automatically mined from image captions and scene graphs. Thus some of them are highly correlated like "huge" and "large", and some of them are very subjective like "busy" and "calm". Such a large and noisy vocabulary of visual concepts is challenging for the mIRT model since current visual concepts are assumed to be independent. It also requires a much longer time to converge when maximizing the ELBO to fit the mIRT model with more concepts. A potential solution is to consider the hierarchical structure of visual concept space and correlations among the concepts and incorporate common-sense knowledge to handle subjective concepts.

More importantly, the competence-aware curriculum can be adapted to other domains that possess compositional structures such as natural language processing. Specifically, in neural machine translation task [4,53], mIRT can be used to model the difficulty and competence of translating different words/phrases and build a curriculum to increase learning speed and data efficiency. mIRT can also be used in the task of semantic parsing [12,36,37] that transforms natural language sentences (*e.g.*, instructions or queries) into logic forms (*e.g.*, lambda-calculus or SQL). The difficulty and competence of different logic predicates can also be estimated by the mIRT model.

Acknowledgements. We thank Yixin Chen from UCLA for helpful discussions. This work reported herein is supported by ARO W911NF1810296, DARPA XAI N66001-17-2-4029, and ONR MURI N00014-16-1-2007.

References

1. Anderson, P., et al.: Bottom-up and top-down attention for image captioning and visual question answering. In: Proceedings of the IEEE Conference on Computer Vision and Pattern Recognition, pp. 6077–6086 (2018)
2. Andreas, J., Rohrbach, M., Darrell, T., Klein, D.: Neural module networks. In: Conference on Computer Vision and Pattern Recognition (CVPR), pp. 39–48 (2015)
3. Andreas, J., Rohrbach, M., Darrell, T., Klein, D.: Learning to compose neural networks for question answering. In: Proceedings of the 2016 Conference of the North American Chapter of the Association for Computational Linguistics: Human Language Technologies (2016)
4. Bahdanau, D., Cho, K., Bengio, Y.: Neural machine translation by jointly learning to align and translate. In: ICLR (2015)
5. Baker, F.B.: The basics of item response theory. In: ERIC (2001)
6. Baker, F.B., Kim, S.H.: Item Response Theory: Parameter Estimation Techniques. CRC Press, Boca Raton (2004)
7. Bengio, Y., Louradour, J., Collobert, R., Weston, J.: Curriculum learning. In: International Conference on Machine Learning (ICML) (2009)
8. Bingham, E., et al.: Pyro: deep universal probabilistic programming. J. Mach. Learn. Res. **20**, 1–6 (2018)
9. Bock, R.D., Aitkin, M.: Marginal maximum likelihood estimation of item parameters: application of an EM algorithm. Psychometrika **46**, 443–459 (1981)
10. Chrupała, G., Kádár, A., Alishahi, A.: Learning language through pictures. In: Association for Computational Linguistics (ACL) (2015)
11. Dasgupta, S., Hsu, D., Poulis, S., Zhu, X.: Teaching a black-box learner. In: ICML (2019)
12. Dong, L., Lapata, M.: Language to logical form with neural attention. In: ACL (2016)
13. Elman, J.L.: Learning and development in neural networks: the importance of starting small. Cognition **48**, 71–99 (1993)
14. Embretson, S.E., Reise, S.P.: Item Response Theory. Psychology Press, New York (2013)
15. Fan, Y., et al.: Learning to teach. In: ICLR (2018)
16. Fazly, A., Alishahi, A., Stevenson, S.: A probabilistic computational model of cross-situational word learning. In: Annual Meeting of the Cognitive Science Society (CogSci) (2010)
17. Gan, C., Li, Y., Li, H., Sun, C., Gong, B.: VQS: linking segmentations to questions and answers for supervised attention in VQA and question-focused semantic segmentation. In: ICCV, pp. 1811–1820 (2017)
18. Gauthier, J., Levy, R., Tenenbaum, J.B.: Word learning and the acquisition of syntactic-semantic over hypotheses. In: Annual Meeting of the Cognitive Science Society (CogSci) (2018)

19. Goyal, Y., Khot, T., Summers-Stay, D., Batra, D., Parikh, D.: Making the V in VQA matter: elevating the role of image understanding in visual question answering. In: Proceedings of the IEEE Conference on Computer Vision and Pattern Recognition, pp. 6904–6913 (2017)
20. Graves, A., Bellemare, M.G., Menick, J., Munos, R., Kavukcuoglu, K.: Automated curriculum learning for neural networks. In: International Conference on Machine Learning (ICML) (2017)
21. Guo, S., et al.: CurriculumNet: weakly supervised learning from large-scale web images. arXiv preprint arXiv:1808.01097 (2018)
22. He, K., Gkioxari, G., Dollár, P., Girshick, R.: Mask R-CNN. In: Conference on Computer Vision and Pattern Recognition (CVPR) (2017)
23. He, K., Zhang, X., Ren, S., Sun, J.: Deep residual learning for image recognition. In: Conference on Computer Vision and Pattern Recognition (CVPR) (2016)
24. Hu, R., Andreas, J., Rohrbach, M., Darrell, T., Saenko, K.: Learning to reason: end-to-end module networks for visual question answering. In: International Conference on Computer Vision (ICCV), pp. 804–813 (2017)
25. Hudson, D.A., Manning, C.D.: Compositional attention networks for machine reasoning. In: International Conference on Learning Representations (ICLR) (2018)
26. Hudson, D.A., Manning, C.D.: GQA: a new dataset for real-world visual reasoning and compositional question answering. In: CVPR (2019)
27. Jiang, L., et al.: Self-paced learning with diversity. In: NIPS (2014)
28. Jiang, L., et al.: Self-paced curriculum learning. In: AAAI (2015)
29. Johnson, J., Hariharan, B., van der Maaten, L., Fei-Fei, L., Lawrence Zitnick, C., Girshick, R.: CLEVR: a diagnostic dataset for compositional language and elementary visual reasoning. In: Conference on Computer Vision and Pattern Recognition (CVPR) (2017)
30. Johnson, J., et al.: Inferring and executing programs for visual reasoning. In: International Conference on Computer Vision (ICCV) (2017)
31. Kingma, D., Ba, J.: Adam: a method for stochastic optimization. In: International Conference on Learning Representations (ICLR) (2015)
32. Krueger, K.A., Dayan, P.: Flexible shaping: how learning in small steps helps. Cognition 110, 380–394 (2009)
33. Kumar, M.P., et al.: Self-paced learning for latent variable models. In: NIPS (2010)
34. Lalor, J.P., Wu, H., Yu, H.: Building an evaluation scale using item response theory. In: Conference on Empirical Methods in Natural Language Processing (EMNLP) (2016)
35. Lalor, J.P., Wu, H., Yu, H.: Learning latent parameters without human response patterns: item response theory with artificial crowds. In: Conference on Empirical Methods in Natural Language Processing (EMNLP) (2019)
36. Liang, C., Berant, J., Le, Q., Forbus, K.D., Lao, N.: Neural symbolic machines: learning semantic parsers on freebase with weak supervision. In: ACL (2016)
37. Liang, C., Norouzi, M., Berant, J., Le, Q., Lao, N.: Memory augmented policy optimization for program synthesis and semantic parsing. In: NIPS (2018)
38. Liu, W., et al.: Iterative machine teaching. In: Proceedings of the 34th International Conference on Machine Learning, vol. 70, pp. 2149–2158. JMLR.org (2017)
39. Malinowski, M., Fritz, M.: A multi-world approach to question answering about real-world scenes based on uncertain input. In: Advances in Neural Information Processing Systems (NeurIPS) (2014)
40. Mansouri, F., Chen, Y., Vartanian, A., Zhu, X., Singla, A.: Preference-based batch and sequential teaching: towards a unified view of models. In: NeurIPS (2019)

41. Mao, J., Gan, C., Kohli, P., Tenenbaum, J.B., Wu, J.: The neuro-symbolic concept learner: interpreting scenes, words, and sentences from natural supervision. In: International Conference on Learning Representations (ICLR) (2019)

42. Misra, I., Girshick, R.B., Fergus, R., Hebert, M., Gupta, A., van der Maaten, L.: Learning by asking questions. In: Conference on Computer Vision and Pattern Recognition (CVPR) (2017)

43. Natesan, P., Nandakumar, R., Minka, T., Rubright, J.D.: Bayesian prior choice in IRT estimation using MCMC and variational Bayes. Front. Psychol. **7**, 1–11 (2016)

44. Pentina, A., Sharmanska, V., Lampert, C.H.: Curriculum learning of multiple tasks. In: Conference on Computer Vision and Pattern Recognition (CVPR), pp. 5492–5500 (2014)

45. Perez, E., Strub, F., de Vries, H., Dumoulin, V., Courville, A.C.: FiLM: visual reasoning with a general conditioning layer. In: AAAI Conference on Artificial Intelligence (AAAI) (2017)

46. Peterson, G.B.: A day of great illumination: B. F. Skinner's discovery of shaping. J. Exp. Anal. Behav. **82**, 317–328 (2004)

47. Platanios, E.A., Stretcu, O., Neubig, G., Póczos, B., Mitchell, T.M.: Competence-based curriculum learning for neural machine translation. In: North American Chapter of the Association for Computational Linguistics (NAACL-HLT) (2019)

48. Reckase, M.D.: The difficulty of test items that measure more than one ability. Appl. Psychol. Meas. **13**, 113–127 (1985)

49. Reckase, M.D.: Multidimensional item response theory models. In: Multidimensional Item Response Theory (2009)

50. Sachan, M., et al.: Easy questions first? A case study on curriculum learning for question answering. In: ACL (2016)

51. Skinner, B.F.: Reinforcement today. Am. Psychol. **47**, 1318–1328 (1958)

52. Spitkovsky, V.I., Alshawi, H., Jurafsky, D.: From baby steps to leapfrog: How less is more in unsupervised dependency parsing. In: Human Language Technologies: The 2010 Annual Conference of the North American Chapter of the Association for Computational Linguistics, pp. 751–759. Association for Computational Linguistics (2010)

53. Sutskever, I., Vinyals, O., Le, Q.V.: Sequence to sequence learning with neural networks. In: Advances in Neural Information Processing Systems, pp. 3104–3112 (2014)

54. Tsvetkov, Y., Faruqui, M., Ling, W., MacWhinney, B., Dyer, C.: Learning the curriculum with Bayesian optimization for task-specific word representation learning. In: ACL (2016)

55. Williams, R.J.: Simple statistical gradient-following algorithms for connectionist reinforcement learning. Mach. Learn. **8**, 229–256 (1992)

56. Wu, L., et al.: Learning to teach with dynamic loss functions. In: NeurIPS (2018)

57. Yi, K., et al.: CLEVRER: collision events for video representation and reasoning. In: ICLR (2020)

58. Yi, K., Wu, J., Gan, C., Torralba, A., Kohli, P., Tenenbaum, J.: Neural-symbolic VQA: disentangling reasoning from vision and language understanding. In: Advances in Neural Information Processing Systems (2018)

59. Zhu, X.: Machine teaching: An inverse problem to machine learning and an approach toward optimal education. In: Twenty-Ninth AAAI Conference on Artificial Intelligence (2015)

60. Zhu, X., Singla, A., Zilles, S., Rafferty, A.N.: An overview of machine teaching. arXiv preprint arXiv:1801.05927 (2018)

Multitask Learning Strengthens Adversarial Robustness

Chengzhi Mao[✉], Amogh Gupta, Vikram Nitin, Baishakhi Ray, Shuran Song, Junfeng Yang, and Carl Vondrick

Columbia University, New York, NY, USA
{mcz,rayb,shurans,junfeng,vondrick}@cs.columbia.edu,
{ag4202,vikram.nitin}@columbia.edu

Abstract. Although deep networks achieve strong accuracy on a range of computer vision benchmarks, they remain vulnerable to adversarial attacks, where imperceptible input perturbations fool the network. We present both theoretical and empirical analyses that connect the adversarial robustness of a model to the number of tasks that it is trained on. Experiments on two datasets show that attack difficulty increases as the number of target tasks increase. Moreover, our results suggest that when models are trained on multiple tasks at once, they become more robust to adversarial attacks on individual tasks. While adversarial defense remains an open challenge, our results suggest that deep networks are vulnerable partly because they are trained on too few tasks.

Keywords: Multitask learning · Adversarial robustness

1 Introduction

Deep networks obtain high performance in many computer vision tasks [18,19, 31,57], yet they remain brittle to adversarial examples. A large body of work has demonstrated that images with human-imperceptible noise [3,11,34,39] can be crafted to cause the model to mispredict. This pervasiveness of adversarial examples exposes key limitations of deep networks, and hampers their deployment in safety-critical applications, such as autonomous driving.

A growing body of research has been dedicated to answering what causes deep networks to be fragile to adversarial examples and how to improve robustness [7,13,21,34,35,41,46,48,49,54]. The investigations center around two factors: the training data and the optimization procedure. For instance, more training data – both labeled and unlabeled – improves robustness [42,51]. It has been theoretically shown that decreasing the input dimensionality of data improves

---

A. Gupta and V. Nitin—Equal Contribution.

Electronic supplementary material The online version of this chapter (https:// doi.org/10.1007/978-3-030-58536-5_10) contains supplementary material, which is available to authorized users.

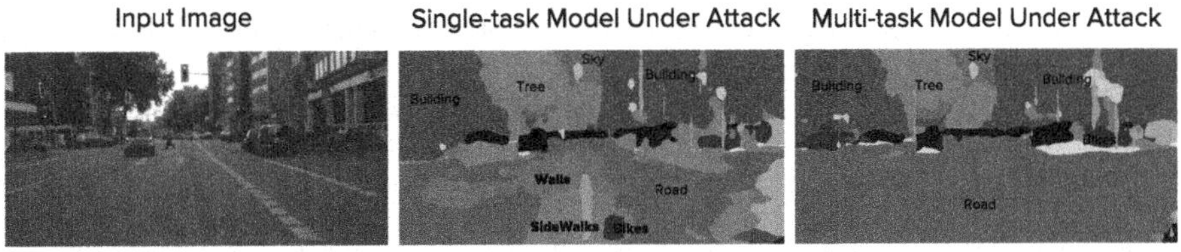

Input Image   Single-task Model Under Attack   Multi-task Model Under Attack

Fig. 1. We find that multitask models are more robust against adversarial attacks. Training a model to solve multiple tasks improves the robustness when one task is attacked. The middle and right column show predictions for single-task and multitask models when one task is adversarially attacked.

robustness [46]. Adversarial training [34] improves robustness by dynamically augmenting the training data using generated adversarial examples. Similarly, optimization procedures that regularize the learning specifically with robustness losses have been proposed [7,55]. This body of work suggests that the fragility of deep networks may stem from the training data and optimization procedure.

In this paper, we pursue a new line of investigation: how learning on multiple tasks affects adversarial robustness. While previous work shows that multitask learning can improve the performance of specific tasks [4,47], we show that it increases robustness too. See Fig. 1. Unlike prior work that trades off performance between natural and adversarial examples [50], our work improves adversarial robustness while also maintaining performance on natural examples.

Using the first order vulnerability of neural networks [46], we theoretically show that increasing output dimensionality – treating each output dimension as an individual task – improves the robustness of the entire model. Perturbations needed to attack multiple output dimensions cancel each other out. We formally quantify and upper bound how much robustness a multitask model gains against a multitask attack with increasing output dimensionality.

We further empirically show that multitask learning improves the model robustness for two classes of attack: both when a single task is attacked or several tasks are simultaneously attacked. We experiment with up to 11 vision tasks on two natural image datasets, Cityscapes [8] and Taskonomy [59]. When all tasks are under attack, multitask learning increases segmentation robustness by up to 7 points and reduces the error of other tasks up to 60% over baselines. We compare the robustness of a model trained for a main task with and without an auxiliary task. Results show that, when the main task is under attack, multitask learning improves segmentation overlap by up to 6 points and reduces the error of the other tasks by up to 23%. Moreover, multitask training is a complementary defense to adversarial training, and it improves both the clean and adversarial performance of the state-of-the-art, adversarially trained, single-task models. Code is available at https://github.com/columbia/MTRobust.

Overall, our experiments show that multitask learning improves adversarial robustness while maintaining most of the the state-of-the-art single-task model performance. While defending against adversarial attacks remains an open problem, our results suggest that current deep networks are vulnerable partly because they are trained for too few tasks.

2 Related Work

We briefly review related work in multitask learning and adversarial attacks.

Multitask Learning: Multitask learning [4,10,15,25,47] aims to solve several tasks at once, and has been used to learn better models for semantic segmentation [28], depth estimation [52], key-point prediction [23], and object detection [29]. It is hypothesized that multitask learning improves the performance of select tasks by introducing a knowledge-based inductive bias [4]. However, multi-objective functions are hard to optimize, where researchers design architectures [20,24,30,33,37] and optimization procedures [5,12,44,58] for learning better multitask models. Our work complements this body of work by linking multitask learning to adversarial robustness.

Adversarial Attacks: Current adversarial attacks manipulate the input [2,6,9, 11,36,45,48,53] to fool target models. While attacking single output models [17, 26] is straightforward, Arnab et al. [2] empirically shows the inherent hardness of attacking segmentation models with dense output. Theoretical insight of this robustness gain, however, is missing in the literature. While past theoretical work showed the hardness of multi-objective optimization [16,43], we leverage this motivation and prove that multitask models are robust when tasks are simultaneously attacked. Our work contributes both theoretical and empirical insights on adversarial attacks through the lens of multitask learning.

Adversarial Robustness: Adversarial training improves models' robustness against attacks, where the training data is augmented using adversarial samples [17,34]. In combination with adversarial training, later works [21,35,54,60] achieve improved robustness by regularizing the feature representations with additional loss, which can be viewed as adding additional tasks. Despite the improvement of robustness, adversarially trained models lose significant accuracy on clean (unperturbed) examples [34,50,60]. Moreover, generating adversarial samples slows down training several-fold, which makes it hard to scale adversarial training to large datasets.

Past work revealed that model robustness is strongly connected to the gradient of the input, where models' robustness is improved by regularizing the gradient norm [7,40,55]. Parseval [7] regularizes the Lipschitz constant—the maximum norm of gradient—of the neural network to produce a robust classifier, but it fails in the presence of batch-normalization layers. [40] decreases the input gradients norm. These methods can improve the model's robustness without compromising clean accuracy. Simon-Gabriel et al. [46] conducted a theoretical analysis of the vulnerability of neural network classifiers, and connected gradient norm and adversarial robustness. Our method enhances robustness by training a multitask model, which complements both adversarial training [34,60] and existing regularization methods [7,38,40,55].

3 Adversarial Setting

The goal of an adversary is to "fool" the target model by adding human-imperceptible perturbations to its input. We focus on untargeted attacks, which are harder to defend against than targeted attacks [14]. We classify adversarial attacks for a multitask prediction model into two categories: adversarial attacks that fool more than one task at once (multitask attacks), and adversarial attacks that fool a specific task (single-task attacks).

3.1 Multitask Learning Objective

Notations. Let $\mathbf{x}$ denote an input example, and $\mathbf{y}_c$ denote the corresponding ground-truth label for task c. In this work, we focus on multitask learning with shared parameters [24,27,30,32,47], where all the tasks share the same "backbone network" $F(\cdot)$ as a feature extractor with task-specific decoder networks $D_c(\cdot)$. The task-specific loss is formulated as:

$$\mathcal{L}_c(\mathbf{x}, \mathbf{y}_c) = \ell(D_c(F(\mathbf{x})), \mathbf{y}_c), \tag{1}$$

where ℓ is any appropriate loss function. For simplicity, we denote $(\mathbf{y}_1, ..., \mathbf{y}_M)$ as $\overline{\mathbf{y}}$, where M is the number of tasks. The total loss for multitask learning is a weighted sum of all the individual losses:

$$\mathcal{L}_{all}(\mathbf{x}, \overline{\mathbf{y}}) = \sum_{c=1}^{M} \lambda_c \mathcal{L}_c(\mathbf{x}, \mathbf{y}_c) \tag{2}$$

For the simplicity of theoretical analysis, we set $\lambda_c = \frac{1}{M}$ for all $c = 1, ..., M$, such that $\sum_{c=1}^{M} \lambda_c = 1$. In our experiments on real-world datasets, we will adjust the λ_c accordingly, following standard practice [27,47].

3.2 Adversarial Multitask Attack Objective

The goal of a multitask attack is to change multiple output predictions together. For example, to fool an autonomous driving model, the attacker may need to deceive both the object classification and depth estimation tasks. Moreover, if we regard each output pixel of a semantic segmentation task as an individual task, adversarial attacks on segmentation models need to flip multiple output pixels, so we consider them as multitask attacks. We also consider other dense output tasks as a variant of multitask, such as depth estimation, keypoints estimation, and texture prediction.

In general, given an input example $\mathbf{x}$, the objective function for multitask attacks against models with multiple outputs is the following:

$$\underset{\mathbf{x}_{adv}}{\arg\max} \, \mathcal{L}_{all}(\mathbf{x}_{adv}, \overline{\mathbf{y}}) \quad \text{s.t.} \quad ||\mathbf{x}_{adv} - \mathbf{x}||_p \le r \tag{3}$$

where the attacker aims to maximize the joint loss function by finding small perturbations within a p-norm bounded distance r of the input example. Intuitively, a multitask attack is not easy to perform because the attacker needs to optimize the perturbation to fool each individual task simultaneously. The robustness of the overall model can be a useful property - for instance, consider an autonomous-driving model trained for both classification and depth estimation. If either of the two tasks is attacked, the other can still be relied on to prevent accidents.

3.3   Adversarial Single-Task Attack Objective

In contrast to a multitask attack, a single-task attack focuses on a selected target task. Compared with attacking all tasks at once, this type of attack is more effective for the target task, since the perturbation can be designed solely for this task without being limited by other considerations. It is another realistic type of attack because some tasks are more important than the others for the attacker. For example, if the attacker successfully subverts the color prediction for a traffic light, the attacker may cause an accident even if the other tasks predict correctly. The objective function for single-task attack is formulated as:

$$\operatorname*{argmax}_{\mathbf{x}_{adv}} \mathcal{L}_c(\mathbf{x}_{adv}, \mathbf{y}_c), \text{s.t.} ||\mathbf{x}_{adv} - \mathbf{x}||_p \leq r \tag{4}$$

For any given task, this single-task attack is more effective than jointly attacking the other tasks. We will empirically demonstrate that multitask learning also improves model robustness against this type of attack in Sect. 5.

4   Theoretical Analysis

We present theoretical insights into the robustness of multitask models. A prevalent formulation of multitask learning work uses shared backbone network with task-specific branches [27,30,32]. We denote the multitask predictor as F and each individual task predictor as F_c. Prior work [46] showed that the norm of gradients captures the vulnerability of the model. We thus measure the multitask models' vulnerability with the same metric. Since we are working with deep networks, we assume all the functions here are differentiable. Details of all proofs are in the supplementary material.

Definition 1. *Given classifier F, input $\mathbf{x}$, output target $\mathbf{y}$, and loss $\mathcal{L}(\mathbf{x}, \mathbf{y}) = \ell(F(\mathbf{x}), \mathbf{y})$, the feasible adversarial examples lie in a p-norm bounded ball with radius r, $B(\mathbf{x}, r) := \{\mathbf{x}_{adv}, ||\mathbf{x}_{adv} - \mathbf{x}|| < r\}$. Then adversarial vulnerability of a classifier over the whole dataset is*

$$\mathbb{E}_{\mathbf{x}}[\Delta\mathcal{L}(\mathbf{x}, \mathbf{y}, r)] = \mathbb{E}_{\mathbf{x}}[\max_{||\delta||_p < r} |\mathcal{L}(\mathbf{x}, \mathbf{y}) - \mathcal{L}(\mathbf{x} + \delta, \mathbf{y})|]$$

$\Delta\mathcal{L}$ captures the maximum change of the output loss from arbitrary input change δ inside the p-norm ball. Intuitively, a robust model should have smaller change of the loss given any perturbation of the input. Given the adversarial noise is imperceptible, i.e., $r \to 0$, we can approximate $\Delta\mathcal{L}$ with a first-order Taylor expansion [46].

Lemma 1. *For a given neural network F that predicts multiple tasks, the adversarial vulnerability is*

$$\mathbb{E}_{\mathbf{x}}[\Delta\mathcal{L}(\mathbf{x},\mathbf{y},r)] \approx \mathbb{E}_{\mathbf{x}}\left[||\partial_{\mathbf{x}}\mathcal{L}_{all}(\mathbf{x},\overline{\mathbf{y}})||_q\right] \cdot ||\delta||_p \propto \mathbb{E}_{\mathbf{x}}\left[||\partial_{\mathbf{x}}\mathcal{L}_{all}(\mathbf{x},\overline{\mathbf{y}})||_q\right]$$

where q is the dual norm of p, which satisfies $\frac{1}{p} + \frac{1}{q} = 1$ and $1 \leq p \leq \infty$. Without loss of generality, let $p = 2$ and $q = 2$. Note that from Eq. 2, we get $\mathcal{L}_{all}(\mathbf{x},\overline{\mathbf{y}}) = \sum_{c=1}^{M}\frac{1}{M}\mathcal{L}_c(\mathbf{x},\mathbf{y}_c)$. Thus we get the following equation:

$$\partial_{\mathbf{x}}\mathcal{L}_{all}(\mathbf{x},\overline{\mathbf{y}}) = \partial_{\mathbf{x}}\sum_{c=1}^{M}\frac{1}{M}\mathcal{L}_c(\mathbf{x},\mathbf{y}_c) = \frac{1}{M}\sum_{c=1}^{M}\partial_{\mathbf{x}}\mathcal{L}_c(\mathbf{x},\mathbf{y}_c) \tag{5}$$

We denote the gradient for task c as $\mathbf{r}_c$, i.e., $\mathbf{r}_c = \partial_{\mathbf{x}}\mathcal{L}_c(\mathbf{x},\mathbf{y}_c)$. We propose the following theory for robustness of different numbers of randomly selected tasks.

Theorem 1. *(**Adversarial Vulnerability of Model for Multiple Correlated Tasks**) If the selected output tasks are correlated with each other such that the covariance between the gradient of task i and task j is $\mathrm{Cov}(\mathbf{r}_i,\mathbf{r}_j)$, and the gradient for each task is i.i.d. with zero mean (because the model is converged), then adversarial vulnerability of the given model is proportional to*

$$\sqrt{\frac{1 + \frac{2}{M}\sum_{i=1}^{M}\sum_{j=1}^{i-1}\frac{\mathrm{Cov}(\mathbf{r}_i,\mathbf{r}_j)}{\mathrm{Cov}(\mathbf{r}_i,\mathbf{r}_i)}}{M}}$$

where M is the number of output tasks selected.

The idea is that when we select more tasks as attack targets, the gradients for each of the individual tasks on average cancels out with each other. We define the joint gradient vector $\mathbf{R}$ as follows:

$$\mathbf{R} = \partial_{\mathbf{x}}\mathcal{L}_{all}(\mathbf{x},\overline{\mathbf{y}}) = \frac{1}{M}\sum_{c=1}^{M}\partial_{\mathbf{x}}\mathcal{L}_c(\mathbf{x},\mathbf{y}_c)$$

The joint gradient is the sum of gradients from each individual task. We then obtain the expectation of the L_2 norm of the joint gradient:

$$\mathbb{E}(||\mathbf{R}||_2^2) = \mathbb{E}\left[||\frac{1}{M}\sum_{i=1}^{M}\mathbf{r}_i||_2^2\right] = \frac{1}{M^2}\mathbb{E}\left[\sum_{i=1}^{M}||\mathbf{r}_i||^2 + 2\sum_{i=1}^{M}\sum_{j=1}^{i}\mathbf{r}_i\mathbf{r}_j\right]$$

$$= \frac{1}{M^2}\left(\sum_{i=1}^{M}\mathbb{E}[\mathrm{Cov}(\mathbf{r}_i,\mathbf{r}_i)] + 2\sum_{i=1}^{M}\sum_{j=1}^{i}\mathbb{E}[\mathrm{Cov}(\mathbf{r}_i,\mathbf{r}_j)]\right)$$

The last equation holds due to the $\mathbf{0}$ mean assumption of the gradient. For further details of the proof, please see the supplementary material.

Corollary 1. *(**Adversarial Vulnerability of Model for Multiple Independent Tasks**) If the output tasks selected are independent of each other, and the gradient for each task is i.i.d. with zero mean, then the adversarial vulnerability of given model is proportional to $\frac{1}{\sqrt{M}}$, where M is the number of independent output tasks selected.*

Based on the independence assumption, all covariances becomes zero. Thus Theorem 1 can be simplified as:

$$\mathbb{E}[\|\partial_{\mathbf{x}}\mathcal{L}_{all}(\mathbf{x}, \overline{\mathbf{y}})\|_2^2] = \mathbb{E}(\|\mathbf{R}\|_2^2) = \frac{1}{M}\mathbb{E}\|\mathbf{r}_i\|^2 = \frac{\sigma^2}{M} \propto \frac{1}{M} \tag{6}$$

Remark 1. By increasing the number of output tasks M, the first order vulnerability [46] of network decreases. In the ideal case, if the model has an infinite number of uncorrelated tasks, then it is impossible to find an adversarial examples that fools all the tasks.

Remark 2. Theorem 1 studies the robustness for multiple **correlated** tasks, which is true for most computer vision tasks. The independent tasks assumption in Corollary 1 is a simplified, idealistic instance of Theorem 1 that upper-bounds the robustness of models under multitask attacks. Together, Theorem 1 and Corollary 1 demonstrate that unless the tasks are 100% correlated (the same task), multiple tasks together are more robust than each individual one.

Our theoretical analysis shows that more outputs, especially if they are less correlated, improve the model's robustness against multitask attacks. Past work shows that segmentation is inherently robust [2,6] compared to classification. Our analysis provides a formal explanation to this inherent robustness because a segmentation model can be viewed as a multitask model (one task per pixel).

5   Experiments

We validate our analysis with empirical results on the Cityscapes and the Taskonomy datasets. We evaluate the robustness of multitask models against two types of attack: multitask attack (Sect. 5.3) and single-task attacks (Sect. 5.4). We also conduct multitask learning experiments on adversarial training and show that they are complementary (Sect. 5.5).

5.1   Datasets

Cityscapes. The Cityscapes dataset [8] consists of images of urban driving scenes. We study three tasks: semantic segmentation, depth estimation, and image reconstruction. We use the full resolution (2048×1024) for analyzing pre-trained state-of-the-art models. We resize the image to 680×340 to train our single task (baseline) and multitask models.[1]

Taskonomy. The Taskonomy dataset [59] consists of images of indoor scenes. We train on up to 11 tasks: semantic segmentation (s), depth euclidean estimation (D), depth zbuffer estimation (d), normal (n), edge texture (e), edge occlusion (E), keypoints 2D (k), keypoints 3D (K), principal curvature (p), reshading (r), and image reconstruction (A). We use the "tiny" version of their dataset splits [1]. We resize the images to 256×256.

---

[1] We use the same dimension for baselines and ours during comparison because input dimension impacts robustness [46].

Attacked Input Images   Ground Truth   Single-Task (s)   Multi-Task (s, d)   Multi-Task (s, d, A)

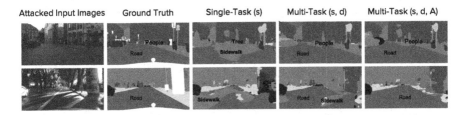

Fig. 2. We show model predictions on Cityscapes under multitask attack. The single-task segmentation model misclassifies the 'road' as 'sidewalk' under attack, while the multitask model can still segment it correctly. The multitask models are more robust than the single-task trained model.

Attacked Input Images   Ground Truth   Single-Task (d)   Multi-Task (s, d)   Multi-Task (s, d, A)

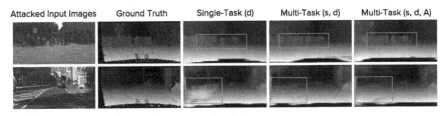

Fig. 3. We show depth predictions of multitask models under multitask attacks. To emphasize the differences, we annotated the figure with red boxes where the errors are particularly noticeable. The multitask trained model outperforms the single-task trained model under attack. (Color figure online)

5.2 Attack Methods

We evaluate the model robustness with L_∞ bounded adversarial attacks, which is a standard evaluation metric for adversarial robustness [34]. We evaluate with four different attacks:

FGSM: We evaluate on the Fast Gradient Sign Method (FGSM) [17], which generates adversarial examples $\mathbf{x}_{adv}$ by $\mathbf{x}_{adv} = \mathbf{x} + \epsilon \cdot \text{sign}(\nabla_{\mathbf{x}} \ell(F(\mathbf{x}), y))$. It is a single step, non-iterative attack.

PGD: Following the attack setup for segmentation in [2], we use the widely used attack PGD (Iteratively FGSM with random start [34]), set the number of iterations of attacks to $\min(\epsilon + 4, \lceil 1.25\epsilon \rceil)$ and step-size $\alpha = 1$. We choose the L_∞ bound ϵ from $\{1, 2, 4, 8, 16\}$ where noise is almost imperceptible. Under $\epsilon = 4$, we also evaluate the robustness using PGD attacks with $\{10, 20, 50, 100\}$ steps, which is a stronger attack compared to 5 steps attack used in [2].

MIM: We also evaluate on MIM attack [11], which adds momentum to iterative attacks to escape local minima and won the NeurIPS 2017 Adversarial Attack Competition.

Houdini: We evaluate the semantic segmentation task with the state-of-the-art Houdini attack [6], which directly attacks the evaluation metric, such as the non-differentiable mIoU criterion (mean Intersection over Union).

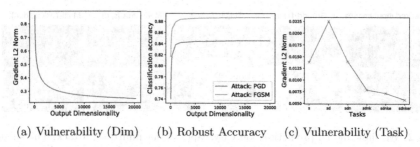

(a) Vulnerability (Dim)   (b) Robust Accuracy   (c) Vulnerability (Task)

Fig. 4. The effect of output dimensionality and number of tasks on adversarial robustness. We analyzed the pre-trained DRN model on Cityscapes (a, b), and a multitask model trained on Taskonomy (c). The x-axis of (a, b) represents the output dimensionality, the x-axis of (c) shows the combination of multiple tasks. The y-axis of (a, c) is the L2 norm of the joint gradient and is proportional to the model's adversarial vulnerability. The y-axis of (b) is classification accuracy. The robust performances for (c) are shown in Fig. 5. Increasing the output dimensionality or number of tasks improves the model's robustness.

We do not use the DAG [53] attack for segmentation because it is an unrestricted attack without controlling L_∞ bound. For all the iterative attacks, the step size is 1.

5.3   Multitask Models Against Multitask Attack

High Output Dimensionality as Multitask. Our experiment first studies the effect of a higher number of output dimensions on adversarial robustness. As an example, we use semantic segmentation. The experiment uses a pre-trained Dilated Residual Network (DRN-105) [56,57] model on the Cityscapes dataset. To obtain the given output dimensionality, we randomly select a subset of pixels from the model output. We mitigate the randomness of the sampling by averaging the results over 20 random samples. Random sampling is a general dimension reduction method, which preserves the correlation and structure for high dimensional, structured data [22]. Figure 4a shows that the model's vulnerability (as measured by the norm of the gradients) decreases as the number of output dimension increases, which validates Theorem 1.

Besides the norm of gradient, we measure the performance under FGSM [17] and PGD [34] adversarial attacks, and show that it improves as output dimensionality increases (Fig. 4b). Notice when few pixels are selected, the robustness gains are faster. This is because with fewer pixels: (1) the marginal gain of the inverse function is larger; and (2) the select pixels are sparse and tend to be far away and uncorrelated to each other. The correlation between the output pixels compounds as more nearby pixels are selected, which slows down the improvements to robustness. The results demonstrate that models with higher output dimension/diversity are inherently more robust against adversarial attacks, consistent with the observation in [2,6] and our Theorem 1.

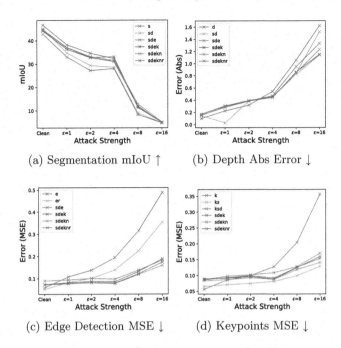

(a) Segmentation mIoU ↑   (b) Depth Abs Error ↓

(c) Edge Detection MSE ↓   (d) Keypoints MSE ↓

Fig. 5. Adversarial robustness against multitask attack on Taskonomy dataset. The x-axis is the attack strength, ranging from no attack (clean) to the strongest attack ($\epsilon = 16$ PGD). For each subfigure, the y-axis shows the performance of one task under multitask attack. ↑ means the higher, the better. ↓ means the lower, the better. The multitask model names are in the legend, we refer to the task by their initials, e.g., 'sde' means the model is trained on segmentation, depth, and edge simultaneously. The blue line is the single-task baseline performance, the other lines are multitask performance. The figures show that it is hard to attack all the tasks in a multitask model simultaneously. Thus multitask models are more robust against multitask attacks. (Color figure online)

Number of Tasks. We now consider the case where the number of tasks increases, which is a second factor that increases output dimensionality. We evaluate the robustness of multitask models on the Cityscapes and Taskonomy datasets. We equally train all the tasks with the shared backbone architecture mentioned in Sect. 3. On Cityscapes, we use DRN-105 model as the architecture for encoder and decoder; on Taskonomy, we use Resnet-18 [59]. Each task has its own decoder. For the Cityscapes dataset, we start with training only the semantic segmentation task, then add the depth estimation and input reconstruction task. For the Taskonomy dataset, following the setup in [47], we start with only semantic segmentation, and add depth estimation, normal, keypoints 2D, edge texture, and reshading tasks to the model one by one. In our figures and tables, we refer to these tasks by the task's first letter.

Figure 4c shows the L2 norm of the joint gradient for many tasks, which measures the adversarial vulnerability. Overall, as we add more tasks, the norm

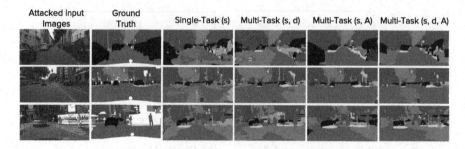

Fig. 6. Performance of single-task attack for multitask models trained on Cityscapes. We show segmentation under attack for single-task and three multitask models. The multitask trained model out-performs the single-task trained model.

of the joint gradient decreases, indicating improvement to robustness [46]. The only exception is the depth estimation task, which we believe is due to the large range of values (0 to $+\infty$) that its outputs take. Empirically, a larger output range leads to a larger loss, which implies a larger gradient value.

We additionally measure the robust performance on different multitask models under multitask attacks. Following the setup in [2], we enumerate the ϵ of the L_∞ attack from 1 to 16. Figure 5 shows the robustness of multitask models using Taskonomy, where the adversarial robustness of multitask models are better than single-task models, even if the clean performance of multitask models may be lower. We also observe some tasks gain more robustness compared to other tasks when they are attacked together, which suggests some tasks are inherently harder to attack. Overall, the attacker cannot simultaneously attack all the tasks successfully, which results in improved overall robustness of multitask models. In Table 1, we observe the same improvement on Cityscapes. Qualitative results are shown in Fig. 2 and Fig. 3. Please see the supplemental material for additional results.

5.4 Multitask Models Against Single-Task Attacks

Following the setup for multitask learning in [27,30,32], we train the multitask models using a main task and auxiliary tasks, where we use $\lambda = 1$ for the main

Table 1. The models' performances under multitask PGD attack and clean images on Cityscapes using DRN-D-105 [57]. The **bold** demonstrate the better performance for each row, underline shows inferior results of multitask learning. The results show that multitask models are overall more robust under multitask attack.

Training Tasks	Baseline s	Multitask sd	sdA	Training Tasks	Baseline d	Multitask sd	sdA
Clean SemSeg ↑	44.77	**46.53**	45.82	Clean Depth ↓	1.82	**1.780**	1.96
PGD SemSeg ↑	15.75	16.01	**16.36**	PGD Depth ↓	6.81	6.08	**5.81**

Table 2. Model's robust performance under $L_\infty = 4$ bounded single-task attacks on Cityscapes. Each column is a DRN-22 model trained on a different combination of tasks, where "s", "d," and "A" denote segmentation, depth, and auto-encoder, respectively. ↑ means the higher, the better. ↓ means the lower, the better. **Bold** in each row, shows the best performance under the same attack. Multitask learning models outperform single-task models except for the <u>underlined</u> ones. While nearly maintaining the performance on clean examples, multitask models are consistently more robust under strong adversarial attacks.

		SemSeg mIoU Score ↑			Depth Abs Error ↓		
	Baseline	Multitask Learning			Baseline	Multitask Learning	
Training Tasks →	s	sd	sA	sdA	d	ds	dA	dAs
λ_a		0.001	0.001	0.001		0.1	0.1	0.01
Clean	48.58	48.61	**49.61**	<u>48.19</u>	1.799	**1.792**	<u>1.823</u>	1.798
FGSM	26.35	<u>26.28</u>	**26.79**	26.71	3.16	3.01	**3.00**	3.24
PGD10	13.04	13.64	**14.76**	14.48	6.96	6.15	**6.03**	6.59
PGD20	11.41	11.98	**12.79**	12.73	8.81	**7.70**	7.64	8.38
PGD50	10.49	10.95	11.68	**11.86**	10.23	**9.07**	9.12	9.81
PGD100	10.15	10.51	11.22	**11.52**	10.8	**9.69**	9.74	10.41
MIM100	9.90	10.17	10.93	**11.24**	12.04	**10.72**	10.97	11.69
Houdini100	5.04	5.14	**6.24**	6.21	-	-	-	-

(Attacks)

task and λ_a for the auxiliary tasks. We then evaluate the robustness of the main task under single-task attacks. On Cityscapes, the main and the auxiliary tasks share 16 layers of an encoding backbone network. The decoding network for each individual task has 6 layers. For all the models, we train for 200 epochs. For adversarial robustness evaluation, we use strong attacks including PGD100 and MIM100 for attacking the segmentation accuracy,[2] and use 100 steps Houdini [6] to attack the non-differentiable mIoU of the Segmentation model directly. We do not use Houdini to attack the depth because the L1 loss for depth is differentiable and does not need any surrogate loss. The results in Table 2 show that multitask learning improves the segmentation mIoU by 1.2 points and the performance of depth estimation by 11% under attack, while maintaining the performance on most of the clean examples. Qualitative results are in Fig. 6.

On the Taskonomy dataset, we conduct experiments on 11 tasks. Following the setup in [47], we use ResNet-18 [19] as the shared encoding network, where each individual task has its own prediction network using the encoded representation. We train single-task models for each of the 11 tasks as baselines. We train a total of 110 multitask models — each main task combined with 10 different auxilliary tasks — for 11 main tasks. We evaluate both the clean performance and adversarial performance. λ_a is either 0.1 or 0.01 based on the tasks. We use PGD attacks bounded with $L_\infty = 4$ with 50 steps, where the step size is 1. The attack performance plateaus for more steps. Figure 7 shows the performance of the main task on both clean and adversarial examples. While maintaining the performance on clean examples (average improvement of 4.7%), multitask learning improves 90/110 the models' performance under attacks, by an average of

---

[2] Suffixed number indicates number of steps for attack.

Training Task 2

	None (Baseline)	SemSeg	DepthZ	Edge2D	Normal	Reshading	Key2D	Key3D	DepthE	AutoE	Edge3D	PCurve	Mean
SemSeg *	13.36	0.00	44.61	4.42	24.48	9.13	17.51	3.14	11.60	5.61	10.18	11.53	14.22
DepthZ (10^{-2})	11.49	59.00	0.00	40.99	-1.10	-8.02	3.22	27.27	8.64	56.65	-6.43	56.18	23.64
Edge2D (10^{-2})	10.67	7.79	12.27	0.00	10.55	6.83	8.81	9.54	8.98	6.85	6.50	5.41	8.35
Normal (10^{-2})	40.93	14.06	-4.75	3.89	0.00	1.04	3.58	2.43	-2.80	12.71	9.42	-0.70	3.89
Reshading (10^{-2})	57.89	15.66	0.18	5.05	2.42	0.00	3.36	7.93	-3.68	-5.37	14.80	0.46	4.08
Key2D (10^{-2})	11.72	7.40	7.08	8.69	10.19	7.13	0.00	6.56	9.41	6.27	8.23	9.88	8.08
Key3D (10^{-2})	49.70	37.72	0.18	-2.19	7.73	15.14	11.81	0.00	-3.04	34.47	-7.48	-6.39	8.80
DepthE (10^{-3})	4.85	27.15	29.95	32.96	11.87	-17.11	24.24	23.07	0.00	23.57	31.19	39.44	22.63
AutoE (10^{-2})	59.31	2.60	-1.62	1.75	-5.08	-0.17	-0.03	-2.37	1.80	0.00	-1.94	-3.68	-0.87
Edge3D (10^{-2})	15.90	8.36	3.58	-2.71	3.38	4.45	1.96	-6.32	3.12	20.69	0.00	6.98	4.35
PCurve (10^{-4})	11.47	22.22	22.38	9.74	19.51	16.14	22.39	9.04	2.81	19.91	9.31	0.00	15.34

(a) Performance Under Attack

Training Task 2

	None (Baseline)	SemSeg	DepthZ	Edge2D	Normal	Reshading	Key2D	Key3D	DepthE	AutoE	Edge3D	PCurve	Mean
SemSeg *	43.19	0.00	7.20	6.92	7.32	7.06	5.21	5.63	3.47	3.03	4.93	3.01	5.38
DepthZ (10^{-2})	2.85	4.15	0.00	-36.05	0.19	12.16	-0.78	-24.91	-17.07	-11.20	-8.26	-64.44	-14.62
Edge2D (10^{-2})	3.38	-15.91	0.05	0.00	-3.64	-1.51	1.60	-4.08	-1.46	-5.62	-5.49	-2.07	-3.81
Normal (10^{-2})	7.00	-2.64	-1.38	-0.13	0.00	0.11	0.10	-2.64	0.82	1.91	0.93	-2.06	-0.50
Reshading (10^{-2})	8.03	0.52	-0.94	1.07	1.57	0.00	-0.18	0.88	1.09	1.71	-0.48	-1.53	0.37
Key2D (10^{-2})	4.16	0.96	6.22	8.67	6.99	0.21	0.00	5.08	7.20	8.01	7.35	6.68	5.74
Key3D (10^{-2})	8.77	3.72	0.96	2.93	1.84	5.17	0.77	0.00	3.18	4.62	4.66	2.19	3.00
DepthE (10^{-3})	6.37	-3.17	6.70	0.35	2.15	8.96	-0.71	-1.53	0.00	6.67	10.32	1.91	3.17
AutoE (10^{-2})	3.47	-4.21	-6.90	-2.27	-3.38	-2.02	-8.96	-8.40	-2.09	0.00	-1.75	-2.40	-4.24
Edge3D (10^{-2})	4.65	-1.00	0.87	-1.69	1.86	-1.64	6.12	0.31	0.82	9.45	0.00	20.35	3.54
PCurve (10^{-4})	10.48	20.19	21.87	20.26	18.47	30.96	26.18	26.26	22.17	25.03	26.19	0.00	23.76

(b) Performance on Clean Examples

Fig. 7. We consider models trained on two tasks. In each matrix, the rows show the first training task and the testing task. The columns show the auxiliary training task. The first column without color shows the absolute value for the baseline model (single-task). The middle colored columns show the relative improvement of multitask models over the single-task model in percentage. The last colored column shows the average relative improvement. We show results for both (a) adversarial and (b) clean performance. Multitask learning improves the performance on clean examples for 70/110 cases, and the performance on adversarial examples for 90/110 cases. While multitask training does not always improve clean performance, we show multitask learning provides more gains for adversarial performance. (Color figure online)

10.23% relative improvement. Our results show that one major advantage of multitask learning, which to our knowledge is previously unknown, is that it improves the model's robustness under adversarial attacks.

5.5   Multitask Learning Complements Adversarial Training

We study whether multitask learning helps adversarial robust training. We use DRN-22 on the Cityscapes dataset, and train both single-task and multitask

Table 3. Adversarial robustness of adversarial training models under $L_\infty = 4$ bounded attacks on Cityscapes. Each column is a model trained on a different combination of tasks. "s", "d", and "A" denote segmentation, depth, and auto-encoder respectively. $\uparrow$ indicates the higher, the better. The $\downarrow$ indicates the lower, the better. **Bold** shows the best performance of the same task for each row. Multitask learning improves both the clean performance and robustness upon single-task learning.

		SemSeg mIoU Score $\uparrow$			Depth Abs Error $\downarrow$		
	Baseline	Multitask Learning			Baseline	Multitask Learning	
Training Tasks $\rightarrow$	s	sd	sA	sdA	d	ds dA	dAs
Clean	41.95	43.27	**43.65**	43.26	2.24	**2.07** 2.15	2.15
PGD50	19.73	**22.08**	20.45	21.93	2.85	**2.61** 2.75	2.67
PGD100	19.63	**21.96**	20.31	21.83	2.85	**2.61** 2.75	2.67
MIM100	19.54	**21.89**	20.20	21.74	2.85	**2.61** 2.75	2.67
Houdini100	17.05	**19.45**	17.36	19.16	-	- -	-

(Attacks)

models for 200 epoch under the same setup. The single-task model follows the standard adversarial training algorithm, where we train the model on the generated single-task (segmentation) adversarial attacks. For the multitask adversarial training, we train it on the generated multitask attack images for both semantic segmentation and the auxiliary task. Details are in the supplementary material. Table 3 shows that multitask learning improves the robust performance of both clean examples and adversarial examples, where segmentation mIoU improves by 2.40 points and depth improves by 8.4%.

6   Conclusion

The widening deployment of machine learning in real-world applications calls for versatile models that solve multiple tasks or produce high-dimensional outputs. Our theoretical analysis explains that versatile models are inherently more robust than models with fewer output dimensions. Our experiments on real-world datasets and common computer vision tasks measure improvements in adversarial robustness under attacks. Our work is the first to connect this vulnerability with multitask learning and hint towards a new direction of research to understand and mitigate this fragility.

Acknowledgements. This work was in part supported by a JP Morgan Faculty Research Award; a DiDi Faculty Research Award; a Google Cloud grant; an Amazon Web Services grant; an Amazon Research Award; NSF grant CNS-15-64055; NSF-CCF 1845893; NSF-IIS 1850069; ONR grants N00014-16-1- 2263 and N00014-17-1-2788. The authors thank Vaggelis Atlidakis, Augustine Cha, Dídac Surís, Lovish Chum, Justin Wong, and Shunhua Jiang for valuable comments.

References

1. https://github.com/StanfordVL/taskonomy/tree/master/data

2. Arnab, A., Miksik, O., Torr, P.H.: On the robustness of semantic segmentation models to adversarial attacks. In: CVPR (2018)
3. Carlini, N., Wagner, D.A.: Towards evaluating the robustness of neural networks. In: 2017 IEEE Symposium on Security and Privacy, pp. 39–57 (2017)
4. Caruana, R.: Multitask learning. Mach. Learn. **28**(1), 41–75 (1997)
5. Chen, Z., Badrinarayanan, V., Lee, C.Y., Rabinovich, A.: Gradnorm: gradient normalization for adaptive loss balancing in deep multitask networks (2017)
6. Cisse, M., Adi, Y., Neverova, N., Keshet, J.: Houdini: fooling deep structured prediction models (2017)
7. Cissé, M., Bojanowski, P., Grave, E., Dauphin, Y., Usunier, N.: Parseval networks: improving robustness to adversarial examples. In: ICML, pp. 854–863 (2017)
8. Cordts, M., et al.: The cityscapes dataset for semantic urban scene understanding. In: Proceedings of the IEEE Conference on Computer Vision and Pattern Recognition (CVPR) (2016)
9. Costales, R., Mao, C., Norwitz, R., Kim, B., Yang, J.: Live trojan attacks on deep neural networks (2020)
10. Doersch, C., Zisserman, A.: Multi-task self-supervised visual learning. arXiv:1708.07860 (2017)
11. Dong, Y., et al.: Boosting adversarial attacks with momentum. In: CVPR, pp. 9185–9193 (2018)
12. Désidéri, J.A.: Multiple-gradient descent algorithm (MGDA) for multiobjective optimization. C.R. Math. **350**, 313–318 (2012)
13. Engstrom, L., et al.: A discussion of adversarial examples are not bugs, they are features. Distill **4**(8) (2019)
14. Engstrom, L., Ilyas, A., Athalye, A.: Evaluating and understanding the robustness of adversarial logit pairing (2018)
15. Evgeniou, T., Pontil, M.: Regularized multi-task learning, pp. 109–117 (2004)
16. Glaßer, C., Reitwießner, C., Schmitz, H., Witek, M.: Approximability and hardness in multi-objective optimization. In: Ferreira, F., Löwe, B., Mayordomo, E., Mendes Gomes, L. (eds.) CiE 2010. LNCS, vol. 6158, pp. 180–189. Springer, Heidelberg (2010). https://doi.org/10.1007/978-3-642-13962-8_20
17. Goodfellow, I.J., Shlens, J., Szegedy, C.: Explaining and harnessing adversarial examples. arXiv:1412.6572 (2014)
18. Gur, S., Wolf, L.: Single image depth estimation trained via depth from defocus cues. In: The IEEE Conference on Computer Vision and Pattern Recognition (CVPR) (2019)
19. He, K., Zhang, X., Ren, S., Sun, J.: Deep residual learning for image recognition. arXiv:1512.03385 (2015)
20. Kaiser, L., et al.: One model to learn them all (2017)
21. Kannan, H., Kurakin, A., Goodfellow, I.J.: Adversarial logit pairing (2018)
22. Keshavan, R.H., Montanari, A., Oh, S.: Matrix completion from noisy entries. In: NIPS (2009)
23. Kocabas, M., Karagoz, S., Akbas, E.: Multiposenet: fast multi-person pose estimation using pose residual network. In: CoRR (2018)
24. Kokkinos, I.: Ubernet: training a universal convolutional neural network for low-, mid-, and high-level vision using diverse datasets and limited memory. arXiv:1609.02132 (2016)
25. Kumar, A., Daume III, H.: Learning task grouping and overlap in multi-task learning (2012)
26. Kurakin, A., Goodfellow, I.J., Bengio, S.: Adversarial examples in the physical world. arXiv:1607.02533 (2017)

27. Lee, T., Ndirango, A.: Generalization in multitask deep neural classifiers: a statistical physics approach (2019)
28. Liu, S., Johns, E., Davison, A.J.: End-to-end multi-task learning with attention. arXiv:1803.10704 (2018)
29. Liu, W., et al.: SSD: single shot multibox detector, pp. 21–37 (2016)
30. Liu, X., He, P., Chen, W., Gao, J.: Multi-task deep neural networks for natural language understanding. In: Proceedings of the 57th Annual Meeting of the Association for Computational Linguistics (2019)
31. Long, J., Shelhamer, E., Darrell, T.: Fully convolutional networks for semantic segmentation. arXiv:1411.4038 (2014)
32. Luong, M.T., Le, Q.V., Sutskever, I., Vinyals, O., Kaiser, L.: Multi-task sequence to sequence learning (2015)
33. Ma, J., Zhao, Z., Yi, X., Chen, J., Hong, L., Chi, E.: Modeling task relationships in multi-task learning with multi-gate mixture-of-experts, pp. 1930–1939 (2018)
34. Madry, A., Makelov, A., Schmidt, L., Tsipras, D., Vladu, A.: Towards deep learning models resistant to adversarial attacks. In: ICLR (2018)
35. Mao, C., Zhong, Z., Yang, J., Vondrick, C., Ray, B.: Metric learning for adversarial robustness (2019)
36. Metzen, J.H., Kumar, M.C., Brox, T., Fischer, V.: Universal adversarial perturbations against semantic image segmentation. In: ICCV (2017)
37. Misra, I., Shrivastava, A., Gupta, A., Hebert, M.: Cross-stitch networks for multi-task learning (2016)
38. Pang, T., Xu, K., Du, C., Chen, N., Zhu, J.: Improving adversarial robustness via promoting ensemble diversity. arXiv:1901.08846 (2019)
39. Papernot, N., McDaniel, P.D., Jha, S., Fredrikson, M., Celik, Z.B., Swami, A.: The limitations of deep learning in adversarial settings. arXiv:1511.07528 (2015)
40. Ross, A.S., Doshi-Velez, F.: Improving the adversarial robustness and interpretability of deep neural networks by regularizing their input gradients. arXiv:1711.09404 (2017)
41. Samangouei, P., Kabkab, M., Chellappa, R.: Defense-GAN: protecting classifiers against adversarial attacks using generative models. arXiv:1805.06605 (2018)
42. Schmidt, L., Santurkar, S., Tsipras, D., Talwar, K., Madry, A.: Adversarially robust generalization requires more data. In: NeurIPS, pp. 5019–5031 (2018)
43. Schutze, O., Lara, A., Coello, C.A.C.: On the influence of the number of objectives on the hardness of a multiobjective optimization problem. IEEE Trans. Evol. Comput. 15(4), 444–455 (2011)
44. Sener, O., Koltun, V.: Multi-task learning as multi-objective optimization (2018)
45. Shen, G., Mao, C., Yang, J., Ray, B.: AdvSPADE: realistic unrestricted attacks for semantic segmentation (2019)
46. Simon-Gabriel, C.J., Ollivier, Y., Bottou, L., Schölkopf, B., Lopez-Paz, D.: First-order adversarial vulnerability of neural networks and input dimension. In: Proceedings of the 36th International Conference on Machine Learning, vol. 97, pp. 5809–5817 (2019)
47. Standley, T., Zamir, A.R., Chen, D., Guibas, L.J., Malik, J., Savarese, S.: Which tasks should be learned together in multi-task learning? arXiv:1905.07553 (2019)
48. Szegedy, C., et al.: Intriguing properties of neural networks. arXiv:1312.6199 (2013)
49. Tramèr, F., Kurakin, A., Papernot, N., Boneh, D., McDaniel, P.D.: Ensemble adversarial training: attacks and defenses. arXiv:1705.07204 (2017)
50. Tsipras, D., Santurkar, S., Engstrom, L., Turner, A., Madry, A.: Robustness may be at odds with accuracy. In: International Conference on Learning Representations (2019)

51. Uesato, J., Alayrac, J., Huang, P., Stanforth, R., Fawzi, A., Kohli, P.: Are labels required for improving adversarial robustness? In: CoRR (2019)
52. Xiao, T., Liu, Y., Zhou, B., Jiang, Y., Sun, J.: Unified perceptual parsing for scene understanding. In: CoRR (2018)
53. Xie, C., Wang, J., Zhang, Z., Zhou, Y., Xie, L., Yuille, A.: Adversarial examples for semantic segmentation and object detection. In: ICCV (2017)
54. Xie, C., Wu, Y., van der Maaten, L., Yuille, A.L., He, K.: Feature denoising for improving adversarial robustness. In: CoRR (2018)
55. Yan, Z., Guo, Y., Zhang, C.: Deep defense: training DNNs with improved adversarial robustness. In: Proceedings of the 32nd International Conference on Neural Information Processing Systems, NIPS 2018, pp. 417–426 (2018)
56. Yu, F., Koltun, V.: Multi-scale context aggregation by dilated convolutions. In: International Conference on Learning Representations (ICLR) (2016)
57. Yu, F., Koltun, V., Funkhouser, T.: Dilated residual networks. In: Computer Vision and Pattern Recognition (CVPR) (2017)
58. Yu, T., Kumar, S., Gupta, A., Levine, S., Hausman, K., Finn, C.: Gradient surgery for multi-task learning (2020)
59. Zamir, A.R., Sax, A., Shen, W.B., Guibas, L., Malik, J., Savarese, S.: Taskonomy: disentangling task transfer learning. In: 2018 IEEE Conference on Computer Vision and Pattern Recognition (CVPR). IEEE (2018)
60. Zhang, H., Yu, Y., Jiao, J., Xing, E.P., Ghaoui, L.E., Jordan, M.I.: Theoretically principled trade-off between robustness and accuracy. arXiv:1901.08573 (2019)

S2DNAS: Transforming Static CNN Model for Dynamic Inference via Neural Architecture Search

Zhihang Yuan[1,2], Bingzhe Wu[1], Guangyu Sun[1,2(✉)], Zheng Liang[1], Shiwan Zhao[3], and Weichen Bi[1]

[1] Peking University, Beijing, China
{yuanzhihang,wubingzhe,gsun,liangzheng,biweichen}@pku.edu.cn
[2] Advanced Institute of Information Technology, Peking University, Hangzhou, China
[3] IBM China Research Laboratory, Beijing, China
zhaosw@cn.ibm.com

Abstract. Recently, dynamic inference has emerged as a promising way to reduce the computational cost of deep convolutional neural networks (CNNs). In contrast to static methods (e.g., weight pruning), dynamic inference adaptively adjusts the inference process according to each input sample, which can considerably reduce the computational cost on "easy" samples while maintaining the overall model performance.

In this paper, we introduce a general framework, S2DNAS, which can transform various static CNN models to support dynamic inference via neural architecture search. To this end, based on a given CNN model, we first generate a CNN architecture space in which each architecture is a multi-stage CNN generated from the given model using some predefined transformations. Then, we propose a reinforcement learning based approach to automatically search for the optimal CNN architecture in the generated space. At last, with the searched multi-stage network, we can perform dynamic inference by adaptively choosing a stage to evaluate for each sample. Unlike previous works that introduce irregular computations or complex controllers in the inference or re-design a CNN model from scratch, our method can generalize to most of the popular CNN architectures and the searched dynamic network can be directly deployed using existing deep learning frameworks in various hardware devices.

Keywords: Dynamic inference · Neural architecture search · CNN

Z. Yuan and B. Wu—Equal contribution.

Electronic supplementary material The online version of this chapter (https://doi.org/10.1007/978-3-030-58536-5_11) contains supplementary material, which is available to authorized users.

A. Vedaldi et al. (Eds.): ECCV 2020, LNCS 12347, pp. 175–192, 2020.
https://doi.org/10.1007/978-3-030-58536-5_11

1 Introduction

In the past years, deep convolutional neural networks (CNNs) have gained great success in many computer vision tasks, such as image classification [12,17,20], object detection [29,32,34], and image segmentation [4,11]. However, the remarkable performance of CNNs always comes with huge computational cost, which impedes their deployment in resource constrained hardware devices. Thus, various methods have been proposed to improve computational efficiency of the CNN inference, including network pruning [10,21,22] and weight quantization [8,31,44]. Most of the previous methods are static approaches, which use fixed computation graphs for all test samples.

Recently, dynamic inference has emerged as a promising alternative to speed up the CNN inference by dynamically changing the computation graph according to each input sample [2,5–7,16,28,39,45]. The basic idea is to allocate less computation for "easy" samples while more computation for "hard" ones. As a result, the dynamic inference can considerably save the computational cost of "easy" samples without sacrificing the overall model performance. Moreover, the dynamic inference can naturally exploit the trade-off between accuracy and computational cost to meet varying requirements (e.g., computational budget) in real-world scenarios.

To enable the dynamic inference of a CNN model, most previous works aim to develop dedicated strategies to dynamically skip some computation operations during the CNN inference according to different input samples. To achieve this goal, these works attempted to add extra controllers in-between the original model to select which computations are executed. For example, well-designed gate-functions were proposed as the controller to select a subset of channels or pixels for the subsequent computation of the convolution layer [5,7,14]. However, these methods lead to irregular computation at channel level or spatial level, which are not efficiently supported by existing software and hardware devices [15, 42,46]. To address this issue, a more aggressive strategy that dynamically skips whole layers was proposed for efficient inference [40,41,45]. Unfortunately, this strategy can only be applied to the CNN model with residual connection [12]. Moreover, the controllers of some methods comes with a considerable complex structure, which cause the increase of the overall computational cost in the inference (see experimental results in Sect. 4).

To mitigate these problems, researchers propose early exiting the "easy" input samples at inference time [1,16,30,39]. A typical solution is to add intermediate prediction layers at multiple layers of a normal CNN model, and then exit the inference when the confidence score of the intermediate classifier is higher than a given threshold. Figure 1a shows the paradigm of these early exiting methods [1,30]. In this paradigm, prediction layers are directly added in-between the original network and the network is split into multiple stages along the layer depth. However, these solutions face the challenge that early classifiers are unable to leverage the semantic-level features produced by the deeper layers. It may cause a significant accuracy drop [16]. As illustrated in Fig. 1a, three

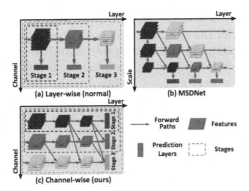

Fig. 1. Three paradigms of early exiting methods. (a) The layer-wise approach splits the network into multiple stages along the layer depth. (b) MSDNet devises a multi-stage CNN in which each stage maintains a feature pyramid. (c) Our proposed channel-wise approach splits the network along the channel width.

prediction layers are added to different depth of the network. Thus, the classifier[1] in the previous stage cannot make use of the semantic-level features produced by the classifier in the late stage.

Huang et al. [16] proposed a novel CNN model, called MSDNet, for solving this issue. The core design of MSDNet is a two-dimensional multi-scale architecture that maintains the coarse and fine level features in every layer as shown in Fig. 1b. Based on this design, MSDNet can leverage the semantic-level features in every prediction layer and achieve the best result. However, MSDNet needs to design specialized network architecture, which cannot generalize to other CNN models and needs massive expertise in architecture design.

To solve the aforementioned issue without designing CNNs from scratch, we propose to transform a given CNN model into a *channel-wise* multi-stage network, which comes with the advantage that the classifier in the early stages can leverage the semantic-level features. Figure 1c intuitively demonstrates the idea behind our method. Different from the normal paradigm in Fig. 1a, our method split the original network into multiple stages along the channel width. The prediction layers are added only to the last convolutional layer, thus all classifiers can leverage the semantic-level features. To reduce the computational cost of the classifiers in the early stages, we propose to cut down the number of channels of each layer in different stages (more details can be found in Sect. 3).

Based on the high-level idea introduced above, we present a general framework called S2DNAS. Given a specific CNN model, the framework can automatically generate the dynamic model following the paradigm showed in Fig. 1c. S2DNAS consists of two components: S2D and NAS. First, the component S2D, which means "static to dynamic", is used to generate a CNN model space based on the given model. This space comprises of different multi-stage CNN networks

---

[1] In this paper, the classifier refers to the whole sub-network in the current stage.

generated from the given model based on the predefined transformations. Then, NAS is used to search for the optimal model in the generated space with the help of reinforcement learning. Specifically, we devise an RNN to decide the setting of each transformation for generating the model. To exploit trade-off between accuracy and computational cost, we design a reward function that can reflect both the classification accuracy and the computational cost inspired by the prior works [13,38,43]. We then use a policy-gradient based algorithm [36] to train the RNN. The RNN will generate better CNN models with reinforcement learning and we can further use the searched model for dynamic inference.

To verify the effectiveness of S2DNAS, we perform extensive experiments by applying our method to various CNN models. With a comparable model accuracy, our method can achieve further computation reduction in contrast to the previous works for dynamic inference.

2 Related Work

Static Method for Efficient CNN Inference. Numerous methods are proposed for improving the efficiency of CNN inference. Two representative research directions are network pruning [9,10,21,22] and quantization [8,25,31,44]. Specifically, network pruning aims to remove redundant weights in a well-trained CNN without sacrificing the model accuracy. In contrast, network quantization aims to reduce the bit-width of both activations and weights. Most works in the above two directions are static, which refers to using the same computation graph for all test samples. Next, we introduce an emerging direction of utilizing dynamic inference for improving the efficiency of CNN inference.

Dynamic Inference. Dynamic inference also refers to adaptive inference in previous works [23,40]. Most previous works aim to develop dedicated strategies to dynamically skip some computation during inference. They attempted to add extra controllers to select which computations are executed [3,5,7,14,24,33,40, 41,45]. Dong et al. [5] proposed to compute the spatial attention using extra convolutional layers then skipping the computation of inactive pixels. Gao et al. [7] proposed to compute the importance of each channel then skipping the computation of those unimportant channels. However, these methods lead to irregular computation at channel level or spatial level, which is not efficiently supported by existing deep learning frameworks and hardware devices. To address this issue, a more aggressive strategy that dynamically skips the whole layers or blocks is proposed [40,41,45]. For example, BlockDrop [45] introduced a policy network to decide which layers should be skipped. Unfortunately, this strategy can only be applied to the CNN model with residual connection. Moreover, these methods introduce extra controllers into the computational graph, the computational cost will remain the same or even increase in some cases. On the other hand, early exiting methods propose to divide a CNN model into multiple stages and exit the inference of "easy" samples in the early stages [1,16,30,39]. The state-of-the-art is MSDNet [16] in which the authors manually design a novel multi-stage network architecture to serve the purpose of dynamic inference.

Neural Architecture Search. Recently, neural architecture search (NAS) has emerged as a promising direction to automatically design the network architecture to meet varying requirements of different tasks [13,26,27,43,47,48]. There are two typical types of works in this research direction, RL-based searching algorithms [47] and differentiable searching algorithms [27]. In this paper, according to the formulation of our specific problem, we choose the RL-based searching algorithm to search for the optimal model in a design space.

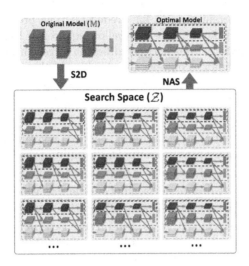

Fig. 2. Overview of S2DNAS. S2D first generates a search space from the original CNN model. Then, NAS searches for the optimal model in the generated space.

3   Our Approach

3.1   Overview of S2DNAS

The overview of S2DNAS is depicted in Fig. 2. At a high level, S2DNAS can be divided into two components, namely, S2D and NAS. Here, S2D means "static-to-dynamic", which is used to generate a search space comprises of dynamic models based on a given static CNN model. Specifically, we define two transformations and then apply the transformations to the original model for generating different dynamic models in the search space. Each of these dynamic models is a multi-stage CNN that can be directly used for dynamic inference. All these generated models form the search space. Once the search space is generated, NAS searches for the optimal model in the space. In what follows, we will give the details of these two components.

3.2   The Details of S2D

Given a CNN model $\mathbb{M}$, the goal of S2D is to generate the search space $\mathcal{Z}$ which consists of different dynamic models transformed from $\mathbb{M}$. Each network in $\mathcal{Z}$ is a multi-stage CNN model in which each stage contains one classifier. These multi-stage CNNs can be generated from $\mathbb{M}$ using two transformations, namely, split and concat. First, we propose split to split the original model along the channel width as Fig. 3 shows. Specifically, we divide the input channels in each layer of the original model into different subsets. And each classifier can use features from different subsets for prediction. The prediction can be done by adding a prediction layer (shown as yellow squares in Fig. 3). Moreover, to enhance the feature interactions between different stages for further performance boost, we propose concat to enforce the classifier in the current stage to reuse the features from previous stages. Next, we will present the details of these two transformations, split and concat. Before that, we first present some basic notations.

Notation. We start with the notation of a normal convolutional layer. Taking the k-th layer of a deep CNN as an example, the input of the k-th layer is denoted as $\mathbf{X}^{(k)} = \{X_1^{(k)}, \cdots, X_C^{(k)}\}$, where C is the number of input channels and $X_i^{(k)}$ is the i-th feature map. We denote the weights of layer k as $w^{(k)} = \{w_1^{(k)}, \cdots, w_O^{(k)}\}$, where O is the number of output channels and $w_i^{(k)} \in \mathbb{R}^{k_c \times k_c \times C}$ ($k_c \times k_c$ is the kernel size). In the following parts, we will present two transformations that can be applied to the original model. The goal of the transformations is to transform a static CNN model $\mathbb{M}$ to a multi-stage model, which can be represented as $a = \{f_1, \cdots, f_s\}$, where f_i is the classifier in the i-th stage. Next, we will introduce the details of the proposed two transformations.

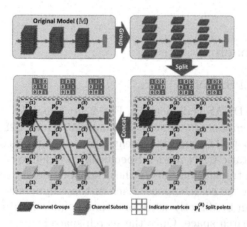

Fig. 3. Illustration of how split and concat are applied to a CNN model. Note that Group is an intermediate step of split for reducing the size of search space (see more details in the main text). (Color figure online)

Split. The `split` transformation is responsible for assigning different subsets of the input channels to the classifiers in different stages. We denote the number of stages as s. A direct way is splitting the input channels into s subsets and allocating the i-th subset to the classifier in the i-th stage. However, this splitting method results in a considerable large search space which poses the obstacle to the subsequent search process (i.e., NAS). In order to reduce the search space generated by this transformation, we propose to first divide the input channels into groups and then assign these groups to different classifiers.

Specifically, we first evenly divide the input channels[2] into G groups thus each group consists of $m = \dfrac{C}{G}$ input channels. Taking the k-th layer as an example, the i-th group consists $\{X^{(k)}_{(i-1)m+1}, \cdots, X^{(k)}_{im}\}$. Once the grouping is finished, these groups are assigned to the classifiers in different stages. Precisely, we use the split points $(p^{(k)}_0, p^{(k)}_1, \cdots, p^{(k)}_{s-1}, p^{(k)}_s)$ to split the groups, here $p^{(k)}_0 = 0$ and $p^{(k)}_s = G$ are two peculiar points, which denote the start and end points. With the split points, we assign the groups between $p^{(k)}_{i-1}$ and $p^{(k)}_i$ to the classifier f_i in the i-th stage. Note that the connection (of the original model $\mathbb{M}$) between different classifiers are removed (see Fig. 3).

Concat. The `concat` transformation is used for enhancing the interaction between different stages. The basic idea is to enable the classifiers in later stages to reuse the features from previous stages. Formally, we use indicator matrices $\{\mathbf{I}^{(k)}\}^L_{k=1}$ to indicate whether to enable the feature reuse at different positions. Here k denotes the k-th layer and L is the depth[3] of the CNN model. The element $m^{(k)}_{ij} \in \mathbf{I}^{(k)}$ indicates whether to reuse the features of the i-th stage in the j-th stage at the k-th layer, i.e., $m^{(k)}_{ij} = 1$ means that the classifier in the j-th stage will concat all the feature maps (of k-th layer) from the i-th stage. Note that we restrict the previous stages from concat the features of the later stages, i.e., $m^{(k)}_{ij} = 0, j < i, \forall k < L$. Moreover, we force the L-th layer (the prediction layer[4]) to concat the features from all the previous stages. We demonstrate a concrete example in Fig. 3 to illustrate how to use the above two transformations to reshape a CNN model.

Architecture Search Space. Based on the above two transformations, we can generate the search space by transforming the original CNN model. Specifically, there are two adjustable settings for the two transformations, splitting points and indicator matrices. Adjusting the splitting points will change the way to assign the feature groups, which is used for the trade-off between accuracy and computational cost of different classifiers. For example, we can assign more features to the early stages for improving the model performance on "easy" samples. Adjusting the indicator matrices accompanies the change of the feature reuse strategy. To reduce the size of the search space, we restrict the feature layers

---

[2] We do not split the first layer.
[3] Omit the batch normalization and pooling layers.
[4] Refer to the last layer of the classifier for prediction.

with the same resolution to use the same split and concat settings in our experiments. Through changing these two settings, we can generate the search space $\mathcal{Z}$ which consists of different multi-stage models. In the following section, we will demonstrate how to search for the optimal model in the generated space.

3.3   The Details of NAS

Once we obtain the search space $\mathcal{Z}$ from the above procedure of S2D, the goal of NAS is to find the optimal model a with high accuracy and low computational cost. Note that the model is jointly determined by the settings of the above two transformations, i.e., the split points and the indicator matrices. With a slight abuse of notation, we also refer the architecture a as these two settings and denote $\mathcal{Z}$ as the space which consists of these different settings. Thus the optimization goal reduces to search for the optimal settings of the proposed transformations which can maximize our predefined metric (see details in the following section).

However, searching the optimal setting is nontrivial due to the huge search space $\mathcal{Z}$. For example, in our experiment on MobileNetV2 [35], the size of the search space is around 10^{11}. Motivated by the recent progress in neural architecture search (NAS) [13,38,47,48], we propose to use a policy gradient based reinforcement learning algorithm for searching. The goal of the algorithm is to optimize the policy π which further proceeds the optimal model. This process can be formulated into a nested optimization problem:

$$\arg\max_{\pi} \mathop{\mathbb{E}}_{a \sim \pi} \left(R(a, W_a^*, \mathcal{D}_{val}) \right)$$
$$\text{s.t. } W_a^* = \arg\min_{W_a} \mathcal{L}(a, W_a, \mathcal{D}_{train}) , \tag{1}$$

where W_a is the corresponding weights of the model a and π is the policy which generates the settings of the transformations. $\mathcal{D}_{val}$ and $\mathcal{D}_{train}$ denote the validation and training datasets, respectively. And R is the reward function for evaluating the quality of the multi-stage model.

To solve the nested optimization problem in Eq. 1, we need to solve two sub-problems, namely, optimizing π when W_a^* is given and optimizing W_a when the architecture a is given. We first present how to optimize the policy π when W_a^* is given.

Optimization of the Transformation Settings. Similar to previous works [47,48], we use a customized recurrent neural network (RNN) to generate the distribution of different transformation settings for each layer of the CNN model. Then a policy gradient based algorithm [36] is used for optimizing the parameters of the RNN to maximize the expected reward, which is defined in Eq. 2. Specifically, the reward in our paper is defined as a weighted product considering both the accuracy and the computational cost:

$$R(a, W_a, \mathcal{D}) = \text{COST}(a, W_a, \mathcal{D})^{\omega} \times \text{ACC}(a, W_a, \mathcal{D}) , \tag{2}$$

where $\text{COST}(a, W_a, \mathcal{D})$ is the average computational cost over the samples of the dataset $\mathcal{D}$ using dynamic inference (i.e. $\frac{1}{|\mathcal{D}|} \sum_i \text{COST}_i \times N_i$, where COST_i is the cost of f_i and N_i is the number of samples exiting at f_i). For a fair comparison with other works of dynamic inference, we use FLOPs[5] as the proxy of the computational cost. ω is a hyper-parameter which can be used for controlling the trade-off between model performance and the computational cost. The $\text{ACC}(a, W_a, \mathcal{D})$ is the accuracy of the multi-stage model on the dataset $\mathcal{D}$ (i.e. $\frac{1}{|\mathcal{D}|} \sum_i \text{ACC}_i \times N_i$, where ACC_i is accuracy of the N_i samples exiting at f_i). Next, we will introduce how to solve the inner optimization problem, i.e., optimizing W_a on the training dataset when the model a is given (Fig. 4).

Optimization of the Multi-stage CNN. The inner optimization problem (i.e., solving for W_a^*) can be solved using the gradient descent algorithm. Specifically, we modify the normal classification loss function (i.e., cross-entropy function) for the case of training multi-stage models. Formally, the loss function is defined as:

$$\mathcal{L} = \sum_{(x,y) \in D_{train}} \sum_{i=1}^{s} \alpha_i \text{CE}(f_i(x, W_a), y) , \tag{3}$$

Here, CE denotes the cross-entropy function, x is the input image and y is the class label. The optimization of the above equation can be regarded as jointly optimizing all the classifiers in different stages. The optimization can be implemented using stochastic gradient descent (SGD) and its variants. We use the optimized W_a for assessing the quality of the model a generated by the RNN, which can be further used for optimizing the RNN. In practice, to reduce the search time, following the previous work [48], we approximate W_a^* by updating it for only several training epochs, without solving the inner optimization problem completely by training the network until convergence.

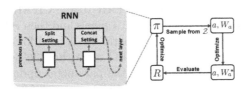

Fig. 4. The process of NAS. The RNN model is responsible for outputting the policy π, i.e., settings of split and concat, which further produces a model a. We can then optimize a for approximating the optimal parameter W_a^*. The computational cost and the accuracy of a is used from evaluating the generated policy π.

---

[5] Here, we regard one multiply-accumulate (MAC) as one floating-point operation (FLOP).

Dynamic Inference of the Searched CNN. Once the optimal multi-stage model $a = \{f_1, \cdots, f_s\}$ is found, we can directly perform dynamic inference using it. Specifically, we set a predefined threshold for each stage. Formally, the threshold of the i-th stage is set to t_i. Then, we can use these thresholds to decide at which stage that the inference should stop. Specifically, given a input sample x, the inference stops at the i-th stage when the i-th classifier outputs a top-1 confidence score $c_i \geq t_i$, here, $c_i = \max(f_i(x))$. Note that the threshold of the classifier should be set before we perform dynamic inference. We use the grid search to find the best threshold on training samples. At first, the correctness and confidences of the samples are obtained by inference them on all the s classifiers. Then we set different thresholds of classifiers from the grid of the thresholds and calculate the rewards on training data set using Eq. 2. The thresholds with the highest reward are choose.

4   Experiments

To verify the effectiveness of S2DNAS, we compare it with different dynamic inference methods on different CNN models. Our experiments have covered a wide range of previous methods of dynamic inference [5,30,39,45]. We also evaluate different aspects of S2DNAS, which are presented in the discussion part.

4.1   Experiment Settings

Model Setup. In our experiments, we conduct experiments on three CNN architectures: ResNet [12], VGG [37], and MobileNetV2 [35][6]. Moreover, to compare with MSDNet, we use the DenseNet-like model and perform S2DNAS on the model.

Searching Details. The CIFAR [19] dataset contains 50k training images and 10k test images. We randomly choose 5k images from the training images as the validation dataset and leave the other 45k images as the training dataset. We use the same input preprocessing for both CIFAR-10 and CIFAR-100. To be specific, the training images are zero-padded with 4 pixels and then randomly cropped to 32×32 resolution. The randomly horizontal flip is used for data augmentation.

For the training of the RNN, the PPO algorithm [36] is used. And we use Adam [18] as the optimizer to perform the parameter update in RNN. For the training of the multi-stage model, we use SGD as the optimizer. The momentum is set to 0.9. The initial learning rate is set to 0.1 and the learning rate is divided by a factor of 10 at 50% and 75% of the total epochs.

For the hyper-parameters of S2DNAS, we set the group number $G = 8$ for every layer. And we set the number of stages $s = 3$. For comparing with MSDNet which contains 5 stages, we set the $s = 5$ for performing S2DNAS on DenseNet-like model. The ω in Eq. 2 is set to -0.06 and all of the α_i in Eq. 3 is set to 1 for all experiments.

---

[6] We use the batch normalization after each convolution layer in VGG and change the stride of the first convolution layer in MobileNetV2 from 2 to 1 for CIFAR.

4.2   Classification Results

In this part, we compare our method with other methods of dynamic inference. To give a comprehensive study of our method, we have covered a wide range of methods, including LCCL [5], BlockDrop [45], Naive [30] and BranchyNet [39]. We conduct experiments on two widely-used image classification benchmarks, CIFAR-10 and CIFAR-100. To show the effectiveness of S2DNAS in reducing the computational cost of CNN models with different architectures, we apply S2DNAS to five typical CNNs with various depth, width, and sub-structures.

The overall results are shown in Table 1. Note that different thresholds (t_i defined in the previous section) lead to different trade-offs between model accuracy and the computational cost. In our experiments, we chose the threshold which leads to the highest reward on the validation dataset. We also provide further results of using different thresholds in the discussion subsection.

As shown in Table 1, for most of the architectures and tasks, our method (denoted as S2DNAS in Table 1) can significantly reduce the computational cost with comparable accuracy with the original CNN model. As mentioned above, we use average FLOPs on the whole test dataset as the metric to measure the computational cost of a given CNN model. For ResNet-20 on CIFAR-10, S2DNAS has reduced the computation cost of the original net from 41M to 16M without the accuracy drop (even with a slight increase as shown in Table 1), which shows a relative cost reduction of 61%.

Our method also shows improvements over other methods for dynamic inference in terms of computational cost reduction. We have reproduced the previous works on these CNN models for comparison. We have also implemented a normal early exiting solution (marked as Naive in Table 1), i.e., directly adding prediction layers (i.e., global average pooling and fully-connected layers) at the intermediate layers of the original models. For example, for ResNet-20 on CIFAR-10, compared with BranchyNet [39], our method has achieved a slight accuracy improvement (from 91.37% to 91.41%) with more computational cost reduction.

One interesting observation is that some methods even cause an increase in computational cost. For example, BlockDrop boosts the FLOPs of the original net about 29%. We infer that this is caused by the controller with high computational cost introduced by BlockDrop in the inference process [45]. We also notice that some of the previous works can not be used for the network without residual connection. For instance, BlockDrop cannot be applied to VGG16-BN. In contrast, our method can generalize to CNN without residual connection. From Table 1, our method can reduce the computational cost of the original VGG16-BN net by 79% with a slight accuracy drop.

Comparison to MSDNet. As mentioned in the introduction section, there is a recent work that proposed a specialized CNN named MSDNet for dynamic inference. Since the method cannot directly be applied to general CNN models, thus for comparison with MSDNet [16], we use the DenseNet-like [17] model[7]. We then

---

[7] Using DenseNet-BC ($d = 100$, $k = 8$) and doubling the growth rate after each transition layer.

Table 1. Evaluations on CNN models with different architectures. The number in the Reduction column denotes the relative cost reduction compared with the original model. Some results are missing because there is no implementation of these CNN models in the reference papers.

Model	Method	CIFAR-10			CIFAR-100		
		FLOPs	Reduction	Accuracy	FLOPs	Reduction	Accuracy
ResNet-20	Original	41M	-	91.25%	41M	-	67.78%
	LCCL	30M	28%	90.95%	40M	1%	68.26%
	BlockDrop	45M	−11%	91.31%	53M	−29%	67.39%
	Naive	34M	18%	91.27%	39M	5%	66.77%
	BranchyNet	33M	20%	91.37%	45M	−9%	67.00%
	S2DNAS	16M	61%	91.41%	25M	39%	67.29%
ResNet-56	Original	126M	-	93.03%	126M	-	71.32%
	LCCL	102M	19%	92.99%	106M	16%	70.33%
	BlockDrop	74M	41%	92.98%	129M	−2%	72.39%
	Naive	68M	46%	92.78%	108M	14%	71.58%
	BranchyNet	73M	42%	92.51%	120M	5%	71.22%
	S2DNAS	37M	71%	92.42%	62M	51%	71.20%
ResNet-110	Original	254M	-	93.57%	254M	-	73.55%
	LCCL	166M	35%	93.44%	210M	17%	72.72%
	BlockDrop	76M	70%	93.00%	153M	40%	73.70%
	Naive	158M	38%	93.13%	217M	15%	73.06%
	BranchyNet	147M	42%	93.33%	243M	5%	73.25%
	S2DNAS	76M	70%	93.39%	113M	56%	73.06%
VGG16-BN	Original	313M	-	93.72%	313M	-	72.93%
	LCCL	269M	14%	92.75%	264M	16%	70.46%
	Naive	185M	41%	93.34%	202M	36%	72.78%
	BranchyNet	162M	48%	93.39%	239M	24%	72.39%
	S2DNAS	66M	79%	93.51%	104M	67%	72.00%
MobileNetV2	Original	91M	-	93.89%	91M	-	74.21%
	LCCL	77M	15%	93.13%	73M	20%	71.11%
	Naive	38M	58%	91.90%	61M	33%	74.03%
	BranchyNet	35M	61%	91.76%	74M	18%	73.71%
	S2DNAS	25M	73%	92.25%	39M	57%	73.50%

apply S2DNAS to it and generate the dynamic models. The results are plotted in Fig. 5. The varying FLOPs metrics of the x-coordinate can be obtained by adjusting the thresholds of each classifier of the dynamic CNN models. As Fig. 5 shows, in most cases, our method can achieve similar accuracy-computation trade-offs. In the case of CIFAR-10, MSDNet outperforms our method when FLOPs is relative to 15M. However, the superiority of MSDNet comes with the cost of manually designing the CNN architecture. In contrast, as Table 1 shows, our method can be applied to various general CNN models.

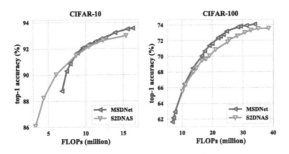

Fig. 5. Comparison to MSDNet.

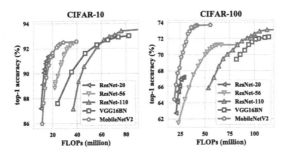

Fig. 6. Trade off accuracy with computational cost by adjusting thresholds of different stages.

4.3 Discussion

Here, we present some discussions on our method for providing further insights.

Trade-off of Accuracy and Computational Cost. A key hyper-parameter of dynamic inference is the threshold setting $t = t_1, \cdots, t_s$, where s is the number of stages. When the model is trained, different threshold settings lead to different trade-offs between the accuracy and the computational cost. To demonstrate how the threshold affects the final model performances, we conduct experiments with different thresholds and plot the results in Fig. 6. All these results show the trend that the increase of computational cost leads to a performance boost. Thus, for practical use, we can set the threshold based on the computational budget of the given hardware device. Moreover, this property also helps to solve the anytime prediction task proposed in the prior work [16].

Difficulty Distribution of Test Dataset. The basic idea of our method is early exiting "easy" samples from the early stages. In this part, we give the statistics of all the samples in the test dataset ($s = 3$, i.e., there are three stages in the trained model). As shown in Table 2, for ResNet-20 on CIFAR-10/100, the inference process of about 50% test samples exits from the first two stages. As a result, S2DNAS can considerably reduce the average computation cost. Further, we observe that the accuracy of the classifier in the first stage[8] is

---

[8] Here, we only consider samples that exit from this stage.

Table 2. Accuracy and fractions of samples in test dataset that exit from each stage.

Dataset	Model	Stage1		Stage2		Stage3	
		Accuracy	Fractions	Accuracy	Fractions	Accuracy	Fractions
CIFAR-10	ResNet-20	98.44%	10.24%	98.89%	41.59%	83.45%	48.17%
	ResNet-56	98.25%	67.50%	89.72%	11.19%	75.36%	21.31%
	ResNet-110	98.43%	61.66%	93.22%	22.28%	74.28%	16.06%
	VGG-16BN	96.54%	87.29%	91.44%	2.22%	68.73%	10.49%
	MobileNetV2	98.62%	50.59%	94.04%	33.21%	68.70%	16.20%
CIFAR-100	ResNet-20	85.27%	58.72%	54.11%	20.68%	29.27%	20.60%
	ResNet-56	97.13%	22.27%	86.64%	29.86%	49.51%	47.87%
	ResNet-110	95.83%	28.04%	85.05%	28.90%	50.19%	43.06%
	VGG-16BN	97.21%	7.18%	90.08%	44.97%	51.20%	47.85%
	MobileNetV2	97.81%	8.68%	90.38%	45.84%	51.85%	45.48%

much higher than the classifier in later stages, which indicates that the classifier can easily classify those samples, i.e., those samples are "easy" samples. This observation also validates the intuition ("easy" samples can be classified using fewer computations) pointed out by some recent works [2,5,7,16].

5   Conclusion

In this paper, we present a general framework called S2DNAS, for transforming various static CNN models into multi-stage models to support dynamic inference. Empirically, our method can be applied to various CNN models to reduce the computational cost, without sacrificing model performance. In contrast to previous methods for dynamic inference, our method comes with two advantages: (1) With our method, we can obtain a dynamic model generated from an existing CNN model instead of manually re-designing a new CNN architecture. (2) The inference of the generated dynamic model does not introduce irregular computations or complex controllers. Thus the generated model can be easily deployed on various hardware devices using existing deep learning frameworks.

These advantages are appealing for deploying a given CNN model into hardware devices with limited computational resources. To be specific, we can first use S2DNAS to transform the given model into the dynamic one then deploy it on the hardware devices. Moreover, our method is orthogonal to previous pruning/quantization methods, which can further reduce the computational cost of the given CNN model. All these properties of our method imply a wide range of application scenarios where the efficient CNN inference is desired.

Acknowledgement. This work is supported by National Natural Science Foundation of China (Grant No. 61832020) and Beijing Academy of Artificial Intelligence (BAAI).

References

1. Berestizshevsky, K., Even, G.: Dynamically sacrificing accuracy for reduced computation: cascaded inference based on softmax confidence. In: Tetko, I.V., Kůrková, V., Karpov, P., Theis, F. (eds.) ICANN 2019. LNCS, vol. 11728, pp. 306–320. Springer, Cham (2019). https://doi.org/10.1007/978-3-030-30484-3_26
2. Bolukbasi, T., Wang, J., Dekel, O., Saligrama, V.: Adaptive neural networks for efficient inference. In: Proceedings of the 34th International Conference on Machine Learning, ICML 2017, Sydney, NSW, Australia, 6–11 August 2017, pp. 527–536 (2017)
3. Cao, S., et al.: SeerNet: predicting convolutional neural network feature-map sparsity through low-bit quantization. In: IEEE Conference on Computer Vision and Pattern Recognition, CVPR 2019, Long Beach, CA, USA, 16–20 June 2019, pp. 11216–11225 (2019)
4. Chen, L.-C., Papandreou, G., Kokkinos, I., Murphy, K., Yuille, A.L.: DeepLab: semantic image segmentation with deep convolutional nets, atrous convolution, and fully connected CRFs. IEEE Trans. Pattern Anal. Mach. Intell. **40**(4), 834–848 (2018)
5. Dong, X., Huang, J., Yang, Y., Yan, S.: More is less: a more complicated network with less inference complexity. In: 2017 IEEE Conference on Computer Vision and Pattern Recognition, CVPR 2017, Honolulu, HI, USA, 21–26 July 2017, pp. 1895–1903 (2017)
6. Figurnov, M., et al.: Spatially adaptive computation time for residual networks. In: 2017 IEEE Conference on Computer Vision and Pattern Recognition, CVPR 2017, Honolulu, HI, USA, 21–26 July 2017, pp. 1790–1799 (2017)
7. Gao, X., Zhao, Y., Dudziak, L., Mullins, R., Xu, C.-Z.: Dynamic channel pruning: feature boosting and suppression. In: 7th International Conference on Learning Representations, ICLR 2019, New Orleans, LA, USA, 6–9 May 2019 (2019)
8. Gupta, S., Agrawal, A., Gopalakrishnan, K., Narayanan, P.: Deep learning with limited numerical precision. In: Proceedings of the 32nd International Conference on Machine Learning, ICML 2015, Lille, France, 6–11 July 2015, pp. 1737–1746 (2015)
9. Han, S., Mao, H., Dally, W.J.: Deep compression: compressing deep neural network with pruning, trained quantization and Huffman coding. In: 4th International Conference on Learning Representations, ICLR 2016, San Juan, Puerto Rico, 2–4 May 2016, Conference Track Proceedings (2016)
10. Han, S., Pool, J., Tran, J., Dally, W.J.: Learning both weights and connections for efficient neural network. In: Advances in Neural Information Processing Systems 28: Annual Conference on Neural Information Processing Systems 2015, 7–12 December 2015, Montreal, Quebec, Canada, pp. 1135–1143 (2015)
11. He, K., Gkioxari, G., Dollár, P., Girshick, R.B.: Mask R-CNN. In: IEEE International Conference on Computer Vision, ICCV 2017, Venice, Italy, 22–29 October 2017, pp. 2980–2988 (2017)
12. He, K., Zhang, X., Ren, S., Sun, J.: Deep residual learning for image recognition. In: 2016 IEEE Conference on Computer Vision and Pattern Recognition, CVPR 2016, Las Vegas, NV, USA, 27–30 June 2016, pp. 770–778 (2016)
13. He, Y., Lin, J., Liu, Z., Wang, H., Li, L.-J., Han, S.: AMC: AutoML for model compression and acceleration on mobile devices. In: Ferrari, V., Hebert, M., Sminchisescu, C., Weiss, Y. (eds.) ECCV 2018. LNCS, vol. 11211, pp. 815–832. Springer, Cham (2018). https://doi.org/10.1007/978-3-030-01234-2_48

14. Hua, W., De Sa, C., Zhang, Z., Suh, G.E.: Channel gating neural networks. CoRR, abs/1805.12549 (2018)
15. Hua, W., Zhou, Y., De Sa, C., Zhang, Z., Suh, G.E.: Boosting the performance of CNN accelerators with dynamic fine-grained channel gating. In: Proceedings of the 52nd Annual IEEE/ACM International Symposium on Microarchitecture, MICRO 2019, Columbus, OH, USA, 12–16 October 2019, pp. 139–150 (2019)
16. Huang, G., Chen, D., Li, T., Wu, F., van der Maaten, L., Weinberger, K.Q.: Multi-scale dense networks for resource efficient image classification. In: 6th International Conference on Learning Representations, ICLR 2018, Vancouver, BC, Canada, 30 April–3 May 2018, Conference Track Proceedings (2018)
17. Huang, G., Liu, Z., van der Maaten, L., Weinberger, K.Q.: Densely connected convolutional networks. In: 2017 IEEE Conference on Computer Vision and Pattern Recognition, CVPR 2017, Honolulu, HI, USA, 21–26 July 2017, pp. 2261–2269 (2017)
18. Kingma, D.P., Ba, J.: Adam: a method for stochastic optimization. In: 3rd International Conference on Learning Representations, ICLR 2015, San Diego, CA, USA, 7–9 May 2015, Conference Track Proceedings (2015)
19. Krizhevsky, A., Hinton, G., et al.: Learning multiple layers of features from tiny images. Technical report. Citeseer (2009)
20. Krizhevsky, A., Sutskever, I., Hinton, G.E.: ImageNet classification with deep convolutional neural networks. In: Advances in Neural Information Processing Systems 25: 26th Annual Conference on Neural Information Processing Systems 2012. Proceedings of a Meeting Held 3–6 December 2012, Lake Tahoe, Nevada, United States, pp. 1106–1114 (2012)
21. LeCun, Y., Denker, J.S., Solla, S.A.: Optimal brain damage. In: Advances in Neural Information Processing Systems 2, [NIPS Conference, Denver, Colorado, USA 27–30 November 1989], pp. 598–605 (1989)
22. Li, H., Kadav, A., Durdanovic, I., Samet, H., Graf, H.P.: Pruning filters for efficient convnets. In: 5th International Conference on Learning Representations, ICLR 2017, Toulon, France, 24–26 April 2017, Conference Track Proceedings (2017)
23. Li, H., Zhang, H., Qi, X., Yang, R., Huang, G.: Improved techniques for training adaptive deep networks. CoRR, abs/1908.06294 (2019)
24. Li,X., Liu, Z., Luo, P., Change Loy, C., Tang, X.: Not all pixels are equal: difficulty-aware semantic segmentation via deep layer cascade. In: 2017 IEEE Conference on Computer Vision and Pattern Recognition, CVPR 2017, Honolulu, HI, USA, 21–26 July 2017, pp. 6459–6468 (2017)
25. Lin, D.D., Talathi, S.S., Annapureddy, V.S.: Fixed point quantization of deep convolutional networks. In: Proceedings of the 33nd International Conference on Machine Learning, ICML 2016, New York City, NY, USA, 19–24 June 2016, pp. 2849–2858 (2016)
26. Liu, C., et al.: Auto-DeepLab: hierarchical neural architecture search for semantic image segmentation. In: IEEE Conference on Computer Vision and Pattern Recognition, CVPR 2019, Long Beach, CA, USA, 16–20 June 2019, pp. 82–92 (2019)
27. Liu, H., Simonyan, K., Yang, Y.: DARTS: differentiable architecture search. In: 7th International Conference on Learning Representations, ICLR 2019, New Orleans, LA, USA, 6–9 May 2019 (2019)

28. Liu, L., Deng, J.: Dynamic deep neural networks: optimizing accuracy-efficiency trade-offs by selective execution. In: Proceedings of the Thirty-Second AAAI Conference on Artificial Intelligence, (AAAI-18), the 30th innovative Applications of Artificial Intelligence (IAAI-18), and the 8th AAAI Symposium on Educational Advances in Artificial Intelligence (EAAI-18), New Orleans, Louisiana, USA, 2–7 February 2018, pp. 3675–3682 (2018)
29. Liu, W., et al.: SSD: single shot multibox detector. In: Leibe, B., Matas, J., Sebe, N., Welling, M. (eds.) ECCV 2016. LNCS, vol. 9905, pp. 21–37. Springer, Cham (2016). https://doi.org/10.1007/978-3-319-46448-0_2
30. Panda, P., Sengupta, A., Roy, K.: Conditional deep learning for energy-efficient and enhanced pattern recognition. In: 2016 Design, Automation & Test in Europe Conference & Exhibition, DATE 2016, Dresden, Germany, 14–18 March 2016, pp. 475–480 (2016)
31. Rastegari, M., Ordonez, V., Redmon, J., Farhadi, A.: XNOR-Net: imagenet classification using binary convolutional neural networks. In: Leibe, B., Matas, J., Sebe, N., Welling, M. (eds.) ECCV 2016. LNCS, vol. 9908, pp. 525–542. Springer, Cham (2016). https://doi.org/10.1007/978-3-319-46493-0_32
32. Redmon, J., Divvala, S.K., Girshick, R.B., Farhadi, A.: You only look once: unified, real-time object detection. In: 2016 IEEE Conference on Computer Vision and Pattern Recognition, CVPR 2016, Las Vegas, NV, USA, 27–30 June 2016, pp. 779–788 (2016)
33. Ren, M., Pokrovsky, A., Yang, B., Urtasun, R.: SBNet: sparse blocks network for fast inference. In: 2018 IEEE Conference on Computer Vision and Pattern Recognition, CVPR 2018, Salt Lake City, UT, USA, 18–22 June 2018, pp. 8711–8720 (2018)
34. Ren, S., He, K., Girshick, R.B., Sun, J.: Faster R-CNN: towards real-time object detection with region proposal networks. In: Advances in Neural Information Processing Systems 28: Annual Conference on Neural Information Processing Systems 2015, Montreal, Quebec, Canada, 7–12 December 2015, pp. 91–99 (2015)
35. Sandler, M., Howard, A.G., Zhu, M., Zhmoginov, A., Chen, L.-C.: MobileNetV 2: inverted residuals and linear bottlenecks. In: 2018 IEEE Conference on Computer Vision and Pattern Recognition, CVPR 2018, Salt Lake City, UT, USA, 18–22 June 2018, pp. 4510–4520 (2018)
36. Schulman, J., Wolski, F., Dhariwal, P., Radford, A., Klimov, O.: Proximal policy optimization algorithms. CoRR, abs/1707.06347 (2017)
37. Simonyan, K., Zisserman, A.: Very deep convolutional networks for large-scale image recognition. In: 3rd International Conference on Learning Representations, ICLR 2015, San Diego, CA, USA, 7–9 May 2015, Conference Track Proceedings (2015)
38. Tan, M., et al.: MnasNet: platform-aware neural architecture search for mobile. In: IEEE Conference on Computer Vision and Pattern Recognition, CVPR 2019, Long Beach, CA, USA, 16–20 June 2019, pp. 2820–2828. Computer Vision Foundation/IEEE, (2019)
39. Teerapittayanon, S., McDanel, B., Kung, H.T.: BranchyNet: fast inference via early exiting from deep neural networks. In: 23rd International Conference on Pattern Recognition, ICPR 2016, Cancún, Mexico, 4–8 December 2016, pp. 2464–2469 (2016)
40. Veit, A., Belongie, S.: Convolutional networks with adaptive inference graphs. In: Ferrari, V., Hebert, M., Sminchisescu, C., Weiss, Y. (eds.) ECCV 2018. LNCS, vol. 11205, pp. 3–18. Springer, Cham (2018). https://doi.org/10.1007/978-3-030-01246-5_1

41. Wang, X., Yu, F., Dou, Z.-Y., Darrell, T., Gonzalez, J.E.: SkipNet: learning dynamic routing in convolutional networks. In: Ferrari, V., Hebert, M., Sminchisescu, C., Weiss, Y. (eds.) ECCV 2018. LNCS, vol. 11217, pp. 420–436. Springer, Cham (2018). https://doi.org/10.1007/978-3-030-01261-8_25
42. Wen, W., Wu, C., Wang, Y., Chen, Y., Li, H.: Learning structured sparsity in deep neural networks. In: Advances in Neural Information Processing Systems 29: Annual Conference on Neural Information Processing Systems 2016, Barcelona, Spain, 5–10 December 2016, pp. 2074–2082 (2016)
43. Wu, B., et al.: FBNet: hardware-aware efficient convnet design via differentiable neural architecture search. In: IEEE Conference on Computer Vision and Pattern Recognition, CVPR 2019, Long Beach, CA, USA, 16–20 June 2019, pp. 10734–10742 (2019)
44. Wu, J., Leng, C., Wang, Y., Hu, Q., Cheng, J.: Quantized convolutional neural networks for mobile devices. In: 2016 IEEE Conference on Computer Vision and Pattern Recognition, CVPR 2016, Las Vegas, NV, USA, 27–30 June 2016, pp. 4820–4828 (2016)
45. Wu, Z., et al.: BlockDrop: dynamic inference paths in residual networks. In: 2018 IEEE Conference on Computer Vision and Pattern Recognition, CVPR 2018, Salt Lake City, UT, USA, 18–22 June 2018, pp. 8817–8826 (2018)
46. Yu, J., Lukefahr, A., Palframan, D.J., Dasika, G.S., Das, R., Mahlke, S.A.: Scalpel: customizing DNN pruning to the underlying hardware parallelism. In: Proceedings of the 44th Annual International Symposium on Computer Architecture, ISCA 2017, Toronto, ON, Canada, 24–28 June 2017, pp. 548–560 (2017)
47. Zoph, B., Le, Q.V.: Neural architecture search with reinforcement learning. In: 5th International Conference on Learning Representations, ICLR 2017, Toulon, France, 24–26 April 2017, Conference Track Proceedings (2017)
48. Zoph, B., Vasudevan, V., Shlens, J., Le, Q.V.: Learning transferable architectures for scalable image recognition. In: 2018 IEEE Conference on Computer Vision and Pattern Recognition, CVPR 2018, Salt Lake City, UT, USA, 18–22 June 2018, pp. 8697–8710 (2018)

Improving Deep Video Compression by Resolution-Adaptive Flow Coding

Zhihao Hu[1], Zhenghao Chen[2], Dong Xu[2], Guo Lu[3]([✉]), Wanli Ouyang[2], and Shuhang Gu[2]

[1] College of Software, Beihang University, Beijing, China
[2] School of Electrical and Information Engineering, The University of Sydney, Sydney, Australia
[3] School of Computer Science and Technology, Beijing Institute of Technology, Beijing, China
sdluguo@gmail.com

Abstract. In the learning based video compression approaches, it is an essential issue to compress pixel-level optical flow maps by developing new motion vector (MV) encoders. In this work, we propose a new framework called Resolution-adaptive Flow Coding (RaFC) to effectively compress the flow maps globally and locally, in which we use multi-resolution representations instead of single-resolution representations for both the input flow maps and the output motion features of the MV encoder. To handle complex or simple motion patterns globally, our frame-level scheme RaFC-frame automatically decides the optimal flow map resolution for each video frame. To cope different types of motion patterns locally, our block-level scheme called RaFC-block can also select the optimal resolution for each local block of motion features. In addition, the rate-distortion criterion is applied to both RaFC-frame and RaFC-block and select the optimal motion coding mode for effective flow coding. Comprehensive experiments on four benchmark datasets HEVC, VTL, UVG and MCL-JCV clearly demonstrate the effectiveness of our overall RaFC framework after combing RaFC-frame and RaFC-block for video compression.

1 Introduction

There is increasing demand for new video compression systems to effectively reduce redundancy in video sequences. The conventional video compression systems are based on hand-designed modules such as block based motion estimation and Discrete Cosine Transform (DCT). Taking advantage of large-scale training datasets and powerful nonlinear modeling capacity of deep neural networks,

---

Z. Hu and Z. Chen—First two authors contributed equally.

Electronic supplementary material The online version of this chapter (https://doi.org/10.1007/978-3-030-58536-5_12) contains supplementary material, which is available to authorized users.

© Springer Nature Switzerland AG 2020
A. Vedaldi et al. (Eds.): ECCV 2020, LNCS 12347, pp. 193–209, 2020.
https://doi.org/10.1007/978-3-030-58536-5_12

the recent deep video compression methods [17, 28, 33] have achieved promising video compression performance (Please refer to Sect. 2 for more details about the related image and video compression methods). Specifically, in the recent end-to-end deep video compression (DVC) framework [17], all modules (e.g., DCT, motion estimation and motion compensation) in the conventional H.264/H.265 codec are replaced with the well-designed neural networks.

In the learning based video compression approaches such as the aforementioned DVC framework, it is a non-trivial task to compress pixel-level optical flow maps. However, such frameworks adopt single representations for both input flow maps and output motion features using a single motion vector (MV) encoder. This cannot effectively handle complex or simple motion patterns in different scenes and fast or slow movement of objects. To this end, in this work we propose a new framework called Resolution-adaptive Flow Coding (RaFC), which can adopt multi-resolution representations for both flow maps and motion features and then automatically decide the optimal resolutions at both frame-level and block-level in order to achieve the optimal rate-distortion trade-off.

At the frame-level, our RaFC-frame scheme can automatically decide the optimal flow map resolution for each video frame in order to effectively handle complex or simple motion patterns globally. As a result, for those frames with complex global motion patterns, high-resolution flow maps containing more detailed optimal flow information are more likely to be selected as the input for the MV encoder. In contrast, for the frames with simple global motion patterns, low-resolution optimal flow maps are generally preferred.

Inspired by the traditional codecs [23, 32], in which the blocks with different sizes are used for motion estimation, we also propose a new scheme RaFC-block, which can decide the optimal resolution for each block based on the rate-distortion (RD) criterion when encoding the motion features. As a result, for the local blocks with complicated motion patterns, our RaFC-block scheme will use high-resolution blocks containing fine motion features. For the blocks within smooth areas, our RaFC-block scheme prefers low-resolution blocks with coarse motion features in order to save bits for encoding their motion features without substantially sacrificing the distortion. In addition, we also propose an overall RaFC framework by combining the two newly proposed schemes RaFC-frame and RaFC-block.

We perform comprehensive experiments on four benchmark datasets HEVC Class E, VTL, UVG and MCL-JCV. The results clearly demonstrate our overall RaFC framework outperforms the baseline algorithms including H.264, H.265 and DVC. Our contributions are summarized as follows:

- To effectively handle complex or simple motion patterns globally, we adopt the multi-resolution representations for the flow maps, in which the optimal resolution at the frame-level can be automatically decided for our method RaFC-Frame based on the RD criterion.
- Using multi-resolution representations for motion features, we additionally propose the RaFC-block method to automatically decide the optimal

resolution at the block-level based on the RD criterion, which can effectively cope with different types of local motion patterns.
- Our overall RaFC framework after combining RaFC-frame and RaFC-block achieves the state-of-the-arts video compression performance on four benchmark datasets including HEVC Class E, VTL, UVG and MCL-JCV.

2 Related Work

2.1 Image Compression

Transform-based image compression methods can efficiently reduce the spatial redundancy. Currently, those approaches (e.g., JPEG [29], BPG [7] and JPEG2000 [24]) are still the most widely used image compression algorithms. Recently, the deep learning based image compression methods [3–6,14,16,21,25–27] have been proposed and achieved the state-of-the-arts performance. The general idea of deep image compression is to transform input images into quantized bit-streams, which can be further compressed through lossless coding algorithms. To achieve this goal, some methods [14,26,27] directly employed recurrent neural networks (RNNs) to compress the images in a progressive manner. Toderici *et al.* [26] firstly introduced a simple RNN-based approach to compress the image and further proposed a method [27], which enhances the performance by progressively compressing reconstructed residual information. Johnston *et al.* [14] also improved Toderici's work by introducing a new objective loss. Other popular approaches use an auto-encoder architecture [5,6,19,25]. Balle *et al.* [5] introduced a continuous and differentiable proxy for the rate-distortion loss and further proposed a variational auto-encoder based compression algorithm [6].

Recently, some methods [6,19] focus on predicting different distribution in different spatial area. And Li *et al.* [16] introduced the importance map to reduce the total binary codes to transmit. All such methods need to transmit the full-resolution feature map to the decoding stage. Our proposed method selects the most optimal resolution at both frame-level and block-level in the encoding side, which saves a lot of bits.

2.2 Video Compression

Traditional video compression algorithms, such as H.264 [32] and H.265 [23], adopted the hand-crafted operations for motion estimation and motion compensation for inter-frame prediction. Even though they can successfully reduce temporal redundancy of video data, those compression algorithms are limited in compression performance as they cannot be jointly optimized.

With the success of deep learning based motion estimation and image compression approaches, some attempts have been made to use neural networks for video compression [8,28,33,34], in which the neural networks are used to replace the modules from the conventional approach. The work in [8] proposed a block based approach, while Tsai *et al.* [28] utilized an auto-encoder approach to compress residual information from H.264. Wu *et al.* [33] predicted and reconstructed video frames by using interpolation. While the above works have achieved remarkable performance, they cannot be trained in an end-to-end fashion, which limits their performance.

Recently, more deep video compression methods [9,11,17,18,22] have been proposed. Lu *et al.* [17] proposed the first end-to-end deep learning video compression (DVC) framework, which replaces all the key components of the traditional video compression codec with deep neural networks. Rippel *et al.* [22] proposed to maintain a state, which contains the past information, compressed motion information and residual information for video compression. Djelouah *et al.* [9] proposed an interpolation based video compression approach, which combines motion compression and image synthesis in a single network. In these works, optical flow information plays an essential role. In order to achieve reasonable compression performance, the state-of-the-art optical flow estimation networks [10,13] have been adopted to provide accurate motion estimation. However, as these optical flow estimation networks were designed for generating accurate full-resolution motion maps, they are not optimal for the video compression task. Recently, Habibian *et al.* [12] proposed a 3D auto-encoder approach without requiring optical flow for motion compensation. However, their algorithm is still limited for capturing fine scale motions.

In contrast to these works, we propose a new framework RaFC to effectively compress optical flow maps, and it can be trained in an end-to-end fashion.

3   Methodology

3.1   System Overview

Figure 1(a) provides an overview of the proposed video compression system. Inspired by the DVC [17] framework, we also use a hybrid coding scheme (e.g., motion coding and residual coding). The overall *coding procedure* is summarized in the following steps.

Motion Coding. We utilize our proposed RaFC method for motion coding. RaFC consists of three modules, motion estimation net, the motion vector (MV) encoder net, and the MV decoder net. The motion estimation net estimates the optical flow V_t between the input frame X_t and the previous reconstructed frame $\hat{X}_{t-1}$ from the decoded frames buffer. Then, the MV encoder net encodes the optical flow maps as motion features/representations M_t, which is further quantized as $\hat{M}_t$ before entropy coding. Finally, the MV decoder net decodes the motion representation $\hat{M}_t$ so that the reconstructed flow map $\hat{V}_t$ is obtained.

Motion Compensation. Based on the reconstructed optical flow map $\hat{V}_t$ from the MV decoder and the reference frame $\hat{X}_{t-1}$, a motion compensation network is employed to obtain the predicted frame $\bar{X}_t$.

Residual Coding. Denote the residual between the original frame X_t and the predicted frame $\bar{X}_t$ by R_t. Like in [17], we adopt a residual encoder network to encode the residual as the latent representation Y_t and then quantized as $\hat{Y}_t$ for

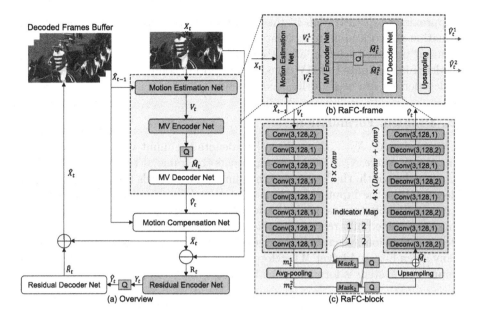

Fig. 1. Overview of our proposed framework and several basic modules used in our pipeline (a), the detailed motion coding modules in our frame-level scheme RaFC-frame (b) and our block-level scheme RaFC-block (c). In RaFC-frame (dashed yellow box), the "Motion Estimation Net" will generate two optical flow maps V_t^1 and V_t^2 with different resolutions and our method automatically select the optimal resolution (see the details in Sect. 3.3(a)). In RaFC-block, the optical flow map V_t (i.e., V_t^1 or V_t^2) is transformed to multi-scale motion features m_t^1 and m_t^2, and we will select the most optimal resolution for each block by using the representations from either m_t^1 or m_t^2 to construct the reorganized motion feature $\hat{M}_t$, which will be used to obtain the reconstructed flow map $\hat{V}_t$ (see the details in Sect. 3.3(b)). In (c), Conv(3,128,2) represents the convolution operation with the kernel size of 3×3, the output channel of 128 and the stride of 2. Each convolution with the stride of 1 is followed by a Leaky ReLU layer. Two masks $Mask_1$ and $Mask_2$ are only used for "Motion Feature Reorganization" and are not used for "Indicator Map Generation" (see Sect. 3.3(b) for more details).

entropy coding. Then the residual decoder network reconstructs the residual $\hat{R}_t$ from the latent representation $\hat{Y}_t$.

Frame Reconstruction. With the predicted frame $\bar{X}_t$ from the motion compensation net and $\hat{R}_t$ obtained from the residual decoder net, the final reconstructed frame for X_t can be obtained by $\hat{X}_t = \bar{X}_t + \hat{R}_t$, which is also sent to the decoded frames buffer and will be used as the reference frame for the next frame X_{t+1}.

Quantization and Bit Estimation. The generated latent representations (e.g., $\hat{Y}_t$) should be quantized before sending to the decoder side. To build an end-to-end optimized system, we follow the method in [6] and add uniform noise to approximate quantization in the training stage. Besides, we use the bitrate estimation network in [6] to estimate the entropy coding bits.

In our proposed scheme, all the components in Fig. 1(a) are included in the encoder side, and only the MV decoder net, motion compensation net and residual decoder net are used in the decoder side.

3.2   Problem Formulation

We use $X = \{X_1, X_2, ..., X_{t-1}, X_t, ...\}$ to denote the input video sequence to be compressed, where $X_t \in \mathbb{R}^{W \times H \times C}$ represents the frame at time step t. W, H, C represent the width, the height and the number of channels (i.e., $C = 3$ for RGB videos). Given the input video sequences, the video encoder will generate the corresponding bitstreams, while the decoder reconstructs the video sequences by using the received bitstreams. To achieve highly efficient compression, the whole video compression system needs to generate high quality reconstructed frames at any given bitrate budget. Therefore, the objective of the learning based video compression system is formulated as follows,

$$RD = R + \lambda D = (\mathbb{H}(\hat{M}_t) + \mathbb{H}(\hat{Y}_t)) + \lambda d(X_t, \hat{X}_t), \tag{1}$$

The term R in Eq. (1) denotes the number of bits used to encode the frame. R is calculated by adding up the number of bits $\mathbb{H}(\hat{M}_t)$ for encoding the flow information and the number of bits $\mathbb{H}(\hat{Y}_t)$ for encoding the residual information. $D = d(X_t, \hat{X}_t)$ denotes the distortion between the input frame and the reconstructed frame, where $d(\cdot)$ represents the metric (mean square error or MS-SSIM [31]) for measuring the difference between two images.

In the traditional video compression system, the rate-distortion optimization (RDO) technique is widely used to select the optimal mode for each coding block. The RDO procedure is formulated as follows,

$$\mathcal{M} = \arg \min_{i \in \mathcal{C}} RD_i \tag{2}$$

where RD_i represents the RD value of the ith mode, and $\mathcal{C}$ represents the candidate modes. The RDO procedure will select the optimal mode $\mathcal{M}$ with the minimum rate-distortion (RD) value to achieve highly efficient video coding.

However, this basic technique is not exploited in the state-of-the-art learning based video compression systems. In this work, we propose the RaFC framework to effectively compress motion information by using multi-resolution representations for the flow maps and motion features. The key idea in our method is to use the RDO technique to select the optimal resolution of optical flow maps or motion features at each block for the current frame.

3.3   Resolution-Adaptive Flow Coding (RaFC)

In this section, we introduce our RaFC scheme for motion compression and present how to select the optimal flow map or motion features by using the RDO technique based on the RD criterion.

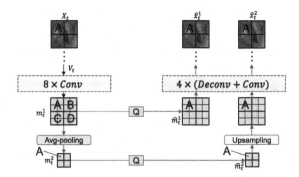

Fig. 2. Generation of the indicator map. The network structures of $8 \times Conv$ and $4 \times (Deconv + Conv)$ are provided in Fig. 1(c). For better illustration, one channel is shown as an example.

(a) Frame-level Scheme RaFC-frame

As shown in Fig. 1(b), given the input frame X_t and its corresponding reference frame $\hat{X}_{t-1}$ from the decoded frames buffer, we utilize the motion estimation network to generate the multi-scale flow maps. Taking advantage of the existing pyramid architecture in Spynet [20] in our work, we generate two flow maps V_t^1 and V_t^2 with the resolutions of $W \times H$ and $\frac{W}{2} \times \frac{H}{2}$, respectively. While more resolutions can be readily used in our RaFC-frame method, we observe that our RaFC-frame scheme based on two-scale optical flow maps has already been able to achieve promising results.

In our proposed frame-level scheme RaFC-frame, the goal is to select the optimal resolution from the multi-scale optical flow maps for the current frame in order to handle complex or simple motion patterns globally. According to the RDO formulation in Eq. (2), we need to calculate the RD values for the two optical flow maps V_t^1 and V_t^2 respectively. The details are provided below.

Calculating the Rate-Distortion (RD) Value. We take the optical flow map V_t^2 as an example to introduce how to calculate the RD value. First, as shown in Fig. 1(b), based on the MV encoder and the MV decoder, we can obtain the reconstructed optical flow map and the corresponding quantized representation $\hat{M}_t^2$. While the resolution of the reconstructed flow map is only $\frac{W}{2} \times \frac{H}{2}$, there is an additional upsampling operation before obtaining $\hat{V}_t^2$, so the resolution of $\hat{V}_t^2$ is also $W \times H$. After going through the subsequent coding procedure, such as the motion compensation unit, the residual encoder unit and the residual decoder unit (see Sect. 3.1 for more details), we arrive at the reconstructed frame $\hat{X}_t^2$ and also obtain the corresponding bitstreams from $\hat{M}_t^2$ and $\hat{Y}_t^2$, for motion information and residual information, respectively. Therefore, based on Eq. (1), we can calculate the RD value for the flow map V_t^2. We can similarly calculate the RD value for the flow map V_t^1. Finally, we select the optimal flow map with the minimum RD value.

After selecting the optimal flow map of the current frame by using the RDO technique in Eq. (2), we can update the network parameters by using the loss function defined in Eq. (1), where $\hat{M}_t$, $\hat{Y}_t$ and $\hat{X}_t$ are obtained based on the selected flow map (i.e., V_t^1 or V_t^2).

(b) Block-level Scheme RaFC-block

Previous learning based video compression systems only use motion features with fix resolution to represent optical flow information. In H.264 and H.265, different block sizes are used for motion estimation. To this end, it is necessary to design an efficient multi-scale motion features in order to handle different types of motion patterns.

As shown in Fig. 1(c), given the optical flow map V_t from one resolution (i.e. V_t can be V_t^1 or V_t^2 from Sect. 3.3(a)), we firstly feed the optical flow map V_t to generate the multi-scale motion features m_t^1 and m_t^2. Here we just use two-resolution motion features as an example, and our approach can be readily used for more resolutions (we use three-resolution motion features in our experiments). Then, the proposed RaFC-block method will select the optimal resolution of the motion features for each block in the reconstructed frame based on the RDO technique. Specifically, we proposed a two-step procedure, which is summarized as follows.

Indicator Map Generation. In Fig. 2, we take an input image with the resolution of 64×64 as an example to introduce how to generate the indicator map with the size of 2×2. After four pairs of convolution layers with the strides of 1 and 2, we can obtain the motion feature m_t^1 with the resolution of 4×4. We divide m_t^1 as 4 blocks A, B, C and D, and each block represents a 2×2 region. Based on m_t^1, we further obtain m_t^2 with the resolution of 2×2 after going through another average pooling layer. Then for each block (A, B, C, or D), we need to decide whether we should choose the 2×2 representation from m_t^1 or the 1×1 representation from m_t^2. The details are provided below.

After quantizing m_t^1 to obtain $\hat{m}_t^1$, we will go through four pairs of deconvolution and convolution layers and the rest coding procedure (e.g. the motion compensation unit, the residual encoder unit and the residual decoder unit), we can obtain the final reconstructed image $\hat{x}_t^1$ with the resolution of 64×64 from $\hat{m}_t^1$. We also quantize m_t^2 as $\hat{m}_t^2$, and go through an additional upsampling layer to reach the same size with $\hat{m}_t^1$. Then after four pairs of deconvolution and convolution layers and the rest coding procedure, we can also obtain $\hat{x}_t^2$ with the resolution of 64×64. We then similarly divide $\hat{x}_t^1$ and $\hat{x}_t^2$ as four blocks A, B, C, and D. For each block in both $\hat{x}_t^1$ and $\hat{x}_t^2$, we can calculate the RD value by using Eq. (1), where the bit rates are calculated by using the corresponding motion features and the residual image at one specific block, and the distortion D is also calculated for this specific block. By choosing the smaller RD value, we can determine which representation of motion feature (i.e., the 2×2 representation from m_t^1 or the 1×1 representation from m_t^2) will be used at each block.

In this way, we can obtain the indicator map which represents the optimal resolution choice at each block. While more advanced approaches can be used

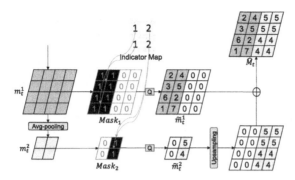

Fig. 3. Motion feature reorganization with the indicator map. For better illustration, one channel is shown as an example.

to decide the indicator map, it is worth mentioning that the aforementioned solution is efficient and achieves promising results (see our results in Sect. 4).

Motion Feature Reorganization. In our approach, we need to reorganize the motion representation based on the indicator map. As shown in Fig. 3, given the indicator map and the quantized features, we first obtain the masked and quantized multi-scale motion features $\tilde{m}_t^1$ and $\tilde{m}_t^2$. The corresponding locations without features, which are also masked at the encoder side, are filled with zeros. Then from bottom to top, $\tilde{m}_t^2$ is first upsampled to the same size of $\tilde{m}_t^1$, which is then added to $\tilde{m}_t^1$. In this way, we can obtain the reorganized motion feature $\hat{M}_t$, which exploits the multi-scale motion representations for better motion compression.

After motion feature reorginzation, we can easily obtain the quantized residual information $\hat{Y}_t$ and the reconstructed frame $\hat{X}_t$ by following the hybrid coding scheme in Fig. 1(a), which includes the motion compensation unit, the residual encoder unit and the residual decoder unit. Then the loss function defined in Eq. (1) will be minimized to update the network parameters.

(c) Our Overall RaFC Framework by Combining both Schemes

The frame-level scheme RaFC-frame selects the optimal resolution of optical flow maps, which is the input of the MV encoder, while the block-level scheme RaFC-block selects the optimal resolution for motion features at each block, which is the output of the MV encoder. Therefore, these two techniques are complementary to each other and can be readily combined.

Specifically, we embed the block-level method RaFC-block into the frame-level method RaFC-frame. For the first input flow map V_t^1, we use the RaFC-block method to decide the optimal indicator map based on the RD criterion at the block level, and then output $\hat{V}_t^1$ based on the reorganized motion feature. After going through the subsequent coding process including the motion compensation unit, the residual encoder unit and the residual decoder unit, we finally obtain the reconstructed frame $\hat{X}_t^1$. Based on the distortion between $\hat{X}_t^1$ and X_t, and the numbers of bits used for encoding both the reorganized motion feature and residual information, we can calculate the RD value. For the second input flow map V_t^2, we perform the same process and calculate the RD value. Finally, we choose the optimal mode with the minimum RD value for encoding motion information of the current frame. Here, the optimal mode includes the selected optical flow map and the corresponding selected resolution of motion features at each block for this selected flow map.

After selecting the optimal mode for encoding the motion information of the current frame, we update all the parameters in our network by minimizing the objective function in Eq. (1), where the distortion and the numbers of bits used to encode the motion features and the residual information are obtained for the selected mode.

4   Experiment

4.1   Experimental Setup

Datasets. We use the Vimeo-90k dataset [35] to train our framework and each clip in this dataset consists of 7 frames with the resolution of 448×256.

For performance evaluation, we use four datasets: HEVC Class E [23], UVG [1], MCL-JCV [30] and VTL [2]. The HEVC Standard Test Sequences have been widely used for evaluating the traditional video compression methods, in which the HEVC class E dataset contains three videos with the resolution of 1280×720. The UVG dataset [1] has seven videos with the resolution of 1920×1080. The MCL-JCV dataset [30] has been widely used for video quality evaluation, which has 30 videos with the resolution of 1920×1080. For the VTL dataset [2], we follow the experimental setting in [9] and use the first 300 frames in each video clip for performance evaluation.

Evaluation Metric. We use PSNR and MS-SSIM [31] to measure the distortion between the reconstructed and ground-truth frames. PSNR is the most widely used metric for measuring compression distortion, while MS-SSIM has been adopted in many recent works to evaluate the subjective visual quality. We use bit per pixel (Bpp) to denote the bitrate cost in the compression procedure.

Implementation Details. We train our model in two stages. At the first stage, we set λ as 2048, and train our model based on mean square error for 2,000,000 steps to obtain a pre-trained model at high bitrate. At the second stage, for different λ values ($\lambda = 256, 512, 1024$ and 2048), we fine-tune the pretrained model for another 500,000 iterations. To achieve better MS-SSIM performance, we additionally fine-tune the models from the second stage for about 80,000 steps by using the MS-SSIM criterion as the distortion term when calculating the RD values.

Our framework is implemented based on Pytorch with CUDA support. In the training phase, we set the batch size as 4. We use the Adam optimizer [15] with the learning rate of $1e-4$ for the first 1,800,000 steps and 1e-5 for the remaining steps. It takes about 6 days to train the proposed model.

In our experiments, motion features ($\hat{m}_t^1$, $\hat{m}_t^2$ and $\hat{m}_t^3$) with three different resolutions are used in our RaFC-block module (note $\hat{m}_t^3$ can be similarly obtained from $\hat{m}_t^2$ as shown in Fig. 2). It is noted that one pixel in $\hat{m}_t^1$, $\hat{m}_t^2$ and $\hat{m}_t^3$ correspond to one block with the resolution of 16×16, 32×32 and 64×64 in the original optical flow map, respectively.

4.2   Experimental Results

The experimental results on different datasets are provided in Fig. 4. In DVC [17], the hyperprior entropy model [6] is used to compress the flow maps. However, other advanced methods like the auto-regressive entropy model [19] can be readily used to compress the flow maps. To this end, we report two results for our RaFC framework, which are denoted as "Ours" and "Ours*". In "Ours", the hyperprior entropy model [6] is incorporated in our RaFC framework in order to fairly compare our RaFC framework with DVC. In "Ours*", the auto-regressive entropy model [19] is incorporated in our RaFC framework to further improve the video compression performance. We use the traditional compression methods H.264 [32], H.265 [23] and the state-of-the-art learning-based compression methods, including DVC [17], AD_ICCV [9], AH_ICCV [12] and CW_ECCV [33] for performance comparison. It is noted that CW_ECCV [33] and AD_ICCV [9] are B-frame based compression methods, while the others are P-frame based compression methods. For H.264 and H.265, we follow the setting in DVC [17] and use FFmpeg with the *default* mode. We use the image compression method [6] to reconstruct the I-frame.

As shown in Fig. 4, our method using the hyperprior entropy model (i.e., "Ours") outperforms the baseline method DVC on all datasets, which demonstrates it is beneficial to use our newly proposed framework RaFC to compress the optical flow maps. In other words, it is necessary to choose the optimal resolutions for the optical flow maps and the corresponding motion features in video compression. When compared with our method using the hyperprior entropy model (i.e., "Ours"), our method using the auto-regressive entropy model (i.e., "Ours*") further improves the results, which demonstrates the effectiveness of the auto-regressive entropy model for flow compression. Our method using the auto-regressive entropy model [6] achieves the best results on all datasets. Specifically, our method (i.e., "Ours*") has about 0.5 dB gain over DVC at 0.1bpp on the UVG dataset. On the MCL-JCV dataset, our approach (i.e., "Ours*") outperforms the interpolation based video compression method AD_ICCV in terms of both PSNR and MS-SSIM. In addition, it also achieves about 0.4dB improvement at 0.2bpp over AD_ICCV on the VTL dataset in terms of PSNR. Although our method is designed for P-frame compression, we can still achieve better compression performance than the B-frame compression methods AD_ICCV and CW_ECCV, which demonstrates the effectiveness of our approach.

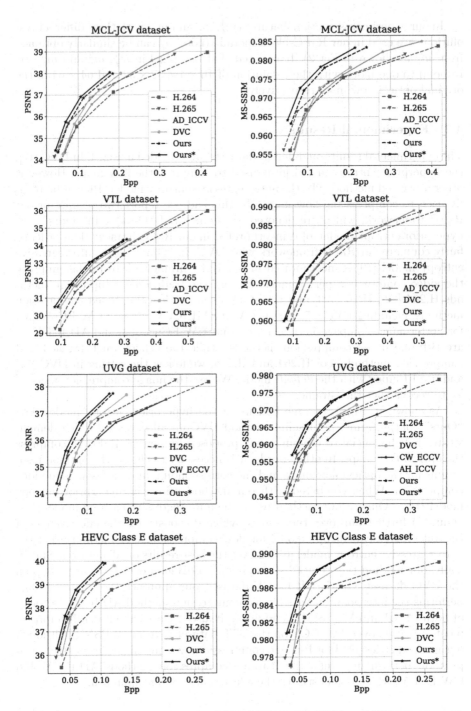

Fig. 4. Experimental results on the MCL-JCV, VTL, UVG and HEVC Class E datasets.

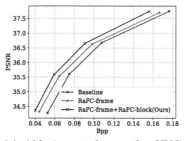 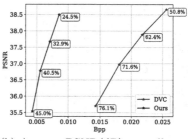

(a) Ablation study on the UVG dataset. DVC is adopted as our baseline method.

(b) Average PSNR(dB) over all predicted frames (i.e., $\bar{X}_t$'s) and the percentage of bits used to encode motion information over the total number of bits at different Bpps on the HEVC Class E dataset.

Fig. 5. Ablation study and model analysis.

Table 1. Percentages of the selected optical flow map resolutions when using our RaFC-frame scheme at different λ values.

	High resolution (i.e., $\hat{V}_t^1$)	Low resolution (i.e., $\hat{V}_t^2$)
$\lambda = 256$	38.89%	61.11%
$\lambda = 512$	45.14%	54.86%
$\lambda = 1024$	57.64%	42.36%
$\lambda = 2048$	63.20%	36.80%

Table 2. Percentages of the selected block resolutions when using our RaFC-block scheme at different λ values.

Block Resolutions	16×16 (i.e., $\hat{m}_t^1$)	32×32 (i.e., $\hat{m}_t^2$)	64×64 (i.e., $\hat{m}_t^3$)
$\lambda = 256$	0.98%	40.55%	58.46%
$\lambda = 512$	27.18%	36.69%	36.11%
$\lambda = 1024$	36.44%	32.27%	31.28%
$\lambda = 2048$	41.91%	31.02%	27.06%

4.3 Ablation Study and Model Analysis

Effectiveness of Different Components. In order to verify the effectiveness of different components in our proposed method, we take the UVG dataset as an example to perform ablation study. In this section, the hyperprior entropy model [6] is used in all methods for fair comparison. As shown in Fig. 5(a), our method RaFC-frame outperforms the baseline DVC algorithm and has achieved 0.5dB improvement when compared with DVC at 0.055bpp. We also observe that our overall framework RaFC by using both RaFC-block scheme and RaFC-frame scheme achieves better result, which indicates that our overall framework combining RaFC-frame and RaFC-block can further improve the performance of RaFC-frame. In other words, it is beneficial to choose the optimal resolution for both the optical flow maps and the corresponding motion representations.

Model Analysis. In Fig. 5(b), we take the HEVC Class E dataset as an example and show the average PSNR results over all predicted frames (i.e. $\bar{X}_t$'s) after motion compensation at different Bpps. When compared with the flow coding

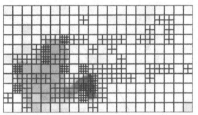

(a) The 6th frame from the HEVC Class E dataset.

(b) The reconstructed optical flow map and the corresponding block selection result by using our method RaFC-block.

Fig. 6. Visualization of the selected block resolutions by using our method RaFC-block.

method in DVC [17], our overall RaFC framework can compress motion information in a much more effective way and save up to 70% bits at the same PSNR when encoding motion information.

Besides, we also report the percentage of bits used to encode motion information over the total number of bits for encoding both motion and residual information at different Bpps when using different λ values. And it is obvious that the percentage drops significantly when comparing our RaFC framework with the baseline DVC method, which indicates our RaFC framework uses less bits to encode flow information.

Resolutions Selection at Various Bit Rates. In our approach, we select the optimal resolution for the optical flow map in RaFC-frame or motion features in RaFC-block. To investigate the effectiveness of our method, we provide the percentage of each selected resolution over the total number resolutions at various bit rates. From Table 1 and Table 2, we observe that low-resolution flow maps and large size blocks take a large portion at lower bit rates (*i.e.*, when λ is small). At higher bit rates (i.e., when λ is large), it is more likely that our methods RaFC-frame and RaFC-block select high resolution flow maps and small block sizes, respectively. This observation is consist with the traditional video compression methods, where large size blocks are often preferred for motion estimation at low bit rates in order to save bits for motion coding.

Visualization of Selected Blocks. In Fig. 6, we visualize the selected blocks with different resolutions by using our method RaFC-block. Figure 6(a) shows the 6th frame of the 1st video from the HEVC Class E dataset and Fig. 6(b) represents the reconstructed optical flow map of this frame and the corresponding block selection result by using our method RaFC-block. It can be observed that the small size blocks are often preferred from areas around the moving object boundaries and large size blocks are always preferred in the smooth areas.

5   Conclusion

In this work, we have proposed a Resolution-adaptive Flow Coding (RaFC) method to efficiently compress the motion information for video compression, which consists of two new schemes RaFC-frame at the frame-level and RaFC-block at the block-level. Our method RaFC-frame can handle complex or simple motion patterns globally by automatically selecting the optimal resolutions from multi-scale flow maps, while our method RaFC-block can cope with different types of motion patterns locally by selecting the optimal resolutions of multi-scale motion features at each block. By performing comprehensive experiments on four benchmark datasets, we show that our RaFC framework outperforms the recent state-of-the-art deep learning based video compression methods. In our future work, we will use the proposed framework for encoding residual information and study more efficient block partitioning strategy.

Acknowledgement. This work was supported by the National Key Research and Development Project of China (No. 2018AAA0101900). The work of Wanli Ouyang was supported by the Australian Medical Research Future Fund MRFAI000085.

References

1. Ultra video group test sequences. http://ultravideo.cs.tut.fi. Accessed 06 Nov 2019
2. Video trace library. http://trace.kom.aau.dk/yuv/index.html. Accessed 06 Nov 2019
3. Agustsson, E., et al.: Soft-to-hard vector quantization for end-to-end learning compressible representations. In: Advances in Neural Information Processing Systems, pp. 1141–1151 (2017)
4. Agustsson, E., Tschannen, M., Mentzer, F., Timofte, R., Gool, L.V.: Generative adversarial networks for extreme learned image compression. In: Proceedings of the IEEE International Conference on Computer Vision, pp. 221–231 (2019)
5. Ballé, J., Laparra, V., Simoncelli, E.P.: End-to-end optimized image compression. International Conference on Learning Representations (ICLR)(2017)
6. Ballé, J., Minnen, D., Singh, S., Hwang, S.J., Johnston, N.: Variational image compression with a scale hyperprior. International Conference on Learning Representations (ICLR) (2018)
7. Bellard, F.: Bpg image format. https://bellard.org/bpg (2015)
8. Chen, Z., He, T., Jin, X., Wu, F.: Learning for video compression. IEEE Trans. Circ. Syst. Video Technol. **30**(2), 566–576 (2019)
9. Djelouah, A., Campos, J., Schaub-Meyer, S., Schroers, C.: Neural inter-frame compression for video coding. In: Proceedings of the IEEE International Conference on Computer Vision, pp. 6421–6429 (2019)
10. Dosovitskiy, A., et al.: Flownet: learning optical flow with convolutional networks. In: Proceedings of the IEEE international conference on computer vision, pp. 2758–2766 (2015)
11. Guo, L., et al.: Content adaptive and error propagation aware deep video compression. Proceedings of the European Conference on Computer Vision (ECCV) (2020)
12. Habibian, A., Rozendaal, T.v., Tomczak, J.M., Cohen, T.S.: Video compression with rate-distortion autoencoders. In: Proceedings of the IEEE International Conference on Computer Vision, pp. 7033–7042 (2019)
13. Hui, T.W., Tang, X., Change Loy, C.: Liteflownet: a lightweight convolutional neural network for optical flow estimation. In: Proceedings of the IEEE Conference on Computer Vision and Pattern Recognition, pp. 8981–8989 (2018)

14. Johnston, N., et al.: Improved lossy image compression with priming and spatially adaptive bit rates for recurrent networks. In: Proceedings of the IEEE Conference on Computer Vision and Pattern Recognition, pp. 4385–4393 (2018)
15. Kingma, D.P., Ba, J.: Adam: a method for stochastic optimization. International Conference for Learning Representations (2015)
16. Li, M., Zuo, W., Gu, S., Zhao, D., Zhang, D.: Learning convolutional networks for content-weighted image compression. In: Proceedings of the IEEE Conference on Computer Vision and Pattern Recognition, pp. 3214–3223 (2018)
17. Lu, G., Ouyang, W., Xu, D., Zhang, X., Cai, C., Gao, Z.: DVC: an end-to-end deep video compression framework. In: Proceedings of the IEEE Conference on Computer Vision and Pattern Recognition, pp. 11006–11015 (2019)
18. Lu, G., Zhang, X., Ouyang, W., Chen, L., Gao, Z., Xu, D.: An end-to-end learning framework for video compression. IEEE Transactions on Pattern Analysis and Machine Intelligence in Press, pp. 1–1 (2020) https://doi.org/10.1109/TPAMI.2020.2988453
19. Minnen, D., Ballé, J., Toderici, G.D.: Joint autoregressive and hierarchical priors for learned image compression. In: Advances in Neural Information Processing Systems, pp. 10771–10780 (2018)
20. Ranjan, A., Black, M.J.: Optical flow estimation using a spatial pyramid network. In: Proceedings of the IEEE Conference on Computer Vision and Pattern Recognition, pp. 4161–4170 (2017)
21. Rippel, O., Bourdev, L.: Real-time adaptive image compression. In: Proceedings of the 34th International Conference on Machine Learning, JMLR. org. **70**, 2922–2930 (2017)
22. Rippel, O., Nair, S., Lew, C., Branson, S., Anderson, A.G., Bourdev, L.: Learned video compression. In: Proceedings of the IEEE International Conference on Computer Vision, pp. 3454–3463 (2019)
23. Sullivan, G.J., Ohm, J.R., Han, W.J., Wiegand, T.: Overview of the high efficiency video coding (hevc) standard. IEEE Trans. Circ. Syst. Video Technol. **22**(12), 1649–1668 (2012)
24. Taubman, D.S., Marcellin, M.W.: Jpeg 2000: standard for interactive imaging. Proc. IEEE **90**(8), 1336–1357 (2002)
25. Theis, L., Shi, W., Cunningham, A., Huszár, F.: Lossy image compression with compressive autoencoders. International Conference for Learning Representations (2017)
26. Toderici, G., et al.: Variable rate image compression with recurrent neural networks. International Conference for Learning Representations (2017)
27. Toderici, G., et al.: Full resolution image compression with recurrent neural networks. In: Proceedings of the IEEE Conference on Computer Vision and Pattern Recognition, pp. 5306–5314 (2017)
28. Tsai, Y.H., Liu, M.Y., Sun, D., Yang, M.H., Kautz, J.: Learning binary residual representations for domain-specific video streaming. In: Thirty-Second AAAI Conference on Artificial Intelligence (2018)
29. Wallace, G.K.: The jpeg still picture compression standard. IEEE Trans. Consum. Electron. **38**(1), xviii–xxxiv (1992)
30. Wang, H., et al.: MCL-JCV: a jnd-based h. 264/avc video quality assessment dataset. In: 2016 IEEE International Conference on Image Processing (ICIP), pp. 1509–1513. IEEE (2016)
31. Wang, Z., Simoncelli, E.P., Bovik, A.C.: Multiscale structural similarity for image quality assessment. In: The Thrity-Seventh Asilomar Conference on Signals, Systems & Computers, 2003. **2**, pp. 1398–1402. IEEE (2003)

32. Wiegand, T., Sullivan, G.J., Bjontegaard, G., Luthra, A.: Overview of the h. 264/avc video coding standard. IEEE Trans. Circ. Syst. Video Technol. **13**(7), 560–576 (2003)
33. Wu, C.Y., Singhal, N., Krahenbuhl, P.: Video compression through image interpolation. In: Proceedings of the European Conference on Computer Vision (ECCV), pp. 416–431 (2018)
34. Xu, M., Li, T., Wang, Z., Deng, X., Yang, R., Guan, Z.: Reducing complexity of HEVC: a deep learning approach. IEEE Trans. Image Process. **27**(10), 5044–5059 (2018)
35. Xue, T., Chen, B., Wu, J., Wei, D., Freeman, W.T.: Video enhancement with task-oriented flow. Int. J. Comput. Vision **127**(8), 1106–1125 (2019)

Motion Capture from Internet Videos

Junting Dong[1], Qing Shuai[1], Yuanqing Zhang[1], Xian Liu[1], Xiaowei Zhou[1], and Hujun Bao[1,2(✉)]

[1] State Key Lab of CAD&CG, Zhejiang University, Hangzhou, China
[2] Zhejiang Lab, Hangzhou, China
bao@cad.zju.edu.cn

Abstract. Recent advances in image-based human pose estimation make it possible to capture 3D human motion from a single RGB video. However, the inherent depth ambiguity and self-occlusion in a single view prohibit the recovery of as high-quality motion as multi-view reconstruction. While multi-view videos are not common, the videos of a celebrity performing a specific action are usually abundant on the Internet. Even if these videos were recorded at different time instances, they would encode the same motion characteristics of the person. Therefore, we propose to capture human motion by jointly analyzing these Internet videos instead of using single videos separately. However, this new task poses many new challenges that cannot be addressed by existing methods, as the videos are unsynchronized, the camera viewpoints are unknown, the background scenes are different, and the human motions are not exactly the same among videos. To address these challenges, we propose a novel optimization-based framework and experimentally demonstrate its ability to recover much more precise and detailed motion from multiple videos, compared against monocular motion capture methods.

Keywords: Motion capture · Human pose estimation

1 Introduction

Human motion capture (MoCap) is a core technology in a variety of applications such as movie production, video game development, sports analysis and interactive entertainment. While there have been some commercial solutions to MoCap, e.g., optical MoCap systems like Vicon, these systems are for professionals but not commodities. The systems are expensive and hard to calibrate. More importantly, the performers need to present in the studio to perform actions, which makes it impossible to collect large-scale motion data for a large population. For example, producing an animated avatar of a celebrity needs to invite the person to the MoCap studio, which is not always feasible especially for amateur productions.

J. Dong and Q. Shuai—Equal contribution.

Electronic supplementary material The online version of this chapter (https://doi.org/10.1007/978-3-030-58536-5_13) contains supplementary material, which is available to authorized users.

ⓒ Springer Nature Switzerland AG 2020
A. Vedaldi et al. (Eds.): ECCV 2020, LNCS 12347, pp. 210–227, 2020.
https://doi.org/10.1007/978-3-030-58536-5_13

Fig. 1. This paper proposes a system for motion capture from a set of Internet videos which record different instances of the same action of a person. The videos were recorded at different times and in different scenes (bottom). Our system synchronizes the videos, recover the camera viewpoints, and reconstruct the motion accurately (top). More video demonstrations are available at https://github.com/zju3dv/iMoCap.

To make human MoCap a commodity, many monocular motion capture algorithms [17,24,51,57] have been developed to recover human motion from single RGB videos. Remarkable progress has been made in past years thanks to the advances in deep learning, public datasets on human bodies, and expressive human models [28,32]. However, all these methods take a single video as input. As 3D reconstruction from a monocular image is inherently ill-posed, it is extremely difficult to recover accurate and detailed motion from a single video. Leveraging multiple views can resolve the ambiguity, but calibrated and synchronized multi-view videos are not common.

Fortunately, we observe that videos of some celebrities doing some specific actions are abundant on the Internet. While those videos were recorded at different times and the motions in these videos are not exactly the same, they encode the same motion characteristics of the person. Compared to a single video, multiple videos provide richer observations about the specific motion. More importantly, the videos are often recorded at different viewpoints which provide multi-view information to help alleviate the 3D ambiguity and self-occlusion issues.

In this paper, we propose to capture human motion from a collection of Internet videos that record different instances of a person's specific performance. However, this new problem brings in many challenges that make existing multi-view MoCap algorithms inapplicable: the human motions are not exactly the same among all videos; the videos are unsynchronized; the camera viewpoints are unknown; and the background scenes can be different. To solve these challenges, we propose an optimization-based framework that simultaneously solves video synchronization, camera calibration, and human motion reconstruction. More specifically, the proposed system initializes per-frame 3D human pose estimation with a learned 3D pose estimator, synchronizes videos by matching frames based on the 3D pose similarity, and jointly optimizes for camera poses and human motions over all the videos. The motions to be recovered are not assumed exactly the same among videos but modeled by a low-rank subspace. Finally, the motion reconstruction and the pose-based video synchronization are iteratively refined. We also show that the video synchronization can be improved by imposing the cycle consistency constraint among multiple videos.

In summary, we make the following contributions:

- We introduce the new task of motion capture from a collection of Internet videos that record different instances of a person's certain action, which is unexplored in the literature to our knowledge.
- We develop a new optimization-based framework to solve this new task. Our technical contributions include pose-based video synchronization, low-rank modeling of motions, and joint optimization for synchronization, camera poses and human motion.
- We show that, compared to using single videos, the joint analysis of multiple videos provides richer information to address occlusion and depth ambiguity, even if the videos record different motion instances.

2   Related Work

Single-view Mocap: There has been remarkable progress on 3D human pose and shape estimation from single images. Many works focus on the skeleton-based 3D human pose estimation, either first estimating 2D pose from images and then lifting it to 3D [7,29,30,36,57], or end-to-end regressing to obtain the 3D pose directly [33,42–45,58]. In addition, a lot of works propose to estimate the 3D pose and shape involving a parametric model of the human body [1,28]. Some early works attempt to use the optimization-based methods [3,16,25,40, 53], which fit the human model to 2D evidence. More recently, many works attempt to directly regress the model from images with a deep network [17,23,24, 31,35,51,54]. However, due to the inherent depth ambiguity of single views, the accuracy of these methods is not comparable with the multi-view reconstruction.

Multi-view Mocap: Markerless multi-view motion capture has been explored in computer vision for many years. The solutions to this problem are mainly divided into two categories: tracking and pose estimation. Most multi-view tracking methods [2,11,15,26,27] fit a human body model, e.g., a triangle mesh or a collection of geometric primitives, to image evidence such as keypoints and silhouette. The main difference between them is the type of image evidence and the way to optimize it. However, these tracking based approaches usually require the initialization of the first frame and easily fall into local optima and tracking failures. Hence, more recent works [4,22,34,41] generally tend to estimate 3D human body based on 2D features detected from images. Burenius et al. [4]

propose to extend the pictorial structure model to 3D and use it to estimate 3D human skeleton from images. Pavlakos et al. [34] propose to use a ConvNet for 2D pose estimation and combine with the 3D pictorial structure model to produce 3D pose estimation. Huang et al. [20] and Joo et al. [22] propose to combine statistical body models with a 2D pose estimator and show impressive results. All the above methods assume the multi-view videos are synchronized with known camera parameters.

There are a few methods [12,13,18,38,49,52,55] which attempt to reconstruct the 3D human motion from multiple uncalibrated and unsynchronized videos. Most methods synchronize the videos using additional information, such as audio [12,18], system time [38], and flashing a light [49], which are unavailable in our scenario. In terms of calibration, many works [12,18,49,52,55] assume that the camera parameters are provided or obtain the camera geometry using structure from motion based on the static background, which is inapplicable in our setting where the scenes are totally different. Some works [13,52] also propose to optimize camera parameters and human poses jointly but they assume the motions among videos are exactly the same.

Video Alignment: When the videos are recording the same event, there are many existing methods to address the temporal alignment problem. Early works [6,46,47,50] generally assume a linear temporal mapping between videos. More recent works propose non-linear solutions based on handcrafted features [48] or learned features [39]. However, for our situation where the videos record similar motions rather than the same event, these approaches are not suitable. Dwibedi et al. [10] propose a self-supervised representation learning method for general video alignment but not tailored for human videos. We will use it as a baseline to evaluate our synchronization component in experiments.

3   Methods

Our goal is to reconstruct human motion from multiple videos. Suppose the videos are synchronized, the cameras are calibrated and the motion is the same in these videos, this problem is reduced to a multi-view 3D pose reconstruction problem, which can be solved by first detecting 2D poses in each view and then lifting them to 3D by triangulation. However, this is not the case in our task, where we need to solve video synchronization and motion reconstruction simultaneously with unknown camera geometry, and the motions are similar but not exactly the same across videos.

To solve this challenging problem, we propose an iterative optimization framework that jointly solves synchronization and reconstruction. The intuition is that, if the 3D pose in each video frame is given, we can synchronize videos based on the 3D poses; and if the videos are synchronized, we can recover 3D poses and camera viewpoints from the corresponding frames using multi-view geometry. Figure 2 presents an overview of our approach. We initialize per-frame 3D poses with a CNN-based estimator and iteratively solve synchronization and

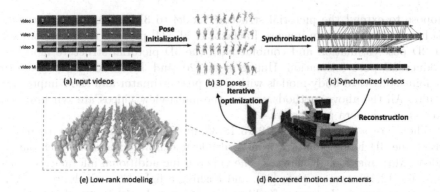

Fig. 2. Overview of our approach. Given multiple Internet videos of an action (a), an off-the-shelf 3D human pose estimator is used to initialize the 3D pose of each frame (b). Then, the 3D poses are used to synchronize all videos (c), from which the human motion and camera parameters are recovered (d) with the motion variation across videos modeled by a low-rank matrix (e). Finally, the optimized pose estimates are used to refine the video synchronization again, and the video synchronization and the motion reconstruction are optimized iteratively.

motion recovery by optimization. In the rest of this section, we first introduce the pose-based video synchronization and then the motion recovery method.

3.1   Pose-Based Video Synchronization

In order to leverage multiple views for pose reconstruction, video synchronization is required, i.e., finding the correspondences of frames between videos. However, this is a challenging task because the appearances are very different among videos due to the different background, clothing, and viewpoints. To address this problem, we propose to synchronize videos directly based on 3D human poses seen in the video frames. The initial poses can be obtained by an off-the-shelf pose estimator [24] and refined after synchronization.

Suppose there are M Internet videos, N_j is the number of frames for video j, and $K_{ij} \in \mathbb{R}^{3 \times J}$ denotes the 3D human pose estimated for the i-th frame of video j. Then, we can measure the likelihood that two frames correspond to each other (a.k.a affinity) based on the similarity between the estimated 3D human poses. Specifically, we compute the Euclidean distance between each pair of 3D poses aligned by the Procrustes method. Then, we map the reciprocal of distance to a value between $[0, 1]$ as the affinity score between two frames. For a pair of videos j_1 and j_2, we construct an affinity matrix $A_{j_1 j_2} \in \mathbb{R}^{N_{j_1} \times N_{j_2}}$ which consists of all affinity scores between frames of two videos. The correspondences to be estimated can be represented as a partial permutation matrix $X_{j_1 j_2} \in \{0, 1\}^{N_{j_1} \times N_{j_2}}$ and efficiently estimated based on $A_{j_1 j_2}$ using an optimal assignment algorithm, e.g., dynamic programming considering the sequential constraint on video frames.

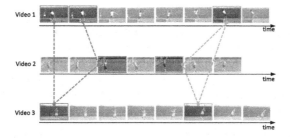

Fig. 3. An illustration of cycle consistency. The green lines denote a set of consistent correspondences and the red lines show a set of inconsistent correspondences.

If we align each pair of videos separately, the resulting correspondences may be inconsistent due to ignoring the cycle consistency constraint. For example, as shown in Fig. 3, the correspondences in green are cycle-consistent since they form a closed cycle and the ones in red are inconsistent. Therefore, we can use the cycle consistency constraint to improve the alignment of multiple videos. To achieve this, we adopt the result in prior work [19] that the cycle consistency is equivalent to a low-rank constraint on the correspondence matrix $\boldsymbol{X}$, which is the concatenation of all pairwise permutation matrix:

$$
\boldsymbol{X} = \begin{pmatrix} \boldsymbol{X}_{11} & \boldsymbol{X}_{12} & \cdots & \boldsymbol{X}_{1M} \\ \boldsymbol{X}_{21} & \boldsymbol{X}_{22} & \cdots & \boldsymbol{X}_{2M} \\ \vdots & \vdots & \ddots & \vdots \\ \boldsymbol{X}_{M1} & \boldsymbol{X}_{M2} & \cdots & \boldsymbol{X}_{MM} \end{pmatrix} \in \mathbb{R}^{N_a \times N_a}.
\tag{1}
$$

N_a is the number of all frames of all videos.

Therefore, we minimize the following objective function to estimate $\boldsymbol{X}$:

$$
f(\boldsymbol{X}) = \|\boldsymbol{A} - \boldsymbol{X}\|_F^2 + \lambda \cdot rank(\boldsymbol{X}),
\tag{2}
$$

where $\boldsymbol{A} \in \mathbb{R}^{N_a \times N_a}$ denotes the concatenation of all $\boldsymbol{A}_{j_1 j_2}$ similar to the form of $\boldsymbol{X}$, λ is the weight of low-rank constraint. This problem can be approximately solved with the convex relaxation algorithms in previous work [9,56]. The relaxed solution $\boldsymbol{X}_{j_1 j_2}$ is usually not a valid permutation matrix but a real matrix with values in $(0, 1)$, which can be regarded as a denoised version of $\boldsymbol{A}_{j_1 j_2}$ with cycle consistency. Finally, to find the frame-to-frame correspondence between video i and video j, we use the dynamic time warping algorithm based on the affinity matrix $\boldsymbol{X}_{j_1 j_2}$.

3.2 Motion Recovery

Even if the videos are synchronized, the problem still cannot be treated as a standard multi-view reconstruction problem for the following two reasons. First, the relative camera poses between videos are unknown and cannot be recovered from structure from motion as the scenes in videos are different. Second, the motions in all videos are not exactly the same. To solve the first issue, we directly register cameras with the human body as the reference and recovery the human motion and camera parameters simultaneously. To address the second issue, we propose to model the motion variation among videos by a low-rank subspace. Before we introduce the methods in detail, we first introduce the representation of human motion.

Motion Representation: For each video, the corresponding 3D human motion is individually represented by a statistical body mesh model SMPL [28] instead of 3D skeleton, since it contains a richer body prior. The SMPL model is parameterized by the pose parameters $\theta \in \mathbb{R}^{72}$, the shape parameters $\beta \in \mathbb{R}^{10}$, and a root translation $\gamma \in \mathbb{R}^3$, and maps a set of parameters to a body mesh denoted by $M(\theta, \beta, \gamma) \in \mathbb{R}^{3 \times N_v}$ with $N_v = 6890$ vertices. A predefined set of 3D body joints $F(\theta, \beta, \gamma) \in \mathbb{R}^{3 \times J}$ can be generated by linear regression from the mesh vertices, where J denotes the number of 3D joints. The SMPL+H model [37] which extends SMPL with hands and SMPL-X model [32] which extends SMPL with face and hands can also be used if the video resolution is sufficient for Open Pose [5] to capture the face and hand motion. Our goal is to recover θ_{ij}, β_{ij}, and γ_{ij}, which denote the pose, shape and translation parameters for each frame i of video j, respectively. Note that we assume the shape parameters β_{ij} remains the same in one video, i.e., $\beta_{ij} = \beta_j$.

SMPL-BA: We attempt to solve camera parameters and SMPL parameters simultaneously by minimizing the reprojection errors of body keypoints detected in video frames, similar to bundle adjustment (BA) in traditional structure from motion. The body keypoints are anchored on the SMPL model. Therefore, we call this procedure SMPL-BA.

Suppose $\boldsymbol{R}_j^c$ and $\boldsymbol{T}_j^c$ denote the rotation and translation of the camera j in the world coordinate system that defines SMPL, respectively. Then, the reprojection error in SMPL-BA can be written as:

$$L_{2d} = \sum_{i,j,z} c_{ijz} \rho \left(\boldsymbol{W}_{ijz} - P\{ \boldsymbol{R}_j^c F(\theta_{ij}, \beta_j, \gamma_{ij})_z + \boldsymbol{T}_j^c \} \right), \tag{3}$$

where $\boldsymbol{W}_{ijz} \in \mathbb{R}^2$ denotes the z-th joint of the estimated 2D pose at i-th frame in video j with corresponding confidence c_{ijz} and P denotes the perspective projection. ρ denotes the Geman-McClure robust error function for suppressing noisy detection.

In (3), the camera poses $\boldsymbol{R}_j^c$ and $\boldsymbol{T}_j^c$ are irrelevant to frame index i. But in practice the camera may move in each video. To address this issue, we assume that the cameras are only allowed to rotate at fixed camera centers, which is a practical assumption, e.g., in sports broadcasting. Then, we propose to compensate for the camera rotation in each video by warping other frames to the first frame using a homography transformation estimated by feature tracking between frames.

Low-Rank Modeling of Motions: When the human motion in each video is not exactly the same, we assume that 3D poses observed in the corresponding frames are very similar which can be approximated by a low-rank matrix:

$$rank(\boldsymbol{\theta}_i) \leq s, \tag{4}$$

where $\boldsymbol{\theta}_i = [\boldsymbol{\theta}_{i1}^T; \boldsymbol{\theta}_{i2}^T; \cdots; \boldsymbol{\theta}_{iM}^T] \in \mathbb{R}^{M \times 72}$ denotes the collection of pose parameters in all videos of frame i and the constant s controls the degree of similarity. Note that each video has its own SMPL parameters. The only constraint that links all videos is the low-rank constraint, which is soft and allows difference among videos.

In addition, we also assume that the 3D trajectories of the root joint of the body should be similar among videos. Suppose the root trajectories in all videos are denoted by $\boldsymbol{\gamma} = [\boldsymbol{\gamma}_1^T; \boldsymbol{\gamma}_2^T; \cdots; \boldsymbol{\gamma}_M^T] \in \mathbb{R}^{M \times 3N}$, where $\boldsymbol{\gamma}_j \in \mathbb{R}^{3N}$ is the trajectory in video j and N is the number of frames. Then, the constraint can be written as:

$$rank(\boldsymbol{\gamma}) \leq s. \tag{5}$$

Intuitively, how much motion variation is allowed depends on the selected rank s. A larger s allows larger variation but imposes less constraint across views. We empirically find that $s = 1$ or 2 works well, which imposes sufficient multi-view constraint while allowing reasonable motion variation across videos.

Objective Function: Combining all discussed above, the final objective function to optimize can be written as:

$$
\begin{aligned}
\min L_{2d} + \lambda_t L_{temp}, \\
\text{s.t.} \, rank(\boldsymbol{\theta}_i) \leq s, i = 1, 2, ..., N, \\
rank(\boldsymbol{\gamma}) \leq s,
\end{aligned}
\tag{6}
$$

where L_{temp} is a temporal smoothing term with weight λ_t to eliminate jittering in motion:

$$L_{temp} = \sum_{i=1}^{N-1} \|\boldsymbol{\theta}_i - \boldsymbol{\theta}_{i+1}\|_F^2. \tag{7}$$

Optimization: To simplify the optimization, we introduce two auxiliary variables $\boldsymbol{Z}_i \in \mathbb{R}^{M \times 72}$ and $\boldsymbol{Y} \in \mathbb{R}^{M \times 3N}$ to decouple the rank constraints with the objective function:

$$
\begin{aligned}
\min L_{2d} + \lambda_t L_{temp} + \lambda_{r_1} \sum_{i=1}^{N} \|\boldsymbol{\theta}_i - \boldsymbol{Z}_i\|_F^2 + \lambda_{r_2} \|\boldsymbol{\gamma} - \boldsymbol{Y}\|_F^2, \\
\text{s.t.} \, rank(\boldsymbol{Z}_i) \leq s, i = 1, 2, \cdots, N, \\
rank(\boldsymbol{Y}) \leq s
\end{aligned}
\tag{8}
$$

where λ_{r_1} and λ_{r_2} are weighting parameters.

The problem in (8) is highly nonconvex. However, reliable initialization allows us to use local optimization to solve this problem. Specifically, we update each variable alternately while the others remain fixed. The pose $\boldsymbol{\theta}_i$, shape $\boldsymbol{\beta}_j$, and translation $\boldsymbol{\gamma}$ parameters of SMPL can be updated with Gradient Descent. It is a standard low-rank approximation problem to update $\boldsymbol{Z}_i$ and $\boldsymbol{Y}$, which can be solved by SVD analytically. The update of $\boldsymbol{R}_j^c$ and $\boldsymbol{T}_j^c$ can be solved with a perspective-n-point (PnP) algorithm that minimizes reprojection errors over all frames of video j.

Initialization: We initialize the SMPL parameters for each frame using a pre-trained neural network [24], which is further refined by minimizing the reprojection error of 2D keypoints for each frame. Next, the videos are initially synchronized based on the initial pose estimates as introduced in Sect. 3.1. Then, a reference video is selected, whose camera coordinate system is regarded as the world frame. Note that the initial SMPL model in each video is defined in the coordinate system of the respective camera. Therefore, the relative camera poses between two videos can be initialized by rigidly aligning the SMPL models, assuming the SMPL pose parameters are the same between videos. When intrinsics are unknown, we set the focal length to be a large constant, approximating a weak-perspective camera model. In this way, the camera poses can be initialized.

3.3   Iterative Optimization

The video synchronization in the first iteration may not be very accurate based on initial pose estimates. Therefore, we propose to refine the synchronization based on the optimized poses. More specifically, the affinity matrix $\boldsymbol{A}$ in Sect. 3.1 is updated with the optimized poses given by the SMPL-BA and the frame correspondences are re-computed using the new affinity matrix. Then, the SMPL-BA is computed again with the updated synchronization. Both synchronization and reconstruction benefit from each other in iterative optimization, which will be experimentally demonstrated in Sect. 4.2.

4   Experiments

4.1   Motion Capture from Internet Videos

There is no existing dataset for our task. Therefore, we collect a new dataset that consists of 20 actions of various actors, such as tennis serves, yoga and Tai Chi. Take tennis serves as an example. We download the publicly available videos of some tennis players from YouTube, and manually crop the videos roughly to obtain a set of video clips of serves for each player. Figure 4 shows the statistics of the number of videos and average number of frames for each action. The dataset is available at https://github.com/zju3dv/iMoCap.

We apply the proposed approach on each action of this dataset to recover the corresponding human motion. Some representative results are visualized in Fig. 6, which shows that the proposed approach is able to recover 3D human motion as well as camera geometry from these videos, even if they were recorded

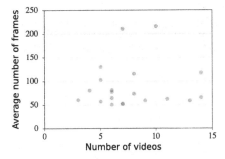
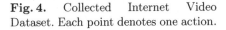

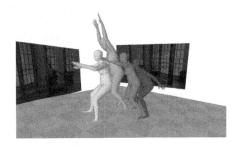

Fig. 4. Collected Internet Video Dataset. Each point denotes one action.

Fig. 5. Trajectory Recovery. Our approach is able to recover the absolute 3D trajectory of human motion. The brightness of human mesh indicates the chronological order.

at different times. These videos record the action from very different viewpoints and therefore provide multi-view constraints to help alleviate the depth ambiguity and self-occlusion issues that often occur for single-view estimation. Consequently, compared to the monocular motion capture algorithm [24], our approach produces much more detailed and faithful motion, as indicated by the circles in Fig. 6. In addition, with the multi-view constraint, our method is also able to recover an accurate 3D trajectory of the body as shown in Fig. 5, which is infeasible for monocular motion capture algorithms. Note that, the proposed approach can be easily extended to hand motion recovery if the 2D hand pose estimation is available as shown in Fig. 1 and 6 (Tai Chi). We find that most of the failure cases are because of failed 2D pose estimation. Also, when the viewpoints of videos are similar, the depth ambiguity cannot be resolved even if multiple videos are used. *More qualitative results and video demonstrations are available in the supplementary material.*

Since the motion in all the videos is not exactly the same, each video has its own SMPL parameters with a low-rank constraint to make the parameters correlated among multiple videos. An alternative is to assume the motions are all the same and use a single model with the same set of SMPL parameters for all videos. We provide a qualitative comparison in Fig. 7. More specifically, we reproject the initial 3D mesh, the 3D mesh reconstructed by our low-rank model, and the 3D mesh reconstructed by the single model to images and compare them in terms of 2D consistency. The results show that the projected mesh using the single model is less consistent with 2D evidence as shown in the red circles, which suggests that the single model cannot model the motion difference among videos. Low-rank modeling is able to capture more detailed motion, such as the curved back of the performer, as shown in the green circles. Overall, low-rank modeling recovers more detailed and natural motion than the results of a single model and initial monocular results.

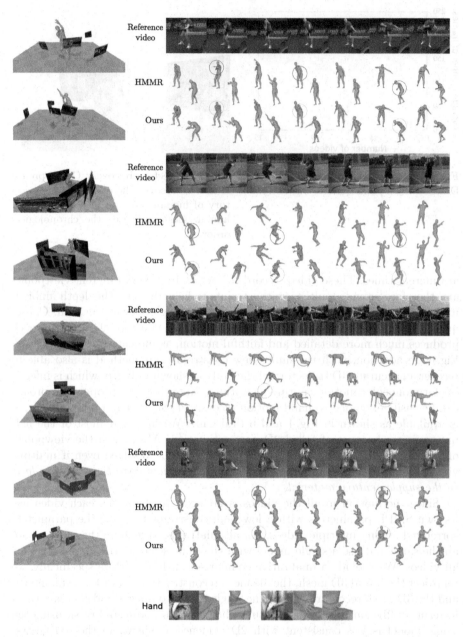

Fig. 6. Results on internet videos of table tennis serves, shotput, yoga, and Tai Chi(with hands motion). The left images present the reconstructed human motion and camera positions visualized in two viewpoints. On the right, we present some frames of the reference video and corresponding motion capture results from HMMR [24] and our method. Red and green cycles emphasize some representative differences between the results from HMMR and our method.

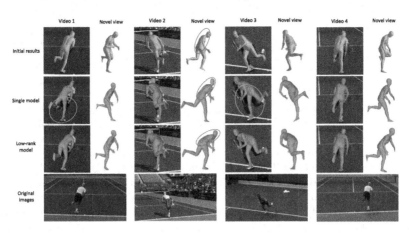

Fig. 7. Effect of low-rank modeling. The projected mesh of a single model is less consistent with 2D evidence (red marks). A low-rank model is able to capture differences among videos and recover more accurate motion such as the curved back of the performer (green marks).

4.2   Quantitative Evaluation

While we have collected a dataset of Internet videos to demonstrate the qualitative performance of our system, *quantitative* evaluation is difficult due to the lack of 3D ground truth, a similar case for most prior work on reconstruction from Internet data. For quantitative analysis, we synthesize a dataset using existing datasets [21] with ground-truth annotations. We select some challenging sequences in the Human3.6M dataset [21] and modify the data to simulate the unsynchronized and uncalibrated scenario. Please refer to Fig. 8 for details. We would like to note that the purpose of evaluation on Human3.6M is not to compare against existing methods in the standard Human3.6M setting, but to provide an ablative analysis of our system when solving the proposed problem.

Video Synchronization: As described in Sect. 3.1, we propose a pose-based video synchronization method to address the different appearances among videos, i.e., background, clothing, and viewpoints. We also impose the cycle consistency constraint to improve the synchronization. Here, we compare with some baselines and use the standard video alignment metric to measure the alignment of two videos. In particular, for each frame of non-reference video v_i, we compute the frame distance between the matched frame and the ground truth position in reference video v_0 and normalize it by the video clip length.

We first propose a simple alternative that uses the DP algorithm to quantize the original affinity matrix directly. The result of this baseline method ('No cycle-consis') is shown in Table 1. The results show that imposing the cycle consistency constraint can reduce the alignment error of video synchronization significantly.

Another baseline is a recent self-supervised representation learning method [10] based on the cycle consistency loss to align videos. Their network is retrained

Fig. 8. Dataset generation for quantitative analysis. We edit the videos in Human3.6M to simulate the unsynchronized scenario. As the dataset is large, we only select a few actions, i.e., SittingDown, Sitting, Smoking, Photo and Phoning. For each action, we first sample N_{s1} frames at equal intervals (**blue lines**) from each video, which results in $N_{s1} - 1$ segments. Then we randomly choose N_{s2} segments and randomly sample N_{s3} frames (**red lines**) from each selected segment. In our experiments, we set $N_{s1} = 150, N_{s2} = 50$ and for each segment N_{s3} is a variable value randomly selected from 1 to the length of the segment. The dataset is available at https://github.com/zju3dv/iMoCap.

on the evaluated sequences and the alignments are obtained by the dynamic time warping algorithm on the features. The results of their method ('TCC') on the dataset are presented in Table 1. The results show that our 3D pose based method outperforms the generic method by a large margin in our problem.

Table 1. Quantitative analysis of synchronization. 'No cycle-consis' denotes our synchronization method without cycle consistency constraint. 'TCC' represents the general video synchronization method [10] based on representation learning.

Method	Synchronization error
Ours	**0.77%**
No cycle-consis	1.19%
TCC [10]	11.24%

Reconstruction: We evaluate the motion reconstruction quantitatively. To evaluate 3D joint error, we use the standard metric, i.e., the mean per joint position error (MPJPE) and the error after rigid alignment (P-MPJPE).

As videos are unsynchronized and uncalibrated, none of the existing multi-view MoCap methods is applicable to the proposed problem. Monocular MoCap methods are the only applicable alternatives. We compare with the state-of-the-art monocular method HMMR [24] and the results are shown in Table 2. Our method significantly reduces the reconstruction error compared to HMMR, which shows the benefit of using multiple videos.

Note that we use a generic 2D pose detector [14] which is not fine-tuned on Human3.6M. With no fine-tuning, we wish to evaluate the generalization ability of our system when applied to unseen and challenging videos. We also report the results with a fine-tuned 2D pose detector [8] in Table 2, which show that using a fine-tuned detector significantly reduces the reconstruction error.

Table 2. Quantitative analysis of reconstruction. 'HMMR' denotes the state-of-the-art monocular motion capture method [24].

Method	MPJPE (mm)	P-MPJPE (mm)
HMMR	109.80	78.26
Ours+generic 2D, 4 videos	76.48	53.34
Ours+fine-tuned 2D, 1 video	80.65	62.58
Ours+fine-tuned 2D, 2 videos	78.45	59.42
Ours+fine-tuned 2D, 3 videos	71.48	53.77
Ours+fine-tuned 2D, 4 videos	**66.53**	**50.33**

Table 3. Quantitative analysis of iterative optimization. 'No iter-opt' denotes our method without iterative optimization of synchronization and reconstruction.

Method	Synchronization error	MPJPE (mm)	P-MPJPE (mm)
Ours	**0.77%**	**66.53**	**50.33**
No iter-opt	0.91%	71.49	54.12

In addition, we validate the influence of the number of videos. We report the reconstruction accuracy with various numbers of videos in Table 2. The results show that more videos improve the accuracy of reconstructed motion. As for Internet videos, while more views generally improve the results, we empirically find that three or four videos are sufficient in most cases.

Iterative Optimization: Our approach iteratively optimizes video synchronization and reconstruction to let them benefit from each other. 'No iter-opt' in Table 3 indicates the result without such an iterative optimization, which shows that iterative optimization reduces both alignment and reconstruction errors.

5 Summary

In this paper, we demonstrated the potential of leveraging multiple Internet videos to recover accurate and detailed human motion, which in a long-term perspective opens up the possibility of collecting high-quality and diverse human motion data for free from existing Internet videos. Unlike standard multi-view motion capture, in this new task the human motions are not exactly the same among all videos; the videos are unsynchronized; the camera viewpoints are unknown; and the background scenes can be different. All these challenges make existing multi-view motion capture algorithms inapplicable. To address all challenges above, we proposed (1) low-rank modeling of motions to handle motion variation among videos; (2) pose-based multi-video synchronization and calibration; and (3) *most importantly* a unified optimization-based framework to solve the entire problem, which doesn't treat synchronization, calibration and motion recovery as separate tasks, but integrates them in a single optimization problem. Both qualitative and quantitative results demonstrated the effectiveness of the proposed approach. Please see the supplementary material for more video demonstrations.

Acknowledgement. The authors would like to acknowledge support from NSFC (No. 61806176) and Fundamental Research Funds for the Central Universities (2019QNA5022).

References

1. Anguelov, D., Srinivasan, P., Koller, D., Thrun, S., Rodgers, J., Davis, J.: Scape: shape completion and animation of people. In: ACM transactions on graphics (TOG), pp. 408–416 (2005)
2. Bo, L., Sminchisescu, C.: Twin gaussian processes for structured prediction. Int. J. Comput. Vis. **87**, 28 (2010). https://doi.org/10.1007/s11263-008-0204-y
3. Bogo, F., Kanazawa, A., Lassner, C., Gehler, P., Romero, J., Black, M.J.: Keep it SMPL: automatic estimation of 3D human pose and shape from a single image. In: Leibe, B., Matas, J., Sebe, N., Welling, M. (eds.) ECCV 2016. LNCS, vol. 9909, pp. 561–578. Springer, Cham (2016). https://doi.org/10.1007/978-3-319-46454-1_34
4. Burenius, M., Sullivan, J., Carlsson, S.: 3D pictorial structures for multiple view articulated pose estimation. In: CVPR, pp. 3618–3625 (2013)
5. Cao, Z., Hidalgo Martinez, G., Simon, T., Wei, S., Sheikh, Y.A.: Open pose: real time multi-person 2D pose estimation using part affinity fields. IEEE Transactions on Pattern Analysis and Machine Intelligence (2019)
6. Caspi, Y., Irani, M.: Spatio-temporal alignment of sequences. IEEE Trans. Pattern Anal. Mach. Intell. **24**(11), 1409–1424 (2002)
7. Chen, C.H., Ramanan, D.: 3D human pose estimation= 2D pose estimation + matching. In: Proceedings of the IEEE Conference on Computer Vision and Pattern Recognition, pp. 7035–7043 (2017)
8. Chen, Y., Wang, Z., Peng, Y., Zhang, Z., Yu, G., Sun, J.: Cascaded Pyramid Network for Multi-Person Pose Estimation. In: Proceedings of the IEEE conference on computer vision and pattern recognition, pp. 7103–7112 (2018)
9. Dong, J., Jiang, W., Huang, Q., Bao, H., Zhou, X.: Fast and robust multi-person 3D pose estimation from multiple views. In: Proceedings of the IEEE Conference on Computer Vision and Pattern Recognition, pp. 7792–7801 (2019)
10. Dwibedi, D., Aytar, Y., Tompson, J., Sermanet, P., Zisserman, A.: Temporal cycle-consistency learning. In: Proceedings of the IEEE Conference on Computer Vision and Pattern Recognition, pp. 1801–1810 (2019)
11. Elhayek, A., et al.: Efficient convnet-based marker-less motion capture in general scenes with a low number of cameras. In: Proceedings of the IEEE Conference on Computer Vision and Pattern Recognition, pp. 3810–3818 (2015)
12. Elhayek, A., Stoll, C., Hasler, N., Kim, K.I., Seidel, H.P., Theobalt, C.: Spatio-temporal motion tracking with unsynchronized cameras. In: 2012 IEEE Conference on Computer Vision and Pattern Recognition, pp. 1870–1877. IEEE (2012)
13. Elhayek, A., Stoll, C., Kim, K.I., Theobalt, C.: Outdoor human motion capture by simultaneous optimization of pose and camera parameters. Comput. Graph. Forum **34**(6), 86–98 (2015)

14. Fang, H.S., Xie, S., Tai, Y.W., Lu, C.: RMPE: regional multi-person pose estimation. In: Proceedings of the IEEE International Conference on Computer Vision, pp. 2334–2343 (2017)
15. Gall, J., Rosenhahn, B., Brox, T., Seidel, H.P.: Optimization and filtering for human motion capture. Int. J. Comput. Vis. **87**(1–2), 75 (2010). https://doi.org/10.1007/s11263-008-0173-1
16. Guan, P., Weiss, A., Balan, A.O., Black, M.J.: Estimating human shape and pose from a single image. In: 2009 IEEE 12th International Conference on Computer Vision, pp. 1381–1388. IEEE (2009)
17. Guler, R.A., Kokkinos, I.: Holopose: holistic 3D human reconstruction in-the-wild. In: Proceedings of the IEEE Conference on Computer Vision and Pattern Recognition, pp. 10884–10894 (2019)
18. Hasler, N., Rosenhahn, B., Thormahlen, T., Wand, M., Gall, J., Seidel, H.P.: Markerless motion capture with unsynchronized moving cameras. In: 2009 IEEE Conference on Computer Vision and Pattern Recognition, pp. 224–231 IEEE (2009)
19. Huang, Q.X., Guibas, L.: Consistent shape maps via semidefinite programming. Comput. Graph. Forum **32**(5), 177–186 (2013)
20. Huang, Y., et al.: Towards accurate marker-less human shape and pose estimation over time. In: 2017 international conference on 3D vision (3DV), pp. 421–430. IEEE (2017)
21. Ionescu, C., Papava, D., Olaru, V., Sminchisescu, C.: Human 3.6m: large scale datasets and predictive methods for 3D human sensing in natural environments. IEEE Trans. Pattern Anal. Mach. Intell. **36**(7), 1325–1339 (2013)
22. Joo, H., Simon, T., Sheikh, Y.: Total capture: a 3D deformation model for tracking faces, hands, and bodies. In: Proceedings of the IEEE conference on computer vision and pattern recognition, pp. 8320–8329 (2018)
23. Kanazawa, A., Black, M.J., Jacobs, D.W., Malik, J.: End-to-end recovery of human shape and pose. In: Proceedings of the IEEE Conference on Computer Vision and Pattern Recognition, pp. 7122–7131 (2018)
24. Kanazawa, A., Zhang, J.Y., Felsen, P., Malik, J.: Learning 3D human dynamics from video. In: Proceedings of the IEEE Conference on Computer Vision and Pattern Recognition, pp. 5614–5623 (2019)
25. Lassner, C., Romero, J., Kiefel, M., Bogo, F., Black, M.J., Gehler, P.V.: Unite the people: Closing the loop between 3d and 2d human representations. In: Proceedings of the IEEE conference on computer vision and pattern recognition, pp. 6050-6059 (2017)
26. Lee, C.S., Elgammal, A.: Coupled visual and kinematic manifold models for tracking. Int. J. Comput. Vis. **87**(1–2), 118 (2010)
27. Li, R., Tian, T.P., Sclaroff, S., Yang, M.H.: 3D human motion tracking with a coordinated mixture of factor analyzers. Int. J. Comput. Vis. **87**(1–2), 170 (2010)
28. Loper, M., Mahmood, N., Romero, J., Pons-Moll, G., Black, M.J.: SMPL: a skinned multi-person linear model. ACM transactions on graphics (TOG). **34**(6), pp. 1–16 (2015)
29. Martinez, J., Hossain, R., Romero, J., Little, J.J.: A simple yet effective baseline for 3D human pose estimation. In: Proceedings of the IEEE International Conference on Computer Vision, pp. 2640–2649 (2017)
30. Moreno-Noguer, F.: 3D human pose estimation from a single image via distance matrix regression. In: Proceedings of the IEEE Conference on Computer Vision and Pattern Recognition, pp. 2823–2832 (2017)

31. Omran, M., Lassner, C., Pons-Moll, G., Gehler, P., Schiele, B.: Neural body fitting: unifying deep learning and model based human pose and shape estimation. In 2018 international conference on 3D vision (3DV), pp. 484–494. IEEE (2018)
32. Pavlakos, G., et al.: Expressive body capture: 3D hands, face, and body from a single image. In: Proceedings of the IEEE Conference on Computer Vision and Pattern Recognition, pp. 10975–10985 (2019)
33. Pavlakos, G., Zhou, X., Daniilidis, K.: Ordinal depth supervision for 3D human pose estimation. In: Proceedings of the IEEE Conference on Computer Vision and Pattern Recognition, pp. 7307–7316 (2018)
34. Pavlakos, G., Zhou, X., Derpanis, K.G., Daniilidis, K.: Harvesting multiple views for marker-less 3D human pose annotations. In: Proceedings of the IEEE conference on computer vision and pattern recognition, pp. 6988–6997 (2017)
35. Pavlakos, G., Zhu, L., Zhou, X., Daniilidis, K.: Learning to estimate 3D human pose and shape from a single color image. In: Proceedings of the IEEE Conference on Computer Vision and Pattern Recognition, pp. 459–468 (2018)
36. Pavllo, D., Feichtenhofer, C., Grangier, D., Auli, M.: 3D human pose estimation in video with temporal convolutions and semi-supervised training. In: Proceedings of the IEEE Conference on Computer Vision and Pattern Recognition, pp. 7753-7762 (2019)
37. Romero, J., Tzionas, D., Black, M.J.: Embodied hands: modeling and capturing hands and bodies together. ACM Trans. Graph. (ToG) **36**(6), 245 (2017)
38. Saini, N., et al.: Markerless outdoor human motion capture using multiple autonomous micro aerial vehicles. In: Proceedings of the IEEE International Conference on Computer Vision, pp. 823–832 (2019)
39. Sermanet, P., et al.: Time-contrastive networks: self-supervised learning from video. In: 2018 IEEE International Conference on Robotics and Automation (ICRA), pp. 1134-1141. IEEE (2018)
40. Sigal, L., Balan, A., Black, M.J.: Combined discriminative and generative articulated pose and non-rigid shape estimation. In: Advances in neural information processing systems, pp. 1337–1344 (2008)
41. Sigal, L., Isard, M., Haussecker, H., Black, M.J.: Loose-limbed people: estimating 3D human pose and motion using non-parametric belief propagation. Int. J. Comput. Vis. **98**(1), 15–48 (2012)
42. Sun, X., Shang, J., Liang, S., Wei, Y.: Compositional human pose regression. In: Proceedings of the IEEE International Conference on Computer Vision, pp. 2602–2611 (2017)
43. Sun, X., Xiao, B., Wei, F., Liang, S., Wei, Y.: Integral human pose regression. In: Proceedings of the European Conference on Computer Vision (ECCV), pp. 529–545 (2018)
44. Tekin, B., Márquez-Neila, P., Salzmann, M., Fua, P.: Learning to fuse 2D and 3D image cues for monocular body pose estimation. In: Proceedings of the IEEE International Conference on Computer Vision, pp. 3941–3950 (2017)
45. Tome, D., Russell, C., Agapito, L.: Lifting from the deep: convolutional 3D pose estimation from a single image. In: Proceedings of the IEEE Conference on Computer Vision and Pattern Recognition, pp. 2500–2509 (2017)
46. Tuytelaars, T., Van Gool, L.: Synchronizing video sequences. In: Proceedings of the 2004 IEEE Computer Society Conference on Computer Vision and Pattern Recognition, 2004. CVPR 2004, pp. 1–1. IEEE (2004)

47. Ukrainitz, Y., Irani, M.: Aligning sequences and actions by maximizing space-time correlations. In: Leonardis, A., Bischof, H., Pinz, A. (eds.) ECCV 2006. LNCS, vol. 3953, pp. 538–550. Springer, Heidelberg (2006). https://doi.org/10.1007/11744078_42

48. Wang, O., Schroers, C., Zimmer, H., Gross, M., Sorkine-Hornung, A.: Videosnapping: interactive synchronization of multiple videos. ACM Trans. Graph. (TOG) **33**(4), 1–10 (2014)

49. Wang, Y., Liu, Y., Tong, X., Dai, Q., Tan, P.: Outdoor markerless motion capture with sparse handheld video cameras. IEEE Trans. Visual. Comput. Graph. **24**(5), 1856–1866 (2017)

50. Wolf, L., Zomet, A.: Wide baseline matching between unsynchronized video sequences. Int. J. Comput. Vis. **68**(1), 43–52 (2006)

51. Xiang, D., Joo, H., Sheikh, Y.: Monocular total capture: posing face, body, and hands in the wild. In: Proceedings of the IEEE Conference on Computer Vision and Pattern Recognition, pp. 10965–10974 (2019)

52. Xu, X., Dunn, E.: Discrete laplace operator estimation for dynamic 3D reconstruction. arXiv preprint arXiv:1908.11044 (2019)

53. Zanfir, A., Marinoiu, E., Sminchisescu, C.: Monocular 3D pose and shape estimation of multiple people in natural scenes-the importance of multiple scene constraints. In: Proceedings of the IEEE Conference on Computer Vision and Pattern Recognition, pp. 2148–2157 (2018)

54. Zanfir, A., Marinoiu, E., Zanfir, M., Popa, A.I., Sminchisescu, C.: Deep network for the integrated 3D sensing of multiple people in natural images. In: Advances in Neural Information Processing Systems, pp. 8410–8419 (2018)

55. Zheng, E., Ji, D., Dunn, E., Frahm, J.M.: Sparse dynamic 3D reconstruction from unsynchronized videos. In: Proceedings of the IEEE International Conference on Computer Vision, pp. 4435–4443 (2015)

56. Zhou, X., Zhu, M., Daniilidis, K.: Multi-image matching via fast alternating minimization. In: Proceedings of the IEEE International Conference on Computer Vision, pp. 4032–4040 (2015)

57. Zhou, X., Zhu, M., Leonardos, S., Derpanis, K.G., Daniilidis, K.: Sparseness meets deepness: 3d human pose estimation from monocular video. In: Proceedings of the IEEE conference on computer vision and pattern recognition, pp. 4966-4975 (2016)

58. Zhou, X., Huang, Q., Sun, X., Xue, X., Wei, Y.: Towards 3D human pose estimation in the wild: a weakly-supervised approach. In: Proceedings of the IEEE International Conference on Computer Vision, pp. 398–407 (2017)

Appearance-Preserving 3D Convolution
for Video-Based Person Re-identification

Xinqian Gu[1,2], Hong Chang[1,2(✉)], Bingpeng Ma[2], Hongkai Zhang[1,2],
and Xilin Chen[1,2]

[1] Key Lab of Intelligent Information Processing of Chinese Academy of Sciences
(CAS), Institute of Computing Technology, CAS, Beijing 100190, China
{xinqian.gu,hongkai.zhang}@vipl.ict.ac.cn, {changhong,xlchen}@ict.ac.cn
[2] University of Chinese Academy of Sciences, Beijing 100049, China
bpma@ucas.ac.cn

Abstract. Due to the imperfect person detection results and posture
changes, temporal appearance misalignment is unavoidable in video-
based person re-identification (ReID). In this case, 3D convolution may
destroy the appearance representation of person video clips, thus it is
harmful to ReID. To address this problem, we propose Appearance-
Preserving 3D Convolution (AP3D), which is composed of two components: an Appearance-Preserving Module (APM) and a 3D convolution
kernel. With APM aligning the adjacent feature maps in pixel level, the
following 3D convolution can model temporal information on the premise
of maintaining the appearance representation quality. It is easy to combine AP3D with existing 3D ConvNets by simply replacing the original
3D convolution kernels with AP3Ds. Extensive experiments demonstrate
the effectiveness of AP3D for video-based ReID and the results on three
widely used datasets surpass the state-of-the-arts. Code is available at:
https://github.com/guxinqian/AP3D.

Keywords: Video-based person re-identification · Temporal
appearance misalignment · Appearance-Preserving 3D Convolution

1 Introduction

Video-based person re-identification (ReID) [11,13,32] plays a crucial role in
intelligent video surveillance system. Compared with image-based ReID [12,28],
the main difference is that the query and gallery in video-based ReID are both
videos and contain additional temporal information. Therefore, how to deal with
the temporal relations between video frames effectively is of central importance
in video-based ReID.

The most commonly used temporal information modeling methods in computer vision include LSTM [10,23], 3D convolution [2,24,29], and Non-local operation [33]. LSTM and 3D convolution are adept at dealing with local temporal
relations and encoding the relative position. Some researchers [2] have demonstrated that 3D convolution is superior to CNN+LSTM on the video classification tasks. In contrast, Non-local operation does not encode the relative position,

© Springer Nature Switzerland AG 2020
A. Vedaldi et al. (Eds.): ECCV 2020, LNCS 12347, pp. 228–243, 2020.
https://doi.org/10.1007/978-3-030-58536-5_14

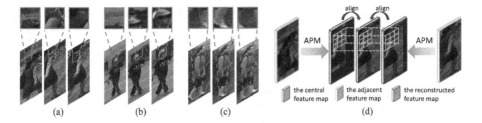

Fig. 1. Temporal appearance misalignment caused by (a) smaller bounding boxes, (b) bigger bounding boxes and (c) posture changes. (d) AP3D firstly uses APM to reconstruct the adjacent feature maps to guarantee the appearance alignment with respect to the central feature map and then performs 3D convolution

but it can model long-range temporal dependencies. These methods are complementary to each other. In this paper, we mainly focus on improving existing 3D convolution to make it more suitable for video-based ReID.

Recently, some researchers [17,19] try to introduce 3D convolution to video-based ReID. However, they neglect that, compared with other video-based tasks, the video sample in video-based ReID consists of a sequence of bounding boxes produced by some pedestrian detector [25,35] (see Fig. 1), not the original video frames. Due to the imperfect person detection algorithm, some resulting bounding boxes are smaller (see Fig. 1(a)) or bigger (see Fig. 1(b)) than the ground truths. In this case, because of the resizing operation before feeding into a neural network, the same spatial positions in adjacent frames may belong to different body parts and the same body parts in adjacent frames may be scaled to different sizes. Even though the detection results are accurate, the misalignment problem may still exist due to the posture changes of the target person (see Fig. 1(c)). Note that one 3D convolution kernel processes the features at the same spatial position in adjacent frames into one value. When temporal appearance misalignment exists, 3D convolution may mixture the features belonging to different body parts in adjacent frames into one feature, which destroys the appearance representations of person videos. Since the performance of video-based ReID highly relies on the appearance representation, so the appearance destruction is harmful. Therefore, it is desirable to develop a new 3D convolution method which can model temporal relations on the premise of maintaining appearance representation quality.

In this paper, we propose Appearance-Preserving 3D convolution (AP3D) to address the appearance destruction problem of existing 3D convolution. As shown in Fig. 1(d), AP3D is composed of an Appearance-Preserving Module (APM) and a 3D convolution kernel. For each central feature map, APM reconstructs its adjacent feature maps according to the cross-pixel semantic similarity and guarantees the temporal appearance alignment between the reconstructed and central feature maps. The reconstruction process of APM can be considered as feature map registration between two frames. As for the problem of asymmetric appearance information (*e.g.*, in Fig. 1(a), the first frame does not contain foot region, thus can not be aligned with the second frame perfectly), Contrastive Attention is proposed to find the unmatched regions between the reconstructed and central feature maps. Then, the learned attention mask is imposed on the reconstructed feature map to avoid error propagation. With APM guaranteeing the appearance alignment, the following 3D convolution can model the spatiotemporal information more effectively and enhance the video representation with higher discriminative ability but no appearance destruction. Consequently, the performance of video-based ReID can be greatly improved. Note that the learning process of APM is unsupervised. In other words, no extra correspondence annotations are required, and the model can be trained only with identification supervision.

The proposed AP3D can be easily combined with existing 3D ConvNets (*e.g.*, I3D [2] and P3D [24]) just by replacing the original 3D convolution kernels with AP3Ds. Extensive ablation studies on two widely used datasets indicate that AP3D outperforms existing 3D convolution significantly. Using RGB information only and without any bells and whistles (*e.g.*, optical flow, complex feature matching strategy), AP3D achieves state-of-the-art results on both datasets.

In summary, the main contributions of our work lie in three aspects: (1) finding that existing 3D convolution is problematic for extracting appearance representation when misalignment exists; (2) proposing an AP3D method to address this problem by aligning the feature maps in pixel level according to semantic similarity before convolution operation; (3) achieving superior performance on video-based ReID compared with state-of-the-art methods.

2   Related Work

Video-Based ReID. Compared with image-based ReID, the samples in video-based ReID contain more frames and additional temporal information. Therefore, some existing methods [4,17,22,32] attempt to model the additional temporal information to enhance the video representations. In contrast, other methods [3,18,21,27] extract video frame features just using image-based ReID model and explore how to integrate or match multi-frame features. In this paper, we try to solve video-based ReID through developing an improved 3D convolution model for better spatiotemporal feature representation.

Temporal Information Modeling. The widely used temporal information modeling methods in computer vision include LSTM [10,23], 3D convolution [2,29], and Non-local operation [33]. LSTM and 3D convolution are adept at modeling local temporal relations and encoding the relative position, while Non-local operation can deal with long-range temporal relations. They are complementary to each other. Zisserman *et al.* [2] has demonstrated that 3D convolution outperforms CNN+LSTM on the video classification task. In this paper, we mainly improve the original 3D convolution to avoid the appearance destruction problem and also attempt to combine the proposed AP3D with some existing 3D ConvNets.

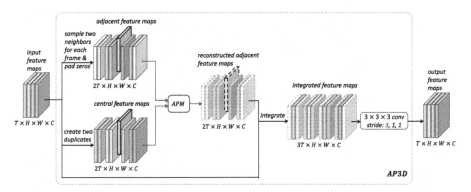

Fig. 2. The overall framework of the proposed AP3D. Each feature map of the input tensor is considered as the central feature map and its two neighbors are sampled as the corresponding adjacent feature maps. APM is used to reconstruct the adjacent feature maps to guarantee the appearance alignment with respect to corresponding central feature maps. Then the following 3D convolution is performed. Note that the temporal stride of 3D convolution kernel is set to its temporal kernel size. In that case, the shape of output tensor is the same as the shape of input tensor

Image Registration. Transforming different images into the same coordinate system is called image registration [1,39]. These images may be obtained at different times, from different viewpoints or different modalities. The spatial relations between these images may be estimated using rigid, affine, or complex deformation models. As for the proposed method, the alignment operation of APM can be considered as feature map registration. Different feature maps are obtained at sequential times and the subject of person is non-rigid.

3   Appearance-Preserving 3D Convolution

In this section, we first illustrate the overall framework of the proposed AP3D. Then, the details of the core module, *i.e.* Appearance-Preserving Module (APM), are explained followed with discussion. Finally, we introduce how to combine AP3D with existing 3D ConvNets.

3.1   The Framework

3D convolution is widely used on video classification task and achieves state-of-the-art performance. Recently, some researchers [17,19] introduce it to video-based ReID. However, they neglect that the performance of ReID tasks is highly dependent on the appearance representation, instead of the motion representation. Due to the imperfect detection results or posture changes, appearance misalignment is unavoidable in video-based ReID samples. In this case, existing 3D convolutions, which process the same spatial position across adjacent

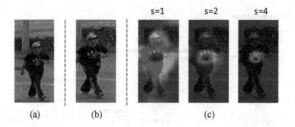

Fig. 3. Visualization of (a) a central frame, (b) its adjacent frame and (c) similarity distribution with different scale factors s on the adjacent feature maps. With a reasonable s, APM can locate the corresponding region on the adjacent feature map w.r.t. the marked position on the central frame accurately (Color figure online)

frames as a whole, may destroy the appearance representation of person videos, therefore they are harmful to ReID.

In this paper, we propose a novel AP3D method to address the above problem. The proposed AP3D is composed of an APM and a following 3D convolution. An example of AP3D with $3 \times 3 \times 3$ convolution kernel is shown in Fig. 2. Specifically, given an input tensor with T frames, each frame is considered as the central frame. We first sample two neighbors for each frame and obtain $2T$ adjacent feature maps in total after padding zeros. Secondly, APM is used to reconstruct each adjacent feature map to guarantee the appearance alignment with corresponding central feature map. Then, we integrate the reconstructed adjacent feature maps and the original input feature maps to form a temporary tensor. Finally, the $3 \times 3 \times 3$ convolution with stride $(3, 1, 1)$ is performed and an output tensor with T frames can be produced. With APM guaranteeing appearance alignment, the following 3D convolution can model temporal relations without appearance destruction. The details of APM are presented in next subsection.

3.2 Appearance-Preserving Module

Feature Map Registration. The objective of APM is reconstructing each adjacent feature map to guarantee that the same spatial position on the reconstructed and corresponding central feature maps belong to the same body part. It can be considered as a graph matching or registration task between each two feature maps. On one hand, since the human body is a non-rigid object, a simple affine transformation can not achieve this goal. On the other hand, existing video-based ReID datasets do not have extra correspondence annotations. Therefore, the process of registration is not that straightforward.

We notice that the middle-level features from ConvNet contain some semantic information [1]. In general, the features with the same appearance have higher cosine similarity, while the features with different appearances have lower cosine similarity [1,13]. As shown in Fig. 3, the red crosses indicate the same position on the central (in Fig. 3(a)) and adjacent (in Fig. 3(b)) frames, but they belong to different body parts. We compute the cross-pixel cosine similarits between the marked position on the central feature map and all positions on the adjacent feature map. After normalization, the similarity distribution is visualized in Fig. 3(c) ($s = 1$). It can be seen that the region with the same appearance is highlighted. Hence, in this paper, we locate the corresponding positions in adjacent frames according to the cross-pixel similarities to achieve feature map registration.

Since the scales of the same body part on the adjacent feature maps may be different, one position on the central feature map may have several corresponding pixels on its adjacent feature map, and vice versa. Therefore, filling the corresponding position on the reconstructed feature map with only the most similar position on the original adjacent feature map is not accurate. To include all pixels with the same appearance, we compute the response y_i at each position on the reconstructed adjacent feature map as a weighted sum of the features x_j at all positions on the original adjacent feature map:

$$y_i = \sum_j \frac{e^{f(c_i, x_j)} x_j}{\sum_j e^{f(c_i, x_j)}}, \tag{1}$$

where c_i is the feature on the central feature map with the same spatial position as y_i and $f(c_i, x_j)$ is defined as the cosine similarity between c_i and x_j with a scale factor $s > 0$:

$$f(c_i, x_j) = s \frac{g(c_i) \cdot g(x_j)}{\|g(c_i)\|\|g(x_j)\|}, \tag{2}$$

where $g(\cdot)$ is a linear transformation that maps the features to a low-dimensional space. The scale factor s is used to adjust the range of cosine similarities. And a big s can make the relatively high similarity even higher while the relatively low similarity lower. As shown in Fig. 3(c), with a reasonable scale factor s, APM can locate the corresponding region on the adjacent feature map precisely. In this paper, We set the scale factor to 4.

Contrastive Attention. Due to the error of pedestrian detection, some regressive bounding boxes are smaller than the ground truths, so some body parts may be lost in the adjacent frames (see Fig. 1(a)). In this case, the adjacent feature maps can not align with the central feature map perfectly. To avoid error propagation caused by imperfect registration, Contrastive Attention is proposed to find the unmatched regions between the reconstructed and central feature maps. Then, the learned attention mask is imposed on the reconstructed feature map. The final response z_i at each position on the reconstructed feature map is defined as:

$$z_i = ContrastiveAtt(c_i, y_i)y_i. \tag{3}$$

Here $ContrastiveAtt(c_i, y_i)$ produces an attention value in $[0, 1]$ accoring to the semantic similarity between c_i and y_i:

$$ContrastiveAtt(c_i, y_i) = sigmoid(w^T(\theta(c_i) \odot \phi(y_i))), \tag{4}$$

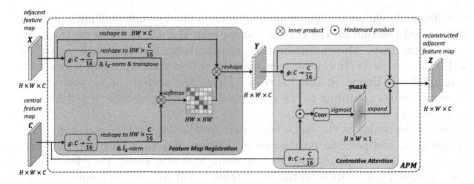

Fig. 4. The illustration of APM. The adjacent feature map is firstly reconstructed by feature map registration. Then a Contrastive Attention mask is multiplied with the reconstructed feature map to avoid error propagation caused by imperfect registration

where w is a learnable weight vector implemented by 1×1 convolution, and $\odot$ is Hadamard product. Since c_i and y_i are from the central and reconstructed feature maps respectively, we use two asymmetric mapping functions $\theta(\cdot)$ and $\phi(\cdot)$ to map c_i and y_i to a shared low-dimension semantic space.

The registration and contrastive attention of APM are illustrated in Fig. 4. All three semantic mappings, *i.e.* g, θ and ϕ, are implemented by 1×1 convolution layers. To reduce the computation, the output channels of these convolution layers are set to $C/16$.

3.3   Discussion

Relations between APM and Non-local. APM and Non-local (NL) operation can be viewed as two graph neural network modules. Both modules consider the feature at each position on feature maps as a node in graph and use weighted sum to estimate the feature. But they have many differences:

(a) NL aims to use spatiotemporal information to enhance feature and its essence is graph convolution or self-attention on a spatiotemporal graph. In contrast, APM aims to reconstruct adjacent feature maps to avoid appearance destruction by the following 3D Conv. Its essence is graph matching or registration between two spatial graphs.

(b) The weights in the weighted sum in NL are used for building dependencies between each pair of nodes only and do not have specific meaning. In contrast, APM defines the weights using cosine similarity with a reasonable scale factor, in order to find the positions with the same appearance on the adjacent feature maps accurately (see Fig. 3).

(c) After APM, the integrated feature maps in Fig. 2 can still maintain spatiotemporal relative relations to be encoded by the following 3D Conv, while NL cannot.

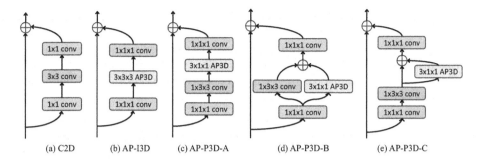

(a) C2D   (b) AP-I3D   (c) AP-P3D-A   (d) AP-P3D-B   (e) AP-P3D-C

Fig. 5. The C2D, AP-I3D and AP-P3D versions of Residual blocks. As for AP-I3D and AP-P3D Residual blocks, only the origional temporal convolution kernels are replaced by AP3Ds

(d) Given a spatiotemporal graph with N frames, the computational complexity of NL is $O(N^2)$, while the computational complexity of APM is only $O(N)$, much lower than NL.

Relations Between Contrastive Attention and Spatial Attention. The Contrastive Attention in APM aims to find the unmatched regions between two frames to avoid error propagation caused by imperfect registration, while the widely used spatial attention [18] in ReID aims to locate more discriminative regions for each frame. As for formulation, Contrastive Attention takes two feature maps as inputs and is imposed on the reconstructed feature map, while Spatial Attention takes one feature map as input and is imposed on itself.

3.4 Combining AP3D with I3D and P3D Blocks

To leverage successful 3D ConvNet designs, we combine the proposed AP3D with I3D [2] and P3D [24] Residual blocks. Transferring I3D and P3D Residual blocks to their AP3D versions just needs to replace the original temporal convolution kernel with AP3D with the same kernel size. The C2D, AP-I3D and AP-P3D versions of Residual blocks are shown in Fig. 5.

4 AP3D for Video-Based ReID

To investigate the effectiveness of AP3D for video-based ReID, we use the 2D ConvNet (C2D) form [13] as our baseline method and extend it into AP3D ConvNet with the proposed AP3D. The details of network architectures are described in Sect. 4.1, and then the loss function we use is introduced in Sect. 4.2.

4.1 Network Architectures

C2D Baseline. We use ResNet-50 [8] pre-trained on ImageNet [26] as the backbone and remove the down-sampling operation of stage$_5$ following [28] to enrich the granularity. Given an input video clip with T frames, it outputs a tensor with shape $T \times H \times W \times 2048$. After spatial max pooling and temporal average pooling, a 2048-dimension feature is produced. Before feeding into the classifier, a BatchNorm [14] operation is used to normalize the feature following [13]. The C2D baseline does not involve any temporal operations except the final temporal average pooling.

AP3D ConvNet. We replace some 2D Residual blocks with AP3D Residual blocks to turn C2D into AP3D ConvNet for spatiotemporal feature learning. Specifically, we investigate replacing one, half of or all Residual blocks in one stage of ResNet, and the results are reported in Sect. 5.4

4.2   Objective Function

Following [30], we combine cross entropy loss and triplet loss [9] for spatiotemporal representation learning. Since cross entropy loss mainly optimizes the features in angular subspace [31], to maintain consistency, we use cosine distance for triplet loss.

5   Experiments

5.1   Datasets and Evaluation Protocol

Datasets. We evaluate the proposed method on three video-based ReID datasets, *i.e.* MARS [37], DukeMTMC-VideoReID [34] and iLIDS-VID [32]. Since MARS and DukeMTMC-VideoReID have fixed train/test splits, for convenience, we perform ablation studies mainly on these two datasets. Besides, we report the final results on iLIDS-VID to compare with the state-of-the-arts.

Evaluation Protocol. We use the Cumulative Matching Characteristics (CMC) and mean Average Precision (mAP) [38] as the evaluation metrics.

5.2   Implementation Details

Training. In the training stage, for each video tracklet, we randomly sample 4 frames with a stride of 8 frames to form a video clip. Each batch contains 8 persons, each person with 4 video clips. We resize all the video frames to 256×128 pixels and use horizontal flip for data augmentation. As for the optimizer, Adam [15] with weight decay 0.0005 is adopted to update the parameters. We train the model for 240 epochs in total. The learning rate is initialized to 3×10^{-4} and multiplied by 0.1 after every 60 epochs.

Testing. In the test phase, for each video tracklet, we first split it into several 32-frame video clips. Then we extract the feature representation for each video clip and the final video feature is the averaged representation of all clips. After feature extraction, the cosine distances between the query and gallery features are computed, based on which the retrieval is performed.

Table 1. Comparison between AP3D and original 3D convolution

Model	Param.	GFLOPs	MARS		Duke-Video	
			top-1	mAP	top-1	mAP
C2D	23.51	16.35	88.9	83.4	95.6	95.1
I3D	27.64	19.37	88.6	83.0	95.4	95.2
AP-I3D	27.68	19.48	**90.1**	84.8	96.2	95.4
P3D-A	24.20	16.85	88.9	83.2	95.0	95.0
AP-P3D-A	24.24	16.90	**90.1**	84.9	96.0	95.3
P3D-B	24.20	16.85	88.8	83.0	95.4	95.3
AP-P3D-B	24.24	16.96	89.9	84.7	**96.4**	**95.9**
P3D-C	24.20	16.85	88.5	83.1	95.3	95.3
AP-P3D-C	24.24	16.90	**90.1**	**85.1**	96.3	95.6

5.3 Comparison with Related Approaches

AP3D vs. Original 3D Convolution. To verify the effectiveness and generalization ability of the proposed AP3D, we implement I3D and P3D residual blocks using AP3D and the original 3D convolution, respectively. Then, we replace one 2D block with 3D block for every 2 residual blocks in $stage_2$ and $stage_3$ of C2D ConvNets, and 5 residual blocks in total are replaced. As shown in Table 1, compared with the C2D baseline, I3D and P3D show close or lower results due to appearance destruction. With APM aligning the appearance representation, the corresponding AP3D versions improve the performance significantly and consistently on both two datasets with few additional parameters and little extra computational complexity. Specifically, AP3D increases about 1% top-1 and 2% mAP over I3D and P3D on MARS dataset. Note that the mAP improvement on DukeMTMC-VideoReID is not as much as that on MARS. One possible explanation is that the bounding boxes of video samples in DukeMTMC-VideoReID dataset are manually annotated and the appearance misalignment is not too serious, so the improvement of AP3D is not very significant.

Compared with other varieties, AP-P3D-C achieves the best performance among most settings. So we conduct the following experiments based on AP-P3D-C (denoted as AP3D for short) if not specifically noted.

AP3D vs. Non-local. Both APM in AP3D and Non-local (NL) are graph-based methods. We insert the same 5 NL blocks into C2D ConvNets and compare AP3D with NL in Table 2. It can be seen that, with fewer parameters and less computational complexity, AP3D outperforms NL on both two datasets.

To compare more fairly, we also implement Contrastive Attention embedded Non-local (CA-NL) and the combination of NL and P3D (NL-P3D). As shown in Table 2, CA-NL achieves the same result as NL on MARS and is still inferior to AP3D. On DukeMTMC-VideoReID, the top-1 of CA-NL is even lower than NL. It is more likely that the Contrastive Attention in APM is designed to avoid

Table 2. Comparison with NL and other temporal information modeling methods

Model	Param.	GFLOPs	MARS		Duke-Video	
			top-1	mAP	top-1	mAP
NL	30.87	21.74	89.6	85.0	96.2	95.6
CA-NL	32.75	21.92	89.6	85.0	95.9	95.6
NL-P3D	31.56	22.17	89.9	84.8	96.2	95.5
AP3D	24.24	16.90	90.1	85.1	96.3	95.6
NL-AP3D	31.60	22.29	**90.7**	**85.6**	**97.2**	**96.1**
Deformable 3D Conv [5]	27.75	19.53	88.5	81.9	95.2	95.0
CNN+LSTM [10]	28.76	16.30	88.7	79.8	95.7	94.6

error propagation caused by imperfect registration. However, the essence of NL is graph convolution on a spatiotemporal graph, not graph registration. So NL can not co-work with Contrastive Attention. Besides, since P3D can not handle appearance misalignment in video-based ReID, NL-P3D shows close results to NL and is inferior to AP3D, too. With APM aligning the appearance, further improvement is achieved by NL-AP3D. This result demonstrates that AP3D and NL are complementary to each other.

AP3D vs. Other Methods for Temporal Information Modeling. We also compare AP3D with Deformable 3D convolution [5] and CNN+LSTM [10]. To compare fairly, the same backbone and hyper-parameters are used. As shown in Table 2, AP3D outperforms these two methods significantly on both two datasets. This comparison further demonstrates the effectiveness of AP3D for learning temporal cues.

5.4   Ablation Study

Effective Positions to Place AP3D Blocks. Table 3 compares the results of replacing a residual block with AP3D block in different stages of C2D ConvNet. In each of these stages, the second last residual block is replaced with the AP3D block. It can be seen that the improvements by placing AP3D block in stage$_2$ and stage$_3$ are similar. Especially, the results of placing only one AP3D block in stage$_2$ or stage$_3$ surpass the results of placing 5 P3D blocks in stage$_{2,3}$. However, the results of placing AP3D block in stage$_1$ or stage$_4$ are worse than the C2D baseline. It is likely that the low-level features in stage$_1$ are insufficient to provide precise semantic information, thus APM in AP3D can not align the appearance representation very well. In contrast, the features in stage$_4$ are insufficient to provide precise spatial information, so the improvement by appearance alignment is also limited. Hence, we only consider replacing the residual blocks in stage$_2$ and stage$_3$.

How Many Blocks Should be Replaced by AP3D? Table 3 also shows the results with more AP3D blocks. We investigate replacing 2 blocks (1 for

Table 3. The results of replacing different numbers of residual blocks in different stages with AP3D block

Model	Stage	Num.	MARS		Duke-Video	
			top-1	mAP	top-1	mAP
C2D			88.9	83.4	95.6	95.1
P3D	stage$_{2,3}$	5	88.5	83.1	95.3	95.3
AP3D	stage$_1$	1	89.0	83.2	95.3	95.1
	stage$_2$	1	89.5	84.0	95.6	**95.4**
	stage$_3$	1	**89.7**	**84.1**	95.9	95.3
	stage$_4$	1	88.8	82.9	95.4	95.0
	stage$_{2,3}$	2	**90.1**	84.7	96.2	95.4
	stage$_{2,3}$	5	**90.1**	**85.1**	**96.3**	**95.6**
	stage$_{2,3}$	10	89.8	84.7	95.9	95.2

Table 4. The results with different backbones

Backbone	Model	MARS		Duke-Video	
		top-1	mAP	top-1	mAP
ResNet-18	C2D	86.9	79.0	93.7	92.9
	P3D	86.9	79.5	93.2	92.9
	AP3D	**88.1**	**80.9**	**94.2**	**93.4**
ResNet-34	C2D	87.5	80.9	94.6	93.6
	P3D	87.6	81.0	94.4	93.7
	AP3D	**88.7**	**82.1**	**95.2**	**94.7**

Table 5. The results of AP3D with/without CA on MARS

Model	w/ CA?	top-1	mAP
I3D	-	88.6	83.0
AP-I3D	✗	89.7	84.7
	✓	**90.1**	**84.8**
P3D	-	88.5	83.1
AP-P3D	✗	89.6	84.8
	✓	**90.1**	**85.1**

each stage), 5 blocks (half of residual blocks in stage$_2$ and stage$_3$) and 10 blocks (all residual blocks in stage$_2$ and stage$_3$) in C2D ConvNet. It can be seen that more AP3D blocks generally lead to higher performance. We argue that more AP3D blocks can perform more temporal communications, which can hardly be realized via the C2D model. As for the results with 10 blocks, the performance drop may lie in the overfitting caused by the excessive parameters.

Effectiveness of AP3D Across Different Backbones. We also investigate the effectiveness and generalization ability of AP3D across different backbones. Specifically, we replace half of the residual blocks in stage$_{2,3}$ of ResNet-18 and ResNet-34 with AP3D blocks. As shown in Table 4, AP3D can improve the results of these two architectures significantly and consistently on both datasets. In particular, AP3D-ResNet-18 is superior to both its ResNet-18 counterparts (C2D and P3D) and the deeper ResNet-34, a model which has almost double the number of parameters and computational complexity, on MARS dataset. This comparison shows that the effectiveness of AP3D does not rely on additional parameters and computational load.

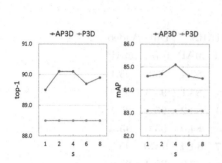

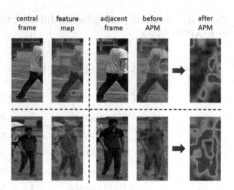

Fig. 6. The results with different s on MARS dataset

Fig. 7. The visualization of the original and the reconstructed feature maps after APM

The Effectiveness of Contrastive Attention. As described in Sect. 3.2, we use Contrastive Attention to avoid error propagation of imperfect registration caused by asymmetric appearance information. To verify the effectiveness, we reproduce AP3D with/without Contrastive Attention (CA) and the experimental results on MARS, a dataset produced by pedestrian detector, are shown in Table 5. It can be seen that, without Contrastive Attention, AP-I3D and AP-P3D can still increase the performance of I3D and P3D baselines by a considerable margin. With Contrastive Attention applied on the reconstructed feature map, the results of AP-I3D and AP-P3D can be further improved.

The Influence of the Scale Factor s. As discussed in Sect. 3.2, the larger the scale factor s, the higher the weights of pixels with high similarity. We show the experimental results with varying s on MARS dataset in Fig. 6. It can be seen that AP3D with different scale factors consistently improves over the baseline and the best performance is achieved when $s = 4$.

5.5 Visualization

We select some misaligned samples and visualize the original feature maps and the reconstructed feature maps in stage$_3$ after APM in Fig. 7. It can be seen that the highlighted regions of the central feature map and the adjacent feature map before APM mainly focus on their own foreground respectively and are misaligned. After APM, the highlighted regions of the reconstructed feature maps are aligned w.r.t.the foreground of the corresponding central frame. It can further validate the alignment mechanism of APM.

5.6 Comparison with State-of-the-Art Methods

We compare the proposed method with state-of-the-art video-based ReID methods which use the same backbone on MARS, DukeMTMC-VideoReID, and iLIDS-VID datasets. The results are summarized in Table 6. Note that these

Table 6. Comparison with state-of-the-arts on MARS, DukeMTMC-VideoReID and iLIDS-VID datasets. 'Flow' denotes optical flow and 'Att.' represents attribute

Method	Modality	MARS		Duke-Video		iLIDS-VID
		top-1	mAP	top-1	mAP	top-1
EUG [34]	RGB	80.8	67.4	83.6	78.3	
DuATM [27]	RGB	81.2	67.7	-	-	-
DRSA [18]	RGB	82.3	65.8	-	-	80.2
TKP [7]	RGB	84.0	73.3	94.0	91.7	-
M3D [17]	RGB	84.4	74.1	-	-	74.0
Snippet [3]	RGB + Flow	86.3	76.1	-	-	85.4
STA [6]	RGB	86.3	80.8	96.2	94.9	-
AttDriven [36]	RGB + Att.	87.0	78.2	-	-	86.3
GLTR [16]	RGB	87.0	78.5	**96.3**	93.7	86.0
VRSTC [13]	RGB	88.5	82.3	95.0	93.5	83.4
NVAN [20]	RGB	90.0	82.8	**96.3**	94.9	-
AP3D	RGB	**90.1**	**85.1**	**96.3**	95.6	**86.7**
NL-AP3D	RGB	**90.7**	**85.6**	**97.2**	96.1	**88.7**

comparison methods differ in many aspects, *e.g.*, using information from different modalities. Nevertheless, using RGB only and with a simple feature integration strategy (*i.e.* temporal average pooling), the proposed AP3D surpasses all these methods consistently on these three datasets. Especially, AP3D achieves 85.1% mAP on MARS dataset. When combined with Non-local, further improvement can be obtained.

6   Conclusion

In this paper, we propose a novel AP3D method for video-based ReID. AP3D consists of an APM and a 3D convolution kernel. With APM guaranteeing the appearance alignment across adjacent feature maps, the following 3D convolution can model temporal information on the premise of maintaining the appearance representation quality. In this way, the proposed AP3D addresses the appearance destruction problem of the original 3D convolution. It is easy to combine AP3D with existing 3D ConvNets. Extensive experiments verify the effectiveness and generalization ability of AP3D, which surpasses start-of-the-art methods on three widely used datasets. As a future work, we will extend AP3D to make it a basic operation in deep neural networks for various video-based recognition tasks.

Acknowledgement. This work is partially supported by Natural Science Foundation of China (NSFC): 61876171 and 61976203.

References

1. Aberman, K., Liao, J., Shi, M., Lischinski, D., Chen, B., Cohen-Or, D.: Neural best-buddies: sparse cross-domain correspondence. ACM Trans. Graph. **37**(4), 69 (2018)
2. Carreira, J., Zisserman, A.: Quo Vadis, action recognition? a new model and the kinetics dataset. In: CVPR (2017)
3. Chen, D., Li, H., Xiao, T., Yi, S., Wang, X.: Video person re-identification with competitive snippet-similarity aggregation and co-attentive snippet embedding. In: CVPR (2018)
4. Chung, D., Tahboub, K., Delp, E.J.: A two stream siamese convolutional neural network for person re-identification. In: Proceedings of the IEEE International Conference on Computer Vision (ICCV) (2017)
5. Dai, J., et al.: Deformable convolutional networks. In: Proceedings of the IEEE International Conference on Computer Vision (ICCV) (2017)
6. Fu, Y., Wang, X., Wei, Y., Huang, T.: STA: Spatial-temporal attention for large-scale video-based person re-identification. In: Proceedings of the AAAI Conference on Artificial Intelligence (AAAI) (2019)
7. Gu, X., Ma, B., Chang, H., Shan, S., Chen, X.: Temporal knowledge propagation for image-to-video person re-identification. In: ICCV (2019)
8. He, K., Zhang, X., Ren, S., Sun, J.: Deep residual learning for image recognition. In: Proceedings of the IEEE Conference on Computer Vision and Pattern Recognition (CVPR) (2016)
9. Hermans, A., Beyer, L., Leibe, B.: In defense of the triplet loss for person re-identification. ArXiv:1703.07737 (2017)
10. Hochreiter, S., Schmidhuber, J.: Long short-term memory. Neural Comput. **9**(8), 1735–1780 (1997)
11. Hou, R., Chang, H., Ma, B., Shan, S., Chen, X.: Temporal complementary learning for video person re-identification. In: ECCV (2020)
12. Hou, R., Ma, B., Chang, H., Gu, X., Shan, S., Chen, X.: Interaction-and-aggregation network for person re-identification. In: CVPR (2019)
13. Hou, R., Ma, B., Chang, H., Gu, X., Shan, S., Chen, X.: VRSTC: occlusion-free video person re-identification. In: CVPR (2019)
14. Ioffe, S., Szegedy, C.: Batch normalization: accelerating deep network training by reducing internal covariate shift. In: ICML (2015)
15. Kingma, D.P., Ba, J.: Adam: a method for stochastic optimization. In: ICLR (2015)
16. Li, J., Wang, J., Tian, Q., Gao, W., Zhang, S.: Global-local temporal representations for video person re-identification. In: ICCV (2019)
17. Li, J., Zhang, S., Huang, T.: Multi-scale 3D convolution network for video based person re-identification. In: AAAI (2019)
18. Li, S., Bak, S., Carr, P., Wang, X.: Diversity regularized spatiotemporal attention for video-based person re-identification. In: CVPR (2018)
19. Liao, X., He, L., Yang, Z., Zhang, C.: Video-based person re-identification via 3D convolutional networks and non-local attention. In: Jawahar, C., Li, H., Mori, G., Schindler, K. (eds.) ACCV 201. Lecture Notes in Computer Science, vol. 11366, pp. 620–634. Springer, Cham (2018). https://doi.org/10.1007/978-3-030-20876-9_39
20. Liu, C.T., Wu, C.W., Wang, Y.C.F., Chien, S.Y.: Spatially and temporally efficient non-local attention network for video-based person re-identification. In: BMVC (2019)

21. Liu, Y., Yan, J., Ouyang, W.: Quality aware network for set to set recognition. In: CVPR (2017)
22. Mclaughlin, N., Rincon, J.M.D., Miller, P.: Recurrent convolutional network for video-based person re-identification. In: CVPR (2016)
23. Ng, Y.H., et al.: Beyond short snippets: deep networks for video classification. In: CVPR (2015)
24. Qiu, Z., Yao, T., Mei, T.: Learning spatio-temporal representation with pseudo-3d residual networks. In: ICCV (2017)
25. Ren, S., He, K., Girshick, R., Sun, J.: Faster R-CNN: Towards real-time object detection with region proposal networks. In: NIPS (2015)
26. Russakovsky, O., et al.: Imagenet large scale visual recognition challenge. Int. J. Comput. Vis. **115**, 211–252 (2015)
27. Si, J., et al.: Dual attention matching network for context-aware feature sequence based person re-identification. In: CVPR (2018)
28. Sun, Y., Zheng, L., Yang, Y., Tian, Q., Wang, S.: Beyond part models: person retrieval with refined part pooling (and a strong convolutional baseline). In: Ferrari, V., Hebert, M., Sminchisescu, C., Weiss, Y. (eds.) ECCV 2018. Lecture Notes in Computer Science, vol. 11208, pp. 501–518. Springer, Cham (2018). https://doi.org/10.1007/978-3-030-01225-0_30
29. Tran, D., Bourdev, L., Fergus, R., Torresani, L., Paluri, M.: Learning spatiotemporal features with 3D convolutional networks. In: ICCV (2015)
30. Wang, G., Yuan, Y., Chen, X., Li, J., Zhou, X.: Learning discriminative features with multiple granularities for person re-identification. In: ACM MM (2018)
31. Wang, H., et al.: CosFace: Large margin cosine loss for deep face recognition. In: CVPR (2018)
32. Wang, T., Gong, S., Zhu, X., Wang, S.: Person re-identification by video ranking. In: Fleet, D., Pajdla, T., Schiele, B., Tuytelaars, T. (eds.) ECCV 2014. LNCS, vol. 8692, pp. 688–703. Springer, Cham (2014). https://doi.org/10.1007/978-3-319-10593-2_45
33. Wang, X., Girshick, R., Gupta, A., He, K.: Non-local neural networks. In: CVPR (2018)
34. Wu, Y., Lin, Y., Dong, X., Yan, Y., Ouyang, W., Yang, Y.: Exploit the unknown gradually: One-shot video-based person re-identification by stepwise learning. In: CVPR (2018)
35. Zhang, H., Chang, H., Ma, B., Wang, N., Chen, X.: Dynamic R-CNN: Towards high quality object detection via dynamic training. In: ECCV (2020)
36. Zhao, Y., Shen, X., Jin, Z., Lu, H., Hua, X.: Attribute-driven feature disentangling and temporal aggregation for video person re-identification. In: CVPR (2019)
37. Zheng, L., et al.: Mars: a video benchmark for large-scale person re-identification. In: Leibe, B., Matas, J., Sebe, N., Welling, M. (eds.) ECCV 2016. LNCS, vol. 9910. Springer, Cham (2016). https://doi.org/10.1007/978-3-319-46466-4_52
38. Zheng, L., et al.: Scalable person re-identification: a benchmark. In: ICCV (2015)
39. Zitová, B., Flusser, J.: Image registration methods: a survey. IVC (2003)

Solving the Blind Perspective-n-Point Problem End-to-End with Robust Differentiable Geometric Optimization

Dylan Campbell[✉], Liu Liu, and Stephen Gould

Australian National University, Australian Centre for Robotic Vision,
Canberra, Australia
{dylan.campbell,liu.liu,stephen.gould}@anu.edu.au

Abstract. Blind Perspective-n-Point (PnP) is the problem of estimating the position and orientation of a camera relative to a scene, given 2D image points and 3D scene points, without prior knowledge of the 2D–3D correspondences. Solving for pose and correspondences simultaneously is extremely challenging since the search space is very large. Fortunately it is a coupled problem: the pose can be found easily given the correspondences and vice versa. Existing approaches assume that noisy correspondences are provided, that a good pose prior is available, or that the problem size is small. We instead propose the first fully end-to-end trainable network for solving the blind PnP problem efficiently and globally, that is, without the need for pose priors. We make use of recent results in differentiating optimization problems to incorporate geometric model fitting into an end-to-end learning framework, including Sinkhorn, RANSAC and PnP algorithms. Our proposed approach significantly outperforms other methods on synthetic and real data.

Keywords: Camera pose estimation · PnP · Implicit differentiation

1 Introduction

The blind Perspective-n-Point (PnP) problem [35] aims to estimate the camera pose from which a set of 2D points were viewed, relative to an unordered 3D point-set. Specifically, the task is to find the rotation and translation that aligns a set of 2D bearing vectors with a set of 3D points, without knowledge of the true 2D–3D correspondences. The camera intrinsic parameters are assumed to be known, which allows 2D points to be expressed as bearing vectors. While a fundamental technique for many computer vision and robotic applications, including augmented reality and visual localization, it remains a challenging problem that has not as yet been satisfactorily solved.

D. Campbell and L. Liu—Equal contribution.

Electronic supplementary material The online version of this chapter (https://doi.org/10.1007/978-3-030-58536-5_15) contains supplementary material, which is available to authorized users.

© Springer Nature Switzerland AG 2020
A. Vedaldi et al. (Eds.): ECCV 2020, LNCS 12347, pp. 244–261, 2020.
https://doi.org/10.1007/978-3-030-58536-5_15

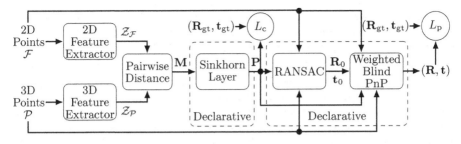

Fig. 1. Network architecture for our end-to-end blind PnP solver. We combine standard neural layers with declarative layers in a bi-level optimization framework to instantiate the traditional camera pose estimation pipeline (feature extraction, feature matching and optimization) in a single neural network. The input is a set of 2D and 3D point coordinates, from which point-wise features are extracted using standard network layers. Feature matching is then performed by computing the pairwise distance between the 2D and 3D point features, and using the Sinkhorn algorithm [42] to obtain a joint probability matrix. Finally, a probability-weighted blind PnP objective function is optimized from a RANSAC initialization to estimate the camera rotation and translation. The key contribution of this work is showing how this optimization procedure may be incorporated into an end-to-end learnable network by the use of declarative layers.

The standard (non-blind) PnP problem [23,32], where 2D–3D correspondences are known, is significantly less complex. It has a closed-form solution for three points [30] and, for a larger number of points, can be embedded in a RANSAC framework [23] to reduce its sensitivity to outliers. However, it is inherently difficult to establish 2D–3D correspondences between modalities. As a result, PnP solvers are typically restricted to applications where both the 2D and 3D data contain visual information, such as structure-from-motion datasets [33]. Even for these, appearance may change seasonally, diurnally, and with weather, and so using geometric rather than visual features may improve generalizability.

Solving the PnP problem without correspondences is much more challenging, because the search space of correspondences and camera poses is very large, the objective function has many local optima, and outliers are prevalent. As a result, it was traditionally the domain of robust geometric algorithms that overcame the search space and non-convexity problems by requiring good pose priors [17,35] or time-consuming global optimization [9,11,12]. Since these techniques were typically iterative, randomized and non-differentiable, this problem has not been amenable to a deep learning solution. Moreover, the geometry of the problem is difficult for a network to learn. However, there is significant opportunity in using a neural network for this problem, since it can effectively recognize patterns in the geometric data and thus reduce the search space and influence of outliers.

Fortunately, the framework of deep declarative networks [25] has recently been proposed, which provides a solution to the problem of including standard neural layers and geometric optimization layers inside the same end-to-end learnable network. This paper applies many of the ideas associated with deep declarative networks by formulating our deep blind PnP solver as a bi-level optimization problem. In this way, we aim to benefit from the pattern recognition capabilities of standard neural networks and the physical models and optimization algorithms used in traditional geometric approaches to the PnP problem.

We focus on the *optimization* part of the traditional camera pose estimation pipeline (feature extraction, matching and optimization) shown in Fig. 1, that is, the remit of the PnP solver itself. To this end, we use an existing network architecture for feature extraction [47] and matching [34]. However, our key insight is that camera pose optimization algorithms, including robust global search techniques such as RANSAC [23] and state-of-the-art nonlinear PnP solvers, can be seamlessly integrated into an end-to-end deep learning framework. Our contributions are: 1) the first fully end-to-end trainable network for solving the blind PnP problem efficiently and globally; 2) the novel deployment of geometric model fitting algorithms as declarative layers inside the network; 3) the novel embedding of non-differentiable robust estimation techniques into the network; and 4) state-of-the-art performance on synthetic and real datasets.

2 Related Work

The majority of camera pose estimation methods assume that a set of 2D–3D correspondences is available and thus a PnP solver [23,32] can be used. Hence, much of the effort in improving the visual localization pipeline focuses on robustly establishing correspondences [38,40] or removing outliers [19,44,47]. These approaches are not appropriate for situations where correspondences cannot be easily obtained, such as when the 3D point-set has no associated visual information. In contrast, we address this problem by deferring the correspondence estimation task until the PnP stage of the pipeline, jointly estimating pose and correspondences. An alternative approach, which does not require explicit correspondences, is learning-based pose regression [27,28,45]. However, Sattler et al. [39] show that this essentially solves an image retrieval task rather than reasoning about 3D structure. Also, the camera is not localized with respect to an explicit 3D map, instead representing the scene implicitly. DSAC and extensions [7,8] can localize with respect to a 3D model, but require many training images from the test scene. In contrast, we eschew visual information to learn generalizable geometric features, and never see the test scene during training.

To solve the blind PnP problem [35] of localizing the camera relative to an explicit unseen 3D model when correspondences cannot be obtained, some approaches seek a local optimum and assume that a good pose prior is available [4,17]. For example, David et al. [17] propose an algorithm that alternates between solving for pose and correspondences, using the Sinkhorn algorithm [42]. To mitigate the pose prior requirement, global search strategies have been proposed, including multiple random starts [17] and probabilistic pose priors [35]. RANSAC [23] can also be used, but becomes intractable for moderately-sized problems. To obviate the need for pose priors and guarantee that a global optimum is found, globally-optimal approaches [9,11,12] use branch-and-bound to systematically reduce the search space. For example, Campbell et al. [12] globally optimize a robust distance between mixture distributions to solve the blind PnP problem. However, these optimal methods are time-consuming and limited to a moderate number of points, unlike our (orders of magnitude) faster approach.

Deep PnP solvers are proposed in existing work [16] (standard PnP) and concurrent work [34] (blind PnP). Due to difficulties inherent in eigendecomposition and outlier filtering, neither approach is end-to-end, despite being highly effective at learning 2D–3D correspondences. The former shows how to avoid unstable eigendecomposition gradients by applying a loss before the pose parameters are estimated, while the latter shows that high-quality 2D–3D correspondence matrices can be learned using optimal transport via the Sinkhorn algorithm [42]. Metric learning can also be used to learn matchable features, as shown for 2D–2D and 3D–3D matching [20]. However, estimating pose from these features or correspondence matrices requires a non-differentiable selection step, such as nearest neighbor search, to reduce the set of correspondences to a tractable size. Different to these approaches, we propose a fully end-to-end trainable blind PnP solver. We directly use an existing ResNet-based [26] feature extraction architecture [47] and a Sinkhorn-based [42] feature matching technique [34] in our network, since our focus is on the joint optimization of all parameters in the camera pose estimation pipeline. Our contribution is orthogonal to these works.

The declarative framework that allows us to incorporate geometric optimization algorithms into a deep network is described in Gould et al. [25]. They present theoretical results and analyses on how to differentiate constrained optimization problems via implicit differentiation. Differentiable convex problems have also been studied recently, including quadratic programs [3] and cone programs [1,2]. In computer vision, the technique has been applied to video classification [21,22], action recognition [14], visual attribute ranking [37], few-shot learning for visual recognition [31], and non-blind PnP in concurrent work [13]. In this work, we show that we can embed geometric model fitting algorithms and non-differentiable robust estimation techniques into a network as declarative layers to solve the blind PnP problem end-to-end.

3   An End-to-End Blind PnP Solver

In this section we present our end-to-end trainable network for solving the blind PnP problem, which we name BPnPNet. We start by formally defining the problem, then provide background on deep declarative networks and show how critical components of a blind PnP solver can be implemented as declarative layers. We then describe our network architecture, loss functions, and learning strategy.

3.1   Problem Formulation

Let $\mathbf{p} \in \mathbb{R}^3$ denote a 3D point and $\mathbf{f} \in \mathbb{R}^3$ denote a unit bearing vector corresponding to a 2D point in the image plane of a calibrated camera. That is, $\|\mathbf{f}\| = 1$ and $\mathbf{f} \propto \mathbf{K}^{-1}[u, v, 1]^\mathsf{T}$, where $\mathbf{K}$ is the matrix of intrinsic camera parameters and (u, v) are the 2D image coordinates. Given a set of bearing vectors $\mathcal{F} = \{\mathbf{f}_i\}_{i=1}^m$ and 3D points $\mathcal{P} = \{\mathbf{p}_i\}_{i=1}^n$, the objective of blind PnP is to find the rotation $\mathbf{R} \in SO(3)$ and translation $\mathbf{t} \in \mathbb{R}^3$ that transforms $\mathcal{P}$ to the coordinate system of $\mathcal{F}$ with the greatest number of one-to-one inlier correspondences, defined by an angular inlier threshold $\theta \in (0, \pi)$. The optimization problem is

$$\underset{\mathbf{R},\mathbf{t},\mathbf{C}}{\text{maximize}} \sum_{i=1}^{m} \sum_{j=1}^{n} \mathbf{C}_{ij} \left(2 \llbracket \angle \left(\mathbf{f}_i, \mathbf{R}\mathbf{p}_j + \mathbf{t} \right) \leqslant \theta \rrbracket - 1 \right)$$
$$\text{subject to } \mathbf{R} \in SO(3), \ \mathbf{t} \in \mathbb{R}^3$$
$$\mathbf{C} \in \mathbb{B}^{m \times n}, \ \mathbf{C} \mathbf{1}^n \in \mathbb{B}^m, \ \mathbf{C}^\mathsf{T} \mathbf{1}^m \in \mathbb{B}^n \tag{1}$$

where $\mathbf{C}$ is a Boolean one-to-one correspondence matrix with at most one non-zero element in each row and column, $\mathbf{C}_{ij}$ is the element at row i and column j, $\mathbb{B} = \{0, 1\}$ is the Boolean domain, $\llbracket \cdot \rrbracket$ is an Iverson bracket, and $\angle(\mathbf{x}, \mathbf{y}) = \arccos(\|\mathbf{x}\|^{-1} \|\mathbf{y}\|^{-1} \mathbf{x}^\mathsf{T} \mathbf{y})$ is a function that returns the angle in $[0, \pi]$ between the vector arguments. This inlier maximization formulation optimizes a robust angular reprojection error. The joint optimization problem can be simplified if either the correspondences or the camera pose is known. If the correspondence matrix $\mathbf{C}$ is known, we have the standard PnP problem. The camera pose can be estimated by minimizing the angular reprojection error

$$\frac{1}{mn} \sum_{i=1}^{m} \sum_{j=1}^{n} \mathbf{C}_{ij} \angle \left(\mathbf{f}_i, \mathbf{R}\mathbf{p}_j + \mathbf{t} \right). \tag{2}$$

If rotation $\mathbf{R}$ and translation $\tilde{\mathbf{t}}$ are known, then $\mathbf{C}$ can be computed as

$$\tilde{\mathbf{C}}_{ij} = \llbracket \angle \left(\mathbf{f}_i, \mathbf{R}\mathbf{p}_j + \mathbf{t} \right) \leqslant \theta \rrbracket \tag{3}$$

followed by the Hungarian algorithm to enforce the one-to-one constraint. If a good pose or correspondence matrix initialization is available, these steps can be alternated to find a good estimate of $\mathbf{R}$, $\mathbf{t}$ and $\mathbf{C}$. This strategy, analogous to the Iterative Closest Point algorithm [6], is taken by the SoftPOSIT algorithm [17]. In contrast, our approach applies this alternation implicitly during training.

3.2 Bi-level Optimization

The conceptual framework that underpins this method is the deep declarative network [25], which interprets training a network as a bi-level optimization problem. According to this view, a network can be composed of multiple imperative and declarative layers. An imperative layer explicitly defines a function for transforming the input to the output, e.g., a convolution layer. In contrast, a declarative layer is implicitly defined in terms of the desired output, formulated as a constrained mathematical optimization problem. Declarative layers are more flexible and general than standard layers, since they admit constraints on the output and decouple the gradient computation from the algorithm used to solve the optimization problem. Crucially, the technique of implicit differentiation enables the back-propagation of gradients through a declarative layer without having to traverse the forward processing function, as shown in Fig. 2.

A bi-level optimization problem [5,43] for end-to-end learning has an upper-level problem solved subject to constraints imposed by a lower-level problem:

$$\text{minimize } L(\mathbf{x}, \mathbf{y})$$
$$\text{subject to } \mathbf{y} \in \arg\min_{\mathbf{u} \in \mathcal{C}} f(\mathbf{x}, \mathbf{u}) \tag{4}$$

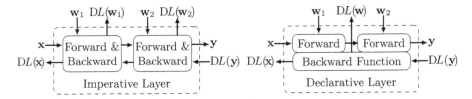

Fig. 2. Comparison of imperative and declarative layers. An imperative layer (left) transforms the input **x** to the output **y** using explicit forward functions, parameterized by network weights **w**. A declarative layer (right) computes the output **y** as a minimizer of an objective function, parameterized by the input **x** and network weights **w**. During learning, the gradient $DL(\mathbf{y})$ of the global loss function with respect to the output is propagated backwards using the chain rule. While the backward function of an imperative layer is tightly coupled with every step of the forward function, effectively unrolling any algorithm applied, the backward function of a declarative layer computes the gradient of the entire layer in one step. The individual forward processing nodes can be recursive or non-differentiable, provided that the objective function optimized by the final forward node is (sub)differentiable in **x**.

where L is a global loss function, f is an objective function, $\mathcal{C}$ is an arbitrary constraint set, and the loss is minimized over all network weights. To solve the bi-level optimization problem (4) by gradient descent, we require the derivative

$$DL(\mathbf{x}, \mathbf{y}) = D_X L(\mathbf{x}, \mathbf{y}) + D_Y L(\mathbf{x}, \mathbf{y}) D\mathbf{y}(\mathbf{x}) \tag{5}$$

in order to back-propagate gradients. The key challenge is to compute $D\mathbf{y}(\mathbf{x})$, for which we use implicit differentiation. We use the notation Df for the derivative of a function $f : \mathbb{R}^n \to \mathbb{R}^m$, an $m \times n$ matrix with entries $(Df(\mathbf{x}))_{ij} = \partial f_i/\partial \mathbf{x}_j$, and we denote the partial derivative over the formal variable X, with all other variables fixed, as $D_X f(\mathbf{x}, \mathbf{y})$. We also use $D_{XY}^2 f$ as shorthand for $D_X (D_Y f)^{\mathsf{T}}$. For completeness, we collect the two results from Gould et al. [25] that are used in this paper. The unconstrained case applies Dini's implicit function theorem [18, p19] to the first-order optimality condition $D_Y f(\mathbf{x}, \mathbf{y}) = \mathbf{0}$.

Lemma 1. *Consider a function* $f : \mathbb{R}^n \times \mathbb{R}^m \to \mathbb{R}$ *and let* $\mathbf{y}(\mathbf{x}) \in \arg\min_{\mathbf{u}} f(\mathbf{x}, \mathbf{u})$. *Assume* $\mathbf{y}(\mathbf{x})$ *exists and that* f *is second-order differentiable in the neighborhood of* $\mathbf{u} = \mathbf{y}(\mathbf{x})$. *Set* $\mathbf{H} = D_{YY}^2 f(\mathbf{x}, \mathbf{y}(\mathbf{x})) \in \mathbb{R}^{m \times m}$ *and* $\mathbf{B} = D_{XY}^2 f(\mathbf{x}, \mathbf{y}(\mathbf{x})) \in \mathbb{R}^{m \times n}$. *Then for non-singular* $\mathbf{H}$ *the derivative of* $\mathbf{y}$ *with respect to* $\mathbf{x}$ *is*

$$D\mathbf{y}(\mathbf{x}) = -\mathbf{H}^{-1}\mathbf{B}. \tag{6}$$

We also require the linear equality constraints case.

Lemma 2. *Consider a function* $f : \mathbb{R}^n \times \mathbb{R}^m \to \mathbb{R}$ *and let* $\mathbf{A} \in \mathbb{R}^{p \times m}$ *and* $\mathbf{d} \in \mathbb{R}^p$ *with* $\mathrm{rank}(\mathbf{A}) = p$ *define a set of* p *under-constrained linear equations* $\mathbf{Au} = \mathbf{d}$. *Also let* $\mathbf{y}(\mathbf{x}) \in \arg\min_{\mathbf{u}} f(\mathbf{x}, \mathbf{u})$ *subject to* $\mathbf{Au} = \mathbf{d}$. *Assume that* $\mathbf{y}(\mathbf{x})$ *exists and that* $f(\mathbf{x}, \mathbf{u})$ *is second-order differentiable in the neighborhood of* $\mathbf{u} = \mathbf{y}(\mathbf{x})$. *Set* $\mathbf{H} = D_{YY}^2 f(\mathbf{x}, \mathbf{y})$ *and* $\mathbf{B} = D_{XY}^2 f(\mathbf{x}, \mathbf{y})$. *Then*

$$D\mathbf{y}(\mathbf{x}) = \left(\mathbf{H}^{-1}\mathbf{A}^{\mathsf{T}}(\mathbf{A}\mathbf{H}^{-1}\mathbf{A}^{\mathsf{T}})^{-1}\mathbf{A}\mathbf{H}^{-1} - \mathbf{H}^{-1}\right)\mathbf{B}. \tag{7}$$

3.3 Declarative Layers for Blind PnP

In this section, we will demonstrate how the theory of bi-level optimization may be applied to the blind PnP problem.

Weighted Blind PnP Layer: This declarative layer operates on the $m \times n$ product set of 2D–3D correspondences, optimizing the lower-level objective function

$$f(\mathbf{P}, \mathbf{r}, \mathbf{t}) = \sum_{i=1}^{m} \sum_{j=1}^{n} \mathbf{P}_{ij} \left(1 - \mathbf{f}_i^{\mathsf{T}} \frac{\mathbf{R_r} \mathbf{p}_j + \mathbf{t}}{\|\mathbf{R_r} \mathbf{p}_j + \mathbf{t}\|} \right) \tag{8}$$

over $\mathbf{r} \in \mathbb{R}^3$ and $\mathbf{t} \in \mathbb{R}^3$, where $\mathbf{P}$ is a fixed joint correspondence probability matrix with $\sum \mathbf{P}_{ij} = 1$, a relaxation of the Boolean correspondence matrix $\mathbf{C}$, and $\mathbf{r}$ is the angle-axis representation of the rotation $\mathbf{R_r}$ such that $\mathbf{R_r} = \exp[\mathbf{r}]_\times$ for the skew symmetric operator $[\cdot]_\times$. The Rodrigues' rotation formula provides an efficient closed-form solution to this exponential map. We use the angle-axis representation to automatically satisfy the constraints on $\mathbf{R}$, that is, $\mathbf{R} \in SO(3)$. We minimize this nonlinear function using the native PyTorch L-BFGS optimizer [10] to find $(\mathbf{r}^\star, \mathbf{t}^\star)$ for a given joint probability matrix $\mathbf{P}$.

Given optimal $(\mathbf{r}^\star, \mathbf{t}^\star)$, corresponding to $\mathbf{y}$ in Lemma 1, we can compute the derivatives $\mathrm{D}\mathbf{r}^\star(\mathbf{P})$ and $\mathrm{D}\mathbf{t}^\star(\mathbf{P})$ using (6). Observe that the gradient computation is agnostic to the choice of optimization algorithm; we do not backpropagate through the L-BFGS iterations that were used to determine $(\mathbf{r}^\star, \mathbf{t}^\star)$. Instead, since the objective function is twice-differentiable, we only require that a (locally) optimal solution be found in order to compute the gradient in one step. While an analytic solution for the gradient can be obtained, it is quite unwieldy. In lieu of this, we use automatic differentiation to compute the necessary Jacobian and Hessian matrices. To be clear, automatic differentiation is applied to the specification of the objective function, not the algorithmic steps used to optimize it, which is distinctly different from standard usage in deep learning.

Importantly, the $m \times n$ product set of correspondences is too large for existing differentiable PnP solvers, such as DLT [16] and DSAC [7]. For $m = n = 1000$, 99.9% of the mn possible correspondences are outliers. Even with the weights $\mathbf{P}$, outliers will dominate the solver. We achieve robustness with top-k RANSAC (see below) and nonlinear optimization. In contrast, the non-robust linear estimate of DLT is unusably poor, and has severe numerical issues [16]. DSAC also fails in this case, because the probability of selecting an inlier hypothesis at random from the product set is vanishingly small. Unlike our declarative layer, DSAC cannot differentiably select hypotheses from the top-k subset.

RANSAC. While Lemma 1 guarantees a local descent direction given the local minimizer $\mathbf{y}$, it is unlikely to be useful for the learning problem (or indeed the inference problem) if $\mathbf{y}$ is a bad estimate. Hence it is helpful for $\mathbf{y}$ to be, on average, a good local optimum—preferably the global optimum. Since the blind PnP objective function is non-convex with many local minima, a standard technique is to apply robust randomized global search such as RANSAC [23]. The declarative framework gives us the opportunity to incorporate this non-differentiable algorithm into an end-to-end learning network. We select a subset of $k = 1.5 \min\{m, n\}$ correspondences from the mn possibilities, choosing those with the highest joint probability in $\mathbf{P}$. We then run RANSAC with the P3P algorithm [24] to find the inlier set, followed by the EPnP algorithm [32] on all inliers to refine the estimate. This robust estimate of the camera pose parameters is used to initialize the nonlinear weighted PnP optimization algorithm.

Since the final processing node in the declarative layer optimizes a twice-differentiable objective function, the non-differentiability of any intermediate computation is irrelevant to the gradient calculation. Note that this procedure for robustly estimating the camera pose parameters has no analytic solution and involves a non-differentiable algorithm. It would not be possible to use standard techniques such as explicit or automatic differentiation to obtain the gradient.

Sinkhorn Layer. We also define a declarative layer for feature matching in order to estimate the joint correspondence probability of the 2D–3D point pairs from a cost matrix $\mathbf{M} \in \mathbb{R}_+^{m \times n}$. This is achieved by encapsulating the Sinkhorn algorithm [42] in a declarative layer, as has been previously demonstrated in the literature [37]. The layer optimizes the lower-level objective function [15]

$$f(\mathbf{M}, \mathbf{P}) = \sum_{i=1}^{m} \sum_{j=1}^{n} (\mathbf{M}_{ij}\mathbf{P}_{ij} + \mu\mathbf{P}_{ij}(\log \mathbf{P}_{ij} - 1)) \tag{9}$$

with respect to $\mathbf{P} \in U(\mathbf{r}, \mathbf{c})$, where the transport polytope

$$U(\mathbf{r}, \mathbf{c}) = \{\mathbf{P} \in \mathbb{R}_+^{m \times n} \mid \mathbf{P}\mathbf{1}^n = \mathbf{r}, \mathbf{P}^{\mathsf{T}}\mathbf{1}^m = \mathbf{c}\} \tag{10}$$

is defined for the prior probability vectors $\mathbf{r} \in \mathbb{R}_+^m$ and $\mathbf{c} \in \mathbb{R}_+^n$ with $\sum \mathbf{r} = 1$ and $\sum \mathbf{c} = 1$, which represent the probability that any given 2D or 3D point has a valid match. In this work, we use uniform priors $\mathbf{r} = \frac{1}{m}\mathbf{1}$ and $\mathbf{c} = \frac{1}{n}\mathbf{1}$.

We run the highly-efficient Sinkhorn algorithm which optimizes this objective function, an entropy-regularized Wasserstein distance, in $O(m^2)$ [15]. This is considerably more efficient than the Hungarian algorithm, which exactly optimizes the Wasserstein distance in $O(m^3)$, while also converging to the Wasserstein distance as $\mu \to 0$. Given optimal $\mathbf{P}^\star$, corresponding to $\mathbf{y}$ in Lemma 2, we can compute the derivative $D\mathbf{P}^\star(\mathbf{M})$ using (7). Unlike the PnP layer, we compute the derivative analytically to ensure memory efficiency. To do so, we need to form the matrices $\mathbf{A}$, $\mathbf{B}$ and $\mathbf{H}$ and perform the necessary inversions. We defer the details to the supplementary material.

The benefits of enclosing this algorithm in a declarative layer include being able to run the algorithm to convergence, rather than fixing the number of iterations, and obviating the need for unrolling the algorithmic steps and maintaining the requisite computation graph, which saves a significant amount of memory. Our implementation is much more memory efficient than that of Santa Cruz et al. [37], reducing $O(m^2n^2)$ memory requirements to $O(mn)$. This allows much larger problems than were considered previously, such as $m = n = 1000$ used in this work. This was achieved by exploiting the block structure of the matrices $\mathbf{A}$ and $\mathbf{H}$ rather than storing them in full, and by computing the vector–Jacobian product rather than the Jacobian itself.

3.4   Network Architecture

Our network architecture is shown in Fig. 1. First, we extract discriminative features from the 2D and 3D point-sets, aiming to recognise patterns in the data that are useful for establishing correspondences. Next, we estimate the correspondence probability for every 2D–3D pair by computing the pairwise distance between features and solving an optimal transport problem. Last, we optimize a weighted blind PnP objective function to obtain the (locally) optimal camera pose, given the data and estimated correspondence probability matrix.

Feature Extraction. To extract discriminative features from the 2D and 3D point-sets, we directly use the point feature extraction model from Yi et al. [47]. This model is a 12-layer ResNet [26], where each layer consists of a perceptron with 128 neurons per point, context normalization, batch normalization and a ReLU nonlinearity, with weights shared between points. Context normalization is the mechanism for sharing information between points, by normalizing with respect to the mean $\boldsymbol{\mu}^l$ and standard deviation σ^l of the feature vectors $\mathbf{z}_i^l$ of every point at the lth layer, and is given by $\text{CN}(\mathbf{z}_i^l) = (\mathbf{z}_i^l - \boldsymbol{\mu}^l)/\sigma^l$.

Before passing the data to the feature extraction networks, we do some initial processing. We convert the homogeneous bearing 3-vectors into inhomogeneous 2-vectors by dividing through by the z coordinate, since this requires fewer network parameters. We also apply a learned 3×3 transformation matrix to the 3D points using the input transform from PointNet [36], to align the points to a canonical orientation. Finally, after obtaining the pointwise feature vectors, we apply L_2 normalization, which is helpful for the ensuing optimization procedure.

Hence the 2D feature extractor encodes a mapping Φ with parameters ϕ from the 2D bearing vector set $\mathcal{F}$ to the feature vector set $\mathcal{Z}_{\mathcal{F}}$, given by $\mathcal{Z}_{\mathcal{F}} = \Phi_\phi(\mathcal{F})$ with $\mathcal{Z}_{\mathcal{F}} = \{\mathbf{z}_{\mathbf{f}_i}\}_{i=1}^m$ and $\mathbf{z}_{\mathbf{f}_i} \in \mathbb{R}^{128}$. The 3D feature extractor encodes a similar mapping, given by $\mathcal{Z}_{\mathcal{P}} = \Psi_\psi(\mathcal{P})$ with $\mathcal{Z}_{\mathcal{P}} = \{\mathbf{z}_{\mathbf{p}_i}\}_{i=1}^n$ and $\mathbf{z}_{\mathbf{p}_i} \in \mathbb{R}^{128}$.

Correspondence Probability. To estimate the probability that a 2D–3D point pair is an inlier correspondence, we compute the pairwise distances between the feature vector sets $\mathcal{Z}_{\mathcal{F}}$ and $\mathcal{Z}_{\mathcal{P}}$ and then solve an optimal transport problem, as was shown to be effective in concurrent work [34]. The elements of the pairwise L_2 distance matrix $\mathbf{M} \in \mathbb{R}_+^{m \times n}$ are computed as $\mathbf{M}_{ij} = \|\mathbf{z}_{\mathbf{f}_i} - \mathbf{z}_{\mathbf{p}_j}\|_2$. We then solve the regularized transport problem [15] using the Sinkhorn algorithm [42] to obtain a joint probability matrix $\mathbf{P}$. The advantage of this approach is that it considers the entirety of $\mathbf{M}$ when estimating the probability $\mathbf{P}_{ij}$, in order to resolve correspondence ambiguities. Finding a jointly optimal solution is critical if the learning goal is to approach the *one-to-one* correspondence matrix $\mathbf{C}$ up to scale. This optimization problem was outlined in Sect. 3.3, where we showed that the gradient computation can be decoupled from the Sinkhorn algorithm using implicit differentiation.

Blind PnP Optimization. Given the joint correspondence probability matrix $\mathbf{P}$, we can now optimize the weighted nonlinear blind PnP objective function (8) to obtain the optimal camera pose $(\mathbf{R}, \mathbf{t})$ for that set of correspondence probabilities. The optimization problem was outlined in Sect. 3.3, where we showed how to find a locally-optimal camera pose using the L-BFGS algorithm, how to ensure it is a good local optimum (on average) using RANSAC, and how to back-propagate through the layer. Hence we have a fully-differentiable way to generate camera pose parameters that are likely to be near the global optimum of the non-convex objective function. At test time, we can either take the network output or the RANSAC output computed within the network; both are evaluated in the experiments.

3.5   Learning from Pose-Labelled Data

Loss Functions. We use two loss functions, one for each component of the coupled problem. The first is a correspondence loss L_c to bring the estimated correspondence matrix $\mathbf{P}$ closer to the ground-truth. The second is a pose loss L_p to encourage the network to generate correspondence matrices that are amenable to our PnP solver. Note that these are distinct, albeit complementary, aims. While a perfect correspondence matrix would generate an accurate camera pose, this is not achievable in practice. Instead, there is a family of correspondence matrices for a fixed suboptimal value of L_c, which will differ considerably in their suitability for the weighted PnP solver.

The correspondence loss L_c arises directly from the problem formulation (1) given the ground-truth rotation $\mathbf{R}_{gt}$ and translation $\mathbf{t}_{gt}$, yielding

$$L_c = \sum_i^m \sum_j^n \mathbf{P}_{ij} \left(1 - 2[\![\angle\left(\mathbf{f}_i, \mathbf{R}_{gt}\mathbf{p}_j + \mathbf{t}_{gt}\right) \leqslant \theta]\!]\right) \tag{11}$$

where $[\![\cdot]\!]$ is an Iverson bracket (indicator function), and θ is the angular inlier threshold. The loss is bounded, since $\sum \mathbf{P}_{ij} = 1$ and so $L_c \in [-1, 1)$, and has the interpretation of maximizing the probability of the inlier correspondences and minimizing the probability of the outlier correspondences, since $\mathbf{P}$ is a joint probability matrix. If the ground-truth correspondence matrix $\mathbf{C}$ is available, this can be used instead of the indicator function. It is also possible to use the (less robust) reprojection error (2), however we found no advantage to this.

The camera pose loss L_p uses standard error measures on rotations and translations, and is given by

$$L_p = L_r + L_t \tag{12}$$

$$L_r = \angle\left(\mathbf{R}, \mathbf{R}_{gt}\right) = \arccos \tfrac{1}{2}\left(\operatorname{trace} \mathbf{R}_{gt}^{\mathsf{T}}\mathbf{R} - 1\right) \tag{13}$$

$$L_t = \|\mathbf{t} - \mathbf{t}_{gt}\|_2 \tag{14}$$

This loss is not bounded, since $L_r \in [0, \pi]$ and $L_t \in [0, \infty)$. The argument of arccos is clamped to between $\pm(1 - \epsilon)$ for $\epsilon = 10^{-7}$ to prevent an infinite gradient at $0°$ and $180°$. Finally, the total loss is given by

$$L = L_c + \gamma_p L_p \tag{15}$$

where γ_p is a hyperparameter that controls the relative influence of L_p.

Learning Strategy. We train the network implemented in PyTorch using the Adam optimizer [29] with a learning rate of 10^{-5} and otherwise default parameters. We use a batch size of 16 and train for 120 epochs (to convergence) with the correspondence loss only ($\gamma_p = 0$), followed by 20–80 epochs with the pose loss as well ($\gamma_p = 1$). This reflects the intuition that the pose loss is more meaningful once the correspondence probability matrix **P** has useful information, having reduced the correspondence search space.

Implementation Details. For the Sinkhorn algorithm, the entropy parameter μ was set to 0.1; for RANSAC, the inlier reprojection error was set to 0.01 and the maximum number of iterations was set to 1000; and for the L-BFGS solver, the line search function was set to strong Wolfe, the maximum number of iterations varied with the number of points to standardize the batch runtime, and the gradient norms were clipped to 100. Ground-truth correspondences were used in training instead of specifying inlier threshold θ. All experiments were run on a single Titan V GPU, and the PyTorch code, including modular Sinkhorn and weighted PnP layers, will be released.

4   Results

Our blind PnP network, named BPnPNet, is evaluated with respect to the baseline algorithms SoftPOSIT [17], RANSAC [23], and GOSMA [12] on synthetic and real data. These are state-of-the-art representative examples of a local blind PnP solver (SoftPOSIT), a global solver (RANSAC), and a globally-optimal solver (GOSMA). For RANSAC, we randomly sample 2D–3D correspondences and use a minimal P3P solver [30].[1] For SoftPOSIT, we provide an initialization using ground-truth pose information, since it is a local solver and therefore requires a good pose prior. The algorithms were stopped early if their runtime for a single point-set pair exceeded 30 s, returning the best pose found so far. This ensured that evaluation time was bounded at four days per algorithm on the datasets tested. Globally-optimal algorithms often exceed this limit, but it is infeasible to evaluate them to convergence on large datasets.

We use the synthetic ModelNet40 dataset [46] and the real-world MegaDepth dataset [33] for evaluation. The former is a CAD mesh model dataset, while the latter is a multi-view photo dataset with COLMAP [41] reconstructions providing the 2D and 3D point-sets. MegaDepth has highly diverse scenes, camera poses, and point distributions. We report quartiles for rotation error (in degrees), translation error, and reprojection error (in degrees), according to (13), (14), and (2) respectively. We denote the first, second (median) and third quartiles as Q1, Q2 and Q3. We also report average runtime for inference (in seconds) and recall at a particular error threshold (as a percentage), that is, the percentage of poses with an error less than that threshold.

---

[1] The probability of choosing a minimal set of 4 true 2D–3D correspondences from the size mn set of all correspondences without replacement is $\prod_{i=0}^{3} \frac{m-i}{(m-i)(n-i)} \approx 10^{-12}$ for $m = n = 1000$ and no outliers. The number of RANSAC iterations required to achieve 90% confidence is thus $\log(1 - 0.9)/\log(1 - 10^{-12}) \approx 2.3 \times 10^{12}$.

Table 1. Results on the ModelNet40 [46] test set. We report quartiles for rotation error (°), translation error and reprojection error (°), and the mean runtime T (s). Note that Ours L_cR is a standard RANSAC baseline: deep 2D–3D feature matching followed by P3P-RANSAC. †Algorithms were run for a maximum of 30 s.

Method	Rotation Error			Translation Error			Reproj. Error			T
	Q1	Q2	Q3	Q1	Q2	Q3	Q1	Q2	Q3	$\bar{x}$
SoftPOSIT [17]	16.1	21.8	28.0	0.33	0.49	0.72	2.82	3.98	5.21	27†
RANSAC [23]	90.8	139	165	0.43	1.15	3.08	4.22	5.87	8.06	30†
GOSMA [12]	10.1	22.1	52.0	0.25	0.46	0.75	1.04	1.62	3.11	30†
Ours L_c	6.08	11.3	18.3	0.34	0.52	0.81	0.56	0.86	1.31	**0.1**
Ours L_cL_p	4.88	9.66	16.0	**0.04**	**0.08**	**0.15**	0.36	0.61	1.03	**0.1**
Ours L_cR	5.49	11.7	20.0	**0.04**	0.09	0.20	0.37	0.70	1.25	**0.1**
Ours L_cL_pR	**3.33**	**8.09**	**15.8**	**0.04**	**0.08**	0.16	**0.28**	**0.52**	1.01	0.1

4.1 Synthetic Data Experiments

In this section, we evaluate our network on the synthetic ModelNet40 dataset [46] and conduct ablation studies. To generate the synthetic data from the mesh models, we uniformly sampled 1000 3D points from each model and generated virtual cameras by drawing Euler rotation angles uniformly from $[0, \pi/4]$ and translations from $[-0.5, 0.5]$, with an offset of 4.5 along the z axis. The points were projected to a 640×480 virtual image with a focal length of 800 and normal noise with $\sigma = 2$ pixels was applied to the 2D points. In this way, we generated training and testing sets of 40000 and 2468 2D–3D point-set pairs respectively, each from the standard train and test splits of ModelNet40. SoftPOSIT was initialized using the mean ground-truth Euler angles and translation, corresponding to a median initial rotation error of 21.5° and translation error of 0.49.

The results on the ModelNet40 test set are shown in Table 1 and Fig. 3. They demonstrate that our network obtains significantly better camera pose results than state-of-the-art local, global and globally-optimal algorithms. The results also include our ablation study, where we compare our model's camera pose output (Ours L_cL_p) with a variant of our model that is learnt without the pose loss (Ours L_c). In all cases, the pose loss improves the results significantly, especially the translation errors. In particular, the recall for rotation errors less than 15° and translation errors less than 0.5 is 72%, an improvement of 25% over using the L_c loss only. We also compare our model's output with the RANSAC pose computed within our network, denoted by R in the results and $(\mathbf{R}_0, \mathbf{t}_0)$ in Fig. 1. The robust RANSAC estimate tends to be more accurate than the model's final output since it is more resistant to errors in the correspondence probability matrix, and so should be used in any applications. Our method is also at least two orders of magnitude faster than the other methods, taking 0.12s on average. Note that Ours L_cR is an example of the standard RANSAC baseline of deep 2D–3D feature matching followed by P3P-RANSAC, without end-to-end

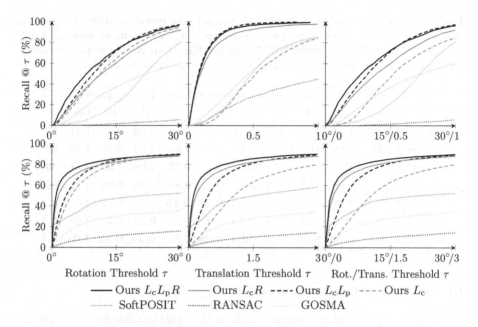

Fig. 3. Recall on the ModelNet40 (top) and MegaDepth (bottom) test sets, with respect to an error threshold τ. R denotes using the RANSAC estimate.

Fig. 4. Qualitative results for the MegaDepth dataset: 3D point-sets projected onto the image using the camera pose found by GOSMA [12] (top) and our method (bottom).

training. Finally, a small reduction of only $2°/0.01$ on the median statistics is observed with weaker pose-only supervision (see supplementary material).

4.2 Real Data Experiments

Here we evaluate our network on the MegaDepth dataset [33]. To generate the splits, we randomly selected landmarks and obtained train and test sets of 40828 and 10795 2D–3D point-set pairs respectively. The landmarks do not overlap across splits; we test how well the network *generalizes* to unseen locations. There

Table 2. Results on the MegaDepth [33] test set. We report quartiles for rotation error (°), translation error and reprojection error (°), and the mean runtime T (s). [†]Algorithms were run for a maximum of 30 s.

Method	Rotation Error			Translation Error			Reproj. Error			T
	Q1	Q2	Q3	Q1	Q2	Q3	Q1	Q2	Q3	$\bar{x}$
SoftPOSIT [17]	1.81	21.4	165	0.24	1.53	6.10	0.92	7.85	24.1	18[†]
RANSAC [23]	66.6	122	155	6.80	15.2	28.2	4.45	8.77	13.3	30[†]
GOSMA [12]	8.69	86.8	145	1.07	5.67	9.34	1.30	13.7	37.1	30[†]
Ours L_c	1.91	4.47	11.4	0.52	1.05	2.34	0.54	1.12	2.81	**0.2**
Ours $L_c L_p$	1.32	3.31	8.84	0.21	0.46	1.08	0.21	0.53	1.64	**0.2**
Ours $L_c R$	0.44	1.55	7.70	0.05	0.18	0.80	0.06	0.16	1.27	**0.2**
Ours $L_c L_p R$	**0.34**	**1.00**	**4.88**	**0.04**	**0.12**	**0.53**	**0.06**	**0.12**	**0.74**	0.2

are between 5–15000 points per set, reflecting the variability of real structure-from-motion data. SoftPOSIT was initialized using the ground-truth, perturbed by up to 10° about a random axis and up to 0.5 m in a random direction.

The results on the MegaDepth test set are shown in Table 2 and Fig. 3, with qualitative results given in Fig. 4. Our approach outperforms the other algorithms by a significant margin, notably doing better than a local optimization algorithm initialized very close to the ground-truth. GOSMA performs poorly on this dataset, despite being optimal, since it rarely converges within the 30 s evaluation limit (often taking minutes). As with the synthetic data, the pose loss improves the results, especially the translation errors. In particular, the recall for rotation errors less than 10° and translation errors less than 1 is 73%, 25% better than without the loss. With the RANSAC estimate, the recall further improves to 82%. Additional results, including outlier analysis, failure cases, and another feature matching approach, are provided in the supplementary material.

5   Conclusion

In this paper, we have proposed the first fully end-to-end trainable network for solving the blind PnP problem. The key insight is that we can back-propagate through a geometric optimization algorithm using the technique of implicit differentiation. This allows us to compute a gradient even when the declarative layer involves non-differentiable RANSAC search and L-BFGS optimization of a nonlinear geometric objective. For such a layer, unrolling the algorithmic steps and computing the gradient with automatic differentiation is not possible, and would not be advisable even if it were due to the memory and computational requirements. Furthermore, we show that our method outperforms state-of-the-art geometric blind PnP solvers by a considerable margin when pose-labelled training data is available. Promisingly, our declarative approach admits the possibility of an unsupervised reprojection error loss, which may be used to fine-tune our pre-trained model to test scene data without the need for ground-truth labels.

Acknowledgements. This work was conducted by the Australian Research Council Centre of Excellence for Robotic Vision (CE140100016), funded by the Australian Government.

References

1. Agrawal, A., Amos, B., Barratt, S., Boyd, S., Diamond, S., Kolter, Z.: Differentiable convex optimization layers. In: Wallach, H., Larochelle, H., Beygelzimer, A., d'Alché Buc, F., Fox, E., Garnett, R. (eds.) Advances in Neural Information Processing Systems 32 (NIPS 2019). Curran Associates, Inc., pp. 9562–9574 (2019)
2. Agrawal, A., Barratt, S., Boyd, S., Busseti, E., Moursi, W.: Differentiating through a cone program. J. Appl. Numer. Optim. **1**(2), 107–115 (2019)
3. Amos, B., Kolter, J.Z.: OptNet: differentiable optimization as a layer in neural networks. In: Precup, D., Teh, Y.W. (eds.) Proceedings of the 34th International Conference on Machine Learning. Proceedings of Machine Learning Research, PMLR, International Convention Centre, Sydney, Australia. **70**, 136–145 (2017)
4. Baka, N., Metz, C., Schultz, C.J., van Geuns, R.J., Niessen, W.J., van Walsum, T.: Oriented Gaussian mixture models for nonrigid 2D/3D coronary artery registration. IEEE Trans. Med. Imag. **33**(5), 1023–1034 (2014). https://doi.org/10.1109/TMI.2014.2300117
5. Bard, J.F.: Practical Bilevel Optimization: Algorithms and Applications. Kluwer Academic Press (1998)
6. Besl, P.J., McKay, N.D.: A method for registration of 3-D shapes. IEEE Trans. Pattern Anal. Mach. Intell. (PAMI) **14**(2), 239–256 (1992)
7. Brachmann, E., et al.: DSAC - differentiable RANSAC for camera localization. In: Proceedings of the 2017 Conference on Computer Vision and Pattern Recognition, Computer Society, pp. 2492–2500. IEEE (2017) https://doi.org/10.1109/CVPR.2017.267
8. Brachmann, E., Rother, C.: Learning less is more - 6D camera localization via 3D surface regression. In: Proceedings of the 2018 Conference on Computer Vision and Pattern Recognition, Computer Society, pp. 4654–4662. IEEE (2018)
9. Brown, M., Windridge, D., Guillemaut, J.Y.: Globally optimal 2D–3D registration from points or lines without correspondences. In: Proceedings of the 2015 International Conference on Computer Vision, pp. 2111–2119 (2015)
10. Byrd, R.H., Lu, P., Nocedal, J., Zhu, C.: A limited memory algorithm for bound constrained optimization. SIAM J. Sci. Comput. **16**(5), 1190–1208 (1995)
11. Campbell, D., Petersson, L., Kneip, L., Li, H.: Globally-optimal inlier set maximisation for camera pose and correspondence estimation. IEEE Trans. Pattern Anal. Mach. Intell. (PAMI) **42**(2), 328–342 (2020). https://doi.org/10.1109/TPAMI.2018.2848650
12. Campbell, D., Petersson, L., Kneip, L., Li, H., Gould, S.: The alignment of the spheres: globally-optimal spherical mixture alignment for camera pose estimation. In: Proceedings of the 2019 Conference on Computer Vision and Pattern Recognition, Computer Society, pp. 11796–11806. IEEE (2019)

13. Chen, B., Parra, A., Cao, J., Li, N., Chin, T.J.: End-to-end learnable geometric vision by backpropagating PnP optimization. In: Proc. of the IEEE Conference on Computer Vision and Pattern Recognition (CVPR), Computer Society, pp. 8100–8109. IEEE (2020)
14. Cherian, A., Fernando, B., Harandi, M., Gould, S.: Generalized rank pooling for action recognition. In: Proceedings of the IEEE Conference on Computer Vision and Pattern Recognition (CVPR), Computer Society, pp. 3222–3231. IEEE (2017)
15. Cuturi, M.: Sinkhorn distances: Light speed computation of optimal transport. In: Burges, C.J.C., Bottou, L., Welling, M., Ghahramani, Z., Weinberger, K.Q. (eds.) Advances in Neural Information Processing Systems (NeurIPS). Curran Associates Inc., pp. 2292–2300 (2013)
16. Dang, Z., Yi, K.M., Hu, Y., Wang, F., Fua, P., Salzmann, M.: Eigendecomposition-free training of deep networks with zero eigenvalue-based losses. In: Ferrari, V., Hebert, M., Sminchisescu, C., Weiss, Y. (eds.) ECCV 2018. LNCS, vol. 11209, pp. 792–807. Springer, Cham (2018). https://doi.org/10.1007/978-3-030-01228-1_47
17. David, P., Dementhon, D., Duraiswami, R., Samet, H.: SoftPOSIT: simultaneous pose and correspondence determination. Int. J. Comput. Vis. (IJCV) **59**(3), 259–284 (2004)
18. Dontchev, A.L., Rockafellar, R.T.: Implicit Functions and Solution Mappings. SSORFE. Springer, New York (2014). https://doi.org/10.1007/978-1-4939-1037-3
19. Enqvist, O., Kahl, F.: Robust optimal pose estimation. In: Forsyth, D., Torr, P., Zisserman, A. (eds.) ECCV 2008. LNCS, vol. 5302, pp. 141–153. Springer, Heidelberg (2008). https://doi.org/10.1007/978-3-540-88682-2_12
20. Fathy, M.E., Tran, Q.-H., Zia, M.Z., Vernaza, P., Chandraker, M.: Hierarchical metric learning and matching for 2D and 3D geometric correspondences. In: Ferrari, V., Hebert, M., Sminchisescu, C., Weiss, Y. (eds.) ECCV 2018. LNCS, vol. 11219, pp. 832–850. Springer, Cham (2018). https://doi.org/10.1007/978-3-030-01267-0_49
21. Fernando, B., Gould, S.: Learning end-to-end video classification with rank-pooling. In: Balcan, M.F., Weinberger, K.Q. (eds.) Proc. of the International Conference on Machine Learning (ICML). PMLR, pp. 1187–1196 (2016)
22. Fernando, B., Gould, S.: Discriminatively learned hierarchical rank pooling networks. Int. J. Comput. Vis. (IJCV) **124**, 335–355 (2017)
23. Fischler, M.A., Bolles, R.C.: Random sample consensus: a paradigm for model fitting with applications to image analysis and automated cartography. Commun. ACM **24**(6), 381–395 (1981)
24. Gao, X.S., Hou, X.R., Tang, J., Cheng, H.F.: Complete solution classification for the perspective-three-point problem. IEEE Trans. Pattern Anal. Mach. Intell. **25**(8), 930–943 (2003)
25. Gould, S., Hartley, R., Campbell, D.: Deep declarative networks: a new hope. Tech. rep., Australian National University (arXiv:1909.04866) (2019)
26. He, K., Zhang, X., Ren, S., Sun, J.: Deep residual learning for image recognition. In: Proceedings of the IEEE Conference on Computer Vision and Pattern Recognition (CVPR), Computer Society, pp. 770–778. IEEE (2016)
27. Kendall, A., Cipolla, R.: Geometric loss functions for camera pose regression with deep learning. In: Proceedings of the 2017 Conference on Computer Vision and Pattern Recognition, Computer Society, pp. 6555–6564. IEEE (2017) https://doi.org/10.1109/CVPR.2017.694

28. Kendall, A., Grimes, M., Cipolla, R.: PoseNet: a convolutional network for real-time 6-DOF camera relocalization. In: Proceedings of the 2015 International Conference on Computer Vision, Computer Society, pp. 2938–2946. IEEE (2015) https://doi.org/10.1109/ICCV.2015.336

29. Kingma, D.P., Ba, J.: Adam: A method for stochastic optimization. In: Bengio, Y., LeCun, Y. (eds.) Proceedings of the International Conference on Learning Representations (ICLR) (2015)

30. Kneip, L., Scaramuzza, D., Siegwart, R.: A novel parametrization of the perspective-three-point problem for a direct computation of absolute camera position and orientation. In: Proceedings of the 2011 Conference on Computer Vision and Pattern Recognition, Computer Society, pp. 2969–2976. IEEE (2011)

31. Lee, K., Maji, S., Ravichandran, A., Soatto, S.: Meta-learning with differentiable convex optimization. In: Proceedings of the IEEE Conference on Computer Vision and Pattern Recognition (CVPR), Computer Society, pp. 10657–10665. IEEE (2019)

32. Lepetit, V., Moreno-Noguer, F., Fua, P.: EPnP: An accurate O(n) solution to the PnP problem. Int. J. Comput. Vis. **81**(2), 155–166 (2009)

33. Li, Z., Snavely, N.: MegaDepth: learning single-view depth prediction from internet photos. In: Proceedings of the IEEE Conference on Computer Vision and Pattern Recognition (CVPR), Computer Society, pp. 2041–2050. IEEE (2018)

34. Liu, L., Campbell, D., Li, H., Zhou, D., Song, X., Yang, R.: Learning 2D–3D correspondences to solve the blind perspective-n-point problem. Tech. rep., Australian National University arXiv:2003.06752 (2019)

35. Moreno-Noguer, F., Lepetit, V., Fua, P.: Pose priors for simultaneously solving alignment and correspondence. In: Forsyth, D., Torr, P., Zisserman, A. (eds.) ECCV 2008. LNCS, vol. 5303, pp. 405–418. Springer, Heidelberg (2008). https://doi.org/10.1007/978-3-540-88688-4_30

36. Qi, C.R., Su, H., Mo, K., Guibas, L.J.: PointNet: deep learning on point sets for 3D classification and segmentation. In: Proceedings of the IEEE Conference on Computer Vision and Pattern Recognition (CVPR), Computer Society, Honolulu, USA, pp. 652–660. IEEE (2017)

37. Santa Cruz, R., Fernando, B., Cherian, A., Gould, S.: Visual permutation learning. IEEE Trans. Pattern Anal. Mach. Intell. (PAMI) **41**(12), 3100–3114 (2019)

38. Sattler, T., Leibe, B., Kobbelt, L.: Efficient effective prioritized matching for large-scale image-based localization. IEEE Trans. Pattern Anal. Mach. Intell. **39**(9), 1744–1756 (2017). https://doi.org/10.1109/TPAMI.2016.2611662

39. Sattler, T., Zhou, Q., Pollefeys, M., Leal-Taixe, L.: Understanding the limitations of CNN-based absolute camera pose regression. In: Proceedings of the IEEE Conference on Computer Vision and Pattern Recognition (CVPR), Computer Society, pp. 3302–3312. IEEE (2019)

40. Schönberger, J.L., Pollefeys, M., Geiger, A., Sattler, T.: Semantic visual localization. In: Proceedings of the IEEE Conference on Computer Vision and Pattern Recognition (CVPR), Computer Society, pp. 6896–6906. IEEE (2018)

41. Schönberger, J.L., Frahm, J.M.: Structure-from-motion revisited. In: Proceedings of the IEEE Conference on Computer Vision and Pattern Recognition (CVPR), Computer Society, pp. 4104–4113. IEEE (2016)

42. Sinkhorn, R.: Diagonal equivalence to matrices with prescribed row and column sums. Am. Math. Mon. **74**(4), 402–405 (1967)

43. von Stackelberg, H., Bazin, D., Urch, L., Hill, R.R.: Market Structure and Equilibrium. Springer (2011) https://doi.org/10.1007/978-3-642-12586-7

44. Svärm, L., Enqvist, O., Kahl, F., Oskarsson, M.: City-scale localization for cameras with known vertical direction. IEEE Trans. Pattern Anal. Mach. Intell. **39**(7), 1455–1461 (2016)
45. Walch, F., Hazirbas, C., Leal-Taixe, L., Sattler, T., Hilsenbeck, S., Cremers, D.: Image-based localization using lstms for structured feature correlation. In: Proceedings of the International Conference on Computer Vision (ICCV), Computer Society, pp. 627–637. IEEE (2017)
46. Wu, Z., Song, S., Khosla, A., Yu, F., Zhang, L., Tang, X., Xiao, J.: 3D ShapeNets: a deep representation for volumetric shapes. In: Proceedings of the IEEE Conference on Computer Vision and Pattern Recognition (CVPR), Computer Society, pp. 1912–1920. IEEE (2015)
47. Yi, K.M., Trulls, E., Ono, Y., Lepetit, V., Salzmann, M., Fua, P.: Learning to find good correspondences. In: Proceedings of the IEEE Conference on Computer Vision and Pattern Recognition (CVPR), Computer Society, pp. 2666–2674. IEEE (2018)

Exploiting Deep Generative Prior for Versatile Image Restoration and Manipulation

Xingang Pan[1(✉)], Xiaohang Zhan[1], Bo Dai[1], Dahua Lin[1], Chen Change Loy[2], and Ping Luo[3]

[1] The Chinese University of Hong Kong, Shatin, Hong Kong
{px117,zx017,bdai,dhlin}@ie.cuhk.edu.hk
[2] Nanyang Technological University, Singapore, Singapore
ccloy@ntu.edu.sg
[3] The University of Hong Kong, Pokfulam, Hong Kong
pluo@cs.hku.hk

Abstract. Learning a good image prior is a long-term goal for image restoration and manipulation. While existing methods like deep image prior (DIP) capture low-level image statistics, there are still gaps toward an image prior that captures rich image semantics including color, spatial coherence, textures, and high-level concepts. This work presents an effective way to exploit the image prior captured by a generative adversarial network (GAN) trained on large-scale natural images. As shown in Fig. 1, the deep generative prior (DGP) provides compelling results to restore missing semantics, *e.g.*, color, patch, resolution, of various degraded images. It also enables diverse image manipulation including random jittering, image morphing, and category transfer. Such highly flexible effects are made possible through relaxing the assumption of existing GAN-inversion methods, which tend to fix the generator. Notably, we allow the generator to be fine-tuned on-the-fly in a progressive manner regularized by feature distance obtained by the discriminator in GAN. We show that these easy-to-implement and practical changes help preserve the reconstruction to remain in the manifold of nature image, and thus lead to more precise and faithful reconstruction for real images. Code is at https://github.com/XingangPan/deep-generative-prior.

1 Introduction

Learning image prior models is important to solve various tasks of image restoration and manipulation, such as *image colorization* [21,36], *image inpainting* [35], *super-resolution* [12,22], and *adversarial defense* [27]. In the past decades, many image priors [13,16,25,26,40] have been proposed to capture certain statistics of

Electronic supplementary material The online version of this chapter (https://doi.org/10.1007/978-3-030-58536-5_16) contains supplementary material, which is available to authorized users.

© Springer Nature Switzerland AG 2020
A. Vedaldi et al. (Eds.): ECCV 2020, LNCS 12347, pp. 262–277, 2020.
https://doi.org/10.1007/978-3-030-58536-5_16

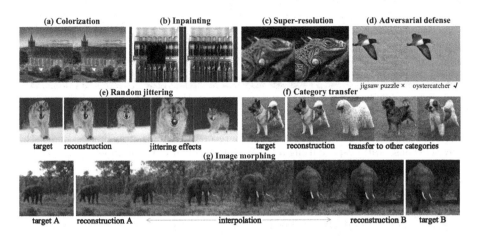

Fig. 1. These image restoration (a)(b)(c)(d) and manipulation (e)(f)(g) effects are achieved by leveraging the rich generative prior of a GAN. The GAN does not see these images during training

natural images. Despite their successes, these priors often serve a dedicated purpose. For instance, markov random field [13,25,40] is often used to model the correlation among neighboring pixels, while dark channel prior [16] and total variation [26] are developed for dehazing and denoising respectively.

There is a surge of interest to seek for more general priors that capture richer statistics of images through deep learning models. For instance, the seminal work on deep image prior (DIP) [30] showed that the structure of a randomly initialized Convolutional Neural Network (CNN) implicitly captures texture-level image prior, thus can be used for restoration by fine-tuning it to reconstruct a corrupted image. SinGAN [28] further shows that a randomly-initialized generative adversarial network (GAN) model is able to capture rich patch statistics after training from a single image. These priors have shown impressive results on some low-level image restoration and manipulation tasks like super-resolution and harmonizing. In both the representative works, the CNN and GAN are trained from a single image of interest from scratch.

In this study, we are interested to go one step further, examining how we could leverage a GAN [14] trained on large-scale natural images for richer priors beyond a single image. GAN is a good approximator for natural image manifold. By learning from large image datasets, it captures rich knowledge on natural images including color, spatial coherence, textures, and high-level concepts, which are useful for broader image restoration and manipulation effects. Specifically, we take a collapsed image (*e.g.*, gray-scale image) as a partial observation of the original natural image, and reconstruct it in the observation space (*e.g.*, grayscale space) with the GAN, the image prior of the GAN would tend to restore the missing semantics (*e.g.*, color) in a faithful way to match natural images. Despite its enormous potentials, it remains a challenging task to exploit a GAN as a prior for general image restoration and manipulation. The key challenge lies

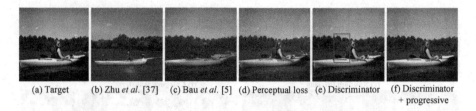

(a) Target   (b) Zhu *et al.* [37]   (c) Bau *et al.* [5]   (d) Perceptual loss   (e) Discriminator   (f) Discriminator
                                                                                                               + progressive

Fig. 2. Comparison of various methods in reconstructing a gray image under the gray-scale observation space using a GAN. Conventional GAN-inversion strategies like (b) [38] and (c) [5] produce imprecise reconstruction for the existing semantics. In this work, we relax the generator so that it can be fine-tuned on-the-fly, achieving more accurate reconstruction as in (d)(e)(f), of which optimization is based on (d) VGG perceptual loss, (e) discriminator feature matching loss, and (f) combined with progressive reconstruction, respectively. We highlight that discriminator is important to preserve the generative prior so as to achieve better restoration for the missing information (*i.e.*, color). The proposed progressive strategy eliminates the 'information lingering' artifacts as in the red box in (e)

in the needs in coping with arbitrary images from different tasks with distinctly different natures. The reconstruction also needs to produce sharp and faithful images obeying the natural image manifold.

An appealing option for our problem is GAN-inversion [2,5,8,38]. Existing GAN-inversion methods typically reconstruct a target image by optimizing over the latent vector, *i.e.*, $\mathbf{z}^* = \arg\min_{\mathbf{z}\in\mathbb{R}^d} \mathcal{L}(\mathbf{x}, G(\mathbf{z};\boldsymbol{\theta}))$, where $\mathbf{x}$ is the target image, G is a fixed generator, $\mathbf{z}$ and $\boldsymbol{\theta}$ are the latent vector and generator parameters, respectively. In practice, we found that this strategy fails in dealing with complex real-world images. In particular, it often results in mismatched reconstructions, whose details (*e.g.*, objects, texture, and background) appear inconsistent with the original images, as Fig. 2 (b)(c) show. On one hand, existing GAN-inversion methods still suffer from the issues of mode collapse and limited generator capacity, affecting their capability in capturing the desired data manifold. On the other hand, perhaps a more crucial limitation is that when a generator is fixed, the GAN is inevitably limited by the training distribution and its inversion cannot faithfully reconstruct unseen and complex images. It is infeasible to carry such assumptions while using a GAN as prior for general image restoration and manipulation.

Despite the gap between the approximated manifold and the real one, the GAN generator still captures rich statistics of natural images. In order to make use of these statistics while avoiding the aforementioned limitation, in this paper we present a relaxed and more practical reconstruction formulation for mining the priors in GAN. Our first reformulation is to allow the generator parameters to be fine-tuned on the target image on-the-fly, *i.e.*, $\boldsymbol{\theta}^*, \mathbf{z}^* = \arg\min_{\boldsymbol{\theta},\mathbf{z}} \mathcal{L}(\mathbf{x}, G(\mathbf{z};\boldsymbol{\theta}))$. This lifts the constraint of confining the reconstruction within the training distribution. Relaxing the assumption with fine-tuning, however, is still not sufficient to ensure good reconstruction quality for arbitrary target images. We found that fine-tuning using a standard loss such as perceptual loss [19] or mean squared error (MSE) in DIP could risk wiping out the originally rich priors. Consequently, the reconstruction may become increasingly unnatural during the reconstruction of a degraded image. Figure 2(d) shows an example, suggesting that a new loss and reconstruction strategy is needed.

Thus, in our second reformulation, we devise an effective reconstruction strategy that consists of two components:

1) *Feature matching loss from the coupled discriminator* - we make full use of the discriminator of a trained GAN to regularize the reconstruction. Note that during training, the generator is optimized to mimic massive natural images via gradients provided by the discriminator. It is reasonable to still adopt the discriminator in guiding the generator to match a single image as the discriminator preserves the original parameter structure of the generator better than other distance metrics. Thus deriving a feature matching loss from the discriminator can help maintain the reconstruction to remain in the natural image space. Although the feature matching loss is not new in the literature [31], its significance to GAN reconstruction has not been investigated before.

2) *Progressive reconstruction* - we observe that a joint fine-tuning of all parameters of the generator could lead to *'information lingering'*, where missing semantics (*e.g.*, color) do not naturally change along with the content when reconstructing a degraded image. This is because the deep layers of the generator start to match the low-level textures before the high-level configurations are aligned. To address this issue, we propose a progressive reconstruction strategy that fine-tunes the generator gradually from the shallowest layers to the deepest layers. This allows the reconstruction to start with matching high-level configurations and gradually shift its focus on low-level details.

Thanks to the proposed techniques that enable faithful reconstruction while maintaining the generator prior, our approach, which we name as Deep Generative Prior (DGP), generalizes well to various kinds of image restoration and manipulation tasks, despite that our method is not specially designed for each task. When reconstructing a corrupted image in a task-dependent observation space, DGP tends to restore the missing information, while keeping existing semantic information unchanged. As shown in Fig. 1(a)(b)(c), color, missing patches, and details of the given images are well restored, respectively. As illustrated in Fig. 1(e)(f), we can manipulate the content of an image by tweaking the latent vector or category condition of the generator. Figure 1(g) shows that image morphing is possible by interpolating between the parameters of two fine-tuned generators and the corresponding latent vectors of these images. To our knowledge, it is the first time these jittering and morphing effects are achieved on a dataset with complex images like ImageNet [10]. We show more interesting examples in the experiments and supplementary material.

2   Related Work

Image Prior. Image priors that describe various statistics of natural images have been widely adopted in computer vision, including markov random fields [13,25,40], dark channel prior [16], and total variation regularizer [26]. Recently, the work of deep image prior (DIP) [30] shows that image statistics are implicitly captured by the structure of CNN, which is also a kind of prior, and could be used to restore corrupted images. SinGAN [28] fine-tunes a randomly initialized GAN on patches of a single image, achieving various image editing or restoration effects. As DIP and SinGAN are trained from scratch, they have limited access to image statistics beyond the input image, which restrains their applicability in tasks such as image colorization. Existing attempts that use a pre-trained GAN as a source of image statistics include [4] and [18], which respectively applies to image manipulation, *e.g.*, editing partial areas of an image, and image restoration, *e.g.*, compressed sensing and super-resolution for human faces. As we will show in our experiments, by using a discriminator based distance metric and a progressive fine-tuning strategy, DGP can better preserve image statistics learned by the GAN and thus allows richer restoration and manipulation effects.

Recently, a concurrent work of multi-code GAN prior [15] also conducts image processing by solving the GAN-inversion problem. It uses multiple latent vectors to reconstruct the target image and keeps the generator fixed, while our method makes the generator image-adaptive by allowing it to be fine-tuned on-the-fly.

Image Restoration and Manipulation. In this paper we demonstrate the effect of applying DGP to multiple tasks of image processing, including image colorization [21], image inpainting [35], super-resolution [12,22], adversarial defence [27], and semantic manipulation [7,38,39]. While many task-specific models and loss functions have been proposed to pursue a better performance on a specific restoration task [12,21,22,27,35,36], there are also works that apply GAN and design task-specific pipelines to achieve various image manipulation effects [4,7,29,31,34,39], such as CycleGAN [39] and StarGAN [7]. In this work we are more interested in uncovering the potential of exploiting the GAN prior as a task-agnostic solution, where we propose several techniques to achieve this goal.

GAN-Inversion. GAN-inversion aims at finding a vector in the latent space that best reconstructs a given image, where the GAN generator is fixed. Previous attempts either optimize the latent vector directly via gradient back-propagation [2,8] or leverage an additional encoder mapping images to latent vectors [11,38]. A more recent approach [5] proposes to add small perturbations to shallow blocks of the generator to ease the inversion task. While these methods could handle datasets with limited complexities or synthetic images sampled by the GAN itself, we empirically found in our experiments they may produce imprecise reconstructions for complex real scenes, *e.g.*, images in the ImageNet [10]. Recently, the work of StyleGAN [20] enables a new way for GAN-inversion by operating in intermediate latent spaces [1], but noticeable mismatches are still observed and the inversion for vanilla GAN (*e.g.*, BigGAN [6]) is still challenging. In this paper, instead of directly applying standard GAN-inversion, we devise a more practical way to reconstruct a given image using the generative prior, which is shown to achieve better reconstruction results.

3   Method

We first provide some preliminaries on DIP and GAN before discussing how we exploit DGP for image restoration and manipulation.

Deep Image Prior. Ulyanov *et al.* [30] show that image statistics are implicitly captured by the structure of CNN. These statistics can be seen as a kind of image prior, which can be exploited in various image restoration tasks by tuning a randomly initialized CNN on the degraded image: $\theta^* = \arg\min_\theta E(\hat{\mathbf{x}}, f(\mathbf{z}; \theta)), \mathbf{x}^* = f(\mathbf{z}; \theta^*)$, where E is a task-dependent distance metric, $\mathbf{z}$ is a randomly chosen latent vector, and f is a CNN with θ being its parameters. $\hat{\mathbf{x}}$ and $\mathbf{x}^*$ are the degraded image and restored image respectively. One limitation of DIP is that the restoration process mainly resorts to existing statistics in the input image, it is thus infeasible to apply DIP on tasks that require more general statistics, such as image colorization [21] and manipulation [38].

Generative Adversarial Networks (GANs). GANs are widely used for modeling complex data such as natural images [9,14,20,33]. In GAN, the underlying manifold of natural images is approximated by the combination of a parametric generator G and a prior latent space $\mathcal{Z}$, so that an image can be generated by sampling a latent vector $\mathbf{z}$ from $\mathcal{Z}$ and applying G as $G(\mathbf{z})$. GAN jointly trains G with a parametric discriminator D in an adversarial manner, where D is supposed to distinguish generated images from real ones.

3.1   Deep Generative Prior

Suppose $\hat{\mathbf{x}}$ is obtained via $\hat{\mathbf{x}} = \phi(\mathbf{x})$, where $\mathbf{x}$ is the original natural image and ϕ is a degradation transform. *e.g.*, ϕ could be a graying transform that turns $\mathbf{x}$ into a grayscale image. Many tasks of image restoration can be regarded as recovering $\mathbf{x}$ given $\hat{\mathbf{x}}$. A common practice is learning a mapping from $\hat{\mathbf{x}}$ to $\mathbf{x}$, which often requires task-specific training for different ϕs. Alternatively, we can also employ statistics of $\mathbf{x}$ stored in some prior, and search in the space of $\mathbf{x}$ for an optimal $\mathbf{x}$ that best matches $\hat{\mathbf{x}}$, viewing $\hat{\mathbf{x}}$ as partial observations of $\mathbf{x}$.

While various priors have been proposed [25,28,30] in the second line of research, in this paper we are interested in studying a more generic image prior, *i.e.*, a GAN generator trained on large-scale natural images for image synthesis. Specifically, a straightforward realization is a reconstruction process based on GAN-inversion, which optimizes the following objective:

$$\mathbf{z}^* = \arg\min_{\mathbf{z} \in \mathbb{R}^d} E(\hat{\mathbf{x}}, G(\mathbf{z}; \theta)), \qquad \mathbf{x}^* = G(\mathbf{z}^*; \theta), \qquad (1)$$

$$= \arg\min_{\mathbf{z} \in \mathbb{R}^d} \mathcal{L}(\hat{\mathbf{x}}, \phi(G(\mathbf{z}; \theta))),$$

where $\mathcal{L}$ is a distance metric such as the L2 distance, G is a GAN generator parameterized by θ and trained on natural images. Ideally, if G is sufficiently powerful that the data manifold of natural images is well captured in G, the above objective will drag $\mathbf{z}$ in the latent space and locate the optimal natural

(a) MSE

(b) Perceptual

(c) Discriminator

(d) Discriminator + progressive

Fig. 3. Comparison of different loss types when fine-tuning the generator to reconstruct the gray image under the gray-scale observation space

image $\mathbf{x}^* = G(\mathbf{z}^*; \boldsymbol{\theta})$, which contains the missing semantics of $\hat{\mathbf{x}}$ and matches $\hat{\mathbf{x}}$ under ϕ. For example, if ϕ is a graying transform, $\mathbf{x}^*$ will be an image with a natural color configuration subject to $\phi(\mathbf{x}^*) = \hat{\mathbf{x}}$. However, in practice it is not always the case.

As the GAN generator is fixed in Eq. (1) and its improved versions, *e.g.*, adding an extra encoder [11,38], these reconstruction methods based on the standard GAN-inversion suffer from an intrinsic limitation, *i.e.*, the gap between the approximated manifold of natural images and the actual one. On one hand, due to issues including mode collapse and insufficient capacity, the generator cannot perfectly grasp the training manifold represented by a dataset of natural images. On the other hand, the training manifold itself is also an approximation of the actual one. Consequently, a sub-optimal $\mathbf{x}^*$ is often retrieved, which often contains significant mismatches to $\hat{\mathbf{x}}$. See Fig. 2 and [5,11] for an illustration.

A Relaxed GAN Reconstruction Formulation. Despite the gap between the approximated manifold and the real one, a well trained GAN generator still covers rich statistics of natural images. In order to make use of these statistics while avoiding the aforementioned limitation, we propose a relaxed GAN reconstruction formulation by allowing parameters $\boldsymbol{\theta}$ of the generator to be moderately fine-tuned along with the latent vector $\mathbf{z}$. Such a relaxation on $\boldsymbol{\theta}$ gives rise to an updated objective:

$$\boldsymbol{\theta}^*, \mathbf{z}^* = \arg\min_{\boldsymbol{\theta}, \mathbf{z}} \mathcal{L}(\hat{\mathbf{x}}, \phi(G(\mathbf{z}; \boldsymbol{\theta}))), \quad \mathbf{x}^* = G(\mathbf{z}^*; \boldsymbol{\theta}^*). \tag{2}$$

We refer to this updated objective as Deep Generative Prior (DGP). With this relaxation, DGP significantly improves the chance of locating an optimal $\mathbf{x}^*$ for $\hat{\mathbf{x}}$, as fitting the generator to a single image is much more achievable than fully capturing a data manifold. Note that the generative prior buried in G, *e.g.*, its ability to output faithful natural images, might be deteriorated during the fine-tuning process. The key to preserve the generative prior lies in the design of a good distance metric $\mathcal{L}$ and a proper optimization strategy.

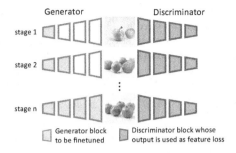

Fig. 4. Progressive reconstruction of the generator can better preserves the consistency between missing and existing semantics in comparison to simultaneous fine-tuning on all the parameters at once. Here the list of images shown in the middle are the outputs of the generator in different fine-tuning stages.

3.2 Discriminator Guided Progressive Reconstruction

To fit the GAN generator to the input image $\hat{\mathbf{x}}$ while retaining a natural output, in this section we introduce a discriminator based distance metric, and a progressive fine-tuning strategy.

Discriminator Matters. As shown in Fig. 3, the choice of distance metric $\mathcal{L}$ significantly affects the optimization of Eq. (2). Existing literature often adopts the Mean-Squared-Error (MSE) [30] or the AlexNet/VGGNet based Perceptual loss [19,38] as $\mathcal{L}$, which respectively emphasize the pixel-wise appearance and the low-level/mid-level texture. However, we empirically found using these metrics in Eq. (2) often cause unfaithful outputs at the beginning of optimization, leading to sub-optimal results at the end. We thus propose to replace them with a discriminator-based distance metric, which measures the L1 distance in the *discriminator feature space*:

$$\mathcal{L}(\mathbf{x}_1, \mathbf{x}_2) = \sum_{i \in \mathcal{I}} \|D(\mathbf{x}_1, i), D(\mathbf{x}_2, i)\|_1, \tag{3}$$

where $\mathbf{x}_1$ and $\mathbf{x}_2$ are two images, corresponding to $\hat{\mathbf{x}}$ and $\phi(G(\mathbf{z}; \boldsymbol{\theta}))$ in Eqs. 1 and 2, and D is the discriminator that is coupled with the generator. $D(\mathbf{x}, i)$ returns the feature of $\mathbf{x}$ at i-block of D, and $\mathcal{I}$ is the index set of used blocks. Compared to the AlexNet/VGGNet based perceptual loss, the discriminator D is trained along with G, instead of being trained for a separate task. D, being a distance metric, thus is less likely to break the parameter structure of G, as they are well aligned during the pre-training. Moreover, we found the optimization of DGP using such a distance metric visually works like an image morphing process. *e.g.*, as shown in Fig. 3, the person on the boat is preserved and all intermediate outputs are all vivid natural images.

Progressive Reconstruction. Typically, we will fine-tune all parameters of $\boldsymbol{\theta}$ simultaneously during the optimization of Eq. (2). However, we observe an

Table 1. Comparison with other GAN-inversion methods, including (a) optimizing latent vector [2,8], (b) learning an encoder [38], (c) a combination of (a)(b) [38], and (d) adding small perturbations to early stages based on (c) [5]. We reported PSNR, SSIM, and MSE. The results are evaluated on the 1k ImageNet validation set

	(a)	(b)	(c)	(d)	Ours
PSNR↑	15.97	11.39	16.46	22.49	**32.89**
SSIM↑	46.84	32.08	47.78	73.17	**95.95**
MSE↓ (×e-3)	29.61	85.04	28.32	6.91	**1.26**

adverse effect of *'information lingering'*, where missing semantics (*e.g.* color) do not shift along with existing context. Taking Fig. 3(c) as an example, the leftmost apple fails to inherit the green color of the initial apple when it emerges. One possible reason is deep blocks of the generator G start to match low-level textures before high-level configurations are aligned. To overcome this problem, we propose a progressive reconstruction strategy for some restoration tasks.

Specifically, as illustrated in Fig. 4, we first fine-tune the shallowest block of the generator, and gradually continue with blocks at deeper depths, so that DGP can control the global configuration at the beginning and gradually shift its attention to details at lower levels. A demonstration of the proposed strategy is included in Fig. 3(d), where DGP splits the apple from one to two at first, then increases the number to five, and finally refines the details of apples. Compared to the non-progressive counterpart, such a progressive strategy better preserves the consistency between missing and existing semantics.

4 Applications

We first compare our method with other GAN inversion methods for reconstruction, and then show the application of DGP in a number of image restoration and image manipulation tasks. We adopt a BigGAN [6] to progressively reconstruct given images based on discriminator feature loss. The BigGAN is pre-trained on the ImageNet training set for conditional image synthesis. For evaluation, we use the ImageNet [10] validation set that has not been observed by BigGAN. To quantitatively evaluate our method on image restoration tasks, we test on 1k images from the ImageNet validation set, where the first image for each class is collected to form the test set. We recommend readers to refer to the supplementary material for implementation details and more qualitative results.

Comparison with Other GAN-inversion Methods. To begin with, we compare with other GAN-inversion methods [2,5,8,38] for image reconstruction. As shown in Table 1, our method achieves a very high PSNR and SSIM scores, outperforming other GAN-inversion methods by a large margin. It can be seen from Fig. 2 that our method achieves more precise reconstruction than conventional GAN-inversion methods like [5,38].

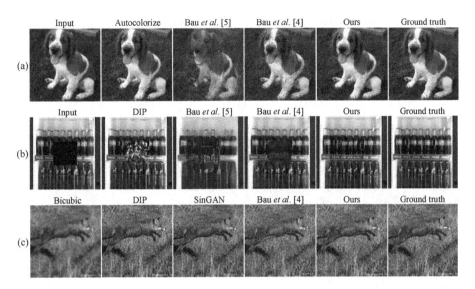

Fig. 5. (a) Colorization. Qualitative comparison of Autocolorize [21], other GAN-inversion methods [4,5], and our DGP. (b) Inpainting. Compared with DIP and [4,5], DGP could preserve the spatial coherence in image inpainting with large missing regions. (c) Super-resolution (×4) on 64 × 64 size images. The comparisons of our method with DIP, SinGAN, and [4] are shown, where DGP produces sharper results

4.1   Image Restoration

Colorization. Image colorization aims at restoring a gray-scale image $\hat{\mathbf{x}} \in \mathbb{R}^{H \times W}$ to a colorful image with RGB channels $\mathbf{x} \in \mathbb{R}^{3 \times H \times W}$. To obtain $\hat{\mathbf{x}}$ from the colorful image $\mathbf{x}$, the degradation transform ϕ is a graying transform that only preserves the brightness of $\mathbf{x}$. By taking this degradation transform to Eq. (2), the goal becomes finding the colorful image $\mathbf{x}^*$ whose gray-scale image is the same as $\hat{\mathbf{x}}$. We optimize Eq. (2) using back-propagation and the progressive discriminator based reconstruction technique in Sect. 3.2.

Figure 5(a) presents the qualitative comparisons with the Autocolorize [21] method. Note that Autocolorize is directly optimized to predict color from gray-scale images while our method does not adopt such task-specific training. Despite so, our method is visually better or comparable to Autocolorize. To evaluate the colorization quality, we report the classification accuracy of a ResNet50 [17] model on the colorized images. The ResNet50 accuracy for Autocolorize [21], Bau *et al.* [5], Bau *et al.* [4], and ours are 51.5%, 56.2%, 56.0%, and 62.8% respectively, showing that DGP outperforms other baselines on this perceptual metric.

Inpainting. The goal of image inpainting is to recover the missing pixels of an image. The corresponding degradation transform is to multiply the original image with a binary mask $\mathbf{m}$: $\phi(\mathbf{x}) = \mathbf{x} \odot \mathbf{m}$, where $\odot$ is Hadamard's product. As before, we put this degradation transform to Eq. (2), and reconstruct target

Table 2. Inpainting evaluation. We reported PSNR and SSIM of the inpainted area. The results are evaluated on the 1k ImageNet validation set

	DIP	Zhu et al. [38]	Bau et al. [5]	Bau et al. [4]	Ours
PSNR↑	14.58	13.70	15.01	14.33	**16.97**
SSIM↑	29.37	33.09	33.95	30.60	**45.89**

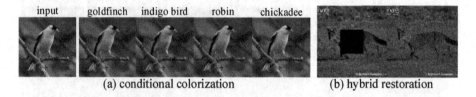

(a) conditional colorization   (b) hybrid restoration

Fig. 6. (a) Colorizing an image under different class conditions. (b) Simultaneously conduct colorization, inpainting, and super-resolution (×2)

Table 3. Super-resolution (×4) evaluation. We reported widely used NIQE [23], PSNR, and RMSE scores. The results are evaluated on the 1k ImageNet validation set. (MSE) and (D) indicate which kind of loss DGP is biased to use

	DIP	SinGAN	Bau et al. [4]	Ours (MSE)	Ours (D)
NIQE↓	6.03	6.28	5.05	5.30	**4.90**
PSNR↑	23.02	20.80	19.89	**23.30**	22.00
RMSE↓	17.84	19.78	25.42	**17.40**	20.09

images with missing boxes. Thanks to the generative image prior of the generator, the missing part tends to be recovered in harmony with the context, as illustrated in Fig. 5(b). In contrast, the absence of a learned image prior would result in messy inpainting results, as in DIP. Quantitative results indicate that DGP outperforms other methods by a large margin, as Table 2 shows.

Super-Resolution. In this task, one is given with a low-resolution image $\hat{x} \in \mathbb{R}^{3 \times H \times W}$, and the purpose is to generate the corresponding high-resolution image $x \in \mathbb{R}^{3 \times fH \times fW}$, where f is the upsampling factor. In this case, the degradation transform ϕ is to downsample the input image by a factor f.

Figure 5(c) and Table 3 show the comparison of DGP with DIP, SinGAN, and Bau et al. [4]. Our method achieves sharper and more faithful super-resolution results than its counterparts. For quantitative results, we could trade off between perceptual quality like NIQE and commonly used PSNR score by using different combination ratios of discriminator loss and MSE loss at the final fine-tuning stage. For instance, when using higher MSE loss, DGP has excellent PSNR and RMSE, and outperforms other counterparts in all the metrics involved. And NIQE could be further improved by biasing towards discriminator loss.

(a) Raccoon   (b) Places   (c) No foreground

Input   DIP   DGP   Input   DIP   DGP

(d) Windows

Fig. 7. Evaluation of DGP on non-ImageNet images, including (a) 'Raccoon', a category outside ImageNet categories, (b) image from Places dataset [37], (c) image without foreground object, and (d) windows. (a)(c)(d) are scratched from Internet

Table 4. Comparison of different loss type and fine-tuning strategy

Task	Metric	MSE	Perceptual	Discriminator	Discriminator+Progressive
Colorization	ResNet50↑	49.1	53.9	56.8	**62.8**
SR	NIQE↓	6.54	6.27	6.06	**4.90**
	PSNR↑	21.24	20.30	21.58	**22.00**

Flexibility of DGP. The generic paradigm of DGP provides more flexibility in restoration tasks. For example, an image of gray-scale bird may have many possibilities when restored in the color space. Since the BigGAN used in our method is a conditional GAN, we could achieve diversity in colorization by using different class conditions when restoring the image, as Fig. 6(a) shows. Furthermore, our method allows hybrid restoration, $i.e.$, jointly conducting colorization, inpainting, and super-resolution. This could naturally be achieved by using a composite of degrade transform $\phi(\mathbf{x}) = \phi_a(\phi_b(\phi_c(\mathbf{x})))$, as shown in Fig. 6(b).

Generalization of DGP. We also test our method on images not belonging to ImageNet. As Fig. 7 shows, DGP restores the color and missed patches of these images reasonably well. Particularly, compared with DIP, DGP fills the missed patches to be well aligned with the context. This indicates that DGP does capture the 'spatial coherence' prior of natural images, instead of memorizing the ImageNet dataset. We scratch a small dataset with 18 images of windows, stones, and libraries to test our method, where DGP achieves 15.34 for PSNR and 41.53 for SSIM, while DIP has only 12.60 for PSNR and 21.12 for SSIM.

Ablation Study. To validate the effectiveness of the proposed discriminator guided progressive reconstruction method, we compare different fine-tuning strategies in Table 4. There is a clear improvement of discriminator feature matching loss over MSE and perceptual loss, and the combination of the pro-

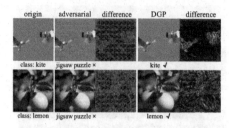

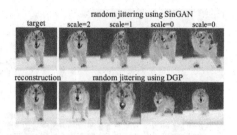

Fig. 8. Adversarial defense. DGP could filter out unnatural perturbations in adversarial samples by reconstruction

Fig. 9. Comparison of random jittering using SinGAN (above) and DGP (below)

Table 5. Adversarial defense evaluation. We reported the classification accuracy of a ResNet50. The results are evaluated on the 1k ImageNet validation set

Method	Clean image	Adversarial	DefenceGAN	DIP	Ours
top1 acc. (%)	74.9	1.4	0.2	37.5	41.3
top5 acc. (%)	92.7	12.0	1.4	61.2	65.9

gressive reconstruction further boosts the performance. Figure 2, Fig. 3, and supplementary material provide qualitative comparisons. The results show that the progressive strategy effectively eliminates the 'information lingering' artifacts.

Adversarial Defense. Adversarial attack methods aim at fooling a CNN classifier by adding a certain perturbation $\Delta\mathbf{x}$ to a target image $\mathbf{x}$ [24]. In contrast, adversarial defense aims at preventing the model from being fooled by attackers. Here we show the potential of DGP in adversarial defense under a black-box attack setting [3].

For adversarial attack, the degradation transform is $\phi(\mathbf{x}) = \mathbf{x} + \Delta\mathbf{x}$, where $\Delta\mathbf{x}$ is the perturbation generated by the attacker. Since calculating $\phi(\mathbf{x})$ is generally not differentiable, here we adopt DGP to directly reconstruct the adversarial image $\hat{\mathbf{x}}$. To prevent $\mathbf{x}^*$ from overfitting to $\hat{\mathbf{x}}$, we stop the reconstruction when the MSE loss reaches 5e-3. We adopt the adversarial transformation networks attacker [3] to produce the adversarial samples[1].

As Fig. 8 shows, the generated adversarial image contains unnatural perturbations, leading to misclassification for a ResNet50 [17]. After reconstructing the adversarial samples using DGP, the perturbations are largely alleviated, and the samples are thus correctly classified. The comparisons of our method with DefenseGAN and DIP are shown in Table 5. DefenseGAN yields poor defense performance due to inaccurate reconstruction. And DGP outperforms DIP, thanks to the learned image prior that produces more natural restored images.

---

[1] We use the code at https://github.com/pfnet-research/nips17-adversarial-attack.

target A   reconstruction A   ←————————interpolation————————→   reconstruction B   target B

(a)

target   reconstruction   transfer to other categories

(b)

Fig. 10. Our method achieves realistic (a) image morphing and (b) category transfer effects

4.2   Image Manipulation

Since DGP enables precise GAN reconstruction while preserving the generative property, it becomes straightforward to apply the fascinating capabilities of GAN to real images, as we will show in this section.

Random Jittering. We show the random jittering effects of DGP, and compare it with SinGAN. Specifically, after reconstructing a target image using DGP, we add Gaussian noise to the latent vector z^* and see how the output changes. As shown in Fig. 9, the dog in the image changes in pose, action, and size, where each variant looks like a natural shift of the original image. For SinGAN, however, the jittering effects seem to preserve some texture, but losing the concept of 'dog'. This is because it cannot learn a valid representation of dog by looking at only one dog. In contrast, in DGP the generator is fine-tuned in a moderate way such that the structure of image manifold is well preserved. Therefore, perturbing z^* corresponds to shifting the image in the natural image manifold.

Image Morphing. The purpose of image morphing is to achieve a visually sound transition from one image to another. Given a GAN generator G and two latent vectors z_A and z_B, morphing between $G(z_A)$ and $G(z_B)$ could naturally be done by interpolating between z_A and z_B. In the case of DGP, however, reconstructing two target images x_A and x_B would result in two generators G_{θ_A} and G_{θ_B}, and the corresponding latent vectors z_A and z_B. Inspired by [32], to morph between x_A and x_B, we apply linear interpolation to both the latent vectors and the generator parameters: $z = \lambda z_A + (1 - \lambda)z_B, \theta = \lambda\theta_A + (1 - \lambda)\theta_B, \lambda \in (0, 1)$, and generate images with the new z and θ. As Fig. 10 shows, our method enables highly photo-realistic image morphing effects.

Category Transfer. In conditional GAN, the class condition controls the content to be generated. So after reconstructing a given image via DGP, we can manipulate its content by tweaking the class condition. Figures 1(f) and 10 present examples of transferring the object category of given images. Our method can transfer the dog and bird to various other categories without changing the pose, size, and image configurations.

5 Conclusion

To summarise, we have shown that a GAN generator trained on massive natural images could be used as a generic image prior, namely deep generative prior (DGP). Embedded with rich knowledge on natural images, DGP could be used to restore the missing information of a degraded image by progressively reconstructing it under the discriminator metric. Meanwhile, such reconstruction strategy addresses the challenge of GAN-inversion, achieving multiple visually realistic image manipulation effects. Our results uncover the potential of a universal image prior captured by a GAN in image restoration and manipulation.

References

1. Abdal, R., Qin, Y., Wonka, P.: Image2stylegan: How to embed images into the stylegan latent space? In: ICCV, pp. 4432–4441 (2019)
2. Albright, M., McCloskey, S.: Source generator attribution via inversion. In: CVPR Workshops (2019)
3. Baluja, S., Fischer, I.: Adversarial transformation networks: learning to generate adversarial examples. arXiv preprint arXiv:1703.09387 (2017)
4. Bau, D., Strobelt, H., Peebles, W., Wulff, J., Zhou, B., Zhu, J.Y., Torralba, A.: Semantic photo manipulation with a generative image prior. ACM Trans. Graph. (TOG) **38**(4), 59 (2019)
5. Bau, D., et al.: Seeing what a gan cannot generate. In: ICCV, pp. 4502–4511 (2019)
6. Brock, A., Donahue, J., Simonyan, K.: Large scale gan training for high fidelity natural image synthesis. In: ICLR (2019)
7. Choi, Y., et al.: Stargan: Unified generative adversarial networks for multi-domain image-to-image translation. In: CVPR (2018)
8. Creswell, A., Bharath, A.A.: Inverting the generator of a generative adversarial network. IEEE Trans. Neural Netw. Learn. Syst. **30**(7), 1967–1974 (2018)
9. Dai, B., Fidler, S., Urtasun, R., Lin, D.: Towards diverse and natural image descriptions via a conditional gan. In: ICCV, pp. 2970–2979 (2017)
10. Deng, J., et al.: Imagenet: a large-scale hierarchical image database. In: CVPR, pp. 248–255 (2009)
11. Donahue, J., Krähenbühl, P., Darrell, T.: Adversarial feature learning. In: ICLR (2017)
12. Dong, C., Loy, C.C., He, K., Tang, X.: Image super-resolution using deep convolutional networks. IEEE Trans. Pattern Anal. Mach. Intell. **38**(2), 295–307 (2015). https://doi.org/10.1109/TPAMI.2015.2439281
13. Geman, S., Geman, D.: Stochastic relaxation, gibbs distributions, and the bayesian restoration of images. IEEE Trans. Pattern Anal. Mach. Intell. **6**, 721–741 (1984)
14. Goodfellow, I., et al.: Generative adversarial nets. In: NIPS, pp. 2672–2680 (2014)
15. Gu, J., Shen, Y., Zhou, B.: Image processing using multi-code GAN prior. In: CVPR (2020)
16. He, K., Sun, J., Tang, X.: Single image haze removal using dark channel prior. IEEE Trans. Pattern Anal. Mach. Intell. **33**(12), 2341–2353 (2010)
17. He, K., Zhang, X., Ren, S., Sun, J.: Deep residual learning for image recognition. In: CVPR, pp. 770–778 (2016)
18. Hussein, S.A., Tirer, T., Giryes, R.: Image-adaptive gan based reconstruction. arXiv preprint arXiv:1906.05284 (2019)

19. Johnson, J., Alahi, A., Fei-Fei, L.: Perceptual losses for real-time style transfer and super-resolution. In: Leibe, B., Matas, J., Sebe, N., Welling, M. (eds.) ECCV 2016. LNCS, vol. 9906, pp. 694–711. Springer, Cham (2016). https://doi.org/10.1007/978-3-319-46475-6_43

20. Karras, T., Laine, S., Aila, T.: A style-based generator architecture for generative adversarial networks. In: CVPR, pp. 4401–4410 (2019)

21. Larsson, G., Maire, M., Shakhnarovich, G.: Learning representations for automatic colorization. In: Leibe, B., Matas, J., Sebe, N., Welling, M. (eds.) ECCV 2016. LNCS, vol. 9908, pp. 577–593. Springer, Cham (2016). https://doi.org/10.1007/978-3-319-46493-0_35

22. Ledig, C., et al.: Photo-realistic single image super-resolution using a generative adversarial network. In: CVPR, pp. 4681–4690 (2017)

23. Mittal, A., Soundararajan, R., Bovik, A.C.: Making a "completely blind" image quality analyzer. IEEE Signal Process. Lett. **20**(3), 209–212 (2012)

24. Nguyen, A., Yosinski, J., Clune, J.: Deep neural networks are easily fooled: high confidence predictions for unrecognizable images. In: CVPR, pp. 427–436 (2015)

25. Roth, S., Black, M.J.: Fields of experts: a framework for learning image priors. In: CVPR, pp. 860–867 (2005)

26. Rudin, L.I., Osher, S., Fatemi, E.: Nonlinear total variation based noise removal algorithms. Phys. D: Nonlinear Phenom. **60**(1–4), 259–268 (1992)

27. Samangouei, P., Kabkab, M., Chellappa, R.: Defense-GAN: protecting classifiers against adversarial attacks using generative models. In: ICLR (2018)

28. Shaham, T.R., Dekel, T., Michaeli, T.: SinGAN: learning a generative model from a single natural image. In: ICCV, pp. 4570–4580 (2019)

29. Shen, Y., Gu, J., Tang, X., Zhou, B.: Interpreting the latent space of GANs for semantic face editing. In: CVPR (2020)

30. Ulyanov, D., Vedaldi, A., Lempitsky, V.: Deep image prior. In: CVPR, pp. 9446–9454 (2018)

31. Wang, T.C., et al.: High-resolution image synthesis and semantic manipulation with conditional GANs. In: CVPR (2018)

32. Wang, X., Yu, K., Dong, C., Tang, X., Loy, C.C.: Deep network interpolation for continuous imagery effect transition. In: CVPR, pp. 1692–1701 (2019)

33. Xiangli, Y., Deng, Y., Dai, B., Loy, C.C., Lin, D.: Real or not real, that is the question. In: International Conference on Learning Representations (2020)

34. Yang, C., Shen, Y., Zhou, B.: Semantic hierarchy emerges in deep generative representations for scene synthesis. arXiv preprint arXiv:1911.09267 (2019)

35. Yeh, R.A., et al.: Semantic image inpainting with deep generative models. In: CVPR, pp. 5485–5493 (2017)

36. Zhang, R., Isola, P., Efros, A.A.: Colorful image colorization. In: Leibe, B., Matas, J., Sebe, N., Welling, M. (eds.) ECCV 2016. LNCS, vol. 9907, pp. 649–666. Springer, Cham (2016). https://doi.org/10.1007/978-3-319-46487-9_40

37. Zhou, B., et al.: Places: a 10 million image database for scene recognition. IEEE Trans. Pattern Anal. Mach. Intell. **40**, 1452–1464 (2017). https://doi.org/10.1109/TPAMI.2017.2723009

38. Zhu, J.Y., Krähenbühl, P., Shechtman, E., Efros, A.A.: Generative visual manipulation on the natural image manifold. In: Leibe, B., Matas, J., Sebe, N., Welling, M. (eds.) ECCV 2016. LNCS, vol. 9909, pp. 597–613. Springer, Cham (2016). https://doi.org/10.1007/978-3-319-46454-1_36

39. Zhu, J.Y., Park, T., Isola, P., Efros, A.A.: Unpaired image-to-image translation using cycle-consistent adversarial networks. In: ICCV (2017)

40. Zhu, S.C., Mumford, D.: Prior learning and gibbs reaction-diffusion. IEEE Trans. Pattern Anal. Mach. Intell. **19**(11), 1236–1250 (1997)

Deep Spatial-Angular Regularization for Compressive Light Field Reconstruction over Coded Apertures

Mantang Guo[1], Junhui Hou[1(✉)], Jing Jin[1], Jie Chen[2], and Lap-Pui Chau[3]

[1] Department of Computer Science, City University of Hong Kong,
Hong Kong, China
{mantanguo2-c,jingjin25-c}@my.cityu.edu.hk
[2] Department of Computer Science,
Hong Kong Baptist University,
Hong Kong, China
jh.hou@cityu.edu.hk, chenjie@comp.hkbu.edu.hk
[3] School of Electrical and Electronics Engineering,
Nanyang Technological University, Nanyang, Singapore
elpchau@ntu.edu.sg

Abstract. Coded aperture is a promising approach for capturing the 4-D light field (LF), in which the 4-D data are compressively modulated into 2-D coded measurements that are further decoded by reconstruction algorithms. The bottleneck lies in the reconstruction algorithms, resulting in rather limited reconstruction quality. To tackle this challenge, we propose a novel learning-based framework for the reconstruction of high-quality LFs from acquisitions via learned coded apertures. The proposed method incorporates the measurement observation into the deep learning framework elegantly to avoid relying entirely on data-driven priors for LF reconstruction. Specifically, we first formulate the compressive LF reconstruction as an inverse problem with an implicit regularization term. Then, we construct the regularization term with an efficient deep spatial-angular convolutional sub-network to comprehensively explore the signal distribution free from the limited representation ability and inefficiency of deterministic mathematical modeling. Experimental results show that the reconstructed LFs not only achieve much higher PSNR/SSIM but also preserve the LF parallax structure better, compared with state-of-the-art methods on both real and synthetic LF benchmarks. In addition, experiments show that our method is efficient and robust to noise, which is an essential advantage for a real camera system. The code is publicly available at https://github.com/angmt2008/LFCA.

Keywords: Light field · Coded aperture · Deep learning · Regularization · Observation model

Electronic supplementary material The online version of this chapter (https://doi.org/10.1007/978-3-030-58536-5_17) contains supplementary material, which is available to authorized users.

1   Introduction

Owing to multi-view and depth information embedded in 4-D light fields (LFs), a large variety of LF based applications have emerged, e.g., image post-refocusing [25], 3-D reconstruction [34], saliency detection [16], view synthesis [6,11,12,37, 45]. Different from earlier LF capturing approaches, i.e., camera gantry [15] and camera array [36], portable micro-lens array-based LF cameras [19,27] are more convenient and cost-effective for capturing a dense LF. By using a micro-lens array placed between the main lens and image sensor, the micro-lens array-based camera records the spatial and angular information of light rays into a multiplexing sensor with only a single shot. However, due to the limited sensor resolution, the projection of the 4-D LF into a 2-D image leads to an inevitable trade-off between spatial and angular resolution. Besides, the large amount of data of captured LFs poses a great challenge to storage and transmission.

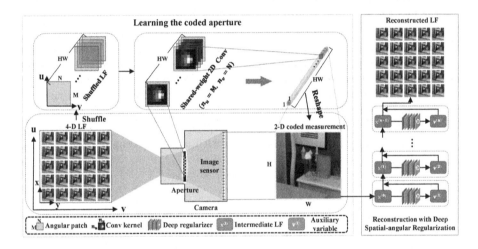

Fig. 1. The pipeline of our deep learning-based compressive LF reconstruction over coded apertures. Our method elegantly incorporates the observation model of coded measurements into deep learning framework. The left side illustrates the acquisition of coded measurements by learning apertures, and the right side shows the reconstruction phase. More details of the reconstruction module are shown in Fig. 2

To preserve the LF resolution and simultaneously reduce the data size of captured LFs, based on a traditional camera, the coded aperture camera was designed, which modulates light rays through the main lens into one or multiple coded measurements with the same size as that of the image sensor. Then, by employing LF reconstruction algorithms, a full 4-D LF can be generated from coded measurements. Earlier LF reconstruction methods [2,17,20,21,24, 39], which either require relatively many measurements or use dictionaries, are limited by the representation ability. Recent deep learning-based methods [7,9, 23] are able to reconstruct LFs from only a few measurements. However, these methods are purely data-driven without taking the special characteristics of

LFs into account. That is, they employ networks for general purposes but not specifically designed for tackling the problem of LF reconstruction from coded measurements, e.g., the plain convolutional layers, the networks used in the fully convolutional network (FCN) [18], and the very deep convolutional network (VDSR) [13], and thus the reconstruction performance is still limited.

In this paper, as shown in Fig. 1, we propose a novel deep learning-based framework to reconstruct high-quality LFs with measurements from adaptively learned coded apertures. First, the coded aperture is modeled and learned by a 2-D convolutional layer with specific configurations, denoted as acquisition layer in the method. Based on the observation model of the coded measurements, we formulate the LF reconstruction from measurements as an inverse problem with an implicit regularization term. Then, we construct the regularization term with a deep spatial-angular convolutional network instead of an deterministic mathematical modeling with a limited representation ability, such that the underlying complex structure can be comprehensively explored. Consequently, the LF reconstruction from coded measurements is solved by training the end-to-end network. Our LF capture and reconstruction method can breakthrough the limitations of conventional optimization method and simultaneously take full advantage of the strong representation ability of deep learning. Note that this paper is focused on developing a novel LF reconstruction method from coded measurements. Together with the experimentally verified robustness of our method against noise as well as the well constructed hardware platforms of coded aperture LF imaging in previous works [3,9,17,20,24], there is no technical barrier to implement the proposed method with a real camera system.

2   Related Work

Based on the inputs, we divide the existing LF reconstruction methods into two categories: sub-aperture image (SAI)-based reconstruction and coded measurement-based reconstruction, which will be reviewed as follows.

2.1   LF Reconstruction from SAIs

For SAI-based LF reconstruction, the input consists of a sparse set of SAIs belonging to a dense LF to be reconstructed. This kind of methods mainly investigate view synthesis to increase the angular resolution of a sparsely-sampled LF, which is always fixed into a regular pattern, e.g., four-corner SAIs [12], borders-diagonal SAIs [30], SAI-pairs [41], or multiplane images [22,44]. Specifically, Shi et al. [30] reconstructed a dense LF from fixed SAI-sampling patterns by exploiting the sparsity in Fourier domain. Yoon et al. [41] proposed a deep learning method to synthesize novel SAI between a pair or stack of SAIs. Besides, input SAIs obtained from different sampling patterns are processed separately by three sub-networks, which is inefficient for LF super-resolution. Kalantari et al. [12] used four-corner SAIs as the input of their algorithm. The quality of reconstructed LFs is rather limited by using a series of hand-crafted features which are extracted from warped images. In addition, it fails to handle the input only with a single SAI as the hand-crafted features cannot be obtained. Such a weakness also exists in the method by Wu *et al.* [38], which reconstructs a dense LF by carrying out super-resolution on angular dimension of each epipolar plane image (EPI) of the input LF. Yeung *et al.* [37] proposed an end-to-end deep learning method to reconstruct a dense LF in a coarse-to-fine manner. However, the method [37] also cannot reconstruct a dense LF from few SAIs. Although the method proposed by Srinivasan *et al.* [31] can reconstruct a dense LF from a single SAI, it is only able to use information from the input SAI, and the coherence among SAIs of LF is lost. Moreover, the quality of reconstructed LF by the method relies heavily on the accuracy of depth estimation.

2.2   LF Reconstruction from Coded Measurements

For coded measurement-based LF reconstruction, earlier methods [2,17,24,26] require relatively many exposures to reconstruct the entire LF. With the introduction of compressive sensing [1], the dense LF can be reconstructed from coded measurements. Two challenges in LF compressive sensing are the design of sensing matrix for projecting LF into proper measurements and the algorithm for inversely reconstructing LF from measurements. Marwah *et al.* [20] designed a sensing matrix by using conventional optimization method with an overcomplete dictionary. Then, the method in [20] formulates LF reconstruction from measurements as a basis pursuit denoise problem which is solved by conventional solvers. Chen *et al.* [3] constructed a dictionary based on perspective shifting of center view of an LF. Miandji *et al.* [21] proposed to aggregate multidimensional dictionary ensemble to encode and decode LFs efficiently in dictionary-based compressive sensing. However, the sensing matrix in these model-based methods has to be carefully designed to make the projection of LF as orthogonal as possible. Furthermore, a large scale of optimization techniques and training data are likely used to improve the representing ability of dictionaries.

Recently, deep learning-based LF reconstruction methods were proposed, which design network architectures to infer LFs from coded measurements by training with a large amount of LF data. Based on the sensing matrix designed by [20], Gupta *et al.* [7] proposed a deep learning-based method for LF compressive sensing. Given coded measurements, the method in [7] employs two plain network branches to generate two coarse LFs and then fuses them together to generate the final LF. Nabati *et al.* [23] improved the sensing matrix which can modulate both color and angular information of an LF into the 2-D coded measurement. Then, the sensing matrix together with the coded measurements are fed into an FCN-based network to reconstruct a dense LF. However, since all these methods use a fixed sensing matrix to modulate LFs, the sensing process is not flexible enough to extract information from LFs. Furthermore, all these reconstruction networks are data-driven models without considering the signal reconstruction principle. Inagaki *et al.* [9] adopted a 1×1 convolutional kernel to simulate the coded aperture process. Then, they employed two sequential sub-networks to reconstruct a dense LF from coded measurements. The first sub-network is constructed by a series of stacked convolutional layers while the second sub-network is a VDSR network [13]. Although all angular information from different angular locations is selectively blended into coded measurements, the reconstruction quality is rather limited due to the plain network architecture.

3   Proposed Method

As shown in Fig. 1, we model the compressive LF reconstruction over coded apertures within an end-to-end deep learning framework. Specifically, the measurements are obtained by modulating 4-D LFs through a learning-based coded aperture. For reconstruction, it is formulated as an inverse problem with a deep spatial-angular regularizer, which elegantly incorporates an LF degradation model into the deep learning framework. In what follows, we demonstrate each component in detail.

3.1   Learning Coded Apertures

In the following, we just consider a single color channel of the LF image in RGB space for simplicity. The other two color channels will be processed in the same manner. The 4-D LF denoted as $\mathscr{L}(u, v, x, y) \in \mathbb{R}^{M \times N \times H \times W}$ can be represented with the two-parallel plane parameterization, where $\{(u, v)|u \in [1, M], v \in [1, N]\}$ and $\{(x, y)|x \in [1, H], y \in [1, W]\}$ are the angular and spatial coordinates, respectively. As shown in Fig. 1, incident light rays are modulated when passing through different aperture positions before converging at the image sensor. Then, the camera captures a 2-D coded measurement $\mathbf{L}_i(x, y) \in \mathbb{R}^{H \times W}$ of the LF. Specifically, the 2-D coded measurement can be formulated as

$$\mathbf{L}_i(x, y) = \sum_{u=1}^{M} \sum_{v=1}^{N} a_i(u, v) \mathscr{L}(u, v, x, y), \tag{1}$$

where $a_i(u, v) \in [0, 1]$ is the transmittance at aperture position (u, v) in the i-th capturing of LF.

Based on the formulation, a coded measurement is the weighted summation of all SAIs. Our method simulates this process by a 2-D convolutional layer with specific configurations. The input of the layer is the entire 4-D LF while the output is 2-D coded measurements corresponding to the input LF. In our simulation, the convolutional operation is carried out on $u - v$ plane, i.e., the angular patch of the input LF, to fuse all $M \times N$ elements in the angular dimension into desired number of elements. We first shuffle the input LF to let the spatial dimension into one axis, i.e., the batch axis during training, to share a same kernel with all angular patches. By setting a proper kernel size, padding and strides, the kernel in the acquisition layer is able to fully cover or slide on the angular patch to obtain desired number of measurements. For example, to capture an LF with angular resolution 7×7 into corresponding 1, 2, and 4 measurements, we set the corresponding kernel size $n_u \times n_v$ to 7×7, 7×6 and 6×6 without padding and one-pixel stride to produce the desired number of measurements.

Besides, we limit the weights into $[0, 1]$ and set bias to 0 in the acquisition layer corresponding to the physical capturing process during training.

Note that the learned aperture can be realized by a typical programmable device, e.g., liquid crystal on silicon (LCoS) display, in a real camera implementation [3,9,17,20,24]. Moreover, there is the color mask which provides an opportunity for modulating both color and angles of incident light rays [23]. Our method can also simulate color LF coded aperture capturing by using three channels in the convolutional layer to modulate the R, G, B channels of the input LF simultaneously. The results are demonstrated in Sect. 4.

3.2   Reconstruction with Deep Spatial-Angular Regularization

The Observation Model. Based on the sensing mechanism in Sect. 3.1, the observation model of coded aperture measurements can be written as

$$l = Ax + \epsilon, \tag{2}$$

where $l = [l_1; l_2; ...; l_k] \in \mathbb{R}^{kHW}$ is the set of k measurements with $l_i \in \mathbb{R}^{HW}$ ($i = 1, 2, ..., k$) being vectorial representation of the i-th measurement L_i, $x \in \mathbb{R}^{HWMN}$ is the vectorial representation of the original LF to be reconstructed, $A \in \mathbb{R}^{kHW \times HWMN}$ denotes the linear degradation/sensing matrix, and $\epsilon \in \mathbb{R}^{kHW}$ denotes the additive noise.

Prior-Driven Solution. Since recovering x from l in Eq. (2) is an ill-posed inverse problem, regularization has to be introduced to constrain the solution space of x. Thus, the problem of LF reconstruction from measurements can be generally cast as

$$\min_{x} \frac{1}{2} \|l - Ax\|_2^2 + \lambda \mathcal{J}(x), \tag{3}$$

where $\mathcal{J}(\cdot)$ is the regularization term, and λ is a positive penalty parameter to balance the two terms.

By introducing an auxiliary variable $v \in \mathbb{R}^{HWMN}$, the optimization problem in Eq. (3) can be decoupled into two sub-problems correspondingly for the data likelihood term and the regularization term [28,33]:

$$\min_{x,v} \frac{1}{2} \|l - Ax\|_2^2 + \lambda \mathcal{J}(v), \quad s.t. \quad x = v. \tag{4}$$

We further convert Eq. (4) into an unconstrained problem by moving the equality constraint into the objective function as a penalty term, i.e.,

$$\min_{x,v} \frac{1}{2} \|l - Ax\|_2^2 + \eta \|x - v\|_2^2 + \lambda \mathcal{J}(v), \tag{5}$$

where $\eta > 0$ is a penalty parameter. Based on the half quadratic splitting method [5,43], the optimization problem in Eq. (5) can be solved by alternatively solving the following two sub-problems until convergence:

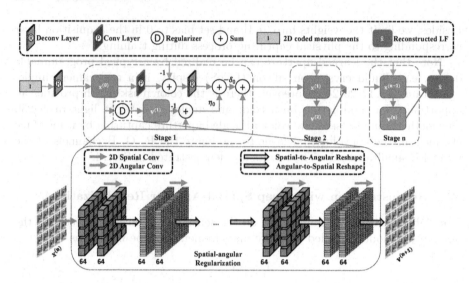

Fig. 2. The architectures of the proposed iterative framework and the deep spatial-angular regularization sub-network

$$\begin{cases} \mathbf{x}^{(t+1)} = \arg\min_{\mathbf{x}} \dfrac{1}{2} \|\mathbf{l} - \mathbf{Ax}\|_2^2 + \eta \left\| \mathbf{x} - \mathbf{v}^{(t)} \right\|_2^2, \\ \mathbf{v}^{(t+1)} = \arg\min_{\mathbf{v}} \eta \left\| \mathbf{x}^{(t+1)} - \mathbf{v} \right\|_2^2 + \lambda \mathcal{J}(\mathbf{v}), \end{cases} \tag{6}$$

where t is the iteration index. For the $\mathbf{x}$-subproblem in Eq. (6), it can be computed with a single step of gradient descent for an inexact solution:

$$\mathbf{x}^{(t+1)} = \mathbf{x}^{(t)} - \delta[\mathbf{A}^{\mathsf{T}}(\mathbf{Ax}^{(t)} - \mathbf{l}) + \eta(\mathbf{x}^{(t)} - \mathbf{v}^{(t)})], \tag{7}$$

where $\delta > 0$ is the parameter controlling the step size. With regard to the $\mathbf{v}$-subproblem, the solution is the proximity term of $\mathcal{J}(\mathbf{v})$ at the point, i.e.,

$$\mathbf{v}^{(t+1)} = \mathcal{D}(\mathbf{x}^{(t+1)}), \tag{8}$$

where $\mathcal{D}(\cdot)$ denotes the proximal operator with respect to a typical regularization $\mathcal{J}(\cdot)$.

Deep Spatial-Angular Regularization. Due to the high-dimensional property and complex geometry structure in LFs, it is difficult to use an explicit regularization term in Eq. (8), which commonly has a limited representation ability, for comprehensively exploring the underlying distribution. To this end, as shown in Fig. 2, we adopt a deep implicit regularization which is constructed by computationally-efficient spatial-angular separable (SAS) convolutional layers [37,40]. The SAS convolution is able to thoroughly detect the dimensional correlations of a 4-D pixel in the LF by alternatively conducting 2-D convolutional operations on spatial and angular planes. Besides, SAS convolution does not significantly increase the number of parameters compared against the 4-D convolution. Furthermore, linear transformations which are implemented by matrices in conventional optimization algorithms, i.e., Eqs. (7) and (8), can be replaced by convolutional layers or networks without impairing the convergence property [5,32,42]. Since the degradation matrix $\mathbf{A}$ and its transpose $\mathbf{A}^\mathsf{T}$ in Eq. (7) are linear projections, we can correspondingly replace them with a convolutional layer as mentioned in Sect. 3.1 and a corresponding deconvolutional layer for inverse projection. Furthermore, in order to preserve the linear property of the transformations, all these layers are not followed by activation functions or bias units. The projection $\mathcal{P}(\cdot)$ conducted by the convolutional layer can be regarded as a linear mapping function which projects $\mathbf{x}$ to $\mathbf{l}$. On the contrary, the inverse projection $\mathcal{R}(\cdot)$ conducted by a deconvolutional layer is regarded as a linear function which projects $\mathbf{l}$ to $\mathbf{x}$, i.e.,

$$\mathbf{l} = \mathcal{P}(\mathbf{x}), \mathbf{x} = \mathcal{R}(\mathbf{l}). \tag{9}$$

Thus, Eq. (8) and Eq. (7) can be respectively rewritten as

$$\begin{cases} \mathbf{v}^{(t+1)} = \mathcal{D}(\mathbf{x}^{(t)}, \theta_d^t), \\ \mathbf{x}^{(t+1)} = \mathbf{x}^{(t)} - \delta_t[\mathcal{R}(\mathcal{P}(\mathbf{x}^{(t)}, \theta_p^t) - \mathbf{l}, \theta_r^t) + \eta_t(\mathbf{x}^{(t)} - \mathbf{v}^{(t+1)})], \end{cases} \tag{10}$$

where θ_r, θ_p, and θ_d are the network parameters which will be learned by the backpropagation algorithm during the training process. In order to enhance the representing ability of the network, parameters in each iterative stage are independently learned without being inherited.

Our iterative framework for reconstructing a 4-D LF from 2-D coded measurements is demonstrated in Fig. 2. Given 2-D coded measurements $\mathbf{l}$, we first use an inverse projection $\mathcal{R}(\mathbf{l}, \theta_r^0)$ to produce an initialization $\mathbf{x}^{(0)}$ for the reconstructed LF. After being processed by the deep regularization $\mathcal{D}(\mathbf{x}^{(0)}, \theta_d^0)$, an optimized LF $\mathbf{x}^{(1)}$ which is better than that from last iterative stage is estimated. Such an iterative stage repeats n times to gradually generate a final reconstructed LF.

3.3   Training and Implementation Details

Training Strategy and Parameter Setting. In our method, the acquisition layer is a 2-D convolutional layer without activation function or bias units for the simulation of coded aperture modulation. The parameters, i.e., the kernel size, padding and stride, in the layer can be flexibly set before training for different simulation tasks. Correspondingly, the parameters in the deconvolutional layer are the same as those in the convolutional layer. It is to ensure that the output and input size of the deconvolutional layer are respectively the same as the input and output size of the convolutional layer. The loss function is the ℓ_1 loss between reconstructed LF and the ground-truth LF. In the regularization sub-network, the kernel size in both spatial and angular convolutional layer is 3×3. The number of feature maps in each layer is 64. The output of each convolutional layer is mapped by a ReLU activation function. Besides, the number of SAS convolutional layers is set to 9 according to our ablation studies in Sect. 4.2. At the training stage, patches of spatial size 32×32 were randomly cropped from LFs contained in the training set. The batch size was set to 5. In order to increase the number of training samples, we randomly cropped 4-D LF patch with size of $M \times N \times 32 \times 32$ from each LF image in the training datasets. The learning rate was initially set to $1e-4$ and reduced to $1e-5$ when the loss stopped decreasing. We chose Adam [14] as the optimizer with $\beta_1 = 0.9$ and $\beta_2 = 0.999$. Our framework was implemented with PyTorch.

Datasets. The training dataset contains both synthetic and real-world LF images. Specifically, there are 100 real-world LF images of size $7 \times 7 \times 376 \times 541$ from Kalantari Lytro [12], 22 synthetic LF images of size $5 \times 5 \times 512 \times 512$ from HCI [8], and 33 synthetic LF images of size $5 \times 5 \times 512 \times 512$ from Inria [29]. The test set contains 30 LF images from Kalantari Lytro [12], 2 LF images from HCI [8] and 4 LF images from Inria [29]. Please refer to the supplementary material for more details of the employed training and testing data.

4 Experiments

In this section, we evaluated the proposed LF reconstruction method by comparing our method with three state-of-the-art methods, followed by a series of comprehensive ablation studies.

4.1 Comparison with State-of-the-Art Methods

We compared with one state-of-the-art deep learning-based LF reconstruction method from coded aperture measurements, i.e., Inagaki *et al.* [9], and two deep learning-based methods from sparsely sampled SAIs, i.e., Kalantari *et al.* [12] and Yeung *et al.* [37]. According to Inagaki *et al.* [9], the performance of traditional compressive sensing-based methods is far below that of deep learning-based ones. Here we omitted the comparison with those methods. Specifically, the detailed experimental settings are listed as follows for fair comparisons:

- all the networks under comparison were re-trained with the same datasets using their source codes with suggested parameters;
- we conducted three tasks, i.e., $1 \to 49$, $2 \to 49$ and $4 \to 49$ ($i \to j$ denotes using i measurements/SAIs to reconstruct an LF with angular resolution j) on real-world dataset, while one task $2 \to 25$ on synthetic dataset;
- we used a same single-channel kernel in our acquisition layer to modulate three color channels of LF, denoted as Ours (Single), for fairly comparing with Inagaki *et al.* [9]. According to the analysis in Sect. 3.1, a three-channel kernel was also trained in another model, denoted as Ours (Multiple);

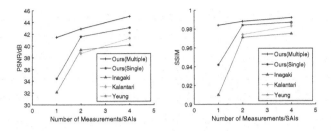

Fig. 3. The quantitative comparisons of all methods on various reconstruction tasks: $1 \rightarrow 49$, $2 \rightarrow 49$, and $4 \rightarrow 49$. Here the PSNR and SSIM values refer to the average of all 30 LFs contained in the test set from Kalantari Lytro [12]. See the supplementary material for the PSNR/SSIM of each LF image

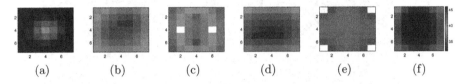

(a)   (b)   (c)   (d)   (e)   (f)

Fig. 4. The average PSNR at each angular position of reconstructed LFs from different methods. The white blocks denote the input SAI positions. From left to right: (a) Inagaki (1) [9], (b) Ours (Single) (1), (c) Kalantari (2) [12], (d) Ours (Single) (2), (e) Yeung (4) [37], (f) Ours (Single) (4). The digits in brackets are the number of input measurements/SAIs for each method

- Kalantari *et al.* [12] cannot handle the task $1 \rightarrow 49$ or $1 \rightarrow 25$ since the hand-craft features cannot be calculated in their algorithm. Besides, Kalantari *et al.* [12] can accept flexible and irregular input patterns. As suggested in [10], we re-trained it with choosing the optimal input patterns for comparison, i.e., the angular coordinates are correspondingly $(4, 2)$ and $(4, 6)$ for task $2 \rightarrow 49$, while $(2, 5)$, $(3, 2)$, $(5, 6)$ and $(6, 3)$ for task $4 \rightarrow 49$;
- and Yeung *et al.* [37] requires the input SAIs with a regular pattern, and it cannot reconstruct an LF from only 1 or 2 SAIs. We thus re-trained it only on the task $4 \rightarrow 49$ with the four-corner SAIs as inputs.

Quantitative Comparison. We calculated the average PSNR and SSIM between the reconstructed LFs and ground-truth ones in Y channel to conduct quantitative comparisons of different methods, as shown in Fig. 3. Besides, the average PSNR at each SAI position of the reconstructed LFs from these methods are shown in Fig. 4. From Figs. 3 and 4, we can draw the following conclusions:

- Ours (Multiple) has better performance than Ours (Single) under all tasks. The possible reason is that the three color channels may have different distributions, and the color channel-tailored aperture can adapt to each channel better than that a common one for three channels;

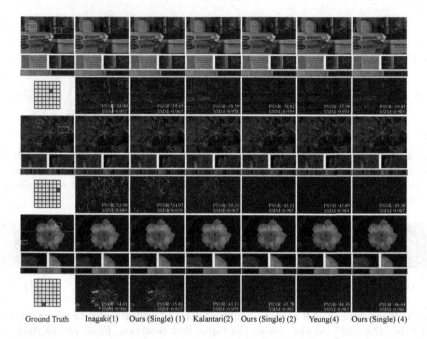

Ground Truth   Inagaki(1)   Ours (Single) (1)   Kalantari(2)   Ours (Single) (2)   Yeung(4)   Ours (Single) (4)

Fig. 5. Visual comparisons of all methods over real LF images under various reconstruction tasks: $1 \rightarrow 49$, $2 \rightarrow 49$ and $4 \rightarrow 49$. The error maps are calculated in gray-scale space. The digits in the brackets are the number of input measurements/SAIs for the methods

- both Ours (Single) and Ours (Multiple) consistently outperform the coded aperture method Inagaki et al. [9] under all tasks. The reason is that Inagaki et al. [9] employs network architecture which is not specifically designed for LF reconstruction, while our algorithm incorporates the observation model for measurements into the deep learning framework elegantly to avoid relying entirely on data-driven priors;
- and our method can preserve a high-quality reconstruction at most aperture positions. Conversely, the quality of SAI in Kalantari et al. [12] and Yeung et al. [37] declines along with the increase of the distance from input SAIs. The possible reason is that they can only extract features from input SAIs, while our method can adaptively condense and utilize information from all aperture positions through measurements.

Visual Comparison of Reconstructed LFs. The visual comparisons of reconstructed LFs from all methods are shown in Figs. 5 and 6, where it can be observed that our method can produce better details than other methods [9,12,37] under all tasks. For Kalantari et al. [12] and Yeung et al. [37], the blurred artifacts and ghost effects exist at occlusion boundaries and high-frequency regions. Additionally, our method can produce better details than Inagaki et al. [9] on synthetic LFs with large disparities. The reason is that the deep spatial-angular regularization in our method can fully exploit dimensional correlations in LFs to achieve a better reconstruction on datasets with large disparities.

Comparison of the LF Parallax Structure. Since the parallax structure among SAIs of an LF is the most valuable information, we conducted experiments to evaluate such a structure embedded in the reconstructed LFs from different methods. First, we compared the EPIs at the bottom of each sub-figure of Figs. 5 and 6, where it can be seen that the EPIs from our method

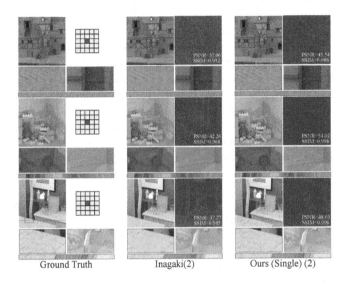

Ground Truth   Inagaki(2)   Ours (Single) (2)

Fig. 6. Visual comparisons of our method against Inagaki [9] over synthetic LF data under the task 2 → 25

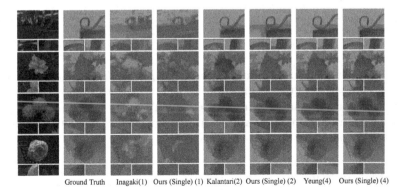

Ground Truth   Inagaki(1)   Ours (Single) (1) Kalantari(2) Ours (Single) (2)   Yeung(4) Ours (Single) (4)

Fig. 7. Comparisons of depth maps estimated with reconstructed LFs by different methods under tasks 1 → 49, 2 → 49 and 4 → 49. The digits in the brackets are the number of input measurements/SAIs for the methods

show clearer and more consistent straight lines compared with those from other methods. Moreover, we conducted more investigations on this important issue. However, due to the lack of standard evaluation metrics, we chose the following two metrics to achieve the target:

(1) It is expected that with a typical LF depth estimation method, the estimated depth maps from LFs with better parallax structure will be more accurate and closer to those from ground-truth LFs. Thus, We applied the same depth estimation algorithm [4] on the reconstructed LFs by different methods and the ground-truth ones. The visualized depth maps illustrated in Fig. 7 show that our method can preserve sharper edges at the occlusion boundaries, which are most similar to those of ground truth, which demonstrates the advantage of our method on preserving the LF parallax structure indirectly.

(2) Considering that the line appearance of EPIs can reflect the parallax structure of LFs, we applied SSIM [35] that measures the similarity between two images by using the structural information on EPIs extracted from the reconstructed LFs, denoted as EPI-SSIM, from different methods to provide a quantitative evaluation towards the LF parallax structure. The average SSIM values listed in Table 1 show that the EPIs from the reconstructed LFs by our method have higher SSIM values, which validates that our reconstructed LFs preserve better parallax structures to some extent.

Table 1. Running time (in second) of different reconstruction methods/average EPI-SSIM of reconstructed LFs by different methods. "–" indicates that the method cannot work on the task

	$1 \rightarrow 49$	$2 \rightarrow 49$	$4 \rightarrow 49$
Ours (Single)	**80.71/0.935**	**80.18/0.981**	80.54/**0.986**
Inagaki et al. [9]	218.94/0.901	179.36/0.968	179.36/0.973
Kalantari et al. [12]	–	84.43/0.972	168.86/0.980
Yeung et al. [37]	–	–	**0.85**/0.974

Comparison of Running Time. We also compared the efficiency of different methods. And the results are shown in Table 1, where it can be observed that our method is faster than all methods except Yeung et al. [37]. Note that all methods were implemented on a desktop with Intel CPU i7–8700 @ 3.70 GHz, 32 GB RAM and NVIDIA GeForce RTX 2080Ti.

4.2   Ablation Study

The Number of Iterative Stages and SAS Convolutional Layers. In our method, the crucial step is alternately updating the $\mathbf{x}^{(t+1)}$ and $\mathbf{v}^{(t+1)}$ from the results of the t-th iterative stage. The number of iterative stages and SAS convolutional layers in the deep regularizer are both key factors to the reconstruction quality. Taking the task $2 \rightarrow 25$ as an example, we separately carried out ablation studies on these two factors.

First, three iterative-stage numbers, i.e., $1, 3$ and 6, were set, while the number of SAS was fixed to 9. The quantitative results shown in Fig. 8(a) indicate that with the number of iterative stages increasing, the quality of reconstructed LFs improves. Besides, the improvement is more obvious from 1 stage to 3 stages but very slight from 3 stages to 6 stages. In practice, the numbers of iterative stages and SAS convolutional layers can be optimally set according to the available computational resources. Then, we set three SAS numbers, i.e., $3, 6$ and 9, and fixed the number of iterative stages to 6. The results are shown in Fig. 8(b). With the number of SAS convolutional layers increasing, the quality of reconstructed LFs also improves. We hence chose 6 stages and 9 SAS convolutional layers to trade-off the quality of LF reconstruction and the computational costs in our method.

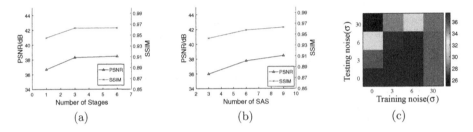

Fig. 8. Comparisons of different numbers of iterative stages (a), different numbers of SAS convolutional layers in deep spatial-angular regularization (b), and different levels of noise (c)

Noisy Measurements. Due to the small size of coded aperture and low-light conditions in practice, the measurements are used to be affected by noise. In order to evaluate the robustness of our method in real applications, we added different levels of Gaussian noise onto the measurements during training and testing. We set three standard deviations for one measurement: $\sigma = 3$, $\sigma = 6$, and $\sigma = 30$ to control the noise level. Here we took the task $2 \rightarrow 25$ on the synthetic LFs as an example. The PSNR value shown in Fig. 8(c) indicate that our method with noisy inputs can preserve the comparable performance on the noise-free case when the testing noise level is lower or slightly higher than that of training. If the noise level is too high (e.g., $\sigma = 30$), the performance declines rapidly.

5   Conclusion and Future Work

We proposed a novel deep learning-based LF reconstruction method from coded aperture measurements, which links the observation model of measurements and deep learning elegantly, making it more physically interpretable. To be specific, we design a deep regularization term with an efficient spatial-angular convolutional sub-network to implicitly and comprehensively explore the signal distribution. Extensive experiments over both real and synthetic datasets demonstrate that our method outperforms state-of-the-art approaches to a significant extent both quantitatively and qualitatively.

Acknowledgements. This work was supported in part by the Hong Kong RGC under Grant 9048123 (CityU 21211518), and in part by the Basic Research General Program of Shenzhen Municipality under Grant JCYJ20190808183003968.

References

1. Ashok, A., Neifeld, M.A.: Compressive light field imaging. In: Three-Dimensional Imaging, Visualization, and Display 2010 and Display Technologies and Applications for Defense, Security, and Avionics IV, vol. 7690, p. 76900Q. International Society for Optics and Photonics (2010)
2. Babacan, S.D., Ansorge, R., Luessi, M., Mataran, P.R., Molina, R., Katsaggelos, A.K.: Compressive light field sensing. IEEE Trans. Image Process. **21**(12), 4746–4757 (2012)
3. Chen, J., Chau, L.P.: Light field compressed sensing over a disparity-aware dictionary. IEEE Trans. Circ. Syst. Video Technol. **27**(4), 855–865 (2015)
4. Chen, J., Hou, J., Ni, Y., Chau, L.P.: Accurate light field depth estimation with superpixel regularization over partially occluded regions. IEEE Trans. Image Process. **27**(10), 4889–4900 (2018)
5. Dong, W., Wang, P., Yin, W., Shi, G., Wu, F., Lu, X.: Denoising prior driven deep neural network for image restoration. IEEE Trans. Pattern Anal. Mach. Intell. **41**(10), 2305–2318 (2018)
6. Guo, M., Zhu, H., Zhou, G., Wang, Q.: Dense light field reconstruction from sparse sampling using residual network. In: Asian Conference on Computer Vision (ACCV), pp. 50–65. Springer (2018)
7. Gupta, M., Jauhari, A., Kulkarni, K., Jayasuriya, S., Molnar, A., Turaga, P.: Compressive light field reconstructions using deep learning. In: IEEE Conference on Computer Vision and Pattern Recognition Workshops (CVPRW), pp. 11–20 (2017)
8. Honauer, K., Johannsen, O., Kondermann, D., Goldluecke, B.: A dataset and evaluation methodology for depth estimation on 4D light fields. In: Lai, S.-H., Lepetit, V., Nishino, K., Sato, Y. (eds.) ACCV 2016. LNCS, vol. 10113, pp. 19–34. Springer, Cham (2017). https://doi.org/10.1007/978-3-319-54187-7_2
9. Inagaki, Y., Kobayashi, Y., Takahashi, K., Fujii, T., Nagahara, H.: Learning to capture light fields through a coded aperture camera. In: European Conference on Computer Vision (ECCV), pp. 418–434 (2018)
10. Jin, J., Hou, J., Chen, J., Zeng, H., Kwong, S., Yu, J.: Deep coarse-to-fine dense light field reconstruction with flexible sampling and geometry-aware fusion. IEEE Trans. Pattern Anal. Mach. Intell. (2020). https://doi.org/10.1109/TPAMI.2020.3026039
11. Jin, J., Hou, J., Yuan, H., Kwong, S.: Learning light field angular super-resolution via a geometry-aware network. In: Thirty-Fourth AAAI Conference on Artificial Intelligence, pp. 11141–11148 (2020)
12. Kalantari, N.K., Wang, T.C., Ramamoorthi, R.: Learning-based view synthesis for light field cameras. ACM Trans. Graph. **35**(6), 193 (2016)

13. Kim, J., Kwon Lee, J., Mu Lee, K.: Accurate image super-resolution using very deep convolutional networks. In: IEEE Conference on Computer Vision and Pattern Recognition (CVPR), pp. 1646–1654 (2016)
14. Kingma, D.P., Ba, J.: Adam: a method for stochastic optimization. arXiv preprint (2014). arXiv:1412.6980
15. Levoy, M., Hanrahan, P.: Light field rendering. In: ACM SIGGRAPH, pp. 31–42 (1996)
16. Li, N., Ye, J., Ji, Y., Ling, H., Yu, J.: Saliency detection on light field. In: IEEE Conference on Computer Vision and Pattern Recognition (CVPR), pp. 2806–2813 (2014)
17. Liang, C.K., Lin, T.H., Wong, B.Y., Liu, C., Chen, H.H.: Programmable aperture photography: multiplexed light field acquisition. In: ACM SIGGRAPH, pp. 1–10 (2008)
18. Long, J., Shelhamer, E., Darrell, T.: Fully convolutional networks for semantic segmentation. In: IEEE Conference on Computer Vision and Pattern Recognition (CVPR), pp. 3431–3440 (2015)
19. Lytro: https://www.lytro.com/ (2016)
20. Marwah, K., Wetzstein, G., Bando, Y., Raskar, R.: Compressive light field photography using overcomplete dictionaries and optimized projections. ACM Trans. Graph. 32(4), 46 (2013)
21. Miandji, E., Hajisharif, S., Unger, J.: A unified framework for compression and compressed sensing of light fields and light field videos. ACM Trans. Graph. 38(3), 1–18 (2019)
22. Mildenhall, B., Srinivasan, P.P., Ortiz-Cayon, R., Kalantari, N.K., Ramamoorthi, R., Ng, R., Kar, A.: Local light field fusion: practical view synthesis with prescriptive sampling guidelines. ACM Trans. Graph. 38(4), 1–14 (2019)
23. Nabati, O., Mendlovic, D., Giryes, R.: Fast and accurate reconstruction of compressed color light field. In: IEEE International Conference on Computational Photography (ICCP), pp. 1–11. IEEE (2018)
24. Nagahara, H., Zhou, C., Watanabe, T., Ishiguro, H., Nayar, S.K.: Programmable aperture camera using LCoS. In: Daniilidis, K., Maragos, P., Paragios, N. (eds.) ECCV 2010. LNCS, vol. 6316, pp. 337–350. Springer, Heidelberg (2010). https://doi.org/10.1007/978-3-642-15567-3_25
25. Ng, R., et al.: Digital Light Field Photography. Stanford University, United States (2006)
26. Qu, W., Zhou, G., Zhu, H., Xiao, Z., Wang, Q., Vidal, R.: High angular resolution light field reconstruction with coded-aperture mask. In: IEEE International Conference on Image Processing (ICIP), pp. 3036–3040. IEEE (2017)
27. RayTrix: 3d light field camera technology. https://raytrix.de/
28. Romano, Y., Elad, M., Milanfar, P.: The little engine that could: regularization by denoising (red). SIAM J. Imaging Sci. 10(4), 1804–1844 (2017)
29. Shi, J., Jiang, X., Guillemot, C.: A framework for learning depth from a flexible subset of dense and sparse light field views. IEEE Trans. Image Process. 28(12), 5867–5880 (2019)
30. Shi, L., Hassanieh, H., Davis, A., Katabi, D., Durand, F.: Light field reconstruction using sparsity in the continuous fourier domain. ACM Trans. Graph. 34(1), 12 (2014)
31. Srinivasan, P.P., Wang, T., Sreelal, A., Ramamoorthi, R., Ng, R.: Learning to synthesize a 4D RGBD light field from a single image. In: IEEE International Conference on Computer Vision (ICCV), vol. 2, p. 6 (2017)

32. Sun, J., et al.: Deep ADMM-Net for compressive sensing MRI. In: Advances in Neural Information Processing Systems (NeurIPS), pp. 10–18 (2016)
33. Venkatakrishnan, S.V., Bouman, C.A., Wohlberg, B.: Plug-and-play priors for model based reconstruction. In: IEEE Global Conference on Signal and Information Processing, pp. 945–948. IEEE (2013)
34. Wang, T.C., Efros, A.A., Ramamoorthi, R.: Depth estimation with occlusion modeling using light-field cameras. IEEE Trans. Pattern Anal. Mach. Intell. **38**(11), 2170–2181 (2016)
35. Wang, Z., Bovik, A.C., Sheikh, H.R., Simoncelli, E.P.: Image quality assessment: from error visibility to structural similarity. IEEE Trans. Image Process. **13**(4), 600–612 (2004)
36. Wilburn, B., et al.: High performance imaging using large camera arrays. In: ACM SIGGRAPH, pp. 765–776 (2005)
37. Wing Fung Yeung, H., Hou, J., Chen, J., Ying Chung, Y., Chen, X.: Fast light field reconstruction with deep coarse-to-fine modeling of spatial-angular clues. In: European Conference on Computer Vision (ECCV), pp. 137–152 (2018)
38. Wu, G., Zhao, M., Wang, L., Dai, Q., Chai, T., Liu, Y.: Light field reconstruction using deep convolutional network on EPI. In: IEEE Conference on Computer Vision and Pattern Recognition (CVPR), pp. 6319–6327 (2017)
39. Yagi, Y., Takahashi, K., Fujii, T., Sonoda, T., Nagahara, H.: PCA-coded aperture for light field photography. In: IEEE International Conference on Image Processing (ICIP), pp. 3031–3035. IEEE (2017)
40. Yeung, H.W.F., Hou, J., Chen, X., Chen, J., Chen, Z., Chung, Y.Y.: Light field spatial super-resolution using deep efficient spatial-angular separable convolution. IEEE Trans. Image Process. **28**(5), 2319–2330 (2018)
41. Yoon, Y., Jeon, H.G., Yoo, D., Lee, J.Y., So Kweon, I.: Learning a deep convolutional network for light-field image super-resolution. In: IEEE International Conference on Computer Vision Workshops (ICCVW), pp. 24–32 (2015)
42. Zhang, J., Ghanem, B.: ISTA-Net: Interpretable optimization-inspired deep network for image compressive sensing. In: IEEE Conference on Computer Vision and Pattern Recognition (CVPR), pp. 1828–1837 (2018)
43. Zhang, K., Zuo, W., Gu, S., Zhang, L.: Learning deep CNN denoiser prior for image restoration. In: IEEE Conference on Computer Vision and Pattern Recognition (CVPR), pp. 3929–3938 (2017)
44. Zhou, T., Tucker, R., Flynn, J., Fyffe, G., Snavely, N.: Stereo magnification: learning view synthesis using multiplane images. ACM Trans. Graph. **37**(4), 1–12 (2018)
45. Zhu, H., Guo, M., Li, H., Wang, Q., Robles-Kelly, A.: Revisiting spatio-angular trade-off in light field cameras and extended applications in super-resolution. IEEE Trans. Visual. Comput. Graph. (2019). https://doi.org/10.1109/TVCG.2019.2957761

Video-Based Remote Physiological Measurement via Cross-Verified Feature Disentangling

Xuesong Niu[1,2(✉)], Zitong Yu[3], Hu Han[1,4], Xiaobai Li[3], Shiguang Shan[1,2,4], and Guoying Zhao[3]

[1] Key Laboratory of Intelligent Information Processing of Chinese Academy of Sciences (CAS), Institute of Computing Technology, CAS, Beijing 100190, China
`xuesong.niu@vipl.ict.ac.cn`, {`hanhu,sgshan`}`@ict.ac.cn`
[2] University of Chinese Academy of Sciences, Beijing 100049, China
[3] Center for Machine Vision and Signal Analysis, University of Oulu, Oulu, Finland
{`zitong.yu,xiaobai.li,guoying.zhao`}`@oulu.fi`
[4] Peng Cheng Laboratory, Shenzhen 518055, China

Abstract. Remote physiological measurements, e.g., remote photoplethysmography (rPPG) based heart rate (HR), heart rate variability (HRV) and respiration frequency (RF) measuring, are playing more and more important roles under the application scenarios where contact measurement is inconvenient or impossible. Since the amplitude of the physiological signals is very small, they can be easily affected by head movements, lighting conditions, and sensor diversities. To address these challenges, we propose a cross-verified feature disentangling strategy to disentangle the physiological features with non-physiological representations, and then use the distilled physiological features for robust multi-task physiological measurements. We first transform the input face videos into a multi-scale spatial-temporal map (MSTmap), which can suppress the irrelevant background and noise features while retaining most of the temporal characteristics of the periodic physiological signals. Then we take pairwise MSTmaps as inputs to an autoencoder architecture with two encoders (one for physiological signals and the other for non-physiological information) and use a cross-verified scheme to obtain physiological features disentangled with the non-physiological features. The disentangled features are finally used for the joint prediction of multiple physiological signals like average HR values and rPPG signals. Comprehensive experiments on different large-scale public datasets of multiple physiological measurement tasks as well as the cross-database testing demonstrate the robustness of our approach.

Electronic supplementary material The online version of this chapter (https://doi.org/10.1007/978-3-030-58536-5_18) contains supplementary material, which is available to authorized users.

A. Vedaldi et al. (Eds.): ECCV 2020, LNCS 12347, pp. 295–310, 2020.
https://doi.org/10.1007/978-3-030-58536-5_18

1   Introduction

Physiological signals, such as average heart rate (HR), respiration frequency (RF) and heart rate variability (HRV), are very important for cardiovascular activity analysis and have been widely used in healthcare applications. Traditional HR, RF and HRV measurements are based on the electrocardiography (ECG) and contact photoplethysmography (cPPG) signals, which require dedicated skin-contact devices for data collection. The usage of these contact sensors may cause discomfort and inconvenience for subjects in healthcare scenarios. Recently, a growing number of physiological measurement techniques have been proposed based on remote photoplethysmography (rPPG) signals, which could be captured from face by ordinary cameras and without any contact. These techniques make it possible to measure HR, RF and HRV remotely and have been developed rapidly [3,9,13,15,16,19,22,25].

The rPPG based physiological measurement is based on the fact that optical absorption of a local tissue varies periodically with the blood volume, which changes accordingly with the heartbeats. However, the amplitude of the light absorption variation w.r.t. blood volume pulse (BVP) is very small, i.e., not visible for human eyes, and thus it can be easily affected by factors such as subject's movements and illumination conditions. To solve this problem, many traditional methods are proposed to remove the non-physiological information using color space projection [3,22] or signal decomposition [15,16,19]. However, these methods are based on assumptions such as linear combination assumption [15,16] or certain skin reflection models [3,22], which may not hold in less-constrained scenarios with large head movement or dim lighting condition.

Besides the hand-crafted traditional methods, there are also some approaches which adopt the strong modeling ability of deep neural networks for remote physiological signal estimation [2,13,17,25]. Most of these methods focus on learning a network mapping from different hand-crafted representations of face videos (e.g., cropped video frames [17,25], motion representation [2] or spatial-temporal map [13]) to the physiological signals. One fact is that, these hand-crafted representations contain not only the information of physiological signals, but also the non-physiological information such as head movements, lighting variations and device noises. The features learned from these hand-crafted representations are usually affected by the non-physiological information. In addition, existing methods use either the rPPG signals [2,25] or the average HR values [13,17] for supervision. However, rPPG signals and average HR describe the subject's physiological status from detailed and general aspects respectively, and making full use of both two physiological signals for training will help the network to dig more representative features.

To address these problems, we propose an end-to-end multi-task physiological measurement network, and train the network using a cross-verified disentangling (CVD) strategy. We first compress the face videos into multi-scale spatial-temporal maps (MSTmaps) to better represent the physiological information in face videos. Although the MSTmaps could highlight the physiological information in face videos, they are still polluted by the non-physiological information such as head movements, lighting conditions and device variations. To automatically disentangle the physiological features with the non-physiological information, we then use a cross-verified disentangling strategy to train our network.

As illustrated in Fig. 1, this disentanglement is realized by designing an autoencoder with two encoders (one for physiological signals and the other for non-physiological information) and using pairwise MSTmaps as input for training. Besides of reconstructing the original MSTmaps, we also cross-generate the pseudo MSTmaps using the physiological and non-physiological features encoded from different original MSTmaps. Intuitively, the physiological and non-physiological features encoded from the pseudo MSTmap are supposed to be similar to the features used to generate the pseudo MSTmap. This principle can be used to guide the encoders to cross-verify the features they are supposed to encode, and thus make sure that physiological encoder only focuses on extracting features for HR and rPPG signal estimation, and the non-physiological encoder to encode other irrelevant information. The physiological features are then used for multi-task physiological measurements, i.e., estimating average HR and rPPG signals synchronously. Results on multiple datasets for different physiological measurement tasks as well as the cross-database testing show the effectiveness of our method.

The contributions of this work include: 1) we propose a novel multi-scale spatial-temporal map to highlight the physiological information in face videos. 2) We propose a novel cross-verified disentangling strategy to distill the physiological features for robust physiological measurements. 3) We propose a multi-task physiological measurement network trained with our cross-verified disentangling strategy, and achieve the state-of-the-art performance on multiple physiological measurement databases as well as the cross-database testing.

2   Related Work

2.1   Video-Based Remote Physiological Measurement

An early study of rPPG based physiological measurement was reported in [20]. After that, many approaches have been reported on this topic. The traditional hand-crafted methods mainly focus on extracting robust physiological signals using different color channels [7,9,15,16] or different regions of interest (ROIs) [5,19]. It has been demonstrated that signal decomposition methods, such as independent component analysis (ICA) [5,15,16], principal components analysis (PCA) [7] and matrix completion [19], are effective to improve the signal-to-noise rate (SNR) of the generated signal. Besides the signal decomposition methods, there are also some approaches aimming to get a robust rPPG signal using reflection model based color space transformation [3,23,24]. Various hand-crafted methods using different signal decomposition methods and skin models lead to improved robustness. However, in less-constrained and complicated scenarios, the assumptions of the signal decomposition methods and skin reflection models may not hold, and the estimation performance will drop significantly.

In addition to these hand-crafted methods, there are also a few approaches aiming to leverage the effectiveness of deep learning to remote physiological measurement. In [2], Chen et al. proposed a motion representation of face video and used a convolution network with attention mechanism to predict the rPPG signals. In [13], Niu et al. utilized a spatial-temporal map as the representation of face video and used a CNN-RNN structure to regress the average HR value. In [17], Spetlik et al. proposed a full convolutional network to estimate the average HR from the cropped faces. In addition to these methods using 2D convolution, Yu et al. [25] proposed a 3D convolution network to regress the rPPG signals from face videos. All these existing methods mainly focused on designing the effective representation of the input face videos [2,13] and using various network structures [17,25] for physiological measurement. However, non-physiological information such as head movements and lighting conditions may have significant impacts on these hand-crafted representations, but few of these methods have considered the non-physiological influences. In addition, these methods only use either rPPG signals or average HR values during training, and did not consider taking advantage of both physiological signals for supervision.

2.2 Disentangle Representation Learning

Disentanglement representation learning is gaining increasing attention in computer vision. In [18], Tran et al. proposed the DR-GAN to disentangle the identity and pose representations for pose-invariant face recognition with adversarial learning. In [10], Liu et al. utilized an auto-encoder model to distill the identity and disentangle the identity features with other facial attributes. In [26], Zhang et al. utilized the walking videos to disentangle the appearance and pose features for gait recognition. Besides the works on recognition tasks, there are also many works using the disentangled representation for image synthesis and editing. In [11], Lu et al. proposed a method for image deblurring by splitting the content and blur features in a blurred image. In [6], Lee et al. proposed an approach for diverse image-to-image translation by embedding images into a domain-invariant content space and a domain specific attribute space. Different from the existing approaches [6,11,18], our method does not need adversarial training, making it more accessible for training. Besides, unlike [10], which requires prior attribute labels for disentanglement, our method does not need the prior non-physiological information for training. In addition, unlike [26] requiring temporal information, our method only needs pairwise MSTmaps for disentanglement. Moreover, this work is the first work leveraging disentangle representation learning to remote physiological measurement.

3 Proposed Method

In this section, we give detailed explanations of the proposed cross-verified feature disentangling strategy for multi-task physiological measurement from face videos. Figure 1 gives an overview of the proposed method, which includes

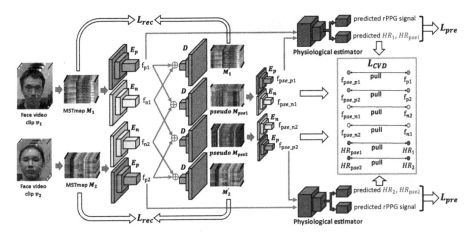

Fig. 1. An overview of our cross-verified feature disentangling (CVD) strategy. Pairwise face video clips are used for training. We first generate the corresponding MSTmaps M_1, M_2 of the input face video clips. Then we feed the MSTmaps into the physiological encoder E_p and non-physiological encoder E_n to get the features. L_{rec} is first used for reconstructing the original MSTmaps. Then we cross-generate the pseudo MSTmaps M_{pse1}, M_{pse2} using features from different original MSTmaps. Disentanglement is realized using L_{CVD} by cross-verifying the encoded features of the original MSTmaps M_1, M_2 and the pseudo MSTmaps M_{pse1}, M_{pse2}. The physiological estimator takes the physiological features $f_{p1}, f_{p2}, f_{pse\_p1}, f_{pse\_p2}$ as input and is optimized by L_{pre}. The modules of the same type in our network use shared weights.

the multi-scale spatial-temporal map generation, the cross-verified disentangling strategy and multi-task physiological measurements.

3.1   Multi-scale Spatial Temporal Map

The amplitude of optical absorption variation of face skin is very small, thus it is important to design a good representation to highlight the physiological information in face videos [2,13]. Considering both the local and the global physiological information in face, we propose a novel multi-scale spatial-temporal map (MSTmap) to represent the facial skin color variations due to heartbeats.

As shown in Fig. 2, we first use an open source face detector SeetaFace[1] to detect the facial landmarks, based on which we define the most informative ROIs for physiological measurement in face, i.e., the forehead and cheek areas. In order to stabilize the landmarks, we also apply a moving average filter to the facial landmark locations across frames. Average pooling is widely used to improve the robustness [7,9,15,16,19,22] against the background and device noises, and the traditional methods usually use the average pixel values of the whole facial ROI for further processing. Different from using the global pooling of the whole

---

[1] https://github.com/seetaface/SeetaFaceEngine.

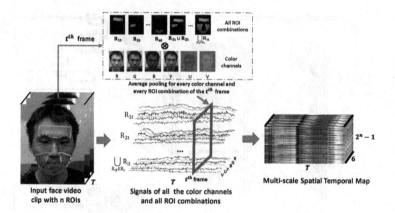

Fig. 2. An illustration of our MSTmap generation from an input face video clip of T frames. For the tth frame, we first detect the n most informative ROIs in face, then calculate the average pixel values for all kinds of ROI combinations and all the color channels in both RGB and YUV color spaces. The average pixel values of different frames are concatenated into temporal sequences, and the temporal sequences of all ROI combinations and color channels are placed into rows. The final multi-scale spatial-temporal map is in the size of $(2^n - 1) \times T \times 6$. (Color figure online)

facial ROI, which may ignore some local information, we generate our MSTmap considering both the local and global facial regions.

Specifically, as shown in Fig. 2, for the tth frame of a video clip, we first get a set of n informative regions of face $R_t = \{R_{1t}, R_{2t}, \cdots, R_{nt}\}$. Then, we calculate the average pixel values of each color channel for all the non-empty subsets of R_t, which are $2^n - 1$ combinations of the elements in R_t. As illustrated in [13], YUV color space is effective in representing the physiological information in face. Therefore, we use both the RGB and YUV color space to calculate the MSTmap. The total number of color channels is 6. For each video clip with T frames, we first obtain $6 \times (2^n - 1)$ temporal sequences of length T. A max-min normalization is applied to all the temporal sequences to scale the temporal series into $[0, 255]$. Then, the $2^n - 1$ temporal signals of each channel are placed into rows, and we can get the final MSTmap with the size of $(2^n - 1) \times T \times 6$ for each video clip.

3.2 Cross-Verified Feature Disentangling

After we generate the MSTmaps as stated in Sect. 3.1, we can further use these informative representations for physiological measurements. One straight forward way is directly using convolution neural networks (CNNs) to regress the physiological signals with the MSTmaps. However, these hand-crafted MSTmaps are usually polluted by non-physiological information such as head movements and illumination conditions. In order to disentangle the physiological feature from the non-physiological information, we use a cross-verified disentangling strategy (CVD) to train the network. An auto-encoder with two encoders (one for physiological features and the other for non-physiological information) is used to learn the features. Pairwise MSTmaps are used as input, and the network is trained with both the original MSTmaps as well as the pseudo MSTmaps cross-decoded using features from different original MSTmaps.

Specifically, as shown in Fig. 1, with pairwise input face video clips v_1, v_2, we first generate the corresponding MSTmaps M_1, M_2. Then, we use the physiological encoder E_p and noise encoder E_n to get the physiological features f_{p1}, f_{p2} and the non-physiological features f_{n1}, f_{n2}. Physiological and non-physiological features encoded from the same MSTmap are first used to reconstruct the original MSTmap to make sure the decoder D can effectively reconstruct the MSTmap. For the input MSTmaps M_1, M_2 and the decoded MSTmaps M_1', M_2', the reconstruction loss is formulated as

$$L_{rec} = \lambda_{rec} \sum_{i=1}^{2} \| M_i - M_i' \|_1 \tag{1}$$

where λ_{rec} is used to balance L_{rec} and other losses.

Then, pseudo MSTmaps M_{pse1}, M_{pse2} are generated with the decoder D using the features from different MSTmaps, i.e., pseudo MSTmap M_{pse1} is generated using f_{p1} and f_{n2}, and M_{pse2} is generated using f_{p2} and f_{n1}. M_{pse1}, M_{pse2} are further fed to E_p and E_n to get the encoded features $f_{pse\_p1}, f_{pse\_n1}, f_{pse\_p2}, f_{pse\_n2}$. Meanwhile, all the physiological features f_{p1}, f_{p2}, f_{pse_p1} and f_{pse_p2} are fed to the physiological estimator for HR and rPPG signal predictions. The detailed architecture of E_p, E_n and D can be found in the supplementary material.

Institutively, taking f_{p1} and f_{pse_p1} as example, if the physiological encoder E_p encodes only the physiological information, f_{p1} is supposed to be the same as f_{pse_p1} since M_{pse1} are generated using f_{p1} and f_{n2} and the physiological representation contained in M_{pse1} should be the same as M_1. This principle can be used to cross-verify the features of the two encoders, and guide the two encoders to focus on the information they are supposed to encode. We consider the physiological features, non-physiological features as well as the predicted HR values $HR_1, HR_{pse1}, HR_2, HR_{pse2}$ with the physiological features, and design our CVD loss L_{CVD} as:

$$L_{CVD} = \lambda_{cvd} \sum_{i=1}^{2} \| f_{pi} - f_{pse\_pi} \|_1 + \lambda_{cvd} \sum_{i=1}^{2} \| f_{ni} - f_{pse\_n(3-i)} \|_1$$
$$+ \sum_{i=1}^{2} \| HR_i - HR_{psei} \|_1 \tag{2}$$

where λ_{cvd} is the balance parameter. With this CVD loss, we can enforce the physiological and non-physiological encoders to obtain features disentangled from each other, and make the physiological encoder to get more representative features for physiological measurements.

3.3 Multi-task Physiological Measurement

The physiological features are further used for multi-task physiological measurements. Typically-considered physiological signals include the average HR values and the rPPG signals. The average HR values can provide general information of the physiological status, while the rPPG signals can give more detailed supervision. Considering that these two signals can provide different aspects of supervision for training, we design our physiological estimator taking both of them into consideration, and expect that this will help the network to learn more robust features and get more accurate predictions.

Specifically, our physiological estimator is a two-head network with one head to regress the rPPG signals and the other to regress the HR values. The detailed architectures can be found in the supplementary material. For the average HR value regression branch, we use a conventional L1 loss function for supervision. For the rPPG signal prediction branch, we use a Pearson correlation based loss to define the similarity between the predicted signal and ground truth, i.e.,

$$L_{rppg} = 1 - \frac{Cov(s_{pre}, s_{gt})}{\sqrt{Cov(s_{pre}, s_{pre})}\sqrt{Cov(s_{gt}, s_{gt})}} \qquad (3)$$

where s_{pre} and s_{gt} are the predicted and ground truth rPPG signals, and $Cov(x, y)$ is the covariance of x and y. Meanwhile, since average HR can also be calculated from the rPPG signal, we add the L_{rppg_hr} to help the average HR estimation using the features of the rPPG branch. The L_{rppg_hr} is formulated as

$$L_{rppg\_hr} = CE(PSD(s_{pre}), HR_{gt}) \qquad (4)$$

where HR_{gt} is the ground-truth HR, and $CE(x, y)$ calculate the cross-entropy loss of the input x and ground-truth y. $PSD(s)$ is the power spectral density of the input signal s. The final loss for physiological measurements is

$$L_{pre} = \|HR_{pre} - HR_{gt}\| + \lambda_{rppg}L_{rppg} + L_{rppg\_hr} \qquad (5)$$

where HR_{pre} and HR_{gt} are the predicted HR and ground-truth HR, and λ_{rppg} is a balancing parameter. The overall loss function of our cross-verified disentangling strategy is

$$L = L_{rec} + L_{CVD} + L_{pre} \qquad (6)$$

4   Experiments

In this section, we provide evaluations of the proposed method including intra-database testing, cross-database testing and the ablation study.

4.1 Databases and Experimental Settings

Databases. We evaluate our method on three widely-used publicly-available physiological measurement databases, i.e., VIPL-HR, OBF, and MMSE-HR.

VIPL-HR [12,13] is a large-scale remote HR estimation database containing 2,378 visible light face videos from 107 subjects. Various less-constrained scenarios, including different head movements, lighting conditions and acquisition devices, are considered in this database. The frame rates of the videos in VIPL-HR vary from 25 fps to 30 fps. Average HR values as well as the BVP signals are provided, and we use the BVP signals as the ground truth rPPG signals. Following the protocol in [13], we conduct a five-fold subject-exclusive cross-validation for the intra-database testing as well as the ablation study for average HR estimation.

OBF [8] is a large-scale database for remote physiological signal analysis. It contains 200 five-minute-long high-quality RGB face videos from 100 subjects. The videos were recorded at 60 fps in OBF, we downsample the frame rate to 30 fps for the convenience of computing. Following [25], we use OBF for evaluations on both HR estimation, RF measuring and HRV analysis. The ground truth RF and HRV features are calculated using the corresponding ECG signals provided. The BVP signals provided are used as the ground truth rPPG signals. We conduct a ten-fold subjective-exclusive cross-validation as [25].

MMSE-HR [19] is a database for remote HR estimation consisting of 102 RGB face videos from 40 subjects recorded at 25 fps. The corresponding average HR values are collected using a biological data acquisition system. Various facial expressions and head movements of the subjects are recorded. The MMSE-HR database is only used for cross-database testing.

Training Details. The detailed architectures of the network can be found in the supplementary material, and all the losses are applied jointly. For all the experiments, the length of face video clip is set to 300 frames, and 6 ROIs as shown in Fig. 2 are considered. For average HR estimation of a 30 s face video as [13], we use a time step of 0.5 s to get all the video clips, and the average of the predicted HRs is regarded as the predicted average HR for the 30 s video. The MSTmaps are resized to 320×320 before being fed to the network for the convenience of computing. Random horizontal and vertical flip of the MSTmaps as well as the data balancing strategy proposed in [14] are used for data augmentation. For all the experiments, we set $\lambda_{rec} = 50$, $\lambda_{cvd} = 10$ and $\lambda_{rppg} = 2$. All the networks are implemented using PyTorch framework[2] and trained with NVIDIA P100. Adam optimizer [4] with an initial learning rate of 0.0005 is used for training. The maximum epoch number for training is set to 70 for experiments on VIPL-HR database and 30 for experiments on OBF database. Code is available.[3]

Evaluation Metrics. Various metrics are used for evaluations. For the task of average HR estimation, we follow previous work [9,13,19,25] and use the metrics including the standard deviation of the error (Std), the mean absolute error

---

[2] https://pytorch.org/.
[3] https://github.com/nxsEdson/CVD-Physiological-Measurement.

Table 1. The HR estimation results by the proposed approach and several state-of-the-art methods on the VIPL-HR database.

Method	Std↓ (bpm)	MAE↓ (bpm)	RMSE↓ (bpm)	r ↑
SAMC [19]	18.0	15.9	21.0	0.11
POS [22]	15.3	11.5	17.2	0.30
CHROM [3]	15.1	11.4	16.9	0.28
I3D [1]	15.9	12.0	15.9	0.07
DeepPhy [2]	13.6	11.0	13.8	0.11
RhythmNet [13]	8.11	5.30	8.14	0.76
Proposed	**7.92**	**5.02**	**7.97**	**0.79**

(MAE), the root mean squared error (RMSE), and the Pearson's correlation coefficients (r). For the evaluation of RF and HRV analysis, following [8, 25], we use Std, RMSE and r as the evaluation of RF and three HRV features, i.e., low frequency (LF), high frequency (HF) and LF/HF.

4.2   Intra-database Testing

Results on Average HR Estimation. We first conduct experiments on VIPL-HR database for average HR estimation using a five-fold subject-exclusive evaluation protocol following [13]. State-of-the-art methods including hand-crafted methods (SAMC [19], POS [22], CHROM [3]) and deep learning based methods (I3D [1], DeepPhy [2], RhythmNet [13]) are used for comparison. We directly take the results of these state-of-the-art methods from [13]. The results of the proposed method and the state-of-the-art methods are given in Table 1.

From the results, we can see that the proposed method achieves promising results with an Std of 7.92 bpm, an MAE of 5.02 bpm, an RMSE of 7.97 bpm and a r of 0.79, which outperform all the state-of-the-art methods including both the hand-crafted and deep learning based methods. In order to further check the correlations between the predicted HRs and the ground-truth HRs, we plot the HR estimation results against the ground truths in Fig. 3(a). From the figure we can see that the predicted HRs and the ground-truth HRs are well correlated in a wide range of HR from 47 bpm to 147 bpm under the less-constrained scenarios of the VIPL-HR database such as large head movements and dim environment. In addition, we also calculate the estimation errors for the large head movement scenario in VIPL-HR and compare the result with RhythmNet [13]. We get an RMSE of 7.44 bpm, which is distinctively better than RhythmNet (9.4 bpm). All the results indicate that the proposed method could effectively distill the physiological information and provide robust physiological measurements.

Besides the average HR estimation for a thirty-second video, we also check the short-time HR estimation performance of the after exercising scenario on the VIPL-HR, in which the subject's HR decreases rapidly. Two examples are given in Fig. 3(b). From the examples, we can see that the proposed approach could

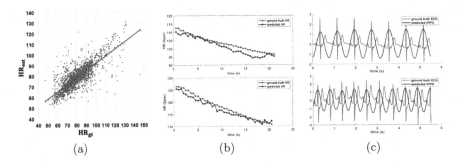

Fig. 3. (a) The scatter plot of the ground truth HR_{gt} and the predicted HR_{est} of all the face videos on VIPL-HR dataset. (b) Two examples of the short-time HR estimation for face videos with significantly decreased HR. (c) Two example curves of the predicted rPPG signals and the ground truth ECG signals used to calculate the HRV features.

follow the trend of HR changes well, which indicates our model is robust in the significant HR changing scenarios.

In addition to experiments on VIPL-HR, we also evaluate the average HR estimation performance on OBF. State-of-the-art methods including ROI_{green} [8], CHROM [3], POS [22] and rPPGNet [25] are used for comparison, and the results of these methods are from [25]. As shown in Table 3, our method also achieves the best performance, indicating the effectiveness of the proposed approach.

Results on RF Measurement and HRV Analysis. We also conduct experiments for RF measurement and HRV analysis on the OBF database. Following [25], we use a 10-fold subject-exclusive protocol for all experiments. RF and three HRV features, i.e., low frequency (LF), high frequency (HF) and LF/HF are considered for evaluations. We compare the proposed method with the state-of-the-art methods including hand-crafted methods ROI_{green} [8], CHROM [3] and POS [22] and the learning based method rPPGNet [25]. The results of ROI_{green} [8], CHROM [3], POS [22] and rPPGNet [25] are taken from [25]. All the results are shown in Table 2.

From the results, we can see that the proposed approach outperforms all the existing state-of-the-art methods by a large margin on all evaluation metrics for RF and all HRV features. These results indicate that our method could not only handle the average HR estimation task but also could give a promising prediction of the rPPG signal for RF measurement and HRV analysis, which show the effectiveness of the proposed method and the potential in many healthcare applications. We further check the predicted rPPG signals of our estimator. Two examples are given in Fig. 3(c). From the results we can see that our method could give an accurate prediction of the interbeat intervals (IBIs), thus can give a robust estimation of RF and HRV features.

Table 2. The results of RF measurement and HRV analysis by the proposed approach and several state-of-the-art methods on the OBF database.

Method	RF(Hz)			LF(u.n)			HF(u.n)			LF/HF		
	Std	RMSE	r	Std	RMSE	r	Std	RMSE	r	Std	RMSE	r
ROI_{green} [8]	0.078	0.084	0.321	0.22	0.24	0.573	0.22	0.24	0.573	0.819	0.832	0.571
CHROM [3]	0.081	0.081	0.224	0.199	0.206	0.524	0.199	0.206	0.524	0.83	0.863	0.459
POS [22]	0.07	0.07	0.44	0.155	0.158	0.727	0.155	0.158	0.727	0.663	0.679	0.687
rPPGNet[25]	0.064	0.064	0.53	0.133	0.135	0.804	0.133	0.135	0.804	0.58	0.589	0.773
Proposed	**0.058**	**0.058**	**0.606**	**0.09**	**0.09**	**0.914**	**0.09**	**0.09**	**0.914**	**0.453**	**0.453**	**0.877**

4.3 Cross-Database Testing

Besides of the intra-database testings on the VIPL-HR and OBF databases, we also conduct cross-database testing on the small-scale HR estimation database MMSE-HR following the protocol of [13]. The model is trained on the VIPL-HR database and directly tested on the MMSE-HR. State-of-the-art method including hand-crafted methods Li2014 [9], CHROM [3], SAMC [19] and deeply learned method RhythmNet [13] are listed for comparison. The results of Li2014 [9], CHROM [3], SAMC [19] and RhythmNet [13] are from [13]. All the results of the proposed approach and the state-of-the-art methods are shown in Table 4.

From the results, we can see that our model gets the best results on all evaluation metrics compared with the state-of-the-art methods. These results indicate that our method could help the network to have a good generalization ability to the unknown scenarios without any prior knowledge, which demonstrates the effectiveness of the proposed approach.

Table 3. The average HR estimation results by the proposed approach and several state-of-the-art methods on the OBF database.

Method	Std↓ (bpm)	RMSE↓ (bpm)	r ↑
ROI_{green} [8]	2.159	2.162	0.99
CHROM [3]	2.73	2.733	0.98
POS [22]	1.899	1.906	0.991
rPPGNet[25]	1.758	1.8	0.992
Proposed	**1.257**	**1.26**	**0.996**

Table 4. The cross-database HR estimation results by the proposed approach and several state-of-the-art methods on the MMSE-HR database.

Method	Std↓ (bpm)	RMSE↓ (bpm)	r ↑
Li2014 [9]	20.02	19.95	0.38
CHROM [3]	14.08	13.97	0.55
SAMC [19]	12.24	11.37	0.71
RhythmNet [13]	6.98	7.33	0.78
Proposed	**6.06**	**6.04**	**0.84**

4.4 Ablation Study

We also provide the results of ablation studies for the proposed method for HR estimation on the VIPL-HR database. All the results are shown in Table 5.

Table 5. The HR estimation results of the ablation study on the VIPL-HR database.

Method	Std↓ (bpm)	MAE↓ (bpm)	RMSE↓ (bpm)	r ↑
MSTmap+HR	10.16	6.39	10.24	0.662
MSTmap+MTL	8.93	5.55	9.03	0.736
STmap+MTL	8.98	5.80	9.01	0.727
STmap+MTL+CVD	8.41	5.34	8.51	0.765
MSTmap+MTL+$N_{movement}$	8.81	5.58	8.96	0.743
MSTmap+MTL+$N_{stdface}$	8.60	5.46	8.72	0.756
MSTmap+MTL+CVD (proposed)	**7.92**	**5.02**	**7.97**	**0.796**

MTL: multi-task learning; **CVD**: cross-verified disentangling

Effectiveness of Multi-task Learning. In order to validate the effectiveness of the multi-task learning, we train our network using just the HR estimation branch (*MSTmap+HR*) as well as using both HR and rPPG estimation branches (*MSTmap+MTL*). The input of the network is our MSTmaps, and the network is trained without the cross-verified disentangling strategy. From the results, we can see that when we use both the rPPG and HR branches for training, the HR estimation results is better than only using the HR estimation branch. The MAE is reduced from 6.39 bpm to 5.55 bpm and the RMSE is reduced from 10.24 bpm to 9.03 bpm. These results indicate that rPPG signals and HR describe different aspects of the subject's physiological status, and learning from both rPPG signals and average HR values can help the network to learn both the general and the detailed features of physiological information, and thus benefit the average HR estimation.

Effectiveness of Multi-scale Spatial Temporal Map. We then test the effectiveness of our MSTmap. We use another effective physiological representation proposed by [13] (STmap) for comparison. Experiments with and without the cross-verified disentangling strategy are conducted. When we train the network without the cross-verified disentangling strategy, using MSTmap as the physiological representation of face video (*MSTmap+MTL*) outperforms using STmap (*STmap+MTL*) as the representation with a comparable RMSE as well as a better MAE of 5.55 bpm and a better Pearson correlation coefficient of 0.736. When we apply the proposed cross-verified disentangling strategy during training, we can see that the model using our MSTmap (*MSTmap+MTL+CVD*) outperforms the model using the STmap (*STmap+MTL+CVD*) for all evaluations. These results indicate that our MSTmap is effective to represent the physiological information in face videos.

Effectiveness of Cross-Verified Disentangling. We further evaluate the effectiveness of our CVD strategy. We first compare the results with and without using the CVD strategy during training. From the results, we can see that no matter what representation of face video we use as input, our

CVD strategy could bring a large improvement to the final results. When using the MSTmap as input, training with cross-verified disentangling strategy ($MSTmap+MTL+CVD$) can reduce the estimation RMSE from 9.03 bpm to 7.97 bpm and the MAE from 5.55 bpm to 5.02 bpm. When we use STmap as the representation ($STmap+MTL+CVD$), our CVD strategy can again bring an improvement of the HR estimation accuracy. These results indicate that our cross-verified disentangling strategy could effectively improve the physiological measurement accuracy.

Besides the experiments with and without the cross-verified disentangling strategy, we also compare the proposed CVD strategy with training the network using pre-defined non-physiological signals for disentanglement. The disentanglement using pre-defined non-physiological signals is implemented by using a decoder with the same architecture as D, and decoded the non-physiological signals with f_n. In [21], two pre-defined non-physiological signals, i.e., the head movements ($N_{movement}$) and the standard deviation of the facial pixel values ($N_{stdface}$), are used to improve the HR estimation accuracy. Following [21], we also use these two non-physiological signals for disentanglement. The results are denoted as $MSTmap+MTL+N_{movement}$ and $MSTmap+MTL+N_{stdface}$.

On one hand, we can see that both pre-defined non-physiological signals could help to reduce the HR estimation errors. When the head movement ($N_{movement}$) is used as the non-physiological signal, the disentanglement achieves a comparable MAE and slightly improves the Std, RMSE and r. When we use the standard deviation of the facial pixel values ($N_{stdface}$) as the non-physiological signals, it achieves better results on all evaluations because $N_{stdface}$ contains more non-physiological information of the face video. These results indicate that the MSTmaps are usually polluted by the non-physiological information, and disentangling strategy is necessary. On the other hand, pre-defined non-physiological signals are only a subset of the non-physiological information of the MSTmaps. Our cross-verified disentangling strategy can help the network to distill the physiological features from data, which are more representative than using the pre-defined non-physiological signals. The results of using the cross-verified disentangling strategy outperform all the disentangling methods using pre-defined physiological signals on all evaluation metrics. All these experiments indicate that our CVD strategy is effective to distill the physiological information and thus benefit the physiological measurements.

5   Conclusions

In this paper, we propose an effective end-to-end multi-task network for multiple physiological measurements using cross-verified disentangling strategy to reduce the influences of non-physiological signals such as head movements, lighting conditions, etc. The input face videos are first compressed into a hand-crafted representation named multi-scale spatial-temporal map to better represent the physiological information in face videos. Then we take pairwise MSTmaps as input and train the network with a cross-verified disentangling strategy to get effective physiological features. The learned physiological features are used for both the average HR estimation and rPPG signals regression. The proposed method achieves state-of-the-art performance in multiple physiological measurement tasks and databases. In our future work, we would like to explore the semi-supervised learning technologies for remote physiological measurement.

Acknowledgment. This work is partially supported by National Key R&D Program of China (grant 2018AAA0102501), Natural Science Foundation of China (grant 61672496), the Academy of Finland for project MiGA (grant 316765), project 6+E (grant 323287), ICT 2023 project (grant 328115), and Infotech Oulu.

References

1. Carreira, J., Zisserman, A.: Quo vadis, action recognition? a new model and the kinetics dataset. In: Proceedings of the IEEE CVPR (2017)
2. Chen, W., McDuff, D.: DeepPhys: video-based physiological measurement using convolutional attention networks. In: Ferrari, V., Hebert, M., Sminchisescu, C., Weiss, Y. (eds.) ECCV 2018. Lecture Notes in Computer Science, vol. 11206, pp. 356–373. Springer, Cham (2018). https://doi.org/10.1007/978-3-030-01216-8_22
3. De Haan, G., Jeanne, V.: Robust pulse rate from chrominance-based rPPG. IEEE Trans. Biomed. Eng. **60**(10), 2878–2886 (2013)
4. Kingma, D.P., Ba, J.: Adam: a method for stochastic optimization. arXiv preprint arXiv:1412.6980 (2014)
5. Lam, A., Kuno, Y.: Robust heart rate measurement from video using select random patches. In: Proceedings of the IEEE ICCV (2015)
6. Lee, H.-Y., Tseng, H.-Y., Huang, J.-B., Singh, M., Yang, M.-H.: Diverse image-to-image translation via disentangled representations. In: Ferrari, V., Hebert, M., Sminchisescu, C., Weiss, Y. (eds.) ECCV 2018. LNCS, vol. 11205, pp. 36–52. Springer, Cham (2018). https://doi.org/10.1007/978-3-030-01246-5_3
7. Lewandowska, M., Ruminski, J., Kocejko, T., Nowak, J.: Measuring pulse rate with a webcam - a non-contact method for evaluating cardiac activity. In: Proceedings of the ComSIS (2011)
8. Li, X., et al.: The OBF database: a large face video database for remote physiological signal measurement and atrial fibrillation detection. In: Proceedings of the IEEE FG (2018)
9. Li, X., Chen, J., Zhao, G., Pietikainen, M.: Remote heart rate measurement from face videos under realistic situations. In: Proceedings of the IEEE CVPR (2014)
10. Liu, Y., Wei, F., Shao, J., Sheng, L., Yan, J., Wang, X.: Exploring disentangled feature representation beyond face identification. In: Proceedings of the IEEE CVPR (2018)
11. Lu, B., Chen, J.C., Chellappa, R.: Unsupervised domain-specific deblurring via disentangled representations. In: Proceedings of the IEEE CVPR (2019)
12. Niu, X., Han, H., Shan, S., Chen, X.: VIPL-HR: A multi-modal database for pulse estimation from less-constrained face video. In: Proceedings of the ACCV (2018)
13. Niu, X., Shan, S., Han, H., Chen, X.: RhythmNet: end-to-end heart rate estimation from face via spatial-temporal representation. IEEE Trans. Image Process. **29**, 2409–2423 (2020)
14. Niu, X., et al.: Robust remote heart rate estimation from face utilizing spatial-temporal attention. In: Proceedings of the IEEE FG (2019)

15. Poh, M.Z., McDuff, D.J., Picard, R.W.: Non-contact, automated cardiac pulse measurements using video imaging and blind source separation. Opt. Express **18**(10), 10762–10774 (2010)
16. Poh, M.Z., McDuff, D.J., Picard, R.W.: Advancements in noncontact, multiparameter physiological measurements using a webcam. IEEE Trans. Biomed. Eng. **58**(1), 7–11 (2011)
17. Spetlik, R., Franc, V., Cech, J., Matas, J.: Visual heart rate estimation with convolutional neural network. In: Proceedings of the BMVC (2018)
18. Tran, L., Yin, X., Liu, X.: Disentangled representation learning GAN for pose-invariant face recognition. In: Proceedings of the IEEE CVPR (2017)
19. Tulyakov, S., Alameda-Pineda, X., Ricci, E., Yin, L., Cohn, J.F., Sebe, N.: Self-adaptive matrix completion for heart rate estimation from face videos under realistic conditions. In: Proceedings of the IEEE CVPR (2016)
20. Verkruysse, W., Svaasand, L.O., Nelson, J.S.: Remote plethysmographic imaging using ambient light. Opt. Express **16**(26), 21434–21445 (2008)
21. Wang, W., den Brinker, A.C., de Haan, G.: Discriminative signatures for remote-PPG. IEEE Trans. Biomed. Eng. **67**(5), 1462–1473 (2020)
22. Wang, W., den Brinker, A.C., Stuijk, S., de Haan, G.: Algorithmic principles of remote PPG. IEEE Trans. Biomed. Eng. **64**(7), 1479–1491 (2017)
23. Wang, W., den Brinker, A.C., Stuijk, S., de Haan, G.: Amplitude-selective filtering for remote-PPG. Biomed. Opt. Express **8**(3), 1965–1980 (2017)
24. Wang, W., Stuijk, S., De Haan, G.: Exploiting spatial redundancy of image sensor for motion robust rPPG. IEEE Trans. Biomed. Eng. **62**(2), 415–425 (2015)
25. Yu, Z., Peng, W., Li, X., Hong, X., Zhao, G.: Remote heart rate measurement from highly compressed facial videos: an end-to-end deep learning solution with video enhancement. In: Proceedings of the IEEE ICCV (2019)
26. Zhang, Z., Tran, L., Yin, X., Atoum, Y., Liu, X.: Gait recognition via disentangled representation learning. In: Proceedings of the IEEE CVPR (2019)

Combining Implicit Function Learning and Parametric Models for 3D Human Reconstruction

Bharat Lal Bhatnagar[1(✉)], Cristian Sminchisescu[2], Christian Theobalt[1], and Gerard Pons-Moll[1]

[1] Max Planck Institute for Informatics, Saarland Informatics Campus, Saarbrucken, Germany
{bbhatnag,theobalt,gpons}@mpi-inf.mpg.de
[2] Google Research, New York, USA
sminchisescu@google.com

Abstract. Implicit functions represented as deep learning approximations are powerful for reconstructing 3D surfaces. However, they can only produce static surfaces that are not controllable, which provides limited ability to modify the resulting model by editing its pose or shape parameters. Nevertheless, such features are essential in building flexible models for both computer graphics and computer vision. In this work, we present methodology that combines detail-rich implicit functions and parametric representations in order to reconstruct 3D models of people that remain controllable and accurate even in the presence of clothing. Given sparse 3D point clouds sampled on the surface of a dressed person, we use an Implicit Part Network (IP-Net) to jointly predict the *outer* 3D surface of the dressed person, the *inner* body surface, and the semantic correspondences to a parametric body model. We subsequently use correspondences to fit the body model to our inner surface and then non-rigidly deform it (under a parametric body + displacement model) to the outer surface in order to capture garment, face and hair detail. In quantitative and qualitative experiments with both full body data and hand scans we show that the proposed methodology generalizes, and is effective even given incomplete point clouds collected from single-view depth images. Our models and code will be publicly released (http://virtualhumans.mpi-inf.mpg.de/ipnet).

Keywords: 3D human reconstruction · Implicit reconstruction · Parametric modelling

1 Introduction

The sensing technology for capturing unstructured 3D point clouds is becoming ubiquitous and more accurate, thus opening avenues for extracting detailed

Electronic supplementary material The online version of this chapter (https://doi.org/10.1007/978-3-030-58536-5_19) contains supplementary material, which is available to authorized users.

A. Vedaldi et al. (Eds.): ECCV 2020, LNCS 12347, pp. 311–329, 2020.
https://doi.org/10.1007/978-3-030-58536-5_19

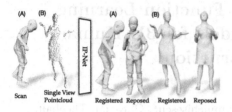

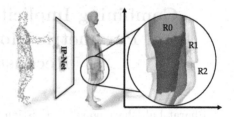

Fig. 1. We combine implicit functions and parametric modeling for detailed and controllable reconstructions from sparse point clouds. IP-Net predictions can be registered with SMPL+D model for control. IP-Net can also register (A) 3D scans and (B) single view point clouds.

Fig. 2. Unlike typical implicit reconstruction methods, IP-Net predicts a double layered surface, classifying the points as lying inside the body (R0), between the body and the clothing (R1) and outside the clothing (R2). IP-Net also predicts part correspondences to the SMPL model.

models from point cloud data. This is important in many 3D applications such as shape analysis and retrieval, 3D content generation, 3D human reconstruction from depth data, as well as mesh registration, which is the workhorse of building statistical shape models [20,27,54]. The problem is extremely challenging as the body can be occluded by clothing, hence identifying body parts given a point cloud is often ambiguous, and reasoning-with (or filling-in) missing data often requires non-local analysis. In this paper, we focus on the reconstruction of human models from sparse or incomplete point clouds, as captured by body scanners or depth cameras. In particular, we focus on extracting detailed 3D representations, including models of the underlying body shape and clothing, in order to make it possible to seamlessly re-pose and re-shape (*control*) the resulting dressed human models. To avoid ambiguity, we refer to static implicit reconstructions as *reconstruction* and our controllable model fit as *registration*. Note that the *registration* involves both reconstruction (explaining the given point cloud geometry) and registration, as it is obtained by deforming a predefined model (Fig. 1).

Learning-based methods are well suited to process sparse or incomplete point clouds, as they can leverage prior data to fill in the missing information in the input, but the choice of output representation limits either the resolution (when working with voxels or meshes), or the surface control (for implicit shape representations [13,14,29,33]). The main limitation of learning an implicit function is that the output is "just" a static surface with *no explicit model to control its pose and shape*. In contrast, parametric body models, such as SMPL [27] allow control, but the resulting meshes are overly-smooth and accurately regressing parameters directly from a point cloud is difficult (see Table 1). Furthermore, the surface of SMPL can not represent clothing, which makes registration difficult. Non-rigidly registering a less constrained parametric model to point clouds using non-linear optimization is possible, but only yields good results when provided with very good initialization close to the data (without local assignment ambiguity) and the point cloud is reasonably complete (see Table 1 and Fig. 4).

The main idea in this paper is to take advantage of the best of both representations (implicit and parametric), and learn to predict body under clothing (including body part labels) in order to make subsequent optimization-based registration feasible. Specifically, we introduce a novel architecture which jointly learns 2 implicit functions for (i) the joint occupancy of the outer (body+clothing) and the inner (body) surfaces and (ii) body part labels. Following recent work [14], we compute a 3-dimensional multi-scale tensor of deep features from the input point cloud, and make predictions at continuous query points. Unlike recent work that only predicts the occupancy of a single surface [13,14,29,33], we jointly learn a continuous implicit function for the inner/outer surface prediction and another classifier for body part label prediction. Our key insight is that since the inner surface (body) can be well approximated by a parametric body model (SMPL), and the predicted body parts constrain the space of possible correspondences, fitting SMPL to the predicted inner surface is very robust. Starting from SMPL fitted to the inner surface, we register it to the outer surface (under an additional displacement model, SMLP+D [6,25]), which in turn allows us to *re-pose and re-shape* the implicitly reconstructed outer surface.

Our experiments show that our implicit network can accurately predict body shape under clothing, the outer surface, and part labels, which makes subsequent parametric model fitting robust. Results on the Renderpeople dataset [1] demonstrate that our tandem of implicit function and parametric fitting yields detailed outer reconstructions, which are controllable, along with an estimation of body shape under clothing. We further achieve comparable performance on body shape under clothing on the BUFF dataset [60] without training on BUFF and without using temporal information. To show that our model can be useful in other domains, we train it on the MANO dataset [42] and show accurate registration using sparse and single view point clouds. Our key contributions can be summarized as follows:

- We propose a unified formulation which combines implicit functions and parametric modelling to obtain high quality controllable reconstructions from partial/ sparse/ dense point clouds of articulated dressed humans.
- Ours is the first approach to jointly reconstruct body shape under clothing along with full dressed reconstruction using a double surface implicit function, in addition to predicting part correspondences to a parametric model.
- Results on a dataset of articulated clothed humans and hands (MANO [42]) show the wide applicability of our approach.

2   Related Work

In this section, we discuss works which extract 3D humans from visual observations using parametric and implicit surface models. We further classify methods in top-down (optimization based) and bottom-up (learning based).

2.1   Parametric Modelling for Humans

Parametric body models factorize deformations into shape and pose [20, 22, 27, 54], soft-tissue [36], and recently even clothing [11, 34, 49], which constraints meshes to the space of humans. Most of current *model based approaches* optimize the pose and shape of SMPL [27] to match *image features*, which are extracted with bottom-up predictors [7, 8, 12, 53, 58]. Alternative methods based on GHUM [54] also exist [57]. The most popular image features are 2D joints [12], or 2D joints and silhouettes [7, 8, 19]. Some work have focused on estimating body shape under clothing [10, 55, 60], or capturing body shape and clothing jointly from scans [35]. These approaches are typically slow, and are susceptible to local-minima.

In contrast, *deep learning based* models predict body model parameters in a feed-forward pass [16, 39, 43] and use bottom-up 2D features for self-supervision [21, 23, 24, 32, 50, 59] during training. These approaches are limited by the shape space of SMPL, can not capture clothing nor surface detail, and lack a feedback loop, which results in miss-alignments between reconstructions and input pixels.

Hybrid methods mitigate these problems by refining feed-forward predictions with optimization at training [57] and/or test time [58], and by predicting displacements on top of SMPL, demonstrating capture of fine details and even *clothing* [6, 11]. However, the initial feed-forward predictions lack surface detail. Predicting normals and displacement maps on a UV-map or geometry image of the surface [9, 40] results in more detail, but predictions are not necessarily aligned with the observations.

Earlier work predicts dense correspondences on a depth map with a random forest and fit a 3D model to them [37, 38, 48]. To predict correspondences from point clouds using CNNs, depth maps can be generated where convolutions can be performed [52]. Our approach differs critically in that i) we do not require generating multiple depth maps, 2) we predict the body shape under clothing which makes subsequent fitting easier, and (ii) our approach can generate complete controllable and detailed surfaces from incomplete point clouds.

2.2   Implicit Functions for Humans

TSDFs [15] can represent the human surface implicitly, which is common in depth-fusion approaches [31, 45]. Such free-form representation has been combined with SMPL [27] to increase robustness and tracking [56]. Alternatively, implicit functions can be parameterized with Gaussian primitives [41, 47]. Since these approaches are not learning based, they can not reconstruct the occluded part of the surface in single view settings.

Voxels discretize the implicit occupancy function, which makes convolution operations possible. CNN based reconstructions using voxels [18, 51, 61] or depth-maps [17, 26, 46] typically produce more details than parametric models, but limbs are often missing. More importantly, unlike our method, the reconstruction quality is limited by the resolution of the voxel grid and increasing the resolution is hard as the memory footprint grows cubically.

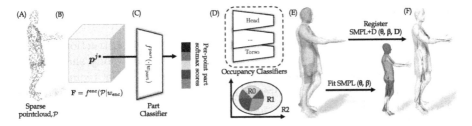

Fig. 3. The input to our method is (A) sparse point cloud $\mathcal{P}$. IP-Net encoder $f^{enc}(\cdot)$ generates an (B) implicit representation of $\mathcal{P}$. IP-Net predicts, for each query point $\boldsymbol{p}^j$, its (C) part label and double layered occupancy. IP-Net uses (D) occupancy classifiers to classify the points as lying inside the body (R0), between the body and the clothing (R1) and outside the body (R2), hence predicting (E) full 3D shape $\mathcal{S}_o$, body shape under clothing $\mathcal{S}_{in}$ and part labels. We register IP-Net predictions with (F) SMPL+D model to make implicit reconstruction controllable for the first time.

Recent methods learn a continuous implicit function representing the object surface directly [13,29,33]. However, these approaches have difficulties reconstructing articulated structures because they use a global shape code, and the networks tend to memorize typical object coordinates [14]. The occupancy can be predicted based on local image features instead [44], which results in medium-scale wrinkles and details, but the approach has difficulties with out of image plane poses, and is designed for image-reconstruction and can not handle point clouds. Recently, IF-Nets [14] have been proposed for 3D reconstruction and completion from point clouds – a mutliscale grid of deep features is first computed from the point cloud, and a decoder network classifies the occupancy based on mutli-scale deep features extracted at continuous point locations. These recent approaches [14,44] make occupancy decisions based on local and global evidence, which results in more robust reconstruction of articulated and fine structures than decoding based on the X-Y-Z point coordinates and a global latent shape code [13,29,30,33]. However, they do not reconstruct shape under clothing and surfaces are not controllable.

2.3   Summary: Implicit vs Parametric Modelling

Parametric models allow control over the surface and never miss body parts, but feed-forward prediction is hard, and reconstructions lack detail. Learning the implicit functions representing the surface directly is powerful because the output is continuous, details can be preserved better, and complex topologies can be represented. However, the output is not controllable, and can not guarantee that all body parts are reconstructed. Naive fitting of a body model to a reconstructed implicit surface often gets trapped into local minimal when the poses are difficult or clothing occludes the body (see Fig. 4). These observations motivate the design of our hybrid method, which retains the benefits of both representations: i) control, ii) detail, iii) alignment with the input point clouds.

3   Method

We introduce IP-Net, a network to generate detailed 3D reconstruction from an unordered sparse point cloud. IP-Net can additionally infer body shape under clothing and the body parts of the SMPL model. Training IP-Net requires supervision on three fronts, i) an outer dressed surface occupancy–directly derived from 3D scans, ii) an inner body surface–we supervise with an optimization based body shape under clothing registration approach and iii) correspondences to the SMPL model–obtained by registering SMPL to scans using custom optimization.

3.1   Training Data Preparation

To generate training data, we require non-rigidly registering SMPL [6,25] to 3D scans and estimating body shape under clothing, which is extremely challenging for the difficult poses in our dataset. Consequently, we first render the scans in multiple views, detect keypoints and joints, and integrate these as viewpoint landmark constraints to regularize registration similarly as in [6,25]. To non-rigidly deform SMPL to scans, we leverage SMPL+D [6,25], which is an extension to SMPL that adds per-vertex free-form displacements on top of SMPL to model deformations due to garments and hair. For the body shape under clothing, we build on top of [60] and propose a similar optimization based approach integrating viewpoint landmarks. Once SMPL+D has been registered to the scans, we transfer body part labels from the SMPL model to the scans. We provide more details in the supplementary. This process to generate training data is fairly robust, but required a lot of engineering to make it work. It also requires rendering multiple views of the scan, and does not work for sparse point clouds or scans without texture.

One of the key contributions of this work is to replace this tedious process with IP-Net, which quickly predicts a double layer implicit surface for body and outer surface, and body part labels to make subsequent registration using SMPL+D easy. We describe our network IP-Net, that infers detailed geometry and SMPL body parts from sparse point clouds next.

3.2   IP-Net: Overview

IP-Net $f(\cdot|w)$ takes in as input a sparse point cloud, $\mathcal{P}$ ($\sim$5k points), from articulated humans in diverse shapes, poses and clothing. IP-Net learns an implicit function to jointly infer outer surface, $\mathcal{S}_o$ (corresponding to full dressed 3D shape) and the inner surface $\mathcal{S}_{in}$ (corresponding to underlying body shape), of the person. Since we intend to register SMPL model to our implicit predictions, IP-Net additionally predicts, for each query point $p^j \in \mathbb{R}^3$, the SMPL body part label $I^j \in \{0, \ldots, N-1\}$ ($N = 14$) . We define I^j as a label denoting the associated body part on the SMPL mesh.

IP-Net: Feature Encoding. Recently, IF-Nets [14] achieve SOTA 3D mesh reconstruction from sparse point clouds. Their success can be attributed to two key insights: using a multi-scale, grid of deep features to represent shape, and predicting occupancy using features extracted at continuous point locations, instead of using the point coordinates. We build our IP-Net encoder $f^{\mathrm{enc}}(\cdot|w_{\mathrm{enc}})$ in the spirit of IF-Net encoder. We denote our multi-scale grid-aligned feature representation as $\mathbf{F} = f^{\mathrm{enc}}(\mathcal{P}|w_{\mathrm{enc}})$ and the features at point $\boldsymbol{p}^j = (x, y, z)$ as $\mathbf{F}^j = \mathbf{F}(x, y, z)$.

IP-Net: Part Classification. Next, we train a multi-class classifier $f^{\mathrm{part}}(\cdot|w_{\mathrm{part}})$ that predicts, for each point $\boldsymbol{p}^j$, its part label (correspondence to nearest SMPL part) conditioned on its feature encoding. More specifically, $f^{\mathrm{part}}(\cdot|w_{\mathrm{part}})$ predicts a per part score vector $\boldsymbol{D}^j \in [0, 1]^N$ at every point $\boldsymbol{p}^j$

$$\boldsymbol{D}^j = f^{\mathrm{part}}(\mathbf{F}^j|w_{\mathrm{part}}). \tag{1}$$

Then, we classify a point with the part label of maximum score

$$I^j = \operatorname*{arg\,max}_{I \in \{0,\ldots,N-1\}} (\boldsymbol{D}^j_I). \tag{2}$$

IP-Net: Occupancy prediction. Previous implicit formulations [14,29,33,44] train a deep neural network to classify points as being inside or outside a *single* surface. In addition, they minimize a classification/ regression loss over sampled points, which biases the network to perform better for parts with large surface area (more points) over smaller regions like hands (less points).

The key distinction between IP-Net and previous implicit approaches is that it classifies points as belonging to 3 different regions: 0-inside the body, 1-between body and clothing and 2-outside. This allows us to recover two surfaces (inner $\mathcal{S}_{in}$ and outer $\mathcal{S}_o$), see Fig. 2 and 3. Furthermore, we use an ensemble of occupancy classifiers $\{f^I(\cdot|w_I)\}_{I=0}^{N-1}$, where each $f^I(\cdot|w_I) : \mathbf{F}^j \mapsto \boldsymbol{o}^j \in [0, 1]^3$ is trained to classify a point $\boldsymbol{p}^j$ with features $\mathbf{F}^j$ into the three regions $o^j \in \{0, 1, 2\}$, $o_j = \arg\max_i \boldsymbol{o}^j_i$. The idea here is to train the ensemble such that $f^I(\cdot|w_I)$ performs best for part I, and predict the final occupancy o^j as a sum weighted by the part classification scores $\boldsymbol{D}^j_I \in \mathbb{R}$ at point $\boldsymbol{p}^j$

$$o^j = \arg\max_i \boldsymbol{o}^j_i, \qquad \boldsymbol{o}^j = \sum_{I=0}^{N-1} \boldsymbol{D}^j_I \cdot f^I(\mathbf{F}^j|w_I), \tag{3}$$

thereby reducing the bias towards larger body parts. After dividing the space in 3 regions the double-layer surface is extracted from the two decision boundaries.

IP-Net: Losses. IP-Net is trained using categorical cross entropy loss for both part-prediction (f^{part}) and occupancy prediction ($\{f^I\}_{I=0}^{N-1}$).

IP-Net: Surface Generation. We use marching cubes [28] on our predicted occupancies to generate a triangulated mesh surface.

We provide more implementation details in the supplementary.

3.3 Registering SMPL to IP-Net Predictions

Implicit based approaches can generate details at arbitrary resolutions but reconstructions are static and not controllable. This makes these approaches unsuitable for re-shaping and re-posing. We propose the first approach to combine implicit reconstruction with parametric modelling which lifts the details from the implicit reconstruction onto the SMPL+D model [6,25] to obtain an editable surface. We describe our registration using IP-Net predictions next. We use SMPL to denote the parametric model constrained to undressed shapes, and SMPL+D (SMPL plus displacements) to represent details like clothing and hair.

Fit SMPL to Implicit Body: We first optimize the SMPL shape, pose and translation parameters $(\boldsymbol{\theta}, \boldsymbol{\beta}, \boldsymbol{t})$ to fit our inner surface prediction $\mathcal{S}_{in}$.

$$E_{\text{data}}(\boldsymbol{\theta}, \boldsymbol{\beta}, \boldsymbol{t}) = \frac{1}{|\mathcal{S}_{in}|} \sum_{v_i \in \mathcal{S}_{in}} d(\boldsymbol{v}_i, \mathcal{M}) + w \cdot \frac{1}{|\mathcal{M}|} \sum_{v_j \in \mathcal{M}} d(\boldsymbol{v}_j, \mathcal{S}_{in}), \qquad (4)$$

where $\boldsymbol{v}_i$ and $\boldsymbol{v}_j$ denote vertices on $\mathcal{S}_{in}$ and SMPL surface $\mathcal{M}$ respectively. $d(\boldsymbol{p}, \mathcal{S})$ computes the distance of point $\boldsymbol{p}$ to surface $\mathcal{S}$. In our experiments we set $w = 0.1$

Additionally, we use the part labels predicted by IP-Net to ensure that correct parts on the SMPL mesh explain the corresponding regions on the inner surface $\mathcal{S}_{in}$. This term is critical to ensure correct registration (see Table 2 and Fig. 5)

$$E_{\text{part}}(\boldsymbol{\theta}, \boldsymbol{\beta}, \boldsymbol{t}) = \frac{1}{|\mathcal{S}_{in}|} \sum_{I=0}^{N-1} \sum_{v_i \in \mathcal{S}_{in}} d(\boldsymbol{v}_i, \mathcal{M}^I) \delta(I^i = I), \qquad (5)$$

where $\mathcal{M}^I$ denotes the surface of the SMPL mesh corresponding to part I and I^i denotes the predicted part label of vertex $\boldsymbol{v}_i$. The final objective can be written as follows

$$E(\boldsymbol{\theta}, \boldsymbol{\beta}, \boldsymbol{t}) = w_{\text{data}} E_{\text{data}} + w_{\text{part}} E_{\text{part}} + w_{\text{lap}} E_{\text{lap}}, \qquad (6)$$

where E_{lap} denotes a Laplacian regularizer. In our experiments we set the blancing weights $w_{\text{data/part/lap}}$ to 100, 10 and 1 respectively based on experimentation.

Register SMPL+D to Full Implicit Reconstruction: Once we obtain the SMPL body parameters $(\boldsymbol{\theta}, \boldsymbol{\beta}, \boldsymbol{t})$ from the above optimization, we jointly optimize the per-vertex displacements $\mathbf{D}$ to fit the outer implicit reconstruction $\mathcal{S}_o$.

$$E_{\text{data}}(\mathbf{D}, \boldsymbol{\theta}, \boldsymbol{\beta}, \boldsymbol{t}) = \frac{1}{|\mathcal{S}_o|} \sum_{v_i \in \mathcal{S}_o} d(\boldsymbol{v}_i, \mathcal{M}) + w \cdot \frac{1}{|\mathcal{M}|} \sum_{v_j \in \mathcal{M}} d(\boldsymbol{v}_j, \mathcal{S}_o) \qquad (7)$$

4   Dataset and Experiments

4.1 Dataset

We train IP-Net on a dataset of 700 scans [2,3] and test on held out 50 scans [1]. We normalize our scans to a bounding box of size 1.6 m. To train IP-Net we need paired data of sparse point clouds (input) and the corresponding outer surface, inner surface and correspondence to SMPL model (output). We generate the sparse point clouds by randomly sampling 5k points on our scans, which we voxelize into a grid of size $128 \times 128 \times 128$ for our input. We use the normalized scans directly as our ground truth dressed meshes and use our method for body shape registration under scan to get the corresponding body mesh $\mathcal{B}$ (see supplementary). For SMPL part correspondences, we manually define 14 parts (left/right forearm, left/right mid-arm, left/right upper-arm, left/right upper leg, left/right mid leg, left/right foot, torso and head) on SMPL mesh and use the fact that our body mesh $\mathcal{B}$, is a template with SMPL-topology registered to the scan; this automatically annotates $\mathcal{B}$ with the part labels. The part label of each query point in $\mathbb{R}^3$, is the label of the nearest vertex on the corresponding body mesh $\mathcal{B}$. Note that part annotations do not require manual effort.

We evaluate the implicit outer surface reconstructions against the GT scans. We use the optimization based approach described in Subsect. 3.1 to obtain ground truth registrations.

Table 1. IP-Net predictions, i.e. the outer/inner surface and correspondences to SMPL are key to high quality SMPL+D registration. We compare the quality (vertex-to-vertex error in cm) of registering to (a) point cloud, (b) implicit reconstruction by IF-Net[14], (c) regressing SMPL+D params and (d) IP-Net predictions. NP* means 'not possible'.

Register SMPL+D	Outer reg.	Inner reg.
(a) Sparse point cloud	14.85	NP*
(b) IF-Net [14]	13.88	NP*
(c) Regress SMPL+D params	32.45	NP*
(d) **IP-Net (Ours)**	**3.67**	**3.32**

Table 2. We compare three possibilities of registering the SMPL model to the implicit reconstruction produced by IP-Net. (a) registering SMPL+D to outer implicit reconstruction, (b) registering SMPL+D using the body prediction and (c) registering SMPL+D using body and part predictions. We report vertex-to-vertex error (cm) between the GT and predicted registered meshes.

	Outer	Inner
(a) Outer only	11.84	11.62
(b) Outer+inner	11.54	11.14
(c) **Outer+inner+parts**	**3.67**	**3.32**

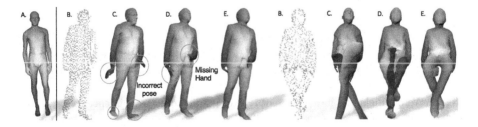

Fig. 4. We compare quality of SMPL+D registration for various alternatives to IP-Net. We show A) colour coded reference SMPL, B) the input point cloud, C) registration directly to sparse PC, D) registration to IFNet [14] prediction and E) registration to IP-Net predictions. It is important to note that poses such as sitting (second set) are difficult to register without explicit correspondences to the SMPL model.

4.2   Outer Surface Reconstruction

For the task of outer surface reconstruction, we demonstrate that IP-Net performs better or on par with state of the art implicit reconstruction methods, Occ.Net [29] and IF-Net [14]. We report the average bi-directional vertex-to-surface error of 9.86 mm, 4.86 mm and 4.95 mm for [14,29] and IP-Net respectively. We show qualitative results in the supplementary. Unlike [14,29] which predict only the outer surface, we infer body shape under clothing and body part labels with the same model.

4.3   Comparison to Baselines

The main contribution of our method is to make implicit reconstructions controllable. We do so by registering SMPL+D model [8,25] to IP-Net outputs: outer surface, inner surface and part correspondences. This raises the following

Table 3. Depth sensors can provide single depth view point clouds. We report registration accuracy (vertex-to-vertex distance in cm) on such data and show that registration using IP-Net predictions is significantly better than alternatives. NP* implies 'not possible'.

Register, single view point cloud	Outer reg.	Inner reg.
Sin. view PC	15.90	NP*
Sin. view PC + IP-Net correspondences (Ours)	14.43	NP*
IP-Net (Ours)	**5.11**	**4.67**

Table 4. An interesting use for IP-Net is to fit the SMPL+D model to sparse point clouds or scans using its part labels. This is useful for scan registration as we can retain the details of the high resolution scan and make it controlable. We report vertex-to-vertex error in cm. See Fig. 6 for qualitative results. NP* implies 'not possible'.

Register **with IP-Net** correspondences	Outer reg.	Inner reg.
Sparse point cloud	13.93	NP*
Scan	3.99	NP*
IP-Net (Ours)	**3.67**	**3.32**

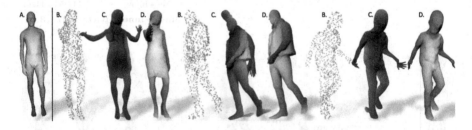

Fig. 5. We highlight the importance of IP-Net predicted correspondences for accurate registration. We show A) color coded SMPL vertices to appreciate registration quality and three sets of comparative results. In each set, we visualize B) the input point cloud, C) registration without using IP-Net correspondences and D) registration with IP-Net correspondences. It can be seen that without correspondences we find problems like 180° flips (dark colors indicate back surface), vertices from torso being used to explain arms etc. These results are quantitatively corroborated in Table 2. (Color figure online)

questions, "Why not a) register SMPL+D directly to the input sparse point cloud?, b) register SMPL+D to the surface generated by an existing reconstruction approach [14]? c) directly regress SMPL+D parameters from the point cloud? and d) How much better is it to register using IP-Net predictions?". Table 1 and Fig. 4 show that option d) (our method) is significantly better than the other baselines (a,b and c). To regress SMPL+D parameters (Option c), we implement a feed forward network that uses a similar encoder as IP-Net, but instead of predicting occupancy and part labels, produces SMPL+D parameters. We notice that the error for this method is dominated by misaligned pose and overall scale of the prediction. If we optimise the global orientation and scale of the predictions, this error is reduced from 32.45 cm to 7.25 cm which is still very high as compared to IP-Net based registration (3.67 cm) which requires no such adjustments. This experiment provides two key insights, i) it is significantly better to make local predictions using implicit functions and later register a parametric model, than to directly regress the parameters of the model and ii) directly registering a parametric model to an existing reconstruction method [14] yields larger errors than registering to IP-Net outputs (13.88 cm vs 3.67 cm).

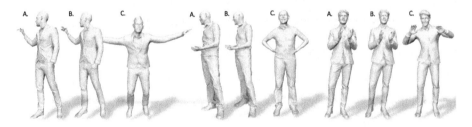

Fig. 6. IP-Net can be used for scan registration. As can be seen from Table 1, registering SMPL+D directly to scan is difficult. We propose to predict the inner body surface and part correspondences for every point on the scan using IP-Net and subsequently register SMPL+D to it. This allows us to retain outer geometric details from the scan while also being able to animate it. We show A) input scan, B) SMPL+D registration using IP-Net, C) scan in a novel pose. See video at [4].

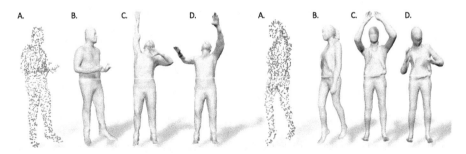

Fig. 7. Implicit predictions by IP-Net can be registered with SMPL+D model and hence reposed. We show, A) input point cloud, B) corresponding SMPL+D registration and C, D) two instances of new poses.

4.4   Body Shape Under Clothing

We quantitatively evaluate our body shape predictions on BUFF dataset [60]. Given a sparse point cloud generated from BUFF scans, IP-Net predicts the inner and outer surfaces along with the correspondences. We use our registration approach, as described in Sect. 3.3 to fit SMPL to our inner surface prediction and evaluate the error as per the protocol described in [60]. It is important to note that the comparison is unfair to our approach on several counts:

1. Our network uses sparse point clouds whereas [60] use 4D scans for their optimization based approach.
2. Our network was not trained on BUFF (noisier scans, missing soles in feet).
3. The numbers reported by [60] are obtained by jointly optimizing the body shape over entire 4D sequence, whereas our network makes a per-frame prediction without using temporal information.

We also compare our method to [55]. We report the following errors (mm): ([60] male: 2.65, female: 2.48), ([55] male: 17.85, female: 18.19) and (Ours male: 3.80, female: 6.17). Note that we did not have gender annotations for training IP-Net and hence generated our training data by registering all the scans to the 'male' SMPL model. This leads to significantly higher errors in estimating the

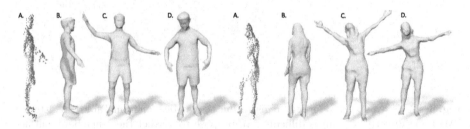

Fig. 8. Single depth view point clouds (A) are becoming increasingly accessible with devices like Kinect. We show our registration using IP-Net (B) and reposing results (C,D) with two novel poses using such data.

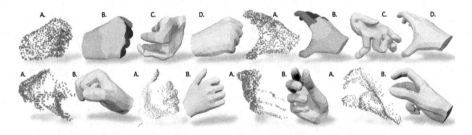

Fig. 9. We extend our idea of predicting implicit correspondences to parametric models to 3D hands. Here, we show results on MANO hand dataset [42]. In the first row we show A) input PC, B) surface and part labels predicted by IP-Net, C) registration without part correspondences, and D) our registration. Registration without part labels is ill-posed and often leads to wrong parts explaining the surface. In the second row we show A) input single-view PC and B) corresponding registrations using IP-Net.

body shape under clothing for 'female' subjects (we think this could be fixed by fitting gender specific models during training data generation). We add subject and sequence wise comparison in the supplementary. We show that our approach can accurately infer body shape under clothing using just a sparse point cloud and is on par with approaches which use much more information.

4.5   Why is Correspondence Prediction Important?

In this experiment, we demonstrate that inner surface reconstruction and part correspondences predicted by IP-Net are key for accurate registration. We discuss three obvious approaches for this registration:

(a) Register SMPL+D directly to the implicit outer surface predicted by IP-Net. This approach is simple and can be used with any other existing implicit reconstruction approaches.
(b) Register SMPL to the inner surface predicted by IP-Net and then non-rigidly register to the outer surface (without leveraging the correspondences).
(c) (Ours) First fit the SMPL model to the inner surface using correspondences and then non-rigidly register SMPL+D model to the implicit outer surface.

We report our results for the aforementioned approaches in Table 2 and Fig. 5. It can clearly be seen (Fig. 5, first set) that the arms of the SMPL model have not snapped to the correct pose. This is to be expected when arms are close to the body and no joint or correspondence information is present. In the second set, we see that vertices from torso are being used to explain the arms while SMPL arms are left hanging out. Third set is the classic case of 180° flipped fitting (dark color indicates back surface). This experiment highlights the importance of inner body surface and part correspondence prediction.

4.6   Why Not Independent Networks for Inner and Outer Surfaces?

IP-Net jointly predicts the inner and the outer surface for a human with clothing. Alternatively, one could train two separate implicit reconstruction networks using an existing SOTA approach. This has a clear disadvantage that one surface cannot reason about another, leading to severe inter-penetrations between the two. We report the average surface area of intersecting mesh faces which is $2000.71\,\mathrm{mm}^2$ for the two independent network approach, whereas with IP-Net the number is $0.65\,\mathrm{mm}^2$, which is four orders of magnitude smaller. We add qualitative results in the supplementary. Our experiment demonstrates that having a joint model for inner and outer surfaces is better.

4.7   Using IP-Net Correspondences to Register Scans

A very powerful use case for IP-Net is scan registration. Current state-of-the art registration approaches [6, 25] for registering SMPL+D to 3D scans are tedious and cumbersome (as described in Sect. 3.1). We provide a simple alternative using IP-Net. We sample points on our scan and generate the voxel grid used by IP-Net as input. We then run our pre-trained network and estimate the inner surface corresponding to the body shape under clothing. We additionally predict correspondences to the SMPL model for *each vertex on the scan*. We then use our registration (Sect. 3.3) to fit SMPL to the inner surface and then non-rigidly register SMPL+D to the scan surface, hence replacing the requirement for accurate 3D joints with IP-Net part correspondences. We show the scan registration and reposing results in Fig. 6 and Table 4. This is a useful experiment that shows that feed-forward IP-Net predictions can be used to replace tedious bottlenecks in scan registration (Fig. 7).

4.8   Registration from Point Clouds Obtained from a Single View

We show that IP-Net can be trained to process sparse point clouds from a single view (such as from Kinect). We show qualitative and quantitative results in Fig. 8 and Table 3, which demonstrate that IP-Net predictions are crucial for successful fitting in this difficult setting. This experiment highlights the general applicability of IP-Net to a variety of input modalities ranging from dense point clouds such as scans to sparse point clouds to single view point clouds.

4.9   Hand Registration

We show the wide applicability of IP-Net by using it for hand registration. We train IP-Net on the MANO hand dataset [42] and show hand registrations to full and single view point cloud in Fig. 9. We report an avg. vertex-to-vertex error of 4.80 mm and 4.87 mm in registration for full and single view point cloud respectively. This experiment shows that the idea of predicting implicit correspondences to a parametric model can be generically applied to different domains.

Limitations of IP-Net. During our experiments we found IP-Net does not perform well with poses that were very different than the training set. We also feel that the reconstructed details can be further improved especially around the face. We encourage the readers to see supplementary for further discussion.

5   Conclusions

Learning implicit functions to model humans has been shown to be powerful but the resulting representations are not amenable to control or reposing which are essential for both animation and inference in computer vision. We have presented methodology to combine expressive implicit function representations and parametric body modelling in order to produce 3D reconstructions of humans that remain controllable even in the presence of clothing.

Given a sparse point cloud representing a human body scan, we use implicit representation obtained using deep learning in order to jointly predict the *outer* 3D surface of the dressed person and the *inner* body surface as well as the semantic body parts of the parametric model. We use the part labels to fit the parametric model to our inner surface and then non-rigidly deform it (under a body prior + displacement model) to the outer surface in order to capture garment, face and hair details. Our experiments demonstrate that 1) predicting a double layer surface is useful for subsequent model fitting resulting in reconstruction improvements of 3 mm and 2) leveraging semantic body parts is *crucial* for subsequent fitting and results in improvements of 8.17 cm. The benefits of our method are paramount for difficult poses or when input is incomplete such as single view sparse point clouds, where the double layer implicit reconstruction and part classification is essential for successful registration. Our method generalizes well to other domains such as 3D hands (as evaluated on the MANO dataset) and even works well when presented with incomplete point clouds from a single depth view, as shown in extensive quantitative and qualitative experiments.

Acknowledgements. We thank Neng Qian, Jiayi Wang and Franziska Mueller for help with MANO experiments, Tribhuvanesh Orekondy for discussions and the reviewers for their feedback. Special thanks to RVH team members [5], their feedback significantly improved the overall writing and readability of this manuscript. We thank Twindom [2] for providing data for this project. This work is funded by the Deutsche Forschungsgemeinschaft (DFG, German Research Foundation) - 409792180 (Emmy Noether Programme, project: Real Virtual Humans) and Google Faculty Research Award.

References

1. https://renderpeople.com/
2. https://web.twindom.com/
3. https://www.treedys.com/
4. http://virtualhumans.mpi-inf.mpg.de/ipnet
5. http://virtualhumans.mpi-inf.mpg.de/people.html
6. Alldieck, T., Magnor, M., Bhatnagar, B.L., Theobalt, C., Pons-Moll, G.: Learning to reconstruct people in clothing from a single RGB camera. In: IEEE/CVF Conference on Computer Vision and Pattern Recognition (CVPR) (2019)
7. Alldieck, T., Magnor, M., Xu, W., Theobalt, C., Pons-Moll, G.: Detailed human avatars from monocular video. In: International Conference on 3D Vision (2018)
8. Alldieck, T., Magnor, M., Xu, W., Theobalt, C., Pons-Moll, G.: Video based reconstruction of 3D people models. In: IEEE Conference on Computer Vision and Pattern Recognition (2018)
9. Alldieck, T., Pons-Moll, G., Theobalt, C., Magnor, M.: Tex2Shape: detailed full human body geometry from a single image. In: IEEE International Conference on Computer Vision (ICCV). IEEE (2019)
10. Bălan, A.O., Black, M.J.: The naked truth: estimating body shape under clothing. In: Forsyth, D., Torr, P., Zisserman, A. (eds.) ECCV 2008. LNCS, vol. 5303, pp. 15–29. Springer, Heidelberg (2008). https://doi.org/10.1007/978-3-540-88688-4_2
11. Bhatnagar, B.L., Tiwari, G., Theobalt, C., Pons-Moll, G.: Multi-garment net: learning to dress 3D people from images. In: IEEE International Conference on Computer Vision (ICCV). IEEE (2019)
12. Bogo, F., Kanazawa, A., Lassner, C., Gehler, P., Romero, J., Black, M.J.: Keep it SMPL: automatic estimation of 3D human pose and shape from a single image. In: Leibe, B., Matas, J., Sebe, N., Welling, M. (eds.) ECCV 2016. LNCS, vol. 9909, pp. 561–578. Springer, Cham (2016). https://doi.org/10.1007/978-3-319-46454-1_34
13. Chen, Z., Zhang, H.: Learning implicit fields for generative shape modeling. In: IEEE Conference on Computer Vision and Pattern Recognition, CVPR 2019, Long Beach, CA, USA, 16–20 June 2019, pp. 5939–5948 (2019)

14. Chibane, J., Alldieck, T., Pons-Moll, G.: Implicit functions in feature space for 3D shape reconstruction and completion. In: IEEE Conference on Computer Vision and Pattern Recognition (CVPR). IEEE (2020)
15. Curless, B., Levoy, M.: A volumetric method for building complex models from range images. In: Proceedings of the 23rd Annual Conference on Computer Graphics and Interactive Techniques, SIGGRAPH 1996, New Orleans, LA, USA, 4–9 August 1996, pp. 303–312. Association for Computing Machinery, New York (1996)
16. Dibra, E., Jain, H., Oztireli, C., Ziegler, R., Gross, M.: Human shape from silhouettes using generative HKS descriptors and cross-modal neural networks. In: IEEE Conference on Computer Vision and Pattern Recognition (2017)
17. Gabeur, V., Franco, J., Martin, X., Schmid, C., Rogez, G.: Moulding humans: nonparametric 3D human shape estimation from single images. In: IEEE International Conference on Computer Vision, ICCV (2019)
18. Gilbert, A., Volino, M., Collomosse, J., Hilton, A.: Volumetric performance capture from minimal camera viewpoints. In: Ferrari, V., Hebert, M., Sminchisescu, C., Weiss, Y. (eds.) ECCV 2018. LNCS, vol. 11215, pp. 591–607. Springer, Cham (2018). https://doi.org/10.1007/978-3-030-01252-6_35
19. Habermann, M., Xu, W., Zollhöfer, M., Pons-Moll, G., Theobalt, C.: LiveCap: real-time human performance capture from monocular video. ACM Trans. Graph. 38(2), 141–1417 (2019). https://doi.org/10.1145/3311970
20. Joo, H., Simon, T., Sheikh, Y.: Total capture: a 3D deformation model for tracking faces, hands, and bodies. In: Proceedings of the IEEE Conference on Computer Vision and Pattern Recognition, pp. 8320–8329 (2018)
21. Kanazawa, A., Black, M.J., Jacobs, D.W., Malik, J.: End-to-end recovery of human shape and pose. In: IEEE Conference on Computer Vision and Pattern Recognition. IEEE Computer Society (2018)
22. Keyang, Z., Bhatnagar, B.L., Pons-Moll, G.: Unsupervised shape and pose disentanglement for 3D meshes. In: The European Conference on Computer Vision (ECCV) (2020)
23. Kolotouros, N., Pavlakos, G., Black, M.J., Daniilidis, K.: Learning to reconstruct 3D human pose and shape via model-fitting in the loop. In: IEEE Conference on Computer Vision and Pattern Recognition (2019)
24. Kolotouros, N., Pavlakos, G., Daniilidis, K.: Convolutional mesh regression for single-image human shape reconstruction. In: IEEE Conference on Computer Vision and Pattern Recognition, CVPR 2019, Long Beach, CA, USA, 16–20 June 2019, pp. 4501–4510 (2019)
25. Lazova, V., Insafutdinov, E., Pons-Moll, G.: 360-degree textures of people in clothing from a single image. In: International Conference on 3D Vision (3DV) (2019)
26. Leroy, V., Franco, J.-S., Boyer, E.: Shape reconstruction using volume sweeping and learned photoconsistency. In: Ferrari, V., Hebert, M., Sminchisescu, C., Weiss, Y. (eds.) ECCV 2018. LNCS, vol. 11213, pp. 796–811. Springer, Cham (2018). https://doi.org/10.1007/978-3-030-01240-3_48
27. Loper, M., Mahmood, N., Romero, J., Pons-Moll, G., Black, M.J.: SMPL: a skinned multi-person linear model. Assoc. Comput. Mach. 34, 248:1–248:16 (2015)
28. Lorensen, W.E., Cline, H.E.: Marching cubes: a high resolution 3D surface construction algorithm. In: SIGGRAPH, pp. 163–169. ACM (1987)
29. Mescheder, L.M., Oechsle, M., Niemeyer, M., Nowozin, S., Geiger, A.: Occupancy networks: learning 3D reconstruction in function space. In: IEEE Conference on Computer Vision and Pattern Recognition, CVPR 2019, Long Beach, CA, USA, 16–20 June 2019, pp. 4460–4470 (2019)

30. Michalkiewicz, M., Pontes, J.K., Jack, D., Baktashmotlagh, M., Eriksson, A.P.: Deep level sets: implicit surface representations for 3D shape inference. CoRR abs/1901.06802 (2019)
31. Newcombe, R.A., Fox, D., Seitz, S.M.: DynamicFusion: reconstruction and tracking of non-rigid scenes in real-time. In: IEEE Conference on Computer Vision and Pattern Recognition, CVPR 2015, Boston, MA, USA, 7–12 June 2015, pp. 343–352 (2015). https://doi.org/10.1109/CVPR.2015.7298631
32. Omran, M., Lassner, C., Pons-Moll, G., Gehler, P., Schiele, B.: Neural body fitting: unifying deep learning and model based human pose and shape estimation. In: International Conference on 3D Vision (2018)
33. Park, J.J., Florence, P., Straub, J., Newcombe, R.A., Lovegrove, S.: DeepSDF: learning continuous signed distance functions for shape representation. In: IEEE Conference on Computer Vision and Pattern Recognition, CVPR 2019, Long Beach, CA, USA, 16–20 June 2019, pp. 165–174 (2019)
34. Patel, C., Liao, Z., Pons-Moll, G.: The virtual tailor: predicting clothing in 3D as a function of human pose, shape and garment style. In: IEEE Conference on Computer Vision and Pattern Recognition (CVPR). IEEE (2020)
35. Pons-Moll, G., Pujades, S., Hu, S., Black, M.: ClothCap: seamless 4D clothing capture and retargeting. ACM Trans. Graph. 36(4), 1–15 (2017)
36. Pons-Moll, G., Romero, J., Mahmood, N., Black, M.J.: Dyna: a model of dynamic human shape in motion. ACM Trans. Graph. 34, 120 (2015)
37. Pons-Moll, G., Taylor, J., Shotton, J., Hertzmann, A., Fitzgibbon, A.: Metric regression forests for human pose estimation. In: British Machine Vision Conference (BMVC). BMVA Press (2013)
38. Pons-Moll, G., Taylor, J., Shotton, J., Hertzmann, A., Fitzgibbon, A.: Metric regression forests for correspondence estimation. Int. J. Comput. Vision 113(3), 163–175 (2015)
39. Popa, A.I., Zanfir, M., Sminchisescu, C.: Deep multitask architecture for integrated 2D and 3D human sensing. In: IEEE Conference on Computer Vision and Pattern Recognition (2017)
40. Pumarola, A., Sanchez, J., Choi, G.P.T., Sanfeliu, A., Moreno-Noguer, F.: 3DPeople: modeling the geometry of dressed humans. CoRR abs/1904.04571 (2019)
41. Rhodin, H., Robertini, N., Casas, D., Richardt, C., Seidel, H.-P., Theobalt, C.: General automatic human shape and motion capture using volumetric contour cues. In: Leibe, B., Matas, J., Sebe, N., Welling, M. (eds.) ECCV 2016. LNCS, vol. 9909, pp. 509–526. Springer, Cham (2016). https://doi.org/10.1007/978-3-319-46454-1_31
42. Romero, J., Tzionas, D., Black, M.J.: Embodied hands: modeling and capturing hands and bodies together. ACM Trans. Graph. (Proc. SIGGRAPH Asia) 36(6), 245:1–245:17 (2017)
43. Rong, Y., Liu, Z., Li, C., Cao, K., Loy, C.C.: Delving deep into hybrid annotations for 3D human recovery in the wild. In: The IEEE International Conference on Computer Vision (ICCV) (2019)
44. Saito, S., Huang, Z., Natsume, R., Morishima, S., Kanazawa, A., Li, H.: PIFu: pixel-aligned implicit function for high-resolution clothed human digitization. CoRR abs/1905.05172 (2019)
45. Slavcheva, M., Baust, M., Cremers, D., Ilic, S.: KillingFusion: non-rigid 3D reconstruction without correspondences. In: 2017 IEEE Conference on Computer Vision and Pattern Recognition, CVPR 2017, Honolulu, HI, USA, 21–26 July 2017, pp. 5474–5483 (2017). https://doi.org/10.1109/CVPR.2017.581
46. Smith, D., Loper, M., Hu, X., Mavroidis, P., Romero, J.: FACSIMILE: fast and accurate scans from an image in less than a second. IEEE International Conference on Computer Vision, ICCV (2019)

47. Stoll, C., Hasler, N., Gall, J., Seidel, H., Theobalt, C.: Fast articulated motion tracking using a sums of Gaussians body model. In: IEEE International Conference on Computer Vision, ICCV 2011, Barcelona, Spain, 6–13 November 2011, pp. 951–958 (2011). https://doi.org/10.1109/ICCV.2011.6126338

48. Taylor, J., Shotton, J., Sharp, T., Fitzgibbon, A.: The vitruvian manifold: inferring dense correspondences for one-shot human pose estimation. In: 2012 IEEE Conference on Computer Vision and Pattern Recognition, pp. 103–110. IEEE (2012)

49. Tiwari, G., Bhatnagar, B.L., Tung, T., Pons-Moll, G.: SIZER: a dataset and model for parsing 3D clothing and learning size sensitive 3D clothing. In: Vedaldi, A., Bischof, H., Brox, Th., Frahm, J.-M. (eds.) European Conference on Computer Vision (ECCV). Springer, Glasgow (2020)

50. Tung, H.Y., Tung, H.W., Yumer, E., Fragkiadaki, K.: Self-supervised learning of motion capture. In: Advances in Neural Information Processing Systems, pp. 5236–5246 (2017)

51. Varol, G., et al.: BodyNet: volumetric inference of 3D human body shapes. In: Ferrari, V., Hebert, M., Sminchisescu, C., Weiss, Y. (eds.) ECCV 2018. LNCS, vol. 11211, pp. 20–38. Springer, Cham (2018). https://doi.org/10.1007/978-3-030-01234-2_2

52. Wei, L., Huang, Q., Ceylan, D., Vouga, E., Li, H.: Dense human body correspondences using convolutional networks. In: Computer Vision and Pattern Recognition (CVPR) (2016)

53. Xiang, D., Joo, H., Sheikh, Y.: Monocular total capture: posing face, body, and hands in the wild. In: The IEEE Conference on Computer Vision and Pattern Recognition (CVPR) (2019)

54. Xu, H., Bazavan, E.G., Zanfir, A., Freeman, W.T., Sukthankar, R., Sminchisescu, C.: GHUM & GHUML: generative 3D human shape and articulated pose models. In: CVPR (2020)

55. Yang, J., Franco, J.-S., Hétroy-Wheeler, F., Wuhrer, S.: Estimation of human body shape in motion with wide clothing. In: Leibe, B., Matas, J., Sebe, N., Welling, M. (eds.) ECCV 2016. LNCS, vol. 9908, pp. 439–454. Springer, Cham (2016). https://doi.org/10.1007/978-3-319-46493-0_27

56. Yu, T., et al.: DoubleFusion: real-time capture of human performances with inner body shapes from a single depth sensor. In: 2018 IEEE Conference on Computer Vision and Pattern Recognition, CVPR 2018, Salt Lake City, UT, USA, 18–22 June 2018, pp. 7287–7296 (2018). https://doi.org/10.1109/CVPR.2018.00761

57. Zanfir, A., Bazavan, E.G., Xu, H., Freeman, B., Sukthankar, R., Sminchisescu, C.: Weakly supervised 3D human pose and shape reconstruction with normalizing flows. In: European Conference on Computer Vision (2020)

58. Zanfir, A., Marinoiu, E., Sminchisescu, C.: Monocular 3D pose and shape estimation of multiple people in natural scenes-the importance of multiple scene constraints. In: Proceedings of the IEEE Conference on Computer Vision and Pattern Recognition, pp. 2148–2157 (2018)

59. Zanfir, A., Marinoiu, E., Zanfir, M., Popa, A.I., Sminchisescu, C.: Deep network for the integrated 3D sensing of multiple people in natural images. In: NIPS (2018)

60. Zhang, C., Pujades, S., Black, M., Pons-Moll, G.: Detailed, accurate, human shape estimation from clothed 3D scan sequences. In: IEEE Conference on Computer Vision and Pattern Recognition (2017)
61. Zheng, Z., Yu, T., Wei, Y., Dai, Q., Liu, Y.: DeepHuman: 3D human reconstruction from a single image. In: The IEEE International Conference on Computer Vision (ICCV) (2019)

Orientation-Aware Vehicle Re-Identification with Semantics-Guided Part Attention Network

Tsai-Shien Chen[1,2], Chih-Ting Liu[1,2], Chih-Wei Wu[1,2],
and Shao-Yi Chien[1,2]

[1] Graduate Institute of Electronic Engineering,
National Taiwan University, Taipei, Taiwan
{tschen,jackieliu,cwwu}@media.ee.ntu.edu.tw, sychien@ntu.edu.tw
[2] NTU IoX Center, National Taiwan University, Taipei, Taiwan

Abstract. Vehicle re-identification (re-ID) focuses on matching images of the same vehicle across different cameras. It is fundamentally challenging because differences between vehicles are sometimes subtle. While several studies incorporate spatial-attention mechanisms to help vehicle re-ID, they often require expensive keypoint labels or suffer from noisy attention mask if not trained with expensive labels. In this work, we propose a dedicated Semantics-guided Part Attention Network (SPAN) to robustly predict part attention masks for different views of vehicles given only image-level semantic labels during training. With the help of part attention masks, we can extract discriminative features in each part separately. Then we introduce Co-occurrence Part-attentive Distance Metric (CPDM) which places greater emphasis on co-occurrence vehicle parts when evaluating the feature distance of two images. Extensive experiments validate the effectiveness of the proposed method and show that our framework outperforms the state-of-the-art approaches.

Keywords: Vehicle re-identification · Spatial attention ·
Semantics-guided learning · Visibility-aware features

1 Introduction

Vehicle re-identification (re-ID) aims to match vehicle images in a camera network. Recently, this task has drawn increasing attention due to practical applications such as urban surveillance and traffic flow analysis. While deep Convolutional Neural Networks (CNN) have shown remarkable performance in vehicle re-ID over the years [22,23,33], various challenges still hinder the performance of vehicle re-ID. One of them is that a vehicle captured from different viewpoints usually has dramatically different visual appearances. On the other hand,

Electronic supplementary material The online version of this chapter (https://doi.org/10.1007/978-3-030-58536-5_20) contains supplementary material, which is available to authorized users.

© Springer Nature Switzerland AG 2020
A. Vedaldi et al. (Eds.): ECCV 2020, LNCS 12347, pp. 330–346, 2020.
https://doi.org/10.1007/978-3-030-58536-5_20

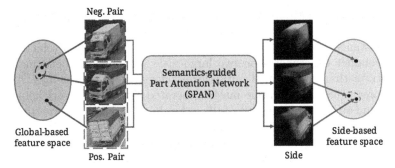

Fig. 1. Concept illustration of Semantics-guided Part Attention Network.
The example images show intra-class difference and inter-class similarity in the vehicle
re-ID problem. It is challenging to separate the negative images merely based on global
feature due to the similar car model and viewpoint. In this example, it is easier to
distinguish two vehicles by the side-based feature. This motivates us to generate the
part (view) attention maps and then emphasize the feature of the co-occurrence vehicle
parts for better re-ID matching.

two different vehicles of the same color and car model are likely to have very
similar appearances. As illustrated in the left part of Fig. 1, it is challenging to
distinguish vehicles by comparing the features extracted from the whole vehicle
images. In such case, the minor differences in specific parts of vehicle such as
decorations or license plates would be a great benefit to identifying two vehicles.
Furthermore, when two vehicles are presented in different orientations, a desired
vehicle re-ID algorithm should be able to focus on the parts (views) that both
appear in the two vehicle images. For example, in the right part of Fig. 1, it is
easier to distinguish the vehicles by comparing their side views. To reach this
idea, we divide it into two steps.

The first step is to extract the feature from specific parts of vehicle images. A
number of work has been proposed to achieve this purpose by learning orientation-
aware features. Nonetheless, existing methods either rely on expensive vehicle
keypoints as guidance to learn an attention mechanism for each part of a vehicle [11,34] or use only viewpoint labels but produce noisy and unsteady attention
outcome which will thus hinder the network to learn subtle differences between
vehicles [42]. In this paper, we introduce the *Semantics-guided Part Attention Network (SPAN)* to generate attention masks for different parts (front, side and rear
views) of a vehicle. As shown in Fig. 1, our SPAN learns to produce meaningful
attention masks. The masks not only help disentangle features of different viewpoints but also improve the interpretability of our learning framework. It is also
worth noting that, instead of expensive keypoints or pixel-level labels for training,
our SPAN requires only *image-level viewpoint labels* which are much easier to be
derived from known camera pose and traffic direction.

For the second step, we design a *Co-occurrence Part-attentive Distance Metric (CPDM)* to better utilize the part features when measuring the distance of
images. The intuition of this metric is that the network should focus on the

parts (views) that both appear in the compared vehicle images. Therefore, the proposed metric allows us to automatically adjust the importance of each part feature distance according to the part visibility in two compared vehicle images.

We conduct experiments on two large-scale vehicle re-ID benchmarks and demonstrate that our method outperforms current state-of-the-arts. Ablation studies prove that the attention masks generated by SPAN extract helpful part features and our CPDM can better utilize the global and part features to improve the re-ID performance. Moreover, qualitative results show that our SPAN can robustly generate meaningful attention maps on vehicles of different types, colors, and orientations. We now highlight our contributions: (1) We propose a Semantics-guided Part Attention Network (SPAN) to generate robust part attention masks which can be used to extract more discriminative features. (2) Our SPAN only needs image-level viewpoint labels instead of expensive keypoints or pixel-level annotations for training. (3) We introduce the Co-occurrence Part-attentive Distance Metric (CPDM) to facilitate vehicle re-ID by focusing on the parts that jointly appear in the compared images. (4) Extensive experiments on public datasets validate the effectiveness of each component and demonstrate that our method performs favorably against state-of-the-art approaches.

2   Related Work

Re-Identification (re-ID). Re-identification studies the problem of identifying identities in different camera views. There are large numbers of studies that focus on re-identifying human [3,15,38,39] and vehicles [20,29,34,42]. Most re-ID methods can be categorized into two types: feature learning and distance metric learning. Feature learning methods [3,15,30,37–39] aim to learn a more discriminative embedding space. Distance metric learning methods [2,4,12,35,41] design distance functions for comparing features of two images. In this work, we design an orientation-aware feature extraction network as well as an orientation-aware distance metric for solving the vehicle re-ID problem.

Vehicle Re-Identification. Vehicle re-ID has received more attention for the past few years due to the releases of large-scale annotated vehicle re-ID datasets. Liu *et al.* [22,23] released a high-quality multi-viewed VeRi-776 dataset. Tang *et al.* [33] proposed a city-scale traffic camera CityFlow dataset. With several datasets, numerous vehicle re-ID methods have been proposed recently. Some methods use CNN model to tackle the vehicle re-ID problem [20,29,33]. However, those methods lack spatial guidance and could be hard to distinguish two similar vehicles with only subtle difference. In contrast, the others adopt the extra information, such as viewpoint or keypoint labels, to generate spatial attentive features. Wang *et al.* [34] and Khorramshahi *et al.* [11] used 20 vehicle keypoints to generate attention maps by categorizing keypoints into four groups which respectively represent front, rear, left or right view of vehicle. Yet, the keypoint information is hard to acquire in real-world scenarios. Also, the keypoint is insufficient to cover all crucial features. Zhou *et al.* [42] proposed a viewpoint-aware attention model to produce attention map for different viewpoints and further generate multi-view features from single view input image. However, due

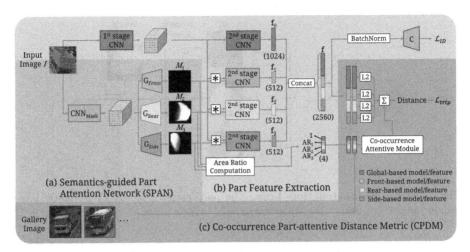

Fig. 2. Architecture of our proposed framework. (a) Semantics-guided Part Attention Network (SPAN) generates the attention masks for each part (view) of a vehicle image. (b) With the attention masks generated by SPAN, Part Feature Extraction produces one global and three part attentive features which are then concatenated into a representative feature. (c) Co-occurrence Part-attentive Distance Metric (CPDM) calculates a weighted feature distance with emphasis on the vehicle parts that appear in both compared images.

to the lack of direct supervision on the generated attention maps, the attention outcomes are noisy and would unfavorably affect the learning of network. In contrast, we design a dedicated network and adopt specific loss functions to supervise the generation of attention maps. Moreover, our network only requires image-level viewpoint labels rather than keypoint labels during training.

Visibility-Aware Features. Utilizing visibility-aware features has gained growing interest considering that there are lots of occluded images in real-world scenarios. Sun *et al.* [31] and Miao *et al.* [5] pre-define several regions among whole images by horizontally or vertically partitioning the images and then produce the confidence score for each region to represent their visibility. However, the visibility of pre-defined region is hard to represent its importance for re-ID matching. For example, the highly visible regions but containing mostly background would be overemphasized while the smaller regions but containing some critical appearances would be neglected. To avoid the issue mentioned above, in this work, we directly use the visibility of specific parts of vehicle to represent its importance. Note that it is only possible when the specific parts are accurately located.

3   Proposed Method

The proposed learning framework for vehicle re-ID consists of three sub-modules as depicted in Fig. 2. First, we learn a Semantics-guided Part Attention Network (SPAN) to predict the attention masks for each part (view) of a vehicle in

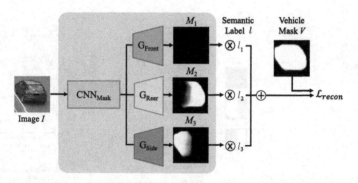

Fig. 3. Mask Reconstruction Loss. The part masks selected by semantic label should jointly reconstruct the whole foreground vehicle mask.

Sect. 3.1. Then, in Sect. 3.2, we apply the attention masks to our main feature extraction network to generate part features in addition to the global features. During both training and inference, the global and part features are combined to evaluate feature distance between two vehicle images with our proposed Co-occurrence Part-attentive Distance Metrics (CPDM) in Sect. 3.3. Last, the overall model learning scheme of our framework is introduced in Sect. 3.4.

3.1 Semantics-Guided Part Attention Network

The goal of our Semantics-guided Part Attention Network (SPAN) is to generate a set of attention masks for different parts (e.g. front, side, and rear view) of a vehicle image. An intuitive approach would be to train a segmentation network with pixel-wise view labels to predict segmented part masks. However, pixel-level annotation is expensive to obtain in real-world data. Instead, we turn to the image-level semantic labels, such as the viewpoint of vehicles which are much easier to be derived from known camera pose and the traffic direction, to learn our attention network. Given a vehicle image I, we define its corresponding semantic label vector as $l \in \mathcal{R}^3$. The semantic label l is encoded from its viewpoint. Its elements represent whether the front, rear or side view of image I are visible or not, respectively. To be more specific, $l_i = 1$ if the ith view is visible, while $l_i = 0$ if it is not. For example, for a vehicle image with the front-side viewpoint, its semantic label vector l will be assigned with $[1, 0, 1]$.

As shown in Fig. 2 (a), our network predicts the attention masks of front, rear and side views M_1, M_2, M_3 with a shared feature extractor CNN_{Mask} and three mask generators G_{Front}, G_{Rear}, and G_{Side}. To ensure our SPAN generating ideal masks, we meticulously design a novel loss function, named mask reconstruction loss, with two auxiliary losses to supervise the learning of network.

Mask Reconstruction Loss. As illustrated in Fig. 3, the main idea of mask reconstruction loss is that the attention masks selected by corresponding semantic labels should jointly reconstruct the foreground mask of a vehicle. For instance, if the image is with rear-side viewpoint, the rear and side masks should jointly reconstruct the whole vehicle foreground mask to the greatest extent possible. To this end, we first need the foreground mask of each vehicle image, which is also automatically generated by our deep segmentation network trained with the preliminary results by GrabCut [28] as the target. The detail of generating foreground masks is shown in the supplementary material; notes that any manually annotated pixel-level label is **not** required here. Thus, with the foreground masks (denoted as V), our mask reconstruction loss can be written as:

$$\mathcal{L}_{recon} = \|V - \sum_{i=1}^{3} (l_i \times M_i)\|_2, \tag{1}$$

which represents the mean square error (MSE) between the foreground mask and the generated mask gated by the semantic label.

Area Constraint Loss. While imposing the mask reconstruction loss, we note that the training is unstable and often leads to undesired results. Take the qualitative result in Fig. 6 "w/o $\mathcal{L}_{area}$" as example, we observe that, for a vehicle image with two visible views, the network only uses single representative mask generator to predict the whole vehicle mask. To prevent network from cheating, we design the area constraint loss to limit the maximum area of each predicted attention mask. Here, we define the area of mask as its L1-norm (sum of all elements) and also define the maximum area ratio of ith view for a semantic label l as $a_{l,i}$. Our area constraint loss can be formulated as:

$$\mathcal{L}_{area} = \sum_{i=1}^{3} \left[\frac{\|M_i\|_1}{\|V\|_1} - a_{l,i} \right]_+, \tag{2}$$

where $\| \cdot \|_1$ represents L1-norm of a given mask. $[\cdot]_+$ is the hinge function since we only penalize the mask with the area ratio (over the whole foreground mask) larger than our expected ratio. For the setting of max area ratio a, the ratio of invisible parts should be 0 intuitively while the ratio of visible part should be 1 for images with merely one visible views. For images with two visible views, the ratio of each view should be set within the range from 0.5 to 1.

Spatial Diversity Loss. In addition to the situation mentioned above, we observe other unfavorable results. Such as the qualitative result in Fig. 6 "w/o $\mathcal{L}_{div}$", for a vehicle image with two visible views, the two corresponding mask generators may predict whole vehicle masks with values of 0.5. Therefore, similar to Li *et al.* [16], we introduce a spatial diversity loss to restrict the overlapped area between masks of different views with the following formulation:

$$\mathcal{L}_{div} = \sum_{(i,j)\in P} [(M_i \cdot M_j) - m_{i,j}]_+, \tag{3}$$

where $m_{i,j}$ is the margin representing the tolerable overlapped area between i^{th} and j^{th} view and P is the set of all view index pairs. For two mutually exclusive

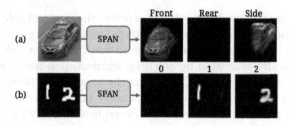

Fig. 4. Illustration of general-purposed SPAN. (a)(b) show the output masks of the vehicle and multi-digit images with different semantic labels respectively. Our proposed SPAN is able to learn to generate the part attention map or localization only given the image-level semantic labels.

views, such as front and rear, the margin is set to 0 intuitively. For two adjacent views, such as front and side, the margin parameter is set to a positive value to tolerate the overlapped situation (e.g. front-side view mirror and headlight could be hard to uniquely assign to either front or side view).

Discussion. SPAN is general-purposed and can be extended to weakly-supervised segmentation which has much weaker supervision setting than regular segmentation because it only requires image-level label for training. It can also well perform on other datasets besides to vehicle images. Figure 4 shows the example results on the multi-digit dataset based on MNIST [14] created by ourselves. For the multi-digit dataset, the semantic label represents which digit is visible in the image. Hence, the network can learn to generate the localization of each digit.

3.2   Part Feature Extraction

With the attention masks generated by our SPAN, we design a part feature extraction module to learn orientation-aware features for vehicle re-ID. As shown in Fig. 2 (b), the module includes two convolution stages. The 1st-stage CNN transforms input images into 1st-stage feature maps. Then, four distinct 2nd-stage CNNs respectively dedicated for extracting global-based, front-based, rear-based and side-based features follow the previous stage. The global-based model simply takes 1st-stage feature map as input and generates the global feature f_0. The other three branches apply the part attention masks to the 1st-stage feature map by element-wise matrix multiplication and then extract part features f_1, f_2 and f_3 by corresponding 2nd-stage CNNs. With one global feature and three part features, unlike previous methods [11,34,42] which embed all part features into one unified vector by additional network, our network simply concatenates them into one representative feature f to best utilize all possible features for vehicles.

3.3   Co-occurrence Part-Attentive Distance Metric

To fully utilize the part features extracted by our SPAN and part feature extraction module, we design the Co-occurrence Part-attentive Distance

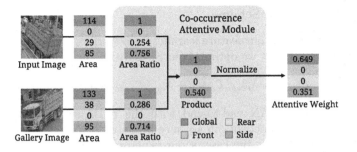

Fig. 5. Co-occurrence Attentive Module. To correctly recognize these two positive images, the feature of side view (co-occurrence view) should be emphasized, while front and rear feature should be relatively neglected. Co-occurrence attentive module is able to re-weigh the importance of each view accordingly.

Metric (CPDM) for both training the CNN and matching images during inference. We note that, in addition to the global feature, the features of the same visible parts on different vehicles are also critical for re-ID. Moreover, the co-occurrence part with greater area ratio often represents higher clarity or is likely to include more key features in the original image. Therefore, we develop the Co-occurrence Attentive Module to re-weigh the importance of different feature distances by comprehensively considering the area ratio of each view in both images. Fig. 5 illustrates an example of Co-occurrence Attentive Module. Given a vehicle image, we first compute the area of global, front, rear and side view by calculating the L1-norm of the attention masks generated by SPAN (the area of global view is defined as the summation of the ones of front, rear and side view). The area ratios of each view are then normalized by the global area. We denote the area ratio of ith view in image I_a as $AR_{a,i}$. For arbitrary two images I_a and I_b, the attentive weight of ith view $w_{(a,b),i}$ can be written as:

$$w_{(a,b),i} = \frac{AR_{a,i} \times AR_{b,i}}{\sum_{i=0}^{3} AR_{a,i} \times AR_{b,i}}. \tag{4}$$

Finally, we use the attentive weights to adjust the weighting for combining feature distances of all global and part features. The final distance $Dist_{(a,b)}$ between two vehicle images I_a and I_b is calculated by:

$$Dist_{(a,b)} = \sum_{i=0}^{3} w_{(a,b),i} \times \|\mathbf{f}_{a,i} - \mathbf{f}_{b,i}\|_2, \tag{5}$$

which is the weighted summation of feature euclidean distances in each view.

Discussion. For two images with completely disjoint views, the attentive weights are all 0 for front, rear, and side views. Hence, the distance between I_a and I_b will be fully determined by their global features $\mathbf{f}_{a,0}$ and $\mathbf{f}_{b,0}$.

3.4   Model Learning Scheme

The learning scheme for our feature learning framework consists of two steps. In the first step (Fig. 2 (a)), we optimize our SPAN with the following loss:

$$\mathcal{L}_{step1} = \lambda_{recon}\mathcal{L}_{recon} + \lambda_{area}\mathcal{L}_{area} + \lambda_{div}\mathcal{L}_{div}. \qquad (6)$$

Instead of training SPAN end-to-end with the re-ID feature extractor network [26], we train this network in advance because SPAN relies on clean viewpoint labels, which is not the case of our experimenting datasets. As a result, we train SPAN with a smaller dataset than the original one but with cleaner viewpoint labels.

In the second stage, we optimize the rest of our network (Fig. 2 (b)(c)) with two common re-ID losses while SPAN is fixed. The first one for metric learning is the triplet loss ($\mathcal{L}_{trip}$) [27], which is calculated based on the weighted distance introduced in Sect. 3.3. The other loss for the discriminative learning is the identity classification loss ($\mathcal{L}_{ID}$) [40]. The overall loss is computed as follows:

$$\mathcal{L}_{step2} = \lambda_{trip}\mathcal{L}_{trip} + \lambda_{ID}\mathcal{L}_{ID}. \qquad (7)$$

During inference, given a query and a gallery image, we extract their features separately by SPAN and the part feature extraction module. The distance of the query and gallery images are then computed by our CPDM for re-ID matching.

4   Experiments

4.1   Datasets and Evaluation Metrics

Our framework is evaluated on two benchmarks, VeRi-776 [22,23] and CityFlow-ReID [33], which are two large-scale vehicle re-ID datasets with multiple viewpoints. VeRi-776 dataset contains 776 different vehicles captured, which is split into 576 vehicles with 37,778 images for training and 200 vehicles with 11,579 images for testing. Wang et al. [34] released the annotated keypoints and viewpoint information for VeRi-776 dataset, which has been widely adopted by other work. In this paper, we only use the viewpoint labels to train our proposed SPAN. CityFlow-ReID is a subset of images sampled from the CityFlow dataset [33]. It consists of 36,935 images of 333 identities in the training set and 18,290 images of another 333 identities in the testing set. However, the viewpoint information of CityFlow-ReID is not available. Thus, we utilize the SPAN pre-trained on VeRi-776 to generate corresponding attention masks. Note that, though VehicleID [19] dataset is also a widely adopted benchmark, it only covers the images with front or rear viewpoint and cannot validate the effectiveness of our method. Hence, we would not use VehicleID in the following experiments.

As in previous vehicle re-ID works, we employ the standard metrics, namely the cumulative matching curve (CMC) and the mean average precision (mAP) [39] to evaluate the results. We report the rank-1 accuracy (R-1) in CMC and the mAP for the testing set in both datasets.

4.2   Implementation Details

For our SPAN (Fig. 2 (a)), we adopt the former four blocks in ResNet-34 [7] (*conv1* to *conv4*) as the feature extractor (CNN_{mask}) to extract the mid-level features which retain more spatial information than those after the last block (*conv5*). Afterwards, the feature map is fed into three generative blocks to generate the part masks. The detailed architecture of SPAN is shown in the supplementary material. This network is trained in advance on a subset of VeRi-776 dataset with balanced images in each viewpoint. For optimizing SPAN with $\mathcal{L}_{step1}$, the coefficients λ_{recon} and λ_{div} are set to 1 and λ_{area} is 0.5.

For our part feature extraction (Fig. 2 (b)), we adopt ResNet-50 [7] as our backbone which is split into two stages. The first four blocks (*conv1* to *conv4*) are in the first stage and the last block (*conv5*) with one fully-connected layer are in the second stage to generate a 1024-d or 512-d feature vector. For optimizing with triplet loss ($\mathcal{L}_{trip}$), we adopt the PK training strategy [8], where we sample $P = 8$ different vehicles and $K = 4$ images for each vehicle in a batch of size 32. In addition, for training identity classification loss ($\mathcal{L}_{ID}$), we adopt a BatchNorm [25] and a fully-connected layer as the classifier [18,25]. The training process lasts for 30,000 iterations with λ_{trip} and λ_{ID} all set to 1 in $\mathcal{L}_{step2}$.

4.3   Ablation Studies and Visualization

In this section, to assess the effectiveness of our Semantics-guided Part Attention Network (SPAN) and Co-occurrence Part-attentive Distance Metric (CPDM), we conduct ablation studies quantitatively on VeRi-776 dataset and visualize the qualitative results of our attention masks compared with the existing methods.

Loss Functions of Our SPAN. We adopt three loss functions to help generating steady and clear attention masks when training SPAN. To evaluate the influence of each loss function, we conduct experiments with multiple combinations of losses and report the re-ID results on VeRi-776 in Table 1 and the corresponding qualitative results of our part attention masks in Fig. 6.

As listed in the first row in Table 1, we show the baseline method which simply transferred the whole vehicle image into a 1024-dim global feature and adopted euclidean distance as the feature distance metric. Except for the baseline method, all other methods in Table 1 adopt CPDM and utilize same architecture in SPAN but trained with different combinations of proposed loss functions. As shown in the second to fourth rows in Table 1 and the corresponding visualized attention masks in Fig. 6, the re-ID performance of those methods are almost the same as the baseline owing to the unfavorable generated attention masks, which cannot benefit the part feature extraction and the following CPDM. Only when simultaneously supervised by proposed three loss functions, our SPAN can generate clear and meaningful attention masks which can further improve the re-ID performance by a large margin as shown in the last row in Table 1.

Table 1. Ablation study of the loss functions for training SPAN (%).

Method	Training Loss			VeRi-776	
	$\mathcal{L}_{recon}$	$\mathcal{L}_{area}$	$\mathcal{L}_{div}$	R-1	mAP
Baseline	-	-	-	92.0	59.1
only $\mathcal{L}_{recon}$	✓	✗	✗	91.8	58.9
w/o $\mathcal{L}_{area}$	✓	✗	✓	92.1	59.2
w/o $\mathcal{L}_{div}$	✓	✓	✗	92.5	59.7
SPAN(Ours)	✓	✓	✓	**93.9**	**68.6**

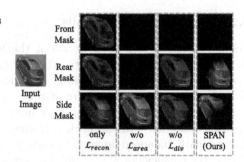

Fig. 6. Results of SPAN training w/ different combinations of losses.

Selection of Hyper-parameters in Loss Functions. There are two hyper-parameters which should be selected for loss functions, including max area ratio a in $\mathcal{L}_{area}$ and margin m in $\mathcal{L}_{div}$. The physical meanings of selection have been discussed in Sect. 3.1. We finally choose $a = 0.7$ for the visible views of two-view images and $m = 0.04$ for two adjacent views based on the experimental results shown in the supplementary material.

Qualitative Results of Part Attention Masks. To verify the robustness of our proposed SPAN, we show some qualitative results of our part attention masks in Fig. 7 (a) and show the comparisons with Wang *et al.* [34] and Zhou *et al.* [42] (VAMI) in Fig. 7 (b)(c), respectively. In Fig. 7 (a), the produced masks from our SPAN can correctly cover all regional features which are belonging to their views while eliminates all redundant information such as features from background or other views. For example, headlights and front bumper are all covered in front mask, while door or background are not. In Fig. 7 (a), the demonstrated vehicles are all in different colors, types (sedan, SUV, pickup, truck, bus, etc.) and orientations, proving that SPAN is robust for various vehicles.

In contrast, the attention masks generated by the previous work [34,42] are more noisy and unsteady. As shown in the left half of Fig. 7 (b), the front mask generated by Wang *et al.* [34] cannot cover the front windshield which possibly contains crucial features such as stickers or patterns. Also, in the right half of Fig. 7 (b), the front face of given vehicle image is not visible but the generated front attention mask fails to shield all features and instead activates on the background. The other example of generating unsteady masks is shown in Fig. 7 (c). Both rear and front masks generated by VAMI [42] fail to consistently embed the rear or front windshield among different vehicle images, which will make the network hard to distinguish two images based on those part features.

Component Analysis of the Proposed Model. Here, we report the re-ID performances to evaluate the effectiveness of each sub-module in our proposed framework in Table 2. The first row demonstrates the baseline model which simply transfers the whole vehicle image into a global feature and uses standard euclidean distance to evaluate the distance between two vehicles. Next, based on

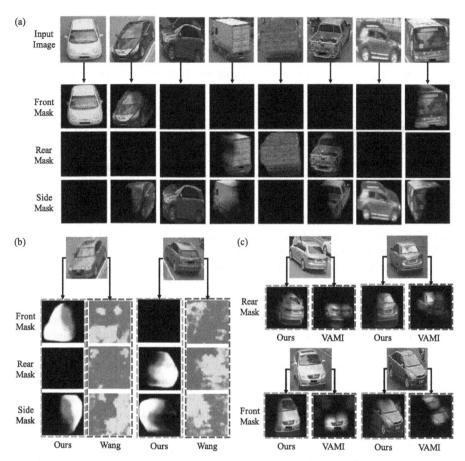

Fig. 7. Qualitative Part Masks. (a) shows some examples of the part masks generated by SPAN. Note that the demonstrated vehicles are all in different colors, types and orientations to verify the robustness of SPAN. (b)(c) show the comparison with Wang *et al.* [34] and Zhou *et al.* [42] (VAMI) respectively. The attention maps generated by their methods are directly from their papers.

our SPAN, we conduct two experiments with different aggregation techniques to combine the global and part features into one vector. The first one utilizes an additional fully-connected layer (FC) to embed the whole features, as shown in the second row in Table 2 (SPAN w/ FC). The other directly concatenates all global and part features, as shown in the third row in Table 2 (SPAN w/ Cat). It shows that compared to the baseline method, the performances are all boosted with the global and part features jointly be utilized. However, concatenating all the features can retain more part information, which achieves better performance for re-ID (from 59.1% to 63.1% than from 59.1% to 60.3% in mAP). Last, we report the results in the last row with the concatenated features and the usage of our CPDM, which is also our final proposed method (**SPAN w/**

Table 2. Ablation studies of the proposed method in terms of R-1 and mAP (%). The effectiveness analysis for each component including the usage of SPAN, feature aggregation methods (Agg.) and distance metrics (Dist.).

Method	Sub-modules			VeRi-776	
	SPAN	Agg.	Dist.	R-1	mAP
Baseline	✗	-	Euc.	92.0	59.1
SPAN w/ FC	✓	FC	Euc.	92.6	60.3
SPAN w/ Cat	✓	Concat.	Euc.	93.0	63.1
SPAN w/ CPDM (Ours)	✓	Concat.	CPDM	**94.0**	**68.9**

CPDM). It shows that with our proposed method, the re-ID performance can outperform the baseline method by a large margin (**9.5%** in mAP), proving that CPDM can better utilize the global and part features to measure the distance between two vehicles by enhancing the importance of the co-occurrence part.

4.4   Comparison with the State-of-the-Arts

We compare our proposed framework with the state-of-the-art vehicle re-ID methods and report the results on VeRi-776 and CityFlow-ReID datasets in Table 3. Note that there are a few of recent works which cannot be fairly compared with ours due to different setting such as the usage of external vehicle re-ID dataset [17], manually annotated bounding boxes for crucial features [6] and large-scale synthetic dataset with various kinds of pixel-level annotations [32]. Therefore those works are not shown in our comparison in Table 3.

Previous vehicle re-ID methods can be mainly summarized into three categories: spatial-attentive feature learning [9,11,20,21,34,42], distance metric learning [1] and embedding learning [24,43]. For spatial-attentive feature learning, proposed methods attempted to guide the network focusing on the regional features which may be useful to distinguish from two vehicles. RAM [20], GRF-GLL [21] and DFFMG [9] simply partitioned the images horizontally and vertically into several regions and extract the corresponding regional features; however, when the given images are in different orientations, the features would fail to consistently attends on same parts of vehicle . To extract orientation-aware features, OIFE [34] and AAVER [11] used extra expensive keypoints information to train their orientation-based region proposal network. Yet, they usually lose some informative information like the sticker on the windshield which is not covered by annotated keypoints. Instead, VAMI [42] used the viewpoint information to generate representative features of each viewpoint and used them to guide the network producing the viewpoint-aware attention maps and features, but the attention outcomes are not steady. To sum up, the unfavorable attention masks generated by existing work would hinder the re-ID performance on the benchmarks. In contrast, our method (**SPAN w/ CPDM**) achieves clear gains

Table 3. Comparison with state-of-the-arts re-ID methods on VeRi-776 and CityFlow-ReID dataset(%). Upper/Lower Group: methods **without/with** spatial-attentive mechanism. All listed scores are from the methods **without** adopting spatial-temporal information [23] or re-ranking [36].

Method	VeRi-776			CityFlow-ReID		
	R-1	R-5	mAP	R-1	R-5	mAP
EALN [24]	84.4	94.1	57.4	–	–	–
MoV1+BS [13]	90.2	96.4	67.6	49.0	63.1	31.3
MTML [10]	92.3	95.7	64.6	48.9	59.7	23.6
OIFE [34]	68.3	89.7	48.0	–	–	–
VAMI [42]	77.0	90.8	50.1	–	–	–
RAM [20]	88.6	94.0	61.5	–	–	–
AAVER [11]	89.0	94.7	61.2	–	–	–
GRF-GGL [21]	89.4	95.0	61.7	–	–	–
DFFMG [9]	–	–	–	48.0	60.0	25.3
SPAN w/ CPDM (Ours)	**94.0**	**97.6**	**68.9**	**59.5**	**61.9**	**42.0**

of **7.2%** and **16.7%** for mAP in VeRi-776 and CityFlow-ReID datasets compared to [21] and [9] respectively, indicating that we can benefit from more meaningful attention masks and better utility of global and part features. Also, our method outperforms other state-of-the-arts in both datasets.

5 Conclusion

In this paper, we present a novel vehicle re-ID feature learning framework including Semantics-guided Part Attention Network (SPAN) and Co-occurrence Part-attentive Distance Metric (CPDM). Our newly-designed SPAN can generate robust and meaningful attention masks on vehicle parts given only the image-level semantic labels for training. This is attributed to the direct supervision by our proposed mask reconstruction loss and two auxiliary losses. With the help of robust attention masks, the part feature extraction network is able to learn a more discriminative representation. Finally, our proposed CPDM can place emphasis on the vehicle parts that co-occurs in two images to better measure the distance between two vehicles. Both qualitative and quantitative results confirm the quality of generated attention masks and the benefit of dedicated part feature extraction and distance metric. Experiments also show that our proposed framework performs favorably against existing vehicle re-ID methods.

Acknowledgment. This research was supported in part by the Ministry of Science and Technology of Taiwan (MOST 108-2633-E-002-001), National Taiwan University (NTU-108L104039), Intel Corporation, Delta Electronics and Compal Electronics.

References

1. Bai, Y., Lou, Y., Gao, F., Wang, S., Wu, Y., Duan, L.Y.: Group-sensitive triplet embedding for vehicle reidentification. IEEE Trans. Multimed. **20**(9), 2385–2399 (2018)
2. Bak, S., Carr, P.: One-shot metric learning for person re-identification. In: IEEE Conference on Computer Vision and Pattern Recognition (CVPR), pp. 2990–2999 (2017)
3. Chen, D., Yuan, Z., Chen, B., Zheng, N.: Similarity learning with spatial constraints for person re-identification. In: IEEE Conference on Computer Vision and Pattern Recognition (CVPR), pp. 1268–1277 (2016)
4. Chen, W., Chen, X., Zhang, J., Huang, K.: Beyond triplet loss: a deep quadruplet network for person re-identification. In: IEEE Conference on Computer Vision and Pattern Recognition (CVPR), pp. 403–412 (2017)
5. Ge, Y., et al.: FD-GAN: pose-guided feature distilling GAN for robust person re-identification. In: Advances in Neural Information Processing Systems, pp. 1222–1233 (2018)
6. He, B., Li, J., Zhao, Y., Tian, Y.: Part-regularized near-duplicate vehicle re-identification. In: Proceedings of the IEEE Conference on Computer Vision and Pattern Recognition, pp. 3997–4005 (2019)
7. He, K., Zhang, X., Ren, S., Sun, J.: Deep residual learning for image recognition. In: IEEE Conference on Computer Vision and Pattern Recognition (CVPR), pp. 770–778 (2016)
8. Hermans, A., Beyer, L., Leibe, B.: In defense of the triplet loss for person re-identification. arXiv 1703.07737 (2017)
9. Huang, P., et al.: Deep feature fusion with multiple granularity for vehicle re-identification. In: IEEE Conference on Computer Vision and Pattern Recognition (CVPR) Workshop, pp. 80–88 (2019)
10. Kanaci, A., Li, M., Gong, S., Rajamanoharan, G.: Multi-task mutual learning for vehicle re-identification. In: IEEE Conference on Computer Vision and Pattern Recognition (CVPR) Workshop, pp. 62–70 (2019)
11. Khorramshahi, P., Kumar, A., Peri, N., Rambhatla, S.S., Chen, J.C., Chellappa, R.: A dual path model with adaptive attention for vehicle re-identification. arXiv 1905.03397 (2019)
12. Koestinger, M., Hirzer, M., Wohlhart, P., Roth, P.M., Bischof, H.: Large scale metric learning from equivalence constraints. In: IEEE Conference on Computer Vision and Pattern Recognition (CVPR), pp. 2288–2295 (2012)
13. Kuma, R., Weill, E., Aghdasi, F., Sriram, P.: Vehicle re-identification: an efficient baseline using triplet embedding. In: 2019 International Joint Conference on Neural Networks (IJCNN), pp. 1–9. IEEE (2019)
14. LeCun, Y., Bottou, L., Bengio, Y., Haffner, P., et al.: Gradient-based learning applied to document recognition. Proc. IEEE **86**(11), 2278–2324 (1998)
15. Li, D., Chen, X., Zhang, Z., Huang, K.: Learning deep context-aware features over body and latent parts for person re-identification. In: IEEE Conference on Computer Vision and Pattern Recognition (CVPR), pp. 384–393 (2017)
16. Li, S., Bak, S., Carr, P., Wang, X.: Diversity regularized spatiotemporal attention for video-based person re-identification. In: IEEE Conference on Computer Vision and Pattern Recognition (CVPR), pp. 369–378 (2018)
17. Liu, C.T., et al.: Supervised joint domain learning for vehicle re-identification. In: Proceedings of CVPR Workshops, pp. 45–52 (2019)

18. Liu, C.T., Wu, C.W., Wang, Y.C.F., Chien, S.Y.: Spatially and temporally efficient non-local attention network for video-based person re-identification (2019)
19. Liu, H., Tian, Y., Yang, Y., Pang, L., Huang, T.: Deep relative distance learning: tell the difference between similar vehicles. In: Proceedings of the IEEE Conference on Computer Vision and Pattern Recognition, pp. 2167–2175 (2016)
20. Liu, X., Zhang, S., Huang, Q., Gao, W.: Ram: a region-aware deep model for vehicle re-identification. In: IEEE International Conference on Multimedia and Expo (ICME), pp. 1–6 (2018)
21. Liu, X., Zhang, S., Wang, X., Hong, R., Tian, Q.: Group-group loss-based global-regional feature learning for vehicle re-identification. IEEE Trans. Image Process. **29**, 2638–2652 (2019)
22. Liu, X., Liu, W., Ma, H., Fu, H.: Large-scale vehicle re-identification in urban surveillance videos. In: IEEE International Conference on Multimedia and Expo (ICME), pp. 1–6 (2016)
23. Liu, X., Liu, W., Mei, T., Ma, H.: A deep learning-based approach to progressive vehicle re-identification for urban surveillance. In: Leibe, B., Matas, J., Sebe, N., Welling, M. (eds.) ECCV 2016. LNCS, vol. 9906, pp. 869–884. Springer, Cham (2016). https://doi.org/10.1007/978-3-319-46475-6_53
24. Lou, Y., Bai, Y., Liu, J., Wang, S., Duan, L.Y.: Embedding adversarial learning for vehicle re-identification. IEEE Trans. Image Process. **28**(8), 3794–3807 (2019)
25. Luo, H., Gu, Y., Liao, X., Lai, S., Jiang, W.: Bag of tricks and a strong baseline for deep person re-identification. In: IEEE Conference on Computer Vision and Pattern Recognition (CVPR) Workshop (2019)
26. Miao, Y., Gowayyed, M., Metze, F.: EESEN: end-to-end speech recognition using deep RNN models and wfst-based decoding. In: IEEE Workshop on Automatic Speech Recognition and Understanding (ASRU), pp. 167–174 (2015)
27. Ristani, E., Tomasi, C.: Features for multi-target multi-camera tracking and re-identification. In: IEEE Conference on Computer Vision and Pattern Recognition (CVPR), pp. 6036–6046 (2018)
28. Rother, C., Kolmogorov, V., Blake, A.: Grabcut: interactive foreground extraction using iterated graph cuts. ACM Trans. Graph. (TOG) **23**(3), 309–314 (2004)
29. Shen, Y., Xiao, T., Li, H., Yi, S., Wang, X.: Learning deep neural networks for vehicle RE-ID with visual-spatio-temporal path proposals. In: Proceedings of the IEEE International Conference on Computer Vision, pp. 1900–1909 (2017)
30. Shi, Z., Hospedales, T.M., Xiang, T.: Transferring a semantic representation for person re-identification and search. In: IEEE Conference on Computer Vision and Pattern Recognition (CVPR), pp. 4184–4193 (2015)
31. Sun, Y., et al.: Perceive where to focus: learning visibility-aware part-level features for partial person re-identification. In: Proceedings of the IEEE Conference on Computer Vision and Pattern Recognition, pp. 393–402 (2019)
32. Tang, Z., et al.: Pamtri: pose-aware multi-task learning for vehicle re-identification using highly randomized synthetic data. In: Proceedings of the IEEE International Conference on Computer Vision, pp. 211–220 (2019)
33. Tang, Z., et al.: Cityflow: a city-scale benchmark for multi-target multi-camera vehicle tracking and re-identification. In: IEEE Conference on Computer Vision and Pattern Recognition (CVPR), pp. 8797–8806 (2019)
34. Wang, Z., et al.: Orientation invariant feature embedding and spatial temporal regularization for vehicle re-identification. In: IEEE International Conference on Computer Vision (ICCV), pp. 379–387 (2017)

35. Yu, H.X., Wu, A., Zheng, W.S.: Cross-view asymmetric metric learning for unsupervised person re-identification. In: IEEE International Conference on Computer Vision (ICCV), pp. 994–1002 (2017)
36. Yu, R., Zhou, Z., Bai, S., Bai, X.: Divide and fuse: a re-ranking approach for person re-identification. arXiv preprint arXiv:1708.04169 (2017)
37. Zhao, H., et al.: Spindle net: person re-identification with human body region guided feature decomposition and fusion. In: IEEE Conference on Computer Vision and Pattern Recognition (CVPR), pp. 1077–1085 (2017)
38. Zhao, L., Li, X., Zhuang, Y., Wang, J.: Deeply-learned part-aligned representations for person re-identification. In: IEEE International Conference on Computer Vision (ICCV), pp. 3219–3228 (2017)
39. Zheng, L., Shen, L., Tian, L., Wang, S., Wang, J., Tian, Q.: Scalable person re-identification: a benchmark. In: IEEE International Conference on Computer Vision (ICCV), pp. 1116–1124 (2015)
40. Zheng, Z., Zheng, L., Yang, Y.: A discriminatively learned cnn embedding for person reidentification. ACM Trans. Multimed. Comput. Commun. Appl. (TOMM) **14**(1), 13 (2018)
41. Zhou, J., Yu, P., Tang, W., Wu, Y.: Efficient online local metric adaptation via negative samples for person re-identification. In: IEEE International Conference on Computer Vision (ICCV), pp. 2420–2428 (2017)
42. Zhou, Y., Shao, L.: Viewpoint aware attentive multi-view inference for vehicle re-identification. In: IEEE Conference on Computer Vision and Pattern Recognition (CVPR), pp. 6489–6498 (2018)
43. Zhu, J., et al.: Vehicle re-identification using quadruple directional deep learning features. IEEE Trans. Intel. Transport. Syst. **21**, 410–420 (2019)

Mining Cross-Image Semantics for Weakly Supervised Semantic Segmentation

Guolei Sun[1], Wenguan Wang[1(✉)], Jifeng Dai[2,3], and Luc Van Gool[1]

[1] ETH Zurich, Zürich, Switzerland
wenguanwang.ai@gmail.com
[2] Sensetime Research, Science Park, Hong Kong
[3] Qing Yuan Research Institute, Shanghai Jiao Tong University, Shanghai, China
https://github.com/GuoleiSun/MCIS_wsss

Abstract. This paper studies the problem of learning semantic segmentation from image-level supervision only. Current popular solutions leverage object localization maps from classifiers as supervision signals, and struggle to make the localization maps capture more complete object content. Rather than previous efforts that primarily focus on *intra-image* information, we address the value of *cross-image* semantic relations for comprehensive object pattern mining. To achieve this, two neural co-attentions are incorporated into the classifier to complimentarily capture cross-image semantic similarities and differences. In particular, given a pair of training images, one co-attention enforces the classifier to recognize the common semantics from co-attentive objects, while the other one, called contrastive co-attention, drives the classifier to identify the unshared semantics from the rest, uncommon objects. This helps the classifier discover more object patterns and better ground semantics in image regions. In addition to boosting object pattern learning, the co-attention can leverage context from other related images to improve localization map inference, hence eventually benefiting semantic segmentation learning. More essentially, our algorithm provides a unified framework that handles well different WSSS settings, *i.e.*, learning WSSS with (1) precise image-level supervision only, (2) extra simple single-label data, and (3) extra noisy web data. It sets new state-of-the-arts on all these settings, demonstrating well its efficacy and generalizability.

Keywords: Semantic segmentation · Weakly supervised learning

1 Introduction

Recently, modern deep learning based semantic segmentation models [6,7], trained with massive manually labeled data, achieve far better performance than

Electronic supplementary material The online version of this chapter (https://doi.org/10.1007/978-3-030-58536-5_21) contains supplementary material, which is available to authorized users.

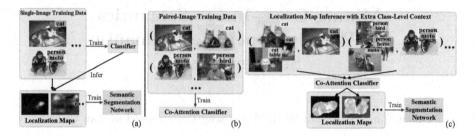

Fig. 1. (a) Current WSSS methods only use single-image information for object pattern discovering. (b–c) Our co-attention classifier leverages cross-image semantics as class-level context to benefit object pattern learning and localization map inference.

before. However, the fully supervised learning paradigm has the main limitation of requiring intensive manual labeling effort, which is particularly expensive for annotating pixel-wise ground-truth for semantic segmentation. Numerous efforts are motivated to develop semantic segmentation with weaker forms of supervision, such as bounding boxes [47], scribbles [38], points [3], image-level labels [48], *etc.* Among them, a prominent and appealing trend is using only image-level labels to achieve weakly supervised semantic segmentation (WSSS), which demands the least annotation efforts and is followed in this work.

To tackle the task of WSSS with only image-level labels, current popular methods are based on network visualization techniques [78,84], which discover discriminative regions that are activated for classification. These methods use image-level labels to train a classifier network, from which class-activation maps are derived as pseudo ground-truths for further supervising pixel-level semantics learning. However, it is commonly evidenced that the trained classifier tends to over-address the most discriminative parts rather than entire objects, which becomes the focus of this area. Diverse solutions are explored, typically adopting: *image-level* operations, such as region hiding and erasing [32,69], *regions growing* strategies that expand the initial activated regions [29,64], and *feature-level* enhancements that collect multi-scale context from deep features [35,71].

These efforts generally achieve promising results, which demonstrates the importance of discriminative object pattern mining for WSSS. However, as shown in Fig. 1(a), they typically use only single-image information for object pattern discovering, ignoring the rich semantic context among the weakly annotated data. For example, with the image-level labels, not only the semantics of each individual image can be identified, the cross-image semantic relations, *i.e.*, two images whether sharing certain semantics, are also given and should be used as cues for object pattern mining. Inspired by this, rather than relying on *intra-image* information only, we further address the value of *cross-image* semantic correlations for complete object pattern learning and effective class-activation map inference (see Fig. 1(b–c)). In particular, our classifier is equipped with a differentiable co-attention mechanism that addresses semantic homogeneity and difference understanding across training *image pairs*. More specifically, two kinds of co-attentions are learned in the classifier. The former one aims to capture cross-image common semantics, which enables the classifier to better ground the common semantic labels over the co-attentive regions. The latter one, called contrastive co-attention, focuses on the rest, unshared semantics, which helps the classifier better separate semantic patterns of different objects. These two co-attentions work in a cooperative and complimentary manner, together making the classifier understand object patterns more comprehensively.

In addition to benefiting object pattern learning, our co-attention provides an efficient tool for precise localization map inference (see Fig. 1(c)). Given a training image, a set of related images (*i.e.*, sharing certain common semantics) are utilized by the co-attention for capturing richer context and generating more accurate localization maps. Another advantage is that our co-attention based classifier learning paradigm brings an efficient data augmentation strategy, due to the use of training image pairs. Overall, our co-attention boosts object discovering during both the classifier's training phase as well as localization map inference stage. This provides the possibility of obtaining more accurate pseudo pixel-level annotations, which facilitate final semantic segmentation learning.

Our algorithm is a unified and elegant framework, which generalizes well different WSSS settings. Recently, to overcome the inherent limitation in WSSS without additional human supervision, some efforts resort to extra image-level supervision from simple single-class data readily available from other existing datasets [37,50], or cheap web-crawled data [20,54,55,70]. Although they improve the performance to some extent, complicated techniques, such as energy function optimization [20,59], heuristic constraints [13,55], and curriculum learning [70], are needed to handle the challenges of domain gap and data noise, restricting their utility. However, due to the use of paired image data for classifier training and object map inference, our method has good tolerance to noise. In addition, our method also handles domain gap naturally, as the co-attention effectively addresses domain-shared object pattern learning and achieves domain adaption as a part of co-attention parameter learning. We conduct extensive experiments on PASCAL VOC 2012 [11], under three WSSS settings, *i.e.*, learning WSSS with (**1**) PASCAL VOC image-level supervision only, (**2**) extra simple single-label data, and (**3**) extra web data. Our algorithm sets state-of-the-art on each case, verifying its effectiveness and generalizability.

2   Related Work

Weakly Supervised Semantic Segmentation. Recently, lots of WSSS methods have been proposed to alleviate labeling cost. Various weak supervision forms have been explored, such as bounding boxes [10,47], scribbles [38], point supervision [3], *etc.* Among them, image-level supervision, due to its less annotation demand, gains most attention and is also adopted in our approach.

Current popular solutions for WSSS with image-level supervision rely on network visualization techniques [78,84], especially the Class Activation Map (CAM) [84], which discovers image pixels that are informative for classification. However, CAM typically only identifies small discriminative parts of objects. Therefore, numerous efforts are made towards expanding the CAM-highlighted regions to the whole objects. In particular, some representative approaches make use of *image-level* hiding and erasing operations to drive a classifier to focus on different parts of objects [32,36,69]. A few ones instead resort to a *regions growing* strategy, *i.e.*, view the CAM-activated regions as initial "seeds" and gradually grow the seed regions until cover the complete objects [2,24,29,64]. Meanwhile, some researchers investigate to directly enhance the activated regions on *feature-level* [33,35,71]. When constructing CAMs, they collect multi-scale context, which is achieved by dilated convolution [71], multi-layer feature fusion [35], saliency-guided iterative training [64], or stochastic feature selection [33]. Some others accumulate CAMs from multiple training phases [25], or self-train a difference detection network to complete the CAMs with trustable information [56]. In addition, a recent trend is to utilize class-agnostic saliency cues to filter out background responses [12,24,33,36,64,69,71] during pseudo ground-truth generation.

Since the supervision provided in above problem setting is so weak, another category of approaches explores to leverage more image-level supervision from other sources. There are mainly two types: (1) exploring simple and single-label examples [37,50] (*e.g.*, images from existing datasets [17,53]); or (2) utilizing near-infinite yet noisy web-sourced image [20,54,55,70] or video [20,34,59] data (also referred as *webly supervised semantic segmentation* [26]). In addition to the common challenge of domain gap between the extra data and target semantic segmentation dataset, the second-type methods need to handle data noise.

Past efforts only consider each image individually, while only few exceptions [12,54] address cross-image information. [54] simply applies off-the-shelf co-segmentation [27] over the web images to generate foreground priors, instead of ours encoding the semantic relations into network learning and inference. For [12], although also exploiting correlations within image pairs, the core idea is to use extra information from a support image to supplement current visual representations. Thus the two images are expected to better contain the same semantics, and unmatched semantics would bring negative influences. In contrast, we view both semantic homogeneity and difference as informative cues, driving our classifier to more explicitly identify the common as well as unshared objects, respectively. Moreover, [12] only utilizes single image to infer the activated objects, but our method comprehensively leverages the cross-image semantics in both classifier training and localization map inference stages. More essentially, our framework is neat and flexible, which is not only able to learn WSSS from clean image-level supervision, but general enough to naturally make use of extra noisy web-crawled or simple single-label data, contrarily to previous efforts which are limited to specific training settings and largely dependent on complicated optimization methods [20,59] or heuristic constraints [55].

Deterministic Neural Attention. Differentiable attention mechanisms enable a neural network to focus more on relevant elements of the input than on irrelevant parts. With their popularity in the field of natural language processing [8,39,43,49,60], attention modeling is rapidly adopted in various computer vision tasks, such as image recognition [14,23,58,65,72], domain adaptation [66,82], human pose estimation [9,63,76], object detection [4] and image generation [75,80,85]. Further, co-attention mechanisms become an essential tool in many vision-language applications and sequential modeling tasks, such as visual question answering [41,44,74,77], visual dialog [73,83], vision-language navigation [67], and video segmentation [42,61], showing its effectiveness in capturing the underlying relations between different entities. Inspired by the general idea of attention mechanisms, this work leverages co-attention to mine semantic relations within training image pairs, which helps the classifier network learn complete object patterns and generate precise object localization maps.

3 Methodology

Problem Setup. Here we follow current popular WSSS pipelines: given a set of training images with image-level labels, a *classification network* is first trained to discover corresponding discriminative object regions. The resulting *object localization maps* over the training samples are refined as pseudo ground-truth masks to further supervise the learning of a *semantic segmentation network*.

Our Idea. Unlike most previous efforts that treat each training image *individually*, we explore cross-image semantic relations as class-level context for understanding object patterns more *comprehensively*. To achieve this, two neural co-attentions are designed. The first one drives the classifier to learn common semantics from the co-attentive object regions, while the other one enforces the classifier to focus on the rest objects for unshared semantics classification.

3.1 Co-attention Classification Network

Let us denote the training data as $\mathcal{I} = \{(I_n, l_n)\}_n$, where I_n is the nth training image, and $l_n \in \{0,1\}^K$ is the associated *ground-truth* image label for K semantic categories. As shown in Fig. 2(a), image pairs, *i.e.*, (I_m, I_n), are sampled from $\mathcal{I}$ for training the classifier. After feeding I_m and I_n into the convolutional embedding part of the classifier, corresponding feature maps, $F_m \in \mathbb{R}^{C \times H \times W}$ and $F_n \in \mathbb{R}^{C \times H \times W}$, are obtained, each with $H \times W$ spatial dimension and C channels.

As in [25,33,34], we can first separately pass F_m and F_n to a *class-aware fully convolutional layer* $\varphi(\cdot)$ to generate *class-aware activation maps*, *i.e.*, $S_m = \varphi(F_m) \in \mathbb{R}^{K \times H \times W}$ and $S_n = \varphi(F_n) \in \mathbb{R}^{K \times H \times W}$, respectively. Then, we apply *global average pooling* (GAP) over S_m and S_n to obtain class score vectors $s_m \in \mathbb{R}^K$ and $s_n \in \mathbb{R}^K$ for I_m and I_n, respectively. Finally, the *sigmoid cross entropy* (CE) loss is used for supervision:

$$
\begin{aligned}
\mathcal{L}_{\text{basic}}^{mn}\big((I_m, I_n), (l_m, l_n)\big) &= \mathcal{L}_{\text{CE}}(s_m, l_m) + \mathcal{L}_{\text{CE}}(s_n, l_n), \\
&= \mathcal{L}_{\text{CE}}\big(\text{GAP}(\varphi(F_m)), l_m\big) + \mathcal{L}_{\text{CE}}\big(\text{GAP}(\varphi(F_n)), l_n\big).
\end{aligned}
\tag{1}
$$

So far the classifier is learned in a standard manner, *i.e.*, only individual-image information is used for semantic learning. One can directly use the activation maps to supervise next-stage semantic segmentation learning, as done in [24, 34]. Differently, our classifier additionally utilizes a co-attention mechanism for further mining cross-image semantics and eventually better localizing objects.

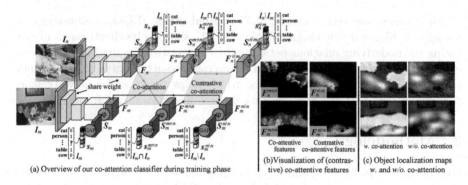

Fig. 2. (a) In addition to mining object semantics from single-image labels, semantic similarities and differences between paired training images are both leveraged for supervising object pattern learning. (b) Co-attentive and contrastive co-attentive features complimentarily capture the shared and unshared objects. (c) Our co-attention classifier is able to learn object patterns more comprehensively. *Zoom-in for details.*

Co-attention for Cross-Image Common Semantics Mining. Our co-attention attends to the two images, *i.e.*, I_m and I_n, simultaneously, and captures their correlations. We first compute the affinity matrix P between F_m and F_n:

$$P = F_m^\top W_P F_n \in \mathbb{R}^{HW \times HW}, \tag{2}$$

where $F_m \in \mathbb{R}^{C \times HW}$ and $F_n \in \mathbb{R}^{C \times HW}$ are flattened into matrix formats, and $W_P \in \mathbb{R}^{C \times C}$ is a learnable matrix. The affinity matrix P stores similarity scores corresponding to all pairs of positions in F_m and F_n, *i.e.*, the $(i, j)^{th}$ element of P gives the similarity between i^{th} location in F_m and j^{th} location in F_n.

Then P is normalized column-wise to derive attention maps across F_m for each position in F_n, and row-wise to derive attention maps across F_n for each position in F_m:

$$A_m = \mathrm{softmax}(P) \in [0,1]^{HW \times HW}, \quad A_n = \mathrm{softmax}(P^\top) \in [0,1]^{HW \times HW}, \tag{3}$$

where softmax is performed column-wise. In this way, A_n and A_m store the co-attention maps in their columns. Next, we can compute attention summaries of F_m (F_n) in light of each position of F_n (F_m):

$$F_m^{m \cap n} = F_n A_n \in \mathbb{R}^{C \times H \times W}, \quad F_n^{m \cap n} = F_m A_m \in \mathbb{R}^{C \times H \times W}, \tag{4}$$

where $F_m^{m \cap n}$ and $F_n^{m \cap n}$ are reshaped into $\mathbb{R}^{C \times W \times H}$. Co-attentive feature $F_m^{m \cap n}$, derived from F_n, preserves the common semantics between F_m and F_n and locate the common objects in F_m. Thus we can expect only the common semantics $l_m \cap l_n$[1] can be safely derived from $F_m^{m \cap n}$, and the same goes for $F_n^{m \cap n}$. Such co-attention based common semantic classification can let the classifier understand the object patterns more completely and precisely.

---

[1] The set operation '∩' is slightly extended here to represent bitwise-and.

To make things intuitive, consider the example in Fig. 2, where I_m contains **Table** and **Person**, and I_n has **Cow** and **Person**. As the co-attention is essentially the affinity computation between all the position pairs between I_m and I_n, only the semantics of the common objects, **Person**, will be preserved in the co-attentive features, *i.e.*, $F_m^{m \cap n}$ and $F_n^{m \cap n}$ (see Fig. 2(b)). If we feed $F_m^{m \cap n}$ and $F_n^{m \cap n}$ into the class-aware fully convolutional layer φ, the generated class-aware activation maps, *i.e.*, $S_m^{m \cap n} = \varphi(F_m^{m \cap n}) \in \mathbb{R}^{K \times H \times W}$ and $S_n^{m \cap n} = \varphi(F_n^{m \cap n}) \in \mathbb{R}^{K \times H \times W}$, are able to locate the common object **Person** in I_m and I_n, respectively. After GAP, the predicted semantic classes (scores) $s_m^{m \cap n} \in \mathbb{R}^K$ and $s_n^{m \cap n} \in \mathbb{R}^K$ should be the common semantic labels $l_m \cap l_n$ of I_m and I_n, *i.e.*, **Person**.

Through co-attention computation, not only the human face, the most discriminative part of **Person**, but also other parts, such as legs and arms, are highlighted in $F_m^{m \cap n}$ and $F_n^{m \cap n}$ (see Fig. 2(b)). When we set the common class labels, *i.e.*, **Person**, as the supervision signal, the classifier would realize that the semantics preserved in $F_m^{m \cap n}$ and $F_n^{m \cap n}$ are related and can be used to recognize **Person**. Therefore, the co-attention, computed across two related images, *explicitly* helps the classifier associate semantic labels and corresponding object regions and better understand the relations between different object parts. It essentially makes full use of the context across training data.

Intuitively, for the co-attention based common semantic classification, the labels $l_m \cap l_n$ shared between I_m and I_n are used to supervise learning:

$$
\begin{aligned}
\mathcal{L}_{\text{co-att}}^{mn}\big((I_m, I_n), (l_m, l_n)\big) &= \mathcal{L}_{\text{CE}}(s_m^{m \cap n}, l_m \cap l_n) + \mathcal{L}_{\text{CE}}(s_n^{m \cap n}, l_m \cap l_n), \\
&= \mathcal{L}_{\text{CE}}\big(\text{GAP}(\varphi(F_m^{m \cap n})), l_m \cap l_n\big) + \\
&\quad \mathcal{L}_{\text{CE}}\big(\text{GAP}(\varphi(F_n^{m \cap n})), l_m \cap l_n\big).
\end{aligned}
\tag{5}
$$

Contrastive Co-attention for Cross-Image Exclusive Semantics Mining. Aside from the co-attention described above that explores cross-image common semantics, we propose a contrastive co-attention that mines semantic differences between paired images. The co-attention and contrastive co-attention complementarily help the classifier better understand the concept of the objects.

As shown in Fig. 2(a), for I_m and I_n, we first derive *class-agnostic co-attentions* from their co-attentive features, *i.e.*, $F_m^{m \cap n}$ and $F_n^{m \cap n}$, respectively:

$$
B_m^{m \cap n} = \sigma(W_B F_m^{m \cap n}) \in [0, 1]^{H \times W}, \quad B_n^{m \cap n} = \sigma(W_B F_n^{m \cap n}) \in [0, 1]^{H \times W}, \tag{6}
$$

where $\sigma(\cdot)$ is the *sigmoid* activation function, and the parameter matrix $W_B \in \mathbb{R}^{1 \times C}$ learns for common semantics collection and is implemented by a convolutional layer with 1×1 kernel. $B_m^{m \cap n}$ and $B_n^{m \cap n}$ are class-agnostic and highlight all the common object regions in I_m and I_n, respectively, based on which we derive contrastive co-attentions:

$$
A_m^{m \setminus n} = 1 - B_m^{m \cap n} \in [0, 1]^{H \times W}, \quad A_n^{n \setminus m} = 1 - B_n^{m \cap n} \in [0, 1]^{H \times W}. \tag{7}
$$

The contrastive co-attention $A_m^{m \setminus n}$ of I_m, as its superscript suggests, addresses those *unshared* object regions that are only of I_m, but not of I_n, and the same goes for $A_n^{n\backslash m}$. Then we get *contrastive co-attentive features*, *i.e.*, unshared semantics in each images:

$$F_m^{m\backslash n} = F_m \otimes A_m^{m\backslash n} \in \mathbb{R}^{C \times H \times W}, \qquad F_n^{n\backslash m} = F_n \otimes A_n^{n\backslash m} \in \mathbb{R}^{C \times H \times W}. \qquad (8)$$

'$\otimes$' denotes element-wise multiplication, where the attention values are copied along the channel dimension. Next, we can sequentially get class-aware activation maps, *i.e.*, $S_m^{m\backslash n} = \varphi(F_m^{m\backslash n}) \in \mathbb{R}^{K \times H \times W}$ and $S_n^{n\backslash m} = \varphi(F_n^{n\backslash m}) \in \mathbb{R}^{K \times H \times W}$, and semantic scores, *i.e.*, $s_m^{m\backslash n} = \text{GAP}(S_m^{m\backslash n}) \in \mathbb{R}^K$ and $s_n^{n\backslash m} = \text{GAP}(S_n^{n\backslash m}) \in \mathbb{R}^K$. For $s_m^{m\backslash n}$ and $s_n^{n\backslash m}$, they are expected to identify the categories of the unshared objects, *i.e.*, $l_m \backslash l_n$ and $l_n \backslash l_m$[2].

Compared with the co-attention that investigates common semantics as informative cues for boosting object patterns mining, the contrastive co-attention addresses complementary knowledge from the semantic differences between paired images. Figure 2(b) gives an intuitive example. After computing the contrastive co-attentions between I_m and I_n (Eq. 7), **Table** and **Cow**, which are unique in their original images, are highlighted. Based on the contrastive co-attentive features, *i.e.*, $F_m^{m\backslash n}$ and $F_n^{n\backslash m}$, the classifier is required to accurately recognize **Table** and **Cow** classes, respectively. When the common objects are filtered out by the contrastive co-attentions, the classifier has a chance to focus more on the rest image regions and mine the unshared semantics more consciously. This also helps the classifier better discriminate the semantics of different objects, as the semantics of common objects and unshared ones are disentangled by the contrastive co-attention. For example, if some parts of **Cow** are wrongly recognized as **Person**-related, the contrastive co-attention will discard these parts in $F_n^{n\backslash m}$. However, the rest semantics in $F_n^{n\backslash m}$ may be not sufficient enough for recognizing **Cow**. This will enforce the classifier to better discriminate different objects.

For the contrastive co-attention based unshared semantic classification, the supervision loss is designed as:

$$\begin{aligned}
\mathcal{L}_{\overline{\text{co-att}}}^{mn}\big((I_m, I_n), (l_m, l_n)\big) &= \mathcal{L}_{\text{CE}}(s_m^{m\backslash n}, l_m \backslash l_n) + \mathcal{L}_{\text{CE}}(s_n^{n\backslash m}, l_n \backslash l_m), \\
&= \mathcal{L}_{\text{CE}}\big(\text{GAP}\big(\varphi(F_m^{m\backslash n})\big), l_m \backslash l_n\big) + \\
&\quad \mathcal{L}_{\text{CE}}\big(\text{GAP}\big(\varphi(F_n^{n\backslash m})\big), l_n \backslash l_m\big).
\end{aligned} \qquad (9)$$

More In-depth Discussion. One can interpret our co-attention classifier from a view of *auxiliary-task learning* [16,45], which is investigated in self-supervised learning field to improve data efficiency and robustness, by exploring auxiliary tasks from inherent data structures. In our case, rather than the task of single-image semantic recognition which has been extensively studied in conventional WSSS methods, we explore two auxiliary tasks, *i.e.*, predicting the common and uncommon semantics from image pairs, for fully mining supervision signals from weak supervision. The classifier is driven to better understand the cross-image semantics by attending to (contrastive) co-attentive features, instead of

---

[2] The set operation '$\backslash$' is slightly extend here, *i.e.*, $l_n \backslash l_m = l_n - l_n \cap l_m$.

only relying on intra-image information (see Fig. 2(c)). In addition, such strategy shares a spirit of *image co-segmentation* [42,62]. Since the image-level semantics of training set are given, the knowledge about some images share or unshare certain semantics should be used as a cue, or supervision signal, to better locate corresponding objects. Our co-attention based learning pipeline also provides an *efficient data augmentation* strategy, due to the use of paired samples, whose amount is near the square of the number of single training images.

3.2   Co-attention Classifier Guided WSSS Learning

Training Co-attention Classifier. The overall training loss for our co-attention classifier ensembles the three terms defined in Eqs. 1, 5, and 9:

$$\mathcal{L} = \sum_{m,n} \mathcal{L}_{\text{basic}}^{mn} + \mathcal{L}_{\text{co-att}}^{mn} + \mathcal{L}_{\overline{\text{co-att}}}^{mn}. \tag{10}$$

The coefficients of different loss terms are set as 1 in our all experiments. During training, to fully leverage the co-attention to mine the common semantics, we sample two images (I_m, I_n) with at least one common class, *i.e.*, $l_m \cap l_n \neq 0$.

Generating Object Localization Maps. Once our image classifier is trained, we apply it over the training data $\mathcal{I} = \{(I_n, l_n)\}_n$ to produce corresponding object localization maps, which are essential for semantic segmentation network training. We explore two different strategies to generate localization maps.

- *Single-round feed-forward prediction*, made over each training image individually. For each training image I_n, running the classifier and directly using its class-aware activation map (*i.e.*, $S_n \in \mathbb{R}^{K \times H \times W}$) as the object localization map L_n, as most previous network visualization based methods [25,34,55] done.
- *Multi-round co-attentive prediction with extra reference information*, which is achieved by considering extra information from other related training images (see Fig. 1(c)). Specifically, given a training image I_n and its associated label vector l_n, we generate its localization map L_n in a *class-wise* manner. For each semantic class $k \in \{1, \cdots, K\}$ labeled for I_n, *i.e.*, $l_{n,k} = 1$ and $l_{n,k}$ is the kth element of l_n, we sample a set of related images $\mathcal{R} = \{I_r\}_r$ from $\mathcal{I}$, which are also annotated with label k, *i.e.*, $l_{r,k} = 1$. Then we compute the co-attentive feature $F_n^{m \cap r}$ from each related image $I_r \in \mathcal{R}$ to I_n, and get the co-attention based class-aware activation map $S_n^{m \cap r}$. Given all the class-aware activation maps $\{S_n^{m \cap r}\}_r$ from $\mathcal{R}$, they are integrated to infer the localization map *only* for class k, *i.e.*, $L_{n,k} = \frac{1}{|\mathcal{R}|} \sum_{r \in \mathcal{R}} S_{n,k}^{m \cap r}$. Here $L_{n,k} \in \mathbb{R}^{H \times W}$ and $S_{n,k}^{(\cdot)} \in \mathbb{R}^{H \times W}$ indicate the feature map at k^{th} channel of $L_n \in \mathbb{R}^{K \times H \times W}$ and $S_n^{(\cdot)} \in \mathbb{R}^{K \times H \times W}$, respectively. '$|\cdot|$' numerates the elements. After inferring the localization maps for all the annotated semantic classes of I_n, we can get L_n.

These two localization map generation strategies are studied in our experiments (Sect. 4.4), and the last one is more favored, as it uses both intra- and inter-image semantics for object inference, and shares a similar data distribution of the training phase. One may notice that the contrastive co-attention is not used here. This is because contrastive co-attentive feature (Eq. 8) is from its original image, which is effective for boosting feature representation learning during classifier training, while contributes little for localization maps inference (with limited cross-image information). Related experiments can be found at Sect. 4.4.

Learning Semantic Segmentation Network. After obtaining high-quality localization maps, we generate pseudo pixel-wise labels for all the training samples $\mathcal{I}$, which can be used to train arbitrary semantic segmentation network. For pseudo groundtruth generation, we follow current popular pipeline [22,24,25,33,34,79], that uses localization maps to extract class-specific object cues and adopts saliency maps [21,40] to get background cues. For the semantic segmentation network, as in [22,25,33,34], we choose DeepLab-LargeFOV [6].

Learning with Extra Simple Single-Label Images. Some recent efforts [37,50] are made towards exploring extra simple single-label images from other existing datasets [17,53] for further boosting WSSS. Though impressive, specific network designs are desired, due to the issue of domain gap between additionally used data and the target complex multi-label dataset, *i.e.*, PASCAL VOC 2012 [11]. Interestingly, our co-attention based WSSS algorithm provides an alternate that addresses the challenge of domain gap naturally. Here we revisit the computation of co-attention in Eq. 2. When I_m and I_n are from different domains, the parameter matrix W_P, in essence, learns to map them into a unified *common semantic space* [46] and the co-attentive features can capture domain-shared semantics. Therefore, for such setting, we learn three different parameter matrixes for W_P, for the cases where I_m and I_n are from (1) the target semantic segmentation domain, (2) the one-label image domain, and (3) two different domains, respectively. Thus the domain adaption is efficiently achieved as a part of co-attention learning. We conduct related experiments in Sect. 4.2.

Learning with Extra Web Images. Another trend of methods [20,26,55,70] address webly supervised semantic segmentation, *i.e.*, leveraging web images as extra training samples. Though cheaper, web data are typically noisy. To handle this, previous arts propose diverse effective yet sophisticated solutions, such as multi-stage training [26] and self-paced learning [70]. Our co-attention based WSSS algorithm can be easily extended to this setting and solve data noise elegantly. As our co-attention classifier is trained with paired images, instead of previous methods only relying on each image individually, our model provides a more robust training paradigm. In addition, during localization map inference, a set of extra related images are considered, which provides more comprehensive and accurate cues, and further improves the robustness. We experimentally demonstrate the effectiveness of our method in such a setting in Sect. 4.3.

3.3   Detailed Network Architecture

Network Configuration. In line with conventions [25,71,81], our image classifier is based on ImageNet [31] pre-trained VGG-16 [57]. For VGG-16 network, the last three fully-connected layers are replaced with three convolutional layers with 512 channels and kernel size 3×3, as done in [25,81]. For the semantic segmentation network, for fair comparison with current top-leading methods [2,25,33,56], we adopt the ResNet-101 [19] version Deeplab-LargeFOV architecture.

Training Phases of the Co-attention Classifier and Semantic Segmentation Network. Our co-attention classifier is fully end-to-end trained by minimizing the loss defined in Eq. 10. The training parameters are set as: initial learning rate (0.001) which is reduced by 0.1 after every 5 epochs, batch size (5), weight decay (0.0002), and momentum (0.9). Once the classifier is trained, we generate localization maps and pseudo segmentation masks over all the training samples (see Sect. 3.2). Then, with the masks, the semantic segmentation network is trained in a standard way [25] using the hyper-parameter setting in [6].

Inference Phase of the Semantic Segmentation Network. Given an *unseen* test image, our segmentation network works in the *standard* semantic segmentation pipeline [6], *i.e.*, directly generating segments without using any other images. Then CRF [30] post-processing is performed to refine predicted masks.

4 Experiment

Overview. Experiments are first conducted over *three* different WSSS settings: **(1)** The most standard paradigm [24,25,56,69] that only allows image-level supervision from PASCAL VOC 2012 [11] (see Sect. 4.1). **(2)** Following [37,50], additional single-label images can be used, yet bringing the challenge of domain gap (see Sect. 4.2). **(3)** Webly supervised semantic segmentation paradigm [26,34,55], where extra web data can be accessed (see Sect. 4.3). Then, in Sect. 4.4, ablation studies are made to assess the effectiveness of essential parts of our algorithm.

Evaluation Metric. In our experiments, the standard intersection over union (IoU) criterion is reported on the val and test sets of PASCAL VOC 2012 [11]. The scores on test set are obtained from official PASCAL VOC evaluation server.

4.1 Experiment 1: Learn WSSS only from PASCAL VOC [11] Data

Experimental Setup: We first conduct experiment following the most standard setting that learns WSSS with only image-level labels [24,25,56,69], *i.e.*, only image-level supervision from PASCAL VOC 2012 [11] is accessible. PASCAL VOC 2012 contains a total of 20 object categories. As in [6,69], augmented training data from [18] are also used. Finally, our model is trained on totally 10,582 samples with only image-level annotations. Evaluations are conducted on the val and test sets, which have 1,449 and 1,456 images, respectively.

Experimental Results: Table 1a compares our approach and current top-leading WSSS methods with image-level supervision, on both PASCAL VOC12 val and test sets. We can observe that our method achieves mIoU scores of 66.2

Table 1. Experimental results for WSSS under three different settings. (a) Standard setting where only PASCAL VOC 2012 images are used (Sect. 4.1). (b) Additional single-label images are used (Sect. 4.2). (c) Additional web-crawled images are used (Sect. 4.3).

Methods	Publication	Val	Test
Using PASCAL VOC data only			
DCSM [68]	ECCV16	44.1	45.1
SEC [29]	ECCV16	50.7	51.7
AFF [51]	ECCV16	54.3	55.5
DCSP [5]	BMVC17	60.8	61.9
CBTS [52]	CVPR17	52.8	53.7
AE-PSL [69]	CVPR17	55.0	55.7
Oh et al. [20]	CVPR17	55.7	56.7
TPL [28]	ICCV17	53.1	53.8
MEFF [15]	CVPR18	-	55.6
GAIN [36]	CVPR18	55.3	56.8
MDC [71]	CVPR18	60.4	60.8
MCOF [64]	CVPR18	60.3	61.2
DSRG [24]	CVPR18	61.4	63.2
PSA [2]	CVPR18	61.7	63.7
SeeNet [22]	NIPS18	63.1	62.8
IRN [1]	CVPR19	63.5	64.8
FickleNet [33]	CVPR19	64.9	65.3
SSDD [56]	ICCV19	64.9	65.5
OAA+ [25]	ICCV19	65.2	66.4
Ours		**66.2**	**66.9**

(a)

Methods	Publication	Val	Test
Using extra simple single-label images			
MCNN [59]	ICCV15	-	36.9
MIL-ILP [50]	CVPR15	32.6	-
MIL-sppxl [50]	CVPR15	36.6	35.8
MIL-bb [50]	CVPR15	37.8	37.0
MIL-seg [50]	CVPR15	42.0	40.6
AttnBN [37]	ICCV19	62.1	63.0
Ours		**67.1**	**67.2**

(b)

Methods	Publication	Val	Test
Using extra noisy web images/videos			
MCNN [59]	ICCV15	38.1	39.8
Shen et al. [54]	BMVC17	56.4	56.9
STC [70]	PAMI17	49.8	51.2
Hong et al. [20]	CVPR17	58.1	58.7
WebS-i1 [26]	CVPR17	51.6	-
WebS-i2 [26]	CVPR17	53.4	55.3
Shen et al. [55]	CVPR18	63.0	63.9
Ours		**67.7**	**67.5**

(c)

and 66.9 on val and test sets respectively, outperforming all the competitors. The performance of our method is 87% of the DeepLab-LargeFOV [6] trained with fully annotated data, which achieved an mIoU of 76.3 on val set. When compared to OAA+ [25], current best-performing method, our approach obtains the improvement of 1.0% on val set. This verifies that the localization maps produced by our co-attention classifier effectively detect more complete semantic regions towards the whole target objects. Note that our network is elegantly trained end-to-end in a single phase. In contrast, many other recent approaches use extra networks [2,25,56] to learn auxiliary information (e.g., integral attention [25], pixel-wise semantic affinity [56], etc.), or adopt multi-step training [1,69,71].

4.2   Experiment 2: Learn WSSS With Extra Simple Single-Label Data

Experimental Setup: Following [37,50], we train our co-attention classifier and segmentation network with PASCAL images and extra single-label images. The extra single-label images are borrowed from the subsets of Caltech-256 [17]

Table 2. Ablation study for different object localization map generate strategies, reported on PASCAL VOC12 val set. See Sect. 4.4 for details.

Method	Inference Mode	Input Image(s)	Val
Basic Classifier	Single-round feed-forward	Test image *only*	61.7
Our Variant	Single-round feed-forward	Test image *only*	64.7
	Multi-round co-attention and contrastive co-attention	Test image and other related images	66.2
Full Model	Multi-round co-attention	Test image and other related images	**66.2**

and ImageNet CLS-LOC [53], and whose annotations are within 20 VOC object categories. There are a total of 20,057 extra single-label images.

Experimental Results: The comparisons are shown in Table 1b. Our method significantly improves the most recent method (*i.e.*, AttnBN [37]) in this setting by 5.0% and 4.2% in val and test sets, respectively. With the fact that objects of the same category but from different domains share similar visual patterns [37], our co-attention provides an end-to-end strategy that efficiently captures the common, cross-domain semantics, and learns domain adaption naturally. Even AttnBN is specifically designed for addressing such setting by knowledge transfer, our method still suppresses it by a large margin. Compared with the setting in Sect. 4.1 where only PASCAL images are used for training, our method obtains improvements on both val and test sets, verifying that it successfully mines knowledge from extra simple single-label data and copes with domain gap well.

4.3 Experiment 3: Learn WSSS with Extra Web-Sourced Data

Experimental Setup: We also conduct experiments using both PASCAL VOC images and webly craweled images as training data. We use the web data provided by [55], which are retrieved from Bing based on class names. The final dataset contains 76,683 images across 20 PASCAL VOC classes.

Experimental Results: Table 1c gives performance comparisons with previous webly supervised segmentation methods. As seen our method outperforms all other approaches and sets new state-of-the-arts with mIoU score of 67.7 and 67.5 on PASCAL VOC 2012 val and test sets, respectively. Among the compared methods, Hong et al.[20] utilize richer information of the temporal dynamics provided by additional large-scale videos. In contrast, although only using static data, our method still outperforms it on the val and test sets by 9.6% and 8.8%, respectively. Compared with Shen et al.[55] using the same web data as ours, our method substantially improves it by a clear margin of 3.6% on the test set.

4.4 Ablation Studies

Inference Strategies. Table 2 shows mIoU scores on PASCAL VOC 2012 val set *w.r.t.* different inference modes (see Sect. 3.2). When using the traditional

Table 3. Ablation study for our co-attention and contrastive co-attention mechanisms for training, reported on PASCAL VOC12 val set. See Sect. 4.4 for details.

Method	(Contrastive) Co-Attention	Training Loss	Val
Basic Classifier	-	$\mathcal{L}_{basic}$ (Eq. 1)	61.7
Our Variant	co-attention *only*	$\mathcal{L}_{basic}$ (Eq. 1)+$\mathcal{L}_{co\text{-}att}$ (Eq. 5)	65.5
Full Model	co-attention +contrastive co-attention	$\mathcal{L}_{basic}$ (Eq. 1)+$\mathcal{L}_{co\text{-}att}$ (Eq. 5)+$\mathcal{L}_{\overline{co\text{-}att}}$ (Eq. 9) $= \mathcal{L}$ (Eq. 10)	**66.2**

inference mode "single-round feed-forward", our method substantially suppresses basic classifier, by improving mIoU score from 61.7 to 64.7. This evidences that co-attention mechanism (trained in an end-to-end manner) in our classifier improves the underlying feature representations and more object regions are identified by the network. We can observe that by using more images to generate localization maps, our method obtains consistent improvement from "Test image *only*" (64.7), to "Test images and other related images" (66.2). This is because more semantic context are exploited during localization map inference. In addition, using contrastive co-attention for localization map inference doesn't boost performance (66.2). This is because the contrastive co-attentive features for one image are derived from the image itself. In contrast, co-attentive features are from the other related image, thus can be effective in the inference stage.

(Contrastive) Co-attention. As seen in Table 3, by only using co-attention (Eq. 5), we already largely suppress the basic classifier (Eq. 1) by 3.8%. When adding additional contrastive co-attention (Eq. 9), we obtain mIoU improvement of 0.7%. Above analysis verify our two co-attentions indeed boost performance.

Number of Related Images for Localization Map Inference. For localization map generation, we use 3 extra related images (Sect. 3.2). Here, we study how the number of reference images affect the performance. From Table 4, it is easily observed that when increasing the number of related images from 0 to 3, the performance

Table 4. Ablation study for using different numbers of related images during object localization map generation, reported on PASCAL VOC12 val set (see Sect. 4.4).

Method	Extra Related Images (#)	Val
Our Variant	0	64.7
	1	65.9
	2	66.0
	4	66.1
	5	66.0
Full Model	3	**66.2**

gets boosted consistently. However, when further using more images, the performance degrades. This can be attributed to the trade-off between useful semantic information and noise brought by related images. From 0 to 3 reference images, more semantic information is used and more integral regions for objects are mined. When further using more related images, useful information reaches its bottleneck and noise, caused by imperfect localization of the classifier, takes over, decreasing performance.

5 Conclusion

This work proposes a co-attention classification network to discover integral object regions by addressing cross-image semantics. With this regard, a co-attention is exploited to mine the common semantics within paired samples, while a contrastive co-attention is utilized to focus on the exclusive and unshared ones for capturing complimentary supervision cues. Additionally, by leveraging extra context from other related images, the co-attention boosts localization map inference. Further, by exploiting additional single-label images and web images, our approach is proven to generalize well under domain gap and data noise. Experiments over three WSSS settings consistently show promising results.

Acknowledgements. This work was partially supported by Zhejiang Lab's Open Fund (No. 2019KD0AB04), Zhejiang Lab's International Talent Fund for Young Professionals, CCF-Tencent Open Fund, and grants from Beijing Academy of Artificial Intelligence (BAAI) (No. BAAI2020ZJ0205).

References

1. Ahn, J., Cho, S., Kwak, S.: Weakly supervised learning of instance segmentation with inter-pixel relations. In: CVPR (2019)
2. Ahn, J., Kwak, S.: Learning pixel-level semantic affinity with image-level supervision for weakly supervised semantic segmentation. In: CVPR (2018)
3. Bearman, A., Russakovsky, O., Ferrari, V., Fei-Fei, L.: What's the point: semantic segmentation with point supervision. In: Leibe, B., Matas, J., Sebe, N., Welling, M. (eds.) ECCV 2016. LNCS, vol. 9911, pp. 549–565. Springer, Cham (2016). https://doi.org/10.1007/978-3-319-46478-7_34
4. Cao, J., Pang, Y., Li, X.: Triply supervised decoder networks for joint detection and segmentation. In: CVPR (2019)
5. Chaudhry, A., Dokania, P.K., Torr, P.H.: Discovering class-specific pixels for weakly-supervised semantic segmentation. In: BMVC (2017)
6. Chen, L.C., Papandreou, G., Kokkinos, I., Murphy, K., Yuille, A.L.: DeepLab: semantic image segmentation with deep convolutional nets, atrous convolution, and fully connected CRFs. TPAMI **40**(4), 834–848 (2017)
7. Chen, L.-C., Zhu, Y., Papandreou, G., Schroff, F., Adam, H.: Encoder-decoder with atrous separable convolution for semantic image segmentation. In: Ferrari, V., Hebert, M., Sminchisescu, C., Weiss, Y. (eds.) ECCV 2018. LNCS, vol. 11211, pp. 833–851. Springer, Cham (2018). https://doi.org/10.1007/978-3-030-01234-2_49
8. Cheng, J., Dong, L., Lapata, M.: Long short-term memory-networks for machine reading. In: EMNLP (2016)
9. Chu, X., Yang, W., Ouyang, W., Ma, C., Yuille, A.L., Wang, X.: Multi-context attention for human pose estimation. In: CVPR (2017)
10. Dai, J., He, K., Sun, J.: BoxSup: exploiting bounding boxes to supervise convolutional networks for semantic segmentation. In: ICCV (2015)
11. Everingham, M., Eslami, S.A., Van Gool, L., Williams, C.K., Winn, J., Zisserman, A.: The pascal visual object classes challenge: a retrospective. IJCV **111**(1), 98–136 (2015)

12. Fan, J., Zhang, Z., Tan, T.: CIAN: cross-image affinity net for weakly supervised semantic segmentation. In: AAAI (2020)
13. Fang, H., Lu, G., Fang, X., Xie, J., Tai, Y., Lu, C.: Weakly and semi supervised human body part parsing via pose-guided knowledge transfer. In: CVPR (2018)
14. Fu, J., et al.: Dual attention network for scene segmentation. In: CVPR (2019)
15. Ge, W., Yang, S., Yu, Y.: Multi-evidence filtering and fusion for multi-label classification, object detection and semantic segmentation based on weakly supervised learning. In: CVPR (2018)
16. Gidaris, S., Singh, P., Komodakis, N.: Unsupervised representation learning by predicting image rotations. In: ICLR (2018)
17. Griffin, G., Holub, A., Perona, P.: Caltech-256 object category dataset (2007)
18. Hariharan, B., Arbeláez, P., Bourdev, L., Maji, S., Malik, J.: Semantic contours from inverse detectors. In: ICCV (2011)
19. He, K., Zhang, X., Ren, S., Sun, J.: Deep residual learning for image recognition. In: CVPR (2016)
20. Hong, S., Yeo, D., Kwak, S., Lee, H., Han, B.: Weakly supervised semantic segmentation using web-crawled videos. In: CVPR (2017)
21. Hou, Q., Cheng, M.M., Hu, X., Borji, A., Tu, Z., Torr, P.: Deeply supervised salient object detection with short connections. TPAMI **41**(4), 815–828 (2019)
22. Hou, Q., Jiang, P., Wei, Y., Cheng, M.M.: Self-erasing network for integral object attention. In: NeurIPS (2018)
23. Hu, J., Shen, L., Sun, G.: Squeeze-and-excitation networks. In: CVPR (2018)
24. Huang, Z., Wang, X., Wang, J., Liu, W., Wang, J.: Weakly-supervised semantic segmentation network with deep seeded region growing. In: CVPR (2018)
25. Jiang, P.T., Hou, Q., Cao, Y., Cheng, M.M., Wei, Y., Xiong, H.K.: Integral object mining via online attention accumulation. In: ICCV (2019)
26. Jin, B., Ortiz Segovia, M.V., Susstrunk, S.: Webly supervised semantic segmentation. In: ICCV (2017)
27. Joulin, A., Bach, F., Ponce, J.: Discriminative clustering for image co-segmentation. In: CVPR (2010)
28. Kim, D., Cho, D., Yoo, D., So Kweon, I.: Two-phase learning for weakly supervised object localization. In: ICCV (2017)
29. Kolesnikov, A., Lampert, C.H.: Seed, expand and constrain: three principles for weakly-supervised image segmentation. In: Leibe, B., Matas, J., Sebe, N., Welling, M. (eds.) ECCV 2016. LNCS, vol. 9908, pp. 695–711. Springer, Cham (2016). https://doi.org/10.1007/978-3-319-46493-0_42
30. Krähenbühl, P., Koltun, V.: Efficient inference in fully connected CRFs with Gaussian edge potentials. In: NeurIPS (2011)
31. Krizhevsky, A., Sutskever, I., Hinton, G.E.: ImageNet classification with deep convolutional neural networks. In: NeurIPS (2012)
32. Kumar Singh, K., Jae Lee, Y.: Hide-and-seek: forcing a network to be meticulous for weakly-supervised object and action localization. In: ICCV (2017)
33. Lee, J., Kim, E., Lee, S., Lee, J., Yoon, S.: FickleNet: weakly and semi-supervised semantic image segmentation using stochastic inference. In: CVPR (2019)
34. Lee, J., Kim, E., Lee, S., Lee, J., Yoon, S.: Frame-to-frame aggregation of active regions in web videos for weakly supervised semantic segmentation. In: ICCV (2019)
35. Lee, S., Lee, J., Lee, J., Park, C.K., Yoon, S.: Robust tumor localization with pyramid grad-cam. arXiv preprint (2018)
36. Li, K., Wu, Z., Peng, K.C., Ernst, J., Fu, Y.: Tell me where to look: guided attention inference network. In: CVPR (2018)

37. Li, K., Zhang, Y., Li, K., Li, Y., Fu, Y.: Attention bridging network for knowledge transfer. In: ICCV (2019)
38. Lin, D., Dai, J., Jia, J., He, K., Sun, J.: ScribbleSup: scribble-supervised convolutional networks for semantic segmentation. In: CVPR (2016)
39. Lin, Z., et al.: A structured self-attentive sentence embedding. In: ICLR (2017)
40. Liu, J.J., Hou, Q., Cheng, M.M., Feng, J., Jiang, J.: A simple pooling-based design for real-time salient object detection. In: CVPR (2019)
41. Lu, J., Yang, J., Batra, D., Parikh, D.: Hierarchical question-image co-attention for visual question answering. In: NeurIPS (2016)
42. Lu, X., Wang, W., Ma, C., Shen, J., Shao, L., Porikli, F.: See more, know more: unsupervised video object segmentation with co-attention Siamese networks. In: CVPR (2019)
43. Luong, M.T., Pham, H., Manning, C.D.: Effective approaches to attention-based neural machine translation. In: EMNLP (2015)
44. Nguyen, D.K., Okatani, T.: Improved fusion of visual and language representations by dense symmetric co-attention for visual question answering. In: CVPR (2018)
45. Odena, A., Olah, C., Shlens, J.: Conditional image synthesis with auxiliary classifier GANs. In: ICML (2017)
46. Pan, B., Cao, Z., Adeli, E., Niebles, J.C.: Adversarial cross-domain action recognition with co-attention. In: AAAI (2020)
47. Papandreou, G., Chen, L.C., Murphy, K.P., Yuille, A.L.: Weakly-and semi-supervised learning of a deep convolutional network for semantic image segmentation. In: ICCV (2015)
48. Pathak, D., Shelhamer, E., Long, J., Darrell, T.: Fully convolutional multi-class multiple instance learning. arXiv preprint (2014)
49. Paulus, R., Xiong, C., Socher, R.: A deep reinforced model for abstractive summarization. In: ICLR (2018)
50. Pinheiro, P.O., Collobert, R.: From image-level to pixel-level labeling with convolutional networks. In: CVPR (2015)
51. Qi, X., Liu, Z., Shi, J., Zhao, H., Jia, J.: Augmented feedback in semantic segmentation under image level supervision. In: Leibe, B., Matas, J., Sebe, N., Welling, M. (eds.) ECCV 2016. LNCS, vol. 9912, pp. 90–105. Springer, Cham (2016). https://doi.org/10.1007/978-3-319-46484-8_6
52. Roy, A., Todorovic, S.: Combining bottom-up, top-down, and smoothness cues for weakly supervised image segmentation. In: CVPR (2017)
53. Russakovsky, O., et al.: ImageNet large scale visual recognition challenge. IJCV 115(3), 211–252 (2015)
54. Shen, T., Lin, G., Liu, L., Shen, C., Reid, I.: Weakly supervised semantic segmentation based on web image co-segmentation. In: BMVC (2017)
55. Shen, T., Lin, G., Shen, C., Reid, I.: Bootstrapping the performance of webly supervised semantic segmentation. In: CVPR (2018)
56. Shimoda, W., Yanai, K.: Self-supervised difference detection for weakly-supervised semantic segmentation. In: ICCV (2019)
57. Simonyan, K., Zisserman, A.: Very deep convolutional networks for large-scale image recognition. arXiv preprint (2014)
58. Sun, M., Yuan, Y., Zhou, F., Ding, E.: Multi-attention multi-class constraint for fine-grained image recognition. In: Ferrari, V., Hebert, M., Sminchisescu, C., Weiss, Y. (eds.) ECCV 2018. LNCS, vol. 11220, pp. 834–850. Springer, Cham (2018). https://doi.org/10.1007/978-3-030-01270-0_49

59. Tokmakov, P., Alahari, K., Schmid, C.: Weakly-supervised semantic segmentation using motion cues. In: Leibe, B., Matas, J., Sebe, N., Welling, M. (eds.) ECCV 2016. LNCS, vol. 9908, pp. 388–404. Springer, Cham (2016). https://doi.org/10.1007/978-3-319-46493-0_24

60. Vaswani, A., et al.: Attention is all you need. In: NeurIPS (2017)

61. Wang, W., Lu, X., Shen, J., Crandall, D.J., Shao, L.: Zero-shot video object segmentation via attentive graph neural networks. In: ICCV (2019)

62. Wang, W., Shen, J.: Higher-order image co-segmentation. IEEE TMM **18**(6), 1011–1021 (2016)

63. Wang, W., Zhu, H., Dai, J., Pang, Y., Shen, J., Shao, L.: Hierarchical human parsing with typed part-relation reasoning. In: CVPR (2020)

64. Wang, X., You, S., Li, X., Ma, H.: Weakly-supervised semantic segmentation by iteratively mining common object features. In: CVPR (2018)

65. Wang, X., Girshick, R., Gupta, A., He, K.: Non-local neural networks. In: CVPR (2018)

66. Wang, X., Li, L., Ye, W., Long, M., Wang, J.: Transferable attention for domain adaptation. In: AAAI (2019)

67. Wang, X., et al.: Reinforced cross-modal matching and self-supervised imitation learning for vision-language navigation. In: CVPR (2019)

68. Shimoda, W., Yanai, K.: Distinct class-specific saliency maps for weakly supervised semantic segmentation. In: Leibe, B., Matas, J., Sebe, N., Welling, M. (eds.) ECCV 2016. LNCS, vol. 9908, pp. 218–234. Springer, Cham (2016). https://doi.org/10.1007/978-3-319-46493-0_14

69. Wei, Y., Feng, J., Liang, X., Cheng, M.M., Zhao, Y., Yan, S.: Object region mining with adversarial erasing: a simple classification to semantic segmentation approach. In: CVPR (2017)

70. Wei, Y.: STC: a simple to complex framework for weakly-supervised semantic segmentation. TPAMI **39**(11), 2314–2320 (2016)

71. Wei, Y., Xiao, H., Shi, H., Jie, Z., Feng, J., Huang, T.S.: Revisiting dilated convolution: a simple approach for weakly-and semi-supervised semantic segmentation. In: CVPR (2018)

72. Woo, S., Park, J., Lee, J.-Y., Kweon, I.S.: CBAM: convolutional block attention module. In: Ferrari, V., Hebert, M., Sminchisescu, C., Weiss, Y. (eds.) ECCV 2018. LNCS, vol. 11211, pp. 3–19. Springer, Cham (2018). https://doi.org/10.1007/978-3-030-01234-2_1

73. Wu, Q., Wang, P., Shen, C., Reid, I., Van Den Hengel, A.: Are you talking to me? Reasoned visual dialog generation through adversarial learning. In: CVPR (2018)

74. Xiong, C., Zhong, V., Socher, R.: Dynamic coattention networks for question answering. In: ICLR (2017)

75. Xu, T., et al.: AttnGAN: fine-grained text to image generation with attentional generative adversarial networks. In: CVPR (2018)

76. Ye, Q., Yuan, S., Kim, T.-K.: Spatial attention deep net with partial PSO for hierarchical hybrid hand pose estimation. In: Leibe, B., Matas, J., Sebe, N., Welling, M. (eds.) ECCV 2016. LNCS, vol. 9912, pp. 346–361. Springer, Cham (2016). https://doi.org/10.1007/978-3-319-46484-8_21

77. Yu, Z., Yu, J., Cui, Y., Tao, D., Tian, Q.: Deep modular co-attention networks for visual question answering. In: CVPR (2019)

78. Zeiler, M.D., Fergus, R.: Visualizing and understanding convolutional networks. In: Fleet, D., Pajdla, T., Schiele, B., Tuytelaars, T. (eds.) ECCV 2014. LNCS, vol. 8689, pp. 818–833. Springer, Cham (2014). https://doi.org/10.1007/978-3-319-10590-1_53

79. Zeng, Y., Zhuge, Y., Lu, H., Zhang, L.: Joint learning of saliency detection and weakly supervised semantic segmentation. In: ICCV (2019)
80. Zhang, H., Goodfellow, I., Metaxas, D., Odena, A.: Self-attention generative adversarial networks. In: ICML (2019)
81. Zhang, X., Wei, Y., Feng, J., Yang, Y., Huang, T.S.: Adversarial complementary learning for weakly supervised object localization. In: CVPR (2018)
82. Zhang, Y., Nie, S., Liu, W., Xu, X., Zhang, D., Shen, H.T.: Sequence-to-sequence domain adaptation network for robust text image recognition. In: CVPR (2019)
83. Zheng, Z., Wang, W., Qi, S., Zhu, S.C.: Reasoning visual dialogs with structural and partial observations. In: CVPR (2019)
84. Zhou, B., Khosla, A., Lapedriza, A., Oliva, A., Torralba, A.: Learning deep features for discriminative localization. In: CVPR (2016)
85. Zhu, Z., Huang, T., Shi, B., Yu, M., Wang, B., Bai, X.: Progressive pose attention transfer for person image generation. In: CVPR (2019)

CoReNet: Coherent 3D Scene Reconstruction from a Single RGB Image

Stefan Popov$^{(\boxtimes)}$, Pablo Bauszat, and Vittorio Ferrari

Google Research, Zürich, Switzerland
spopov@google.com, pablo.bauszat@gmail.com, vittoferrari@google.com

Abstract. Advances in deep learning techniques have allowed recent work to reconstruct the shape of a single object given only one RBG image as input. Building on common encoder-decoder architectures for this task, we propose three extensions: (1) ray-traced skip connections that propagate local 2D information to the output 3D volume in a physically correct manner; (2) a hybrid 3D volume representation that enables building translation equivariant models, while at the same time encoding fine object details without an excessive memory footprint; (3) a reconstruction loss tailored to capture overall object geometry. Furthermore, we adapt our model to address the harder task of reconstructing multiple objects from a single image. We reconstruct all objects jointly in one pass, producing a coherent reconstruction, where all objects live in a single consistent 3D coordinate frame relative to the camera and they do not intersect in 3D space. We also handle occlusions and resolve them by hallucinating the missing object parts in the 3D volume. We validate the impact of our contributions experimentally both on synthetic data from ShapeNet as well as real images from Pix3D. Our method improves over the state-of-the-art single-object methods on both datasets. Finally, we evaluate performance quantitatively on multiple object reconstruction with synthetic scenes assembled from ShapeNet objects.

1 Introduction

3D reconstruction is key to genuine scene understanding, going beyond 2D analysis. Despite its importance, this task is exceptionally hard, especially in its most general setting: from one RGB image as input. Advances in deep learning techniques have allowed recent work [6,7,10,25,30,33,35,44,46,47] to reconstruct the shape of a single object in an image.

In this paper, we first propose several improvements for the task of reconstructing a single object. As in [10,49], we build a neural network model which takes a RGB input image, encodes it and then decodes it into a reconstruction of the full volume of the scene. We then extract object meshes in a second stage. We extend this simple model with three technical contributions: (1) Ray-traced

Electronic supplementary material The online version of this chapter (https://doi.org/10.1007/978-3-030-58536-5_22) contains supplementary material, which is available to authorized users.

A. Vedaldi et al. (Eds.): ECCV 2020, LNCS 12347, pp. 366–383, 2020.
https://doi.org/10.1007/978-3-030-58536-5_22

Fig. 1. 3D reconstructions from a single RGB image, produced by our model. **Left:** Coherent reconstruction of multiple objects in a synthetic scene (shown from a view matching the input image, and another view). We reconstruct all objects in their correct spatial arrangement in a common coordinate frame, enforce space exclusion, and hallucinate occluded parts. **Right:** Reconstructing an object in a real-world scene. The top image shows the reconstruction overlaid on the RGB input. The bottom row shows the input next to two other views of the reconstruction.

skip connections as a way to propagate local 2D information to the output 3D volume in a physically correct manner (Sect. 3.3). They lead to sharp reconstruction details because visible object parts can draw information directly from the image; (2) A hybrid 3D volume representation that is both regular and implicit (Sect. 3.1). It enables building translation equivariant 3D models using standard convolutional blocks, while at the same time encoding fine object details without an excessive memory footprint. Translation equivariance is important for our task, since objects can appear anywhere in space; (3) A reconstruction training loss tailored to capture overall object geometry, based on a generalization of the intersection-over-union metric (IoU) (Sect. 3.4). Note that our model reconstructs objects at the pose (translation, rotation, scale) seen from the camera, as opposed to a canonical pose in many previous works [7,10,25,49].

We validate the impact of our contributions experimentally on synthetic data from ShapeNet [4] (Sect. 4.1) as well as real images from Pix3D [41] (Sect. 4.2). The experiments demonstrate that (1) our proposed ray-traced skip connections and IoU loss improve reconstruction performance considerably; (2) our proposed hybrid volume representation enables to reconstruct at resolutions higher than the one used during training; (3) our method improves over the state-of-the-art single-object 3D reconstruction methods on both ShapeNet and Pix3D datasets.

In the second part of this paper, we address the harder task of reconstructing scenes consisting of spatial arrangements of multiple objects. In addition to the shape of individual objects at their depicted pose, we also predict the semantic class of each object. We focus on coherent reconstruction in this scenario, where we want to (1) reconstruct all objects and the camera at their correct relative pose in a single consistent 3D coordinate frame, (2) detect occlusions and resolve them fully, hallucinating missing parts (*e.g.* a chair behind a table), and (3) ensure that each point in the output 3D space is occupied by at most one object (space exclusion constraint). We achieve this through a relatively simple modification of our single-object pipeline. We predict a probability distribution over semantic classes at each point in the output 3D space and we make the final mesh extraction step aware of this.

The technical contributions mentioned above for the single-object case are even more relevant for reconstructing scenes containing multiple objects. Ray-traced skip connections allow the model to propagate occlusion boundaries and object contact points detected on the 2D image into 3D, and to also understand the depth relations among objects locally. The IoU loss teaches our model to output compact object reconstructions that do not overlap in 3D space. The hybrid volume representation provides a fine discretization resolution, which can compensate for the smaller fraction of the scene volume allocated to each object in comparison to the single object case.

We experimentally study our method's performance on multiple object reconstruction with synthetic scenes assembled from ShapeNet objects (Sect. 4.3). We validate again the impact of our technical contributions, and study the effect of the degree of object occlusion, distance to the camera, number of objects in the scene, and their semantic classes. We observe that ray-traced skip connections and the IoU loss bring larger improvements than in the single object case. We show that our model can handle multiple object scenes well, losing only a fraction of its performance compared to the single object case.

Finally, we study the effect of image realism on reconstruction performance both in the single-object (Sect. 4.1) and multi-object (Sect. 4.3) cases. We render our images with either (1) local illumination against uniform background like most previous works [7,10,25,33,44,46] or (2) a physically-based engine [32], adding global illumination effects, such as shadows and reflections, non-trivial background, and complex lighting from an environment map and finite extent light sources. We publicly release these images, our models, and scene layouts [1].

2   Related Work

Single Object Reconstruction. In the last few years there has been a surge of methods for reconstructing the 3D shape of one object from a single RGB image. Many of them [7,10,47,49,50] employ voxel grids in their internal representation, as they can be handled naturally by convolutional neural networks. Some works have tried to go beyond voxels: (1) by using a differentiable voxels-to-mesh operation [21]; (2) by producing multiple depth-maps and/or silhouettes from fixed viewpoints that can be subsequently fused [33,35,38,51]; (3) by operating on point clouds [8,24], cuboidal primitives [30,44], and even directly on meshes [5,46]. A recent class of methods [6,25,31] use a continuous volume representation through implicit functions. The model receives a query 3D point as part of its input and returns the occupancy at that point.

We build on principles from these works and design a new type of hybrid representation that is both regular like voxels and continuous like implicit functions

(Sect. 3.1). We also address more complex reconstruction tasks: we reconstruct objects in the pose as depicted in the image and we also tackle scenes with multiple objects, predicting the semantic class of each object. Finally, we experiment with different levels of rendering realism.

Multi-object Reconstruction. IM2CAD [15] places multiple CAD models from a database in their appropriate position in the scene depicted in the input image. It only reconstructs the pose of the objects and copies over their whole CAD models, without trying to reconstruct their particular 3D shapes as they appear in the input image. 3D-RCNN [18] learns a per-class linear shape basis from a training dataset of 3D models. It then uses a render-and-compare approach to fit the coefficients of this basis to objects detected in the test image. This method only outputs 3D shapes that lie on a simple linear subspace spanning the training samples. Instead our model can output arbitrary shapes, and the mapping between image appearance and shape is more complex as it is modeled by a deep neural network. Tulsiani et al. [43] first detects object proposals [52] and then reconstructs a pose and a voxel grid for each, based on local features for the proposal and a global image descriptor. Mesh-RCNN [9] extends Mask-RCNN [13] to predict a 3D mesh for each detected object in an image. It tries to predict the objects positions in the image plane correctly, but it cannot resolve the fundamental scale/depth ambiguity along the Z-axis.

All four methods [9,15,18,43] first detect objects in the 2D image, and then reconstruct their 3D shapes independently. Instead, we reconstruct all objects jointly and without relying on a detection stage. This allows us to enforce space exclusion constraints and thus produce a globally coherent reconstruction.

The concurrent work [29] predicts the 3D pose of all objects jointly (after 2D object detection). Yet, it still reconstructs their 3D shape independently, and so the reconstructions might overlap in 3D space. Moreover, in contrast to [15, 18,29,43] our method is simpler as it sidesteps the need to explicitly predict per-object poses, and instead directly outputs a joint coherent reconstruction.

Importantly, none of these works [9,15,18,29] offers true quantitative evaluation of 3D shape reconstruction on multiple object scenes. One of the main reasons for this is the lack of datasets with complete and correct ground truth data. One exception is [43] by evaluating on SunCG [39], which is now banned. In contrast, we evaluate our method fully, including the 3D shape of multiple objects in the same image. To enable this, we create two new datasets of scenes assembled from pairs and triplets of ShapeNet objects, and we report performance with a full scene evaluation metric (Sect. 4.3).

Finally, several works tackle multiple object reconstruction from an RGB-D image [28,39], exploiting the extra information that depth sensors provides.

Neural Scene Representations. Recent works [23,26,27,34,36,37] on neural scene representations and neural rendering extract latent representations of the scene geometry from images and share similar insights to ours. In particular, [16,45] use unprojection, a technique to accumulate latent scene information from multiple views, related to our ray-traced skip connections. Others [34,37] can also reconstruct (single-object) geometry from one RGB image.

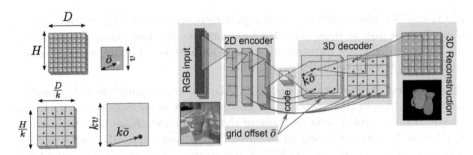

Fig. 2. Left: 2D slice of the output grid (yellow points) and a decoder layer grid (blue points). The output grid is offset by $\bar{o}$ from the origin. The decoder grid, which has k times lower resolution, by $k\bar{o}$. **Right:** Side-cut of our model's architecture. Ray-traced skip connections (red) propagate data from the encoder to the decoder, $\bar{o}$ is appended to the channels of select decoder layers (green). (Color figure online)

3 Proposed Approach

For simplicity and compactness of exposition, we present directly our full method, which can reconstruct multiple objects in the same image. Our reconstruction pipeline takes a single RGB image as input and outputs a set of meshes – one for each object in the scene. It is trained to jointly predict the object shapes, their pose relative to the camera, and their class label.

At the core of our pipeline is a neural network model that receives a single RGB image and a set of volume query points as input and outputs a probability distribution over C possible classes at each of these points. One of the classes is *void* (*i.e.* empty space), while the rest are object classes, such as chair and table. Predicting normalized distributions creates competition between the classes and forces the model to learn about space exclusion. For single object models, we use two classes (foreground and *void*, $C = 2$).

To create a mesh representation, we first reconstruct a fine discretization of the output volume (Sect. 3.5). We query the model repeatedly, at different locations in 3D space, and we integrate the obtained outputs. We then apply marching cubes [20] over the discretization, in a way that enforces the space exclusion constraint. We jitter the query points randomly during training. For single object models, we treat all meshes as parts of one single output object.

3.1 3D Volume Representation

We want our model to reconstruct the large 3D scene volume at a fine spatial resolution, so we can capture geometric details of individual objects, but without an excessive memory footprint. We also want it to be *translation equivariant*: if the model sees *e.g.* chairs only in one corner of the scene during training, it should still be able to reconstruct chairs elsewhere. This is especially important in a multi-object scenario, where objects can appear anywhere in space.

A recent class of models [6,25,31] address our first requirement through an implicit volume representation. They input a compact code, describing the volume contents, and a query point. They output the occupancy of the volume at that point. These models can be used for reconstruction by conditioning on a code extracted from an image, but are not translation equivariant by design.

Models based on a voxel grid representation [7,10] are convolutional in nature, and so address our translation equivariance requirement, but require excessive memory to represent large scenes at fine resolutions (cubic in the number of voxels per dimension).

We address both requirements with a new hybrid volume representation and a model architecture based on it. Our model produces a multinomial distribution over the C possible classes on a regular grid of points. The structure of the grid is fixed (*i.e.* fixed resolution $W \times H \times D$ and distance between points v), but we allow the grid to be placed at an arbitrary spatial offset $\bar{o}$, smaller than v (Fig. 2). The offset value is an input to our model, which then enables fine-resolution reconstruction (see below).

This representation combines the best of voxel grids and implicit volumes. The regular grid structure allows to build a fully convolutional model that is translation equivariant by design, using only standard 3D convolution building blocks. The variable grid offset allows to reconstruct regular samplings of the output volume at any desired resolution (multiple of the model grid's resolution), while keeping the model memory footprint constant. To do this, we call our model repeatedly with different appropriately chosen grid offsets during inference (Sect. 3.5) and integrate the results into a single, consistent, high-resolution output. We sample the full output volume with random grid offsets $\bar{o}$ during training.

3.2 Core Model Architecture

We construct our model on top of an encoder-decoder skeleton (Fig. 2). A custom decoder transforms the output of a standard ResNet-50 [14] encoder into a $W \times H \times D \times C$ output tensor – a probability distribution over the C possible classes for each point in the output grid. The decoder operations alternate between upscaling, using transposed 3D convolutions with stride larger than 1, and data mixing while preserving resolution, using 3D convolutions with stride 1.

We condition the decoder on the grid offset $\bar{o}$. We further create *ray-traced skip connections* that propagate information from the encoder to the decoder layers in a physically accurate manner in Sect. 3.3. We inject $\bar{o}$ and the ray-traced skip connections before select data mixing operations.

3.3 Ray Traced Skip Connections

So far we relied purely on the encoder to learn how to reverse the physical process that converts a 3D scene into a 2D image. This process is well understood however [12,32] and many of its elements have been formalized mathematically. We propose to inject knowledge about it into the model, by connecting each pixel in the input image to its corresponding frustum in the output volume (Fig. 3).

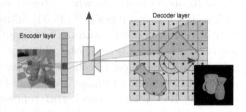

Fig. 3. Pixels in the 2D encoder embed local image information, which ray-traced skip connections propagate to all 3D decoder grid points in the corresponding frustum.

We assume for now that the camera parameters are known. We can compute the 2D projection on the image plane of any point in the 3D output volume and we use this to build ray-traced skip connections. We choose a source 2D encoder layer and a target 3D decoder layer. We treat the $W_e \times H_e \times C_e$ encoder layer as a $W_e \times H_e$ image with C_e channels, taken by our camera. We treat the $W_d \times H_d \times D_d \times C_d$ decoder layer as a $W_d \times H_d \times D_d$ grid of points. We project the decoder points onto the encoder image, then sample it at the resulting 2D coordinates, and finally carry the sampled data over to the 3D decoder. This creates skip connections in the form of rays that start at the camera image plane, pass through the camera pinhole and end at the decoder grid point (Fig. 3). We connect several of the decoder layers to the encoder in this manner, reducing the channel count beforehand to $0.75 \cdot C_d$ by using 1×1 convolutions.

Decoder Grid Offset. An important detail is how to choose the parameters of the decoder layer's grid. The resolution is determined by the layer itself (*i.e.* $W_d \times H_d \times D_d$). It has k times lower resolution than the output grid (by design). We choose $v_d = kv$ for distance between the grid points and $\bar{o}_d = k\bar{o}$ for grid offset (Fig. 2). This makes the decoder grid occupy the same space as the final output grid and respond to changes in the offset $\bar{o}$ in a similar way. In turn, this aids implicit volume reconstruction in Sect. 3.5 with an additional parallax effect.

Obtaining Camera Parameters. Ray-traced skip connections rely on known camera parameters. In practice, the intrinsic parameters are often known. For individual images, they can be deduced from the associated metadata (*e.g.* EXIF in JPEGs). For 3D datasets such as Pix3D [41] and Matterport3D [3] they are usually provided. When not available, we can assume default intrinsic parameters, leading to still plausible 3D reconstructions (*e.g.* correct relative proportions but wrong global object scale). The extrinsic parameters in contrast are usually unknown. We compensate for this by reconstructing relative to the *camera* rather than in world space, resulting in an identity extrinsic camera matrix.

3.4   IoU Training Loss

The output space of our model is a multinomial distribution over the C possible classes (including *void*), for each point in 3D space. This is analog to multi-class recognition in 2D computer vision and hence we could borrow the categorical cross-entropy loss common in those works [13,14]. In our case, most space in the output volume in empty, which leads to most predicted points having the *void* label. Moreover, as only one object can occupy a given point in space, then all but one of the C values at a point will be 0. This leads to even more sparsity. A better loss, designed to deal with extreme class imbalance, is the *focal loss* [22].

Both categorical cross-entropy and the focal loss treat points as a batch of independent examples and average the individual losses. They are not well suited for 3D reconstruction, as we care more about overall object geometry, not independent points. The 3D IoU metric is better suited to capture this, which inspired us to create a new *IoU loss*, specifically aiming to minimize it. Similar losses have been successfully applied to 2D image segmentation problems [2,40].

We generalize IoU, with support for continuous values and multiple classes:

$$IoU_g(g,p) = \frac{\sum\limits_{i \in G} \sum\limits_{c=1}^{C-1} \min(g_{ic}, p_{ic}) \cdot \mu(g_{ic})}{\sum\limits_{i \in G} \sum\limits_{c=1}^{C-1} \max(g_{ic}, p_{ic}) \cdot \mu(g_{ic})}, \quad \mu(g_{ic}) = \begin{cases} 1, & \text{if } g_{ic} = 1 \\ \frac{1}{C-1}, & \text{if } g_{ic} = 0 \end{cases} \quad (1)$$

where i loops over the points in the grid, c – over the $C - 1$ non-void classes, $g_{ic} \in \{0, 1\}$ is the one-hot encoding of the ground truth label, indicating whether point i belongs to class c, and $p_{ic} \in [0, 1]$ is the predicted probability. $\mu(g_{ic})$ balances for the sparsity due to multiple classes, as $C - 1$ values in the ground truth one-hot encoding will be 0.

With two classes (*i.e.* $C = 2$) and binary values for p and g, IoU_g is equivalent to the intersection-over-union measure. The max operator acts like logical *and*, min like logical *or*, and $\mu(g_{ic})$ is always one. In the case where there is a single object class to be reconstructed we use $1 - IoU_g$ as a loss (Sect. 4.1). With multiple objects, we combine IoU_g with categorical cross entropy into a product (Sect. 4.3).

3.5   Mesh Reconstruction

Our end-goal is to extract a set of meshes that represent the surface of the objects in the scene. To do this, we first reconstruct an arbitrary fine discretization of the volume, with a resolution that is an integer multiple n of the model's output resolution. We call the model n^3 times, each time with a different offset $\bar{o} \in \left\{ \frac{0+0.5}{n}v, \frac{1+0.5}{n}v, \ldots, \frac{n-1+0.5}{n}v \right\}^3$ and we interleave the obtained grid values. The result is a $nW \times nH \times nD \times C$ discretization of the volume.

We then extract meshes. We break the discretization into C slices of shape $nW \times nH \times nD$, one for each class. We run marching cubes [20] with threshold 0.5 on each slice independently and we output meshes, except for the slice

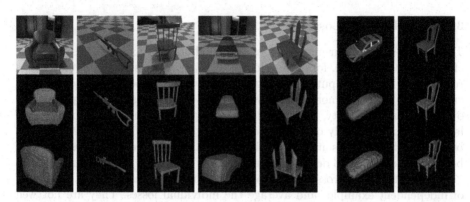

Fig. 4. Single object experiments (Sect. 4.1). **Left:** Scenes reconstructed by h_7, shown from two different viewpoints. Our model handles thin structures and hallucinates invisible back-facing object parts. **Right:** Scenes reconstructed by y_1. Despite the low resolution of y_1 (32^3, second row), we reconstruct high-quality meshes (first row) by sampling y_1 with 4^3 grid offsets (see Sect. 3).

corresponding to the *void* class. The 0.5 threshold enforces space exclusion, since at most one value in a probability distribution can be larger than 0.5.

4   Experiments

We first present experiments on single object reconstruction on synthetic images from ShapeNet [4] (Sect. 4.1) and on real images from Pix3D [41] (Sect. 4.2). Then we evaluate performance on multiple object reconstruction in Sect. 4.3.

4.1   Single Object Reconstruction on ShapeNet

Dataset. We use ShapeNet [4], following the setting of [7]. We consider the same 13 classes, train on 80% of the object instances and test on 20% (the official ShapeNet-trainval and ShapeNet-test splits). We normalize and center each object in the unit cube and render it from 24 random viewpoints, with two levels of photorealism (see inset figure on the right): *low realism*, using local illumination on a uniform background, with no secondary effects such as 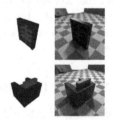 shadows and reflections; and *high realism*, with full global illumination using PBRT's renderer [32], against an environment map background, and with a ground plane. The low realism setting is equivalent to what was used in previous works [7,25,46].

Default Settings. Unless specified otherwise, we train and evaluate at the same grid resolution (128^3), and we use the same camera parameters in all scenes. We evaluate intersection-over-union as a volumetric metric, reporting mean over the

Table 1. Reconstruction performance in % for **(a)** our single object experiments on the left, and **(b)** our multiple object experiments on the right.

(a)

id	skip conn.	loss	rea-lism	IoU mean	IoU glob.	F@1%
h_1	No	focal	low	50.8	52.0	45.0
h_2	No	IoU	low	53.0	53.9	47.8
h_3	Yes	Xent	low	54.1	55.2	52.9
h_4	Yes	focal	low	56.6	57.5	54.4
h_5	Yes	IoU	low	**57.9**	58.7	**57.5**
h_6	Yes	Focal	high	58.1	58.4	57.3
h_7	Yes	IoU	high	**59.1**	59.3	**59.5**

(b)

id	data	skip conn.	loss	rea-lism	IoU mean	IoU glob.
m_1	pairs	no	focal	high	34.9	46.4
m_2	pairs	no	IoU	high	33.1	43.4
m_3	pairs	yes	focal	low	40.4	49.7
m_4	pairs	yes	IoU	low	41.8	50.6
m_5	pairs	yes	Xent	high	30.0	43.5
m_6	pairs	yes	focal	high	42.7	52.4
m_7	pairs	yes	IoU	high	**43.1**	52.7
m_8	tripl.	yes	focal	high	43.0	49.1
m_9	tripl.	yes	IoU	high	**43.9**	49.8
m_{10}	single	yes	focal	high	43.4	53.9
m_{11}	single	yes	IoU	high	46.9	56.4

classes (mIoU) as well as the global mean over all object instances. We also evaluate the predicted meshes with the F@1%-score [42] as a surface metric. As commonly done [6,25,31], we pre-process the ground-truth meshes to make them watertight and to remove hidden and duplicate surfaces. We sample all meshes uniformly 100K points, then compute F-score for each class, and finally report the average over classes.

Reconstruction Performance. We report the effect of hyper parameters on performance in Table 1(a) and show example reconstructions in Fig. 4. Ray-traced skip connections improve mIoU by about 5% and F@1% by 10%, in conjunction with any loss. Our IoU loss performs best, followed by focal and categorical cross entropy (Xent). Somewhat surprisingly, results are slightly better on images with high realism, even though they are visually more complex. Shadows and reflections might be providing additional reconstruction cues in this case. Our best model for low realism images is h_5 and for high realism it is h_7.

Comparison to State-of-the-Art. We compare our models to state-of-the art single object reconstruction methods [7,25,46,49]. We start with an exact comparison to ONN [25]. For this we use the open source implementation provided by the authors to train and test their model on our low-realism images train and test sets. We then use our evaluation procedure on their output predictions. As Table 2 shows, ONN achieves 52.6% mIoU on our data with our evaluation (and 51.5% with ONN's evaluation procedure). This number is expectedly lower than the 57.1% reported in [25] as we ask ONN to reconstruct each shape at the pose depicted in the input image, instead of the canonical pose. From ONN's perspective, the training set contains 24 times more different shapes, one for each rendered view of an object. Our best model for low-realism renderings h_5 outperforms ONN on every class and achieves 57.9% mIoU. ONN's perfor-

Table 2. Comparison to state of the art. The first three rows compare ONN [25] to our models h_2 and h_5, all trained on our data. The next three rows are taken from [25] and report performance of 3D-R2N2 [7], Pix2Mesh [46], and ONN [25] on their data.

model	mIoU	airplane	bench	cabinet	car	chair	display	lamp	loudspeaker	rifle	sofa	table	telephone	vessel
ONN^*	52.6	45.8	45.1	43.8	54.0	58.5	55.4	39.5	57.0	48.0	68.0	50.7	68.3	49.9
h_2	53.0	46.9	44.3	44.7	56.4	57.4	53.8	35.9	58.1	53.4	67.2	49.7	70.9	49.9
h_5	57.9	53.0	50.8	50.9	57.3	63.0	57.2	42.1	60.8	64.6	70.6	55.5	73.1	54.0
3D-R2N2	49.3	42.6	37.3	66.7	66.1	43.9	44.0	28.1	61.1	37.5	62.6	42.0	61.1	48.2
Pix2Mesh	48.0	42.0	32.3	66.4	55.2	39.6	49.0	32.3	59.9	40.2	61.3	39.5	66.1	39.7
ONN	57.1	57.1	48.5	73.3	73.7	50.1	47.1	37.1	64.7	47.4	68.0	50.6	72.0	53.0

mance is comparable to h_2, our best model that, like ONN, does not use skip connections.

We then compare to 3D-R2N2 [7], Pix2Mesh [46], and again ONN [25], using their mIoU as reported by [25] (Table 2). Our model h_5 clearly outperforms 3D-R2N2 (+8.6%) and Pix2Mesh (+9.9%). It also reaches a slightly better mIoU than ONN (+0.8%), while reconstructing in the appropriate pose for each input image, as opposed to a fixed canonical pose. We also compare on the Chamfer Distance surface metric, implemented exactly as in [25]. We obtain 0.15, which is better than 3D-R2N2 (0.278), Pix2Mesh (0.216), and ONN (0.215), all compared with the same metric (as reported by [25]).[1]

Finally, we compare to Pix2Vox [49] and its extension Pix2Vox++ [50] (concurrent work to ours). For a fair comparison we evaluate our h_5 model on a 32^3 grid of points, matching the 32^3 voxel grid output by [49,50]. We compare directly to the mIoU they report. Our model h_5 achieves 68.9% mIoU in this case, +2.8% higher than Pix2Vox (66.1% for their best Pix2Vox-A model) and +1.9% higher than Pix2Vox++ (67.0% for their best Pix2Vox++/A model).

Reconstructing at High Resolutions. Our model can perform reconstruction at a higher resolution than the one used during training (Sect. 3.1). We study this here by reconstructing at 2× and 4× higher resolution. We train one model (y_1) using a 32^3 grid and one (y_2) using a 64^3 grid, with ray-traced skip connections, images with low realism, and focal loss. We then reconstruct a 128^3 discretization from each model, by running inference multiple times at different grid offsets (64 and 8 times, respectively, Sect. 3.5). At test time, we always measure performance on the 128^3 reconstruction, regardless of training resolution.

---

[1] Several other works [8,11], including very recent ones [5,51], report Chamfer Distance and not IoU. They adopt subtly different implementations, varying the underlying point distance metric, scaling, point sampling, and aggregation across points. Thus, they report different numbers for the same works, preventing direct comparison.

Figure 4 shows example reconstructions. We compare performance to h_4 from Table 1, which was trained with same settings but at native grid resolution 128^3. Our first model (trained on 32^3 and producing 128^3) achieves 53.1% mIoU. The second model (trained on 64^3 and producing 128^3) gets to 56.1%, comparable to h_4 (56.6%). This demonstrates that we can reconstruct at substantially higher resolution than the one used during training.

4.2   Single Object Reconstruction on Pix3D

We evaluate the performance of our method on real images using the Pix3D dataset [41], which contains 10069 images annotated with 395 unique 3D models from 9 classes (bed, bookcase, chair, desk, misc, sofa, table, tool, wardrobe). Most of the images are of indoor scenes, with complex backgrounds, occlusion, shadows, and specular highlights.

The images often contain multiple objects, but Pix3D provides annotations for exactly one of them per image. To deal with this discrepancy, we use a single object reconstruction pipeline. At test time, our method looks at the whole image, but we only reconstruct the volume inside the 3D box of the object annotated in the ground-truth. This is similar to how other methods deal with this discrepancy at test time[2].

Generalization Across Domains (ShapeNet to Pix3D). We first perform experiments in the same settings as previous works [7,41,49], which train on synthetic images of ShapeNet objects. As they do, we focus on chairs. We train on the high-realism synthetic images from Sect. 4.1. For each image we crop out the chair and paste it over a random background from OpenImages [17,19], a random background from SunDB [48], and a white background. We start from a model pre-trained on ShapeNet (h_7, Sect. 4.1) and continue training on this data.

We evaluate on the 2894 Pix3D images with chairs that are neither occluded nor truncated. We predict occupancy on a 32^3 discretization of 3D space. This is the exact same setting used in [7,41,49,50]. Our model achieves 29.7% IoU, which is higher than Pix2Vox [49] (28.8%, for their best Pix2Vox-A model), the Pix3D method [41] (28.2%, for their best 'with pose' model), 3D-R2N2 [7] (13.6%, as reported in [41]), and the concurrent work Pix2Vox++ [50] (29.2% for their best Pix2Vox++/A model).

This setting is motivated by the fact that most real-world images do not come with annotations for the ground-truth 3D shape of the objects in them. Therefore, it represents the common scenario of training from synthetic data with available 3D supervision.

Fine Tuning on Real Data from Pix3D. We now consider the case where we do have access to a small set of real-world images with ground-truth 3D shapes

---

[2] Pix2Vox [49] and Pix2Vox++ [50] crop the input image before reconstruction, using the 2D projected box of the ground-truth object. MeshRCNN [9] requires the ground-truth object 3D center as input. It also crops the image through the ROI pooling layers, using the 2D projected ground-truth box to reject detections with IoU < 0.3.

Fig. 5. Qualitative results on Pix3D. For each example, the large image shows our reconstruction overlaid on the RGB input. The smaller images show the RGB input, and our reconstruction viewed from two additional viewpoints.

for training. For this we use the S_1 and S_2 train/test splits of Pix3D defined in [9]. There are no images in common between the test and train splits in both S_1 and S_2. Furthermore, in S_2 also the set of object instances is disjoint between train and test splits. In S_1 instead, some objects are allowed to be in both the splits, albeit with a different pose and against a different background.

We train two models, one for S_1 and one for S_2. In both cases, we start from a model pre-trained on ShapeNet (h_7) and we then continue training on the respective Pix3D train set. On average over all 9 object classes, we achieve 33.3% mIoU on the test set of S_1, and 23.6% on the test set of S_2, when evaluating at 128^3 discretization of 3D space (Fig. 5).

As a reference, we compare to a model trained only on ShapeNet. As above, we start from h_7 and we augment with real-world backgrounds. We evaluate performance on all 9 object classes on the test splits of S_1 and S_2. This leads to 20.9% mIoU for S1 and 20.0% for S_2. This confirms that fine-tuning on real-world data from Pix3D performs better than training purely on synthetic data.

4.3   Multiple Object Reconstruction

Datasets and Settings. We construct two datasets by assembling objects from ShapeNet. The first is *ShapeNet-pairs*, with several pairs of object classes: bed-pillow, bottle-bowl, bottle-mug, chair-table, display-lamp, guitar-piano, motorcycle-car. The second is *ShapeNet-triplets*, with bottle-bowl-mug and chair-sofa-table. We randomly sample the object instances participating in each combination from ShapeNet, respecting its official trainval and test splits. For each

Fig. 6. Pairs and triplets reconstructed by m_7 and m_9, shown from the camera and from one additional viewpoint. Our model hallucinates the occluded parts and reconstructs all objects in their correct spatial arrangement, in a common coordinate frame.

image we generate, we random sample two/three object instances, place them at random locations on the ground plane, with random scale and rotation, making sure they do not overlap in 3D, and render the scene from a random camera viewpoint (yaw and pitch). We construct the same number of scenes for every pair/triplet for training and testing. Note how the objects' scales and rotations, as well as the camera viewpoints, vary between the train and test splits and between images within a split (but their overall distribution is the same). Like the single-object case, the object instances are disjoint in the training and test splits. In total, for pairs we generate 365'600 images on trainval and 91'200 on test; for triplets we make 91'400 on trainval and 22'000 on test.

We perform experiments varying the use of ray-traced skip connections, the image realism, and the loss. Besides categorical cross entropy (*Xent*) and focal loss, we also combine Xent and IoU_g (1) into a product. The IoU part pushes the model to reconstruct full shapes, while Xent pushes it to learn the correct class for each 3D point. We train on the train and val splits together, and test on the test split, always with grid resolution 128^3.

Reconstruction Performance. Table 1(b) summarizes our results, including also multi-class reconstruction of images showing a single ShapeNet object for reference (bottom row, marked 'single'). We show example reconstructions in Fig. 6. On ShapeNet-pairs, using ray-traced skip connections improves mIoU substantially (by 8–10%), in conjunction with any loss function. The improvement is twice as large than in the single object case (Table 1), confirming that ray-traced skip connections indeed help more for multiple objects. They allow the model to propagate occlusion boundaries and object contact points detected on the 2D image into 3D, and also to understand the depth relations among objects locally. When using skip connections, our IoU loss performs best, followed closely by the focal loss. The cross-entropy loss underperforms in comparison (−13% mIoU). As with the single-object case, results are slightly better on higher image realism.

Importantly, we note that performance for pairs/triplets is only mildly lower than for the easier single-object scenario. To investigate why, we compare the single-object models m_{11} and h_7 from Table 1. They differ in the number of classes they handle (14 for m_{11}, 2 for h_7) but have otherwise identical settings. While the difference in their mean IoUs is 12% (46.9% *vs.* 59.1%), their global IoUs are close (56.4% *vs.* 59.3%). Hence, our model is still good at reconstructing the overall shapes of objects, but makes some mistakes in assigning the right class.

Finally, we note that reconstruction is slightly better overall for triplets rather than for pairs. This is due to the different classes involved. On pairs composed of the same classes appearing in the triplets, results are better for pairs.

In conclusion, these results confirm that we are able to perform 3D reconstruction in the harder multiple object scenario.

Occlusion and Distance. In Fig. 7, we break down the performance (mIoU) of m_7 by the degree of object occlusion (blue), and also by the object depth for unoccluded objects (i.e. distance to the camera, green). The performance gracefully degrades as occlusion increases, showing that our model can handle it well. Interestingly, the performance remains steady with increasing depth, which correlates to object size in the image. This shows that our model reconstructs far-away objects about as well as nearby ones.

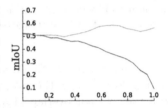

Fig. 7. mIoU *vs.* object occlusion and depth. (Color figure online)

Generalizations. In the suppl. material we explore even more challenging scenarios, where the number of objects varies between training and test images, and where the test set contains combintations of classes not seen during training.

5   Conclusions

We made three contributions to methods for reconstructing the shape of a single object given one RBG image as input: (1) ray-traced skip connections that propagate local 2D information to the output 3D volume in a physically correct manner; (2) a hybrid 3D volume representation that enables building translation equivariant models, while at the same time producing fine object details with limited memory; (3) a reconstruction loss tailored to capture overall object geometry. We then adapted our model to reconstruct multiple objects. By doing so jointly in a single pass, we produce a coherent reconstruction with all objects in one consistent 3D coordinate frame, and without intersecting in 3D space. Finally, we validated the impact of our contributions on synthetic data from ShapeNet as well as real images from Pix3D, including a full quantitative evaluation of 3D shape reconstruction of multiple objects in the same image.

References

1. https://github.com/google-research/corenet
2. Berman, M., Triki, A.R., Blaschko, M.B.: The Lovász-Softmax loss: a tractable surrogate for the optimization of the intersection-over-union measure in neural networks. In: CVPR (2018)
3. Chang, A.X., et al.: Matterport3D: learning from RGB-D data in indoor environments. In: 2017 International Conference on 3D Vision (2017)
4. Chang, A.X., et al.: ShapeNet: an information-rich 3D model repository. CoRR abs/1512.03012 (2015). http://arxiv.org/abs/1512.03012
5. Chen, Z., Tagliasacchi, A., Zhang, H.: BSP-Net: generating compact meshes via binary space partitioning. In: CVPR (2020)
6. Chen, Z., Zhang, H.: Learning implicit fields for generative shape modeling. In: CVPR (2019)
7. Choy, C.B., Xu, D., Gwak, J.Y., Chen, K., Savarese, S.: 3D-R2N2: a unified approach for single and multi-view 3D object reconstruction. In: Leibe, B., Matas, J., Sebe, N., Welling, M. (eds.) ECCV 2016. LNCS, vol. 9912, pp. 628–644. Springer, Cham (2016). https://doi.org/10.1007/978-3-319-46484-8_38
8. Fan, H., Su, H., Guibas, L.J.: A point set generation network for 3D object reconstruction from a single image. In: CVPR (2017)
9. Gkioxari, G., Malik, J., Johnson, J.: Mesh R-CNN. In: ICCV (2019)
10. Girdhar, R., Fouhey, D.F., Rodriguez, M., Gupta, A.: Learning a predictable and generative vector representation for objects. In: Leibe, B., Matas, J., Sebe, N., Welling, M. (eds.) ECCV 2016. LNCS, vol. 9910, pp. 484–499. Springer, Cham (2016). https://doi.org/10.1007/978-3-319-46466-4_29
11. Groueix, T., Fisher, M., Kim, V.G., Russell, B.C., Aubry, M.: A Papier-Mâché approach to learning 3D surface generation. In: CVPR (2018)
12. Hartley, R.I., Zisserman, A.: Multiple View Geometry in Computer Vision. Cambridge University Press, Cambridge (2000). ISBN 0521623049
13. He, K., Gkioxari, G., Dollár, P., Girshick, R.B.: Mask R-CNN. In: ICCV (2017)
14. He, K., Zhang, X., Ren, S., Sun, J.: Deep residual learning for image recognition. In: CVPR (2016)
15. Izadinia, H., Shan, Q., Seitz, S.M.: IM2CAD. In: CVPR, pp. 2422–2431 (2017)
16. Kar, A., Häne, C., Malik, J.: Learning a multi-view stereo machine. In: NIPS (2017)
17. Krasin, I., et al.: OpenImages: a public dataset for large-scale multi-label and multi-class image classification (2017). Dataset https://g.co/dataset/openimages
18. Kundu, A., Li, Y., Rehg, J.M.: 3D-RCNN: instance-level 3D object reconstruction via render-and-compare. In: CVPR (2018)
19. Kuznetsova, A., et al.: The Open Images Dataset V4: unified image classification, object detection, and visual relationship detection at scale. arXiv preprint arXiv:1811.00982 (2018)
20. Lewiner, T., Lopes, H., Vieira, A.W., Tavares, G.: Efficient implementation of marching cubes' cases with topological guarantees. J. Graph. GPU Game Tools 8(2), 1–15 (2003)
21. Liao, Y., Donné, S., Geiger, A.: Deep marching cubes: learning explicit surface representations. In: CVPR (2018)
22. Lin, T., Goyal, P., Girshick, R.B., He, K., Dollár, P.: Focal loss for dense object detection. In: ICCV (2017)
23. Lombardi, S., Simon, T., Saragih, J., Schwartz, G., Lehrmann, A., Sheikh, Y.: Neural volumes: learning dynamic renderable volumes from images. ACM Trans. Graph. 38(4), 65:1–65:14 (2019)

24. Mandikal, P., Navaneet, K.L., Agarwal, M., Radhakrishnan, V.B.: 3D-LMNet: latent embedding matching for accurate and diverse 3D point cloud reconstruction from a single image. In: BMVC (2018)
25. Mescheder, L.M., Oechsle, M., Niemeyer, M., Nowozin, S., Geiger, A.: Occupancy networks: learning 3D reconstruction in function space. In: CVPR (2019)
26. Nguyen-Phuoc, T., Li, C., Balaban, S., Yang, Y.: RenderNet: a deep convolutional network for differentiable rendering from 3D shapes. In: NIPS (2018)
27. Nguyen-Phuoc, T., Li, C., Theis, L., Richardt, C., Yang, Y.L.: HoloGAN: unsupervised learning of 3D representations from natural images. In: ICCV (2019)
28. Nicastro, A., Clark, R., Leutenegger, S.: X-Section: cross-section prediction for enhanced RGB-D fusion. In: ICCV (2019)
29. Nie, Y., Han, X., Guo, S., Zheng, Y., Chang, J., Zhang, J.J.: Total3DUnderstanding: joint layout, object pose and mesh reconstruction for indoor scenes from a single image. In: IEEE/CVF Conference on Computer Vision and Pattern Recognition (CVPR) (2020)
30. Niu, C., Li, J., Xu, K.: Im2Struct: recovering 3D shape structure from a single RGB image. In: CVPR (2018)
31. Park, J.J., Florence, P., Straub, J., Newcombe, R.A., Lovegrove, S.: DeepSDF: learning continuous signed distance functions for shape representation. In: CVPR (2019)
32. Pharr, M., Jakob, W., Humphreys, G.: Physically Based Rendering: From Theory to Implementation, 3rd edn. Morgan Kaufmann Publishers Inc., San Francisco (2016)
33. Richter, S.R., Roth, S.: Matryoshka networks: predicting 3D geometry via nested shape layers. In: CVPR (2018)
34. Saito, S., Huang, Z., Natsume, R., Morishima, S., Kanazawa, A., Li, H.: PIFu: pixel-aligned implicit function for high-resolution clothed human digitization. In: ICCV (2019)
35. Shin, D., Fowlkes, C.C., Hoiem, D.: Pixels, voxels, and views: a study of shape representations for single view 3D object shape prediction. In: CVPR (2018)
36. Sitzmann, V., Thies, J., Heide, F., Niessner, M., Wetzstein, G., Zollhofer, M.: DeepVoxels: learning persistent 3D feature embeddings. In: CVPR (2019)
37. Sitzmann, V., Zollhöfer, M., Wetzstein, G.: Scene representation networks: continuous 3D-structure-aware neural scene representations. In: NIPS (2019)
38. Soltani, A.A., Huang, H., Wu, J., Kulkarni, T.D., Tenenbaum, J.B.: Synthesizing 3D shapes via modeling multi-view depth maps and silhouettes with deep generative networks. In: CVPR (2017)
39. Song, S., Yu, F., Zeng, A., Chang, A.X., Savva, M., Funkhouser, T.: Semantic scene completion from a single depth image. In: CVPR (2017)
40. Sudre, C.H., Li, W., Vercauteren, T., Ourselin, S., Jorge Cardoso, M.: Generalised dice overlap as a deep learning loss function for highly unbalanced segmentations. In: Cardoso, M.J., et al. (eds.) DLMIA/ML-CDS 2017. LNCS, vol. 10553, pp. 240–248. Springer, Cham (2017). https://doi.org/10.1007/978-3-319-67558-9_28
41. Sun, X., et al.: Pix3D: dataset and methods for single-image 3D shape modeling. In: IEEE Conference on Computer Vision and Pattern Recognition (CVPR) (2018)
42. Tatarchenko, M., Richter, S.R., Ranftl, R., Li, Z., Koltun, V., Brox, T.: What do single-view 3D reconstruction networks learn? In: CVPR (2019)
43. Tulsiani, S., Gupta, S., Fouhey, D.F., Efros, A.A., Malik, J.: Factoring shape, pose, and layout from the 2D image of a 3D scene. In: CVPR (2018)
44. Tulsiani, S., Su, H., Guibas, L.J., Efros, A.A., Malik, J.: Learning shape abstractions by assembling volumetric primitives. In: CVPR (2017)

45. Tung, H.F., Cheng, R., Fragkiadaki, K.: Learning spatial common sense with geometry-aware recurrent networks. In: CVPR (2019)
46. Wang, N., Zhang, Y., Li, Z., Fu, Y., Liu, W., Jiang, Y.-G.: Pixel2Mesh: generating 3D mesh models from single RGB images. In: Ferrari, V., Hebert, M., Sminchisescu, C., Weiss, Y. (eds.) ECCV 2018. LNCS, vol. 11215, pp. 55–71. Springer, Cham (2018). https://doi.org/10.1007/978-3-030-01252-6_4
47. Wu, J., Zhang, C., Xue, T., Freeman, W.T., Tenenbaum, J.B.: Learning a probabilistic latent space of object shapes via 3D generative-adversarial modeling. In: NIPS (2016)
48. Xiao, J., Hays, J., Ehinger, K., Oliva, A., Torralba, A.: SUN database: large-scale scene recognition from abbey to zoo. In: CVPR, pp. 3485–3492 (2010)
49. Xie, H., Yao, H., Sun, X., Zhou, S., Zhang, S.: Pix2Vox: context-aware 3D reconstruction from single and multi-view images. In: ICCV (2019)
50. Xie, H., Yao, H., Zhang, S., Zhou, S., Sun, W.: Pix2Vox++: multi-scale context-aware 3D object reconstruction from single and multiple images. IJCV **128**, 2919–2935 (2020)
51. Yao, Y., Schertler, N., Rosales, E., Rhodin, H., Sigal, L., Sheffer, A.: Front2Back: single view 3D shape reconstruction via front to back prediction. In: CVPR (2020)
52. Zitnick, C.L., Dollár, P.: Edge boxes: locating object proposals from edges. In: Fleet, D., Pajdla, T., Schiele, B., Tuytelaars, T. (eds.) ECCV 2014. LNCS, vol. 8693, pp. 391–405. Springer, Cham (2014). https://doi.org/10.1007/978-3-319-10602-1_26

Layer-Wise Conditioning Analysis in Exploring the Learning Dynamics of DNNs

Lei Huang[1](✉), Jie Qin[1], Li Liu[1], Fan Zhu[1], and Ling Shao[1,2]

[1] Inception Institute of Artificial Intelligence (IIAI), Abu Dhabi, UAE
{lei.huang,jie.qin,li.liu,fan.zhu,ling.shao}@inceptioniai.org
[2] Mohamed bin Zayed University of Artificial Intelligence, Abu Dhabi, UAE

Abstract. Conditioning analysis uncovers the landscape of an optimization objective by exploring the spectrum of its curvature matrix. This has been well explored theoretically for linear models. We extend this analysis to deep neural networks (DNNs) in order to investigate their learning dynamics. To this end, we propose layer-wise conditioning analysis, which explores the optimization landscape with respect to each layer independently. Such an analysis is theoretically supported under mild assumptions that approximately hold in practice. Based on our analysis, we show that batch normalization (BN) can stabilize the training, but sometimes result in the false impression of a local minimum, which has detrimental effects on the learning. Besides, we experimentally observe that BN can improve the layer-wise conditioning of the optimization problem. Finally, we find that the last linear layer of a very deep residual network displays ill-conditioned behavior. We solve this problem by only adding one BN layer before the last linear layer, which achieves improved performance over the original and pre-activation residual networks.

Keywords: Conditioning analysis · Normalization · Residual network

1 Introduction

Deep neural networks (DNNs) have been extensively used in various domains [26]. Their success depends heavily on the improvement of training techniques [15,17,22], e.g., fine weight initialization [12,14,17,39], normalization of internal representations [22,46], and well-designed optimization methods [24,49]. It is believed that these techniques are well connected to the curvature of the loss [25, 38,39]. Analyzing this curvature is thus essential in determining various learning behaviors of DNNs.

Electronic supplementary material The online version of this chapter (https://doi.org/10.1007/978-3-030-58536-5_23) contains supplementary material, which is available to authorized users.

A. Vedaldi et al. (Eds.): ECCV 2020, LNCS 12347, pp. 384–401, 2020.
https://doi.org/10.1007/978-3-030-58536-5_23

In the interest of optimization, conditioning analysis uncovers the landscape of an optimization objective by exploring the spectrum of its curvature matrix. This has been well explored for linear models both in terms of regression [28] and classification [44], where the convergence condition of the optimization problem is controlled by the maximum eigenvalue of the curvature matrix [27,28], and the learning time of the model is lower-bounded by its condition number [27,28]. However, in the context of deep learning, the conditioning analysis suffers from several barriers: (1) the model is over-parameterized and whether the direction with respect to small/zero eigenvalues contributes to the optimization progress is unclear [34,37]; (2) the memory and computational costs are extremely high [11,37].

This paper aims to bridge the gap between the theoretical analyses developed by the optimization community and the empirical techniques used for training DNNs, in order to better understand the learning dynamics of DNNs. We propose a layer-wise conditioning analysis, where we analyze the optimization landscape with respect to each layer independently by exploring the spectra of their curvature matrices. The motivation behind our layer-wise conditioning analysis is based on the recent success of second curvature approximation techniques in DNNs [1,3,31,32,41]. We show that the maximum eigenvalue and the condition number of the block-wise Fisher information matrix (FIM) can be characterized based on the spectrum of the covariance matrix of the input and output-gradient, under mild assumptions, which makes evaluating optimization behavior practical in DNNs. Another theoretical base is the recently proposed proximal back-propagation [7,10,50] where the original optimization problem can be approximately decomposed into multiple independent sub-problems with respect to each layer [50]. We provide the connection between our analysis and the proximal back-propagation [10].

Based on our layer-wise conditioning analysis, we show that batch normalization (BN) [22] can adjust the magnitude of the layer activations/gradients, and thus stabilizes the training. However, this kind of stabilization can drive certain layers into a particular state, referred to as **weight domination**, where the gradient update is feeble. This sometimes has detrimental effects on the learning (Sect. 4.1). We also experimentally observe that BN can improve the layer-wise conditioning of the optimization problem. Furthermore, we find that the unnormalized network has several small eigenvalues in the layer curvature matrix, which are mainly caused by the so-called **dying neurons** (Sect. 4.2), while this behavior is almost entirely absent in batch normalized networks.

We further analyze the ignored difficulty in training very deep residual networks [15]. Using our layer-wise conditioning analysis, we show that the difficulty mainly arises from the ill-conditioned behavior of the last linear layer. We solve this problem by only adding one BN layer before the last linear layer, which achieves improved performance over the original [15] and pre-activation [16] residual networks (Sect. 5).

2   Preliminaries

Optimization Objective. Consider a true data distribution $p_*(\mathbf{x}, \mathbf{y}) = p(\mathbf{x})p(\mathbf{y}|\mathbf{x})$ and the sampled training sets $\mathcal{D} \sim p_*(\mathbf{x}, \mathbf{y})$ of size N. We focus on a supervised learning task aiming to learn the conditional distribution $p(\mathbf{y}|\mathbf{x})$ using the model $q(\mathbf{y}|\mathbf{x})$, where $q(\mathbf{y}|\mathbf{x})$ is represented as a function $f_\theta(\mathbf{x})$ parameterized by θ. From an optimization perspective, we aim to minimize the empirical risk, averaged over the sample loss represented as $\ell(\mathbf{y}, f_\theta(\mathbf{x}))$ in training sets $\mathcal{D}$: $\mathcal{L}(\theta) = \frac{1}{N} \sum_{i=1}^{N}(\ell(\mathbf{y}^{(i)}, f_\theta(\mathbf{x}^{(i)})))$.

Gradient Descent. In general, the gradient descent (GD) update seeks to iteratively reduce the loss $\mathcal{L}$ by $\theta_{t+1} = \theta_t - \eta \frac{\partial \mathcal{L}}{\partial \theta}$, where η is the learning rate. For large-scale learning, stochastic gradient descent (SGD) is extensively used to approximate the gradients $\frac{\partial \mathcal{L}}{\partial \theta}$ with a mini-batch gradient. In theory, the convergence behaviors (*e.g.*, the number of iterations required for convergence to a stationary point) depend on the Lipschitz constant C_L of the gradient function of $\mathcal{L}$, which characterizes the global smoothness of the optimization landscape. In practice, the Lipschitz constant is either unknown for complicated functions or too conservative to characterize the convergence behaviors [5]. Researchers thus turn to the local smoothness, characterized by the Hessian matrix $\mathbf{H} = \frac{\partial \mathcal{L}^2}{\partial \theta \partial \theta}$ under the condition that $\mathcal{L}$ is twice differentiable.

Approximate Curvature Matrices. The Hessian describes the local curvature of the optimization landscape. Such curvature information intuitively guides the design of second-order optimization algorithms [5,37], where the update direction is adjusted by multiplying the inverse of a pre-conditioned matrix $\mathbf{G}$ as: $\frac{\partial \hat{\mathcal{L}}}{\partial \theta} = \mathbf{G}^{-1} \frac{\partial \mathcal{L}}{\partial \theta}$. $\mathbf{G}$ is a positive definite matrix that approximates the Hessian and is expect to sustain the its positive curvature. The second moment matrix of sample gradient: $\mathbf{M} = \mathbb{E}_\mathcal{D}(\frac{\partial \ell}{\partial \theta} \frac{\partial \ell}{\partial \theta}^T)$ is usually used as the pre-conditioned matrix [29,36]. Besides, Pascanu and Bengio [35] showed that the FIM: $\mathbf{F} = \mathbb{E}_{p(\mathbf{x}), q(\mathbf{y}|\mathbf{x})}(\frac{\partial \ell}{\partial \theta} \frac{\partial \ell}{\partial \theta}^T)$ can be viewed as a pre-conditioned matrix when performing the natural gradient descent algorithm [35]. Fore more analyses on the connections among $\mathbf{H}$, $\mathbf{F}$, $\mathbf{M}$ please refer to [5,30]. In this paper, we refer to the analysis of the spectrum of the (approximate) curvature matrices as *conditioning analysis*.

Conditioning Analysis for Linear Models. Consider a linear regression model with a scalar output $f_\mathbf{w}(\mathbf{x}) = \mathbf{w}^T\mathbf{x}$, and mean square error loss $\ell = (y - f_\theta(\mathbf{x}))^2$. As shown in [27,28], the learning dynamics in such a quadratic surface are fully controlled by the spectrum of the Hessian matrix $\mathbf{H} = \mathbb{E}_\mathcal{D}(\mathbf{x}\mathbf{x}^T)$. There are two statistical momentums that are essential for evaluating the convergence behaviors of the optimization problem. One is the maximum eigenvalue of the curvature matrix λ_{max}, and the other is the condition number of the curvature matrix, denoted by $\kappa = \frac{\lambda_{max}}{\lambda_{min}}$, where λ_{min} is the minimum nonzero eigenvalue of the curvature matrix. Specifically, λ_{max} controls the upper bound and the optimal learning rate (*e.g.*, the optimal learning rate is $\eta = \frac{1}{\lambda_{max}(\mathbf{H})}$

and the training will diverge if $\eta \geq \frac{2}{\lambda_{max}(\mathbf{H})}$). κ controls the iterations required for convergence (e.g., the lower bound of the iteration is $\kappa(\mathbf{H})$ [28]). If $\mathbf{H}$ is an identity matrix that can be obtained by whitening the input, the GD can converge within only one iteration. It is easy to extend the solution of linear regression from a scalar output to a vectorial output $f_{\mathbf{W}}(\mathbf{x}) = \mathbf{W}^T \mathbf{x}$. In this case, the Hessian is represented as

$$\mathbf{H} = \mathbb{E}_{\mathcal{D}}(\mathbf{x}\mathbf{x}^T) \otimes \mathbf{I}, \tag{1}$$

where $\otimes$ indicates the Kronecker product [13] and $\mathbf{I}$ denotes the identity matrix. For the linear classification model with cross entropy loss, the Hessian is approximated by [44]:

$$\mathbf{H} = \mathbb{E}_{\mathcal{D}}(\mathbf{x}\mathbf{x}^T) \otimes \mathbf{S}. \tag{2}$$

$\mathbf{S} \in \mathbb{R}^{c \times c}$ is defined by $\mathbf{S} = \frac{1}{c}(\mathbf{I}_c - \frac{1}{c}\mathbf{1}_c)$, where c is the number of classes and $\mathbf{1}_c \in \mathbb{R}^{c \times c}$ denotes a matrix in which all entries are 1. Equation 2 assumes the Hessian does not significantly change from the initial region to the optimal region [44].

3   Layer-Wise Conditioning Analysis for DNNs

Considering a multilayer perceptron (MLP), $f_\theta(\mathbf{x})$ can be represented as a layer-wise linear and nonlinear transformation, as follows:

$$\mathbf{h}_k = \mathbf{W}_k\mathbf{x}_{k-1}, \quad \mathbf{x}_k = \phi(\mathbf{h}_k), \quad k = 1, \ldots, K, \tag{3}$$

where $\mathbf{x}_0 = \mathbf{x}$ and the learnable parameters $\theta = \{\mathbf{W}_k \in \mathbb{R}^{d_k \times d_{k-1}}, k = 1, \ldots, K\}$. To simplify the denotation, we set $\mathbf{x}_K = \mathbf{h}_K$ as the output of the network $f_\theta(\mathbf{x})$.

A conditioning analysis on the full curvature matrix for DNNs is difficult due to the high memory and computational costs [11,34]. We thus seek to analyze an approximation of the curvature matrix. One successful example in second-order optimization over DNNs is approximating the FIM using the Kronecker product (K-FAC) [1,3,31,41]. In the K-FAC approach, there are two assumptions: (1) weight-gradients in different layers are assumed to be uncorrelated; (2) the input and output-gradient in each layer are approximated as independent, so the full FIM can be represented as a block diagonal matrix, $\mathbf{F} = diag(F_1, \ldots, F_K)$, where F_k is the sub-FIM (the FIM with respect to the parameters in a certain layer) and computed as:

$$F_k = \mathbb{E}_{p(\mathbf{x}), \; q(\mathbf{y}|\mathbf{x})}((\mathbf{x}_k\mathbf{x}_k^T) \otimes (\frac{\partial \ell}{\partial \mathbf{h}_k}^T \frac{\partial \ell}{\partial \mathbf{h}_k})) \approx \mathbb{E}_{p(\mathbf{x})}(\mathbf{x}_k\mathbf{x}_k^T) \otimes \mathbb{E}_{q(\mathbf{y}|\mathbf{x})}(\frac{\partial \ell}{\partial \mathbf{h}_k}^T \frac{\partial \ell}{\partial \mathbf{h}_k}). \tag{4}$$

$\mathbf{x}_k$ denotes the layer input, and $\frac{\partial \ell}{\partial \mathbf{h}_k}$ denotes the layer output-gradient. We note that Eq. 4 is similar to Eqs. 1 and 2, and all of them depend on the covariance matrix of the (layer) input. The main difference is that, in Eq. 4, the covariance of output-gradient is considered and its value changes over different optimization regions, while in Eqs. 1 and 2, the covariance of output-gradient is constant.

Based on this observation, we propose layer-wise conditioning analysis, *i.e.*, we analyze each sub-FIM F_k's spectrum $\lambda(F_k)$ independently. We expect the spectra of sub-FIMs: $\{\lambda(F_k)\}_{k=1}^{K}$ to effectively reveal that of the full FIM: $\lambda(\mathbf{F})$, at least in terms of analyzing the learning dynamics of the DNNs. Specifically, we analyze the maximum eigenvalue $\lambda_{max}(F_k)$ and condition number $\kappa(F_k)$[1]. Based on the conclusion on the conditioning analysis of linear models shown in Sect. 2, there are two remarkable properties that can be used to implicitly uncover the landscape of the optimization problem:

- *Property 1*: $\lambda_{max}(F_k)$ indicates the magnitude of the weight-gradient in each layer, which shows the steepness of the landscape w.r.t.different layers.
- *Property 2*: $\kappa(F_k)$ indicates how easy it is to optimize the corresponding layer.

Discussion. One concern is the validity of the assumptions the K-FAC approximation is based on. Note that [30,31] have provided some empirical evidence to support their effectiveness in approximating the full FIM with block diagonal sub-FIMs. [23,43] also exploited similar assumptions to derive the mean&variance of eigenvalues (and maximum eigenvalue) of the full FIM, which is calculated using information from layer inputs and output-gradients. Here, we argue that the assumptions required for our analysis are weaker than those of the K-FAC approximation, since we only care about whether or not the spectra of sub-FIMs can accurately reveal the spectrum of full FIM. We conduct experiments to analyze the training dynamics of the unnormalized ('Plain') and

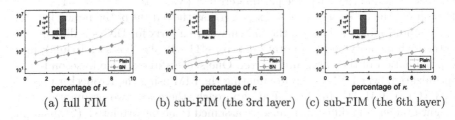

(a) full FIM   (b) sub-FIM (the 3rd layer)   (c) sub-FIM (the 6th layer)

Fig. 1. Conditioning analysis for unnormalized ('Plain') and normalized networks ('BN'). We show the maximum eigenvalue λ_{max} and the generalized condition number κ_p for comparison between the full FIM $\mathbf{F}$ and sub-FIMs $\{F_k\}$. The experiments are performed on an 8-layer MLP with 24 neurons in each layer, for MNIST classification. The input image is center-cropped and resized to 12×12 to remove uninformative pixels. We report the corresponding spectrum at random initialization [27]. Here, we report the results of the 3rd and 6th layers in (b) and (c), respectively. We have similar observations for other layers (See *SM*).

---

[1] We evaluate the general condition number with respect to the percentage: $\kappa_p = \frac{\lambda_{max}}{\lambda_p}$, where λ_p is the pd-th eigenvalue (in descending order) and d is the number of eigenvalues, *e.g.*, $\kappa_{100\%}$ is the original definition of the condition number.

batch normalized [22] ('BN') networks, by looking at the spectra of full curvature matrix and sub-curvature matrices. Figure 1 shows the results based on an
8-layer MLP with 24 neurons in each layer. By observing the results from the full
FIM (Fig. 1(a)), we find that: (1) the unnormalized network suffers from gradient
vanishing (the maximum eigenvalue is around $1e^{-5}$), while the batch normalized
network has an appropriate magnitude of gradient (the maximum eigenvalue is
around 1); (2) 'BN' has better conditioning than 'Plain', which suggests batch
normalization (BN) can improve the conditioning of the network, as observed in
[11,38]. We also obtain a similar conclusion when observing the results from the
sub-FIMs (Fig. 1(b) and (c)). This experiment demonstrates that our layer-wise
conditioning analysis has the potentiality to uncover the training dynamics of
the networks if the full conditioning analysis can. We also conduct experiments
on MLPs with different layers and neurons, and further analyze the spectrum
of the second moment matrix of sample gradient $\mathbf{M}$ (please refer to the *Supplementary Materials (SM)* for details). We have the same observations as in the
first experiment.

Furthermore, we find that investigating $\{\lambda(F_k)\}_{k=1}^{K}$ is more beneficial for
diagnosing the problems behind training DNNs than investigating $\lambda(\mathbf{F})$, *e.g.*, it
enables the gradient vanishing/explosion to be located with respect to a specific
layer from $\{\lambda_{max}(F_k)\}_{k=1}^{K}$, but not $\lambda_{max}(\mathbf{F})$. For example, we know that the 8-
layer unnormalized MLP described in Fig. 1 suffers from difficulty in training, but
we cannot accurately diagnose the problem by only investigating the spectrum
of the full FIM. However, by looking into the layer inputs and output-gradients,
we find that this MLP suffers from exponentially decreased magnitudes of inputs
(forward) and output-gradients (backward). This can be resolved this by using
a better initialization with appropriate variance [14] or using BN [22]. We further elaborate on how to use the layer-wise conditioning analysis to 'debug' the
training of DNNs in the subsequent sections.

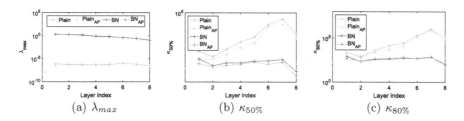

Fig. 2. Validation in approximating the sub-FIMs. The experimental setups are the
same as in Fig. 1. We compare maximum eigenvalue λ_{max} and generalized condition
number κ_p of the sub-FIMs (solid lines) and the approximated ones (dashed lines).

3.1 Efficient Computation

We denote the covariance matrix of the layer input as $\Sigma_{\mathbf{x}} = \mathbb{E}_{p(\mathbf{x})}(\mathbf{x}\mathbf{x}^T)$ and the
covariance matrix of the layer output-gradient as $\Sigma_{\nabla \mathbf{h}} = \mathbb{E}_{q(\mathbf{y}|\mathbf{x})}(\frac{\partial \ell}{\partial \mathbf{h}}^T \frac{\partial \ell}{\partial \mathbf{h}})$. The

condition number and maximum eigenvalue of the sub-FIM F can be derived based on the spectrum of $\Sigma_{\mathbf{x}}$ and $\Sigma_{\nabla\mathbf{h}}$, as shown in the following proposition.

Proposition 1. *Given $\Sigma_{\mathbf{x}}$, $\Sigma_{\nabla\mathbf{h}}$ and $F = \Sigma_{\mathbf{x}} \otimes \Sigma_{\nabla\mathbf{h}}$, we have: 1) $\lambda_{max}(F) = \lambda_{max}(\Sigma_{\mathbf{x}}) \cdot \lambda_{max}(\Sigma_{\nabla\mathbf{h}})$; and 2) $\kappa(F) = \kappa(\Sigma_{\mathbf{x}}) \cdot \kappa(\Sigma_{\nabla\mathbf{h}})$.*

The proof is shown in the *SM*. Proposition 1 provides an efficient way to calculate the maximum eigenvalue and condition number of sub-FIM F by computing those of $\Sigma_{\mathbf{x}}$ and $\Sigma_{\nabla\mathbf{h}}$. In practice, we use the empirical distribution $\mathbb{D}$ to approximate the expected distribution $p(\mathbf{x})$ and $q(\mathbf{y}|\mathbf{x})$ when calculating $\Sigma_{\mathbf{x}}$ and $\Sigma_{\nabla\mathbf{h}}$, since this is very efficient and can be performed with only one forward and backward pass, as has been shown in FIM approximation [1,31].

Note that Proposition 1 depends on the second assumption of Eq. 4. We experimentally demonstrate the effectiveness of such an approximation in Fig. 2, finding that the maximum eigenvalue and the condition number of the sub-FIMs match well with the approximated ones.

3.2 Connection to Proximal Back-Propagation

Carreira-Perpinan and Wang [7] proposed to use auxiliary coordinates to redefine the optimization object $\mathcal{L}(\theta)$ with equality constraints imposed on each neuron. They solved the constrained optimization by adding a quadratic penalty as:

$$\widetilde{\mathcal{L}}(\theta, \mathbf{z}) = \mathcal{L}(\mathbf{y}, f_K(\mathbf{W}_K, \mathbf{z}_{K-1})) + \sum_{k=1}^{K-1} \frac{\lambda}{2}\|\mathbf{z}_k - f_k(\mathbf{W}_k, \mathbf{z}_{k-1}))\|^2, \qquad (5)$$

where $f_k(\cdot, \cdot)$ is a function with respect to each layer. As shown in [7], the solution for minimizing $\widetilde{\mathcal{L}}(\theta, \mathbf{z})$ converges to the solution for minimizing $\mathcal{L}(\theta)$ as $\lambda \to \infty$, under mild conditions. Furthermore, the proximal propagation [10] and the following back-matching propagation [50] reformulate each sub-problem independently with a backward order, minimizing each layer object $\mathcal{L}_k(\mathbf{W}_k, \mathbf{z}_{k-1}; \hat{\mathbf{z}}_k)$, given the target signal $\hat{\mathbf{z}}_k$ from the upper layer, as follows:

$$\begin{cases} \mathcal{L}(\mathbf{y}, f_K(\mathbf{W}_K, \mathbf{z}_{K-1})), & for\ k = K \\ \frac{1}{2}\|\hat{\mathbf{z}}_k - f_k(\mathbf{W}_k, \mathbf{z}_{k-1}))\|^2, & for\ k = K-1, ..., 1. \end{cases} \qquad (6)$$

It has been shown that the produced $\mathbf{W}_k$ using gradient update w.r.t. $\mathcal{L}(\theta)$ equals to the $\mathbf{W}_k$ produced by the back-matching propagation (Procedure 1 in [50]) with one-step gradient update w.r.t. Eq. 6, given an appropriate step size. Note that the target signal $\hat{\mathbf{z}}_k$ is obtained by back-propagation, which means the loss $\mathcal{L}(\theta)$ would be smaller if $f_k(\mathbf{W}_k, \mathbf{z}_{k-1})$ is more close to $\hat{\mathbf{z}}_k$. The loss $\mathcal{L}(\theta)$ will be reduced more efficiently, if the sup-optimization problems in Eq. 6 are well-conditioned. Please refer to [10,50] for more details. If we view the auxiliary variable as the pre-activation in a specific layer, the sub-optimization problem in each layer is formulated as:

$$\begin{cases} \mathcal{L}(\mathbf{y}, \mathbf{W}_K \mathbf{z}_{K-1}), & for\ k = K \\ \frac{1}{2}\|\hat{\mathbf{z}}_k - \mathbf{W}_k \mathbf{z}_{k-1}\|^2, & for\ k = K-1, ..., 1. \end{cases} \qquad (7)$$

It is clear that the sub-optimization problems with respect to $\mathbf{W}_k$ are actually linear classification (for k = K) or regression (for $k = 1, \ldots, K-1$) models. Their conditioning analysis is thoroughly characterized in Sect. 2.

This connection suggests: (1) the quality (conditioning) of the full optimization problem $\mathcal{L}(\theta)$ is well correlated to its sub-optimization problems shown in Eq. 7, whose local curvature matrix can be well explored; (2) We can diagnose the ill behaviors of a DNN by speculating its spectra with respect to certain layers.

4 Exploring Batch Normalized Networks

Let x denote the input for a given neuron in one layer of a DNN. Batch normalization (BN) [22] standardizes the neuron within m mini-batch data by:

$$BN(x^{(i)}) = \gamma \frac{x^{(i)} - \mu}{\sqrt{\sigma^2 + \epsilon}} + \beta, \tag{8}$$

where $\mu = \frac{1}{m} \sum_{i=1}^{m} x^{(i)}$ and $\sigma^2 = \frac{1}{m} \sum_{i=1}^{m} (x^{(i)} - \mu)^2$ are the mean and variance, respectively. The learnable parameters γ and β are used to recover the representation capacity. BN is a ubiquitously employed technique in various architectures [15,19,22,48] due to its ability in stabilizing and accelerating training. Here, we explore how BN stabilizes and accelerates training based on our layer-wise conditioning analysis.

4.1 Stabilizing Training

From the perspective of a practitioner, two phenomena relate to the instability in training a DNN: (1) the training loss first increases significantly and then diverges; or (2) the training loss hardly changes, compared to the initial condition. The former is mainly caused by weights with large updates (*e.g.*, exploded gradients or optimization with a large learning rate). The latter is caused by weights with few updates (vanished gradients or optimization with a small learning rate). In the following theorem, we show that the unnormalized rectifier neural network is very likely to encounter both phenomena.

Theorem 1. *Given a rectifier neural network (Eq. 3) with nonlinearity $\phi(\alpha\mathbf{x}) = \alpha\phi(\mathbf{x})$ ($\alpha > 0$), if the weight in each layer is scaled by $\widehat{\mathbf{W}}_k = \alpha_k \mathbf{W}_k$ ($k = 1, \ldots, K$ and $\alpha_k > 0$), we have the scaled layer input: $\widehat{\mathbf{x}}_k = (\prod_{i=1}^{k} \alpha_i)\mathbf{x}_k$. Assuming that $\frac{\partial \mathcal{L}}{\partial \widehat{\mathbf{h}}_K} = \mu \frac{\partial \mathcal{L}}{\partial \mathbf{h}_K}$, we have the output-gradient: $\frac{\partial \mathcal{L}}{\partial \widehat{\mathbf{h}}_k} = \mu(\prod_{i=k+1}^{K} \alpha_i)\frac{\partial \mathcal{L}}{\partial \mathbf{h}_k}$, and weight-gradient: $\frac{\partial \mathcal{L}}{\partial \widehat{\mathbf{W}}_k} = (\mu \prod_{i=1, i \neq k}^{K} \alpha_i)\frac{\partial \mathcal{L}}{\partial \mathbf{W}_k}$, for all $k = 1, \ldots, K$.*

The proof is shown in the *SM*. From Theorem 1, we observe that the scaled factor α_k of the weight in layer k will affect all other layers' weight-gradients. Specifically, if all $\alpha_k > 1$ ($\alpha_k < 1$), the weight-gradient will increase (decrease) exponentially for one iteration. Moreover, such an exponentially increased weight-gradient will be sustained and amplified in the subsequent iteration, due to the increased magnitude of the weight caused by updating. That is why the unnormalized neural network will diverge, once the training loss increases over a few continuous iterations. We show that such instability can be relieved by BN, based on the following theorem.

Theorem 2. *Under the same condition as Theorem 1, for the normalized network with $\mathbf{h}_k = \mathbf{W}_k \mathbf{x}_{k-1}$ and $\mathbf{s}_k = BN(\mathbf{h}_k)$, we have: $\widehat{\mathbf{x}}_k = \mathbf{x}_k$, $\frac{\partial \mathcal{L}}{\partial \widehat{\mathbf{h}}_k} = \frac{1}{\alpha_k} \frac{\partial \mathcal{L}}{\partial \mathbf{h}_k}$, $\frac{\partial \mathcal{L}}{\partial \widehat{\mathbf{W}}_k} = \frac{1}{\alpha_k} \frac{\partial \mathcal{L}}{\partial \mathbf{W}_k}$, for all $k = 1, \ldots, K$.*

The proof is shown in the *SM*. From Theorem 2, the scaled factor α_k of the weight will not affect other layers' activations/gradients. The magnitude of the weight-gradient is inversely proportional to the scaled factor. Such a mechanism will stabilize the weight growth/reduction, as shown in [22,47]. Note that the behaviors when stabilizing training (Theorem 2) also apply for other activation normalization methods [2,18,45]. We note that the scale-invariance of BN in stabilizing training has been analyzed in previous work [2]. Different to their analyses on the normalization layer itself, we provide an explicit formulation of weight-gradients and output-gradients in a network, which is more important when characterizing the learning dynamics of DNNs.

Empirical Analysis. We further conduct experiments to show how the activation/gradient is affected by initialization in unnormalized DNNs (indicated as 'Plain') and batch normalized DNNs (indicated as 'BN'). We train a 20-layer MLP, with 256 neurons in each layer, for MNIST classification. The nonlinearity is ReLU. We use the full gradient descent[2], and report the results based on the best training loss among learning rates in $\{0.05, 0.1, 0.5, 1\}$. In Fig. 3(a) and (b), we observe that the magnitude of the layer input (output-gradient) of 'Plain' for random initialization [27] suffers from exponential decrease during forward pass (backward pass). The main reason for this is that the weight has a small magnitude, based on Theorem 1. This problem can be relieved by He-initialization [14], where the magnitude of the input/output-gradient is stable across layers (Fig. 3(c) and (d)). We observe that BN can well preserve the magnitude of the input/output-gradient across different layers for both initialization methods.

Weight Domination. It was shown the scale-invariant property of BN has an implicit early stopping effect on the weight matrices [2], helping to stabilize learning towards convergence. Here, we show that this layer-wise 'early stopping' sometimes results in the false impression of a local minimum, which has detrimental effects on the learning, since the network does not well learn the

---

[2] We also perform SGD with a batch size of 1024, and further perform experiments on convolutional neural networks (CNNs) for CIFAR-10 and ImageNet. The results are shown in *SM*, in which we have the same observation as the full gradient descent.

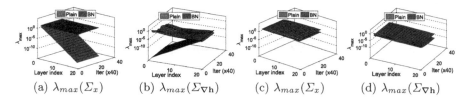

(a) $\lambda_{max}(\Sigma_x)$ (b) $\lambda_{max}(\Sigma_{\nabla h})$ (c) $\lambda_{max}(\Sigma_x)$ (d) $\lambda_{max}(\Sigma_{\nabla h})$

Fig. 3. Analysis of the magnitude of the layer input (indicated by $\lambda_{max}(\Sigma_x)$) and layer output-gradient (indicated by $\lambda_{max}(\Sigma_{\nabla h})$). The experiments are performed on a 20-layer MLP with 256 neurons in each layer, for MNIST classification. The results of (a) (b) are under random initialization [27], while (c) (d) He-initialization [14].

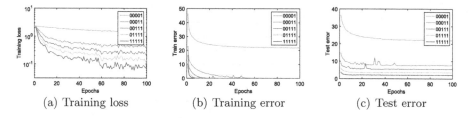

(a) Training loss   (b) Training error   (c) Test error

Fig. 4. Exploring the effectiveness of weight domination. We run the experiments on a 5-layer MLP with BN and the number of neuron in each layer is 256. We simulate weight domination in a given layer by blocking its weight updates. We denote '0' in the legend as the state of weight domination (the first digit represents the first layer).

representation in the corresponding layer. For illustration, we provide a rough definition termed *weight domination*, with respect to a given layer.

Definition 1. *Let $\mathbf{W}_k$ and $\frac{\partial \mathcal{L}}{\partial \mathbf{W}_k}$ be the weight matrix and its gradient in layer k. If $\lambda_{max}(\frac{\partial \mathcal{L}}{\partial \mathbf{W}_k}) \ll \lambda_{max}(\mathbf{W}_k)$, where $\lambda_{max}(\cdot)$ indicates the maximum singular value of a matrix, we refer to layer k has a state of **weight domination**.*

Weight domination implies a smoother gradient with respect to the given layer. This is a desirable property for linear models (the distribution of the input is fixed), where the optimization objective targets to arrive the stationary points with smooth (zero) gradient. However, weight domination is not always desirable for a given layer of a DNN, since such a state of one layer is possibly caused by the increased magnitude of the weight matrix or decreased magnitude of the layer input (the non-convex optimization in Eq. 7), not necessary driven by the optimization objective itself. Although BN ensures a stable distribution of layer inputs, a network with BN still has the possibility that the magnitude of the weight in a certain layer is significantly increased. We experimentally observe this phenomenon, as shown in the *SM*. A similar phenomenon is also observed in [47], where BN results in large updates of the corresponding weights.

Weight domination sometimes harms the learning of the network, because this state limits its ability to learn the representation in the corresponding layer. To investigate this, we conduct experiments on a 5-layer MLP and show the

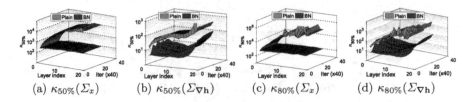

(a) $\kappa_{50\%}(\Sigma_x)$ (b) $\kappa_{50\%}(\Sigma_{\nabla h})$ (c) $\kappa_{80\%}(\Sigma_x)$ (d) $\kappa_{80\%}(\Sigma_{\nabla h})$

Fig. 5. Analysis on the condition number of the layer input ($\kappa_p(\Sigma_x)$) and layer output-gradient ($\kappa_p(\Sigma_{\nabla h})$). The experimental setups are the same as in Fig. 3.

results in Fig. 4. We observe that the network with weight domination in certain layers, can still decrease the loss, but has degenerated performance. We also conduct experiments on CNNs for CIFAR-10 datasets, shown in the *SM*.

4.2 Improved Conditioning

One motivation behind BN is that whitening the input can improve the conditioning of the optimization [22] (*e.g.*, the Hessian will be an identity matrix under the condition that $\mathbb{E}_\mathcal{D}(\mathbf{x}\mathbf{x}^T) = \mathbf{I}$ for a linear model, based on Eq. 1, and thus can accelerate training [9,21]. However, such a motivation is seldom validated by either theoretical or empirical analysis on the context of DNNs [9,38]. Furthermore, it only holds under the condition that BN is placed before the linear layer, while, in practice, BN is typically placed after the linear layer, as recommended in [22]. In this section, we will empirically explore this motivation using our layer-wise conditioning analysis for the scenario of training DNNs.

We first experimentally observe that BN not only improves the conditioning of the layer input's covariance matrix, but also improves the conditioning of the output-gradient's covariation, as shown in Fig. 5. It has been shown that centered data is more likely to be well-conditioned [20,28,33,40]. This suggests that placing BN after the linear layer can improve the conditioning of the output-gradient, because centering the activation, with the gradient back-propagating through such a transformation [22], also centers the gradient.

We also observe that the unnormalized network ('Plain') has several small eigenvalues. For further exploration, we monitor the output of each neuron in each layer, and find that 'Plain' has some neurons that are not activated (zero output of ReLU) for all training examples. We refer to these neurons as *dying neurons*. We also observe that 'Plain' has some neurons that are always activated for every training example, which we refer to as *full neurons*. This observation is most obvious in the initial iterations. The number of dying/full neurons increases as the layer number increases (Please refer to *SM* for details). We conjecture that the dying neurons causes 'Plain' to have numerous small/zero eigenvalues. In contrast, batch normalized networks have no dying/full neurons, because the centering operation ensures that half the examples get activated. This further suggests that placing BN before the nonlinear activation can improve the conditioning.

5   Training Very Deep Residual Networks

Residual networks [15] have significantly relieved the difficulty of training deep networks by their introduction of the residual connection, which makes training networks with hundreds or even thousands of layers possible. However, residual networks also suffer from degenerated performance when the model is extremely deep (*e.g.*, the 1202-layer residual network has worse performance than the 110-layer one), as shown in [15]. He *et al.* [15] argued that this is from over-fitting, not optimization difficulty. Here, we show that a very deep residual network may also suffer difficulty in optimization.

We perform experiments on CIFAR-10 with residual networks, following the same experimental setup as in [15], except that we run the experiments on one GPU. We vary the network depth, ranging in $\{56, 110, 230, 1202\}$, and show the training loss in Fig. 6(a). We observe the residual networks have an increased loss in the initial iterations, which is amplified for deeper networks. Later, the training gets stuck in a state of randomly guessing (the loss stays at $\ln 10$). Although the networks can escape such a state with enough iterations, they suffer from degenerated training performance, especially if they are very deep.

Analysis of Learning Dynamics. To explore why residual networks have such a mysterious behavior, we perform the layer-wise conditioning analysis on the last linear layer (before the cross entropy loss). We monitor the maximum eigenvalue of the covariance matrix $\lambda_{\Sigma_\mathbf{x}}$, the maximum eigenvalue of the second moment matrix of the weight-gradient $\lambda_{\Sigma_{\frac{\partial \mathcal{L}}{\partial \mathbf{W}}}}$, and the norm of the weight ($\|\mathbf{W}\|_2$).

We observe that the initial increase in loss is mainly caused by the large magnitude of $\lambda_{\Sigma_\mathbf{x}}$[3] (Fig. 6(b)), which results in a large magnitude for $\lambda_{\Sigma_{\frac{\partial \mathcal{L}}{\partial \mathbf{W}}}}$ (Fig. 6(c)), and thus a large magnitude for $\|\mathbf{W}\|_2$ (Fig. 6(d)). The increased $\|\mathbf{W}\|_2$ further facilities the increase of the loss. However, the learning objective is to decrease the loss, and thus it should decrease the magnitude of $\mathbf{W}$ or $\mathbf{x}$ (based on Eq. 7) in this case. Apparently, $\mathbf{W}$ is harder to adjust, because the landscape of its loss surface is controlled by $\mathbf{x}$, and all the values of $\mathbf{x}$ are non-negative with large magnitude. The network thus tries to decrease $\mathbf{x}$ based on the given learning objective. We experimentally find that the learnable parameters of BN have a large number of negative values, which causes the ReLUs (positioned after the residual adding operation) deactivated. Such a dynamic results in a significant reduction in the magnitude of $\lambda_{\Sigma_\mathbf{x}}$. The small $\mathbf{x}$ and large $\mathbf{W}$ drive the last linear layer of the network into the state of weight domination, and make the network display a random guess behavior. Although the residual network can escape such a state with enough iterations, the weight domination hinders optimization and results in degenerated training performance.

---

[3] The large magnitude of $\lambda_{\Sigma_\mathbf{x}}$ is caused mainly by the addition of multiple residual connections from the previous layers with ReLU output.

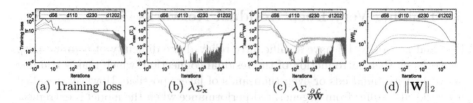

(a) Training loss   (b) $\lambda_{\Sigma_\mathbf{x}}$ (c) $\lambda_{\Sigma \frac{\partial \mathcal{L}}{\partial \mathbf{W}}}$ (d) $\|\mathbf{W}\|_2$

Fig. 6. Analysis on the last linear layer in residual networks for CIFAR-10 classification. We vary the depth ranging in $\{56, 110, 230, 1202\}$ and analyze the results over the course of training. We show (a) the training loss; (b) the maximum eigenvalue of the input's covariance matrix; (c) the maximum eigenvalue of the second moment matrix of the weight-gradient; and (d) the F2-norm of the weight. Note that both the x- and y-axes are in log scale.

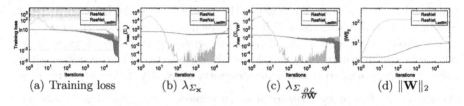

(a) Training loss   (b) $\lambda_{\Sigma_\mathbf{x}}$ (c) $\lambda_{\Sigma \frac{\partial \mathcal{L}}{\partial \mathbf{W}}}$ (d) $\|\mathbf{W}\|_2$

Fig. 7. Analysis of how ResNet$_{LastBN}$ solves the ill-conditioned problem of its last linear layer on the 1202-layer network for CIFAR-10 classification.

5.1 Proposed Solution

Based on the above analysis, it is essential to reduce the large magnitude of $\lambda(\Sigma_\mathbf{x})$. We propose a simple solution and add one BN layer before the last linear layer to normalize its input. We refer to this residual network as 'ResNet$_{LastBN}$', and the original one as 'ResNet'. We also conduct an analysis on the last linear layer of ResNet$_{LastBN}$, providing a comparison between ResNet and ResNet$_{LastBN}$ on the 1202-layer in Fig. 7. We observe that ResNet$_{LastBN}$ can be steadily trained. It does not reach the state of weight domination or encounter a large magnitude of $\mathbf{x}$ in the last linear layer.

We try a similar solution where a constant is divided before the linear layer, and we find it also benefits the training. However, the main disadvantage of this solution is that the value of the constant has to be finely tuned on networks with different depths. We also try putting one BN before the average pooling, which has similar effects as putting it before the last linear layer. We note that Bjorck et al. [4] proposed to train a 110-layer residual network with only one BN layer, which is placed before the average pooling. They showed that this achieves good results. However, we argue that this does not hold for very deep networks. We perform an experiment on the 1202-layer residual network, and find that the model always fails in training with various hyper-parameters.

ResNet$_{LastBN}$, a simple revision of ResNet, achieves significant improvement in performance for very deep residual networks. Figure 8(a) and (b) show

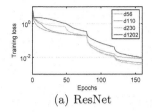 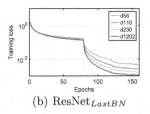

(a) ResNet   (b) ResNet$_{LastBN}$

Fig. 8. Training loss comparison between (a) ResNet and (b) ResNet$_{LastBN}$ with different depth on CIFAR-10. We evaluate the training loss with respect to the epochs.

Table 1. Comparison of test error (%) on CIFAR-10. The results are shown in the format of 'mean ± std' computed over five random seeds.

Method	Depth-56	Depth-110	Depth-230	Depth-1202
ResNet [15]	7.52 ± 0.30	6.89 ± 0.52	7.35 ± 0.64	9.42 ± 3.10
PreResNet [16]	6.89 ± 0.09	6.25 ± 0.08	6.12 ± 0.21	6.07 ± 0.10
PreResNet$_{v1}$	6.75 ± 0.26	6.37 ± 0.24	6.32 ± 0.21	7.89 ± 0.58
ResNet$_{LastBN}$	**6.50 ± 0.22**	**6.10 ± 0.09**	**5.94 ± 0.18**	**5.68 ± 0.14**

the training loss of ResNet and ResNet$_{LastBN}$, respectively, on the CIFAR-10 dataset. We observe that ResNet, with a depth of 1202, appears to have degenerated training performance, especially in the initial phase. Note that, as the depth increases, ResNet obtains worse training performance in the first 80 epochs (before the learning rate is reduced), which coincides with our previous analysis. ResNet$_{LastBN}$ obtains nearly the same training loss for the networks with different depths in the first 80 epochs. Moreover, ResNet$_{LastBN}$ shows lower training loss with increasing depth. Comparing Fig. 8(b) to (a), we observe that ResNet$_{LastBN}$ has better training loss than ResNet for all depths (*e.g.*, at a depth of 56, the loss of ResNet is 0.081, while for ResNet$_{LastBN}$ it is 0.043.).

Table 1 shows the test errors. We observe that ResNet$_{LastBN}$ achieves better test performance with increasing depth, while ResNet has degenerated performance. Compared to ResNet, ResNet$_{LastBN}$ has consistently improved performance over different depths. Particularly, ResNet$_{LastBN}$ reduces the absolute test error of ResNet by 1.02%, 0.79%, 1.41% and 3.74% at depths of 56, 110, 230 and 1202, respectively. Due to ResNet$_{LastBN}$'s optimization efficiency, the training performance is likely improved if we amplify the regularization of the training. Thus, we set the weight decay to 0.0002 and double the training iterations, finding that the 1202-layer ResNet$_{LastBN}$ achieves a test error of 4.79 ± 0.12. We also train a 2402-layer network. We observe that *ResNet* cannot converge, while ResNet$_{LastBN}$ achieves a test error of 5.04 ± 0.30.

We further perform the experiment on CIFAR-100 and use the same experimental setup as CIFAR-10. Table 2 shows the test errors. ResNet$_{LastBN}$ reduces the absolute test error of ResNet by 0.78%, 1.25%, 3.45% and 4.98% at depths of 56, 110, 230 and 1202, respectively. We also validate the effectiveness of ResNet$_{LastBN}$ on the large-scale ImageNet classification, with 1000 classes [8]. ResNet$_{LastBN}$ has better optimization efficiency and achieves better test performance, compared to ResNet. Please refer to the *SM* for more details.

5.2   Revisiting the Pre-activation Residual Network

We note that He *et al.* [16] tried to improve the optimization and generalization of the original residual network [15] by re-arranging the activation functions (using the pre-activation). By looking into the implementation of [16], we find that it also uses an extra BN layer before the last average pooling. It is interesting to investigate which component in [16] (*e.g.*, the pre-activation or the extra BN layer) benefits the optimization behaviors, using our analysis. Here, we denote 'PreResNet' as the pre-activation residual network [16], and denote 'PreResNet$_{v1}$' as the PreResNet without the extra BN layer. We use the conditioning analysis on the last linear layer of the 1202-layer network (see *SM* for details). We observe that: (1) PreResNet$_{v1}$ also gets stuck in the weight domination state with its last linear layer, even though it escapes this states faster than ResNet; (2) PreResNet, like our proposed ResNet$_{LastBN}$, does not suffer the ill-conditioned problem in its last linear layer. These observations suggest that the pre-activation can relieve the ill-conditioned problem to some degree, but more importantly, the extra BN layer is key to improving the optimization efficiency of PreResNet [16] for very deep networks.

Table 2. Comparison of test error (%) on CIFAR-100. The results are shown in the format of 'mean ± std', computed over five random seeds.

Method	Depth-56	Depth-110	Depth-230	Depth-1202
ResNet [15]	29.60 ± 0.41	28.3 ± 1.09	29.25 ± 0.44	30.49 ± 4.44
PreResNet [16]	29.29 ± 0.44	27.58 ± 0.12	26.72 ± 0.33	26.23 ± 0.26
PreResNet$_{v1}$	29.60 ± 0.21	28.54 ± 0.26	27.92 ± 0.34	30.07 ± 2.04
ResNet$_{LastBN}$	**28.82 ± 0.38**	**27.05 ± 0.23**	**25.80 ± 0.10**	**25.51 ± 0.27**

We report the test errors of PreResNet and PreResNet$_{v1}$ in Tables 1 and 2. We find that 'PreResNet' generally has better test performance than PreResNet$_{v1}$, especially for very deep networks (*e.g.*, the 1202-layer one). This supports our arguments that the extra BN layer is the key component of PreResNet [16] for very deep networks. Interestingly, we further observe that our proposed ResNet$_{LastBN}$ is consistently better than PreResNet [16] over different layers and datasets. This demonstrates the effectiveness of our proposed architecture. We believe that our analysis method can be further used to improve residual architectures by looking into the intermediate (inner) layers of networks.

6   Conclusion and Future Work

We proposed a layer-wise conditioning analysis to investigate the learning dynamics of DNNs. Such an analysis is theoretically derived under mild assumptions that approximately hold in practice. Based on our layer-wise conditioning analysis, we showed how batch normalization stabilizes training and improves the conditioning of the optimization problem. We further found that very deep residual networks still suffer difficulty in optimization, which is caused by the ill-conditioned state of the last linear layer. We remedied this by adding only one BN layer before the last linear layer.

We believe there are many potential applications of our method, *e.g.*, investigating the training dynamics of other normalization methods (layer normalization [2] and instance normalization [42]) and comparing them to BN. We also believe it would be interesting to analyze the training dynamics of GANs [6] using our method. We expect our method to provide new insights for analyzing and understanding training techniques for DNNs.

References

1. Ba, J., Grosse, R., Martens, J.: Distributed second-order optimization using Kronecker-factored approximations. In: ICLR (2017)
2. Ba, J., Kiros, R., Hinton, G.E.: Layer normalization. CoRR abs/1607.06450 (2016)
3. Bernacchia, A., Lengyel, M., Hennequin, G.: Exact natural gradient in deep linear networks and its application to the nonlinear case. In: NeurIPS (2018)
4. Bjorck, J., Gomes, C., Selman, B.: Understanding batch normalization. In: NeurIPS (2018)
5. Bottou, L., Curtis, F.E., Nocedal, J.: Optimization methods for large-scale machine learning. SIAM Rev. **60**(2), 223–311 (2018)
6. Brock, A., Donahue, J., Simonyan, K.: Large scale GAN training for high fidelity natural image synthesis. In: ICLR (2019)
7. Carreira-Perpinan, M., Wang, W.: Distributed optimization of deeply nested systems. In: AISTATS (2014)
8. Deng, J., Dong, W., Socher, R., Li, L.J., Li, K., Fei-Fei, L.: ImageNet: a large-scale hierarchical image database. In: CVPR (2009)
9. Desjardins, G., Simonyan, K., Pascanu, R., kavukcuoglu, K.: Natural neural networks. In: NeurIPS (2015)
10. Frerix, T., Möllenhoff, T., Möller, M., Cremers, D.: Proximal backpropagation. In: ICLR (2018)
11. Ghorbani, B., Krishnan, S., Xiao, Y.: An investigation into neural net optimization via Hessian eigenvalue density. In: ICML (2019)
12. Glorot, X., Bengio, Y.: Understanding the difficulty of training deep feedforward neural networks. In: Proceedings of the Thirteenth International Conference on Artificial Intelligence and Statistics, AISTATS 2010 (2010)
13. Grosse, R.B., Martens, J.: A Kronecker-factored approximate Fisher matrix for convolution layers. In: ICML (2016)
14. He, K., Zhang, X., Ren, S., Sun, J.: Delving deep into rectifiers: surpassing human-level performance on ImageNet classification. In: ICCV (2015)

15. He, K., Zhang, X., Ren, S., Sun, J.: Deep residual learning for image recognition. In: CVPR (2016)
16. He, K., Zhang, X., Ren, S., Sun, J.: Identity mappings in deep residual networks. In: Leibe, B., Matas, J., Sebe, N., Welling, M. (eds.) ECCV 2016. LNCS, vol. 9908, pp. 630–645. Springer, Cham (2016). https://doi.org/10.1007/978-3-319-46493-0_38
17. Hinton, G.E., Salakhutdinov, R.R.: Reducing the dimensionality of data with neural networks. Science **313**, 504–507 (2006)
18. Hoffer, E., Banner, R., Golan, I., Soudry, D.: Norm matters: efficient and accurate normalization schemes in deep networks. In: NeurIPS (2018)
19. Huang, G., Liu, Z., Weinberger, K.Q.: Densely connected convolutional networks. In: CVPR (2017)
20. Huang, L., Liu, X., Liu, Y., Lang, B., Tao, D.: Centered weight normalization in accelerating training of deep neural networks. In: ICCV (2017)
21. Huang, L., Yang, D., Lang, B., Deng, J.: Decorrelated batch normalization. In: CVPR (2018)
22. Ioffe, S., Szegedy, C.: Batch normalization: accelerating deep network training by reducing internal covariate shift. In: ICML (2015)
23. Karakida, R., Akaho, S., Amari, S.: Universal statistics of Fisher information in deep neural networks: mean field approach. In: AISTATS (2019)
24. Kingma, D.P., Ba, J.: Adam: a method for stochastic optimization. CoRR abs/1412.6980 (2014)
25. Kohler, J., Daneshmand, H., Lucchi, A., Zhou, M., Neymeyr, K., Hofmann, T.: Towards a theoretical understanding of batch normalization. arXiv preprint arXiv:1805.10694 (2018)
26. LeCun, Y., Bengio, Y., Hinton, G.E.: Deep learning. Nature **521**, 436–444 (2015)
27. LeCun, Y., Bottou, L., Orr, G.B., Müller, K.-R.: Efficient BackProp. In: Orr, G.B., Müller, K.-R. (eds.) Neural Networks: Tricks of the Trade. LNCS, vol. 1524, pp. 9–50. Springer, Heidelberg (1998). https://doi.org/10.1007/3-540-49430-8_2
28. LeCun, Y., Kanter, I., Solla, S.A.: Second order properties of error surfaces: learning time and generalization. In: NeurIPS (1990)
29. Martens, J.: Deep learning via Hessian-free optimization. In: ICML, pp. 735–742 (2010)
30. Martens, J.: New perspectives on the natural gradient method. CoRR abs/1412.1193 (2014)
31. Martens, J., Grosse, R.: Optimizing neural networks with Kronecker-factored approximate curvature. In: ICML (2015)
32. Martens, J., Sutskever, I., Swersky, K.: Estimating the Hessian by back-propagating curvature. In: ICML (2012)
33. Montavon, G., Müller, K.-R.: Deep Boltzmann machines and the centering trick. In: Montavon, G., Orr, G.B., Müller, K.-R. (eds.) Neural Networks: Tricks of the Trade. LNCS, vol. 7700, pp. 621–637. Springer, Heidelberg (2012). https://doi.org/10.1007/978-3-642-35289-8_33
34. Papyan, V.: The full spectrum of deep net Hessians at scale: dynamics with sample size. CoRR abs/1811.07062 (2018)
35. Pascanu, R., Bengio, Y.: Revisiting natural gradient for deep networks. In: ICLR (2014)
36. Roux, N.L., Manzagol, P., Bengio, Y.: Topmoumoute online natural gradient algorithm. In: NeurIPS, pp. 849–856 (2007)
37. Sagun, L., Evci, U., Güney, V.U., Dauphin, Y.N., Bottou, L.: Empirical analysis of the Hessian of over-parametrized neural networks. CoRR abs/1706.04454 (2017)

38. Santurkar, S., Tsipras, D., Ilyas, A., Madry, A.: How does batch normalization help optimization? In: NeurIPS (2018)
39. Saxe, A.M., McClelland, J.L., Ganguli, S.: Exact solutions to the nonlinear dynamics of learning in deep linear neural networks. In: ICLR (2014)
40. Schraudolph, N.N.: Accelerated gradient descent by factor-centering decomposition. Technical report (1998)
41. Sun, K., Nielsen, F.: Relative Fisher information and natural gradient for learning large modular models. In: ICML (2017)
42. Ulyanov, D., Vedaldi, A., Lempitsky, V.S.: Instance normalization: the missing ingredient for fast stylization. CoRR abs/1607.08022 (2016)
43. Wei, M., Stokes, J., Schwab, D.J.: Mean-field analysis of batch normalization. arXiv:1903.02606 (2019)
44. Wiesler, S., Ney, H.: A convergence analysis of log-linear training. In: NeurIPS (2011)
45. Wu, S., Li, G., Deng, L., Liu, L., Xie, Y., Shi, L.: L1-norm batch normalization for efficient training of deep neural networks. CoRR (2018)
46. Wu, Y., He, K.: Group normalization. In: Ferrari, V., Hebert, M., Sminchisescu, C., Weiss, Y. (eds.) ECCV 2018. LNCS, vol. 11217, pp. 3–19. Springer, Cham (2018). https://doi.org/10.1007/978-3-030-01261-8_1
47. Yang, G., Pennington, J., Rao, V., Sohl-Dickstein, J., Schoenholz, S.S.: A mean field theory of batch normalization. In: ICLR (2019)
48. Zagoruyko, S., Komodakis, N.: Wide residual networks. In: BMVC (2016)
49. Zeiler, M.D.: ADADELTA: an adaptive learning rate method. CoRR abs/1212.5701 (2012)
50. Zhang, H., Chen, W., Liu, T.Y.: On the local Hessian in back-propagation. In: NeurIPS (2018)

RAFT: Recurrent All-Pairs Field Transforms for Optical Flow

Zachary Teed[✉] and Jia Deng

Princeton University, Princeton, USA
{zteed,jiadeng}@cs.princeton.edu

Abstract. We introduce Recurrent All-Pairs Field Transforms (RAFT), a new deep network architecture for optical flow. RAFT extracts per-pixel features, builds multi-scale 4D correlation volumes for all pairs of pixels, and iteratively updates a flow field through a recurrent unit that performs lookups on the correlation volumes. RAFT achieves state-of-the-art performance. On KITTI, RAFT achieves an F1-all error of 5.10%, a 16% error reduction from the best published result (6.10%). On Sintel (final pass), RAFT obtains an end-point-error of 2.855 pixels, a 30% error reduction from the best published result (4.098 pixels). In addition, RAFT has strong cross-dataset generalization as well as high efficiency in inference time, training speed, and parameter count. Code is available at https://github.com/princeton-vl/RAFT.

1 Introduction

Optical flow is the task of estimating per-pixel motion between video frames. It is a long-standing vision problem that remains unsolved. The best systems are limited by difficulties including fast-moving objects, occlusions, motion blur, and textureless surfaces.

Optical flow has traditionally been approached as a hand-crafted optimization problem over the space of dense displacement fields between a pair of images [13,21,50]. Generally, the optimization objective defines a trade-off between a *data* term which encourages the alignment of visually similar image regions and a *regularization* term which imposes priors on the plausibility of motion. Such an approach has achieved considerable success, but further progress has appeared challenging, due to the difficulties in hand-designing an optimization objective that is robust to a variety of corner cases.

Recently, deep learning has been shown as a promising alternative to traditional methods. Deep learning can side-step formulating an optimization problem and train a network to directly predict flow. Current deep learning methods [20,22,25,42,48] have achieved performance comparable to the best traditional methods while being significantly faster at inference time. A key question

Electronic supplementary material The online version of this chapter (https:// doi.org/10.1007/978-3-030-58536-5_24) contains supplementary material, which is available to authorized users.

A. Vedaldi et al. (Eds.): ECCV 2020, LNCS 12347, pp. 402–419, 2020.
https://doi.org/10.1007/978-3-030-58536-5_24

for further research is designing effective architectures that perform better, train more easily and generalize well to novel scenes.

We introduce Recurrent All-Pairs Field Transforms (RAFT), a new deep network architecture for optical flow. RAFT enjoys the following strengths:

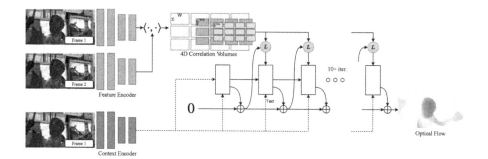

Fig. 1. RAFT consists of 3 main components: (1) A feature encoder that extracts per-pixel features from both input images, along with a context encoder that extracts features from only I_1. (2) A correlation layer which constructs a 4D $W \times H \times W \times H$ correlation volume by taking the inner product of all pairs of feature vectors. The last 2-dimensions of the 4D volume are pooled at multiple scales to construct a set of multi-scale volumes. (3) An *update operator* which recurrently updates optical flow by using the current estimate to look up values from the set of correlation volumes.

- *State-of-the-art accuracy:* On KITTI [18], RAFT achieves an F1-all error of 5.10%, a 16% error reduction from the best published result (6.10%). On Sintel [11] (final pass), RAFT obtains an end-point-error of 2.855 pixels, a 30% error reduction from the best published result (4.098 pixels).
- *Strong generalization:* When trained only on synthetic data, RAFT achieves an end-point-error of 5.04 pixels on KITTI [18], a 40% error reduction from the best prior deep network trained on the same data (8.36 pixels).
- *High efficiency:* RAFT processes 1088×436 videos at 10 frames per second on a 1080Ti GPU. It trains with 10X fewer iterations than other architectures. A smaller version of RAFT with 1/5 of the parameters runs at 20 frames per second while still outperforming all prior methods on Sintel.

RAFT consists of three main components: (1) a feature encoder that extracts a feature vector for each pixel; (2) a correlation layer that produces a 4D correlation volume for all pairs of pixels, with subsequent pooling to produce lower resolution volumes; (3) a recurrent GRU-based *update operator* that retrieves values from the correlation volumes and iteratively updates a flow field initialized at zero. Figure 1 illustrates the design of RAFT.

The RAFT architecture is motivated by traditional optimization-based approaches. The feature encoder extracts per-pixel features. The correlation layer computes visual similarity between pixels. The update operator mimics the steps of an iterative optimization algorithm. But unlike traditional approaches, features and motion priors are not handcrafted but learned—learned by the feature encoder and the update operator respectively.

The design of RAFT draws inspiration from many existing works but is substantially novel. First, RAFT maintains and updates a single fixed flow field at high resolution. This is different from the prevailing coarse-to-fine design in prior work [22,23,42,48,49], where flow is first estimated at low resolution and upsampled and refined at high resolution. By operating on a single high-resolution flow field, RAFT overcomes several limitations of a coarse-to-fine cascade: the difficulty of recovering from errors at coarse resolutions, the tendency to miss small fast-moving objects, and the many training iterations (often over 1M) typically required for training a multi-stage cascade.

Second, the update operator of RAFT is recurrent and lightweight. Many recent works [22,24,25,42,48] have included some form of iterative refinement, but do not tie the weights across iterations [22,42,48] and are therefore limited to a fixed number of iterations. To our knowledge, IRR [24] is the only deep learning approach [24] that is recurrent. It uses FlowNetS [15] or PWC-Net [42] as its recurrent unit. When using FlowNetS, it is limited by the size of the network (38M parameters) and is only applied up to 5 iterations. When using PWC-Net, iterations are limited by the number of pyramid levels. In contrast, our update operator has only 2.7M parameters and can be applied 100+ times during inference without divergence.

Third, the update operator has a novel design, which consists of a convolutional GRU that performs lookups on 4D multi-scale correlation volumes; in contrast, refinement modules in prior work typically use only plain convolution or correlation layers.

We conduct experiments on Sintel [11] and KITTI [18]. Results show that RAFT achieves state-of-the-art performance on both datasets. In addition, we validate various design choices of RAFT through extensive ablation studies.

2   Related Work

Optical Flow as Energy Minimization. Optical flow has traditionally been treated as an energy minimization problem which imposes a tradeoff between a *data* term and a *regularization* term. Horn and Schnuck [21] formulated optical flow as a continuous optimization problem using a variational framework, and were able to estimate a dense flow field by performing gradient steps. Black and Anandan [9] addressed problems with oversmoothing and noise sensitivity by introducing a robust estimation framework. TV-L1 [50] replaced the quadratic penalties with an L1 data term and total variation regularization, which allowed for motion discontinuities and was better equipped to handle outliers. Improvements have been made by defining better matching costs [10,45] and regularization terms [38].

Such continuous formulations maintain a single estimate of optical flow which is refined at each iteration. To ensure a smooth objective function, a first order

Taylor approximation is used to model the data term. As a result, they only work well for small displacements. To handle large displacements, the coarse-to-fine strategy is used, where an image pyramid is used to estimate large displacements at low resolution, then small displacements refined at high resolution. But this coarse-to-fine strategy may miss small fast-moving objects and have difficulty recovering from early mistakes. Like continuous methods, we maintain a single estimate of optical flow which is refined with each iteration. However, since we build correlation volumes for all pairs at both high resolution and low resolution, each local update uses information about both small and large displacements. In addition, instead of using a subpixel Taylor approximation of the data term, our update operator learns to propose the descent direction.

More recently, optical flow has also been approached as a discrete optimization problem [13,35,47] using a global objective. One challenge of this approach is the massive size of the search space, as each pixel can be reasonably paired with thousands of points in the other frame. Menez et al. [35] pruned the search space using feature descriptors and approximated the global MAP estimate using message passing. Chen et al. [13] showed that by using the distance transform, solving the global optimization problem over the full space of flow fields is tractable. DCFlow [47] showed further improvements by using a neural network as a feature descriptor, and constructed a 4D cost volume over all pairs of features. The 4D cost volume was then processed using the Semi-Global Matching (SGM) algorithm [19]. Like DCFlow, we also constructed 4D cost volumes over learned features. However, instead of processing the cost volumes using SGM, we use a neural network to estimate flow. Our approach is end-to-end differentiable, meaning the feature encoder can be trained with the rest of the network to directly minimize the error of the final flow estimate. In contrast, DCFlow requires their network to be trained using an embedding loss between pixels; it cannot be trained directly on optical flow because their cost volume processing is not differentiable.

Direct Flow Prediction. Neural networks have been trained to directly predict optical flow between a pair of frames, side-stepping the optimization problem completely. Coarse-to-fine processing has emerged as a popular ingredient in many recent works [8,20,22–24,42,48,49,51]. In contrast, our method maintains and updates a single high-resolution flow field.

Iterative Refinement for Optical Flow. Many recent works have used iterative refinement to improve results on optical flow [22,25,39,42,48] and related tasks [28,29,44,52]. Ilg et al. [25] applied iterative refinement to optical flow by stacking multiple FlowNetS and FlowNetC modules in series. SpyNet [39], PWC-Net [42], LiteFlowNet [22], and VCN [48] apply iterative refinement using coarse-to-fine pyramids. The main difference of these approaches from ours is that they do not share weights between iterations.

More closely related to our approach is IRR [24], which builds off of the FlownetS and PWC-Net architecture but shares weights between refinement networks. When using FlowNetS, it is limited by the size of the network (38M parameters) and is only applied up to 5 iterations. When using PWC-Net, iterations are limited by the number of pyramid levels. In contrast, we use a much simpler refinement module (2.7M parameters) which can be applied for 100+ iterations during inference without divergence. Our method also shares similar-ites with Devon [31], namely the construction of the cost volume without warping and fixed resolution updates. However, Devon does not have any recurrent unit. It also differs from ours regarding large displacements. Devon handles large displacements using a dilated cost volume while our approach pools the correlation volume at multiple resolutions.

Our method also has ties to TrellisNet [5] and Deep Equilibrium Models (DEQ) [6]. Trellis net uses depth tied weights over a large number of layers, DEQ simulates an infinite number of layers by solving for the fixed point directly. TrellisNet and DEQ were designed for sequence modeling tasks, but we adopt the core idea of using a large number of weight-tied units. Our update operator uses a modified GRU block [14], which is similar to the LSTM block used in TrellisNet. We found that this structure allows our update operator to more easily converge to a fixed flow field.

Learning to Optimize. Many problems in vision can be formulated as an optimization problem. This has motivated several works to embed optimization problems into network architectures [3,4,32,43,44]. These works typically use a network to predict the inputs or parameters of the optimization problem, and then train the network weights by backpropagating the gradient through the solver, either implicitly [3,4] or unrolling each step [32,43]. However, this technique is limited to problems with an objective that can be easily defined.

Another approach is to learn iterative updates directly from data [1,2]. These approaches are motivated by the fact that first order optimizers such as Primal Dual Hybrid Gradient (PDHG) [12] can be expressed as a sequence of iterative update steps. Instead of using an optimizer directly, Adler et al. [1] proposed building a network which mimics the updates of a first order algorithm. This approach has been applied to inverse problems such as image denoising [26], tomographic reconstruction [2], and novel view synthesis [17]. TVNet [16] imple-mented the TV-L1 algorithm as a computation graph, which enabled the training the TV-L1 parameters. However, TVNet operates directly based on intensity gradients instead of learned features, which limits the achievable accuracy on challenging datasets such as Sintel.

Our approach can be viewed as learning to optimize: our network uses a large number of update blocks to emulate the steps of a first-order optimization algorithm. However, unlike prior work, we never explicitly define a gradient with respect to some optimization objective. Instead, our network retrieves features from correlation volumes to propose the descent direction.

3   Approach

Given a pair of consecutive RGB images, I_1, I_2, we estimate a dense displacement field $(\mathbf{f}^1, \mathbf{f}^2)$ which maps each pixel (u, v) in I_2 to its corresponding coordinates $(u', v') = (u + f^1(u), v + f^2(v))$ in I_2. An overview of our approach is given in Fig. 1. Our method can be distilled down to three stages: (1) feature extraction,

(2) computing visual similarity, and (3) iterative updates, where all stages are differentiable and composed into an end-to-end trainable architecture.

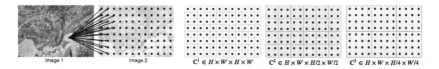

Fig. 2. Building correlation volumes. Here we depict 2D slices of a full 4D volume. For a feature vector in I_1, we take the inner product with all pairs in I_2, generating a 4D $W \times H \times W \times H$ volume (each pixel in I_2 produces a 2D response map). The volume is pooled using average pooling with kernel sizes $\{1, 2, 4, 8\}$.

3.1   Feature Extraction

Features are extracted from the input images using a convolutional network. The feature encoder network is applied to both I_1 and I_2 and maps the input images to dense feature maps at a lower resolution. Our encoder, g_θ outputs features at 1/8 resolution $g_\theta : \mathbb{R}^{H \times W \times 3} \mapsto \mathbb{R}^{H/8 \times W/8 \times D}$ where we set $D = 256$. The feature encoder consists of 6 residual blocks, 2 at 1/2 resolution, 2 at 1/4 resolution, and 2 at 1/8 resolution (more details in the supplemental material).

We additionally use a context network. The context network extracts features only from the first input image I_1. The architecture of the context network, h_θ is identical to the feature extraction network. Together, the feature network g_θ and the context network h_θ form the first stage of our approach, which only need to be performed once.

3.2   Computing Visual Similarity

We compute visual similarity by constructing a full correlation volume between all pairs. Given image features $g_\theta(I_1) \in \mathbb{R}^{H \times W \times D}$ and $g_\theta(I_2) \in \mathbb{R}^{H \times W \times D}$, the correlation volume is formed by taking the dot product between all pairs of feature vectors. The correlation volume, $\mathbf{C}$, can be efficiently computed as a single matrix multiplication.

$$\mathbf{C}(g_\theta(I_1), g_\theta(I_2)) \in \mathbb{R}^{H \times W \times H \times W}, \qquad C_{ijkl} = \sum_h g_\theta(I_1)_{ijh} \cdot g_\theta(I_2)_{klh} \qquad (1)$$

Correlation Pyramid: We construct a 4-layer pyramid $\{\mathbf{C}^1, \mathbf{C}^2, \mathbf{C}^3, \mathbf{C}^4\}$ by pooling the last two dimensions of the correlation volume with kernel sizes 1, 2, 4, and 8 and equivalent stride (Fig. 2). Thus, volume $\mathbf{C}^k$ has dimensions $H \times W \times H/2^k \times W/2^k$. The set of volumes gives information about both large and small displacements; however, by maintaining the first 2 dimensions (the I_1 dimensions) we maintain high resolution information, allowing our method to recover the motions of small fast-moving objects.

Correlation Lookup: We define a lookup operator $L_{\mathbf{C}}$ which generates a feature map by indexing from the correlation pyramid. Given a current estimate of optical flow $(\mathbf{f}^1, \mathbf{f}^2)$, we map each pixel $\mathbf{x} = (u, v)$ in I_1 to its estimated correspondence in I_2: $\mathbf{x}' = (u + f^1(u), v + f^2(v))$. We then define a local grid around $\mathbf{x}'$

$$\mathcal{N}(\mathbf{x}')_r = \{\mathbf{x}' + \mathbf{dx} \mid \mathbf{dx} \in \mathbb{Z}^2, \|\mathbf{dx}\|_1 \leq r\} \qquad (2)$$

as the set of integer offsets which are within a radius of r units of $\mathbf{x}'$ using the L1 distance. We use the local neighborhood $\mathcal{N}(\mathbf{x}')_r$ to index from the correlation volume. Since $\mathcal{N}(\mathbf{x}')_r$ is a grid of real numbers, we use bilinear sampling.

  We perform lookups on all levels of the pyramid, such that the correlation volume at level k, $\mathbf{C}^k$, is indexed using the grid $\mathcal{N}(\mathbf{x}'/2^k)_r$. A constant radius across levels means larger context at lower levels: for the lowest level, $k = 4$ using a radius of 4 corresponds to a range of 256 pixels at the original resolution. The values from each level are then concatenated into a single feature map.

Efficient Computation for High Resolution Images: The all pairs correlation scales $O(N^2)$ where N is the number of pixels, but only needs to be computed once and is constant in the number of iterations M. However, there exists an equivalent implementation of our approach which scales $O(NM)$ exploiting the linearity of the inner product and average pooling. Consider the cost volume at level m, $\mathbf{C}^m_{ijkl}$, and feature maps $g^{(1)} = g_\theta(I_1)$, $g^{(2)} = g_\theta(I_2)$:

$$\mathbf{C}^m_{ijkl} = \frac{1}{2^{2m}} \sum_p^{2^m} \sum_q^{2^m} \langle g^{(1)}_{i,j}, g^{(2)}_{2^m k+p, 2^m l+q} \rangle = \langle g^{(1)}_{i,j}, \frac{1}{2^{2m}} (\sum_p^{2^m} \sum_q^{2^m} g^{(2)}_{2^m k+p, 2^m l+q}) \rangle$$

which is the average over the correlation response in the $2^m \times 2^m$ grid. This means that the value at $\mathbf{C}^m_{ijkl}$ can be computed as the inner product between the feature vector $g_\theta(I_1)_{ij}$ and $g_\theta(I_2)$ pooled with kernel size $2^m \times 2^m$.

  In this alternative implementation, we do not precompute the correlations, but instead precompute the pooled image feature maps. In each iteration, we compute each correlation value on demand—only when it is looked up. This gives a complexity of $O(NM)$.

  We found empirically that precomputing all pairs is easy to implement and not a bottleneck, due to highly optimized matrix routines on GPUs—even for 1088×1920 videos it takes only 17% of total inference time. Note that we can always switch to the alternative implementation should it become a bottleneck.

3.3   Iterative Updates

Our update operator estimates a sequence of flow estimates $\{\mathbf{f}_1, ..., \mathbf{f}_N\}$ from an initial starting point $\mathbf{f}_0 = \mathbf{0}$. With each iteration, it produces an update direction $\Delta\mathbf{f}$ which is applied to the current estimate: $\mathbf{f}_{k+1} = \Delta\mathbf{f} + \mathbf{f}_{k+1}$.

  The update operator takes flow, correlation, and a latent hidden state as input, and outputs the update $\Delta\mathbf{f}$ and an updated hidden state. The architecture of our update operator is designed to mimic the steps of an optimization algorithm. As such, we used tied weights across depth and use bounded activations to encourage convergence to a fixed point. The update operator is trained to perform updates such that the sequence converges to a fixed point $\mathbf{f}_k \rightarrow \mathbf{f}^*$.

Initialization: By default, we initialize the flow field to 0 everywhere, but our iterative approach gives us the flexibility to experiment with alternatives. When applied to video, we test *warm-start* initialization, where optical flow from the previous pair of frames is forward projected to the next pair of frames with occlusion gaps filled in using nearest neighbor interpolation.

Inputs: Given the current flow estimate $\mathbf{f}^k$, we use it to retrieve correlation features from the correlation pyramid as described in Sect. 3.2. The correlation features are then processed by 2 convolutional layers. Additionally, we apply 2 convolutional layers to the flow estimate itself to generate flow features. Finally, we directly inject the input from the context network. The input feature map is then taken as the concatenation of the correlation, flow, and context features.

Update: A core component of the update operator is a gated activation unit based on the GRU cell, with fully connected layers replaced with convolutions:

$$z_t = \sigma(\text{Conv}_{3\times3}([h_{t-1}, x_t], W_z)) \tag{3}$$

$$r_t = \sigma(\text{Conv}_{3\times3}([h_{t-1}, x_t], W_r)) \tag{4}$$

$$\tilde{h}_t = \tanh(\text{Conv}_{3\times3}([r_t \odot h_{t-1}, x_t], W_h)) \tag{5}$$

$$h_t = (1 - z_t) \odot h_{t-1} + z_t \odot \tilde{h}_t \tag{6}$$

where x_t is the concatenation of flow, correlation, and context features previously defined. We also experiment with a separable ConvGRU unit, where we replace the 3×3 convolution with two GRUs: one with a 1×5 convolution and one with a 5×1 convolution to increase the receptive field without significantly increasing the size of the model.

Flow Prediction: The hidden state outputted by the GRU is passed through two convolutional layers to predict the flow update $\Delta\mathbf{f}$. The output flow is at $1/8$ resolution of the input image. During training and evaluation, we upsample the predicted flow fields to match the resolution of the ground truth.

Upsampling: The network outputs optical flow at $1/8$ resolution. We upsample the optical flow to full resolution by taking the full resolution flow at each pixel to be the convex combination of a 3×3 grid of its coarse resolution neighbors. We use two convolutional layers to predict a $H/8 \times W/8 \times (8 \times 8 \times 9)$ mask and perform softmax over the weights of the 9 neighbors. The final high resolution flow field is found by using the mask to take a weighted combination over the neighborhood, then permuting and reshaping to a $H \times W \times 2$ dimensional flow field. This layer can be directly implemented in PyTorch using the `unfold` function.

3.4   Supervision

We supervised our network on the l_1 distance between the predicted and ground truth flow over the full sequence of predictions, $\{\mathbf{f}_1, ..., \mathbf{f}_N\}$, with exponentially increasing weights. Given ground truth flow $\mathbf{f}_{gt}$, the loss is defined as

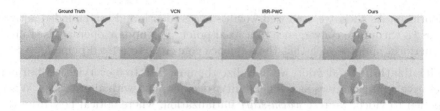

Fig. 3. Flow predictions on the Sintel test set.

$$\mathcal{L} = \sum_{i=1}^{N} \gamma^{i-N} ||\mathbf{f}_{gt} - \mathbf{f}_i||_1 \tag{7}$$

where we set $\gamma = 0.8$ in our experiments.

4   Experiments

We evaluate RAFT on Sintel [11] and KITTI [18]. Following previous works, we pretrain our network on FlyingChairs [15] and FlyingThings [33], followed by dataset specific finetuning. Our method achieves state-of-the-art performance on both Sintel (both clean and final passes) and KITTI. Additionally, we test our method on 1080p video from the DAVIS dataset [37] to demonstrate that our method scales to videos of very high resolutions.

Implementation Details: RAFT is implemented in PyTorch [36]. All modules are initialized from scratch with random weights. During training, we use the AdamW [30] optimizer and clip gradients to the range $[-1, 1]$. Unless otherwise noted, we evaluate after 32 flow updates on Sintel and 24 on KITTI. For every update, $\Delta\mathbf{f} + \mathbf{f}_k$, we only backpropgate the gradient through the $\Delta\mathbf{f}$ branch, and zero the gradient through the $\mathbf{f}_k$ branch as suggested by [20].

Training Schedule: We train RAFT using two 2080Ti GPUs. We pretrain on FlyingThings for 100k iterations with a batch size of 12, then train for 100k iterations on FlyingThings3D with a batch size of 6. We finetune on Sintel for another 100k by combining data from Sintel [11], KITTI-2015 [34], and HD1K [27] similar to MaskFlowNet [51] and PWC-Net+ [41]. Finally, we finetune on KITTI-2015 for an additionally 50k iterations using the weights from the model finetuned on Sintel. Details on training and data augmentation are provided in the supplemental material. For comparison with prior work, we also include results from our model when finetuning only on Sintel and only on KITTI.

4.1   Sintel

We train our model using the FlyingChairs→FlyingThings schedule and then evaluate on the Sintel dataset using the *train* split for validation. Results are shown in Table 1 and Fig. 3, and we split results based on the data used for

Fig. 4. Flow predictions on the KITTI test set.

training. C + T means that the models are trained on FlyingChairs(C) and FlyingThings(T), while +ft indicates the model is finetuned on Sintel data. Like PWC-Net+ [41] and MaskFlowNet [51] we include data from KITTI and HD1K when finetuning. We train 3 times with different seeds, and report results using the model with the median accuracy on the clean pass of Sintel (train).

When using C+T for training, our method outperforms all existing approaches, despite using a significantly shorter training schedule. Our method achieves an average EPE (end-point-error) of 1.43 on the Sintel (train) clean pass, which is a 29% lower error than FlowNet2. These results demonstrates good cross dataset generalization. One of the reasons for better generalization is the structure of our network. By constraining optical flow to be the product of a series of identical update steps, we force the network to learn an update operator which mimics the updates of a first-order descent algorithm. This constrains the search space, reduces the risk of over-fitting, and leads to faster training and better generalization.

When evaluating on the Sintel (test) set, we finetune on the combined clean and final passes of the training set along with KITTI and HD1K data. Our method ranks 1st on both the Sintel clean and final passes, and outperforms all prior work by 0.9 pixels (36%) on the clean pass and 1.2 pixels (30%) on the final pass. We evaluate two versions of our model, Ours (two-frame) uses zero initialization, while Ours (warp-start) initializes flow by forward projecting the flow estimate from the previous frame. Since our method operates at a single resolution, we can initialize the flow estimate to utilize motion smoothness from past frames, which cannot be easily done using the coarse-to-fine model.

4.2   KITTI

We also evaluate RAFT on KITTI and provide results in Table 1 and Fig. 4. We first evaluate cross-dataset generalization by evaluating on the KITTI-15 (train) split after training on Chairs(C) and FlyingThings(T). Our method outperforms prior works by a large margin, improving EPE (end-point-error) from 8.36 to 5.04, which shows that the underlying structure of our network facilitates generalization. Our method ranks 1st on the KITTI leaderboard among all optical flow methods.

4.3   Ablations

We perform a set of ablation experiments to show the relative importance of each component. All ablated versions are trained on FlyingChairs(C) + FlyingThings(T). Results of the ablations are shown in Table 2. In each section of the table, we test a specific component of our approach in isolation, the settings which are used in our final model is underlined. Below we describe each of the experiments in more detail.

Table 1. Results on Sintel and KITTI datasets. We test the generalization performance on Sintel(train) after training on FlyingChairs(C) and FlyingThing(T), and outperform all existing methods on both the clean and final pass. The bottom two sections show the performance of our model on public leaderboards after dataset specific finetuning. S/K includes methods which use only Sintel data for finetuning on Sintel and only KITTI data when finetuning on KITTI. +S+K+H includes methods which combine KITTI, HD1K, and Sintel data when finetuning on Sintel. Ours (warm-start) ranks 1st on both the Sintel clean and final passes, and 1st among all flow approaches on KITTI. ([1]FlowNet2 originally reported results on the disparity split of Sintel, 3.54 is the EPE when their model is evaluated on the standard data [22]. [2][23] finds that HD1K data does not help significantly during Sintel finetuning and reports results without it.)

Training data	Method	Sintel (train)		KITTI-15 (train)		Sintel (test)		KITTI-15 (test)
		Clean	Final	F1-epe	F1-all	Clean	Final	F1-all
–	FlowFields [7]	–	–	–	–	3.75	5.81	15.31
–	FlowFields++ [40]	–	–	–	–	2.94	5.49	14.82
S	DCFlow [47]	–	–	–	–	3.54	5.12	14.86
S	MRFlow [46]	–	–	–	–	2.53	5.38	12.19
C + T	HD3 [49]	3.84	8.77	13.17	24.0	–	–	–
	LiteFlowNet [22]	2.48	4.04	10.39	28.5	–	–	–
	PWC-Net [42]	2.55	3.93	10.35	33.7	–	–	–
	LiteFlowNet2 [23]	2.24	3.78	8.97	25.9	–	–	–
	VCN [48]	2.21	3.68	8.36	25.1	–	–	–
	MaskFlowNet [51]	2.25	3.61	–	23.1	–	–	–
	FlowNet2 [25]	2.02	3.54[1]	10.08	30.0	3.96	6.02	–
	Ours (small)	2.21	3.35	7.51	26.9	–	–	–
	Ours (2-view)	**1.43**	**2.71**	**5.04**	**17.4**	–	–	–
C + T + S/K	FlowNet2 [25]	(1.45)	(2.01)	(2.30)	(6.8)	4.16	5.74	11.48
	HD3 [49]	(1.87)	(1.17)	(1.31)	(4.1)	4.79	4.67	6.55
	IRR-PWC [24]	(1.92)	(2.51)	(1.63)	(5.3)	3.84	4.58	7.65
	VCN [48]	(1.66)	(2.24)	(1.16)	(4.1)	2.81	4.40	**6.30**
	ScopeFlow [8]	–	–	–	–	3.59	4.10	6.82
	Ours (2-view, bilinear)	(1.09)	(1.53)	(1.07)	(3.9)	2.77	3.61	**6.30**
	Ours (warm-start, bilinear)	(1.10)	(1.61)	–	–	**2.42**	**3.39**	–
C + T + S + K + H	LiteFlowNet2[b] [23]	(1.30)	(1.62)	(1.47)	(4.8)	3.45	4.90	7.74
	PWC-Net+ [41]	(1.71)	(2.34)	(1.50)	(5.3)	3.45	4.60	7.72
	MaskFlowNet [51]	–	–	–	–	2.52	4.17	6.10
	Ours (2-view)	(0.76)	(1.22)	(0.63)	(1.5)	1.94	3.18	**5.10**
	Ours (warm-start)	(0.77)	(1.27)	–	–	**1.61**	**2.86**	–

Architecture of Update Operator: We use a gated activation unit based on the GRU cell. We experiment with replacing the convolutional GRU with a set of 3 convolutional layers with ReLU activation. We achieve better performance by using the GRU block, likely because the gated activation makes it easier for the sequence of flow estimates to converge.

Weight Tying: By default, we tied the weights across all instances of the update operator. Here, we test a version of our approach where each update operator

Table 2. Ablation experiments. Settings used in our final model are underlined. See Sect. 4.3 for details.

Experiment	Method	Sintel (train)		KITTI-15 (train)		Parameters
		Clean	Final	F1-epe	F1-all	
Reference Model (bilinear upsampling), Training: 100k(C) → 60k(T)						
Update Op.	ConvGRU	1.63	2.83	5.54	19.8	4.8M
	Conv	2.04	3.21	7.66	26.1	4.1M
Tying	Tied Weights	1.63	2.83	5.54	19.8	4.8M
	Untied Weights	1.96	3.20	7.64	24.1	32.5M
Context	Context	1.63	2.83	5.54	19.8	4.8M
	No Context	1.93	3.06	6.25	23.1	3.3M
Feature Scale	Single-Scale	1.63	2.83	5.54	19.8	4.8M
	Multi-Scale	2.08	3.12	6.91	23.2	6.6M
Lookup Radius	0	3.41	4.53	23.6	44.8	4.7M
	1	1.80	2.99	6.27	21.5	4.7M
	2	1.78	2.82	5.84	21.1	4.8M
	4	1.63	2.83	5.54	19.8	4.8M
Correlation Pooling	No	1.95	3.02	6.07	23.2	4.7M
	Yes	1.63	2.83	5.54	19.8	4.8M
Correlation Range	32px	2.91	4.48	10.4	28.8	4.8M
	64px	2.06	3.16	6.24	20.9	4.8M
	128px	1.64	2.81	6.00	19.9	4.8M
	All-Pairs	1.63	2.83	5.54	19.8	4.8M
Features for Refinement	Correlation	1.63	2.83	5.54	19.8	4.8M
	Warping	2.27	3.73	11.83	32.1	2.8M
Reference Model (convex upsampling), Training: 100k(C) → 100k(T)						
Upsampling	Convex	1.43	2.71	5.04	17.4	5.3M
	Bilinear	1.60	2.79	5.17	19.2	4.8M
Inference Updates	1	4.04	5.45	15.30	44.5	5.3M
	3	2.14	3.52	8.98	29.9	5.3M
	8	1.61	2.88	5.99	19.6	5.3M
	32	1.43	2.71	5.00	17.4	5.3M
	100	1.41	2.72	4.95	17.4	5.3M
	200	1.40	2.73	4.94	17.4	5.3M

learns a separate set of weights. Accuracy is better when weights are tied and the parameter count is significantly lower.

Context: We test the importance of context by training a model with the context network removed. Without context, we still achieve good results, outperforming all existing works on both Sintel and KITTI. But context is helpful. Directly injecting image features into the update operator likely allows spatial information to be better aggregated within motion boundaries.

Feature Scale: By default, we extract features at a single resolution. We also try extracting features at multiple resolutions by building a correlation volume at each scale separately. Single resolution features simplifies the network architecture and allows fine-grained matching even at large displacements.

Lookup Radius: The lookup radius specifies the dimensions of the grid used in the lookup operation. When a radius of 0 is used, the correlation volume is retrieved at a single point. Surprisingly, we can still get a rough estimate of flow when the radius is 0, which means the network is learning to use 0'th order information. However, we see better results as the radius is increased.

Correlation Pooling: We output features at a single resolution and then perform pooling to generate multiscale volumes. Here we test the impact when this pooling is removed. Results are better with pooling, because large and small displacements are both captured.

Correlation Range: Instead of all-pairs correlation, we also try constructing the correlation volume only for a local neighborhood around each pixel. We try a range of 32 pixels, 64 pixels, and 128 pixels. Overall we get the best results when the all-pairs are used, although a 128px range is sufficient to perform well on Sintel because most displacements fall within this range. That said, all-pairs is still preferable because it eliminates the need to specify a range. It is also more convenient to implement: it can be computed using matrix multiplication allowing our approach to be implemented entirely in PyTorch.

Features for Refinement: We compute visual similarity by building a correlation volume between all pairs of pixels. In this experiment, we try replacing the correlation volume with a warping layer, which uses the current estimate of optical flow to warp features from I_2 onto I_1 and then estimates the residual displacement. While warping is still competitive with prior work on Sintel, correlation performs significantly better, especially on KITTI.

Features for Refinement: We compute visual similarity by building a correlation volume between all pairs of pixels. In this experiment, we try replacing the correlation volume with a warping layer, which uses the current estimate of optical flow to warp features from I_2 onto I_1 and then estimates the residual displacement. While warping is still competitive with prior work on Sintel, correlation performs significantly better, especially on KITTI.

Upsampling: RAFT outputs flow fields at 1/8 resolution. We compare bilinear upsampling to our learned upsampling module. The upsampling module produces better results, particularly near motion boundaries.

Inference Updates: Although we unroll 12 updates during training, we can apply an arbitrary number of updates during inference. In Table 2 we provide numerical results for selected number of updates, and test an extreme case of 200 to show that our method doesn't diverge. Our method quickly converges, surpassing PWC-Net after 3 updates and FlowNet2 after 6 updates, but continues to improve with more updates.

4.4   Timing and Parameter Counts

Inference time and parameter counts are shown in Fig. 5. Accuracy is determined by performance on the Sintel(train) final pass after training on FlyingChairs and FlyingThings $(C + T)$. In these plots, we report accuracy and timing after 10 iterations, and we time our method using a GTX 1080Ti GPU. Parameters counts for other methods are taken as reported in their papers, and we report times when run on our hardware. RAFT is more efficient in terms of parameter count, inference time, and training iterations. Ours-S uses only 1M parameters, but outperforms PWC-Net and VCN which are more than 6x larger. We provide an additional table with numerical values for parameters, timing, and training iterations in the supplemental material.

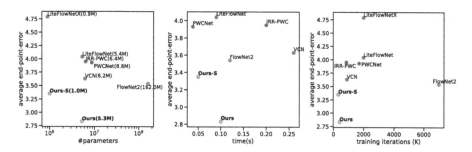

Fig. 5. Plots comparing parameter counts, inference time, and training iterations vs. accuracy. Accuracy is measured by the EPE on the Sintel(train) final pass after training on C + T. *Left*: Parameter count vs. accuracy compared to other methods. RAFT is more parameter efficient while achieving lower EPE. *Middle*: Inference time vs. accuracy timed using our hardware *Right:* Training iterations vs. accuracy (taken as product of iterations and GPUs used).

Fig. 6. Results on 1080p (1088 × 1920) video from DAVIS (550 ms per frame).

4.5   Video of Very High Resolution

To demonstrate that our method scales well to videos of very high resolution we apply our network to HD video from the DAVIS [37] dataset. We use 1080p (1088×1920) resolution video and apply 12 iterations of our approach. Inference takes 550 ms for 12 iterations on 1080p video, with all-pairs correlation taking 95ms. Figure 6 visualizes example results on DAVIS.

5   Conclusions

We have proposed RAFT—Recurrent All-Pairs Field Transforms—a new end-to-end trainable model for optical flow. RAFT is unique in that it operates at a single resolution using a large number of lightweight, recurrent update operators. Our method achieves state-of-the-art accuracy across a diverse range of datasets, strong cross dataset generalization, and is efficient in terms of inference time, parameter count, and training iterations.

Acknowledgments. This work was partially funded by the National Science Foundation under Grant No. 1617767.

References

1. Adler, J., Öktem, O.: Solving ill-posed inverse problems using iterative deep neural networks. Inverse Probl. **33**(12), 124007 (2017)
2. Adler, J., Öktem, O.: Learned primal-dual reconstruction. IEEE Trans. Med. Imag. **37**(6), 1322–1332 (2018)
3. Agrawal, A., Amos, B., Barratt, S., Boyd, S., Diamond, S., Kolter, J.Z.: Differentiable convex optimization layers. In: Advances in Neural Information Processing Systems, pp. 9558–9570 (2019)
4. Amos, B., Kolter, J.Z.: OptNet: differentiable optimization as a layer in neural networks. In: Proceedings of the 34th International Conference on Machine Learning, vol. 70, pp. 136–145. JMLR. org (2017)
5. Bai, S., Kolter, J.Z., Koltun, V.: Trellis networks for sequence modeling. arXiv preprint arXiv:1810.06682 (2018)
6. Bai, S., Kolter, J.Z., Koltun, V.: Deep equilibrium models. In: Advances in Neural Information Processing Systems, pp. 688–699 (2019)
7. Bailer, C., Taetz, B., Stricker, D.: Flow fields: dense correspondence fields for highly accurate large displacement optical flow estimation. In: Proceedings of the IEEE International Conference on Computer Vision, pp. 4015–4023 (2015)
8. Bar-Haim, A., Wolf, L.: ScopeFlow: dynamic scene scoping for optical flow. In: Proceedings of the IEEE/CVF Conference on Computer Vision and Pattern Recognition, pp. 7998–8007 (2020)
9. Black, M.J., Anandan, P.: A framework for the robust estimation of optical flow. In: 1993 (4th) International Conference on Computer Vision, pp. 231–236. IEEE (1993)
10. Brox, T., Bregler, C., Malik, J.: Large displacement optical flow. In: 2009 IEEE Conference on Computer Vision and Pattern Recognition, pp. 41–48. IEEE (2009)

11. Butler, D.J., Wulff, J., Stanley, G.B., Black, M.J.: A naturalistic open source movie for optical flow evaluation. In: Fitzgibbon, A., Lazebnik, S., Perona, P., Sato, Y., Schmid, C. (eds.) ECCV 2012. LNCS, vol. 7577, pp. 611–625. Springer, Heidelberg (2012). https://doi.org/10.1007/978-3-642-33783-3_44
12. Chambolle, A., Pock, T.: A first-order primal-dual algorithm for convex problems with applications to imaging. J. Math. Imag. Vis. **40**(1), 120–145 (2011)
13. Chen, Q., Koltun, V.: Full flow: optical flow estimation by global optimization over regular grids. In: Proceedings of the IEEE Conference on Computer Vision and Pattern Recognition, pp. 4706–4714 (2016)
14. Cho, K., Van Merriënboer, B., Bahdanau, D., Bengio, Y.: On the properties of neural machine translation: encoder-decoder approaches. arXiv preprint arXiv:1409.1259 (2014)
15. Dosovitskiy, A., et al.: FlowNet: learning optical flow with convolutional networks. In: Proceedings of the IEEE International Conference on Computer Vision, pp. 2758–2766 (2015)
16. Fan, L., Huang, W., Gan, C., Ermon, S., Gong, B., Huang, J.: End-to-end learning of motion representation for video understanding. In: Proceedings of the IEEE Conference on Computer Vision and Pattern Recognition, pp. 6016–6025 (2018)
17. Flynn, J., et al.: DeepView: high-quality view synthesis by learned gradient descent (2019)
18. Geiger, A., Lenz, P., Stiller, C., Urtasun, R.: Vision meets robotics: the kitti dataset. Int. J. Robot. Res. **32**(11), 1231–1237 (2013)
19. Hirschmuller, H.: Stereo processing by semiglobal matching and mutual information. IEEE Trans. Pattern Anal. Mach. Intell. **30**(2), 328–341 (2007)
20. Hofinger, M., Bulò, S.R., Porzi, L., Knapitsch, A., Kontschieder, P.: Improving optical flow on a pyramidal level. In: ECCV (2020)
21. Horn, B.K., Schunck, B.G.: Determining optical flow. In: Techniques and Applications of Image Understanding, vol. 281, pp. 319–331. International Society for Optics and Photonics (1981)
22. Hui, T.W., Tang, X., Change Loy, C.: LiteflowNet: a lightweight convolutional neural network for optical flow estimation. In: Proceedings of the IEEE Conference on Computer Vision and Pattern Recognition, pp. 8981–8989 (2018)
23. Hui, T.W., Tang, X., Loy, C.C.: A lightweight optical flow CNN-revisiting data fidelity and regularization. arXiv preprint arXiv:1903.07414 (2019)
24. Hur, J., Roth, S.: Iterative residual refinement for joint optical flow and occlusion estimation. In: Proceedings of the IEEE Conference on Computer Vision and Pattern Recognition, pp. 5754–5763 (2019)
25. Ilg, E., Mayer, N., Saikia, T., Keuper, M., Dosovitskiy, A., Brox, T.: FlowNet 2.0: evolution of optical flow estimation with deep networks. In: Proceedings of the IEEE Conference on Computer Vision and Pattern Recognition, pp. 2462–2470 (2017)
26. Kobler, E., Klatzer, T., Hammernik, K., Pock, T.: Variational networks: connecting variational methods and deep Learning. In: Roth, V., Vetter, T. (eds.) GCPR 2017. LNCS, vol. 10496, pp. 281–293. Springer, Cham (2017). https://doi.org/10.1007/978-3-319-66709-6_23
27. Kondermann, D., et al.: The HCI benchmark suite: stereo and flow ground truth with uncertainties for urban autonomous driving. In: Proceedings of the IEEE Conference on Computer Vision and Pattern Recognition Workshops, pp. 19–28 (2016)

28. Li, X., Wu, J., Lin, Z., Liu, H., Zha, H.: Recurrent squeeze-and-excitation contex-taggregation net for single image deraining. In: Ferrari, V., Hebert, M., Sminchis-escu, C., Weiss, Y. (eds.) ECCV 2018. LNCS, vol. 11211, pp. 262–277. Springer, Cham (2018). https://doi.org/10.1007/978-3-030-01234-2_16

29. Liang, Z., et al.: Learning for disparity estimation through feature constancy. In: Proceedings of the IEEE Conference on Computer Vision and Pattern Recognition, pp. 2811–2820 (2018)

30. Loshchilov, I., Hutter, F.: Decoupled weight decay regularization. arXiv preprint arXiv:1711.05101 (2017)

31. Lu, Y., Valmadre, J., Wang, H., Kannala, J., Harandi, M., Torr, P.: Devon: deformable volume network for learning optical flow. In: The IEEE Winter Conference on Applications of Computer Vision, pp. 2705–2713 (2020)

32. Lv, Z., Dellaert, F., Rehg, J.M., Geiger, A.: Taking a deeper look at the inverse compositional algorithm. In: Proceedings of the IEEE Conference on Computer Vision and Pattern Recognition, pp. 4581–4590 (2019)

33. Mayer, N., et al.: A large dataset to train convolutional networks for disparity, optical flow, and scene flow estimation. In: Proceedings of the IEEE Conference on Computer Vision and Pattern Recognition, pp. 4040–4048 (2016)

34. Menze, M., Geiger, A.: Object scene flow for autonomous vehicles. In: Proceedings of the IEEE Conference on Computer Vision and Pattern Recognition, pp. 3061–3070 (2015)

35. Menze, M., Heipke, C., Geiger, A.: Discrete optimization for optical flow. In: Gall, J., Gehler, P., Leibe, B. (eds.) GCPR 2015. LNCS, vol. 9358, pp. 16–28. Springer, Cham (2015). https://doi.org/10.1007/978-3-319-24947-6_2

36. Paszke, A., et al.: Automatic differentiation in PyTorch (2017)

37. Pont-Tuset, J., Perazzi, F., Caelles, S., Arbeláez, P., Sorkine-Hornung, A., Van Gool, L.: The 2017 davis challenge on video object segmentation. arXiv preprint arXiv:1704.00675 (2017)

38. Ranftl, R., Bredies, K., Pock, T.: Non-local total generalized variation for optical flow estimation. In: Fleet, D., Pajdla, T., Schiele, B., Tuytelaars, T. (eds.) ECCV 2014. LNCS, vol. 8689, pp. 439–454. Springer, Cham (2014). https://doi.org/10.1007/978-3-319-10590-1_29

39. Ranjan, A., Black, M.J.: Optical flow estimation using a spatial pyramid network. In: Proceedings of the IEEE Conference on Computer Vision and Pattern Recognition, pp. 4161–4170 (2017)

40. Schuster, R., Bailer, C., Wasenmüller, O., Stricker, D.: Flowfields++: accurate optical flow correspondences meet robust interpolation. In: 2018 25th IEEE International Conference on Image Processing (ICIP), pp. 1463–1467. IEEE (2018)

41. Sun, D., Yang, X., Liu, M.Y., Kautz, J.: Models matter, so does training: an empirical study of cnns for optical flow estimation. arXiv preprint arXiv:1809.05571 (2018)

42. Sun, D., Yang, X., Liu, M.Y., Kautz, J.: PWC-Net: CNNS for optical flow using pyramid, warping, and cost volume. In: Proceedings of the IEEE Conference on Computer Vision and Pattern Recognition, pp. 8934–8943 (2018)

43. Tang, C., Tan, P.: BA-Net: dense bundle adjustment network. arXiv preprint arXiv:1806.04807 (2018)

44. Teed, Z., Deng, J.: DeepV2D: video to depth with differentiable structure from motion. arXiv preprint arXiv:1812.04605 (2018)

45. Weinzaepfel, P., Revaud, J., Harchaoui, Z., Schmid, C.: DeepFlow: large displacement optical flow with deep matching. In: Proceedings of the IEEE International Conference on Computer Vision, pp. 1385–1392 (2013)

46. Wulff, J., Sevilla-Lara, L., Black, M.J.: Optical flow in mostly rigid scenes. In: Proceedings of the IEEE Conference on Computer Vision and Pattern Recognition, pp. 4671–4680 (2017)
47. Xu, J., Ranftl, R., Koltun, V.: Accurate optical flow via direct cost volume processing. In: Proceedings of the IEEE Conference on Computer Vision and Pattern Recognition, pp. 1289–1297 (2017)
48. Yang, G., Ramanan, D.: Volumetric correspondence networks for optical flow. In: Advances in Neural Information Processing Systems, pp. 793–803 (2019)
49. Yin, Z., Darrell, T., Yu, F.: Hierarchical discrete distribution decomposition for match density estimation. In: Proceedings of the IEEE Conference on Computer Vision and Pattern Recognition, pp. 6044–6053 (2019)
50. Zach, C., Pock, T., Bischof, H.: A duality based approach for realtime TV-L^1 optical flow. In: Hamprecht, F.A., Schnörr, C., Jähne, B. (eds.) DAGM 2007. LNCS, vol. 4713, pp. 214–223. Springer, Heidelberg (2007). https://doi.org/10.1007/978-3-540-74936-3_22
51. Zhao, S., Sheng, Y., Dong, Y., Chang, E.I., Xu, Y., et al.: MaskflowNet: asymmetric feature matching with learnable occlusion mask. In: Proceedings of the IEEE/CVF Conference on Computer Vision and Pattern Recognition, pp. 6278–6287 (2020)
52. Zhou, H., Ummenhofer, B., Brox, T.: DeepTAM: deep tracking and mapping. In: Ferrari, V., Hebert, M., Sminchisescu, C., Weiss, Y. (eds.) ECCV 2018. LNCS, vol. 11220, pp. 822–838. Springer, Cham (2018). https://doi.org/10.1007/978-3-030-01270-0_50

Domain-Invariant Stereo Matching Networks

Feihu Zhang[1]([✉]), Xiaojuan Qi[2], Ruigang Yang[3], Victor Prisacariu[1],
Benjamin Wah[4], and Philip Torr[1]

[1] University of Oxford, Oxford, England
feihu.zhang@eng.ox.ac.uk
[2] University of Hong Kong, Pok Fu Lam, Hong Kong
[3] Baidu Research, Beijing, China
[4] Chinese University of Hong Kong, Sha Tin, China

Abstract. State-of-the-art stereo matching networks have difficulties in generalizing to new unseen environments due to significant domain differences, such as color, illumination, contrast, and texture. In this paper, we aim at designing a domain-invariant stereo matching network (DSMNet) that generalizes well to unseen scenes. To achieve this goal, we propose i) a novel "domain normalization" approach that regularizes the distribution of learned representations to allow them to be invariant to domain differences, and ii) an end-to-end trainable structure-preserving graph-based filter for extracting robust structural and geometric representations that can further enhance domain-invariant generalizations. When trained on synthetic data and generalized to real test sets, our model performs significantly better than all state-of-the-art models. It even outperforms some deep neural network models (*e.g.* MC-CNN [61]) fine-tuned with test-domain data. The code is available at https://github.com/feihuzhang/DSMNet.

1 Introduction

Stereo reconstruction is a fundamental problem in computer vision, robotics and autonomous driving. It aims to estimate 3D geometry by computing disparities between matching pixels in a stereo image pair. Recently, many end-to-end deep neural network models (*e.g.* [5,19,63]) have been developed for stereo matching that achieve impressive accuracy on several datasets or benchmarks.

However, state-of-the-art stereo matching networks (supervised [5,19,63] and unsupervised [51,68]) cannot generalize well to unseen data without fine-tuning or adaptation. Their difficulties lie in the large domain differences (such as color, illumination, contrast and texture). As illustrated in Fig. 1, the pre-trained models on one specific dataset produce poor results on other unseen scenes.

Electronic supplementary material The online version of this chapter (https://doi.org/10.1007/978-3-030-58536-5_25) contains supplementary material, which is available to authorized users.

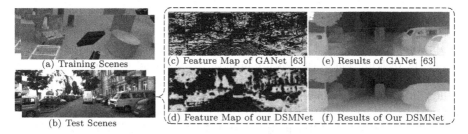

Fig. 1. Visualization of the feature maps and disparity results. GANet [63] is used for comparisons. The features used for matching (outputs of the feature extraction networks) are visualized in (c) and (d). Models are trained on synthetic data (Sceneflow [32]) and tested on novel real scenes (KITTI [33]). The feature maps from GANet has many artifacts (*i.e.* noise). Our DSMNet mainly captures the structure and shape information as robust features, and there is no distortions or artifacts in the feature map. It can produce accurate disparity estimations in the novel test scenes.

Domain adaptation and transfer learning methods (*e.g.* [3,12,51]) attempt to transfer or adapt from one source domain to another new domain. Typically, a large number of stereo images from the new domain are required for the adaptation. However, these cannot be easily obtained in many real scenarios. Yet, we still need a good method for disparity estimation even without data from the new domain for adaptation.

We focus on the more challenging but crucial domain generalization [1] problem that assumes no access to target information for adaptation or fine-tuning. Namely, we are trying to design a model that can generalize well to unseen data without any re-training or adaptation. The difficulties in developing such a domain-invariant stereo matching network (DSMNet) come from the significant domain differences (Fig. 1(a)–(b)) which can be roughly categorized as i) image-level styles (*e.g.* color, illumination), ii) local variations (*e.g.* contrast), iii) texture patterns, details and noise conditions and iv) other complicated domain shifts (*e.g.* uncommon/non-linear contents). They can be approximated by:

$$f(p) = \alpha_I(\alpha_p \cdot \phi(p) + \beta_p) + \beta_I. \tag{1}$$

Here, p is the feature of each pixel (e.g. RGB). Without domain shifts, $f(p) = p$ for different datasets. In practice, domain shifts are varying in different datasets.

The i) image-level style differences can be represented as α_I and β_I. The ii) local variations (e.g. contrasts) are α_p. β_p represents the iii) image details/noise. Pixels of an image have the same α_I and β_I. The local shifts α_p and β_p are varying in different regions/pixels. And ϕ is the expression of iv) other uncommon domain differences that cannot be easily formulated as specific models.

Figure 1 visualizes the features learned by state-of-the-art stereo matching model [63]. Such domain differences make the learned features unstable, distorted and noisy, leading to many wrong matching results (Fig. 1(e)) when applied to the novel test data (Fig. 1(c)).

In this paper, we propose two novel trainable neural network layers for constructing the DSMNet for cross-domain generalization without fine-tuning or adaptation. The proposed novel **domain normalization (DN)** layer fully regulates the distribution of the feature in both the image-level spatial (height and width) and the pixel-level channel dimensions. It can therefore reduce the domain shifts/differences of i) image-level styles (α_I and β_I in Eq. (1)) and ii) local contrast variations (α_p in Eq. (1)) between different datasets/scenes. Our non-local **structure-preserving graph-based filtering (SGF)** layer can further smooth and reduce the iii) domain-sensitive local details/noise (β_p in Eq. (1)). It also helps capture more robust structural and geometric representations (*e.g.* shape and structure, as in Fig. 1(d)) that are more robust to iv) many other complicated domain differences (ϕ in Eq. (1)) for stereo reconstruction.

We formulate our method as an end-to-end deep neural network and train it only with synthetic data. In experiments, without any fine-tuning or adaptation on the real test data, our DSMNet far outperforms: 1) almost all state-of-the-art stereo matching models (*e.g.* GANet [63]) trained on the same synthetic dataset, 2) most of the traditional methods (*e.g.* Cosfter filter, SGM [14] *et al.*), 3) most of the unsupervised/self-supervised models trained on the target test domains. Our model even surpasses some of the fine-tuned (on the target domains) supervised neural network models (*e.g.* MC-CNN [61], content-CNN [31], DispNetC [32]). Also, it doesn't sacrifice fine-tuned accuracy for generalization. After fine-tuning on the target scenes, it can achieve state-of-the-art accuracy (*e.g.* on KITTI benchmark). Moreover, our method can be easily extended to the optical flow task. It also significantly improves the generalization abilities of the optical flow networks (*e.g.* FlowNew2 [17], PwcNet [48]).

2 Related Work

2.1 Deep Neural Networks for Stereo Matching

In recent years, deep neural networks have seen great success in stereo matching [5,19,32,44,63]. These models can be categorized into three types: 1) learning better features for traditional stereo matching algorithms, 2) correlation-based deep neural networks, 3) cost-volume based stereo matching networks.

In the first category, deep neural networks have been used to compute patch-wise similarity scores as the matching costs [61,64]. The costs are then fed into the traditional cost aggregation and disparity computation/refinement methods [14] to get the final disparity maps. The models are, however, limited by the traditional matching cost aggregation step and often produce wrong predictions in occluded regions, large textureless/reflective regions and around object edges.

DispNetC [32], a typical method in the second category, computes the correlations by warping between stereo views and attempts to predict the per-pixel disparity by minimizing a regression training loss. Many other sate-of-the-art methods, including iResNet [28], CRL [38], SegStereo [57], EdgeStereo [47], HD^3 [60], and MADNet [51], are all based on color or feature correlations between the left and right views for disparity estimation.

The recently developed cost-volume based models explicitly learn feature extraction, cost volume, and regularization function all end to end. Examples include GC-Net [19], PSM-Net [5], StereoNet [20], AnyNet [55], GANet [63] and EMCUA [36]. They all utilize a similarity cost as the third dimension to build the 4D cost volume in which the real geometric context is maintained.

Others, like [13], combine the correlation and cost volume strategies.

The common feature of these models is that they all require a large number of training samples with ground truths. More importantly, a model trained on one domain cannot generalize well to new scenes without fine-tuning or retraining.

2.2 Adaptation and Self-supervised Learning

Self-supervised Learning: A recent trend of training stereo matching networks in an unsupervised manner relies on image reconstruction losses that are achieved by warping left and right views [67,68]. However, they cannot solve the occlusions and reflective regions where there is no correspondence between the left and the right views. Also, they cannot generalize well to other new domains.

Domain Adaptation: Some methods pre-train the models on synthetic data and then explore the cross-domain knowledge to adapt [12,39] for a new domain. Others focus on the online or offline adaptations [41,49–51]. For example, MAD-Net [51] is proposed to adapt the pre-trained model online and in real time. But, it has poor accuracy even after the adaptation. Moreover, the domain adaptation approaches require a large number of stereo images from the target domain for adaptations. However, these cannot be easily obtained in many real scenarios. And, in this case, we still need a good method for disparity estimation even without data from the new domain for adaptation.

2.3 Cross-Domain Generalization

In contrast to domain adaptation, domain generalization [1,11] is a much harder problem that assumes no access to target information for adaptation or fine-tuning. There are many approaches that explore the idea of domain-invariant feature learning. Previous approaches focus on developing data-driven strategies to learn invariant features from different source domains [11,22,34]. Some recent methods utilize meta-learning that takes variations in multiple source domains to generalize to novel test distributions [1,23]. Other approaches [24,25] employ an invariant adversarial network to learn domain-invariant representations for image recognition. Choy *et al.* [7] develop a universal feature learning framework for visual correspondences using deep metric learning.

In contrast to the above approaches, there are methods that try to improve the batch or instance normalization in order to improve the generalization and robustness for style transfer or image recognition [35,37].

In summary, for stereo matching, work is seldom done to improve the generalization ability of the end-to-end deep neural network models, especially when developing the domain-invariant stereo matching networks.

3   Proposed DSMNet

To address the challenges of domain shifts (Eq. (1)), we propose 1) a novel domain normalization (DN) to remove the influence of the image-level domain shifts (α_I and β_I: *e.g.* color, style, illuminance) and the local contrast variations (α_p in Eq. (1)), as well as 2) the trainable structure-preserving graph-based filtering (SGF) layer to smooth the domain-sensitive local noise/details (β_p) and capture the structural and geometric context as robust features for domain-invariant stereo reconstruction.

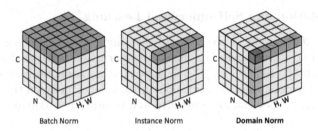

Fig. 2. Normalization methods. Each subplot shows a feature map tensor, with N as the batch axis, C as the channel axis, and (H, W) as the spatial axes. The blue elements in set S are normalized by the same mean and variance. The proposed domain normalization consists of image-level normalization (blue, Eq. (2)) and pixel-level normalization of each C-channel feature vector (green, Eq. (4)). (Color figure online)

3.1   Domain Normalization

Batch Normalization (BN) has become the default feature normalization operation for constructing end-to-end deep stereo matching networks [5,19,32, 47,51,63]. Although it can reduce the internal covariate shift effects in training deep networks, it is domain-dependent and has negative influence on the cross-domain generalization ability.

BN normalizes the features as follows:

$$\hat{x}_i = \frac{1}{\sigma}(x_i - \mu_i).$$

(2)

Here x and $\hat{x}$ are the input and output features, respectively, and i indexes elements in a tensor (*i.e.* feature maps, as illustrated in Fig. 2) of size $N \times C \times H \times W$ (N: batch size, C: channels, H: spatial height, W: spatial width). μ_i and σ_i are the corresponding channel-wise mean and standard deviation (std) and are computed by:

$$\mu_i = \frac{1}{m}\sum_{k \in S_i} x_k, \quad \sigma_i = \sqrt{\frac{1}{m}\sum_{k \in S_i}(x_k - \mu_i)^2 + \epsilon},$$

(3)

where S_i is the set of elements in the same channel as element i (Fig. 2), and ϵ is a small constant to avoid dividing by zeros.

Mean μ and standard deviation σ are computed per batch in the training phase, and the accumulated values of the training set are utilized for inference. However, different domains may have different μ and σ caused by color shifts, contrast, and illumination. (Fig. 1(a)–(b)). Thus μ and σ computed for one dataset are not transferable to others.

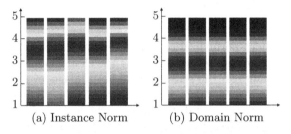

(a) Instance Norm   (b) Domain Norm

Fig. 3. Norm (α_p of Eq. (1)) distributions of the features for different datasets (left to right: synthetic SceneFlow, KITTI, Middlebury, CityScapes and ETH 3D). The output of the feature extraction network is used for the study. The norm (α_p) of the feature vector at each pixel is counted. Instance normalization can only reduce the image-level differences, but does not normalize the C-channel feature vectors at pixel level.

Instance Normalization (IN) [35,40] overcomes the dependency on dataset statistics by normalizing each sample separately, where elements in S_i are confined to be from the same sample as illustrated in Fig. 2. In theory, IN is domain-invariant, and normalization across the spatial dimensions (H, W) reduces image-level style variations.

However, matching of stereo views is realized at the pixel level by finding an accurate correspondence for each pixel using its C-channel feature vector. Any inconsistence of the feature norm and scaling will significantly influence the matching cost and similarity measurements.

Figure 3 illustrates that IN cannot regulate the norm distribution of pixel-wise feature vectors that vary in datasets/domains.

We propose in Fig. 2 our **domain-invariant normalization (DN)**. Our method normalizes features along the spatial axis (H, W) to induce style-invariant representations similar to IN as well as along the channel dimension (C) to enhance the local invariance.

Our DN is realized as follows:

$$\hat{x}'_i = \frac{\hat{x}_i}{\sqrt{\sum_{i \in S'_i} |\hat{x}_i|^2 + \epsilon}}, \tag{4}$$

where S'_i (green region in Fig. 2) includes C elements from the same example (N axis) and the same spatial location (H, W axis). $\hat{x}_i$ is computed as Eqs. (2) and (3) with elements in S_i from the same channel and sample (blue in Fig. 2).

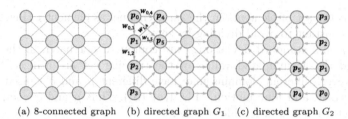

(a) 8-connected graph   (b) directed graph G_1 (c) directed graph G_2

Fig. 4. Illustration of the graph construction. The 8-way connected graph is separated into two directed graphs G_1 and G_2.

In our DN, besides normalization across spatial dimension, we also employ L_2 normalization to normalize features along the channel axis. They collaborate with each other to address the sensitivity to both image-level domain shift (α_I and β_I in Eq. (1)) and the local contrast variations (α_p). As illustrated in Fig. 3, it helps regulate the norm (α_p) distribution of the features in different datasets and improves the robustness to local contrast variations.

Finally, the trainable per-channel scale γ and shift β are added to enhance the discriminative representation ability as BN and IN. The final formulation is:

$$y_i = \gamma_i \, \hat{x}_i' + \beta_i. \tag{5}$$

3.2 Structure-Preserving Graph-Based Filtering

We propose a trainable Structure-preserving Graph-based Filter (SGF) that exploits contextual information and avoid solely memorizing local domain-sensitive texture patterns, details or noise (see Fig. 1(c)) for robust stereo matching.

Our inspiration comes from traditional graph-based filters that are remarkably effective in employing structural and geometric information for structure-preserving texture and detail removing/smoothing [62], denoising [6,62], as well as depth-aware estimation and enhancement [29,58].

Formulation. For a 2D image/feature map I, we construct an 8-connected graph by connecting pixel $\mathbf{p}$ to its eight neighborhoods (see Fig. 4). To avoid loops and achieve fast information aggregation over the graph, we split it into two reverse directed graphs G_1, G_2 (see Fig. 4(b) and 4(c)).

We assign weight ω_e to each edge $e \in G$, and a feature (or color) vector $C(\mathbf{p})$ to each node $\mathbf{p} \in G$. We also allow $\mathbf{p}$ to propagate information to itself with weight $\omega_e(\mathbf{p}, \mathbf{p})$. For graph G_i ($i = 0, 1$), our SGF is defined as follows:

$$C_i^A(\mathbf{p}) = \frac{\sum\limits_{\mathbf{q} \in G_i} W(\mathbf{q}, \mathbf{p}) \cdot C(\mathbf{q})}{\sum\limits_{\mathbf{q} \in G_i} W(\mathbf{q}, \mathbf{p})}, \qquad W(\mathbf{q}, \mathbf{p}) = \sum\limits_{l_{\mathbf{q}, \mathbf{p}} \in G_i} \prod\limits_{e \in l_{\mathbf{q}, \mathbf{p}}} \omega_e. \tag{6}$$

Here, $l_{\mathbf{q}, \mathbf{p}}$ is a feasible path from $\mathbf{q}$ to $\mathbf{p}$. Note that $e(\mathbf{q}, \mathbf{q})$ is included in the path and counts for the start node $\mathbf{q}$. Unlike traditional geodesic filters, we consider all valid paths from source node $\mathbf{q}$ to target node $\mathbf{p}$. The propagation weight along path $l_{\mathbf{q},\mathbf{p}}$ is the product of all edge weights ω_e along the path. Here weight $W(\mathbf{q}, \mathbf{p})$ is defined as the sum of the weights of all feasible paths from $\mathbf{q}$ to $\mathbf{p}$, which determines how much information is diffused to $\mathbf{p}$ from $\mathbf{q}$.

For the edge weight $\omega_{(\mathbf{q},\mathbf{p})}$, we define it in a self-regularized manner as follows:

$$\omega_e(\mathbf{q}, \mathbf{p}) = \frac{\mathbf{x_p}^T \mathbf{x_q}}{\|\mathbf{x_p}\|_2 \cdot \|\mathbf{x_q}\|_2}, \tag{7}$$

where $\mathbf{x_p}$ and $\mathbf{x_q}$ represent the feature vectors of $\mathbf{p}$ and $\mathbf{q}$, respectively.

Compared to other local filters, such as Gaussian filter, median filter, and bilateral filter that can only propagate information in a local region determined by the filter kernel size, our SGF allows the propagation of long-range information over the whole image. More importantly, the filtering weights is defined as a spatial accumulation along all feasible paths in a graph. Similar to Geodesic filter [29] and tree filter [46,59], this path-based filtering kernel helps better preserve the structures of the feature maps.

For stable training and to avoid extreme values, we further add a normalization constraint to the weights associated with $\mathbf{p}$ in the graph G_i as:

$$\sum_{\mathbf{q} \in N_\mathbf{p}} \omega_{e(\mathbf{q},\mathbf{p})} = 1. \tag{8}$$

Here, $N_\mathbf{p}$ is the set of the connected neighbors of $\mathbf{p}$ (including itself), and $e(\mathbf{q}, \mathbf{p})$ is the directed edge connecting $\mathbf{q}$ and $\mathbf{p}$. For example, in Fig. 4(b), for node $\mathbf{p}_0$, $\omega_{e(\mathbf{p}_0,\mathbf{p}_0)} = 1$; and for node $\mathbf{p}_4$, $\omega_{0,4} + \omega_{1,4} + \omega_{e(\mathbf{p}_4,\mathbf{p}_4)} = 1$.

If Eq. (8) holds, we can further derive $\sum_{\mathbf{q} \in G_i} W(\mathbf{q}, \mathbf{p}) = 1$[1]. Equation (6) can then be simplified as follows:

$$C_i^A(\mathbf{p}) = \sum_{\mathbf{q} \in G_i} W(\mathbf{q}, \mathbf{p}) \cdot C(\mathbf{q}), \qquad W(\mathbf{q}, \mathbf{p}) = \sum_{l_{\mathbf{q},\mathbf{p}} \in G_i} \prod_{e \in l_{\mathbf{q},\mathbf{p}}} \omega_e. \tag{9}$$

Such a transformation not only increases the robustness in training but also reduces the computational costs.

Linear Implementation. Equation (9) can be realized as an iterative linear aggregation, where the node is sequentially updated following the direction of the graph (e.g. top to bottom, then left to right in G_1). In each step, $\mathbf{p}$ is updated as:

$$C_i^A(\mathbf{p}) = \omega_{e(\mathbf{p},\mathbf{p})} \cdot C(\mathbf{p}) + \sum_{\mathbf{q} \in N_\mathbf{p}, \mathbf{q} \neq \mathbf{p}} \omega_{e(\mathbf{q},\mathbf{p})} \cdot C_i^A(\mathbf{q})$$
$$s.t. \sum_{\mathbf{q} \in N_\mathbf{p}} \omega_{e(\mathbf{q},\mathbf{p})} = 1. \tag{10}$$

Finally, we repeat the aggregation process for G_1 and G_2 where the updated representation with G_1 is used as the input for aggregation with G_2. The aggregation of Eq. (10) is a linear process with time complexity of $O(n)$ (with n nodes in the graph). During training, backpropagation can be realized by reversing the propagation which is also a linear process (refer to the supplementary material).

---

[1] Proof is in the supplementary material: https://github.com/feihuzhang/DSMNet

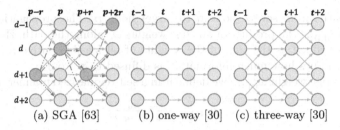

Fig. 5. Special cases of our graph-based filter. (a) Semi-global aggregation (SGA) layer [63]. The dark blue node represents the maximum of each column. (b) and (c) are the affinity-based spatial propagations [30]. They aggregate from column t to $t+1$. (Color figure online)

Relations to Existing Approaches. We show that the recently proposed semi-global aggregation (SGA) layer [63] and affinity-based propagation approach [30] are special cases of our SGF (Eq. (9)). In addition, we compare it with non-local strategies, [54,56], graph neural networks [65] and the attention mechanism [16].

a) Semi-global Aggregation (SGA) [63] is proposed as a differentiable approximation of SGM [14] and can be presented as follows:

$$C_{\mathbf{r}}^A(\mathbf{p}, d) = \text{sum} \begin{cases} \omega_0(\mathbf{p}, \mathbf{r}) \cdot C(\mathbf{p}, d) \\ \omega_1(\mathbf{p}, \mathbf{r}) \cdot C_{\mathbf{r}}^A(\mathbf{p} - \mathbf{r}, d) \\ \omega_2(\mathbf{p}, \mathbf{r}) \cdot C_{\mathbf{r}}^A(\mathbf{p} - \mathbf{r}, d-1) & s.t. \sum_{i=0,1,2,3,4} \omega_i(\mathbf{p}, \mathbf{r}) = 1 \quad (11) \\ \omega_3(\mathbf{p}, \mathbf{r}) \cdot C_{\mathbf{r}}^A(\mathbf{p} - \mathbf{r}, d+1) \\ \omega_4(\mathbf{p}, \mathbf{r}) \cdot \max_i C_{\mathbf{r}}^A(\mathbf{p} - \mathbf{r}, i) \end{cases}$$

The aggregations are in four directions, namely $\mathbf{r} = \{(0,1), (0,-1), (1,0), (-1,0)\}$. Taking the right to left propagation ($\mathbf{r} = (0,1)$) as an example, we can construct a propagation graph in Fig. 5(a). The y-coordinate represents disparity d, and the x-coordinate is the indexes of the pixels/nodes. Compared to our graph in Fig. 4(b), edges connecting top and bottom nodes are removed, and the maximum of each column is densely connected to every node of the next column (red edges). Equation (11) can then be realized by our SGF of Eq. (9). Here, $(\mathbf{p} - \mathbf{r}, d \pm 1)$ are the neighborhood nodes of $\mathbf{p}$, and $\omega_{0,...4}$ are the corresponding edge weights.

b) Affinity-based Spatial Propagation in [30] can be achieved as:

$$C^A(\mathbf{p}, d) = \left(1 - \sum_{\mathbf{q} \in N_{\mathbf{p}}, \mathbf{q} \neq \mathbf{p}} \omega_{e(\mathbf{q}, \mathbf{p})}\right) C(\mathbf{p}) + \sum_{\mathbf{q} \in N_{\mathbf{p}}, \mathbf{q} \neq \mathbf{p}} \omega_{e(\mathbf{q}, \mathbf{p})} C^A(\mathbf{q}), \quad (12)$$

where $\omega_{e(\mathbf{q}, \mathbf{p})}$ are the learned affinities. $1 - \sum_{\mathbf{q} \in N_{\mathbf{p}}} \omega_{e(\mathbf{q}, \mathbf{p})}$ is equal to our weight $\omega_{e(\mathbf{p}, \mathbf{p})}$ for $\mathbf{p}$. The graphs for filtering can be constructed as in Fig. 5(b) and (c) for the one-way and three-way propagations [30], respectively.

c) Non-local Strategies, Graph Neural Networks and Attentions [16,45,54,56, 65] can be used for non-local feature aggregation. But, they are implemented

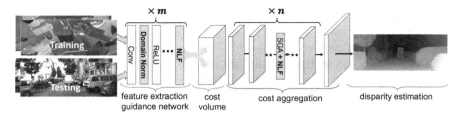

Fig. 6. Overview of the network architecture. Synthetic data are used for training, while using data from other new domains (*e.g.* real KITTI dataset) for testing. The backbone of GANet [63] is used as the baseline. The proposed DN layer is used after each convolutional layer in the feature extraction and guidance network. Several SGF layers are implemented for both feature extraction and cost aggregation.

without spatial and structural awareness. Existing attentions and GNNs used in image segmentation task only consider the feature similarity for aggregation which treat pixel locations equally. In geometric problem (*e.g.* stereo matching), spatial proximity is crucial for learning accurate depths since pixels in the same object/class (with similar features) must be spatially close enough to have similar depth values. Therefore, these similarity/affinity based attentions and non-local networks will easily smooth out depth edges and thin structures (as illustrated in the supplementary material). Our SGF utilizes both the feature affinity and the spatial proximity for non-local graph-based filtering. It spatially aggregates the features along the paths which can better preserve the structure of the disparity maps. More importantly, Our graph filter has lower (linear) complexity in both memory requirement and computation since it is realized by the linear spatial propagation and the weight matrix is only $5 \times N$.

3.3   Network Architecture

As illustrated in Fig. 6, we utilize the backbone of GANet as the baseline architecture. The LGA layer in [63] is removed since it's domain-dependent and captures a lot of local patterns that are very sensitive to domain shifts.

We replace the original batch normalization layer by our proposed domain normalization layer for feature extraction. For the feature extraction network, we utilize a total of seven proposed filtering layers. For 3D cost aggregation of the cost volume, two SGF layers are further added for cost volume filtering in each channel/depth. *Details of the architecture are in the supplementary.*

4   Experimental Results

In our experiments, we train our method only with synthetic data and test it on four real datasets to evaluate its domain generalization ability. During training, we use disparity regression [19] for disparity prediction, and the smooth L_1 loss to compute the errors for back-propagation (the same as in [5,63]). All the models are optimized with Adam ($\beta_1 = 0.9$, $\beta_2 = 0.999$). We train with a batch size of 8 on four GPUs using 288×624 random crops from the input images. The maximum of the disparity is set as 192. We train the model on the synthetic dataset for 10 epochs with a constant learning rate of 0.001. All other training settings are kept the same as those in [63].

4.1 Datasets

KITTI stereo 2012 [10] and 2015 [33] datasets provide about 400 image pairs of outdoor driving scenes for training, where the disparity labels are transformed from Velodyne LiDAR points. The **Cityscapes** dataset [8] provides a large amount of high-resolution ($1k \times 2k$) stereo images collected from city driving scenes. The disparity labels are pre-computed by SGM [14] which is not accurate enough for training deep neural network models. The **Middlebury** stereo dataset [42] is designed for indoor scenes with higher resolution (up to $2k \times 3k$). But it provides no more than 50 image pairs that are not enough to train robust deep neural networks. In addition, **ETH 3D** dataset [43] provides 27 pairs of gray images for training.

These existing real datasets are all limited by their small quantity or poor ground-truth labels, making them insufficient for training. Hence, we use them as test sets for evaluating our models' cross-domain generalization ability.

We mainly use synthetic data to train our domain-invariant models. The existing Scene Flow synthetic dataset [32] contains 35k training image pairs with a resolution of 540×960. This dataset has a limited number of the outdoor driving scenes that provide stereo pairs with a few settings of the camera baselines and image resolutions. We use CARLA [9] to generate a new supplementary synthetic dataset (with 20k stereo pairs) with more diverse settings, including two kinds of image resolutions (720×1080 and 1080×1920), three different focal lengths, and five different camera baselines (in a range of 0.2–1.5 m). This supplementary dataset[1] can significantly improve the diversity of the training set.

The two advantages in using synthetic data are that it can avoid all the difficulties of labeling a large amount of real data, and that it can eliminate the negative influence of wrong depth values in real datasets.

4.2 Ablation Study

We evaluate the performance of our DSMNet with numerous settings, including different architectures, normalization strategies and numbers (0–9) of the proposed SGF layers. As listed in Table 1, the full-setting DSMNet far outperforms the baseline in accuracy by 3% on the KITTI and 8% on the Middlebury datasets. Our proposed domain normalization improves the accuracy by about 1.5%, and the SGF layers contribute another 1.4% on the KITTI dataset.

---

[1] Available at https://github.com/feihuzhang/DSMNet.

Table 1. Ablation study. Models are trained on synthetic data (SceneFlow). Threshold error rates (%) are used for evaluations.

Normalization	SGF		Backbone	Middlebury	KITTI
	Feature	Cost volume		2-pixel	3-pixel
BN			Ours	30.3	9.4
DN			Ours	27.1	7.9
DN	+3		Ours	24.2	7.1
DN	+7		Ours	22.9	6.8
DN	+9		Ours	22.4	6.8
DN	+7	+2	Ours	**21.8**	**6.5**
BN			PSMNet	39.5	16.3
BN			GANet	32.2	11.7
DN	+7	+2	PSMNet	26.1	8.5
DN	+7	+2	GANet	23.7	7.3

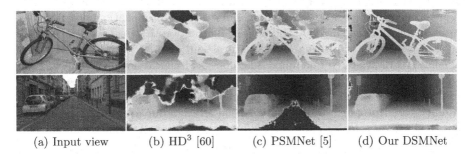

(a) Input view   (b) HD³ [60]   (c) PSMNet [5]   (d) Our DSMNet

Fig. 7. Comparisons with state-of-the-art models. Models are trained on synthetic data and evaluated on high-resolution real datasets (Middlebury and CityScapes). Our DSMNet can produce much more accurate disparity estimation.

Moreover, our proposed layers are generic and could be seamlessly integrated into other deep stereo matching models. Here, we replace our backbone model with GANet [63] and PSMNet [5]. The accuracies are improved by 4–8% on KITTI dataset and 8–13% on Middlebury dataset for coss-domain evaluations compared with the original PSMNet and GANet.

4.3   Component Analysis and Comparisons

To further validate the superiorities of the proposed layers, we compare each of them with other related normalization and attention/affinity strategies.

Normalization Strategies. Table 2 compares our domain normalization with batch normalization [18], instance normalization [53], and the recently proposed adaptive batch-instance normalization [35]. There are also some adaptive BNs

Table 2. Comparisons with Existing Normalization and Filtering/Attention Strategies

Normalization	Middlebury (full)	KITTI	Affinity/Attention	Middlebury (full)	KITTI
Batch norm	29.1	7.3	Attention [16]	25.2	5.9
Instance norm	27.1	6.4	Denoising [56]	25.9	6.1
Adaptive norm [35]	28.2	6.8	Affinity [30]	23.1	5.2
Our domain norm	20.1	4.1	Our graph filter	20.1	4.1

[26,27] for domain adaptation, but they require the full access to the target dataset and cannot be used for more challenging domain generalization task. We keep all other settings the same as our DSMNet and only replace the normalization method for training and evaluation. Our domain normalization is superior to others for domain-invariant stereo matching because it can fully regulate the distribution of the feature vectors and remove both image-level and local contrast differences for cross-domain generalization.

Attentions and Non-local Approaches. Finally, we compare our SGF with attention and non-local networks, including affinity-based propagation [30], non-local neural network denoising [56], and non-local attention [16] (in Table 2). Our SGF layer is better for capturing the structural and geometric context for robust domain-invariant stereo matching. The non-local neural network denoising [56] and non-local attention [16] do not have spatial constraints that usually lead to smoothness of the depth edges (as shown in the supplementary material). Affinity-based propagations [30] are special cases of our proposed SGF and are not as effective in feature and cost volume aggregations for stereo matching.

4.4 Cross-Domain Evaluations

In this section, we compare our DSMNet with state-of-the-art stereo matching models by training with synthetic data and evaluating on real test sets.

Comparisons with State-of-the-Art Models. In Table 3 and Fig. 7, we compare our DSMNet with other state-of-the-art neural network models on the four real datasets. All models are trained on synthetic data (either SceneFlow or a mixture of SceneFlow and Carla). We find that DSMNet far outperforms the state-of-the-art models by 3–30% in error rates for all these datasets. It is also far better than traditional algorithms, like SGM [14], costfilter [15] and patchmatch [2].

Evaluation on the KITTI Benchmark. Table 4 presents the performance of our DSMNet on the KITTI benchmark [33]. Our model far outperforms most of the unsupervised/self-supervised models trained on the KITTI domain. It is even better than supervised stereo matching networks (including, MC-CNN [61], content-CNN [31], and DispNetC [32]) trained or fine-tuned on the KITTI dataset. When compared with other fine-tuned state-of-the-art models (*e.g.* PSMNet [5], HD³ [60], GANet-deep [63]), our DSMNet (without fine-tuning) produces more accurate object boundaries (Fig. 8).

Table 3. Evaluations on the KITTI, Middlebury, and ETH 3D validation datasets. Threshold error rates (%) are used.

Models	KITTI		Middlebury			ETH3D	Carla
	2012	2015	Full	Half	Quarter		
CostFilter [15]	21.7	18.9	57.2	40.5	17.6	31.1	41.1
PatchMatch [2]	20.1	17.2	50.2	38.6	16.1	24.1	30.1
SGM [14]	7.1	7.6	38.1	25.2	10.7	12.9	20.2
Training set	SceneFlow						
HD^3 [60]	23.6	26.5	50.3	37.9	20.3	54.2	35.7
gwcnet [13]	20.2	22.7	47.1	34.2	18.1	30.1	33.2
PSMNet [5]	15.1	16.3	39.5	25.1	14.2	23.8	25.9
GANet [63]	10.1	11.7	32.2	20.3	11.2	14.1	18.8
Our DSMNet	**6.2**	**6.5**	**21.8**	**13.8**	**8.1**	**6.2**	**9.8**
Training set	SceneFlow + Carla						
HD^3 [60]	19.1	19.5	47.3	35.2	19.5	45.2	–
gwcnet [13]	17.2	18.1	45.2	31.8	17.2	29.4	–
PSMNet [5]	10.3	11.0	35.5	23.7	13.8	20.3	–
GANet [63]	7.2	7.6	31.9	19.7	11.4	13.5	–
Our DSMNet	**3.9**	**4.1**	**20.1**	**13.6**	**8.2**	**6.0**	–

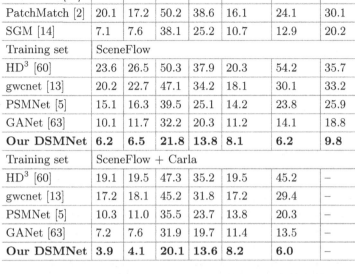

(a) Input view   (b) MC-CNN [61]   (c) PSMNet [5]

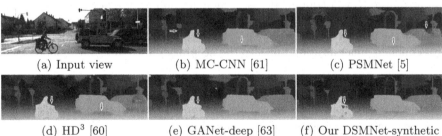

(d) HD^3 [60]   (e) GANet-deep [63]   (f) Our DSMNet-synthetic

Fig. 8. Comparisons with the fine-tuned state-of-the-art models. Our model is trained only with synthetic data. All others are fine-tuned on the KITTI target scenes. As pointed by arrows, our DSMNet can produce more accurate object boundaries.

4.5 Fine-Tuning

In this section, we show the best performance of our DSMNet when fine-tuned on the target domain. We fine-tune the model pre-trained on synthetic data for a further 700 epochs using the KITTI 2015 training set. The learning rate begins at 0.001 for the first 300 epochs and decreases to 0.0001 for the rest. The results of the test set are submitted to KITTI 2015 benchmark for evaluations.

Table 5 compares the results of the fine-tuned DSMNet and those of other state-of-the-art DNN models. We find that DSMNet outperforms most of the

Table 4. Cross-domain evaluation on KITTI 2015 benchmark (all area). DSM-Net is trained only with synthetic data.

Models	Training set	Error rate (%)
Our DSMNet	**Synthetic**	**3.71**
MC-CNN-acrt [61]	Kitti-gt	3.89
DispNetC [32]	Kitti-gt	4.34
Content-CNN [31]	Kitti-gt	4.54
MADNet-finetune [51]	Kitti-gt	4.66
Weak Supervise [52]	Kitti-gt	4.97
MADNet [51]	Kitti (no gt)	8.23
OASM-Net [21]	Kitti (no gt)	8.98
Unsupervised [68]	Kitti (no gt)	9.91

Table 5. In-domain (after fine-tuning) evaluation (error rates: %) on the KITTI 2015 benchmark

Models	Non-occluded	All area
GANet + Our SGF	**1.58**	**1.77**
GANet-deep [63]	1.63	1.81
DSMNet-finetune	1.71	1.90
AcfNet [66]	1.72	1.89
GANet-15 [63]	1.73	1.93
HD³ [60]	1.87	2.02
gwcnet-g [13]	1.92	2.11
PSMNet [5]	2.14	2.32
GCNet [19]	2.61	2.87

recent models (including PSMNet [5], HD³ [60], GwcNet [13] and GANet-15 [63]) by a noteworthy margin. This implies that DSMNet can achieve the same accuracy by fine-tuning on one specific dataset, without sacrificing accuracy to improve its generalization ability.

We also separately test the effectiveness of our SGF layer. Using the current best "GANet-deep" [63] (including the Local Guided Aggregation layer) as the baseline, we add five filtering layers for feature extraction. All other settings are kept the same as the original GANet. After training on synthetic data and fine-tuning on the KITTI training dataset, the new model got a new state-of-the-art accuracy (1.58%) and **ranked No. 1** on KITTI 2015 benchmark (non-occluded area, by the time of submission). This shows that our SGF can improve not only cross-domain generalization but also the accuracy on the test domains.

Table 6. Evaluations of the optical flow networks for cross-domain generalization

Original models	Error rates (%)	Improved models	Error rates (%)
FlowNet2 [63]	34.1	Domain-invaraint FlowNet2	16.2
PwcNet [48]	16.9	Domain-invaraint PwcNet	11.2

5   Extension for Optical Flow

Similar to stereo matching, optical flow is also based on pixel-to-pixel similarity measurement for dense correspondence matching between two different images. Therefore, our domain-invariant matching network can be easily extended to the optical flow task. We use FlowNet2 [17] and PwcNet [48] as baselines and employ our DN and graph filtering to realize the domain-invariant optical flow networks. The models are trained on synthetic FlyingThings3D [32] and MPI

Sintel [4] datasets and evaluated on real flow dataset (KITTI 2015). As shown in Table 6 and Fig. 9, Accuracies are significantly improved by 5.7–17% in the cross-domain evaluations. This further demonstrates the effectiveness of our proposed domain-invariant network.

(a) Synthetic training data

(b) Real test view   (c) FlowNet2 [17]   (d) Domain-invariant FlowNet2

Fig. 9. Performance illustration with optical flow. Models are trained only with synthetic data. (a) Example of the synthetic training data (MPI Sintel [4]), (b) the real test view from KITTI 2015 dataset, (c) the result (top) and the error map (bottom) of the original FlowNet2, (d) the result (top) and the error map (bottom) of the domain-invariant FlowNet2 powered by our DSMNet.

6   Conclusion

In this paper, we proposed two end-to-end trainable neural network layers for our domain-invariant stereo matching network. Our novel domain normalization can fully regulate the distribution of learned features to address significant domain shifts, and our SGF can capture more robust non-local structural and geometric features for accurate disparity estimation in cross-domain situations. We have verified our model on four real datasets and shown its superior accuracy when compared to other state-of-the-art models in the cross-domain generalization.

Acknowledgement. Research is supported by Baidu, the ERC grant ERC-2012-AdG 321162-HELIOS, EPSRC grant Seebibyte EP/M013774/1 and EPSRC/MURI grant EP/N019474/1. We would also like to acknowledge the Royal Academy of Engineering.

References

1. Balaji, Y., Sankaranarayanan, S., Chellappa, R.: MetaReg: towards domain generalization using meta-regularization. In: Advances in Neural Information Processing Systems, pp. 998–1008 (2018)
2. Bleyer, M., Rhemann, C., Rother, C.: PatchMatch stereo-stereo matching with slanted support windows. In: British Machine Vision Conference (BMVC), pp. 1–11 (2011)
3. Bousmalis, K., Silberman, N., Dohan, D., Erhan, D., Krishnan, D.: Unsupervised pixel-level domain adaptation with generative adversarial networks. In: Proceedings of the IEEE Conference on Computer Vision and Pattern Recognition (CVPR), pp. 3722–3731 (2017)

4. Butler, D.J., Wulff, J., Stanley, G.B., Black, M.J.: A naturalistic open source movie for optical flow evaluation. In: Fitzgibbon, A., Lazebnik, S., Perona, P., Sato, Y., Schmid, C. (eds.) ECCV 2012. LNCS, vol. 7577, pp. 611–625. Springer, Heidelberg (2012). https://doi.org/10.1007/978-3-642-33783-3_44

5. Chang, J.R., Chen, Y.S.: Pyramid stereo matching network. In: Proceedings of the IEEE Conference on Computer Vision and Pattern Recognition (CVPR), pp. 5410–5418 (2018)

6. Chen, X., Kang, S.B., Yang, J., Yu, J.: Fast patch-based denoising using approximated patch geodesic paths. In: Proceedings of the IEEE Conference on Computer Vision and Pattern Recognition (CVPR), pp. 1211–1218 (2013)

7. Choy, C.B., Gwak, J., Savarese, S., Chandraker, M.: Universal correspondence network. In: Advances in Neural Information Processing Systems, pp. 2414–2422 (2016)

8. Cordts, M., et al.: The cityscapes dataset for semantic urban scene understanding. In: Proceedings of the IEEE Conference on Computer Vision and Pattern Recognition (CVPR), pp. 3213–3223 (2016)

9. Dosovitskiy, A., Ros, G., Codevilla, F., Lopez, A., Koltun, V.: CARLA: an open urban driving simulator. arXiv preprint arXiv:1711.03938 (2017)

10. Geiger, A., Lenz, P., Urtasun, R.: Are we ready for autonomous driving? The KITTI vision benchmark suite. In: Proceedings of the IEEE Conference on Computer Vision and Pattern Recognition (CVPR), pp. 3354–3361. IEEE (2012)

11. Ghifary, M., Bastiaan Kleijn, W., Zhang, M., Balduzzi, D.: Domain generalization for object recognition with multi-task autoencoders. In: Proceedings of the IEEE International Conference on Computer Vision (ICCV), pp. 2551–2559 (2015)

12. Guo, X., Li, H., Yi, S., Ren, J., Wang, X.: Learning monocular depth by distilling cross-domain stereo networks. In: Proceedings of the European Conference on Computer Vision (ECCV), pp. 484–500 (2018)

13. Guo, X., Yang, K., Yang, W., Wang, X., Li, H.: Group-wise correlation stereo network. In: Proceedings of the IEEE Conference on Computer Vision and Pattern Recognition (CVPR), pp. 3273–3282 (2019)

14. Hirschmuller, H.: Stereo processing by semiglobal matching and mutual information. IEEE Trans. Pattern Anal. Mach. Intell. **30**(2), 328–341 (2008)

15. Hosni, A., Rhemann, C., Bleyer, M., Rother, C., Gelautz, M.: Fast cost-volume filtering for visual correspondence and beyond. IEEE Trans. Pattern Anal. Mach. Intell. **35**(2), 504–511 (2013)

16. Huang, Z., Wang, X., Huang, L., Huang, C., Wei, Y., Liu, W.: CCNet: criss-cross attention for semantic segmentation. In: Proceedings of the IEEE International Conference on Computer Vision (ICCV), pp. 603–612 (2019)

17. Ilg, E., Mayer, N., Saikia, T., Keuper, M., Dosovitskiy, A., Brox, T.: FlowNet 2.0: evolution of optical flow estimation with deep networks. In: Proceedings of the IEEE Conference on Computer Vision and Pattern Recognition (CVPR), pp. 2462–2470 (2017)

18. Ioffe, S., Szegedy, C.: Batch normalization: accelerating deep network training by reducing internal covariate shift. arXiv preprint arXiv:1502.03167 (2015)

19. Kendall, A., et al.: End-to-end learning of geometry and context for deep stereo regression. CoRR, abs/1703.04309 (2017)

20. Khamis, S., Fanello, S.R., Rhemann, C., Kowdle, A., Valentin, J.P.C., Izadi, S.: StereoNet: guided hierarchical refinement for real-time edge-aware depth prediction. CoRR, abs/1807.08865 (2018)

21. Li, A., Yuan, Z.: Occlusion aware stereo matching via cooperative unsupervised learning. In: Jawahar, C.V., Li, H., Mori, G., Schindler, K. (eds.) ACCV 2018. LNCS, vol. 11366, pp. 197–213. Springer, Cham (2019). https://doi.org/10.1007/978-3-030-20876-9_13

22. Li, D., Yang, Y., Song, Y.Z., Hospedales, T.M.: Deeper, broader and artier domain generalization. In: Proceedings of the IEEE International Conference on Computer Vision (ICCV), pp. 5542–5550 (2017)

23. Li, D., Yang, Y., Song, Y.Z., Hospedales, T.M.: Learning to generalize: meta-learning for domain generalization. In: Proceedings of the Thirty-Second AAAI Conference on Artificial Intelligence (2018)

24. Li, H., Jialin Pan, S., Wang, S., Kot, A.C.: Domain generalization with adversarial feature learning. In: Proceedings of the IEEE Conference on Computer Vision and Pattern Recognition (CVPR), pp. 5400–5409 (2018)

25. Li, Y., et al.: Deep domain generalization via conditional invariant adversarial networks. In: Proceedings of the European Conference on Computer Vision (ECCV), pp. 624–639 (2018)

26. Li, Y., Wang, N., Shi, J., Hou, X., Liu, J.: Adaptive batch normalization for practical domain adaptation. Pattern Recogn. 80, 109–117 (2018)

27. Li, Y., Wang, N., Shi, J., Liu, J., Hou, X.: Revisiting batch normalization for practical domain adaptation. arXiv preprint arXiv:1603.04779 (2016)

28. Liang, Z., et al.: Learning for disparity estimation through feature constancy. In: Proceedings of the IEEE Conference on Computer Vision and Pattern Recognition (CVPR), pp. 2811–2820 (2018)

29. Liu, M.Y., Tuzel, O., Taguchi, Y.: Joint geodesic upsampling of depth images. In: Proceedings of the IEEE Conference on Computer Vision and Pattern Recognition (CVPR), pp. 169–176 (2013)

30. Liu, S., De Mello, S., Gu, J., Zhong, G., Yang, M.H., Kautz, J.: Learning affinity via spatial propagation networks. In: Advances in Neural Information Processing Systems, pp. 1520–1530 (2017)

31. Luo, W., Schwing, A.G., Urtasun, R.: Efficient deep learning for stereo matching. In: Proceedings of the IEEE Conference on Computer Vision and Pattern Recognition (CVPR), pp. 5695–5703 (2016)

32. Mayer, N., et al.: A large dataset to train convolutional networks for disparity, optical flow, and scene flow estimation. In: Proceedings of the IEEE Conference on Computer Vision and Pattern Recognition (CVPR), pp. 4040–4048 (2016)

33. Menze, M., Geiger, A.: Object scene flow for autonomous vehicles. In: Proceedings of the IEEE Conference on Computer Vision and Pattern Recognition (CVPR), pp. 3061–3070 (2015)

34. Motiian, S., Piccirilli, M., Adjeroh, D.A., Doretto, G.: Unified deep supervised domain adaptation and generalization. In: Proceedings of the IEEE International Conference on Computer Vision (ICCV), pp. 5715–5725 (2017)

35. Nam, H., Kim, H.E.: Batch-instance normalization for adaptively style-invariant neural networks. In: Advances in Neural Information Processing Systems, pp. 2558–2567 (2018)

36. Nie, G.Y., et al.: Multi-level context ultra-aggregation for stereo matching. In: Proceedings of the IEEE Conference on Computer Vision and Pattern Recognition (CVPR), pp. 3283–3291 (2019)

37. Pan, X., Luo, P., Shi, J., Tang, X.: Two at once: enhancing learning and generalization capacities via IBN-Net. In: Proceedings of the European Conference on Computer Vision (ECCV), pp. 464–479 (2018)

38. Pang, J., Sun, W., Ren, J.S., Yang, C., Yan, Q.: Cascade residual learning: a two-stage convolutional neural network for stereo matching. In: IEEE International Conference on Computer Vision Workshops (ICCVW) (2017)
39. Pang, J., et al.: Zoom and learn: generalizing deep stereo matching to novel domains. In: Proceedings of the IEEE Conference on Computer Vision and Pattern Recognition (CVPR), pp. 2070–2079 (2018)
40. Park, T., Liu, M.Y., Wang, T.C., Zhu, J.Y.: Semantic image synthesis with spatially-adaptive normalization. In: Proceedings of the IEEE Conference on Computer Vision and Pattern Recognition (CVPR), pp. 2337–2346 (2019)
41. Poggi, M., Pallotti, D., Tosi, F., Mattoccia, S.: Guided stereo matching. In: Proceedings of the IEEE Conference on Computer Vision and Pattern Recognition (CVPR), pp. 979–988 (2019)
42. Scharstein, D., et al.: High-resolution stereo datasets with subpixel-accurate ground truth. In: Jiang, X., Hornegger, J., Koch, R. (eds.) GCPR 2014. LNCS, vol. 8753, pp. 31–42. Springer, Cham (2014). https://doi.org/10.1007/978-3-319-11752-2_3
43. Schops, T., et al.: A multi-view stereo benchmark with high-resolution images and multi-camera videos. In: Proceedings of the IEEE Conference on Computer Vision and Pattern Recognition (CVPR), pp. 3260–3269 (2017)
44. Seki, A., Pollefeys, M.: SGM-Nets: semi-global matching with neural networks. In: Proceedings of the IEEE Conference on Computer Vision and Pattern Recognition (CVPR), pp. 6640–6649 (2017)
45. Shi, L., Zhang, Y., Cheng, J., Lu, H.: Non-local graph convolutional networks for skeleton-based action recognition. arXiv preprint arXiv:1805.07694 (2018)
46. Song, L., et al.: Learnable tree filter for structure-preserving feature transform. In: Advances in Neural Information Processing Systems, pp. 1709–1719 (2019)
47. Song, X., Zhao, X., Fang, L., Hu, H.: EdgeStereo: an effective multi-task learning network for stereo matching and edge detection. arXiv preprint arXiv:1903.01700 (2019)
48. Sun, D., Yang, X., Liu, M.Y., Kautz, J.: PWC-Net: CNNs for optical flow using pyramid, warping, and cost volume. In: Proceedings of the IEEE Conference on Computer Vision and Pattern Recognition (CVPR), pp. 8934–8943 (2018)
49. Tonioni, A., Poggi, M., Mattoccia, S., Di Stefano, L.: Unsupervised adaptation for deep stereo. In: The IEEE International Conference on Computer Vision (ICCV) (2017)
50. Tonioni, A., Rahnama, O., Joy, T., Stefano, L.D., Ajanthan, T., Torr, P.H.: Learning to adapt for stereo. In: Proceedings of the IEEE Conference on Computer Vision and Pattern Recognition (CVPR), pp. 9661–9670 (2019)
51. Tonioni, A., Tosi, F., Poggi, M., Mattoccia, S., Stefano, L.D.: Real-time self-adaptive deep stereo. In: Proceedings of the IEEE Conference on Computer Vision and Pattern Recognition (CVPR), pp. 195–204 (2019)
52. Tulyakov, S., Ivanov, A., Fleuret, F.: Weakly supervised learning of deep metrics for stereo reconstruction. In: Proceedings of the IEEE International Conference on Computer Vision (ICCV), pp. 1339–1348 (2017)
53. Ulyanov, D., Vedaldi, A., Lempitsky, V.: Instance normalization: the missing ingredient for fast stylization. arXiv preprint arXiv:1607.08022 (2016)
54. Wang, X., Girshick, R., Gupta, A., He, K.: Non-local neural networks. In: Proceedings of the IEEE Conference on Computer Vision and Pattern Recognition (CVPR), pp. 7794–7803 (2018)
55. Wang, Y., et al.: Anytime stereo image depth estimation on mobile devices. arXiv preprint arXiv:1810.11408 (2018)

56. Xie, C., Wu, Y., van der Maaten, L., Yuille, A.L., He, K.: Feature denoising for improving adversarial robustness. In: Proceedings of the IEEE Conference on Computer Vision and Pattern Recognition (CVPR), pp. 501–509 (2019)
57. Yang, G., Zhao, H., Shi, J., Deng, Z., Jia, J.: SegStereo: exploiting semantic information for disparity estimation. arXiv preprint arXiv:1807.11699 (2018)
58. Yang, Q.: A non-local cost aggregation method for stereo matching. In: Proceedings of the IEEE Conference on Computer Vision and Pattern Recognition (CVPR), pp. 1402–1409. IEEE (2012)
59. Yang, Q.: Stereo matching using tree filtering. IEEE Trans. Pattern Anal. Mach. Intell. **37**(4), 834–846 (2014)
60. Yin, Z., Darrell, T., Yu, F.: Hierarchical discrete distribution decomposition for match density estimation. In: Proceedings of the IEEE Conference on Computer Vision and Pattern Recognition (CVPR), pp. 6044–6053 (2019)
61. Zbontar, J., LeCun, Y.: Computing the stereo matching cost with a convolutional neural network. In: Proceedings of the IEEE Conference on Computer Vision and Pattern Recognition (CVPR), pp. 1592–1599 (2015)
62. Zhang, F., Dai, L., Xiang, S., Zhang, X.: Segment graph based image filtering: fast structure-preserving smoothing. In: Proceedings of the IEEE International Conference on Computer Vision (ICCV), pp. 361–369 (2015)
63. Zhang, F., Prisacariu, V., Yang, R., Torr, P.H.: GA-Net: guided aggregation net for end-to-end stereo matching. In: Proceedings of the IEEE Conference on Computer Vision and Pattern Recognition (CVPR), pp. 185–194 (2019)
64. Zhang, F., Wah, B.W.: Fundamental principles on learning new features for effective dense matching. IEEE Trans. Image Process. **27**(2), 822–836 (2018)
65. Zhang, S., Yan, S., He, X.: LatentGNN: learning efficient non-local relations for visual recognition. arXiv preprint arXiv:1905.11634 (2019)
66. Zhang, Y., et al.: Adaptive unimodal cost volume filtering for deep stereo matching. In: Proceedings of the Thirty-Second AAAI Conference on Artificial Intelligence (2020)
67. Zhong, Y., Dai, Y., Li, H.: Self-supervised learning for stereo matching with self-improving ability. arXiv preprint arXiv:1709.00930 (2017)
68. Zhou, C., Zhang, H., Shen, X., Jia, J.: Unsupervised learning of stereo matching. In: Proceedings of the IEEE International Conference on Computer Vision (ICCV), pp. 1567–1575 (2017)

DeepHandMesh: A Weakly-Supervised Deep Encoder-Decoder Framework for High-Fidelity Hand Mesh Modeling

Gyeongsik Moon[1], Takaaki Shiratori[2], and Kyoung Mu Lee[1(✉)]

[1] ECE & ASRI, Seoul National University, Seoul, Korea
{mks0601,kyoungmu}@snu.ac.kr
[2] Facebook Reality Labs, Pittsburgh, USA
tshiratori@fb.com

Abstract. Human hands play a central role in interacting with other people and objects. For realistic replication of such hand motions, high-fidelity hand meshes have to be reconstructed. In this study, we firstly propose DeepHandMesh, a weakly-supervised deep encoder-decoder framework for high-fidelity hand mesh modeling. We design our system to be trained in an end-to-end and weakly-supervised manner; therefore, it does not require groundtruth meshes. Instead, it relies on weaker supervisions such as 3D joint coordinates and multi-view depth maps, which are easier to get than groundtruth meshes and do not dependent on the mesh topology. Although the proposed DeepHandMesh is trained in a weakly-supervised way, it provides significantly more realistic hand mesh than previous fully-supervised hand models. Our newly introduced penetration avoidance loss further improves results by replicating physical interaction between hand parts. Finally, we demonstrate that our system can also be applied successfully to the 3D hand mesh estimation from general images. Our hand model, dataset, and codes are publicly available(https://mks0601.github.io/DeepHandMesh/).

1 Introduction

Social interactions are vital to humans: every day, we spend a large amount of time on interactions and communications with other people. While facial motion and speech play a central role in communication, important non-verbal information is also communicated via body motion, especially hand and finger motion, to emphasize our speech, clarify our ideas, and convey emotions. Modeling and replicating detailed hand geometry and motion is essential to enrich experience in various applications, including remote communications in virtual/augmented reality and digital storytelling such as movies and video games.

Electronic supplementary material The online version of this chapter (https://doi.org/10.1007/978-3-030-58536-5_26) contains supplementary material, which is available to authorized users.

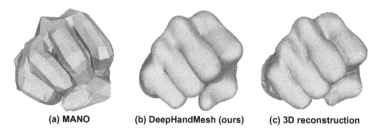

Fig. 1. Qualitative result comparison between (a) MANO [26], (b) our DeepHandMesh, and (c) 3D reconstruction [5].

A pioneering work of hand geometry modeling is MANO by Romero et al. [26], which consists of linear models of identity- and pose-dependent correctives with linear blend skinning (LBS) as an underlying mesh deformation algorithm. The model is learned in a fully-supervised manner by minimizing the per-vertex distance between output and groundtruth meshes that are obtained by registering a template mesh to 3D hand scans [2,11].

Although MANO has been widely used for hand pose and geometry estimation [1,3,9], there exist limitations. First, their method requires groundtruth hand meshes (*i.e.*, the method requires per-vertex supervision to train the linear model). As the hand contains many self-occlusions and self-similarities, existing mesh registration methods [2,11] sometimes fail. To obtain the best quality of groundtruth hand meshes, Romero et al. [26] manually inspected each registered mesh and discarded failed ones from the training data, which requires extensive manual labor. Second, its fidelity is limited. As MANO uses the hand parts of SMPL [19], its resolution is low (*i.e.*, 778 vertices). This low resolution could limit the expressiveness of the reconstructed hand meshes. Also, MANO consists of linear models, optimized by the classical optimization framework. As recent deep neural networks (DNNs) that consist of many non-linear modules show noticeable performance in many computer vision and graphics tasks, utilizing the DNNs with recent deep learning optimization techniques can give more robust and stable results. Finally, it does not consider physical interaction between hand parts. A model without consideration of the physical interaction could result in implausible hand deformation, such as penetration between hand parts.

In this paper, we firstly present *DeepHandMesh*, a weakly-supervised deep encoder-decoder framework for high-fidelity hand mesh modeling, that produces high-fidelity hand meshes from single images. Unlike existing methods such as MANO that require mesh registration for per-vertex supervision (*i.e.*, full supervision), DeepHandMesh utilizes only 3D joint coordinates and multi-view depth maps for supervision (*i.e.*, weak supervision). Therefore, our method avoids expensive data pre-processing such as registration and manual inspection. In addition, obtaining the 3D joint coordinates and depth maps is much easier compared with the mesh registration. The 3D joint coordinates can be obtained from powerful state-of-the-art multi-view 3D human pose estimation methods [17], and the depth maps can be rendered from 3D reconstruction [5] based on the solid mathematical theory about epipolar geometry. Furthermore, these are independent of topology of a hand model, allowing us to use hand meshes with various topology and to be free from preparing topology-specific data such as registered meshes for each topology. To achieve high-fidelity hand meshes, Deep-HandMesh is based on a DNN and optimized with recent deep learning optimization techniques, which provides more robust and stable results. We also use a high-resolution hand model to benefit from the expressiveness of the DNN. Our DeepHandMesh can replicate realistic hand meshes with details such as creases and skin bulging, as well as holistic hand poses. In addition, our newly designed penetration avoidance loss further improves results by enabling our system to replicate physical interaction between hand parts. Figure 1 shows that the proposed DeepHandMesh provides significantly more realistic hand meshes than the existing fully-supervised hand model (*i.e.*, MANO [26]).

As learning a high-fidelity hand model only via weak supervisions is a challenging problem, we assume a personalized environment (*i.e.*, assume the same subject in the training and testing stage). We discuss the limitations of the assumption and future research directions in the later section. To demonstrate the effectiveness of DeepHandMesh for practical purposes, we combine our Deep-HandMesh with 3D pose estimation to build a model-based 3D hand mesh estimation system from a single image, as shown in Fig. 2, and train it on a public dataset captured from general environments. The experimental results show that our DeepHandMesh can be applied to 3D high-fidelity hand mesh estimation from general images in real-time (*i.e.*, 50 fps).

Our contributions can be summarized as follows.

- We firstly propose a deep learning-based weakly-supervised encoder-decoder framework (DeepHandMesh) that is trained in an end-to-end, weakly-supervised manner for high-fidelity hand mesh modeling. Our proposed Deep-HandMesh does not require labor-intensive manual intervention, such as mesh registration.
- Our weakly-supervised DeepHandMesh provides significantly more realistic hand meshes than previous fully-supervised hand models. In addition, we newly introduce a penetration avoidance loss, which can make Deep-HandMesh firstly reproduce physical interaction between hand parts.
- We show that our framework can be applied to practical purposes, such as 3D hand mesh estimation from general images in real-time.

2   Related Works

3D Hand Pose Estimation. 3D hand pose estimation methods can be categorized into depth map-based and RGB-based ones according to their input. Early depth map-based methods are mainly based on a generative approach, which fits

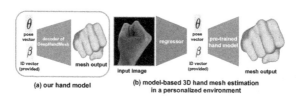

(a) our hand model

(b) model-based 3D hand mesh estimation in a personalized environment

Fig. 2. (a) The hand model outputs meshes from the hand model parameters. Our main goal is to train a high-fidelity hand model in a weakly-supervised way. (b) The model-based 3D hand mesh estimation system outputs inputs of the hand model and use a pre-trained hand model to produce final hand meshes.

a pre-defined hand model to the input depth map by minimizing hand-crafted cost functions [28,32] using particle swarm optimization [28], iterative closest point [31], or their combination [25]. Most of recent depth map-based methods are based on a discriminative approach, which directly localizes hand joints from an input depth map. Tompson et al. [33] utilized a neural network to localize hand joints by estimating 2D heatmaps for each hand joint. Ge et al. [6] extended this method by estimating multi-view 2D heatmaps. Moon et al. [20] designed a 3D CNN that takes a voxel representation of a hand as input and outputs a 3D heatmap for each joint. Wan et al. [34] proposed a self-supervised system, which can be trained from only an input depth map.

The powerful performance of the recent CNN makes 3D hand pose estimation methods work well on RGB images. Zimmermann et al. [39] proposed a DNN that learns an implicit 3D articulation prior. Mueller et al. [22] used an image-to-image translation model to generate synthetic hand images for more effective training of a pose prediction model. Cai et al. [4] and Iqbal et al. [12] implicitly reconstruct depth map from an input RGB image and estimate 3D hand joint coordinates from it. Spurr et al. [30] and Yang et al. [35] proposed variational auto-encoders (VAEs) that learn a latent space of a hand skeleton and appearance.

3D Hand Shape Estimation. Panteleris et al. [23] fitted a pre-defined hand model by minimizing reprojection errors of 2D joint locations w.r.t. hand landmarks detected by OpenPose [29]. Ge et al. [7] proposed a graph convolution-based network which directly estimates vertices of a hand mesh. Many recent methods are based on the MANO hand model. They train their new encoders and use a pre-trained MANO model as a decoder to generate hand meshes. Baek et al. [1] trained their network to estimate input vectors of the MANO model using neural renderer [14]. Boukhayma et al. [3] proposed a network that takes a single RGB image and estimates pose and shape vectors of MANO. Their network is trained by minimizing the distance of the estimated hand joint locations and groundtruth. Recently, Zimmermann et al. [40] proposed a marker-less captured 3D hand pose and mesh dataset.

3D Hand Model. MANO [26] is the most widely used hand model. It takes pose and shape vectors (*i.e.*, relative rotation of hand joint w.r.t. its parent joint

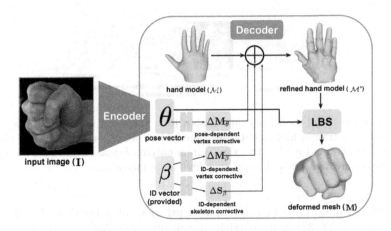

Fig. 3. Overall pipeline of the proposed DeepHandMesh.

and principal component analysis coefficients of hand shape space, respectively) as inputs and outputs deformed mesh using LBS and per-vertex correctives. It is trained from registered hand meshes in a fully-supervised way by minimizing the per-vertex distance between the output and the groundtruth hand meshes. Recently, Kulon et al. [16] proposed a hand model that takes a mesh latent code and outputs a hand mesh using mesh convolution. To obtain the groundtruth meshes, they registered their new high-resolution hand model to 3D joint coordinates of Panoptic dome dataset [29]. They also compute a distribution of valid poses from the hand meshes registered to ∼1000 scans from the MANO dataset. They use this distribution to sample groundtruth hand meshes and train their hand model in a fully-supervised way using per-vertex mesh supervision.

All the above 3D hand models rely on mesh supervision (*i.e.*, trained by minimizing the per-vertex distance between output and groundtruth hand mesh) during training. In contrast, our DeepHandMesh is trained in a weakly-supervised setting, which does not require any groundtruth hand meshes. Although ours is trained without mesh supervision, it successfully reconstructs significantly more high-fidelity hand meshes, including creases and skin bulging, compared with previous hand models. Also, our DeepHandMesh is the first hand model that can replicate physical interaction between hand parts. This is a significant advancement compared with previous hand models.

3   Hand Model

Our hand model is defined as $\mathcal{M} = \{\bar{\mathbf{M}}, \mathbf{S}; \mathbf{W}, \mathcal{H}\}$. $\bar{\mathbf{M}} = [\bar{\mathbf{m}}_1, \ldots, \bar{\mathbf{m}}_V]^T \in \mathbb{R}^{V \times 3}$ denotes vertex coordinates of a zero-pose template hand mesh, where $\bar{\mathbf{m}}_v$ is 3D coordinates of vth vertex of $\bar{\mathbf{M}}$. V denotes the number of vertices. $\mathbf{S} \in \mathbb{R}^{J \times 3}$ means the translation vector of each hand joint from its parent joint, where J is the number of joints. $\mathbf{W} \in \mathbb{R}^{V \times J}$ denotes skinning weights for LBS. Finally, $\mathcal{H}$ denotes a hand joint hierarchy. Our template hand model is prepared by artists. The parameters on the right of the semicolon do not change during training. Thus, we omit them hereafter for simplicity.

4 Encoder

4.1 Hand Pose Vector

The encoder takes a single RGB image of a hand $\mathbf{I}$ and estimates its hand pose vector $\theta \in \mathbb{R}^{N_{\mathrm{P}}}$, where $N_{\mathrm{P}} = 28$ denotes the degrees of freedom (DOFs) of it. Among all the DOFs of the hand joint rotation $3J$, we selected N_{P} DOFs based on the prior knowledge of human hand anatomical property and the hand models of [36,38]. For the enabled DOFs, the estimated hand pose vector is used as a relative Euler angle w.r.t. its parent joint. We set all the disabled DOFs to zero and fixed them during the optimization.

4.2 Network Architecture

Our encoder consists of ResNet-50 [10] and two fully-connected layers. The ResNet extracts a hand image feature from the input RGB image $\mathbf{I}$. Then, the extracted feature is passed to the two fully-connected layers, which outputs the hand pose vector θ. The hidden activation size of the fully-connected layers is 512, and the ReLU activation function is used after the first fully-connected layer. To ensure θ in the range of $(-\pi, \pi)$, we apply a hyperbolic tangent activation function at the output of the second fully-connected layer and multiply it by π.

5 Decoder

5.1 Hand Model Refinement

To replicate details on the hand model, we designed the decoder to estimate three correctives from a pre-defined identity vector $\beta \in \mathbb{R}^{N_{\mathrm{I}}}$ and an estimated hand pose vector θ, inspired by [19,26], as shown in Fig. 3. As the proposed Deep-HandMesh assumes a personalized environment (*i.e.*, assumes the same subject in the training and testing stage), we pre-define β as a $N_{\mathrm{I}} = 32$ dimensional randomly initialized normal Gaussian vector for each subject. β is fixed during training and testing. Note that DeepHandMesh does not require a personalized hand model to be given. Rather, it personalizes an initial hand mesh for a training subject during training.

The first corrective is identity-dependent skeleton corrective $\varDelta \mathbf{S}_{\beta} \in \mathbb{R}^{J \times 3}$. As hand shape and size vary for each person, 3D joint locations can be different for each person. To personalize $\mathbf{S}$ to a training subject, we build two fully-connected layers in our decoder and estimate $\varDelta \mathbf{S}_{\beta}$ from the pre-defined identity code β.

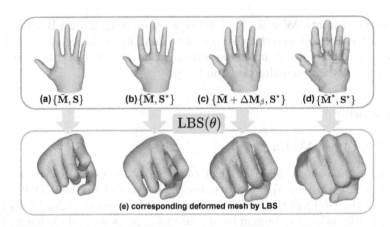

Fig. 4. (a)–(d): Visualized hand model refined by different combinations of correctives. (e): Deformed hand model using LBS.

The hidden activation size of the fully-connected layer is 256. The estimated $\Delta\mathbf{S}_\beta$ is added to $\mathbf{S}$, yielding $\mathbf{S}^*$. Figure 4 (b) shows the effect of $\Delta\mathbf{S}_\beta$.

The second corrective is identity-dependent per-vertex corrective $\Delta\mathbf{M}_\beta \in \mathbb{R}^{V\times 3}$. In addition to the 3D joint locations, hand shape such as finger thickness is also different for each person. To cope with the shape difference, we build two fully-connected layers and estimate $\Delta\mathbf{M}_\beta$ from the identity code β. The hidden activation size of the fully-connected layer is 256. The estimated $\Delta\mathbf{M}_\beta$ is added to $\bar{\mathbf{M}}$. Figure 4 (c) shows the effect of $\Delta\mathbf{M}_\beta$.

The last corrective is pose-dependent per-vertex corrective $\Delta\mathbf{M}_\theta \in \mathbb{R}^{V\times 3}$. When making a pose (*i.e.*, θ varies), local deformation of hand geometry such as skin bulging and crease appearing/disappearing also occurs. To recover such phenomena, we build two fully-connected layers to estimate $\Delta\mathbf{M}_\theta$ from the hand pose vector θ. The hidden activation size of the fully-connected layer is 256. The estimated $\Delta\mathbf{M}_\theta$ is added to $\bar{\mathbf{M}}$. For stable training, we do not back-propagate gradient from $\Delta\mathbf{M}_\theta$ through θ. Figure 4 (d) shows the effect of $\Delta\mathbf{M}_\theta$.

The final refined hand model $\mathcal{M}^*$ is obtained as follows:

$$\bar{\mathbf{M}}^* = \bar{\mathbf{M}} + \Delta\mathbf{M}_\theta + \Delta\mathbf{M}_\beta, \ \mathbf{S}^* = \mathbf{S} + \Delta\mathbf{S}_\beta,$$
$$\mathcal{M}^* = \{\bar{\mathbf{M}}^*, \mathbf{S}^*\}.$$

5.2   Hand Model Deformation

We first perform 3D rigid alignment from the hand model space to the dataset space for the global alignment using the wrist and finger root positions. Then, we use the LBS algorithm to holistically deform our hand model. LBS is a widely used algorithm to deform a mesh according to linear combinations of joint rigid transformation [19,26]. Specifically, each vertex $\mathbf{m}_v$ of a deformed hand mesh

$\mathbf{M} \in \mathbb{R}^{V \times 3}$ is obtained as follows:

$$\mathbf{m}_v = (\mathbf{I}_3, \mathbf{0}) \cdot \sum_{j=1}^{J} w_{v,j} \mathbf{T}_j(\theta, \mathbf{S}^*; \mathcal{H}) \begin{pmatrix} \bar{\mathbf{m}}_v^* \\ 1 \end{pmatrix} \tag{1}$$
$$= \mathrm{LBS}(\theta, \mathbf{S}^*, \bar{\mathbf{m}}_v^*), v = 1, \ldots, V,$$

where $\mathbf{T}_j(\theta, \mathbf{S}^*; \mathcal{H}) \in \mathrm{SE}(3)$ denotes transformation matrix for joint j. It encodes the rotation and translation from the zero pose to the target pose, constructed by traversing the hierarchy $\mathcal{H}$ from the root to j. $w_{v,j}$ and $\bar{\mathbf{m}}_v^*$ denote jth joint of vth vertex skinning weight from $\mathbf{W}$ and vth vertex coordinate from $\bar{\mathbf{M}}^*$, respectively. The visualization of a deformed mesh is shown in Fig. 4 (e).

6   Training DeepHandMesh

We use four loss functions to train DeepHandMesh. The **Pose loss** and **Depth map loss** are responsible for the weak supervision. The **Penetration loss** helps to reproduce physical interaction between hand parts and the **Laplacian loss** acts as a regularizer to make output hand meshes smooth.

Pose Loss. We perform forward kinematics from the estimated hand pose vector θ and refined skeleton $\mathbf{S}^*$ to get the 3D coordinates of the hand joints $\mathbf{P} = [\mathbf{p}_1, \ldots, \mathbf{p}_J]^T \in \mathbb{R}^{J \times 3}$. We minimize $L1$ distance between the estimated and the groundtruth coordinates. The pose loss is defined as follows: $L_{\mathrm{pose}} = \frac{1}{J} \sum_{j=1}^{J} \|\mathbf{p}_j - \mathbf{p}_j^*\|_1$, where $*$ indicates the groundtruth.

Depth Map Loss. We render 2D depth maps $\mathcal{D} = (\mathbf{D}_1, \ldots, \mathbf{D}_{C_{\mathrm{out}}})$ of $\mathbf{M}$ from randomly selected C_{out} target views, and minimize Smooth_{L1} distance [8] between the rendered and the groundtruth depth maps following Ge et al. [7]. To make the depth map loss differentiable, we use Neural Renderer [14]. The depth map loss is defined as follows: $L_{\mathrm{depth}} = \frac{1}{C_{\mathrm{out}}} \sum_{c=1}^{C_{\mathrm{out}}} \delta_c (\mathrm{Smooth}_{L1}(\mathbf{D}_c, \mathbf{D}_c^*))$, where $*$ indicates the groundtruth. δ_c is a binary map whose pixel value of each grid is one if it is foreground (*i.e.*, a depth value is defined in $\mathbf{D}_c$ and $\mathbf{D}_c^*$), and zero otherwise.

Penetration Loss. To penalize penetration between hand parts, we introduce two penetration avoidance regularizers. We consider the fingers as rigid hand parts and the palm as a non-rigid hand part. The regularizers are designed for each of the rigid and non-rigid parts.

For the rigid parts (*i.e.*, fingers), we use a regularizer similar to that in Wan et al. [34], which represents each rigid part with a combination of spheres. Specifically, we compute a pair of the center and radius of spheres $\{s_k^{p(j),j} = (c_k^{p(j),j}, r_k^{p(j),j})\}_{k=1}^{K}$ between joint j and its parent joint $p(j)$, where $K = 10$ denotes the number of spheres between the adjacent joints. The center $c_k^{p(j),j}$ is computed by linearly interpolating $\bar{\mathbf{p}}_{p(j)}$ and $\bar{\mathbf{p}}_j$, where $c_1^{p(j),j} = \bar{\mathbf{p}}_{p(j)}$ and $c_K^{p(j),j} = \bar{\mathbf{p}}_j$. $\bar{\mathbf{p}}_j$ denotes the 3D coordinate of hand joint j obtained from forward

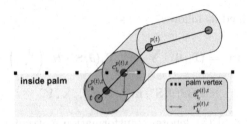

Fig. 5. Visualized example of penetration between a finger and palm.

kinematics using $\theta = \mathbf{0}$ and $\mathbf{S}^*$. Each radius $r_k^{p(j),j}$ is obtained by calculating the distance between $c_k^{p(j),j}$ and the closest vertex in $\mathrm{LBS}(\mathbf{0}, \mathbf{S}^*, \bar{\mathbf{M}}^*)$. Given these spheres, the penetration avoidance term between the rigid hand parts $L_{\mathrm{penet}}^{\mathrm{r}}$ is defined as follows:

$$L_{\mathrm{penet}}^{\mathrm{r}} = \sum_{\substack{k,k' \\ j \neq j', p(j') \\ j' \neq p(j)}} \max(r_k^{p(j),j} + r_{k'}^{p(j'),j'} - ||c_k^{p(j),j} - c_{k'}^{p(j'),j'}||_2, 0), \qquad (2)$$

which indicates that the distances of any pairs of the spheres except the ones associated with adjacent joints are enforced to be greater than the sum of the radii of the paired spheres. This prevents overlap between the spheres, thus avoiding penetration between the rigid parts.

However, $L_{\mathrm{penet}}^{\mathrm{r}}$ does not help prevent penetration at the non-rigid hand part (*i.e.*, the palm). The underlying assumption of $L_{\mathrm{penet}}^{\mathrm{r}}$ is that surface geometry can be approximated by many spheres. While this assumption holds for the fingers due to the cylindrical shape, it does not often hold for the palm, *i.e.*, the spheres along the joints in the palm cannot approximate the palm surface particularly when pose-dependent corrective replicating skin bulging is applied. Additionally, $L_{\mathrm{penet}}^{\mathrm{r}}$ does not produce surface deformation, e.g., finger-palm collision often makes large deformation to the palm surface. $L_{\mathrm{penet}}^{\mathrm{r}}$ does not help replicate such deformation.

To address those limitations, we propose a new penetration avoidance term $L_{\mathrm{penet}}^{\mathrm{nr}}$ for the non-rigid hand part. For this, we only consider penetration between fingertips and palm as illustrated in Fig. 5. Among $\mathbf{M}$, vertices whose most dominant joint in the skinning weight $\mathbf{W}$ is the palm are considered as ones for the palm $\mathbf{M}_\gamma$. Then, the distance between $c_k^{p(t),t}$ and $\mathbf{M}_\gamma$ is calculated, where t is one of fingertip joints. Among the distances, the shortest one is denoted as $d_k^{p(t),t}$. If there exists l_t where $d_{l_t}^{p(t),t}$ is smaller than $r_{l_t}^{p(t),t}$, we consider that $c_{l_t}^{p(t),t}$ penetrates $\mathbf{M}_\gamma$. If there are more than one l_t, we use the one closest to the $p(t)$, which is considered as a starting point of penetration. Based on human hand anatomical property, we can conclude that the spheres from l_t to the fingertip $\{s_k^{p(t),t}\}_{k=l_t}^{K}$ are penetrating $\mathbf{M}_\gamma$. Then, we enforce $\{d_k^{p(t),t}\}_{k=l_t}^{K}$ to be the same as

$\{r_k^{p(t),t}\}_{k=l_t}^K$. The penetration avoidance term for the non-rigid hand part $L_{\text{penet}}^{\text{nr}}$ is defined as follows:

$$L_{\text{penet}}^{\text{nr}} = \sum_t g(t), \qquad (3)$$

$$\text{where } g(t) = \begin{cases} \sum_{k=l_t}^K |d_k^{p(t),t} - r_k^{p(t),t}|, & \text{if } l_t \text{ exists} \\ 0, & \text{otherwise.} \end{cases} \qquad (4)$$

The final penetration avoidance loss function is defined as follows: $L_{\text{penet}} = L_{\text{penet}}^{\text{r}} + \lambda_{\text{nr}} L_{\text{penet}}^{\text{nr}}$, where $\lambda_{\text{nr}} = 5$.

Laplacian Loss. To preserve local geometric structure of the deformed mesh based on the mesh topology, we add a Laplacian regualarizer [18] as follows: $L_{\text{lap}} = \frac{1}{V} \sum_{v=1}^V \left(\mathbf{m}_v - \frac{1}{||\mathcal{N}(v)||} \sum_{v' \in \mathcal{N}(v)} \mathbf{m}_{v'} \right)$, where $\mathcal{N}(v)$ denotes neighbor vertices of $\mathbf{m}_v$.

Our DeepHandMesh is trained in an end-to-end manner. Note that although our DeepHandMesh is trained without per-vertex mesh supervision, it can be trained with a single regularizer L_{lap}. The total loss function L is defined as follows: $L = L_{\text{pose}} + L_{\text{depth}} + L_{\text{penet}} + \lambda_{\text{lap}} L_{\text{lap}}$, where $\lambda_{\text{lap}} = 5$.

7   Implementation Details

PyTorch [24] is used for implementation. The ResNet in the encoder is initialized with the publicly released weights pre-trained on the ImageNet dataset [27], and the weights of the remaining part are initialized by Gaussian distribution with zero mean and $\sigma = 0.01$. The weights are updated by the Adam optimizer [15] with a mini-batch size of 32. The number of rendering views is $C_{\text{out}} = 6$. We use 256×256 as the size of $\mathbf{I}$ and depth maps of $\mathcal{D}$. We observed that changing C_{out} and resolution of $\mathbf{I}$ and depth maps of $\mathcal{D}$ does not affect much the quality of the resulting mesh. The number of vertices in our hand model is 12,553. We train our DeepHandMesh for 35 epochs with a learning rate of 10^{-4}. The learning rate is reduced by a factor of 10 at the 30th and 32nd epochs. We used four NVIDIA Titan V GPUs for training, which took 9 h. Both the encoder and decoder of our DeepHandMesh run at 100 fps, yielding real-time performance (50 fps).

8   Experiment

8.1   Dataset

We used the same data capture studio with Moon et al. [21]. The experimental image data was captured by 80 calibrated cameras capable of synchronously capturing images with 4096×2668 pixels at 30 frames per second. All cameras lie on the front, side, and top hemisphere of the hand and are placed at a distance of about one meter from it. During capture, each subject was instructed to make a pre-defined set of 40 hand motions and 15 conversational gestures. We pre-processed the raw video data by performing multi-view 3D hand pose

(a) without L_{pose} **(b) with** L_{pose} **(c) 3D recon.**   **(d) without** L_{penet} **(e) with** L_{penet} **(f) 3D recon.**

Fig. 6. (a)–(c): Deformed hand mesh trained without and with L_{pose}, and corresponding 3D reconstruction [5]. (d)–(f): Deformed hand mesh trained without and with L_{penet}, and corresponding 3D reconstruction [5].

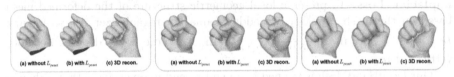

(a) without L_{penet} **(b) with** L_{penet} **(c) 3D recon.**   **(a) without** L_{penet} **(b) with** L_{penet} **(c) 3D recon.**   **(a) without** L_{penet} **(b) with** L_{penet} **(c) 3D recon.**

Fig. 7. Deformed hand mesh trained without and with L_{penet}, and corresponding 3D reconstruction [5].

estimation [17] and multi-view 3D reconstruction [5]. We split our dataset into training and testing sets. The training set 404 K images per subject with the 40 pre-defined hand poses, and the test set 80 K images per subject with the 15 conversational gestures. There are four subjects (one female and three males), and we show more detailed description and various examples of our dataset in the supplementary material.

8.2   Ablation Study

Effect of Each Loss Function. To investigate the effect of each loss function, we visualize test results from models trained with different combinations of loss functions in Fig. 6. In the figure, (a), (b), (d), and (e) are the results of our DeepHandMesh, and (c) and (f) are the results of 3D reconstruction [5], respectively.

The model trained without L_{pose} (a) gives wrong joint locations. Also, there are severe artifacts at occluded hand regions (*e.g.*, the black area on the palm region) because of skin penetration. This is because L_{depth} cannot backpropagate gradients through occluded areas. In contrast, L_{pose} can give gradients at the invisible regions, which makes more stable and accurate results, as shown in (b). The model trained without L_{penet} (d) cannot prevent penetration between fingers and palm. However, L_{penet} penalizes this, and the fingertip locations are placed more plausibly, and the palm vertices are deformed according to the physical interaction between the fingers and palm, as shown in (e). Figure 7 additionally shows the effectiveness of the proposed L_{penet}.

Effect of Identity-Dependent Correctives. To demonstrate the effectiveness of our identity-dependent corrective (*i.e.*, $\Delta \mathbf{S}_\beta$ and $\Delta \mathbf{M}_\beta$), we visualize how

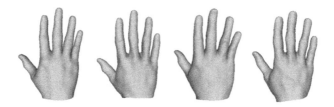

Fig. 8. Visualized hand models of zero pose from different subjects.

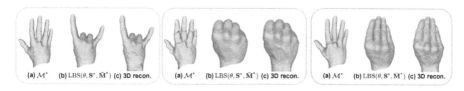

Fig. 9. (a) Refined hand model, (b) deformed hand mesh, and (c) 3D reconstruction [5] from various hand poses of a one subject.

our DeepHandMesh handles different identities in Fig. 8. The figures are drawn by setting $\theta = \mathbf{0}$ to normalize hand pose. As the figures show, our identity-dependent corrective successfully personalizes the initial hand model to each subject by adjusting the hand bone lengths and skin.

Effect of Pose-Dependent Corrective. To demonstrate the effectiveness of our pose-dependent per-vertex corrective $\Delta\mathbf{M}_\theta$, we visualize the hand meshes of different poses in Fig. 9. All the hand meshes are from the same subject to normalize identity. For each hand pose, (a) shows the hand model after model refinement with zero pose. (b) shows deformed (a) using LBS, and (c) shows 3D reconstruction meshes. As the figure shows, our pose-dependent correctives successfully recover details according to the poses. Note that in (b), we approximately reproduced local deformation based on the blood vessels.

8.3   Comparison with State-of-the-Art Methods

We compare our DeepHandMesh with widely used hand model MANO [26] on our dataset. For comparison, we train a model whose encoder is the same one as ours, and decoder is the pre-trained MANO model. The pre-trained MANO model is fixed during the training, and we use the same loss functions as ours. We pre-defined identity code β for each subject and estimate the shape vector of MANO from the code using two fully-connected layers to compare both models in the personalized environment. Figure 10 shows the proposed DeepHandMesh provides significantly more realistic hand mesh from various hand poses and identities. In the last row, MANO suffers from the unrealistic physical interaction between hand parts such as finger penetration and flat palm skin. In contrast, our DeepHandMesh does not suffer from finger penetration and can replicate physical interaction between finger and palm skin. Table 1 shows the 3D joint

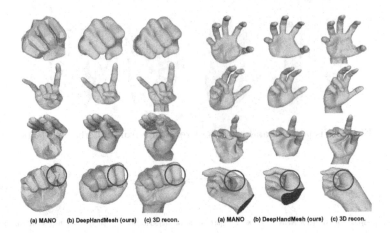

Fig. 10. Estimated hand mesh comparison from various hand poses and subjects with the state-of-the-art method. The red circles in the last row show physical interaction between hand parts. (Color figure online)

Table 1. 3D joint distance error $\mathbf{P}_{err}$ and mesh vertex error $\mathbf{M}_{err}$ comparison between MANO and DeepHandMesh on test set consists of unseen hand poses.

Methods	$\mathbf{P}_{err}$ (mm)	$\mathbf{M}_{err}$ (mm)
MANO	13.81	8.93
DeepHandMesh (Ours)	**9.86**	**6.55**

coordinate distance error and mesh vertex error from the closest point on the 3D reconstruction meshes for unseen hand poses, indicating that our DeepHandMesh outperforms MANO on the unseen hand pose images. For more comparisons, we experimented with lower-resolution hand mesh in the supplementary material.

We found that comparisons between DeepHandMesh and MANO with publicly available 3D hand datasets [37,39] were difficult because DeepHandMesh assumes a personalized environment (*i.e.*, assumes the same subject in training and testing stages). However, we believe the qualitative and quantitative comparisons in Fig. 10 and Table 1 still show the superiority of the proposed DeepHandMesh.

8.4   3D Hand Mesh Estimation from General Images

To demonstrate a use case of DeepHandMesh for general images, we developed a model-based 3D hand mesh estimation system based on DeepHandMesh. Figure 11 shows that our model-based 3D hand mesh estimation system generates realistic hand meshes without mesh supervision from the test set of the RHD [39]. For this, we first pre-trained DeepHandMesh, and replaced its encoder with a randomly initialized one that has exactly the same architecture with our

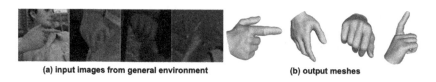

(a) input images from general environment   (b) output meshes

Fig. 11. 3D hand mesh estimation results from general images.

encoder, as illustrated in Fig. 2. We trained the new encoder on the training set of RHD, by minimizing L_{pose}. The RHD dataset 44 K images synthesized by animating the 3D human models. During the training, the decoder is fixed, which is a similar training strategy with that of MANO-based 3D hand mesh estimation methods [1,3]. As our DeepHandMesh assumes a personalized environment, we used a groundtruth bone length to adjust a bone length of the output 3D joint coordinates. The inputs of the decoder are joint rotations and identity code without any image appearance information like MANO; therefore, the decoder can easily generalize to general images, although it is trained on the data captured from the controlled environment.

9   Discussion

Our DeepHandMesh assumes a personalized environment. Future work should consider cross-identity hand mesh modeling by estimating the Gaussian identity code. However, training cross-identity hand mesh model in a weakly-supervised way is very hard. As MANO is trained in a fully-supervised way, they could perform principal component analysis (PCA) on the groundtruth hand meshes in zero-pose and model the identity as coefficients of the principal components. On the other hand, there is no groundtruth mesh under the weakly-supervised setting, therefore performing PCA on meshes is not possible. Generative models (*e.g.*, VAE) can be designed to learn a latent space of identities from registered meshes like [13]; however, training a generative model in a weakly supervised way without registered meshes also remains challenging. We believe the extension of DeepHandMesh to handle cross-identity in a weakly-supervised setting could be an interesting future direction.

10   Conclusion

We presented a novel and powerful weakly-supervised deep encoder-decoder framework, DeepHandMesh, for high-fidelity hand mesh modeling. In contrast to the previous hand models [16,26], DeepHandMesh is trained in a weakly-supervised setting; therefore, it does not require groundtruth hand mesh. Our model successfully generates more realistic hand mesh compared with the previous fully-supervised hand models. The newly introduced penetration avoidance loss makes the result even more realistic by replicating physical interactions between hand parts.

Acknowledgments. This work was partially supported by the Next-Generation Information Computing Development Program (NRF-2017M3C4A7069369) and the Visual Turing Test project (IITP-2017-0-01780) funded by the Ministry of Science and ICT of Korea.

References

1. Baek, S., Kim, K.I., Kim, T.K.: Pushing the envelope for RGB-based dense 3D hand pose estimation via neural rendering. In: CVPR (2019)
2. Bogo, F., Romero, J., Loper, M., Black, M.J.: FAUST: dataset and evaluation for 3D mesh registration. In: CVPR (2014)
3. Boukhayma, A., de Bem, R., Torr, P.H.: 3D hand shape and pose from images in the wild. In: CVPR (2019)
4. Cai, Y., Ge, L., Cai, J., Yuan, J.: Weakly-supervised 3D hand pose estimation from monocular RGB images. In: ECCV (2018)
5. Galliani, S., Lasinger, K., Schindler, K.: Massively parallel multiview stereopsis by surface normal diffusion. In: ICCV (2015)
6. Ge, L., Liang, H., Yuan, J., Thalmann, D.: Robust 3D hand pose estimation in single depth images: from single-view CNN to multi-view CNNs. In: CVPR (2016)
7. Ge, L., et al.: 3D hand shape and pose estimation from a single RGB image. In: CVPR (2019)
8. Girshick, R.: Fast R-CNN. In: ICCV (2015)
9. Hasson, Y., et al.: Learning joint reconstruction of hands and manipulated objects. In: CVPR (2019)
10. He, K., Zhang, X., Ren, S., Sun, J.: Deep residual learning for image recognition. In: CVPR (2016)
11. Hirshberg, D.A., Loper, M., Rachlin, E., Black, M.J.: Coregistration: simultaneous alignment and modeling of articulated 3D shape. In: Fitzgibbon, A., Lazebnik, S., Perona, P., Sato, Y., Schmid, C. (eds.) ECCV 2012. LNCS, vol. 7577, pp. 242–255. Springer, Heidelberg (2012). https://doi.org/10.1007/978-3-642-33783-3_18
12. Iqbal, U., Molchanov, P., Breuel Juergen Gall, T., Kautz, J.: Hand pose estimation via latent 2.5D heatmap regression. In: ECCV (2018)
13. Jiang, Z.H., Wu, Q., Chen, K., Zhang, J.: Disentangled representation learning for 3D face shape. In: CVPR (2019)
14. Kato, H., Ushiku, Y., Harada, T.: Neural 3D mesh renderer. In: CVPR (2018)
15. Kingma, D.P., Ba, J.: Adam: a method for stochastic optimization. In: ICLR (2014)
16. Kulon, D., Wang, H., Güler, R.A., Bronstein, M., Zafeiriou, S.: Single image 3D hand reconstruction with mesh convolutions. In: BMVC (2019)
17. Li, W., et al.: Rethinking on multi-stage networks for human pose estimation. arXiv preprint arXiv:1901.00148 (2019)
18. Liu, S., Chen, W., Li, T., Li, H.: Soft rasterizer: differentiable rendering for unsupervised single-view mesh reconstruction. In: ICCV (2019)
19. Loper, M., Mahmood, N., Romero, J., Pons-Moll, G., Black, M.J.: SMPL: a skinned multi-person linear model. ACM TOG **34**(6), 1–6 (2015)
20. Moon, G., Yong Chang, J., Mu Lee, K.: V2V-PoseNet: voxel-to-voxel prediction network for accurate 3D hand and human estimation from a single depth map. In: CVPR (2018)
21. Moon, G., Yu, S.I., Wen, H., Shiratori, T., Lee, K.M.: InterHand2.6M: a dataset and baseline for 3D interacting hand pose estimation from a single RGB image. In: ECCV (2020)

22. Mueller, F., et al.: GANerated hands for real-time 3D hand tracking from monocular RGB. In: CVPR (2018)
23. Panteleris, P., Oikonomidis, I., Argyros, A.: Using a single RGB frame for real time 3D hand pose estimation in the wild. In: WACV (2018)
24. Paszke, A., et al.: Automatic differentiation in PyTorch (2017)
25. Qian, C., Sun, X., Wei, Y., Tang, X., Sun, J.: Realtime and robust hand tracking from depth. In: CVPR (2014)
26. Romero, J., Tzionas, D., Black, M.J.: Embodied hands: modeling and capturing hands and bodies together. ACM TOG **36**(6), 245 (2017)
27. Russakovsky, O., et al.: ImageNet large scale visual recognition challenge. IJCV **1153**(3), 211–252 (2015)
28. Sharp, T., et al.: Accurate, robust, and flexible real-time hand tracking. In: ACM Conference on Human Factors in Computing Systems (2015)
29. Simon, T., Joo, H., Matthews, I., Sheikh, Y.: Hand keypoint detection in single images using multiview bootstrapping. In: CVPR (2017)
30. Spurr, A., Song, J., Park, S., Hilliges, O.: Cross-modal deep variational hand pose estimation. In: CVPR (2018)
31. Tagliasacchi, A., Schröder, M., Tkach, A., Bouaziz, S., Botsch, M., Pauly, M.: Robust articulated-ICP for real-time hand tracking. In: Computer Graphics Forum (2015)
32. Tang, D., Taylor, J., Kohli, P., Keskin, C., Kim, T.K., Shotton, J.: Opening the black box: hierarchical sampling optimization for estimating human hand pose. In: ICCV (2015)
33. Tompson, J., Stein, M., Lecun, Y., Perlin, K.: Real-time continuous pose recovery of human hands using convolutional networks. ACM TOG **33**(5), 1 (2014)
34. Wan, C., Probst, T., Gool, L.V., Yao, A.: Self-supervised 3D hand pose estimation through training by fitting. In: CVPR (2019)
35. Yang, L., Yao, A.: Disentangling latent hands for image synthesis and pose estimation. In: CVPR (2019)
36. Yuan, S., Ye, Q., Stenger, B., Jain, S., Kim, T.K.: BigHand2.2M benchmark: hand pose dataset and state of the art analysis. In: CVPR (2017)
37. Zhang, J., Jiao, J., Chen, M., Qu, L., Xu, X., Yang, Q.: 3D hand pose tracking and estimation using stereo matching. In: ICIP (2017)
38. Zhou, X., Wan, Q., Zhang, W., Xue, X., Wei, Y.: Model-based deep hand pose estimation. In: IJCAI (2016)
39. Zimmermann, C., Brox, T.: Learning to estimate 3D hand pose from single RGB images. In: ICCV (2017)
40. Zimmermann, C., Ceylan, D., Yang, J., Russell, B., Argus, M., Brox, T.: FreiHAND: a dataset for markerless capture of hand pose and shape from single RGB images. In: ICCV (2019)

Content Adaptive and Error Propagation Aware Deep Video Compression

Guo Lu[1,2], Chunlei Cai[2], Xiaoyun Zhang[2], Li Chen[2(✉)], Wanli Ouyang[3], Dong Xu[3], and Zhiyong Gao[2]

[1] Beijing Institute of Technology, Beijing, China
[2] School of Electronic Information and Electrical Engineering,
Shanghai Jiao Tong University, Shanghai, China
hilichen@sjtu.edu.cn
[3] School of Electrical and Information Engineering, The University of Sydney,
Sydney, Australia

Abstract. Recently, learning based video compression methods attract increasing attention. However, the previous works suffer from error propagation due to the accumulation of reconstructed error in inter predictive coding. Meanwhile, the previous learning based video codecs are also not adaptive to different video contents. To address these two problems, we propose a content adaptive and error propagation aware video compression system. Specifically, our method employs a joint training strategy by considering the compression performance of multiple consecutive frames instead of a single frame. Based on the learned long-term temporal information, our approach effectively alleviates error propagation in reconstructed frames. More importantly, instead of using the handcrafted coding modes in the traditional compression systems, we design an online encoder updating scheme in our system. The proposed approach updates the parameters for encoder according to the rate-distortion criterion but keeps the decoder unchanged in the inference stage. Therefore, the encoder is adaptive to different video contents and achieves better compression performance by reducing the domain gap between the training and testing datasets. Our method is simple yet effective and outperforms the state-of-the-art learning based video codecs on benchmark datasets without increasing the model size or decreasing the decoding speed.

1 Introduction

With the increasing amount of video content, it is a huge challenge to store and transmit videos. In literature, a large number of algorithms [30,39] have

G. Lu and C. Cai—Authors contributed equally.

Electronic supplementary material The online version of this chapter (https://doi.org/10.1007/978-3-030-58536-5_27) contains supplementary material, which is available to authorized users.

A. Vedaldi et al. (Eds.): ECCV 2020, LNCS 12347, pp. 456–472, 2020.
https://doi.org/10.1007/978-3-030-58536-5_27

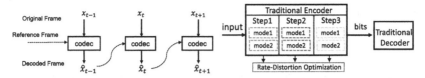

(a) The error propagation issue in the video compression system.

(b) Adaptive mode selection in the traditional video compression system.

Fig. 1. Two motivations of our proposed method.

been proposed to improve the video compression performance. However, all the traditional video compression algorithms [30,39] depend on the hand-designed techniques and highly engineered modules without considering the power of end-to-end learning systems.

Recently, a few learning based image and video compression methods [5,8, 9,19,20,23,24,40] have been proposed. For example, Lu *et al.* [20] proposed an end-to-end video compression system by replacing all the key components in the traditional video compression methods with neural networks.

However, the current state-of-the-art learning based video compression algorithms [20,22,40] still have two drawbacks. First, the error propagation problem is not considered in the training procedure of learning based video compression systems. As shown in Fig. 1(a), the previously decoded frame $\hat{x}_{t-1}$ in the coding procedure will be used as the reference frame to compress the current frame x_t. Since the video compression is a lossy procedure, the previously decoded frame $\hat{x}_{t-1}$ inevitably has reconstruction error, which will be propagated to the subsequent frames because of the inter-frame predictive coding scheme. As the encoding procedure continues, the error will be accumulated frame by frame, which will decrease the compression performance significantly. However, the current approaches [20,40] train the codecs by only minimizing the distortion between the current frame x_t and the decoded frame $\hat{x}_t$, but ignore the influence of $\hat{x}_t$ on the subsequent encoding process for frame x_{t+1} and so on. Therefore, it is critical to build an error propagation aware training strategy for the deep video compression system.

Second, the current learning based encoders [20,40] are not *adaptive* to different video content as the traditional codecs. As shown in Fig. 1(b), the encoder in H.264 [39] or H.265 [30] selects different coding modes (*e.g.*, the size of coding unit) for videos with different contents. In contrast, once the training procedure is finished, the parameters in the learning based encoder are fixed, thus the encoder cannot adapt to different contents in videos and may not be optimal for the current video frame. Furthermore, considering the domain gap due to resolutions or motion magnitudes between the training and testing datasets, the learned encoder may achieve inferior performance for the videos with some specific contents, such as videos with complex motion scenes. To achieve content adaptive coding, it is necessary to update the encoder in the inference stage for the learning based video compression system.

In this paper, we propose a content adaptive and error propagation aware deep video compression method. Our method is a P-frame video compression method and is proposed for low-latency applications. Specifically, to alleviate error accumulation, the video compression system is optimized by minimizing the rate-distortion cost from multiple consecutive frames instead of that from a single frame only. This joint training strategy exploits the long-term information in the coding procedure, therefore the learning based video codec not only achieves high compression performance for the current frame but also guarantees that the decoded current frame is also useful for the coding procedure of the subsequent frames. Furthermore, we propose an online encoder updating scheme to improve the video compression performance. Instead of using the hand-crafted modes in H.264/H.265, the parameters of the encoder will be updated based on the rate-distortion objective for *each* video frame. Our scheme enables adaption of the encoder according to different video contents while keeping the decoder unchanged. Experimental results demonstrate the superiority of the proposed method over the traditional codecs. Our approach is simple yet effective and outperforms the state-of-the-art method [20] without increasing the model size or computational complexity in the decoder side.

The contributions of our work can be summarized as follows,

1. An error propagation aware (EPA) training strategy is proposed by considering more temporal information to alleviate error accumulation for the learning based video compression system.
2. We achieve content adaptive video compression in the inference stage by allowing the online update of the video encoder.
3. The proposed method does not increase the model size or computational complexity of the decoder and outperforms the state-of-the-art learning based video codecs.

2   Related Work

2.1   Image Compression

Traditional image compression methods [1,4,29,35] use hand-crafted techniques, such as discrete cosine transform (DCT)[7] and discrete wavelet transform [28] to reduce the spatial redundancy. Recently, learning based image compression approaches attracted increasing attention [5,6,8,9,14,19,23–25,31–33]. In [8], a CNN based end-to-end image compression framework is proposed by considering both the rate and distortion terms. Furthermore, to obtain the accurate probability model of each symbol, Ballé *et al.* [9] estimate the hyperprior for the compressed features and improves the performance of entropy coding. Since the image compression methods rely on intra prediction, the error propagation reduction issue is not exploited in the existing image compression work.

2.2   Video Compression

The traditional video compression methods [30,39] follow the classical block based hybrid coding framework, which uses motion-compensated prediction and transform coding. Although each module is well-designed, the traditional video compression systems cannot benefit from the power of deep neural networks.

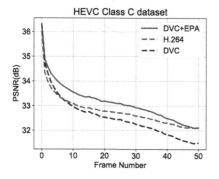

Fig. 2. The PSNR values of the reconstructed frames from different algorithms.

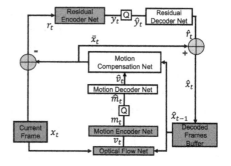

Fig. 3. The architecture of DVC in [20].

Recently, more and more end-to-end frameworks [12,13,16,17,20,26,34,40] were proposed for video compression. In [40], the video compression task was formulated as frame interpolation, in which the motion information is compressed by using the traditional image compression method [4]. Lu *et al.* [20] proposed a fully end-to-end video compression system. Their approach follows the hybrid coding framework and uses neural networks to implement all components in video compression. In [26], an end-to-end video compression framework was proposed and the corresponding motion and residual information are jointly compressed. Habibian *et al.* [16] used a 3D auto-encoder to build a video compression framework and employed an auto-regressive prior as the entropy model. In [15], the residual information is computed in the latent space and the proposed framework can directly decode the motion and blending coefficients. It is worth mentioning that the methods [13,15,40] are based on frame interpolation and designed for B-frame compression.

We would like to highlight that the learning based video codecs in these methods [12,13,15,16,20,26,34,40] are optimized by minimizing the distortion of a single frame without considering the error propagation problem for videos. More importantly, their encoders are not adaptive to different video contents. Although these methods achieve comparable or even better performance than H.264, we believe that the capability of the existing network architecture is not fully exploited, and the video compression performance can be further improved by using our proposed methods.

3   Motivations Related to Learning Based Video Compression System

3.1   Error Propagation

Error propagation is a common issue in the video compression systems, mainly due to the inter-prediction. In Fig. 2, we provide the PSNRs of the reconstructed frames from the H.264 algorithm [27] and the learning based video codec DVC [20]. It is obvious that the PSNR drops when the time step increases. A possible explanation is that video compression is a lossy procedure and the encoding procedure of the current frame relies on the previous reconstructed frame, which is distorted and thus the error propagates to the subsequent frames. Let us take the DVC model as an example. The PSNR value of the 5th reconstructed frame is 33.52 dB, while the PSNR value of the 6th reconstructed frame is 33.37 dB(0.15 dB drop). Furthermore, as the time step increases, the PSNR of the 50th reconstructed frame is only 31.50 dB.

Although error propagation is inevitable for such a predictive coding framework, it is possible and beneficial to alleviate the error propagation issue and further improve the compression performance (see the curve DVC+EPA in Fig. 2).

3.2   The Content Adaptive Coding Scheme

To improve the compression performance, the traditional video encoders [30,39] use the rate-distortion costs to select the optimal mode for the current frame. For example, the encoder prefers to use a large block size for homogeneous regions while a small block size is adopted for complex regions. To this end, the encoder will calculate the rate-distortion cost for each mode in the coding procedure. In contrast, the current learning based video compression systems [20,40] do not employ the content adaptive coding scheme. In other words, the rate-distortion technique is no longer exploited in the inference stage. Therefore, the compressed features are not optimal for the current frame.

More importantly, the encoders are optimized by the rate-distortion optimization (RDO) technique in the training dataset, due to the domain gap between the training and testing datasets in terms of resolution or motion magnitudes, the learned encoders may be far from optimal for the testing dataset. For example, the average motion magnitude between neighboring frames in the training dataset is in the range of [1, 8] pixels [41]. However, the motion in some testing datasets (*e.g.*, the HEVC Class C dataset) is much larger and more complex. The experimental results in [20] also indicate that the compression performance on the HEVC Class C dataset decreases when compared with other datasets.

4   Proposed Method

4.1   Introduction of the DVC Framework

In this paper, we use the framework in [20] as our baseline algorithm to demonstrate the effectiveness of our new approach. In [20], the deep video compression

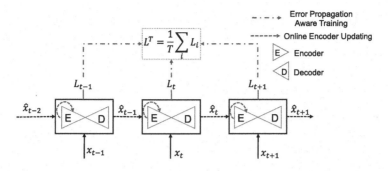

Fig. 4. The proposed content adaptive and error propagation aware deep video compression method.

(DVC) framework follows the classical hybrid coding approach and designs two auto-encoder style networks to compress the motion and residual information, respectively. The architecture of DVC is shown in Fig. 3. The modules with **green color** (*i.e.,* the optical flow net, the motion encoder net and the residual encoder net) represent the **Encoder**. The other modules (*i.e.,* MV decoder, motion compensation net and residual decoder net) represent the **Decoder**. Here, we use Φ_E and Φ_D to represent the trainable parameters in the Encoder and Decoder, respectively. In the inference stage, the parameters in both Encoder and Decoder are fixed. In the coding procedure, we first estimate the motion information v_t between the current frame x_t and previous reconstructed frame $\hat{x}_t$. The motion information will be compressed by the auto-encoder style network and the reconstructed optical flow $\hat{v}_t$ will be used for the motion compensation network. Then we obtain the predicted frame $\bar{x}_t$ and the corresponding residual information r_t. Finally, we use the residual compression network to compress the residual information and obtain the final reconstructed frame $\hat{x}_t$ based on $\bar{x}_t$ and the reconstructed residual $\hat{r}_t$.

The DVC model is optimized by minimizing the following rate-distortion (RD) trade-off,

$$L_t = \lambda D_t + R_t = \lambda d(x_t, \hat{x}_t) + [H(\hat{y}_t) + H(\hat{m}_t)] \tag{1}$$

L_t is the loss function for the current time step t. $d(\cdot, \cdot)$ is the distortion metric between x_t and $\hat{x}_t$. $\hat{y}_t$ and $\hat{m}_t$ are the compressed latent representations from residual and motion information, respectively. $H(\hat{y}_t)$ and $H(\hat{m}_t)$ are the corresponding number of bits used for compressing these latent representations. It is noticed that the whole network is optimized to minimize the rate-distortion criterion for the current time step t.

However, this scheme ignores two critical ***dependencies*** for learning based video compression. First, the compression system, including the encoder and decoder, ignores the potential influence from the reconstruction error of $\hat{x}_t$ to the next frame x_{t+1} in the training procedure and thus leads to error propagation. Second, the encoder itself is fixed and does not depend on the current frame

x_t, which deteriorates the compression performance in the inference stage. In the next section, we will introduce how to address these two issues in video compression.

4.2   The Error Propagation Aware Training Strategy

To alleviate error accumulation in video compression, we propose an error propagation aware training strategy. Specifically, we design a joint training strategy to train the video codec by using the information from different time steps in one video clip and combines all the information to optimize the learned codec for better video compression performance.

The proposed training procedure is shown in Fig. 4. For the current frame x_t, the corresponding reconstructed frame after the encoding and decoding procedure is $\hat{x}_t$. Given x_t and $\hat{x}_t$, we can calculate the RD cost L_t. Then $\hat{x}_t$ will be used as the reference frame in the encoding procedure of x_{t+1}, and we obtain the reconstructed frame $\hat{x}_{t+1}$ and the RD cost L_{t+1}. As the coding procedure continues, the reconstructed error will propagate to the subsequent frames. Meanwhile, we also obtain a series of RD costs, which measure the compression performance at the current time step.

Then, we propose a new objective function by considering the compression performance for both the current frame and the subsequent frames that rely on the current reconstructed frame. Therefore, the loss function is formulated as follows,

$$L^T = \frac{1}{T} \sum_t L_t = \frac{1}{T} \sum_t \{\lambda d(x_t, \hat{x}_t) + [H(\hat{y}_t) + H(\hat{m}_t)]\} \tag{2}$$

where T is the time interval(*i.e.*, the number of frames used in training procedure) and set as 5 in our experiments, L^T represents the error propagation aware loss function. Therefore, our new training objective will optimize the video codec by employing the objectives from multiple time steps.

As shown in Fig. 2, the video codec DVC with an error propagation aware (DVC+EPA) training strategy significantly reduces error accumulation. For example, the proposed method has 0.61 dB (32.11 dB vs. 31.50 dB) improvement over the baseline DVC algorithm [20] for the 50th frame, and the gain becomes larger when the time step increases.

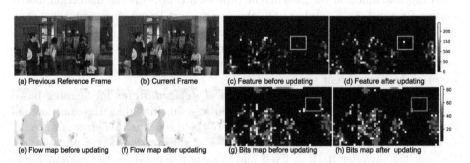

(a) Previous Reference Frame   (b) Current Frame   (c) Feature before updating   (d) Feature after updating

(e) Flow map before updating   (f) Flow map after updating   (g) Bits map before updating   (h) Bits map after updating

Fig. 5. Visual comparison before and after using our online encoder updating scheme.

4.3   The Online Encoder Updating Scheme

To optimize the encoder for each frame and mitigate the domain gap between training and testing data, we propose an online encoder updating scheme in the inference stage. Our method will update the encoder according to the input image while keeping the decoder unchanged. In other words, we use the training dataset to obtain a general decoder and employ the testing dataset to update the CNN parameters of the encoder. Based on the training strategy described in the previous section, we can obtain the learned encoder(E) and decoder(D). For the given original frame x_t and the reference frame $\hat{x}_{t-1}$, the objective L_t at the current frame is obtained according to Eq. (1). Then, the parameters Φ_D in the decoder are fixed while the parameters Φ_E are updated by minimizing L_t. After several iterations, we obtain the content adaptive encoder, which is optimal for the current frame x_t. Finally, the updated encoder based on testing data and the learned decoder from training data is employed for the actual compression procedure. In our implementation, the maximum iteration number is set to 10. To reduce computational complexity, we will compare L_t between two consecutive iterations and stop the optimization procedure once the loss becomes stable.

In contrast to other low-level vision tasks, the ground-truth frame for video compression is available at the encoder side. As a result, we can update the encoder by using the original frame as long as the decoder remains unchanged.

In Fig. 5, we provide the visual results before and after the online updating procedure. It is observed that the output feature from the residual encoder (Fig. 5(c)) and (Fig. 5(d)) has changed after the updating procedure which is optimized for the current frame. More importantly, as shown in Fig. 5(e) and Fig. 5(f), the optical flow map after the updating process contains more details, which is beneficial for accurate prediction. For example, based on the optical flow map in Fig. 5(e), the PSNR of the warped frame is 33.40 dB, while the corresponding PSNR of the warped frame is 34.13 dB based on the updated optical flow map in Fig. 5(f). Furthermore, for the estimated bits map shown in Fig. 5(g) and Fig. 5(h), it is observed that the bits map after the updating process allocates fewer bits for the background region. The experimental results show that the coding bits drop from 0.056 bpp to 0.051 bpp after the online encoder updating procedure. However, the reconstructed frame has better visual quality after the online updating procedure (36.47 dB vs. 36.40 dB).

5   Experiments

5.1   Experimental Setup

Datasets. In the training stage, we use the Vimeo-90k dataset[41]. Vimeo-90k is a widely used dataset for low-level vision tasks [21,37]. It is also used in the recent learning based video compression tasks [15,20]. To evaluate the compression performance of different methods, we employ the four widely used datasets in our experiments.

Specifically, for the HEVC Common Test Sequences [30], we use Class B, Class C, Class D and Class E in our experiments. We don't include the video sequences from the HEVC Class A dataset since it requires more than 11Gb memory for evaluation, which exceeds the capacity of our 1080Ti machine. More details about the other testing datasets including Video Trace Library(VTL) [3], Ultra Video Group(UVG) [2] and MCL-JCV [36], are provided in the supplementary material.

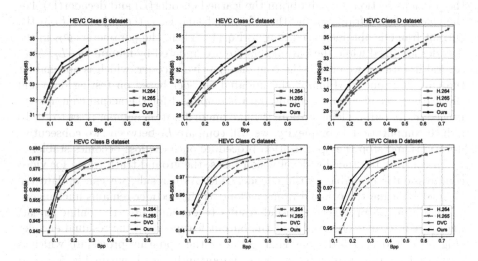

Fig. 6. Comparison between our proposed method with the learning based video codec in [20], H.264 [39] and H.265 [30] at the fixed GoP setting.

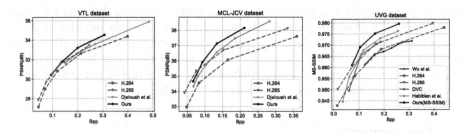

Fig. 7. Comparison between our proposed method and the learning based video codecs [15,16,40] at the fixed GoP setting.

Implementation Details. We train four models with different λ values (256, 512, 1024, 2048) in Eq. (1). To generate the I-frame/key-frame for video compression, we use the learning based image compression method in [9], in which the corresponding λ in the image codec are empirically set to 1024, 2048, 4096 and 12000, respectively.

In our implementation, we use DVC [20] as the baseline method. In the training stage, the whole network is first optimized by using the loss in Eq. (1), then is fine-tuned based on the error propagation aware loss in Eq. (2). The corresponding batch sizes are set to 4 and 1, respectively. The resolution of the training images is 256×256. We use Adam optimizer [18] and the initial learning rate is set as $1e - 4$ for the first 2M steps, and the learning rate is then set to $1e - 5$ for the remaining 0.5M steps. In the inference stage, the encoder is also optimized by using Adam optimizer [18] to achieve content adaptive encoding. The proposed method is implemented based on Tensorflow. It takes about 5 d to train the whole network by using two GTX 1080Ti GPUs.

In our experiments, we use the PSNR and MS-SSIM [38] to measure the distortion between the original frame and the reconstructed frame. The bits per pixel(bpp) represents the coding bits in the compression procedure. The bpp values are estimated from the theoretical values based on the probability of the latent space values.

5.2   Comparison with the State-of-the-art Methods

Evaluation Setting. To make fair comparison with the state-of-the-art learning based video compression methods and the traditional video codecs H.264/H.265, we follow the existing evaluation protocols in [15,20,40] to perform extensive experiments. Specifically, all the existing learning based methods [15,20,40] use the *fixed GoP setting*. For example, the GoP size for the UVG dataset is set to 12 while the corresponding GoP size is set to 10 for the HEVC Common Test Sequences [20,40]. And the corresponding GoP size for H.265/H.264 in these works is also fixed to 12 or 10. We follow the same settings and provide the experimental results in Fig. 6 and Fig. 7.

For the common testing cases of the traditional video codecs, the GoP size is usually not fixed. To further evaluate the performance of the learning based video codec and the traditional video codec (*e.g.*, H.265), we do not impose any restriction on the GoP size in the codec. Specifically, we adopt *veryfast* mode in FFmpeg with the *default Setting*.[1] We evaluate the compression performance for all the video frames on the HEVC Class B, Class C and Class D datasets. The experimental results are provided in Fig. 8.

Baseline Algorithms. The learning based codecs in Wu *et al.* [40] and Djelouah *et al.* [15] are based on frame interpolation and designed for B-frame video compression, while the methods in [16,20] are for P-frame based video compression. Since the B-frame based compression methods employ two reference frames, the coding performance is generally better than P-frame based compression method [30]. We use the P-frame based compression method DVC [20] as our baseline algorithm and we also demonstrate that the proposed method outperforms all

---

[1] ffmpeg -pix_fmt yuv420p -s WxH -r 50 -i video.yuv -c:v libx265 -preset veryfast -tune zerolatency -x265-params "qp=Q" output.mkv; Q is the quantization parameter. W and H are the height and width of the yuv video.

the learning based methods, including the B-frame based compression methods [15,40].

Quantitative Evaluation at the *fixed GoP setting*. As shown in Fig. 6, we provide the compression performance of different methods on the HEVC Common Test Sequences. When compared with the baseline DVC [20] algorithm, our proposed method significantly improves the compression performance. For example, our proposed method has about 1 dB improvement on the HEVC Class C dataset at 0.3 bpp. It is also observed that the proposed method outperforms the H.264 algorithm and is comparable with H.265 in terms of PSNR. The BDBR and BD-PSNR results when compared with H.264 are provided in Table 1. The experimental results on Class E are provided in the supplementary material.

Table 1. The BDBR and BD-PSNR results of different algorithms when compared with H.264. Negative values in BDBR represent the bitrate saving.

Dataset	BDBR(%)			BD-PSNR(dB)		
	H.265	DVC	Ours	H.265	DVC	Ours
Class B	−32.0	−27.9	−41.7	0.78	0.71	1.12
Class C	−20.8	−3.5	−25.9	0.91	0.13	1.18
Class D	−12.3	−6.2	−25.1	0.57	0.26	1.25

We also provide the experimental results when the distortion is evaluated by MS-SSIM. As shown in Fig. 6, our approach outperforms H.265 in terms of MS-SSIM. One possible explanation is that the traditional codecs [30,39] use the block based coding scheme, which inevitably generates the block artifacts.

In Fig. 7, we evaluate the compression performance on the MCL-JCV, VTL and UVG datasets. We compare our proposed method with the recent learning based method [15], which utilizes B-frame based compression scheme. As shown in Fig. 7, although we only use one reference frame, the proposed method still achieves better compression performance on the VTL dataset.

In Fig. 7, we also compare the proposed method with another state-of-the-art learning based video compression method [16] on the UVG dataset. For fair comparison with [16], we also use MS-SSIM as the loss function to optimize the network. The experimental results demonstrate that the proposed approach outperforms [16] by a large margin.

Quantitative Evaluation at the *default setting*. In this section, we also compare the results when the traditional codecs use variable GoP sizes. As shown in Fig. 8, our method outperforms the previous DVC algorithm [20] by a large margin, especially for the HEVC Class C dataset. A possible explanation is that error propagation is more severe as the GoP size becomes larger, which means our proposed scheme will bring more improvements. Although the proposed method cannot outperform H.265 at the default setting, the compression performance of these two methods is generally comparable. Considering that the traditional video codecs exploit other coding techniques, such as multiple reference frames or adaptive quantization parameters, which are not used by the current learning based video compression systems, it is possible to further improve the performance of learning based video codec in the future.

5.3   Ablation Study

The Error Propagation Aware Training Scheme. To demonstrate the effectiveness of our proposed error propagation aware training strategy, we compare the compression performance of different methods in Fig. 9. Specifically, the brown line represents the DVC algorithm [20], while the green line represents the DVC algorithm with the error propagation aware (EPA) training strategy.

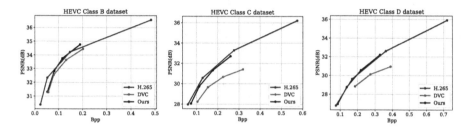

Fig. 8. Evaluation results for all video frames on the HEVC Class B, Class C and Class D at the default setting.

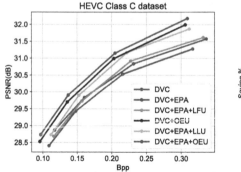

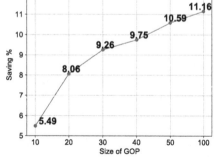

Fig. 9. Ablation study.

Fig. 10. The bitrate saving when comparing DVC+EPA with DVC [20] at different GoP sizes.

Table 2. BDBR(%) performance at different time intervals (*i.e.*, T in Eq. (2)).

T	2	3	4	5	6
BDBR	−0.42	−2.12	−3.68	−5.59	−5.61

It is noticed that the proposed training scheme improves the performance by 0.2 dB on the HEVC Class C dataset(GoP = 20), which demonstrates that the proposed scheme can alleviate error accumulation by exploiting temporal neighboring frames in the training stage.

In practical applications, the GoP size for video compression is usually set as 50 or larger to reduce the bandwidth. And error accumulation is more severe as the GoP size increases. In Fig. 10, we investigate the effectiveness of our newly proposed error propagation aware training scheme when the GoP sizes are set as different numbers. We use BDBR [10] to measure the bitrate saving when compared with the baseline DVC method [20]. Specifically, the proposed scheme saves 5.49% bitrate when the GoP size is set to 10 and saves up to 10.59% bitrate when the GoP size is set to 50. The experimental results demonstrate that the proposed method has achieved better compression performance for video sequences with the large GoP size.

To further investigate the proposed error propagation aware training strategy, we provide the compression results when the method is optimized by using different time intervals T. As shown in Table 2, the proposed scheme saves more bitrates when T increases. For example, the proposed training scheme saves 2.12% bitrate when setting $T = 3$, while the corresponding bitrate saving is 5.59% when setting $T = 5$. One explanation is that we can use long-term temporal information when T increases, which effectively alleviates error accumulation. In our experiments, we set T to 5 by default.

The Online Encoder Updating Scheme. To demonstrate the effectiveness of our proposed online encoder updating (OEU) scheme in the inference stage, we compare the compression performance of the baseline algorithm with or without using our updating scheme. In Fig. 9, the proposed online encoder updating scheme (DVC+OEU, the blue line) significantly improves the compression performance by more than 0.5 dB. Besides, the red line represents the full model of our proposed method, which achieves the best compression performance by using both the online updating scheme and the error propagation aware training strategy.

In [11], Campos et al. adaptively refined the latent representations of the learning based image codecs for better compression performance. Furthermore, we provide the experimental result for the latent features updating (LFU) scheme, where $\hat{m}_t$ and $\hat{y}_t$ are updated and the encoder itself is fixed. The corresponding RD curve (DVC+EPA+LFU) is depicted by the cyan line. Compared with our proposed training scheme (DVC+EPA), we observe that the performance can be further improved by optimizing the latent representation at a high bitrate. However, it is obvious that adaptively optimizing the whole encoder (red line in Fig. 9) achieves better performance. A possible explanation is that updating the encoder provides a larger search range and thus it is more likely to obtain an optimal encoder for the current frame.

Besides, we also provide the compression results when only partial neural networks are updated in the inference stage. Specifically, we use the last layers updating (LLU) scheme, where only the last layers in the residual encoder and motion encoder are updated according to the rate-distortion technique.

The experimental results are denoted by the yellow line (DVC+EPA+LLU) in Fig. 9. It is observed that the partial updating strategy is also useful for video compression. However, the performance is inferior to the proposed approach, where all components in the encoder are updated.

5.4 Discussion

Computational Complexity. In this paper, we use an adaptive encoder in the inference stage to improve compression performance. Since the online rate-distortion optimization scheme is required at the encoder side, it will increase the computational complexity. However, it is noticed that the numbers of iterations for different video sequences are different.

For the video sequences with simple motion scenes, such as the HEVC Class B dataset, the encoder learned from the training dataset is already near-optimal and it only requires 3 iterations to obtain the optimal parameters. For the videos with complex motion scenes, such as the HEVC Class C dataset, more iterations are required to learn the optimal encoder. However, we also obtain a larger improvement($\sim$1 dB). And the corresponding encoding speed of our approach is 1.4 fps while the speed of baseline DVC is 7.1 fps when using 10 iterations. More performance improvement for some test sequences can also be observed by using more iterations. It is noted that the runtime of our approach is evaluated on one Nvidia 1080Ti GPU and we use the plain Tensorflow operations without any specific optimization.

More importantly, a lot of applications, such as video-on-demand applications, are not sensitive to the computational complexity at the encoder side. Considering that our approach is generic and boosts the compression performance without increasing the decoding time, it is feasible to integrate the proposed techniques with other learning based video codecs, such as [15,40], to further improve the compression performance.

Entropy Coding. We use the same entropy coding methods as in the DVC baseline, in which the entropy coding methods in [8,9] are employed for motion coding and residual coding, respectively. While the advanced entropy coding methods may partially alleviate the domain gap, the existing advanced entropy models like [24] usually adopt the autoregressive prior technique, which increases the runtime in the decoder side significantly. In contrast, our approach keeps the decoder unchanged without increasing the computational complexity. More importantly, our online encoder updating (OEU) scheme not only improves the internal entropy coding module but also optimizes the whole encoder (including motion estimation, motion compensation, etc.), which is more effective for video compression.

6 Conclusion

In this paper, we have proposed a content adaptive and error propagation aware deep video compression method. Our approach alleviates error accumulation in the training stage and achieves content adaptive coding by using the online encoder updating scheme in the inference stage. The proposed method is fairly simple yet effective and improves compression performance without increasing the model size or decreasing the decoding speed. The experimental results show that the compression performance of our proposed method outperforms the state-of-the-art learning based video compression methods.

Acknowledgment. This work was supported in part by National Natural Science Foundation of China (61771306) Natural Science Foundation of Shanghai(18ZR1418100),111 plan (B07022), Shanghai Key Laboratory of Digital Media Processing and Transmissions(STCSM 18DZ2270700). Dong Xu was partially supported by the Australian Research Council (ARC) Future Fellowship under Grant FT180100116. Wanli Ouyang was supported by SenseTime, the Australian Research Council Grant DP200103223, and Australian Medical Research Future Fund MRFAI000085.

References

1. Bellard, F.: BPG image format. http://bellard.org/bpg/. Accessed 30 Oct 2018
2. Ultra video group test sequences. http://ultravideo.cs.tut.fi. Accessed 30 Oct 2018
3. Video trace library (VTL) dataset. http://trace.kom.aau.dk/. Accessed 30 Oct 2018
4. Webp. https://developers.google.com/speed/webp/. Accessed 30 Oct 2018
5. Agustsson, E., et al.: Soft-to-hard vector quantization for end-to-end learning compressible representations. In: NIPS, pp. 1141–1151 (2017)
6. Agustsson, E., Tschannen, M., Mentzer, F., Timofte, R., Gool, L.V.: Generative adversarial networks for extreme learned image compression. In: 2019 IEEE/CVF International Conference on Computer Vision, ICCV 2019, pp. 221–231. IEEE (2019)
7. Ahmed, N., Natarajan, T., Rao, K.R.: Discrete cosine transform. IEEE Trans. Comput. **100**(1), 90–93 (1974)
8. Ballé, J., Laparra, V., Simoncelli, E.P.: End-to-end optimized image compression. In: Proceedings of the 5th International Conference on Learning Representations, ICLR (2017)
9. Ballé, J., Minnen, D., Singh, S., Hwang, S.J., Johnston, N.: Variational image compression with a scale hyperprior. In: Proceedings of the 6th International Conference on Learning Representations, ICLR (2018)
10. Bjontegaard, G.: Calculation of average PSNR differences between RD-curves. VCEG-M33 (2001)
11. Campos, J., Meierhans, S., Djelouah, A., Schroers, C.: Content adaptive optimization for neural image compression. In: IEEE CVPR Workshops 2019 (2019)
12. Chen, Z., He, T., Jin, X., Wu, F.: Learning for video compression. IEEE Trans. Circuits Syst. Video Techn. **30**(2), 566–576 (2020). https://doi.org/10.1109/TCSVT.2019.2892608
13. Cheng, Z., Sun, H., Takeuchi, M., Katto, J.: Learning image and video compression through spatial-temporal energy compaction. In: Proceedings of the IEEE Conference on Computer Vision and Pattern Recognition, CVPR, pp. 10071–10080 (2019)
14. Choi, Y., El-Khamy, M., Lee, J.: Variable rate deep image compression with a conditional autoencoder. In: 2019 IEEE/CVF International Conference on Computer Vision, ICCV 2019, pp. 3146–3154. IEEE (2019)

15. Djelouah, A., Campos, J., Schaub-Meyer, S., Schroers, C.: Neural inter-frame compression for video coding. In: The IEEE International Conference on Computer Vision (ICCV) (2019)
16. Habibian, A., van Rozendaal, T., Tomczak, J.M., Cohen, T.: Video compression with rate-distortion autoencoders. In: 2019 IEEE/CVF International Conference on Computer Vision, ICCV 2019, pp. 7032–7041. IEEE (2019)
17. Hu, Z., Chen, Z., Xu, D., Lu, G., Ouyang, W., Gu, S.: Improving deep video compression by resolution-adaptive flow coding. In: ECCV (2020)
18. Kingma, D.P., Ba, J.: Adam: a method for stochastic optimization. arXiv preprint arXiv:1412.6980 (2014)
19. Li, M., Zuo, W., Gu, S., Zhao, D., Zhang, D.: Learning convolutional networks for content-weighted image compression. In: CVPR (2018)
20. Lu, G., Ouyang, W., Xu, D., Zhang, X., Cai, C., Gao, Z.: DVC: an end-to-end deep video compression framework. In: Proceedings of the IEEE Conference on Computer Vision and Pattern Recognition, CVPR, pp. 11006–11015 (2019)
21. Lu, G., Ouyang, W., Xu, D., Zhang, X., Gao, Z., Sun, M.T.: Deep kalman filtering network for video compression artifact reduction. In: ECCV (2018)
22. Lu, G., Zhang, X., Ouyang, W., Chen, L., Gao, Z., Xu, D.: An end-to-end learning framework for video compression. IEEE Trans. Pattern Anal. Mach. Intell. **PP**, 1 (2020)
23. Mentzer, F., Agustsson, E., Tschannen, M., Timofte, R., Van Gool, L.: Conditional probability models for deep image compression. In: CVPR, p. 3, no. 2 (2018)
24. Minnen, D., Ballé, J., Toderici, G.D.: Joint autoregressive and hierarchical priors for learned image compression. In: Advances in Neural Information Processing Systems, pp. 10771–10780 (2018)
25. Rippel, O., Bourdev, L.: Real-time adaptive image compression. In: ICML (2017)
26. Rippel, O., Nair, S., Lew, C., Branson, S., Anderson, A.G., Bourdev, L.D.: Learned video compression. In: 2019 IEEE/CVF International Conference on Computer Vision, ICCV 2019, pp. 3453–3462. IEEE (2019)
27. Schwarz, H., Marpe, D., Wiegand, T.: Overview of the scalable video coding extension of the H.264/AVC standard. IEEE Trans. Circuits Syst. Video Technol. **17**(9), 1103–1120 (2007)
28. Shensa, M.J.: The discrete wavelet transform: wedding the a trous and Mallat algorithms. IEEE Trans. Signal Process. **40**(10), 2464–2482 (1992)
29. Skodras, A., Christopoulos, C., Ebrahimi, T.: The JPEG 2000 still image compression standard. IEEE Signal Process. Mag. **18**(5), 36–58 (2001)
30. Sullivan, G.J., Ohm, J.R., Han, W.J., Wiegand, T., et al.: Overview of the high efficiency video coding (HEVC) standard. TCSVT **22**(12), 1649–1668 (2012)
31. Theis, L., Shi, W., Cunningham, A., Huszár, F.: Lossy image compression with compressive autoencoders. In: Proceedings of the 5th International Conference on Learning Representations, ICLR (2017)
32. Toderici, G., et al.: Variable rate image compression with recurrent neural networks. In: Proceedings of the 4th International Conference on Learning Representations, ICLR (2016)
33. Toderici, G., et al.: Full resolution image compression with recurrent neural networks. In: CVPR, pp. 5435–5443 (2017)
34. Tsai, Y.H., Liu, M.Y., Sun, D., Yang, M.H., Kautz, J.: Learning binary residual representations for domain-specific video streaming. In: Thirty-Second AAAI Conference on Artificial Intelligence (2018)
35. Wallace, G.K.: The JPEG still picture compression standard. IEEE Trans. Consum. Electron. **38**(1), xviii–xxxiv (1992)

36. Wang, H., et al.: MCL-JCV: a JND-based H.264/AVC video quality assessment dataset. In: 2016 IEEE International Conference on Image Processing (ICIP), pp. 1509–1513. IEEE (2016)
37. Wang, X., Chan, K.C., Yu, K., Dong, C., Change Loy, C.: EDVR: video restoration with enhanced deformable convolutional networks. In: Proceedings of the IEEE Conference on Computer Vision and Pattern Recognition Workshops (2019)
38. Wang, Z., Simoncelli, E., Bovik, A., et al.: Multi-scale structural similarity for image quality assessment. In: ASILOMAR Conference on Signals systems and Computers, vol. 2, pp. 1398–1402. IEEE (2003). 1998
39. Wiegand, T., Sullivan, G.J., Bjontegaard, G., Luthra, A.: Overview of the H.264/AVC video coding standard. TCSVT **13**(7), 560–576 (2003)
40. Wu, C.Y., Singhal, N., Krahenbuhl, P.: Video compression through image interpolation. In: ECCV (2018)
41. Xue, T., Chen, B., Wu, J., Wei, D., Freeman, W.T.: Video enhancement with task-oriented flow. Int. J. Comput. Vision **127**(8), 1106–1125 (2019)

Towards Streaming Perception

Mengtian Li[1(✉)], Yu-Xiong Wang[1,2], and Deva Ramanan[1,3]

[1] CMU, Pittsburgh, USA
mtli@cs.cmu.edu
[2] UIUC, Urbana, USA
[3] Argo AI, Pittsburgh, USA
https://www.cs.cmu.edu/ mengtial/proj/streaming

Abstract. Embodied perception refers to the ability of an autonomous agent to perceive its environment so that it can (re)act. The responsiveness of the agent is largely governed by latency of its processing pipeline. While past work has studied the algorithmic trade-off between latency and accuracy, there has not been a clear metric to compare different methods along the Pareto optimal latency-accuracy curve. We point out a discrepancy between standard offline evaluation and real-time applications: by the time an algorithm finishes processing a particular image frame, the surrounding world has changed. To these ends, we present an approach that coherently integrates latency and accuracy into a single metric for real-time online perception, which we refer to as "streaming accuracy". The key insight behind this metric is to jointly evaluate the output of the entire perception stack at every time instant, forcing the stack to consider the amount of streaming data that should be ignored while computation is occurring. More broadly, building upon this metric, we introduce a meta-benchmark that systematically converts any image understanding task into a streaming perception task. We focus on the illustrative tasks of object detection and instance segmentation in urban video streams, and contribute a novel dataset with high-quality and temporally-dense annotations. Our proposed solutions and their empirical analysis demonstrate a number of surprising conclusions: (1) there exists an optimal "sweet spot" that maximizes streaming accuracy along the Pareto optimal latency-accuracy curve, (2) asynchronous tracking and future forecasting naturally emerge as internal representations that enable streaming image understanding, and (3) dynamic scheduling can be used to overcome temporal aliasing, yielding the paradoxical result that latency is sometimes minimized by sitting idle and "doing nothing".

1 Introduction

Embodied perception refers to the ability of an autonomous agent to perceive its environment so that it can (re)act. A crucial quantity governing the responsiveness of the agent is its reaction time. Practical applications, such as self-driving

Electronic supplementary material The online version of this chapter (https://doi.org/10.1007/978-3-030-58536-5_28) contains supplementary material, which is available to authorized users.

© Springer Nature Switzerland AG 2020
A. Vedaldi et al. (Eds.): ECCV 2020, LNCS 12347, pp. 473–488, 2020.
https://doi.org/10.1007/978-3-030-58536-5_28

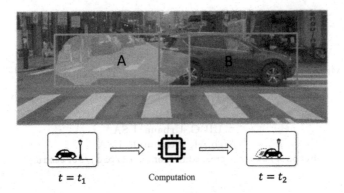

Fig. 1. Latency is inevitable in a real-world perception system. The system takes a snapshot of the world at t_1 (the car is at location A), and when the algorithm finishes processing this observation, the surrounding world has already changed at t_2 (the car is now at location B, and thus there is a mismatch between prediction A and ground truth B). If we define streaming perception as a task of continuously reporting back the current state of the world, then how should one evaluate vision algorithms under such a setting? We invite the readers to watch a video on the project website that compares a standard frame-aligned visualization with our latency-aware visualization [Link].

vehicles or augmented reality and virtual reality (AR/VR), may require reaction time that rivals that of humans, which is typically 200 ms (ms) for visual stimuli [14]. In such settings, low-latency algorithms are imperative to ensure safe operation or enable a truly immersive experience.

Historically, the computer vision community has not particularly focused on algorithmic latency. This is one reason why a disparate set of techniques (and conference venues) have been developed for robotic vision. Interestingly, latency has been well studied recently (*e.g.*, fast but not necessarily state-of-the-art accurate detectors such as [16,18,25]). But it has still been primarily explored in an *offline* setting. Vision-for-online-perception imposes quite different latency demands as shown in Fig. 1, because by the time an algorithm finishes processing a particular image frame—say, after 200 ms—the surrounding world has changed! This forces perception to be ultimately predictive of the future. In fact, such predictive forecasting is a fundamental property of human vision (*e.g.*, as required whenever a baseball player strikes a fast ball [22]). So we argue that streaming perception should be of interest to general computer vision researchers.

Contribution (Meta-Benchmark). To help explore embodied vision in a truly online streaming context, we introduce a general meta-benchmark that systematically converts *any* image understanding task into a streaming image understanding task. Our key insight is that streaming perception requires understanding the state of the world at all time instants—*when a new frame arrives, streaming algorithms must report the state of the world even if they have not done processing the previous frame*. Within this meta-benchmark, we introduce an approach to measure the real-time performance of perception systems. The approach is as simple as querying the state of the world at all time instants, and the quality of the response is measured by the *original* task metric. Such an approach naturally merges latency and accuracy into a single metric. Therefore, the trade-off between accuracy versus latency can now be measured quantitatively. Interestingly, our meta-benchmark naturally evaluates the perception stack *as a whole*. For example, a stack may include detection, tracking, and forecasting modules. Our meta-benchmark can be used to directly compare such modular stacks to end-to-end black-box algorithms [19]. In addition, our approach addresses the issue that overall latency of concurrent systems is hard to evaluate (*e.g.*, latency cannot be simply characterized by the runtime of a single module).

Contribution (Analysis). Motivated by perception for autonomous vehicles, we instantiate our meta-benchmark on the illustrative tasks of object detection and instance segmentation in urban video streams. Accompanied with our streaming evaluation is a novel dataset with high-quality, high-frame-rate, and temporally-dense annotations of urban videos. Our evaluation on these tasks demonstrates a number of surprising conclusions. (1) Streaming perception is significantly more challenging than offline perception. Standard metrics like object-detection average precision (AP) dramatically drop (from 38.0 to 6.2), indicating the need for the community to focus on such problems. (2) Decision-theoretic scheduling, asynchronous tracking, and future forecasting naturally *emerge* as internal representations that enable accurate streaming image understanding, recovering much of the performance drop (boosting performance to 17.8). With simulation, we can verify that infinite compute resources modestly improves performance to 20.3, implying that our conclusions are fundamental to streaming processing, no matter the hardware. (3) It is well known that perception algorithms can be tuned to trade off accuracy versus latency. Our analysis shows that there exists an optimal "sweet spot" that uniquely maximizes streaming accuracy. This provides a different perspective on such well-explored trade-offs. (4) Finally, we demonstrate the effectiveness of decision-theoretic reasoning that dynamically schedules which frame to process at what time. Our analysis reveals the paradox that latency is minimized by sometimes sitting idle and "doing nothing"! Intuitively, it is sometimes better to wait for a fresh frame rather than to begin processing one that will soon become "stale".

2   Related Work

Latency Evaluation. Latency is a well-studied subject in computer vision. One school of research focuses on reducing the FLOPS of backbone networks [12,28], while another school focuses on reducing the runtime of testing time algorithms [16,18,25]. We follow suit and create a latency-accuracy plot under our experiment setting (Fig. 2). While such a plot is suggestive of the trade-off for offline data processing (*e.g.*, archived video footage), it fails to capture the fact that *when the algorithm finishes processing, the surrounding world has already* *changed.* Therefore, we believe that existing plots do not reveal the streaming performance of these algorithms. Aside from computational latency, prior work has also investigated algorithmic latency [21], evaluated by running algorithms on a video in the *offline* fashion and measuring how many frames are required to detect an object after it appears. In comparison, our evaluation is done in the more realistic online real-time setting, and applies to any single image understanding task, instead of just object detection.

Real-Time Evaluation. There has not been much prior effort to evaluate vision algorithms in the real-time fashion in the research community. Notable exceptions include work on real-time tracking and real-time simultaneous localization and mapping (SLAM). First, the VOT2017 tracking benchmark specifically included a real-time challenge [15]. Its benchmark toolkit sends out frames at 20 FPS to participants' trackers and asks them to report back results before the next frame arrives. If the tracker fails to respond in time, the last reported result is used. This is equivalent to applying zero-order hold to trackers' outputs. In our benchmarks, we adopt a similar zero-order hold strategy, but extend it to a broader context of arbitrary image understanding tasks and allow for a more delicate interplay between detection, tracking, and forecast-

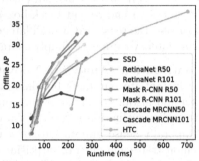

Fig. 2. Prior art routinely explores the trade-off between detection accuracy versus runtime. We generate the above plot by varying the input resolution of each detection network. We argue that such plots are exclusive to offline processing and fail to capture latency-accuracy trade-offs in streaming perception. AP stands for average precision, and is a standard metric for object detection [17].

ing. Second, the literature on real-time SLAM also considers benchmark evaluation under a "hard-enforced" real-time requirement [4,8]. Our analysis suggests that hard-enforcement is too stringent of a formulation; algorithms should be allowed to run longer than the frame rate, but should still be scored on their ability to report the state of the world (*e.g.*, localized map) at frame rate.

Progressive and Anytime Algorithms. There exists a body of work on progressive and anytime algorithms that can generate outputs with lower latency. Such work can be traced back to classic research on intelligent planning under resource constraints [3] and flexible computation [11], studied in the context of AI with bounded rationality [26]. Progressive processing [30] is a paradigm that splits up an algorithm into sequential modules that can be dynamically scheduled. Often, scheduling is formulated as a decision-theoretic problem under resource constraints, which can be solved in some cases with Markov decision processes (MDPs) [29,30]. Anytime algorithms are capable of returning a solution at any point in time [29]. Our work revisits these classic computation paradigms in the context of streaming perception, specifically demonstrating that classic visual tasks (like tracking and forecasting) naturally emerge in such bounded resource settings.

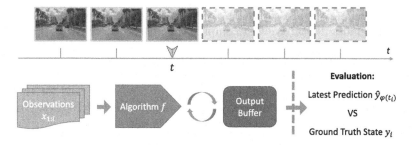

Fig. 3. Our proposed streaming perception evaluation. A streaming algorithm f is provided with (timestamped) observations up until the current time t and refreshes an output buffer with its latest prediction of the current state of the world. At the same time, the benchmark constantly queries the output buffer for estimates of world states. Crucially, f must consider the amount of streaming observations that should be ignored while computation is occurring.

3   Proposed Evaluation

In the previous section, we have shown that existing latency evaluation fails to capture the streaming performance. To address this issue, here we propose a new method of evaluation. Intuitively, a streaming benchmark no longer evaluates a function, but a piece of executable code over a continuous time frame. The code has access to a sensor input buffer that stores the most recent image frame. The code is responsible for maintaining an output buffer that represents the up-to-date estimate of the state of the world (*e.g.*, a list of bounding boxes of objects in the scene). The benchmark examines this output buffer, comparing it with a ground truth stream of the actual world state (Fig. 3).

3.1   Formal Definition

We model a data stream as a set of sensor observations, ground-truth world states, and timestamps, denoted respectively as $\{(x_i, y_i, t_i)\}_{i=1}^{T}$. Let f be a streaming algorithm to be evaluated. At any *continuous* time t, the algorithm f is provided with observations (and timestamps) that have appeared so far:

$$\{(x_i, t_i) | t_i \leq t\} \qquad \text{[accessible input at time } t] \qquad (1)$$

We allow the algorithm f to generate an output prediction at *any time*. Let s_j be the timestamp that indicates when a particular prediction $\hat{y}_j$ is produced. The subscript j indexes over the N outputs generated by f over the entire stream:

$$\{(\hat{y}_j, s_j)\}_{j=1}^{N} \qquad \text{[all outputs by } f] \qquad (2)$$

Note that this output stream is *not* synchronized with the input stream, and N has no direct relationship with T. Generally speaking, we expect algorithms to run slower than the frame rate ($N < T$).

We benchmark the algorithm f by comparing its most recent output at time t_i to the ground-truth y_i. We first compute the index of the most recent output:

$$\varphi(t) = \arg \max_j s_j < t \qquad \text{[real-time constraint]} \qquad (3)$$

This is equivalent to the benchmark applying a *zero-order hold* for the algorithm's outputs to produce continuous estimation of the world states. Given an arbitrary single-frame loss L, the benchmark formally evaluates:

$$L_{\text{streaming}} = L(\{(y_i, \hat{y}_{\varphi(t_i)})\}_{i=1}^{T}) \qquad \text{[evaluation]} \qquad (4)$$

By construction, the streaming loss above can be applied to *any* single-frame task that computes a loss over a set of ground truth and prediction pairs.

3.2   Emergent Tracking and Forecasting

At first glance, "instant" evaluation may seem unreasonable: the benchmark at time t queries the state at time t. Although x_t is made available to the algorithm, any finite-time algorithm cannot make use of it to generate its prediction. For example, if the algorithm takes time Δt to perform its computation, then to make a prediction at time t, it can only use data before time $t - \Delta t$. We argue that this is the *realistic* setting for streaming perception, both in biological and robotic systems. Humans and autonomous vehicles must react to the instantaneous state of the world when interacting with dynamic scenes. Such requirements strongly suggest that perception should be inherently predictive of the future. Our benchmark similarly "forces" algorithms to reason and forecast into the future, to compensate for the mismatch between the last processed observation and the present.

One may also wish to take into account the inference time of downstream actuation modules (that say, need to optimize a motion plan that will be executed given the perceived state of the world). It is straightforward to extend our benchmark to require algorithms to generate a forecast of the world state when the downstream module finishes its processing. For example, at time t the benchmark queries the state of the world at time $t + \eta$, where $\eta > 0$ represents the inference time of the downstream actuation module.

In order to forecast, the algorithms need to reason temporally through tracking (in the case of object detection). For example, constant velocity forecasting requires the tracks of each object over time in order to compute the velocity. Generally, there are two categories of trackers—post-hoc association [2] and template-based visual tracking [20]. In this paper, we refer them in short as "association" and "tracking", respectively. Association of previously computed detections can be made extremely lightweight with simple linking of bounding boxes (*e.g.*, based on the overlap). However, association does not make use of the image itself as done in (visual) tracking. We posit that trackers may produce better streaming accuracy for scenes with highly unpredictable motion. As part of emergent solutions to our streaming perception problem, we include both association and tracking in our experiments in the next section.

(a) Single GPU model   (b) Infinite GPU model

Fig. 4. Two computation models considered in our evaluation. Each block represents an algorithm running on a device and its length indicates its runtime.

Finally, it is natural to seek out an end-to-end system that directly optimizes streaming perception accuracy. We include one such method in Appendix C.2 to show that tracking and forecasting-based representations may also emerge from gradient-based learning.

3.3 Computational Constraints

Because our metric is runtime dependent, we need to specify the computational constraints to enable a fair comparison between algorithms. We first investigate a single GPU model (Fig. 4a), which is used for existing latency analysis in prior art. In the single GPU model, only a single GPU job (*e.g.*, detection or visual tracking) can run at a time. Such a restriction avoids multi-job interference and memory capacity issues. Note that a reasonable number of CPU jobs are allowed to run concurrently with the GPU job. For example, we allow bounding box association and forecasting modules to run on the CPU in Fig. 7.

Nowadays, it is common to have multiple GPUs in a single system. We investigate an *infinite* GPU model (Fig. 4b), with no restriction on the number of GPU jobs that can run concurrently. We implement this infinite computation model with *simulation*, described in the next subsection.

3.4 Challenges for Practical Implementation

While our benchmark is conceptually simple, there are several practical hurdles. First, we require high-frame-rate ground truth annotations. However, due to high annotation cost, most existing video datasets are annotated at rather sparse frame rates. For example, YouTube-VIS is annotated at 6 FPS, while the video data rate is 30 FPS [27]. Second, our evaluation is *hardware dependent*—the same algorithm on different hardware may yield different streaming performance. Third, stochasticity in actual runtimes yields stochasticity in the streaming performance. Note that the last two issues are also prevalent in *existing* offline runtime analyses. Here we present high-level ideas for the solutions and leave additional details to Appendix A.2 & A.3.

Pseudo Ground Truth. We explore the use of pseudo ground truth labels as a surrogate to manual high-frame-rate annotations. The pseudo labels are obtained by running state-of-the-art, *arbitrarily expensive* offline algorithms on each frame of a benchmark video. While the absolute performance numbers (when benchmarked on ground truth and pseudo ground truth labels) differ, we find that the rankings of algorithms are remarkably stable. The Pearson correlation coefficient of the scores of the two ground truth sets is 0.9925, suggesting that the real score is literally a linear function of the pseudo score. Moreover, we find that offline pseudo ground truth could also be used to self-supervise the training of streaming algorithms.

Simulation. While streaming performance is hardware dependent, we now demonstrate that the benchmark can be evaluated on simulated hardware. In simulation, the benchmark assigns a runtime to each module of the algorithm, instead of measuring the wall-clock time. Then based on the assigned runtime, the simulator generates the corresponding output timestamps. The assigned runtime to each module provides a layer of abstraction on the hardware.

The benefit of simulation is to allow us to assess the algorithm performance on non-existent hardware, *e.g.*, a future GPU that is 20% faster or infinite GPUs in a single system. Simulation also allows our benchmark to inform practitioners about computation platforms necessary to obtain a certain level of accuracy.

Runtime-Induced Variance. Due to algorithmic choice and system scheduling, different runs of the same algorithm may end up with different runtimes. This variation across runs also affects the overall streaming performance. Fortunately, we empirically find that such variance causes a standard deviation of up to 0.5% under our experiment setting. Therefore, we omit variance report in our experiments.

4   Solutions and Analysis

In this section, we instantiate our meta-benchmark on the illustrative task of object detection. While we show results on streaming detection, several key ideas also generalize to other tasks. An instantiation on instance segmentation can be found in Appendix A.6. We first explain the setup and present the solutions and analysis. For the solutions, we first consider single-frame detectors, and then add forecasting and tracking one by one into the discussion. We focus on the most effective combination of detectors, trackers, and forecasters which we have evaluated, but include additional methods in Appendix C.

4.1   Setup

We extend the publicly available video dataset Argoverse 1.1 [5] with our own annotations for streaming evaluation, which we name Argoverse-HD (High-frame-rate Detection). It contains diverse urban outdoor scenes from two US cities. We select Argoverse for its embodied setting (autonomous driving) and its high-frame-rate sensor data (30 FPS). We focus on the task of 2D object detection for our streaming evaluation. Under this setting, the state of the world y_t is a list of bounding boxes of the objects of interest. While Argoverse has multiple sensors, we only use the center RGB camera for simplicity. We collect

Dataset	AP	AP_L	AP_M	AP_S	AP_{50}	AP_{75}
MS COCO	37.6	50.3	41.4	20.7	59.8	40.5
Argoverse-HD (Ours)	30.6	52.4	33.1	12.2	52.3	31.2

Fig. 5. Comparison between our dataset and MS COCO [17]. Top shows an example image from Argoverse 1.1 [5], overlaid with our dense 2D annotation (at 30 FPS). Bottom presents results of Mask R-CNN [10] (ResNet 50) evaluated on the two datasets. AP_L, AP_M and AP_S denote AP for large, medium and small objects respectively. AP_{50}, AP_{75} denote AP with IoU (Intersection over Union) thresholds at 0.5 and 0.75 respectively. We first observe that the APs are roughly comparable, showing that our annotation is reasonable in evaluating object detection performance. Second, we see a significant drop in AP_S from COCO to ours, suggesting that the detection of small objects is more challenging in our setting. For self-driving vehicle applications, those small objects are important to identify when the ego-vehicle is traveling at a high speed or making unprotected turns.

our own annotations since the dataset does not provide dense 2D annotations[1]. For the annotations, we follow MS COCO [17] class definitions and format. For example, we include the "iscrowd" attribute for ambiguous cases where each instance cannot be identified, and therefore the algorithms will not be wrongfully penalized. We use only a subset of 8 classes (from 80 MS COCO classes) that are directly relevant to autonomous driving: person, bicycle, car, motorcycle, bus, truck, traffic light, and stop sign. This definition allows us to evaluate off-the-shelf models trained on MS COCO. No training is involved in the following experiments unless otherwise specified. All numbers are computed on the validation set, which contains 24 videos ranging from 15–30 s each (the total number of frames is 15k). Figure 5 shows a comparison of our annotation with that of MS COCO. Additional comparison with other related datasets can be found in Appendix A.4. All output timing is measured on a single Geforce GTX 1080 Ti GPU (a Tesla V100 counterpart is provided in Appendix A.7).

4.2 Detection-Only

Table 1 includes the main results of using just detectors for streaming perception. We first examine the case of running a state-of-the-art detector—Hybrid Task

---

[1] It is possible to derive 2D annotations from the provided 3D annotations, but we find that such derived annotations are highly imprecise.

Table 1. Performance of existing detectors for streaming perception. The number after @ is the input scale (the full resolution is 1920 × 1200). * means using GPU for image pre-processing as opposed to using CPU in the off-the-shelf setting. The last column is the mean runtime of the detector for a single frame in milliseconds (mask branch disabled if applicable). The first baseline is to run an accurate detector (row 1), and we observe a significant drop of AP in the online real-time setting (row 2). Another commonly adopted baseline for embodied perception is to run a fast detector (row 3–4), whose runtime is smaller than unit time interval (33 ms for 30 FPS streams). Neither of these baselines achieves good performance. Searching over a wide suite of detectors and input scales, we find that the optimal solution is Mask R-CNN (ResNet 50) operating at 0.5 input scale (row 5–6). In addition, our scheduling algorithm (Algorithm 1) boosts the performance by 1.0/2.3 for AP/AP_L (row 7). In the hypothetical infinite GPU setting, a more expensive detector yields better trade-off (input scale switching from 0.5 to 0.75, almost doubling the runtime), and it further boosts the performance to 14.4 (row 8), which is the optimal solution achieved by just running the detector. Simulation suggests that 4 GPUs suffice to maximize streaming accuracy for this solution

ID	Method	Detector	AP	AP_L	AP_M	AP_S	AP_{50}	AP_{75}	Runtime
1	Accurate (Offline)	HTC @ s1.0	38.0	64.3	40.4	17.0	60.5	38.5	700.5
2	Accurate	HTC @ s1.0	6.2	9.3	3.6	0.9	11.1	5.9	700.5
3	Fast	RetinaNet R50 @ s0.2	5.5	14.9	0.4	0.0	9.9	5.6	36.4
4	Fast*	RetinaNet R50 @ s0.2	6.0	18.1	0.5	0.0	10.3	6.3	**31.2**
5	Optimized	Mask R-CNN R50 @ s0.5	10.6	21.2	6.3	0.9	22.5	8.8	77.9
6	Optimized*	Mask R-CNN R50 @ s0.5	**12.0**	**24.3**	**7.9**	**1.0**	**25.1**	**10.1**	56.7
7	+ Scheduling (Algorithm 1)	Mask R-CNN R50 @ s0.5	13.0	26.6	9.2	1.1	26.8	11.1	56.7
8	+ Infinite GPUs	Mask R-CNN R50 @ s0.75	14.4	24.3	11.3	2.8	30.6	12.1	92.7

Cascade (HTC) [6], both in the offline and the streaming settings. The AP drops significantly in the streaming setting. Such a result is not entirely surprising due to its high runtime (700 ms). A commonly adopted strategy for real-time applications is to run a detector that is within the frame rate. We point out that this strategy may be problematic, since such a hard-constrained time budget results in poor accuracy for challenging tasks (Table 1 row 3–4). In addition, we find that many existing network implementations are optimized for throughput rather than latency, reflecting the bias of the community for offline versus online processing! For example, image pre-processing (*e.g.*, resizing and normalizing) is often done on CPU, where it can be pipelined with data pre-fetching. By moving it to GPU, we save 21ms in latency (for an input of size 960 × 600).

In our benchmarks, it is a *choice* for the streaming algorithm to decide when and what to process. Figure 6 compares a straight-forward schedule with our dynamic schedule (Algorithm 1). Such subtlety is the result of temporal quantization. While spatial quantization has been studied in computer vision [10], temporal quantization in the streaming setting has not been well explored. Notably, it is difficult to pre-compute the optimal schedule because of the stochasticity of actual runtimes. Our proposed scheduling policy (Algorithm 1) minimizes the expected temporal mismatch of the output stream and the data stream, thus

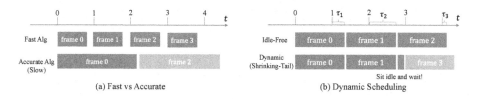

Fig. 6. Algorithm scheduling for streaming perception with a single GPU. (a) A fast detector finishes processing the current frame before the next frame arrives. An accurate (and thus slow) detector does not process every frame due to high latency. In this example, frame 1 is skipped. Note that the goal of streaming perception is not to process every frame but to produce accurate state estimations in a timely manner. (b) A straight-forward schedule for slow algorithms (runtime > unit time interval) is to always process the latest available frame upon the completion of the previous processing (idle-free). However, the latest available frame might be stale, and we find that it might be better to sit idle and wait (our dynamic schedule, Algorithm 1). In this illustration, when the algorithm finishes processing frame 1, Algorithm 1 determines that frame 2 is stale and decides to wait for frame 3 by comparing the tails τ_2 and τ_3.

Algorithm 1. Shrinking-tail policy

1: Given finishing time s and algorithm runtime r in the unit of frames (assuming $r > 1$), this policy returns whether the algorithm should wait for the next frame
2: Define tail function $\tau(t) = t - \lfloor t \rfloor$
3: **return** $[\tau(s + r) < \tau(s)]$ (Iverson bracket)

---

increasing the overall streaming performance. Empirically, we find that it raises the AP for the detector (Table 1 row 7). We provide *theoretical reasoning showing its superiority* and results for a wide suite of detectors in Appendix B.1. Note that Algorithm 1 is by construction task agnostic (not specific to object detection).

4.3   Forecasting

Now we expand our solution space to include forecasting methods. We experimented with both constant velocity models and first-order Kalman filters. We find good performance with the latter, given a small modification to handle asynchronous sensor measurements (Fig. 7). The classic Kalman filter [13] operates on uniform time steps, coupling prediction and correction updates at each step. In our case, we perform correction updates only when a sensor measurement is available, but predict at every step. Second, due to frame-skipping, the Kalman filter should be time-varying (the transition and the process noise depend on the length of the time interval, details can be found in Appendix B.2). Association for bounding boxes across frames is required to update the Kalman filter, and we apply IoU-based greedy matching. For association and forecasting, the computation involves only bounding box coordinates and therefore is very lightweight (< 2ms on CPU). We find that such overhead has little influence on the overall AP. The results are summarized in Table 2.

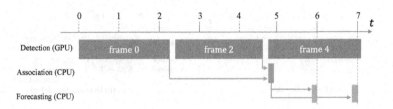

Fig. 7. Scheduling for association and forecasting. Association takes place immediately after a new detection result becomes available, and it links the bounding boxes in two consecutive detection results. Forecasting takes place right before the next time step and it uses an asynchronous Kalman filter to produce an output as the estimation of the current world state. By default, the prediction step also updates internal states in the Kalman filter and is always called before the update step. In our case, we perform multiple update-free predictions (green blocks) until we receive a frame result. (Color figure online)

Table 2. Streaming perception with joint detection, association, and forecasting. Association is done by IoU-based greedy matching, while forecasting is done by an asynchronous Kalman filter. First, we observe that forecasting greatly boosts the performance (from Table 1 row 7's 13.0 to row 1's 16.7). Also, with forecasting compensating for algorithm latency, it is now desirable to run a more expensive detector (row 2). Searching again over a large suite of detectors after adding forecasting, we find that the optimal detector is still Mask R-CNN (ResNet 50), but at input scale 0.75 instead of 0.5 (runtime 93 ms and 57 ms)

ID	Method	AP	AP_L	AP_M	AP_S	AP_{50}	AP_{75}
1	Detection + Scheduling + Association + Forecasting	16.7	39.9	14.9	1.2	31.2	16.0
2	+ Re-optimize Detection (s0.5 → s0.75)	17.8	33.3	16.3	3.2	35.2	16.5
3	+ Infinite GPUs	20.3	38.5	19.9	4.0	39.1	18.9

Streamer (Meta-detector) Note that our dynamic scheduler (Algorithm 1) and asynchronous Kalman forecaster can be applied to *any* off-the-shelf detector, regardless of its underlying latency (or accuracy). This means that we can assemble these modules into a *meta-detector* – which we call Streamer – that converts any detector into a streaming detection system that reports real-time detections at an arbitrary framerate. Appendix B.4 evaluates the improvement in streaming AP across 80 different settings (8 detectors × 5 image scales × 2 compute models), which vary from 4% to 80% with an average improvement of 33%.

4.4   Visual Tracking

Visual tracking is an alternative for low-latency inference, due to its faster speed than a detector. For our experiments, we adopt the state-of-the-art multi-object tracker [1] (which is second place in the MOT'19 challenge [7] and is open sourced), and modify it to only track previously identified objects to make it

Table 3. Streaming perception with joint detection, visual tracking, and forecasting. We see that initially visual trackers do not outperform simple association (Table 2) with the corresponding setting in the single GPU case. But that is reversed if the tracker can be optimized to run faster (2x) while maintaining the same accuracy (row 6). Such an assumption is not unreasonable given the fact that the tracker's job is as simple as updating locations of previously detected objects

ID	Method	AP	AP_L	AP_M	AP_S	AP_{50}	AP_{75}
1	Detection + Visual Tracking	12.0	29.7	11.2	0.5	23.3	11.3
2	+ Forecasting	13.7	38.2	14.2	0.5	24.6	13.6
3	+ Re-optimize Detection (s0.5 → s0.75)	16.5	31.0	14.5	2.8	33.4	14.8
4	+ Infinite GPUs w/o Forecasting	14.4	24.2	11.2	2.8	30.6	12.0
5	+ Forecasting	20.1	38.3	19.7	3.9	38.9	18.7
6	Detection + Simulated Fast Tracker (2x) + Forecasting + Single GPU	19.8	39.2	20.2	3.4	38.6	18.1

faster than the base detector (see Appendix B.3). This tracker is built upon a two-stage detector and for our experiment, we try out the configurations of Mask R-CNN with different backbones and with different input scales. Also, we need a scheduling scheme for this detection plus tracking setting. For simplicity, we only explored running detection at fixed strides of 2, 5, 15, and 30. For example, stride 30 means that we run the detector once and then run the tracker 29 times, with the tracker getting reset after each new detection. Table 3 row 1 contains the best configuration over backbone, input scale, and detection stride.

5   Discussion

Streaming Perception Remains a Challenge. Our analysis suggests that streaming perception involves careful integration of detection, tracking, forecasting, and dynamic scheduling. While we present several strong solutions for streaming perception, the gap between the streaming performance and the offline performance remains significant (20.3 versus 38.0 in AP). This suggests that there is considerable room for improvement by building a better detector, tracker, forecaster, or even an end-to-end model that blurs boundary of these modules.

Formulations of Real-Time Computation. Common folk wisdom for real-time applications like online detection requires that detectors run within the sensor frame rate. Indeed, classic formulations of anytime processing require algorithms to satisfy a "contract" that they will finish under a compute budget [29]. Our analysis suggests that this view of computation might be too myopic as evidenced by contemporary robotic systems [24]. Instead, we argue that the sensor rate and compute budget should be seen as design choices that can be tuned to optimize a downstream task. Our streaming benchmark allows for such a global perspective.

Generalization to Other Tasks. By construction, our meta-benchmark and dynamic scheduler (Algorithm 1) are not restricted to object detection. We illustrate such generalization with an additional task of instance segmentation

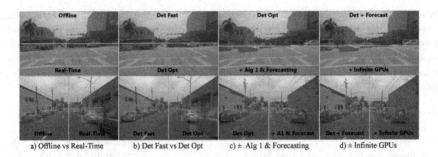

a) Offline vs Real-Time   b) Det Fast vs Det Opt   c) ± Alg 1 & Forecasting   d) ± Infinite GPUs

Fig. 8. Qualitative results. Video results can be found on the project website [Link].

a) Pseudo ground truth   b) Real-time latency   c) Instance mask forecasting

Fig. 9. Generalization to instance segmentation. (a) The offline pseudo ground truth we adopt for evaluation is of high quality. (b) A similar latency pattern can be observed for instance segmentation as in object detection. (c) Forecasting for instance segmentation can be implemented as forecasting the bounding boxes and then warping the masks accordingly.

(Fig. 9). However, there are several practical concerns that need to be addressed. Densely annotating video frames for instance segmentation is almost prohibitively expensive. Therefore, we adopt offline pseudo ground truth (Sect. 3.4) to evaluate streaming performance. Another concern is that the forecasting module is task-specific. In the case of instance segmentation, we implement it as forecasting the bounding boxes and then warping the masks accordingly. Please refer to Appendix A.6 for the complete streaming instance segmentation benchmark.

6   Conclusion and Future Work

We introduce a meta-benchmark for systematically converting any image understanding task into a streaming perception task that naturally trades off computation between multiple modules (*e.g.*, detection versus tracking). We instantiate this meta-benchmark on tasks of object detection and instance segmentation. In general, we find online perception to be dramatically more challenging than its offline counterpart, though significant performance can be recovered by incorporating forecasting. We use our analysis to develop a simple meta-detector that converts any detector (with any internal latency) into a streaming perception system that can operate at any frame rate dictated by a downstream task (such as a motion planner). We hope that our analysis will lead to future endeavor in this under-explored but crucial aspect of real-time embodied perception. For example, streaming benchmarks can be used to motivate attentional processing; by spending more compute only on spatially [9] or temporally [23] challenging regions, one may achieve even better efficiency-accuracy tradeoffs.

Acknowledgements. This work was supported by the CMU Argo AI Center for Autonomous Vehicle Research and was supported by the Defense Advanced Research Projects Agency (DARPA) under Contract No. HR001117C0051. Annotations for ArgoVerse-HD were provided by Scale AI.

References

1. Bergmann, P., Meinhardt, T., Leal-Taixé, L.: Tracking without bells and whistles. In: ICCV (2019)
2. Bewley, A., Ge, Z., Ott, L., Ramos, F., Upcroft, B.: Simple online and realtime tracking. In: ICIP (2016)
3. Boddy, M., Dean, T.L.: Deliberation scheduling for problem solving in time-constrained environments. Artif. Intell. **67**(2), 245–285 (1994)
4. Cadena, C., et al.: Past, present, and future of simultaneous localization and mapping: toward the robust-perception age. IEEE Trans. Robot. **67**(2), 245–286 (2016)
5. Chang, M.F., et al.: Argoverse: 3D tracking and forecasting with rich maps. In: CVPR (2019)
6. Chen, K., et al.: Hybrid task cascade for instance segmentation. In: CVPR (2019)
7. Dendorfer, P., et al.: CVPR19 tracking and detection challenge: how crowded can it get? arXiv:1906.04567 (2019)
8. Engel, J., Koltun, V., Cremers, D.: Direct sparse odometry. TPAMI **1**(2), 4 (2017)
9. Gao, M., Yu, R., Li, A., Morariu, V.I., Davis, L.S.: Dynamic zoom-in network for fast object detection in large images. In: CVPR (2018)
10. He, K., Gkioxari, G., Dollár, P., Girshick, R.B.: Mask R-CNN. In: ICCV (2017)
11. Horvitz, E.J.: Computation and action under bounded resources. Ph.D. thesis, Stanford University (1990)
12. Howard, A.G., et al.: MobileNets: efficient convolutional neural networks for mobile vision applications. arXiv:1704.04861 (2017)
13. Kalman, R.E.: A new approach to linear filtering and prediction problems. Trans. ASME-J. Basic Eng. **82**(Series D), 35–45 (1960)
14. Kosinski, R.J.: A literature review on reaction time. Clemson Univ. **10** (2008)
15. Kristan, M., et al.: The visual object tracking VOT2017 challenge results (2017)
16. Lin, T.Y., Goyal, P., Girshick, R., He, K., Dollár, P.: Focal loss for dense object detection. In: ICCV (2017)
17. Lin, T.Y., et al.: Microsoft COCO: common objects in context. In: ECCV (2014)
18. Liu, W., et al.: SSD: Single shot multibox detector. In: ECCV (2016)
19. Luc, P., Couprie, C., LeCun, Y., Verbeek, J.: Predicting future instance segmentations by forecasting convolutional features. In: ECCV (2018)
20. Lukezic, A., Vojír, T., Zajc, L.C., Matas, J., Kristan, M.: Discriminative correlation filter with channel and spatial reliability. In: CVPR (2017)
21. Mao, H., Yang, X., Dally, W.J.: A delay metric for video object detection: what average precision fails to tell. In: ICCV (2019)
22. McLeod, P.: Visual reaction time and high-speed ball games. Perception **16**(1), 49–59 (1987)

23. Mullapudi, R.T., Chen, S., Zhang, K., Ramanan, D., Fatahalian, K.: Online model distillation for efficient video inference. In: ICCV (2019)
24. Quigley, M., et al.: ROS: an open-source robot operating system. In: ICRA Workshop on Open Source Software, Kobe, Japan (2009)
25. Redmon, J., Divvala, S., Girshick, R., Farhadi, A.: You only look once: unified, real-time object detection. In: CVPR (2016)
26. Russell, S.J., Wefald, E.: Do the Right Thing: Studies in Limited Rationality. MIT Press, Cambridge (1991)
27. Yang, L., Fan, Y., Xu, N.: Video instance segmentation. In: ICCV (2019)
28. Zhang, X., Zhou, X., Lin, M., Sun, J.: ShuffleNet: an extremely efficient convolutional neural network for mobile devices. In: CVPR (2018)
29. Zilberstein, S.: Using anytime algorithms in intelligent systems. AI Mag. **17**(3), 73 (1996)
30. Zilberstein, S., Mouaddib, A.I.: Optimal scheduling of progressive processing tasks. Int. J. Approx. Reason. **25**(3), 169–186 (2000)

Towards Automated Testing and Robustification by Semantic Adversarial Data Generation

Rakshith Shetty[1(✉)], Mario Fritz[2], and Bernt Schiele[1]

[1] Max Planck Institute for Informatics, Saarland Informatics Campus, Saarbrücken, Germany
{rshetty,schiele}@mpi-inf.mpg.de
[2] CISPA Helmholtz Center for Information Security, Saarbrücken, Germany
fritz@cispa.saarland

Abstract. Widespread application of computer vision systems in real world tasks is currently hindered by their unexpected behavior on unseen examples. This occurs due to limitations of empirical testing on finite test sets and lack of systematic methods to identify the breaking points of a trained model. In this work we propose semantic adversarial editing, a method to synthesize plausible but difficult data points on which our target model breaks down. We achieve this with a differentiable object synthesizer which can change an object's appearance while retaining its pose. Constrained adversarial optimization of object appearance through this synthesizer produces rare/difficult versions of an object which fool the target object detector. Experiments show that our approach effectively synthesizes difficult test data, dropping the performance of YoloV3 detector by more than 20 mAP points by changing the appearance of a single object and discovering failure modes of the model. The generated semantic adversarial data can also be used to robustify the detector through data augmentation, consistently improving its performance in both standard and out-of-dataset-distribution test sets, across three different datasets.

1 Introduction

Performance evaluation of computer vision systems is predominantly done by empirical evaluation on a fixed test set, often drawn from a similar distribution as the training data. However, due to limited sample size a fixed test set only captures a small portion of errors the model would make on diverse data seen during real-world deployment. This discrepancy manifests as poor out-of-dataset-distribution (OODD) generalization [15,27,28], vulnerabilities to input noise [14,24] and adversarial perturbations [12]. In this work, we propose automated testing through semantic adversarial editing which synthesizes difficult

Electronic supplementary material The online version of this chapter (https://doi.org/10.1007/978-3-030-58536-5_29) contains supplementary material, which is available to authorized users.

© Springer Nature Switzerland AG 2020
A. Vedaldi et al. (Eds.): ECCV 2020, LNCS 12347, pp. 489–506, 2020.
https://doi.org/10.1007/978-3-030-58536-5_29

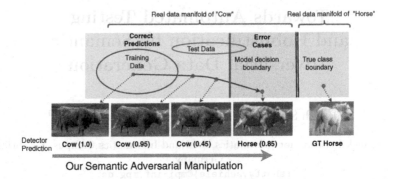

Fig. 1. Standard testing paradigms only covers a small portion of errors models make in the real world due to sample size limitation. We propose semantic adversarial testing to find targeted failure cases through continuous optimization of object appearance to cross the model's decision boundary, while remaining within the true class boundary.

cases, targeted for a particular model, exposing its weaknesses. The error cases synthesized by our model often have atypical appearance and outside the distribution covered by fixed size datasets, however still within the true class boundary to human observers. Apart from its usefulness for testing, our semantic adversarial data can also be used to robustify the target model and improve its performance on OODD data.

To create reliable test data, we need to ensure that the generated sample is consistent with its label. At the same time, the created test data needs to be difficult, ideally capturing the different failure modes of the target model. Simply gathering more data is expensive and inefficient as the process is not targeted to the model. Our approach to meet both these criteria is to start from a real data point, and to make constrained semantic edits through a differentiable synthesizer model. The synthesis process is adversarially optimized to produce semantic changes which fool the target model. By only editing the appearances of individual objects with their pose and the scene held intact, we keep the changes minimal and realistic. An example is seen in Fig. 1 where the appearance of "cow" is edited to change the detector prediction to "horse".

Our key insight to constrain the semantic adversarial objects to be label-consistent is by limiting the range of synthesized appearance to be a combination of real ones. We first select a set of guiding templates by sampling instances of the same class from the real data. Then a new appearance is synthesized for the target object optimized to fool the detector, while staying within the convex hull spanned by the appearance of guiding instances. Since changing pose realistically is a much harder task, requiring reasoning over both object and the context, we keep it fixed. Our synthesizer network disentangles the object's pose from its appearance thus allowing editing the appearance without affecting the pose.

Since our semantic adversarial object synthesis process is fully differentiable, we can mine new errors for a target model by directly optimizing the appearance to fool it. We demonstrate this by creating hard test data for the YoloV3 object detector [29]. The same mechanism can also be used to generate hard training data for the detector. The synthesized examples are hard positive examples, often lying close to detectors class boundaries. Our experiments on three dataset, COCO, BDD100k and Pascal, show that using the generated data to fine-tune the detector model improves the model performance and generalization to data distribution shift. To summarize, main contributions of our work are:

- We propose the first method for automatized testing of computer vision models finding new error cases by synthesizing semantic adversarial examples.
- We design an object synthesizer network which disentangles object shape and appearance. This is achieved through a novel binary part segmentation bottleneck which scales better to the diverse object classes.
- We propose a novel mechanism to semantically change the object appearance to fool detectors, while keeping the appearance within the class boundaries as verified by a human study. Experiments show that our semantic adversary editing the appearance of a single object drops the detector performance by 20 mAP points and helps find new vulnerabilities of the model.
- Utility of our generated data is further shown by using it for training the YoloV3 detector. Experiments on three datasets show that the generated data helps improve the detector performance and generalization to OODD data.

2   Related Work

Our work connects to four lines of research: robustness testing, semantic adversarial attacks, data augmentation for object detection and generative models. This section will discuss these connections and how our work differs.

Robustness of Vision Models. CNN based computer vision models generalize poorly when tested on OODD test data. This includes data with various noise [14, 24], translation and scaling [3], rotations [13], out of context objects [31,34], or test set resampling [27,28]. With simulated data and 3D rendering, models can also be fooled by unusual object poses [2] and lighting [21]. Difficult natural "adversarial" examples where ImageNet classifiers fail was collected in [15,32], but through manual and expensive process. In our work we focus on real data and manipulate object appearances to efficiently synthesize error cases for detectors.

Semantic Adversaries. Deep models have also been attacked with semantic adversarial examples. [6,9] shows we can fool image classifiers by applying the right adversarial translations and rotations to an image. This is generalized to adversarial spatial deformations in [1,43]. Attempts to semantically change the object appearance have been limited to parametric color distortions [16] and using a generative model for faces [36] and digits [37]. Our work moves beyond the prior research in both scale and scope. We generate semantic adversarial objects by optimizing both appearance and position of the objects and use it to attack detectors on three large datasets with diverse object classes.

Data Augmentation for Object Detection. A related research area is the data augmentation for object detection which focus on altering individual

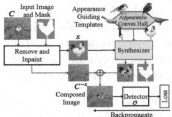
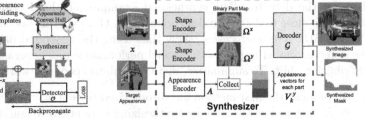

Fig. 2. Our overall pipeline for creating the semantic adversaries

Fig. 3. Synthesizer architecture to generate objects with disentangled appearance and pose latents

objects [7,8,38,41,42]. An early work [42] uses an adversary to partially mask objects to create hard occlusions. Objects are transferred onto new backgrounds for data augmentation with a cut-paste mechanism in [8]. [7] refines this by also heeding to the context for picking a location to paste objects. Yet, this does not take the object pose into account. [38] takes the cut and paste approach further by training a network to predict worst case position, rotation and scale of the added object to fool the detector. Our work, in contrast, resynthesizes the appearance of entire objects to fool the targeted detector, while preserving original context and pose. Thus our synthesis process allows wider range of semantic changes compared to occlusions, and better preserves realism and image context compared to cut-and-paste approaches. We compare our approach to a recent work on data augmenting the object detector by switching instances of objects [41]. While [41] circumvents the context issue by switching instances in-place through shape matching, it does not allow generating targeted hard examples.

Unsupervised Disentangling of Appearence and Pose. Our synthesizer architecture is based on unsupervised generative models for disentangling object appearance and pose [17,19,22,35]. Most similar to our design is the model in [22]. In[22], two encoders are used to create latent vectors of pose and appearance, with a Gaussian keypoint bottleneck regulating the pose encoding to carry only spatial information. The key difference in our work is we propose binary segmentation maps as the bottleneck, which scales better to large number of diverse object classes seen in our experiments on COCO dataset.

3   Synthesizing Semantic Adversarial Objects

Our main goal is to efficiently synthesize hard/error cases for an object detector from data manifold. We achieve this goal by starting from a real data point and adversarially editing its appearance through a synthesizer network to fool the target detector. This is a continuous optimization problem efficiently solvable through gradient descent. Additionally, we also need to make sure the synthesized sample is realistic and matches the original label. This is achieved first by only editing appearance of selected objects while retaining its pose, ensuring that the object instance fits well to the image context. Additionally, we constrain the space of appearances allowed during optimization to keep label consistency.

Our solution, shown in Fig. 2, consists of two key contributions. First, we build an object synthesizer which disentangles object's pose and appearance, thus allowing us to generate various appearances for an object while keeping its original pose. This is enabled by a binary part segmentation bottleneck, which scales better to diverse object classes, a key requirement to scale to detection datasets like COCO. Second, we propose a novel optimization formulation wherein the latent appearance codes in the synthesizer are constrained to the convex hull of guiding templates. Under this constraint, the appearance is optimized to find the adversarial appearance for an object instance to fool the target detector.

3.1   Synthesizer Design

To achieve disentanglement between pose and appearance we propose a modular architecture consisting of an appearance encoder producing latent codes representing the object appearance, a shape encoder producing a binary part segmentation of the object and a decoder which utilizes both the parts and the appearance vectors to synthesize the object. Note that the whole model is learned with only self-supervision, by learning to autoencode objects in the dataset. The overall architecture of synthesizer is shown in Fig. 3. While this architecture is inspired by recent works [17,22], our solution differs in two crucial aspects, the type of bottleneck and the architecture of the decoder. To understand this difference, let us walk through the process of synthesizing an object given an instance x with the target shape and an instance y with target appearance.

Shape Encoder. A representation of the input object shape is first extracted by the shape encoder. This is a CNN with Unet [30] structure which maps the input image into a $K \times M \times N$ dimensional tensor Z, where K is the number of parts and M,N are the spatial dimensions. Z_{kij} represents the likelihood of the kth part being present at locations ij. To disentangle shape and appearance, we need to restrict Z to only carry information about the spatial layout of the object instance. Prior works [17,22] do this by approximating Z with 2D Gaussians. While this works for classes like "person" whose parts fit well with gaussian shapes, we find that it does not work well on diverse object classes with complex sub-parts like "bicycle", "bus", and so on. Instead, we solve this by bottlenecking the information in Z by converting it to a spatial probability distribution and sampling binary masks from it. Specifically, we obtain the part-probability distribution as $P_{kij} = \text{softmax}_k[Z_{ij}]$ and sample binary part maps $\Omega_{kij} = \text{gumbel\_softmax}_k[P_{ij}]$ from it. Here we use gumbel softmax approximation [18,23] to sample from the multinomial distribution P_{kij} in-order to keep the sampling process differentiable.

Appearance Encoder. Object appearance is encoded with a CNN, which maps the input image to a tensor A of dimensions $D \times M \times N$. This spatial appearance map is reduced to K appearance codes $V = [V_1 \cdots V_k]$, one for each part, by averaging A over the part activations $V_k = \sum_{ij} P_{kij} A_{ij}$.

Decoder Network. Now using the appearance vector V^y extracted from image y and the binary part segmentation Ω^x extracted from image x, the decoder network $\mathcal{G}$ synthesizes the desired object and its segmentation mask. The appearance vectors V^y_k are first projected onto their corresponding binary part activation map to reconstruct the spatial appearance map $\widetilde{A}^y = V^y \Omega^x$. Our decoder architecture, in contrast to [22], utilizes spatially adaptive normalization layers [25] to input the appearance code at different resolutions to produce the four channel output (image + mask). We find that this helps better preserve the smaller appearance details in generated images as compared to inputting the appearance codes at the first layer. Full network configuration is provided in the supplementary Sect. 1.

Training the Synthesizer. We train the Synthesizer by learning to autoencode objects and to transfer appearance to other instances, similar to prior works [22]. Additionally we use an adversarial discriminator $\mathcal{D}$, to improve the sharpness of the generated images. When autoencoding, the model is trained end-to-end with l_1 reconstruction loss for the image and cross-entropy loss for the segmentation mask. Paired training data for learning to transfer appearance is created in two ways. First, we apply simple affine transformations to object instances x to obtain $T(x)$, creating paired data. Now the appearance can be transferred by reconstructing x using shape $\Omega^{T(x)}$ and appearance V^x encodings and vice-versa. Secondly, the model is trained to transfer appearance to a random instance y of the same class by using the discriminator real/fake loss and cyclic reconstruction loss [45]. Precisely, given shape code Ω^y and V^x, we generate a hybrid object xy and use the discriminator $\mathcal{D}$ to evaluate realism and provide a training signal. We also re-encode xy to obtain appearance code V^{xy}, and use it to reconstruct the original image as $\widetilde{x} = \mathcal{G}(\Omega^x, V^{xy})$. Apart from the reconstruction losses, we

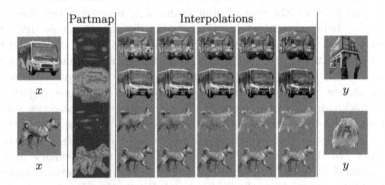

Fig. 4. Appearance interpolations with Our (even rows) and the Gaussian bottleneck model (odd rows). The objects are generated using the shape code from x and by interpolating the appearance vectors from x and y. More examples are in the supplementary.

also impose additional constraints on the appearance and shape latent codes to provide intermediate supervision. For example, $\Omega^{T(x)}$ should be same as $T(\Omega^x)$ since an affine transformed input image should lead to an affine transformed part-map. Equations for these training losses are given below.

$$L_{\mathrm{r}} = |x - \mathcal{G}(\Omega^x, \mathcal{V}^x)| + |T(x) - \mathcal{G}(\Omega^{T(x)}, \mathcal{V}^x)| + |x - \mathcal{G}(\Omega^x, V^{xy})| \qquad (1)$$

$$L_{\mathrm{d}} = \mathcal{D}(\mathcal{G}(\Omega^x, \mathcal{V}^x)) + \mathcal{D}(\mathcal{G}(\Omega^{T(x)}, \mathcal{V}^x)) + \mathcal{D}(\mathcal{G}(\Omega^y, \mathcal{V}^x)) \qquad (2)$$

$$L_{\mathrm{a}} = \|\mathcal{V}^x - \mathcal{V}^{T(x)}\| + \|\mathcal{V}^x - \mathcal{V}^{xy}\| \qquad (3)$$

$$L_{\mathrm{p}} = -P^{T(x)} \log(T(P^x)) \qquad (4)$$

Figure 4 compares the appearance transfer produced by our model trained on COCO dataset and a baseline model with identical structure, except using 2D Gaussians to bottleneck the shape encoding. We see big difference in quality of the generated images especially for objects like bus and dog. This performance gap can be understood by looking at the part representations extracted using the two methods also shown in Fig. 4. We see that while Gaussian part maps are very crude approximations, our binary part maps captures detailed shape information, enabling better reconstruction and interpolation of appearance.

Fig. 5. Intermediate steps when optimizing the appearance to fool the detector.

3.2 Synthesizing Semantic Adversaries

Now that we have a synthesizer which can effectively change appearance of a target object x using an appearance guiding template, let us leverage it to produce semantic adversaries to fool an object detector. We start by extracting the shape (Ω^x) and appearance ($\mathcal{V}^x$) representations for the target instance x occurring in image C, which we wish to edit to fool the detector $\mathcal{O}$. Instance x is removed from C using the ground-truth box and an object removal in-painter from [33] to obtain canvas image C^{-x}. A new version of object x is synthesized as $\mathcal{G}(\Omega^x, \mathcal{V}^x)$ and is pasted in place of the original to get the composed image. We denote this as $C^{-x} + \mathcal{G}(\Omega^x, \mathcal{V}^x)$. This process is illustrated in Fig. 2.

A simple way to fool the detector would be to adversarially optimize the appearance vector $\mathcal{V}^x$ until the object detector fails on the generated image $\mathcal{G}(\Omega^x, \mathcal{V}^x)$. However, in unconstrained optimization the appearance vectors often move into areas where synthesizer produces unrealistic images, which also fools the detector. We overcome this with a novel scheme which keeps the adversarially optimized $\mathcal{V}^x$ from going far from the synthesizer's input distribution. We first sample a set $I = \{i_1, \cdots i_n\}$ of n guiding templates belonging to the same class and extract appearance codes for each of them $\mathcal{V}^I = \{\mathcal{V}_k^{i_1}, \cdots \mathcal{V}_k^{i_n}\}$. Now the appearance vector for the generated object is optimized to fool the detector while constraining it to remain within the convex hull spanned by $\mathcal{V}^I$.

$$\mathcal{V}_k^{adv} = \left\{ \sum_{j=1}^n \alpha_1^j V_1^{i_j}, \cdots \sum_{j=1}^n \alpha_k^j V_k^{i_j} \right\} \tag{5}$$

$$\max_{(a_1^1, \cdots a_k^n)} \mathcal{L}_{det} \left[\mathcal{O} \left(C^{-x} + G \left(\Omega^x, \mathcal{V}_k^{adv} \right) \right) \right] \tag{6}$$

Here $\alpha_k^j = \text{softmax}_n(a_k^j)$, with $\{a_k^1 \cdots a_k^n\}$ being the interpolation co-efficients for part k and $\mathcal{L}_{det}$ is the detector loss function which we maximize. There are total of $n \times k$ interpolation coefficients which are optimized to find the adversary. Having independent part coefficients allows mixing and matching appearances from different templates for each part, and thus allowing richer appearances space to be explored through optimization. Further, since we only manipulate the latent appearance codes, the adversary cannot directly manipulate pixels to produce noisy patterns to fool the detector, but must instead rely on semantic changes. Detector loss is usually a sum of classification, objectness and box regression losses. We discard the box regression losses, as they are not directly affected by appearance and often leads to unstable behavior in optimization. Hence the detector loss becomes $\mathcal{L}_{det} = \lambda L_{obj} + (1 - \lambda)L_{cls}$, where $\lambda \in [0, 1]$ is a co-efficient controlling how much the adversary focusses on causing missed detection versus misclassification. Spatial perturbations like position or scale of the object can be easily incorporated into our formulation by inserting a parametrized affine transformation matrix before pasting the object onto canvas image, allowing position and appearance to be jointly optimized to fool the detector.

Figure 5 depicts the adversarial appearance optimization steps to fool the YoloV3 detector. First row shows the synthesized bird changing from a reconstruction in the zeroth step to a different color by the fourth step, causing detector confidence to drop. More optimization leads to a bigger failure in the detector where the brown head and the yellow circle in the body of the synthesized bird causes the detector to see the object as a "dog" and a "frisbee". The second row shows a case where the appearance of the "ball" is slowly changed to camouflage with the background and cause a missed detection. These examples show that our method makes large semantic changes to the object appearance which fools the detector while looking plausible to human eye (empirically verified in Sect. 4.2).

4 Experiments and Results

We evaluate our semantic adversary for two applications, as a diagnostic tool to find failure modes in the detector and as a hard data generation mechanism to improve the performance of these detectors. We measure the effectiveness of the semantic adversary in terms of the detector performance on generated adversarial test set. We verify label consistency of the semantic adversary by a human study where observers verify if the original class is preserved after adversarial editing. We also qualitatively examine the synthesized error cases and find different mechanisms which cause detector failures. Data augmentation experiments are run on three different datasets, COCO, VOC and BDD100k, and we measure the benefit of the generated adversarial data for improving model performance on both standard test, as well as generalization to out-of-dataset-distribution. First, we describe the experimental setup and datasets, followed by the analysis on effectiveness of the semantic adversary for diagnostics and data augmentation.

4.1 Setup and Datasets

We conduct our data augmentation experiments on three datasets – COCO [20], PascalVOC [10](VOC) and BDD100k [44]. COCO and VOC contains both indoor and outdoor images with common objects like person, car, table etc. While COCO has 120k training images with 80 classes, VOC is smaller with 20 classes and 14k training data (combining 2007 + 2012 splits). BDD100k is a large scale driving dataset with 100k street scenes captured from a car driving around major US cities, with annotation of objects like person, car, traffic light and so on. The object synthesizer and removal inpainter are both trained on the COCO dataset, due to availability of instance segmentation masks needed to extract the object patches. Since all the classes in VOC and 9/10 classes in BDD100k are part of COCO (except "rider" class), COCO trained model can be used to synthesize adversarial objects on these datasets. The synthesizer operates at 128×128 resolution. The generated objects are scaled to match the target box.

We use the YoloV3 [29] model as the target detector, as it is a popular single staged detector with fast runtime, making adversarial attack experiments run quicker. We train our baseline model from scratch using the implementation available in [39], using all the standard data augmentation methods including color jittering and rotation. However, to keep the synthesis single resolution, all our detector models are trained on single fixed resolution (416×416 on COCO and VOC, 704×1248 on BDD100k) as opposed to multi-scale training used in YoloV3, yielding a lower baseline performance. All the improvement reported from training on our synthesized data is in-complimentary to the standard augmentations. The evaluation is also performed at these fixed resolutions. The models on BDD100k and VOC are trained after initializing from a trained COCO model. This ensures that these models have already been exposed to the instances from the COCO dataset. When training on synthetic data, we start from the pre-trained model and fine-tune the last two layers in case of COCO and BDD100k and last three layers in case of Pascal. For fair comparison we also further fine-tune the pre-trained model using the exact same configuration, but only with real data to obtain the *Base-FT* model in all three datasets.

Apart from evaluating on i.i.d test sets, we also measure the generalization to OODD data. This tests our hypothesis that semantic adversarial data improves the model robustness to OODD samples, since our adversarial data often contains atypical objects, from the tail of appearance distribution. To do this, we test the COCO trained model on the UnRel [26] and VOC test sets. The models are tested on the overlapping 29 classes in UnRel and all 20 classes in VOC. UnRel data contains objects in unusual relationships and contexts and will measure if the model generalizes to rare cases. Similarly VOC trained models are also tested on UnRel, for the overlapping 14 classes. The BDD100k models are tested on D2-City [4], with driving images from Chinese cities.

4.2   Semantic Adversary for Automated Testing

To quantify the effectiveness of the semantic adversary, we create adversarial test sets using the COCO training images by optimizing the appearance of selected objects in each image to fool the detector. Objects are selected at random as long as they are not too small/large (≥ 32 pixels and $\leq 30\%$ of the image area). We do this with three variants of our approach. First only optimizes the appearance of one object instance. The second variant optimizes both the position and the appearance of the same object. In the third variant two random objects are chosen from each image and their position and appearance are adversarially optimized. Each of these test sets contain 37k images. Object detector is run on these three sets and performance is measured using mean average precision (mAP@0.5)

Quantitative Analysis. The results are reported in Table 1. We see that all the semantic adversaries drop the performance of the model significantly, with mAP dropping from 81.2 on the corresponding real data to 62.4 with just optimizing the appearance. Optimizing the position and scale of the object along with appearance further degrades the detector performance, with mAP dropping to 59.5. When we adversarially modify two objects jointly, the detector performance drops again to 46.5 mAP, making it a 57% drop in detector performance. To understand this performance drop, we look at the effect on the detector's confidence for each object instance. We consider it a success if the detector's confidence drops after adding the semantic adversary. Table 1 presents the success rate on the edited as well as untouched objects in the same image. Firstly, we see that all three strategies drop the detector's confidence on more

Table 1. Overall and instance-level detector performance under semantic advesarial editing. *Co-occuring* refers to the other untouched objects in the image.

Table 2. Human study results on the label correctness of semantic adversarial editing.

Optimize	n_obj	mAP ↓	Success rate by instance type ↑		
			Edited	Co-occuring	Combined
Real data	0	81.2	-	-	-
Appear	1	62.4	74.99	58.62	61.10
Pos + Appear	1	59.5	**77.69**	59.30	62.13
Pos + Appear	2	**46.5**	77.10	**61.54**	**65.82**

Instance	Label correctness
Real	99%
Random label	11%
SemAdv (appearance)	93%

Fig. 6. Qualitative examples of the failure cases discovered by our semantic adversary. Green boxes are correct detections, **purple** boxes indicate missed detections and red boxes show the misclassified objects. Only relevant detections are marked (Color figureonline).

than 74% of the semantically edited instances. Interestingly, about 60% of the untouched co-occurring instances are also negatively affected. This is often due to the contextual changes caused by the misclassification of the edited instances or minor occlusions produced by the edited instance. We note again that our semantic adversarial attacks are efficient, performed in just 10 steps of gradient descent. To put this in context, adversarial color jittering [16] takes about 200 trials to attack (success in 50% of cases) a simpler classification model on a smaller CIFAR-10 dataset.

We also compare the effectiveness of our semantic adversary to a standard L_∞ norm adversarial attack. For fair comparison we also restrict the L_∞ attacker to change pixels within the bounding box of a single object. The experiments show that our single object semantic adversarial attack (mA P= 59.5) is roughly equivalent in strength to a L_∞ norm attack with $\epsilon = 8/255$ (mAP = 58.7). Full results and details are presented in the supplementary material.

Human Study. A natural question at this point is if the semantic adversarial samples are within the true class boundary. To answer this we turn to human observers. We conduct a study where a human judge is presented with an image and asked if the object highlighted with the box belongs to the specified class. If they consider the label correct, they are asked to also rate how typical the object appearance is from 1 to 5, with 1 corresponding to very unusual and 5 corresponding to very typical appearance. The study is conducted on a mix of 250 real and semantically edited instances each, with each instance rated by three independent observers. We also introduce additional 10% samples where labels are shuffled. This is done in-order to verify the work of the human annotators. More details including exact instructions and interface is presented in the supplementary Sect. 3. Table 2 shows the label correctness results as judged by majority vote of three humans. As expected the label correctness is very high on real instances and is very low on label-shuffled instances. We also see that in 93% of cases, humans agree that semantic adversary preserves the label of the object instance. Performance drop on semantic adversary is small for human observers compared to the significant drop by object detectors seen before. The typicality rating provided by the human judges on real and semantically edited instances shown in Fig. 7 helps understand this gap. While most of the real samples have a typicality rating of 4 or 5 (very typical), semantic adversary have lower rating between 2–3. These results show that the semantic adversary generates atypical examples which are still correctly detectable by humans, but are hard for our detectors. This is further supported by lower performance of the detector on less typical real samples (Accuracy>70% on typicality rating = 5 and accuracy 35–40% on typicality rating between 1–3). Further details are in supplementary Sect. 3.

Qualitative Analysis. Examining the cases where the semantic adversary fools the detector reveals that the adversary causes missed detections and misclassification through four main mechanisms listed below and illustrated in Fig. 6. All examples are from the strategy with editing single object and position.

- **Camouflaging** - Semantic adversary often causes missed detections by changing the appearance of the object to blend with the background. First row of Fig. 6 illustrates a frisbee, a stop-sign and a hydrant being camouflaged.
- **Occlusion** - Second row shows cases where the semantic adversary causes missed detections by moving to partially occlude some co-occurring objects.
- **Appearance** - In many instances, the appearance of the object is altered to include small visual features which trigger misclassification by the detector. These can be seen in the third row in Fig. 6. We see that "cow" is changed to "horse" based on color change and person is misclassified as a dog due to a small change in hue. These cases indicate that detectors often rely on false correlations of low-level textures or colors to certain classes, and often fail when these textures are altered, as also shown for classifiers in recent work [11].

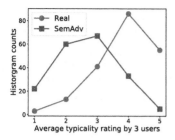

Fig. 7. Comparing the typicality rating between real data and semantic adversary.

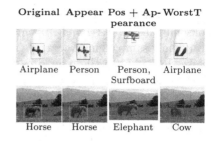

Original Appear Pos + Ap- WorstT
                  pearance

Airplane Person Person, Airplane
                  Surfboard

Horse Horse Elephant Cow

Fig. 8. Comparing various adversarial strategies and the worst-case template baseline.

- **Contextual Appearance** - Last row shows examples where with a change in an object appearance, the contextual evidence overrides the visual features, causing misclassification. Eg. in the first image, a dog changed to white color is misclassified as a sheep as there are other sheep present nearby. Similarly, a falling person is mistaken as an airplane, and a surfboard as a boat.

We note that despite a few generation artifacts, with the COCO dataset being hard for current GAN models, these samples look plausible to human eye and we would not make the same predictions as the detector. This makes it a useful tool to explore the breaking points of a trained detector.

4.3   Semantic Adversary for Data Augmentation

Apart from being a useful diagnostic tool, semantic adversaries can be used to generate training data. By targeting the detector, we can create tailored hard positives for the model, and thus get the most benefit when added to the model training set. We generate the training data with a similar process as in the previous section: first selecting an eligible object from each training image, adversarially optimizing its appearance and adding it back to the training set. The model is then fine-tuned with a combination of the original and the synthesized adversarial data for 50 epochs, and performance on the standard test sets and OODD data is measured. We now present this data augmentation results on the three datasets.

COCO Dataset. Table 3 shows the data augmentation results on the COCO dataset. Comparing the *Baseline* and *Base+FT* models we see that the further fine-tuning the last two layers of the model improves the performance a bit on COCO and VOC test sets, while reducing a bit on UnRel (39.0 vs 38.8). Comparing this with the basic semantic adversary augmented model *SA-Rand-App*, which only edits object appearance, we see a bigger improvement on all three test sets. Table 3 also shows *Base+FreeAdv*, an adversarial baseline which allows for free manipulation of appearance vector under l_∞ constraint, without the convex hull constraint. Its poor performance compared to *SA-Rand-App* shows that the unconstrained attack does not work well as often the adversarial sample looks unrealistic. *SA-Rand-App* model generates the semantic adversaries using randomly sampled instances for guidance. We can further target the model weaknesses by sampling the templates from hard instances for the detector, i.e. setting the probability of picking an instance inversely proportional to the detectors confidence on it. The model trained this way, *SA-App*, further improves the performance a bit on COCO and significantly on UnRel (39.6 vs 39.2). Moreover, the *SA* model which jointly optimizes position and appearance gets even better results, improving over *SA-App* in COCO and VOC test sets. Now, we compare our approach to a simpler baseline, where an object is replaced with the template which increases the detector loss the most. While this often fools the detector, it also places instances which do not fit with the image context, as seen in the examples in Fig. 8. Thus, the *Base-WorstT* model using this data in training performs worse than *SA-App* on all test sets.

We can increase the benefit of semantic adversaries by editing more objects in the image, creating harder data. The *SA#2* model which edits appearance and position of two objects improves on COCO (+0.3 mAP) and UnRel (+0.4 mAP) test sets compared to *SA* editing single objects. Since our adversarial data is adaptive to the model, we can continue generating harder examples attacking the newly trained model. *SA#2x2* does this, further training the *SA#2* model using the adversarial data generated by attacking *SA#2*. This second iteration helps and *SA#2x2* still improves. By repeating this four times, we get the *SA#2x4* model which outperforms the baseline on all three test sets, COCO (+1.0 mAP), VOC (+1.3 mAP) and UnRel (+1.8 mAP). The gain is larger on OODD test sets, VOC and UnRel, indicating that training with semantic adversary improves the robustness of the model to input distribution changes. We also compare our approach to recent data augmentation approaches PSIS [41] and AutoAug [5]. For PSIS, we use the data provided by the authors [40] to fine-tune our baseline model same as before. While the PSIS data improves over the baseline, it falls short compared to our *SA#2x4* model in all test sets. Table 3 also shows that our approach is complimentary to AutoAug [5], which applies augmentation policies on the entire image. While auto-augment improves the baseline performance, our *SA#2* model improves even more when combined with AutoAug.

PascalVOC Dataset. Results in Table 4 for data augmentation on VOC dataset show that, semantic adversarial data improves performance here as well. The *SAx5* model, which edits appearance and position of a single object, is better than the baseline on both VOC (+1 mAP) and the UnRel (+2.1 mAP) test sets, again with bigger gains on the OODD data. *SA#2x5* which creates two adversarial objects underperforms *SAx5*, since VOC images often have only a single object, which causes the *SA#2x5* to add too many out-of-context objects in its generation.

BDD100k Dataset. On BDD100k (see Table 5), we found that adversary often caused drastic appearance changes when fooling the classifier. Since BDD100k has only 10 classes, the class boundaries are well separated and fooling the classifier needs large unrealistic appearance changes. Instead, optimizing to only

Table 3. Data augmentation results on COCO dataset. Metric used is mAP@0.5

Model	Obj	COCO	VOC	UnRel
Baseline	-	46.1	66.4	39.0
Base+FT	-	46.2	66.9	38.8
Base+FreeAdv	1	45.8	65.3	37.9
Base+WorstT	1	46.2	66.8	39.2
SA-Rand-App	1	46.5	67.1	39.2
SA-App	1	46.6	67.0	39.6
SA	1	46.7	67.3	39.4
SA#2	2	46.9	67.4	39.8
SA#2 ×2	2	47.0	67.4	40.4
SA#2 ×4	2	**47.1**	**67.7**	**40.8**
Base+AutoAug[5]	-	47.0	67.6	40.4
SA#2+AutoAug[5]	2	47.8	68.1	41.5
PSIS [41]	-	46.7	67.5	39.8

Table 4. Data augmentation results on VOC using semantic adversary.

Model	Obj	VOC	UnRel
Base+FT	0	74.0	42.9
Base+WorstT	1	73.7	43.4
SA ×5	1	**75.0**	**45.0**
SA#2 ×5	2	74.0	44.3

Table 5. Data augmentation results on BDD100k dataset.

Model	λ	BDD	D2City
Base+FT	-	50.7	34.7
SA-App	0.5	50.8	34.6
SA-App	1.0	**51.4**	**35.1**
SA	1.0	51.2	35.0

reduce the objectness score (setting $\lambda = 1$) leads to more realistic synthesis. This is seen when comparing *SA-App* models with $\lambda = 0.5$ and $\lambda = 1.0$. The model with $\lambda = 1.0$ performs much better on both the BDD and the OODD D2-city test sets, while also improving over the fine-tuned baseline. Additionally, we see in Table 5 that the *SA-App* performs better than *SA* which optimizes position, showing that it is better to edit objects in-place in structured scenes in BDD.

5 Conclusions

We present a method for automatic test case generation through semantic adversarial optimization of object appearances. Our approach can synthesize new OODD hard examples which cause failures in the target detector, while remaining realistic to human eye. Analysis of the synthesized data shows the different failure modes discovered by the process includes camouflaging, occlusions and appearance changes. Our adversarial data is also useful for data augmentation, consistently improving the detector on standard and OODD test sets, in three datasets. We hope that our work will facilitate future approaches to test models beyond finite datasets and hence develop more reliable performance metrics.

References

1. Alaifari, R., Alberti, G.S., Gauksson, T.: ADef: an iterative algorithm to construct adversarial deformations. In: Proceedings of the International Conference on Learning Representations (ICLR) (2019)
2. Alcorn, M.A., et al.: Strike (with) a pose: neural networks are easily fooled by strange poses of familiar objects. In: Proceedings of the IEEE Conference on Computer Vision and Pattern Recognition (CVPR), June 2019

3. Azulay, A., Weiss, Y.: Why do deep convolutional networks generalize so poorly to small image transformations? J. Mach. Learn. Res. (JMLR) **20**(184), 1–25 (2019)

4. Che, Z., et al.: D2-city: a large-scale dashcam video dataset of diverse traffic scenarios. arXiv preprint arXiv:1904.01975 (2019)

5. Cubuk, E.D., Zoph, B., Mane, D., Vasudevan, V., Le, Q.V.: Autoaugment: learning augmentation strategies from data. In: Proceedings of the IEEE Conference on Computer Vision and Pattern Recognition, pp. 113–123 (2019)

6. Dumont, B., Maggio, S., Montalvo, P.: Robustness of rotation-equivariant networks to adversarial perturbations. arXiv preprint arXiv:1802.06627 (2018)

7. Dvornik, N., Mairal, J., Schmid, C.: Modeling visual context is key to augmenting object detection datasets. In: Ferrari, V., Hebert, M., Sminchisescu, C., Weiss, Y. (eds.) ECCV 2018. LNCS, vol. 11216, pp. 375–391. Springer, Cham (2018). https://doi.org/10.1007/978-3-030-01258-8_23

8. Dwibedi, D., Misra, I., Hebert, M.: Cut, paste and learn: Surprisingly easy synthesis for instance detection. In: Proceedings of the IEEE International Conference on Computer Vision (ICCV), pp. 1301–1310 (2017)

9. Engstrom, L., Tran, B., Tsipras, D., Schmidt, L., Madry, A.: Exploring the landscape of spatial robustness. In: Chaudhuri, K., Salakhutdinov, R. (eds.) Proceedings of the International Conference on Machine Learning (ICML), vol. 97, pp. 1802–1811. PMLR, Long Beach, California, USA, 09–15 June 2019

10. Everingham, M., Van Gool, L., Williams, C.K.I., Winn, J., Zisserman, A.: The PASCAL visual object classes challenge 2012 (VOC2012) Results. http://www.pascal-network.org/challenges/VOC/voc2012/workshop/index.html

11. Geirhos, R., Rubisch, P., Michaelis, C., Bethge, M., Wichmann, F.A., Brendel, W.: Imagenet-trained CNNs are biased towards texture; increasing shape bias improves accuracy and robustness. In: Proceedings of the International Conference on Learning Representations (ICLR) (2019)

12. Goodfellow, I., et al.: Generative adversarial nets. In: Advances in Neural Information Processing Systems (NeurIPS) (2014)

13. Hamdi, A., Ghanem, B.: Towards analyzing semantic robustness of deep neural networks. In: Proceedings of the IEEE International Conference on Computer Vision Workshops (ICCV Workshops) (2019)

14. Hendrycks, D., Dietterich, T.: Benchmarking neural network robustness to common corruptions and perturbations. In: Proceedings of the International Conference on Learning Representations (ICLR) (2019)

15. Hendrycks, D., Zhao, K., Basart, S., Steinhardt, J., Song, D.: Natural adversarial examples. arXiv preprint arXiv:1907.07174 (2019)

16. Hosseini, H., Poovendran, R.: Semantic adversarial examples. In: Proceedings of the IEEE Conference on Computer Vision and Pattern Recognition Workshops (CVPR Workshops), pp. 1614–1619 (2018)

17. Jakab, T., Gupta, A., Bilen, H., Vedaldi, A.: Unsupervised learning of object landmarks through conditional image generation. In: Bengio, S., Wallach, H., Larochelle, H., Grauman, K., Cesa-Bianchi, N., Garnett, R. (eds.) Advances in Neural Information Processing Systems (NeurIPS), pp. 4016–4027. Curran Associates, Inc. (2018)

18. Jang, E., Gu, S., Poole, B.: Categorical reparameterization with gumbel-softmax. In: Proceedings of the International Conference on Learning Representations (ICLR) (2016)

19. Li, Y.J., Lin, C.S., Lin, Y.B., Wang, Y.C.F.: Cross-dataset person re-identification via unsupervised pose disentanglement and adaptation. In: Proceedings of the IEEE International Conference on Computer Vision (ICCV), pp. 7919–7929 (2019)

20. Lin, T.-Y., et al.: Microsoft COCO: common objects in context. In: Fleet, D., Pajdla, T., Schiele, B., Tuytelaars, T. (eds.) ECCV 2014. LNCS, vol. 8693, pp. 740–755. Springer, Cham (2014). https://doi.org/10.1007/978-3-319-10602-1_48

21. Liu, H.T.D., Tao, M., Li, C.L., Nowrouzezahrai, D., Jacobson, A.: Beyond pixel norm-balls: parametric adversaries using an analytically differentiable renderer. In: Proceedings of the International Conference on Learning Representations (ICLR) (2019)

22. Lorenz, D., Bereska, L., Milbich, T., Ommer, B.: Unsupervised part-based disentangling of object shape and appearance. In: 2019 IEEE/CVF Conference on Computer Vision and Pattern Recognition (CVPR), pp. 10947–10956 (2019)

23. Maddison, C.J., Mnih, A., Teh, Y.W.: The concrete distribution: a continuous relaxation of discrete random variables (2016)

24. Michaelis, C., et al.: Benchmarking robustness in object detection: autonomous driving when winter is coming. arXiv preprint arXiv:1907.07484 (2019)

25. Park, T., Liu, M.Y., Wang, T.C., Zhu, J.Y.: Semantic image synthesis with spatially-adaptive normalization. In: Proceedings of the IEEE Conference on Computer Vision and Pattern Recognition (CVPR) (2019)

26. Peyre, J., Laptev, I., Schmid, C., Sivic, J.: Weakly-supervised learning of visual relations. In: Proceedings of the IEEE International Conference on Computer Vision (ICCV) (2017)

27. Recht, B., Roelofs, R., Schmidt, L., Shankar, V.: Do cifar-10 classifiers generalize to cifar-10? arXiv preprint arXiv:1806.00451 (2018)

28. Recht, B., Roelofs, R., Schmidt, L., Shankar, V.: Do ImageNet classifiers generalize to ImageNet? In: Chaudhuri, K., Salakhutdinov, R. (eds.) Proceedings of the International Conference on Machine Learning (ICML), vol. 97, pp. 5389–5400. PMLR, Long Beach, California, USA, 09–15 June 2019

29. Redmon, J., Farhadi, A.: Yolov3: an incremental improvement. arXiv preprint arXiv:1804.02767 (2018)

30. Ronneberger, O., Fischer, P., Brox, T.: U-Net: convolutional networks for biomedical image segmentation. In: Navab, N., Hornegger, J., Wells, W.M., Frangi, A.F. (eds.) MICCAI 2015. LNCS, vol. 9351, pp. 234–241. Springer, Cham (2015). https://doi.org/10.1007/978-3-319-24574-4_28

31. Rosenfeld, A., Zemel, R., Tsotsos, J.K.: The elephant in the room. arXiv preprint arXiv:1808.03305 (2018)

32. Shankar, V., Dave, A., Roelofs, R., Ramanan, D., Recht, B., Schmidt, L.: A systematic framework for natural perturbations from videos. In: Proceedings of the International Conference on Machine Learning Workshops (ICML Workshop) (2019)

33. Shetty, R., Fritz, M., Schiele, B.: Adversarial scene editing: automatic object removal from weak supervision. In: Advances in Neural Information Processing Systems (NeurIPS) (2018)

34. Shetty, R., Schiele, B., Fritz, M.: Not using the car to see the sidewalk-quantifying and controlling the effects of context in classification and segmentation. In: Proceedings of the IEEE Conference on Computer Vision and Pattern Recognition (CVPR), pp. 8218–8226 (2019)

35. Siarohin, A., Lathuilière, S., Tulyakov, S., Ricci, E., Sebe, N.: Animating arbitrary objects via deep motion transfer. In: Proceedings of the IEEE Conference on Computer Vision and Pattern Recognition (CVPR), pp. 2372–2381 (2019)

36. Song, Y., Shu, R., Kushman, N., Ermon, S.: Constructing unrestricted adversarial examples with generative models. In: Bengio, S., Wallach, H., Larochelle, H., Grauman, K., Cesa-Bianchi, N., Garnett, R., (eds.) Advances in Neural Information Processing Systems (NeurIPS), pp. 8312–8323. Curran Associates, Inc. (2018)

37. Stutz, D., Hein, M., Schiele, B.: Disentangling adversarial robustness and generalization. In: Proceedings of the IEEE Conference on Computer Vision and Pattern Recognition (CVPR), pp. 6976–6987 (2019)
38. Tripathi, S., Chandra, S., Agrawal, A., Tyagi, A., Rehg, J.M., Chari, V.: Learning to generate synthetic data via compositing. In: Proceedings of the IEEE Conference on Computer Vision and Pattern Recognition (CVPR), pp. 461–470 (2019)
39. Ultralytics: pytorch implementation of YoloV3 (2019). https://github.com/ultralytics/yolov3. Accessed 11 Nov 2019
40. Wang, H.: Implentation of data augmentation for object detection via progressive and selective instance-switching (2019). https://github.com/Hwang64/PSIS. Accessed 11 Nov 2019
41. Wang, H., Wang, Q., Yang, F., Zhang, W., Zuo, W.: Data augmentation for object detection via progressive and selective instance-switching. arXiv preprint arXiv:1906.00358 (2019)
42. Wang, X., Shrivastava, A., Gupta, A.: A-fast-RCNN: hard positive generation via adversary for object detection. In: Proceedings of the IEEE Conference on Computer Vision and Pattern Recognition (CVPR), pp. 2606–2615 (2017)
43. Xiao, C., Zhu, J.Y., Li, B., He, W., Liu, M., Song, D.: Spatially transformed adversarial examples. In: Proceedings of the International Conference on Learning Representations (ICLR) (2018)
44. Yu, F., et al.: BDD100k: a diverse driving dataset for heterogeneous multitask learning. In: Proceedings of the IEEE Conference on Computer Vision and Pattern Recognition (CVPR), pp. 2636–2645 (2020)
45. Zhu, J.Y., Park, T., Isola, P., Efros, A.A.: Unpaired image-to-image translation using cycle-consistent adversarial networks. In: Proceedings of the IEEE International Conference on Computer Vision (ICCV) (2017)

Adversarial Generative Grammars
for Human Activity Prediction

A. J. Piergiovanni[1]([⊠]), Anelia Angelova[1], Alexander Toshev[1],
and Michael S. Ryoo[1,2]

[1] Robotics at Google, Mountain View, USA
ajpiergi@google.com, anelia@google.com, toshev@google.com, mryoo@google.com
[2] Stony Brook University, New York, USA

Abstract. In this paper we propose an adversarial generative grammar model for future prediction. The objective is to learn a model that explicitly captures temporal dependencies, providing a capability to forecast multiple, distinct future activities. Our adversarial grammar is designed so that it can learn stochastic production rules from the data distribution, jointly with its latent non-terminal representations. Being able to select *multiple* production rules during inference leads to different predicted outcomes, thus efficiently modeling many plausible futures. The adversarial generative grammar is evaluated on the Charades, MultiTHUMOS, Human3.6M, and 50 Salads datasets and on two activity prediction tasks: future 3D human pose prediction and future activity prediction. The proposed adversarial grammar outperforms the state-of-the-art approaches, being able to predict much more accurately and further in the future, than prior work. **Code will be open sourced.**

1 Introduction

Future prediction in videos is one of the most challenging visual tasks. Accurately predicting future activities or human pose has many important applications, e.g., in video analytics and robot action planning. Prediction is particularly hard because it is not a deterministic process as multiple potential 'futures' are possible, especially for predicting real-valued output vectors with non-unimodal distribution. Given these challenges, we address the important question of how the sequential dependencies in the data should be modeled and how multiple possible long-term future outcomes can be predicted at any given time.

Electronic supplementary material The online version of this chapter (https://doi.org/10.1007/978-3-030-58536-5_30) contains supplementary material, which is available to authorized users.

A. Vedaldi et al. (Eds.): ECCV 2020, LNCS 12347, pp. 507–523, 2020.
https://doi.org/10.1007/978-3-030-58536-5_30

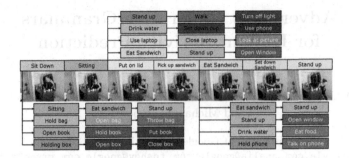

Fig. 1. The Adversarial Generative Grammar predicts future activities in videos and can generate many other plausible ones.

We propose an *Adversarial Generative Grammar* (AGG) model for future prediction. The model is a differentiable form of a regular grammar trained with adversarial sampling of various possible futures, which is able to output real-valued predictions (e.g., 3D human pose) or semantic prediction (e.g., activity classes). Learning sequences of actions or other sequential processes with the production rules of a grammar is valuable, as it imposes temporal structural dependencies and captures relationships between latent states. Each (learned) production rule of a grammar model is able to take a state representation and transition to a different future state. Using multiple rules allows the model to capture multiple branching possibilities (Fig. 1). This capability makes the grammar learning unique, different from previous sequential models including many recurrent neural network (RNN) models.

The main technical contribution of this work is the introduction of adversarial learning approach for differentiable grammar models. This is essential, as the adversarial process allows the grammar model to produce multiple candidate future sequences that follow a similar distribution to sequences seen in the data. A brute force implementation of differentiable grammar learning would need to enumerate all possible rules and generate multiple sequence branches (exponential growth in time) to consider multiple futures. Our adversarial stochastic sampling process allows for much more memory- and computationally-efficient learning without such enumeration. Additionally, unlike other techniques for future generation (e.g., autoregressive RNNs), we show the adversarial grammar is able to learn longer sequences, can handle multi-label settings, and predict much further into the future.

To our knowledge, AGG is the first approach of adversarial grammar learning. It enables qualitatively and quantitatively better solutions - ones able to successfully produce multiple feasible long-term future predictions for real-valued outputs. The proposed approach is driven entirely by the structure imposed from learning grammar rules and adversarial losses – i.e., no direct supervised loss is used for the grammar model training.

The proposed approach is evaluated on different future activity prediction tasks: (i) on future action prediction – multi-class classification and multi-class multi-label problems and (ii) on 3D human pose prediction, which predicts the 3D joint positions of the human body in the future. The proposed method is tested on four challenging datasets: Charades, MultiTHUMOS, 50 Salads, and Human3.6M. It outperforms previous state-of-the-art methods, including RNNs, LSTMs, GRUs, grammar and memory based methods.

2 Related Work

Grammar Models for Visual Data. The notion of grammars in computational science was introduced by [4] for description of language, and has found a widespread use in natural language understanding. In the domain of visual data, grammars are used to parse images of scenes [13,38,39]. In their position paper, [39] present a comprehensive grammar-based language to describe images, and propose MCMC-based inference. More recently, a recursive neural net based approach was applied to parse scenes by [29]. However, these previous works either use a traditional symbolic grammar formulation or use a neural network without explicit representation of grammar. In the context of temporal visual data, grammars have been applied to activity recognition and parsing [23,24,27,32] but not to prediction or generation. [25] used traditional stochastic grammar to predict activities, but only within 3 s.

Generative Models for Sequences. Generative Adversarial Networks (GANs) are a very powerful mechanism for data generation by an underlying learning of the data distribution through adversarial sampling [12]. GANs have been very popular for image generation tasks [2,6,16,33]. Prior work on using GANs for improved sequences generation [8,14,37] has also been successful. Fraccaro et al. [10] proposed a stochastic RNN which enables generation of different sequences from a given state. However, to our knowledge, no prior work explored end-to-end adversarial training of formal grammar as we do. Qi et al. [26] showed a grammar could be used for future prediction, and our work builds on this by learning the grammar structure differntiably from data.

Differentiable Rule Learning. Previous approaches that address differentiable rule or grammar learning are most aligned to our work [34]. Unlike the prior work, we are able to handle larger branching factors and demonstrate successful results in real-valued output spaces, benefiting from the adversarial learning.

Future Pose Prediction. Previous approaches for human pose prediction [11,15,31] are relatively scarce. The dominant theme is the use of recurrent models (RNNs or GRUs/LSTMs) [11,22]. Tang et al. [31] use attention models specifically to target long-term predictions, up to 1 s in the future. Jain et al. [17] propose a structural RNN which learns the spatio-temporal relationship of pose joints. The above models, contrary to ours, cannot produce multiple futures, making them limited for long-term anticipation. These results are only within short-term horizons and the produced sequences often 'interpolate' actual data examples. Although our approach is more generic and is not limited to just pose forecasting, we show that it is able to perform successfully too on this task, outperforming others.

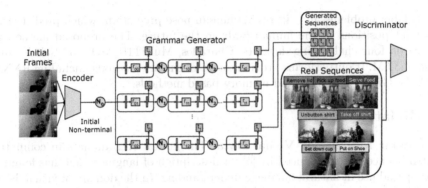

Fig. 2. Overview of the adversarial grammar model. The initial non-terminal is produced by an encoder based on the input video. The grammar then generates multiple possible sequences from the non-terminal. The generated and real sequences are used to train the adversarial discriminator, evaluating whether the generated sequences match the distribution of real sequences.

Video Prediction. Our approach is also related to the video prediction literature [1,5,9,20], but more in-depth survey is beyond the scope of this work.

3   Approach

We first introduce a differentiable form of a formal grammar, where its production rules are implemented with fully-differentiable functions to be applied to non-terminals and terminals represented with latent vectors (Sect. 3.3). Unlike traditional grammar induction with symbolic representations, our approach allows joint learning of latent representations and differentiable functions with the standard back-propagation. Next, we present the adversarial grammar learning approach that actually enables training of such functions and representations without spending an exponential amount of memory and computation (Sect. 3.4). Our adversarial grammar is trained to generate multiple candidate future sequences. This enables robust future prediction, which, more importantly, can easily generate multiple realistic futures.

We note that the proposed approach, based on stochastic sequence learning, is driven entirely by the adversarial losses which help model the data distribution over long sequences. That is, while direct supervised losses can be used, we implement our approach with adversarial losses only, which learn the underlying distribution. All experiments below demonstrate the success of this approach, despite being more challenging.

3.1   Preliminaries

A formal regular grammar is represented as the tuple $(\mathcal{N}, \mathcal{T}, \mathcal{P}, N_0)$ where $\mathcal{N}$ is a finite non-empty set of non-terminals, $\mathcal{T}$ is a finite set of terminals (or output symbols, e.g., here actions), $\mathcal{P}$ is a set of production rules, and N_0 is the starting non-terminal symbol, $N_0 \in \mathcal{N}$. Production rules in a regular grammar are of the form $A \rightarrow aB$, $A \rightarrow b$, and $A \rightarrow \epsilon$, where $A, B \in \mathcal{N}$, $a, b \in \mathcal{T}$, and ϵ is the empty string. Autoregressively applying production rules to the non-terminal generates a sequence of terminals. Note that we only implement rules of form $A \rightarrow aB$ in our grammar, allowing it to generate sequences infinitely and we represented N as a real-valued vector.

Our objective is to learn such non-terminals $\mathcal{N}$ and terminals $\mathcal{T}$ as latent vector representations directly from training data, and model the production rules $\mathcal{P}$ as a (differentiable) generative neural network function. That is, the goal is to learn a nonlinear function G that maps a non-terminal to a *set* of (non-terminal, terminal) pairs; here G is a neural network with learnable parameters.

$$G : \mathcal{N} \rightarrow \{(\mathcal{N}, \mathcal{T})\} \tag{1}$$

Note that this is a mapping from a single non-terminal to multiple (non-terminal, terminal) pairs. The selection of different rules enables modeling of multiple different sequences, generating different future outcomes, unlike existing deterministic models (e.g., RNNs).

The learned production rules allow modeling of the transitions between continuous events in time, for example 3D human pose or activities, which can naturally spawn into many possible futures at different points similarly to switching between rules in a grammar. For example, an activity corresponding to 'walking' can turn into 'running' or 'stopping' or continuing the 'walking' behaviour.

More formally, for any latent non-terminal $N \in \mathcal{N}$, the grammar production rules are generated by applying the function G (a sequence of fully connected layers), to N as:

$$G(N) = \{(N_i, t_i)\}_{i=1:K}, \tag{2}$$

where each pair corresponds to a particular production rule for this non-terminal:

$$N \rightarrow t_1 N_1$$
$$N \rightarrow t_2 N_2 \ldots \tag{3}$$
$$N \rightarrow t_K N_K, \text{ where } N_1, N_2, \ldots N_K \in \mathcal{N}, t_1, t_2, \ldots t_K \in \mathcal{T}, \text{ for } K \text{ rules.}$$

This function is applied recursively to obtain a number of output sequences, similar to prior recurrent methods (e.g., RNNs such as LSTMs and GRUs). However, in RNNs, the learned state is required to abstract multiple potential possibilities into a single representation, as the mapping from the state representation to the next representation is deterministic. As a result, when learning from sequential data with multiple possibilities, standard RNNs tend to learn states as a mixture of multiple sequences instead of learning more discriminative states. By learning explicit production rules, our states lead to more salient and distinct predictions which can be exploited for learning long-term, complex output tasks with multiple possibilities, as shown later in the paper.

3.2   Learning the Starting Non-terminal

Given an initial input data sequence (e.g., a short video or pose sequences), we learn to generate its corresponding starting non-terminal N_0 (i.e., root node). This is used as input to G so as to generate a sequence of terminal symbols starting from the given non-terminal. Concretely, given an input sequence X, a function s (a CNN) is learned which gives the predicted starting non-terminal:

$$N_0 = s(X). \tag{4}$$

Notice that the function $s(X)$ serves as a jointly-learned blackbox parser that is able to estimate the non-terminal corresponding to the current state of the model, allowing future sequence generation to start from such non-terminal.

3.3   Grammar Learning

Given a starting non-terminal, the function G is applied recursively to obtain the possible sequences where j is an index in the sequence and i is one of the possible rules:

$$\begin{cases} G(N_0) = \{(N_i^1, t_i^1)\}_i, & j = 0 \\ G(N^j) = \{(N_i^{j+1}, t_i^{j+1})\}_i, & \text{for } j > 0 \end{cases} \tag{5}$$

For example, suppose W is the non-terminal that encodes the activity for 'walking' sequences. Let *walking* denote the terminal of a grammar. An output of the rule $W \to walkingW$ will be able to generate a sequence of continual 'walking' behavior. Additional rules, e.g., $W \to stoppingU$, $W \to runningV$, can be learned, allowing for the activity to switch to 'stopping' or 'running' (with the non-terminals U, V respectively learning to generate their corresponding potential futures, e.g. 'sitting down', or 'falling'). Clearly, for real valued outputs, such as 3D human pose, the number and dimensionality of the non-terminals required will be larger. We also note that the non-terminals act as a form of memory, capturing the current state with the Markov property.

To accomplish the above task, G (in Eq. 2) has a special structure. G takes an input of $N \in \mathcal{N}$, then using several nonlinear transformations (e.g., fully connected layers with activation functions), maps N to a binary vector r corresponding to a set of rules: $r = f_R(N)$. Here, r is a vector with the size $|\mathcal{P}|$ whose elements specify the probability of each rule given input non-terminal. We learn $|\mathcal{P}|$ rules which are shared globally, but only a (learned) subset are selected for each non-terminal as the other rule probabilities are zero. This is conceptually similar to using memory with recurrent neural network methods [36], but the main difference is that the rule vectors are used to build grammar-like rule structures which are more advantageous in explicitly modeling of temporal dependencies.

In order to generate multiple outputs, the candidate rules, r are followed by the Gumbel-Softmax function [18,21], which allows for stochastic selection of a rule. This function is differentiable and samples a single rule from the candidate rules based on the learned rule probabilities. The probabilities are learned to model the likelihood of each generated sequence, and this formulation allows the 'branching' of sequence predictions as the outcome of the Gumbel-Softmax function differs every time, following the probability distribution.

For each given rule r, two nonlinear functions $f_T(r)$ and $f_N(r)$ are then learned, so that they output the resulting terminal and non-terminal for the rule r: $N_{new} = f_N(r)$, $t_{new} = f_T(r)$. These functions are both implemented as a sequence of fully-connected layers followed by a non-linear activation function (e.g., softmax or sigmoid depending on the task). The schematic of G is visualized in Fig. 2, and more details on the functions are provided in the later sections.

The non-terminals and terminals are modeled as sets of high dimensional vectors with pre-specified size and are learned jointly with the rules (all are tunable parameters and naturally more complex datasets require larger capacity). For example, for a C-class classification problem, the terminals are represented as C-dimensional vectors matching the one-hot encoding for each class.

Difference to stochastic RNNs. Standard recurrent models have a deterministic state given some input, while the grammar is able to generate multiple potential next non-terminals (i.e., states). This is particularly important for multi-modal state distributions. Stochastic RNNs (e.g., [10]) address this by allowing the next state to be stochastically generated, but this is difficult to control, as the next state now depends on a random value. In the grammar model, the next non-terminal is sampled randomly, but from a set of fixed candidates while following the learned probability distribution. By maintaining a set of candidates, the next state can be selected randomly or by some other method (e.g., greedily taking most probable, beam search, etc.), giving more control over the generated sequences.

3.4 Adversarial Grammar Learning

The function G generates a set of (non-terminal, terminal) pairs, which is applied recursively to the non-terminals, resulting in new production rules and the next sets of (non-terminal, terminal) pairs. Note that in most cases, each rule generates a different non-terminal, thus sampling G many times will lead to a variety of generated sequences. As a result, an exponential number of sequences will need to be generated during training, to cover the possible sequences, and enumerating all possible sequences is computationally prohibitive beyond $k = 2$.[1] This restricts the tasks that can be addressed to ones with lower dimensional outputs because of memory limits. When $k = 1$, i.e. when there is no branching, we have an RNN-like model, unable to generate multiple possible future sequences (we also tested this in ablation experiments below).

Stochastic Adversarial Sampling. We address this problem by using stochastic adversarial rule sampling. Given the non-terminals, which effectively contain

---

[1] For a branching factor of k rules per non-terminal with a sequence of length L, there are in k^L terminals and non-terminals (for $k = 2$, $L = 10$ we have ~1000 and for $k = 3$ ~60,000).

a number of potential 'futures', we use *an adversarial-based sampling*, similar to GAN approaches [12], which learns to sample the most likely rules for the given input (Fig. 2). The use of a discriminator network allows the model to generate realistic sequences that may not exactly match the ground truth (but are still realistic) without being penalized.

Generator: We use the function G, which is the function modeling the learned grammar described above, as the *generator function*.

Discriminator: We build an additional *discriminator function* D. Following standard GAN training, the discriminator function returns a binary prediction which discriminates examples from the data distribution vs. generated ones. Note that the adversarial process is designed to ultimately generate terminals, i.e., the final output sequence for the model. D is defined as:

$$p = D(t, n) \tag{6}$$

where $t = t_0 t_1 t_2 \ldots t_L$ is the input sequence of terminals, $n = N_0 N_1 N_2 \ldots N_L$ is the sequence of non-terminals (L is the length of the sequence) and $p \in [0, 1]$ and reflects when the input sequence of terminals is from the data distribution or not. Note that our discriminator is also conditioned on the non-terminal sequence ($n = N_0 N_1 N_2 \ldots N_L$), thus the distribution of non-terminals is learned implicitly as well.

The discriminator function D is implemented as follows: given an input sequence of non-terminals and terminals, we apply several 1D convolutional layers to the terminals and non-terminals, then concatenate their representations followed by a fully-connected layer to produce the binary prediction (see the supp. material).

Adversarial Generative Grammar (AGG). The discriminator and generator (grammar) functions are trained to work jointly, generating sequences which match the data distribution. The optimization objective is defined as:

$$\min_G \max_D \; E_{x \sim p_{data}(x)}[\log D(x)] + \\ E_{z \sim s(X)}[\log(1 - D(G(z)))] \tag{7}$$

where $p_{data}(x)$ is the real data distribution and $G(z)$ is the generated sequence from an initial state based on a sequence of frames (X). That is, the fist part of the loss works on sequences of actions or human pose, whereas the second works over generated sequences ($s(X)$ is the video embedding, or starting non-terminal).

Alternatively, the sequences generated by G could be compared to the ground truth to compute a loss during training (e.g., maximum likelihood estimation), however, doing so requires enumerating many possibilities in order learn multiple, distinct possible sequences. Without such enumeration, the model converges to a mixture representing possible sequences from the data distribution. By using the adversarial training of G, our model is able to generate sequences that match the distribution observed in the dataset. This allows for computationally feasible learning of longer, higher-dimensional sequences.

Architecture Details. The functions G, f_N and f_t, f_R are implemented as networks using several fully-connected layers. The detailed architectures depend on the task and dataset, and we provide them in the supplemental material. For the pose forecasting, the function s is implemented as a two-layer GRU module [3] followed by a 1x1 convolutional layer with D_N outputs to produce the starting non-terminal. For activity prediction, s is implemented as two sequential temporal 1D convolutional layers which produce the starting non-terminal.

4 Experiments

We conduct experiments on two sets of problems for future prediction: future 3D human pose forecasting and future activity prediction. The experiments are done on four public datasets and demonstrate strong performance of the proposed approach over the state-of-the-art and the ability to produce multiple future outcomes, to handle multi-label datasets, and to predict further in the future than prior work.

Table 1. Evaluation of future pose for specific activity classes. Results are Mean Angle Error (lower is better). Human3.6M dataset.

Methods	Walking							
	80 ms	160 ms	320 ms	400 ms	560 ms	640 ms	720 ms	1000 ms
ERD [11]	0.77	0.90	1.12	1.25	1.44	1.45	1.46	1.44
LSTM-3LR [11]	0.73	0.81	1.05	1.18	1.34	1.36	1.37	1.36
Res-GRU [22]	0.27	0.47	0.68	0.76	0.90	0.94	0.99	1.06
Zero-velocity [22]	0.39	0.68	0.99	1.15	1.35	1.37	1.37	1.32
MHU [31]	0.32	0.53	0.69	0.77	0.90	0.94	0.97	1.06
Ours	**0.25**	**0.43**	**0.65**	**0.75**	**0.79**	**0.85**	**0.92**	**0.96**

Methods	Greeting							
	80 ms	160 ms	320 ms	400 ms	560 ms	640 ms	720 ms	1000 ms
ERD [11]	0.85	1.09	1.45	1.64	1.93	1.89	1.92	1.98
LSTM-3LR [11]	0.80	0.99	1.37	1.54	1.81	1.76	1.79	1.85
Res-GRU [22]	**0.52**	**0.86**	1.30	1.47	1.78	1.75	1.82	1.96
Zero-velocity [22]	0.54	0.89	1.30	1.49	1.79	1.74	1.77	1.80
MHU [31]	0.54	0.87	1.27	**1.45**	1.75	1.71	1.74	1.87
Ours	**0.52**	**0.86**	**1.26**	1.45	**1.58**	**1.69**	**1.72**	**1.79**

Methods	Taking photo							
	80 ms	160 ms	320 ms	400 ms	560 ms	640 ms	720 ms	1000 ms
ERD [11]	0.70	0.78	0.97	1.09	1.20	1.23	1.27	1.37
LSTM-3LR [11]	0.63	0.64	0.86	0.98	1.09	1.13	1.17	1.30
Res-GRU [22]	0.29	0.58	0.90	1.04	1.17	1.23	1.29	1.47
Zero-velocity [22]	0.25	0.51	0.79	0.92	1.03	**1.06**	**1.13**	1.27
MHU [31]	0.27	0.54	0.84	0.96	1.04	1.08	1.14	1.35
Ours	**0.24**	**0.50**	**0.76**	**0.89**	**0.95**	1.08	1.15	**1.24**

4.1   Datasets

MultiTHUMOS: The MultiTHUMOS dataset [35] is a well-established video understanding dataset for multi-class activity prediction. It contains 400 videos spanning about 30 h of video and 65 action classes.

Charades: Charades [28] is a challenging video dataset containing longer-duration activities recorded in home environments. Charades is a multi-class multi-label dataset in which multiple activities are often co-occurring. We use it to demonstrate the ability of the model to handle complex data. It contains 9858 videos of 157 action classes.

Human3.6M: The Human 3.6M dataset [15] is a popular benchmark for future pose prediction. It has 3.6 million 3D human poses of 15 activities. The goal is to predict the future 3D locations of 32 joints in the human body.

50 Salads: The 50 Salads [30] is a video dataset of 50 salad preparation sequences (518,411 frames total) with an average length of 6.4 min per video. It has been used recently for future activity prediction [7,19], making it suitable for the evaluation of our method.

Table 2. Evaluation of future pose for short-term and long-term prediction horizons. Measured with Mean Angle Error (lower is better) on Human3.6M. No predictions beyond 1 s are available for prior work.

Method	80 ms	160 ms	320 ms	560 ms	640 ms	720 ms	1 s	2 s	3 s	4 s
ERD [11]	0.93	1.07	1.31	1.58	1.64	1.70	1.95	–	–	–
LSTM-3LR [11]	0.87	0.93	1.19	1.49	1.55	1.62	1.89	–	–	–
Res-GRU [22]	0.40	0.72	1.09	1.45	1.52	1.59	1.89	–	–	–
Zero-vel. [22]	0.40	0.71	1.07	1.42	1.50	1.57	1.85	–	–	–
MHU-MSE [31]	0.39	0.69	1.04	1.40	1.49	1.57	1.89	–	–	–
MHU [31]	0.39	0.68	1.01	1.34	1.42	**1.49**	1.80	–	–	–
AGG (Ours)	**0.36**	**0.65**	**0.98**	**1.27**	**1.40**	**1.49**	**1.74**	**2.25**	**2.70**	**2.98**

4.2   Human Pose Forecasting

We first evaluate the approach on forecasting 3D human pose, a real valued structured-output problem. This is a challenging task [11,17] but is of high importance, e.g., for motion planning in robotics. It also showcases the use of the Adversarial Grammar, as using the standard grammar is not feasible due to the memory and computation constraints for this real-valued dataset.

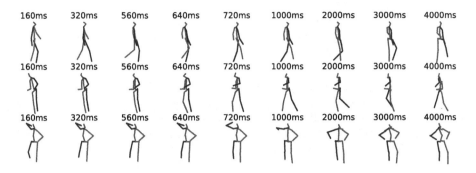

Fig. 3. Example results for 3D pose predictions. Top: walking, middle: greeting, bottom: posing.

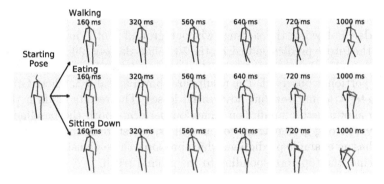

Fig. 4. Starting from a neutral pose, the grammar is able to generate multiple sequences by selecting different rules. Top: a walking sequence, middle: eating, bottom: sitting.

Human 3.6M Dataset. We conduct experiments on the well established future pose prediction benchmark Human3.6M [15]. We here predict the future 3D locations of 32 joints in the human body. We use quaternions to represent each joint location, allowing for a more continuous joint representation space. We also predict differences, rather than absolute positions, which we found leads to more stable learning. Previous work demonstrated prediction results up to a second on this dataset. This work can generate future sequences for longer horizons, 4 s in the future.

We compare against the state-of-the-art methods on the Human 3.6M benchmark [11,15,17,22,31] using the Mean Angle Error (MAE) metric as introduced by [17]. Table 1 shows results on several activities and Table 2 shows average MAE for all activities compared to the state-of-the-art methods, consistent with the protocol in prior work. As seen from the tables, our work outperforms prior work. Furthermore, we are able to generate results at larger time horizons of four seconds in the future. In Fig. 3, we show some predicted future poses for several different activities, confirming the results reflect the characteristics of the actual behaviors. In Fig. 4, we show the ability of the adversarial grammar to generate different sequences from a given starting state. Here, given the same starting state, we select different rules, which lead to different sequences corresponding to walking, eating or sitting.

4.3   Activity Forecasting in Videos

We further test the method for video activity anticipation, where the goal is to predict future activities at various time-horizons, using an initial video sequence as input. We predict future activities on three video understanding datasets MultiTHUMOS [35], Charades [28] and 50-salads [30] using the standard evaluation protocols per dataset. We also predict from 1 to 45 s in the future on MultiTHUMOS and Charades, which is much further into the future than prior approaches.

50 Salads. Following the setting 'without ground truth' in [19] and [7], we evaluate the future prediction task on the 50 Salads dataset [30]. As per standard evaluation protocol, we report prediction on portions of the video when 20% and 30% portion is observed. The results are shown in Table 3, where Grammar-only denotes training without adversarial losses. The results confirm that our approach allows better prediction which outperforms both the baseline, which is already a strong grammar model, as well as, the state-of-the-art approaches. Figure 5 has an example prediction, which proposes three plausible continuations of the recipe, the top corresponding to the ground truth.

Table 3. Results on 50 Salads without ground-truth observations. The proposed work outperforms the grammar baselines and the state-of-the-art.

Observation	20%				30%			
Prediction	10%	20%	30%	50%	10%	20%	30%	50%
Nearest-Neighbor [7]	19.0	16.1	14.1	10.4	21.6	15.5	13.5	13.9
RNN [7]	30.1	25.4	18.7	13.5	30.8	17.2	14.8	9.8
CNN [7]	21.2	19.0	16.0	9.9	29.1	20.1	17.5	10.9
TCA [19]	32.5	27.6	21.3	16.0	35.1	27.1	22.1	15.6
Grammar (from [7])	24.7	22.3	19.8	12.7	29.7	19.2	15.2	13.1
Grammar only	39.2	32.1	24.8	19.3	38.4	29.5	25.5	18.5
AGG (Ours)	**39.5**	**33.2**	**25.9**	**21.2**	**39.5**	**31.5**	**26.4**	**19.8**

MultiTHUMOS. We here present our future prediction results on the MultiTHUMOS dataset [35][2]. We use a standard evaluation metric: we predict the

---

[2] Note that most of the previous works used the MultiTHUMOS dataset and the Charades dataset for per-frame activity categorization; our works showcases a long-term activity forecasting capability, instead.

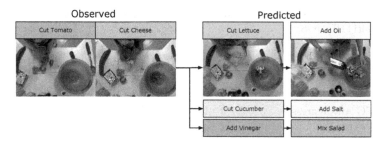

Fig. 5. Example sequence from 50-salads showing the observed frames and the next two predictions.

activities occurring T seconds in the future and compute the mean average precision (mAP) between the predictions and ground truth. As the grammar model is able to generate multiple, different future sequences, we also report the maximum mAP the model could obtain by selecting the best of 10 different future predictions. We compare the predictions at 1, 2, 5, 10, 20, 30 and 45 s into the future. As little work has explored long-term future activity prediction (with the exception of [35] which predicts within a second), we compare against four different baseline methods: (i) repeating the activity prediction of the last seen frame, (ii) using a fully connected layer to predict the next second (applied autoregressively), (iii) using a fully-connected layer to directly predict activities at various future times, and (iv) an LSTM applied autoregressively to future activity predictions.

Table 4 shows activity prediction accuracy for the MultiTHUMOS dataset. In the table, we also report our approach when limited to generating a single outcome ('AGG-single'), to be consistent to previous methods which are not able to generate more than one outcome. We also compare to grammar without adversarial learning, trained by pruning the exponential amount of future sequences to fit into the memory ('Grammar only').

As seen, our approach outperforms alternative methods. We observe that the gap to other approaches widens further in the future: 3.9 mAP for the LSTM vs 11.2 of ours at 45 sec. in the future, as the autoregressive predictions of an LSTM become noisy. Due to the structure of the grammar model, we are able to generate better long-term predictions. We also find that by predicting multiple futures and taking the max improves performance, confirming that the grammar model is generating different sequences, some of which more closely match the ground truth (see also Fig. 6).

Charades. Table 5 shows the future activity prediction results on Charades, using the same protocol as MultiTHUMOS. Similar to our MultiTHUMOS experiments, we observe that the adversarial grammar model provides more accurate future prediction than previous work, outperforming the grammar-only model in most cases. While the grammar-only model performs slightly better at 10 and 20 s, it is not computationally feasible for real-valued tasks due to the

Table 4. Prediction mAP for future activities (higher is better) from 1 ss to 45 s in the future. MultiTHUMOS.

Method	1 s	2 s	5 s	10 s	20 s	30 s	45 s
Random	2.6	2.6	2.6	2.6	2.6	2.6	2.6
Last Predicted Action	16.5	16.0	15.1	12.7	8.7	5.8	5.9
FC Autoregressive	17.9	17	14.5	7.7	4.5	4.2	4.7
FC Direct	13.7	9.8	11.0	7.3	8.0	5.5	8.2
LSTM (Autoregressive)	16.5	15.7	12.5	6.8	4.1	3.2	3.9
Grammar only	18.7	18.6	13.5	12.8	10.5	8.2	8.5
AGG-single (Ours)	19.3	19.6	13.1	13.6	11.7	10.4	**11.4**
AGG (Ours)	**22.0**	**19.9**	**15.5**	**14.4**	**13.3**	**10.8**	**11.4**

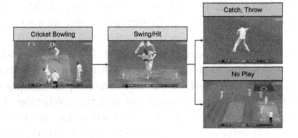

Fig. 6. Example video and activity sequence from MultiTHUMOS (a cricket game). The adversarial grammar is able to learn two possible sequences: a hit/play and no play, instead of picking only the most likely one.

memory constraint. We note that Charades is more challenging than others on both recognition and prediction. Figure 1 shows a true sequence and several other sequences generated by the adversarial grammar. As Charades contains many different possible sequences, generating multiple futures is beneficial.

Ablation Study. We conduct additional experiments to examine the importance of learning grammar with multiple possibilities (i.e., branching). Table 6 compares the models with and without the branching capability. These models use the exact same network architecture as our full models, while the only difference is that they do not generate multiple possible sequences for its learning. That is, they just become standard RNNs, constrained to have our grammar structure. We are able to observe that the ability to consider multiple possibilities during the learning is important, and that our adversarial training is beneficial. Note that we restricted these models to only generate one sequence with the highest likelihood during the inference for fair comparison.

Table 5. Prediction accuracy for future activities for 45 s in the future on the Charades dataset.

Method	1 s	2 s	5 s	10 s	20 s	30 s	45 s
Random	2.4	2.4	2.4	2.4	2.4	2.4	2.4
Last Predicted Action	15.1	13.8	12.8	10.2	7.6	6.2	5.7
FC Autoregressive	13.5	14.0	12.6	6.7	3.7	3.5	5.1
FC Direct	15.2	14.5	12.2	9.1	6.6	6.5	5.5
LSTM (Autoregressive)	12.6	12.7	12.4	10.8	7.0	6.1	5.4
Grammar only	15.7	14.8	12.9	**11.2**	**8.5**	6.6	8.5
AGG-single (Ours)	15.9	15.0	13.1	10.5	7.4	6.2	8.8
AGG (Ours)	**17.0**	**15.9**	**13.4**	10.7	7.8	**7.2**	**9.8**

Table 6. Ablation of our grammar learning on Charades.

Method	1 s	5 s	45 s
Grammar only - no branching	12.2	8.4	3.8
Grammar only	15.7	12.9	8.5
Adversarial Grammar (AGG) - no branching	14.2	12.5	5.5
Adversarial Grammar (AGG)	**15.9**	**13.1**	**8.8**

5   Conclusion

We proposed a differentiable adversarial generative grammar which shows strong performance for future prediction of human pose and activities. Because of the structure we impose for learning grammar-like rules for sequences and learning in adversarial fashion, the model is able to generate multiple sequences that follow the distribution seen in data. One challenge is evaluating future predictions when the ground truth only contains one of many potentially valid sequences. In the future, other forms of evaluation, such as asking humans to rate a generated sequence, could be explored.

References

1. Babaeizadeh, M., Finn, C., Erhan, D., Campbell, R.H., Levine, S.: Stochastic variational video prediction. arXiv preprint arXiv:1710.11252 (2017)
2. Brock, A., Donahue, J., Simonyan, K.: Large scale GAN training for high fidelity natural image synthesis. In: ICLR (2019)
3. Cho, K., et al.: Learning phrase representations using RNN encoder-decoder for statistical machine translation. In: EMNLP (2014)
4. Chomsky, N.: Three models for the description of language. IRE Trans. Inf. Theor. **2**(3), 113–124 (1956)
5. Denton, E., Fergus, R.: Stochastic video generation with a learned prior. arXiv preprint arXiv:1802.07687 (2018)

6. Denton, E.L., Soumith Chintala, R.F.: Deep generative image models using a Laplacian pyramid of adversarial networks. In: Advances in Neural Information Processing Systems (NeurIPS) (2015)
7. Farha, Y.A., Richard, A., Gall, J.: When will you do what? - anticipating temporal occurrences of activities. In: CVPR (2018)
8. Fedus, W., Goodfellow, I., Dai, A.: Maskgan: better text generation via filling in the _. In: ICLR (2018)
9. Finn, C., Goodfellow, I., Levine, S.: Unsupervised learning for physical interaction through video prediction. In: Advances in Neural Information Processing Systems (NeurIPS), pp. 64–72 (2016)
10. Fraccaro, M., Sønderby, S.K., Paquet, U., Winther, O.: Sequential neural models with stochastic layers. In: Advances in Neural Information Processing Systems, pp. 2199–2207 (2016)
11. Fragkiadaki, K., Levine, S., Felsen, P., Malik, J.: Recurrent network models for human dynamics. In: ICCV (2015)
12. Goodfellow, I., et al.: Generative adversarial nets. In: Advances in Neural Information Processing Systems (NeurIPS) (2014)
13. Han, F., Zhu, S.C.: Bottom-up/top-down image parsing with attribute grammar. IEEE Trans. Pattern Anal. Mach. Intell. 31(1), 59–73 (2008)
14. Hu, Z., Yang, Z., Liang, X., Salakhutdinov, R., Xing, E.P.: Toward controlled generation of text. In: ICML (2017)
15. Ionescu, C., Papava, D., Olaru, V., Sminchisescu, C.: Human3.6M: large scale datasets and predictive methods for 3D human sensing in natural environments. IEEE Trans. Pattern Anal. Mach. Intell. 36, 1325–1339 (2014)
16. Isola, P., Zhu, J.Y., Zhou, T., Efros, A.A.: Image-to-image translation with conditional adversarial networks. In: CVPR (2017)
17. Jain, A., Zamir, A.R., Savarese, S., Saxena, A.: Structural-RNN: deep learning on spatio-temporal graphs. In: CVPR (2016)
18. Jang, E., Gu, S., Poole, B.: Categorical reparameterization with gumbel-softmax. In: ICLR (2017)
19. Ke, Q., Fritz, M., Schiele, B.: Time-conditioned action anticipation in one shot. In: CVPR (2019)
20. Lee, A.X., Zhang, R., Ebert, F., Abbeel, P., Finn, C., Levine, S.: Stochastic adversarial video prediction. arXiv preprint arXiv:1804.01523 (2018)
21. Maddison, C.J., Mnih, A., Teh, Y.W.: The concrete distribution: a continuous relaxation of discrete random variables. In: ICLR (2017)
22. Martinez, J., Black, M., Romero, J.: On human motion prediction using recurrent neural networks. In: CVPR (2017)
23. Moore, D., Essa, I.: Recognizing multitasked activities from video using stochastic context-free grammar. In: Proceedings of AAAI Conference on Artificial Intelligence (AAAI), pp. 770–776 (2002)
24. Pirsiavash, H., Ramanan, D.: Parsing videos of actions with segmental grammars. In: CVPR, pp. 612–619 (2014)
25. Qi, S., Huang, S., Wei, P., Zhu, S.C.: Predicting human activities using stochastic grammar. In: Proceedings of the IEEE International Conference on Computer Vision, pp. 1164–1172 (2017)
26. Qi, S., Jia, B., Zhu, S.C.: Generalized earley parser: bridging symbolic grammars and sequence data for future prediction. arXiv preprint arXiv:1806.03497 (2018)

27. Ryoo, M.S., Aggarwal, J.K.: Recognition of composite human activities through context-free grammar based representation. In: 2006 IEEE Computer Society Conference on Computer Vision and Pattern Recognition (CVPR 2006), vol. 2, pp. 1709–1718. IEEE (2006)
28. Sigurdsson, G.A., Varol, G., Wang, X., Farhadi, A., Laptev, I., Gupta, A.: Hollywood in homes: crowdsourcing data collection for activity understanding. In: Leibe, B., Matas, J., Sebe, N., Welling, M. (eds.) ECCV 2016. LNCS, vol. 9905, pp. 510–526. Springer, Cham (2016). https://doi.org/10.1007/978-3-319-46448-0_31
29. Socher, R., Lin, C.C., Manning, C., Ng, A.Y.: Parsing natural scenes and natural language with recursive neural networks. In: Proceedings of the 28th International Conference on Machine Learning (ICML-11), pp. 129–136 (2011)
30. Stein, S., McKenna, S.J.: Combining embedded accelerometers with computer vision for recognizing food preparation activities. In: Proceedings of the 2013 ACM International Joint Conference on Pervasive and Ubiquitous Computing, pp. 729–738. ACM (2013)
31. Tang, Y., Ma, L., Liu, W., Zheng, W.S.: Long-term human motion prediction by modeling motion context and enhancing motion dynamic. In: IJCAI (2018)
32. Vo, N.N., Bobick, A.F.: From stochastic grammar to bayes network: Probabilistic parsing of complex activity. In: Proceedings of the IEEE Conference on Computer Vision and Pattern Recognition, pp. 2641–2648 (2014)
33. Wang, T.C., Liu, M.Y., Zhu, J.Y., Tao, A., Kautz, J., Catanzaro, B.: High-resolution image synthesis and semantic manipulation with conditional gans. In: CVPR (2018)
34. Yang, F., Yang, Z., Cohen, W.W.: Differentiable learning of logical rules for knowledge base reasoning. In: Advances in Neural Information Processing Systems (NeurIPS) (2017)
35. Yeung, S., Russakovsky, O., Jin, N., Andriluka, M., Mori, G., Fei-Fei, L.: Every moment counts: Dense detailed labeling of actions in complex videos. Int. J. Comput. Vis. (IJCV) **126**, 1–15 (2015)
36. Yogatama, D., et al.: Memory architectures in recurrent neural network language models. In: ICLR (2018)
37. Yu, L., Zhang, W., J. Wang, Yu, Y.: Seqgan: sequence generative adversarial nets with policy gradient. In: Proceedings of AAAI Conference on Artificial Intelligence (AAAI) (2017)
38. Zhao, Y., Zhu, S.C.: Image parsing with stochastic scene grammar. In: Advances in Neural Information Processing Systems, pp. 73–81 (2011)
39. Zhu, S.C., Mumford, D.: A stochastic grammar of images. Foundations and Trends® in Computer Graphics and Vision, vol. 2 (2007)

GDumb: A Simple Approach that Questions Our Progress in Continual Learning

Ameya Prabhu[1(✉)], Philip H. S. Torr[1], and Puneet K. Dokania[1,2]

[1] University of Oxford, Oxford, UK
[2] Five AI Ltd., Oxford, UK
{ameya,phst,puneet}@robots.ox.ac.uk

Abstract. We discuss a general formulation for the Continual Learning (CL) problem for classification—a learning task where a stream provides samples to a learner and the goal of the learner, depending on the samples it receives, is to continually upgrade its knowledge about the old classes and learn new ones. Our formulation takes inspiration from the open-set recognition problem where test scenarios do not necessarily belong to the training distribution. We also discuss various quirks and assumptions encoded in recently proposed approaches for CL. We argue that some oversimplify the problem to an extent that leaves it with very little practical importance, and makes it extremely easy to perform well on. To validate this, we propose GDumb that (1) greedily stores samples in memory as they come and; (2) at test time, trains a model from scratch using samples only in the memory. We show that even though GDumb is not specifically designed for CL problems, it obtains state-of-the-art accuracies (often with large margins) in almost all the experiments when compared to a multitude of recently proposed algorithms. Surprisingly, it outperforms approaches in CL formulations for which they were specifically designed. This, we believe, raises concerns regarding our progress in CL for classification. Overall, we hope our formulation, characterizations and discussions will help in designing realistically useful CL algorithms, and GDumb will serve as a strong contender for the same.

1 Introduction

A fundamental characteristic of natural intelligence is its ability to continually learn new concepts while updating information about the old ones. Realizing that very objective in machines is precisely the motivation behind continual learning (CL). While current machine learning (ML) algorithms can achieve excellent performance given any single task, learning new (or even related) tasks continually is extremely difficult for them as, in such scenarios, they are prone to the

Electronic supplementary material The online version of this chapter (https://doi.org/10.1007/978-3-030-58536-5_31) contains supplementary material, which is available to authorized users.

A. Vedaldi et al. (Eds.): ECCV 2020, LNCS 12347, pp. 524–540, 2020.
https://doi.org/10.1007/978-3-030-58536-5_31

phenomenon called catastrophic forgetting [1,2]. Significant attention has been paid recently to this problem [3–8] and a diverse set of approaches have been proposed in the literature (refer [9] for an overview). However, these approaches impose different sets of simplifying constraints to the CL problem and propose tailored algorithms for the same. Sometimes these constraints are so rigid that they even break the notion of learning continually, for example, one such constraint would be knowing a priori the subset of labels a given input might take. In addition, these approaches are never tested exhaustively on useful scenarios. Keeping this observation in mind, we suggest that it is of paramount importance to understand the caveats in these simplifying assumptions, understand why these simplified forms are of little practical usability, and shift our focus on a more *general* and practically useful form of continual learning formulation to help progress the field.

To this end, we first provide a *general* formulation of CL for classification. Then, we investigate popular variants of existing CL algorithms, and categorize them based on the simplifying assumptions they impose over the said general formulation. We discuss how each of them impose constraints either over the growing nature of the label space, the size of the label space, or over the resources available. One of the primary drawbacks of these restricted settings is that algorithms tailored towards them fail miserably when exposed to a slightly different variant of CL, making them extremely specific to a particular situation. We would also like to emphasize that there is no explicit consensus among researchers regarding which formulation of CL is the most appropriate, leading to a diverse experimental scenarios, none of which actually mimic the general form of CL problem one would face when exposed to the real-world.

Then, we take a step back and design an extremely simple algorithm with almost no simplifying assumptions compared to the recent approaches. We call this approach GDumb (Greedy Sampler and Dumb Learner). As the name suggest, the two core components of our approach are a *greedy sampler* and a *dumb learner*. Given a memory budget, the sampler greedily stores samples from a data-stream while making sure that the classes are balanced, and, *at inference*, the learner (neural network) is trained from scratch (hence dumb) using all the samples stored in the memory. When tested on a variety of scenarios on which various recent works have proposed highly tuned algorithms, GDumb surprisingly provides state-of-the-art results with large margins in almost all the cases.

The fact that GDumb, even though not designed to handle the intricacies in the challenging CL problems, outperforms recently proposed algorithms in their own experimental set-ups, is alarming. It raises concerns relating to the popular and widely used assumptions, evaluation metrics, and also questions the efficacy of various recently proposed algorithms for continual learning.

2   Problem Formulation, Assumptions, and Trends

To provide a *general* and practically useful view of CL problem, we begin with the following example. Imagine a robot walking in a living room solving a task that requires it to identify all the objects it encounters. In this setting, the robot will be identifying known objects that it has learned in the past, will be learning about a few *unknown* objects by asking an oracle to provide labels for them, and, at the same time, will be updating its information about the known objects if the new instances of them provided extra cues useful for the task. In a nutshell, the robot begins with some partial information about the world and keeps on improving its knowledge about it as it explores new parts of the world.

Inspired by this example, a realistic formulation of continual learning for classification would be where there is a stream of training samples or data accessible to a learner, each sample comprising a two-tuple $(\mathbf{x}_t, \mathbf{y}_t)$, where t represents the timestamp or the sample index. Let $\mathcal{Y}_t = \cup_{i=1}^{t}\mathbf{y}_i$ be the set of labels seen until time t, then it is trivial to note that $\mathcal{Y}_{t-1} \subseteq \mathcal{Y}_t$. This formulation implies that the stream might give us a sample that either belongs to a new class or to the old ones. Under this setting, at any *given* t, the objective is to provide a mapping $f_{\theta_t} : \mathbf{x} \to \mathbf{y}$ that can accurately map a sample $\mathbf{x}$ to a label $\mathbf{y} \in \mathcal{Y}_t \cup \bar{\mathbf{y}}$, where $\bar{\mathbf{y}}$ indicates that the sample does not belong to any of the learned classes. Notice, addition of this extra label $\bar{\mathbf{y}}$ assumes that while training, there is incomplete knowledge about the world and a test sample might come from outside the training distribution. Interestingly, it connects an instance of CL very well with the well known open-set classification problem [10]. However, in CL, the learner, with the help of an oracle (*e.g.*, active learning), could improve its knowledge about the world by learning the semantics of samples inferred as $\bar{\mathbf{y}}$.

2.1 Simplifying Assumptions in Continual Learning

The above discussed formulation is general in the sense that it does not put any constraints whatsoever on the growing nature of the label space, nature of test samples, and size of the output space $|\lim_{t\to\infty} \mathcal{Y}_t|$. It does not put any restrictions on the resources (compute and storage) one might pick to get a reliable mapping $f_{\theta_t}(.)$ either, however, the lack of information about the nature and the size of the output space makes the problem extremely hard. This has compelled almost all the work in this direction to impose additional simplifying constraints or assumptions. These assumptions are so varied that it is difficult to compare one CL algorithm with another as a slight variation in the assumption might change the complexity of the problem dramatically. For better understanding, below we discuss all the popular assumptions, highlight their drawbacks, and categorize various recently proposed CL algorithms depending on the simplifying assumptions they make. One assumption common to all is that the test samples always belong to the training distribution.

Disjoint Task Formulation: This formulation is being used in almost all the recent works [4–8] whereby the assumption made is that at a particular *duration* in time, the data-stream will provide samples specific to a task, in a pre-defined order of *tasks*, and the aim is to learn the mapping by learning each task at a time sequentially. In particular, let $\mathcal{Y} = \lim_{t\to\infty} \mathcal{Y}_t$ be the set of labels that the stream might present until it runs out of samples. Recall, in the general

CL formulation, the size of $\mathcal{Y}$ is unknown and the samples can be presented in any order. This label space $\mathcal{Y}$ is then divided into different disjoint subsets (could be a random or an informed split), where each label subset $\mathcal{Y}_i$ represents a task and the *sharp* transition between these sets is called *task boundaries*. Let there be m splits (typically the split is balanced with nearly equal number of classes) then $\mathcal{Y} = \cup_i^m \mathcal{Y}_i$, and $\mathcal{Y}_i \cap \mathcal{Y}_j = \emptyset, \forall i \neq j$. An easy and widely used example is to divide ten digits of MNIST into 5 disjoint tasks where each task comprises of the samples from two consecutive digits and the stream is *controlled* to provide samples for each task in a pre-defined order, say $\{0, 1\}, \cdots, \{8, 9\}$. This formulation simplifies the general CL problem to a great extent as the unknown growing nature of the label space is now being restricted and is known. It provides a very strong prior to the learner and helps in deciding both the space budget and the family of functions $f_\theta(.)$ to learn.

Task-Incremental *v/s* **Class-Incremental**: To further make the training and the inference easier, a popular choice of CL formulation is the task-incremental continual learning (TI-CL) [7] where, along with the disjoint task assumption, the task information (or id) is also passed by an oracle during training and inference. Thus, instead of a two-tuple, a three-tuple $(\mathbf{x}, \mathbf{y}, \alpha)$ is given where $\alpha \in \mathbb{N}$ represents the task identifier. This formulation is also known as *multi-head* and is an extremely simplified form of the continual learning problem [8]. For instance, in the above mentioned MNIST example, at inference, if the input is $(\mathbf{x}, \alpha = 3)$, it implies that the sample either belongs to class 4, or to 5. Knowing this subset of labels a-prior dramatically reduces the label space during training and inference, and is relatively impractical to know in real-world scenarios. Whereas, in a class-incremental formulation (CI-CL) [4,8], also known as the *single-head*, we do not have any such information about the task id.

Online CL *v/s* ***Offline CL***: Note, the disjoint task formulation placed a restriction on the growing nature of the label space and inherently restricted the size of it, however, it did not put any constraints on the learner itself. Therefore, the learning paradigm may store task-specific samples coming from the stream depending on the space budget and then use them to update the parameters. Under this setting, in the *online* CL formulation, even though the learner is allowed to store samples as they come, they are not allowed to use a sample more than once for parameter update. Thus, the learner can not use the same sample (unless it is in the memory) multiple times at different iterations of the learning process. In contrast, *offline* CL allows unrestricted access to the *entire* dataset corresponding to a particular task (not to the previous ones) and one can use this dataset to learn the mapping by revisiting the samples again and again while performing multiple passes over the data [4].

Memory Based CL: As mentioned earlier, we only have access to all/subset of samples corresponding to the *current* task. This restriction makes it extremely hard for the model to perform well, in particular, on CI-CL setting as the absence of samples from the previous tasks makes it difficult to learn to distinguish samples from the current and the previous tasks due to catastrophic forgetting.

Table 1. Here we categorize various recently proposed CL approaches depending on the underlying simplifying assumptions they impose.

Form.	CI-CL	Online	Disjoint	Papers	Regularize	Memory	Distill	Param iso
A	✓	✓	✓	MIR[11], GMED[12]	×	✓	×	×
				LwM[13], DMC[14]	×	×	✓	×
				SDC [15]	✓	×	×	×
B	✓	×	✓	BiC[16], iCARL[4] UCIR[17], EEIL[18] IL2M[19], WA[20] PODNet[21], MCIL[22]	×	✓	✓	×
				RPS-Net[23], iTAML[24]	×	✓	✓	✓
				CGATE[25]	×	✓	×	✓
				RWALK[8]	✓	✓	×	×
C	×	×	✓	PNN[26], DEN[27]	×	×	×	✓
				DGR [28]	×	✓	×	×
				LwF[3]	×	×	✓	×
				P&C[29]	×	×	✓	✓
				APD[30]	✓	×	×	✓
				VCL[31]	✓	✓	×	×
				MAS[32], IMM[33] SI[5], Online-EWC[29] EWC[6]	✓	×	×	×
D	×	✓	✓	TinyER[34], HAL[35]	×	✓	×	×
				GEM[7], AGEM[36]	✓	✓	×	×
E	✓	✓	×	GSS[37]	×	✓	×	×

Very little forgetting is normally observed in TI-CL as the given task identifier works as the indicator of task boundary, thus the model does not have to learn to differentiate labels among tasks. To reduce forgetting, a common practice, inspired by the complementary learning systems theory [38, 39], is to store a subset of samples from each task and use them while training on the current task. There primarily are two components under this setting: a *learner* and a *memorizer* (or sampler). The learner has the goal of obtaining representations which generalize beyond current task. The memorizer, on the other hand, deals with remembering (storing) a collection of episode-like memories from the previous tasks. In recent approaches [4,7,8], the learner is modeled by a neural network and the memorizer is modeled by memory slots which store samples previously encountered.

2.2 Recent Trends in Continual Learning

Typically, continual learning approaches are categorized by ways they tackle forgetting such as (1) regularization-based, (2) replay (or memory)-based, (3) distillation-based, and (4) parameter-isolation based (for details refer [9]). However, they do vary in terms of simplifying assumptions they encode, and we argue that keeping track of these assumptions is extremely important for fair comparisons, and also to understand the limitations of each of them. Since all these algorithms in some sense use combinations of the above discussed simplifying assumptions, to give a bird's eye view over all the recently proposed approaches, we categorize them in Table 1 depending on the simplifying assumptions they make. For example, Table 1 indicates that RWalk [8] is an approach designed for a CL formulation that is offline, class-incremental, and assumes sharp task boundaries. Algorithmically, it is regularization based and uses memory. Note, one can potentially modify these approaches to apply to other settings as well. For example, the same RWalk can also be used without memory, or can be applied on task-incremental offline formulation. However, we focus on the formulation these methods were originally designed for. We now discuss some high-level problems associated with the simplifying assumptions these approaches make.

Most models, metrics, classifiers, and samplers for CL inherently encode disjoint task (or sharp task boundary) assumption into their design, hence fail to generalize even with slight deviation from the this formulation. Similarly, popular metrics like forgetting and intransigence [7,8] are designed with this specific formulation encoded in their formal definition, and break with simple modifications like blurry boundaries (class-based, instead of sample-based, definitions of forgetting would appear as classes mix because of blurred boundaries). Moving to TI-CL v/s CI-CL, these are two extreme cases where CI-CL (single-head) faces scaling issues as there is no restriction on the size of $|\lim_{t\to\infty} \mathcal{Y}_t|$, and TI-CL (multi-head) imposes a fixed, coherent two-level hierarchy among classes with oracle labels. This formulation is unrealistic in the sense that it does not allow dynamic contexts [40].

Lastly, Offline CL v/s Online CL is normally defined depending on whether an algorithm is allowed to revisit a sample repeatedly (unless it is in the memory) during the training process or not. The intention here is to make the continual learning algorithm fast enough so that it can learn *quickly* from a single (or few) sample without having the need of revisiting it. This distinction makes sense if we imagine a data stream spitting samples very fast, then the learner has to adapt itself very quickly. Therefore, the algorithm must provide an acceptable trade-off between space (number of samples to store) and time (training complexity) budgets. However, because of the lack of proper definition and evaluation schemes, there are algorithms doing very well on one end (use a sample only once), however, performing very poorly on the other end (very expensive learning process). For example, GEM [7], a widely known online CL algorithm, uses a sample only once, however, solves a quadratic program for parameter updates which is very time consuming. Therefore, without proper metrics or procedures to quantify how well various CL algorithms balance both space and time complexities, categorizing them into offline vs online might lead to wrong conclusions.

3   Greedy Sampler and Dumb Learner (GDumb)

We now propose a simple approach that does not put any restrictions, as discussed above, over the growing nature of the label space, task boundaries, online vs offline, and the ordering of the samples in which the data-stream provides

---

Algorithm 1. Greedy Balancing Sampler

1: **Init:** counter $C_0=\{\}$, $\mathcal{D}_0=\{\}$ with capacity k. Online samples arrive from $t=1$
2:
3: **function** SAMPLE($x_t, y_t, \mathcal{D}_{t-1}, \mathcal{Y}_{t-1}$)   ▷ Input: New sample and past state
4:     $k_c = \frac{k}{|\mathcal{Y}_{t-1}|}$
5:     **if** $y_t \notin \mathcal{Y}_{t-1}$ or $C_{t-1}[y_t] < k_c$ **then**
6:         **if** $\sum_i C_i >= k$ **then**   ▷ If memory is full, replace
7:             $y_r = argmax(C_{t-1})$ ▷ Select largest class, break ties randomly
8:             $(x_i, y_i) = \mathcal{D}_{t-1}.random(y_r)$ ▷ Select random sample from class y_r
9:             $\mathcal{D}_t = (\mathcal{D}_{t-1} - (x_i, y_i)) \cup (x_t, y_t)$
10:            $C_t[y_r] = C_{t-1}[y_r] - 1$
11:        **else**   ▷ If memory has space, add
12:            $\mathcal{D}_t = \mathcal{D}_{t-1} \cup (x_t, y_t)$
13:        **end if**
14:        $\mathcal{Y}_t = \mathcal{Y}_{t-1} \cup y_t$
15:        $C_t[y_t] = C_{t-1}[y_t] + 1$
16:    **end if**
17:    **return** $\mathcal{D}_t$
18: **end function**

---

them. Thus, can easily be applied to all the CL formulations discussed in Table 1. The only requirement is to be allowed to store some episodic memories. We emphasize that we do not claim that our approach solves the general CL problem. Rather, we experimentally show that our simple approach, that does not encode anything specific to the challenging CL problem at hand, is surprisingly effective compared to other approaches over all the formulations discussed previously, and also exposes important shortcomings with recent formulations and algorithms.

As illustrated in Fig. 1, our approach comprises of two key components: a *greedy balancing sampler* and a *learner*. Given a memory budget, say k samples, the sampler greedily stores samples from the data-stream (max k samples) with the constraint to asymptotically balance class distribution (Algorithm 1). It is greedy in the sense that whenever it encounters a new class, the sampler simply creates a new bucket for that class and starts removing samples from the old ones, in particular, from the one with a maximum number of samples. Any tie is broken randomly, and a sample is also removed randomly assuming that each sample is equally important. Note, this sampler does not rely on task boundaries or any information about the number of samples in each class.

Let the set of samples greedily stored by the sampler in the memory at any instant in time be $\mathcal{D}_t$ (a dataset with $\leq k$ samples). Then, the objective of the learner, a deep neural network in our experiments, is to learn a mapping $f_{\theta_t} : \mathbf{x} \rightarrow \mathbf{y}$, where $(\mathbf{x}, \mathbf{y}) \in \mathcal{D}_t$. This way, using a small dataset that the sampler has stored, the learner learns to classify all the labels seen until time t. Let $\mathcal{Y}_t$ represents the set of labels in $\mathcal{D}_t$. Then, at inference, given a sample $\mathbf{x}$, the prediction is made as

$$\hat{y} = \arg\max \mathbf{p} \odot \mathbf{m}, \qquad (1)$$

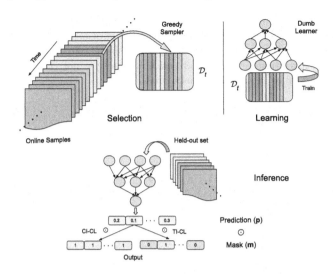

Fig. 1. Our approach (GDumb): The sampler greedily stores samples while balancing the classes. When asked, the learner trains a network from scratch on memory $\mathcal{D}_t$ provided by the sampler. If a mask **m** is given at inference, GDumb classifies on the subset of labels provided by the mask. Depending on the mask, GDumb's inference can vary between two extremes: CI (class-incremental) and TI (task-incremental) formulations.

where, **p** is the softmax probabilities over all the classes in $\mathcal{Y}_t$, $\mathbf{m} \in \{0,1\}^{|\mathcal{Y}_t|}$ is a user-defined mask, and $\odot$ denotes the Hadamard product. Note, our prediction procedure allows us to mask any combination of labels at inference. When **m** consists of all ones, the inference is exactly the same as that of *single-head* or class-incremental, and when the masking is done depending on the subset of classes in a particular task, it is exactly the same as *multi-head* or task-incremental. Since our sampler does not put any restrictions on the flow of the samples from the data-stream, and our learner does not require any task-boundaries, our overall approach puts minimal restrictions on the general continual learning problem. We would also like to emphasize that we do not use the class $\bar{\mathbf{y}}$ as discussed in our general formulation in Sect. 2, we leave that for future work. However, our objective does encapsulate all the recently proposed CL formulations with minimal possible assumptions, allowing us to provide a fair comparison.

4   Experiments

We now compare GDumb with various existing algorithms for several recently proposed CL formulations. As shown in Table 1, there broadly are five such formulations $\{A, B, \cdots, E\}$. Since even within a formulation there can be sub-categories depending on the resources used, we further enumerate them and present a more detailed categorization, keeping fair comparisons in mind, in Table 2. For example, B1 and B2 belong to the same formulation B, however,

Table 2. Various CL formulations we considered in this work to evaluate GDumb. These formulations differ in terms of simplifying assumptions (refer Table 1) and also in terms of resources used. We ensure that selected benchmarks are diverse, covering all popular categorizations. Note, in B3 and D, memory is not constant– it increases over tasks uniformly by (+size) for xtasks times.

Form.	Designed in	Model (Dataset)	memory (k)	Metric	CI-CL	Online	Disjoint
A1	[11]	MLP-400 (MNIST); ResNet18 (CIFAR10)	300, 500; 200, 500, 1000	Acc. (at end)			
A2	[12]	MLP-400 (MNIST); ResNet18 (CIFAR10)	500; 500	Acc. (at end)	✓	✓	✓
A3	[41]	MLP-400 (MNIST); ResNet18 (CIFAR10)	500; 1000	Acc. (at end)			
B1	[42]; [23]	MLP-400 (MNIST); ResNet18 (SVHN)	4400	Acc. (at end)			
B2	[4]	ResNet32 (CIFAR100)	2000	Acc. (avg in t)	✓	✗	✓
B3	[21]	ResNet32 (CIFAR100); ResNet18 (ImageNet100)	1000-2000 (+20) x50	Acc. (avg in t)			
C1	[42]	MLP-400 (MNIST)	4400	Acc. (at end)	✗	✗	✓
C2	[9]	Many (TinyImageNet)	4500,9000	Acc. (at end)			
D	[36]	ResNet-18-S (CIFAR10)	0-1105 (+65) x17	Acc. (at end)	✗	✓	✓
E	[37]	MLP-100 (MNIST); ResNet-18 (CIFAR10)	300; 500	Acc. (at end)	✓	✓	✗

they differ in terms of architectures, datsets, and memory sizes used in their respective papers. Therefore, in total, we pick 10 different formulations, most of them having multiple architectures and datasets (refer Appendix B for details).

Implementation details: GDumb uses the same fixed training settings, with *no* hyperparameter tuning whatsoever, in all the CL formulations. This is possible because of the fact that GDumb does not impose any simplifying assumptions. All results measure accuracy (fraction of correctly classifications) evaluated on the held-out test set. For all the formulations, GDumb uses an SGD optimizer, fixed batch size of 16, learning rates $[0.05, 0.0005]$, an SGDR [45] schedule with $T_0 = 1$, $T_{mult} = 2$ and warm start of 1 epoch. Early stopping with patience of 1 cycle of SGDR, along with standard data augmentation is used. GDumb uses cutmix [46] with $p = 0.5$ and $\alpha = 1.0$ for regularization on all datasets except MNIST. The training set-up comprises of an Intel i7 4790, 32 GB RAM and a single GTX 1070 GPU. All results are averaged over 3 random seeds, each with different class-to-task assignments. In formulations B2 and B3, we strictly follow class order specified in iCARL [4] and PODNet [21]. Our pytorch implementation is publicly available at: https://github.com/drimpossible/GDumb.

Table 3. (CI-Online-Disjoint) Performance on formulation A1.

Method	MNIST	
k	(300)	(500)
MLP-100		
FSS-Clust [37]	75.8 ± 1.7	83.4 ± 2.6
GSS-Clust [37]	75.7 ± 2.2	83.9 ± 1.6
GSS-IQP [37]	75.9 ± 2.5	84.1 ± 2.4
GSS-Greedy [37]	82.6 ± 2.9	84.8 ± 1.8
GDumb (Ours)	**88.9 ± 0.6**	**90.0 ± 0.4**
MLP-400		
GEN [43]	-	75.5 ± 1.3
GEN-MIR [11]	-	81.6 ± 0.9
ER [44]	-	82.1 ± 1.5
GEM [7]	-	86.3 ± 1.4
ER-MIR [11]	-	87.6 ± 0.7
GDumb (Ours)	-	**91.9 ± 0.5**
(A1)		

Method	CIFAR10		
k	(200)	(500)	(1000)
GEM [7]	16.8 ± 1.1	17.1 ± 1.0	17.5 ± 1.6
iCARL [4]	28.6 ± 1.2	33.7 ± 1.6	32.4 ± 2.1
ER [44]	27.5 ± 1.2	33.1 ± 1.7	41.3 ± 1.9
ER-MIR [11]	29.8 ± 1.1	40.0 ± 1.1	47.6 ± 1.1
ER5 [11]	-	-	42.4 ± 1.1
ER-MIR5 [11]	-	-	49.3 ± 0.1
GDumb (Ours)	**35.0 ± 0.6**	**45.8 ± 0.9**	**61.3 ± 1.7**
(A1)			

4.1   Results and Discussions

Class Incremental Online CL with Disjoint Tasks (Form. A): The first sub-category under this formulation is A1, which follows exactly the same setting, on Split-MNIST and Split-CIFAR10, as presented in MIR [11]. Results are shown in Table 3. We observe that on both MNIST and CIFAR10 for all choices of k, GDumb outperforms all the approaches by a large margin. For example, in the case of MNIST with memory size $k = 500$, GDumb outperforms ER-MIR [11] by around 4.3%. Similarly, on CIFAR10 with memory sizes of 200, 500, and 1K, our approach outperforms current approaches by nearly 5%, 6% and 11%, respectively, convincingly achieving state-of-the-art results. Note, increasing the memory size from 200 to 1K in CIFAR10 increases the performance of GDumb by 26.3% (expected as GDumb is trained only on memory), whereas, this increase is only 18% in the case of ER-MIR [11]. Similar or even much worst improvements are noticed in other recent approaches, suggesting they might not be utilizing the memory samples efficiently.

We now benchmark our approach on Split-MNIST and Split-CIFAR10 as detailed by parallel works GMED [12] (sub-category A2) and ARM [41] (sub-category A3). We present results in Table 4 and show that GDumb outperforms parallel works like HAL [35], QMED [12], ARM [41] in addition to outperforming recent works GSS [37], MIR [11], and ADI [47], consistently across datasets. It outperforms the best alternatives in QMED [12] by over 4% and 10% on MNIST and CIFAR10, respectively, and in ARM [41] by over 5% and 13% on MNIST and CIFAR10 datasets, respectively. Results from ARM [41] indicate—(i) GDumb consistently outperforms other experience replay approaches and (ii) experience replay methods obtain much better performance than generative replay with much smaller memory footprint.

Table 4. (CI-Online-Disjoint) Performance on formulations A2 (left) and A3 (right).

Method	MNIST	CIFAR-10
k	(500)	(500)
Fine tuning	18.8 ± 0.6	18.5 ± 0.2
AGEM [36]	29.0 ± 5.3	18.5 ± 0.6
BGD [48]	13.5 ± 5.1	18.2 ± 0.5
GEM [7]	87.2 ± 1.3	20.1 ± 1.4
GSS-Greedy [37]	84.2 ± 2.6	28.0 ± 1.3
HAL [35]	77.9 ± 4.2	32.1 ± 1.5
ER [44]	81.0 ± 2.3	33.3 ± 1.5
MIR [11]	84.9 ± 1.7	34.5 ± 2.0
GMED (ER) [12]	82.7 ± 2.1	35.0 ± 1.5
GMED (MIR) [12]	87.9 ± 1.1	35.5 ± 1.9
GDumb (Ours)	**91.9 ± 0.5**	**45.8 ± 0.9**

(A2)

Method	MNIST		CIFAR10	
	Memory	Accuracy	Memory	Accuracy
Finetune	0	18.8 ± 0.5	0	15.0 ± 3.1
GEN [28]	4.58	79.3 ± 0.6	34.5	15.3 ± 0.5
GEN-MIR [11]	4.31	82.1 ± 0.3	38.0	15.3 ± 1.2
LwF [3]	1.91	33.3 ± 2.5	4.38	19.2 ± 0.3
ADI [47]	1.91	55.4 ± 2.6	4.38	24.8 ± 0.9
ARM [41]	1.91	56.2 ± 3.5	4.38	26.4 ± 1.2
ER [44]	0.39	83.2 ± 1.9	3.07	41.3 ± 1.9
ER-MIR [11]	0.39	85.6 ± 2.0	3.07	47.6 ± 1.1
iCarl [4] (5 iter)	-	-	3.07	32.4 ± 2.1
GEM [7]	0.39	86.3 ± 0.1	3.07	17.5 ± 1.6
GDumb (ours)	**0.39**	**91.9 ± 0.5**	**3.07**	**61.3 ± 1.7**

(A3)

Class Incremental Offline CL with Disjoint Tasks (Form. B):

We proceed next to offline CI-CL formulations. We first compare our proposed approach with 12 popular methods on sub-category B1. Results are presented in Table 5 (left). Our approach outperforms all memory-based methods like GEM, RtF, DGR by over 5% on MNIST. We outperform the recent RPS-Net [23] and OvA-INN [49] by over 1%, and are as good as iTAML [24], on MNIST. On SVHN, we outperform recently proposed methods like RPS-Net by 4.5% and far exceeding methods like GEM. Note, we achieve the same accuracy as the best offline CL method iTAML [24] despite using an extremely simple approach in online fashion.

We now discuss two very interesting sub-categories B2 (as in iCARL [4]) and B3 (from a very recent work PODNet [21]). The primary difference between B2 and B3 merely lies in the number of classes per task. However, as will be seen, this minor difference changes the complexity of the problem dramatically. In the case of B2, CIFAR100 is divided into 20 tasks, whereas, B3 starts with a network trained on 50 classes and then learns one class per task incrementally (leading to 50 new tasks). Performance of GDumb on B2 and B3 formulations are shown in Table 5 (center) and Table 5 (right), respectively. An interesting observation here is that GDumb which performed nearly 20% worse than BiC and iCARL in B2, performs over 10–15% better than BiC, UCIR and iCARL in B3. This drastic shift against previous results might suggest that having higher number of classes per task and less number of tasks might give added advantage to scaling/bias correction type approaches, which otherwise would quickly deteriorate over greater timesteps. Furthermore, we note that our simple baseline narrowly outperforms PODNet (CNN) on both CIFAR100 and ImageNet100 datasets.

Table 5. (CI-Offline-Disjoint) Performance on B1, B2, and B3.

Method	MNIST	SVHN
	No memory	
MAS [32]	19.5 ± 0.3	17.3
SI [5]	19.7 ± 0.1	17.3
EWC [6]	19.8 ± 0.1	18.2
Online EWC [29]	19.8 ± 0.04	18.5
LwF [3]	24.2 ± 0.3	-
	k=4400	
DGR [28]	91.2 ± 0.3	-
DGR+Distill	91.8 ± 0.3	-
GEM [7]	92.2 ± 0.1	75.6
RtF [50]	92.6 ± 0.2	-
RPS-Net [23]	96.2	88.9
OvA-INN [49]	96.4	-
iTAML [24]	97.9	**94.0**
GDumb (Ours)	**97.8 ± 0.2**	93.4 ± 0.4
(B1)		

Method	CIFAR100	
	Acc. (Avg)	Acc. (last)
	No memory	
Finetune	17.8 ± 0.72	5.9 ± 0.15
SI [5]	23.6 ± 1.90	13.3 ± 1.14
MAS [32]	24.7 ± 1.76	10.4 ± 0.80
EWC [6]	25.4 ± 1.99	9.5 ± 0.83
RWALK [8]	25.6 ± 1.92	11.1 ± 2.14
LwF [3]	32.3 ± 1.92	14.1 ± 0.87
DMC [14]	45.0 ± 1.96	23.8 ± 1.90
	k=2000	
GDumb (Ours)	45.2 ± 1.70	24.1 ± 0.97
DMC++ [14]	56.8 ± 0.86	-
iCARL [4]	58.8 ± 1.90	42.9 ± 0.79
EEIL [18]	63.4 ± 1.6	-
BiC [16]	63.8	46.9
(B2)		

Method	CIFAR100	ImageNet100
iCaRL [4]	44.2 ± 1.0	54.97
BiC [16]	47.1 ± 1.5	46.49
UCIR (NME) [17]	48.6 ± 0.4	55.44
UCIR (CNN) [17]	49.3 ± 0.3	57.25
PODNet (CNN) [21]	58.0 ± 0.5	62.08
GDumb (CNN)	58.4 ± 0.8	**62.86**
PODNet (NME) [21]	**61.4 ± 0.7**	-
(B3)		

Task Incremental Offline CL with Disjoint Tasks (Form. C): We now proceed to compare the performance of GDumb in task incremental formulation. Recall, GDumb does not put any restrictions such as task vs class incremental, or online vs offline. However, in the case of GDumb, we use masking (subset of labels in a task) over softmax probabilities at test time to mimic TI-CL formulation.

Table 6 (left) shows the results on C1, a very widely used and most popular offline TI-CL (or multi-head) formulation for CL on Split-MNIST. We now move to C2 on Split-TinyImagenet (offline TI-CL formulation as in [9]). Note, in this particular formulation we used a different architecture called DenseNet-100-BC [54]. Results are presented in Table 6 (middle). We observe that for $k = 9000$, we outperform all 10 approaches including GEM and iCARL by margins of atleast 7%. When the memory is halved to $k = 4500$, we perform slightly better than GEM and nearly 3% worse than iCARL. Since we used different architecture, we do not claim that we would notice similar improvements had we trained GDumb using the networks used in respective papers. However, these results are still encouraging as the approaches we compare against are trained in TI-CL manner and GDumb is always trained in CI-CL manner (much more difficult).

Task Incremental Online CL with Disjoint Tasks (Form. D): We now compare GDumb with 12 TI-CL tuned online approaches with small memory (not a favourable setting for GDumb as it relies totally on the samples in the memory) and detail the results in Table 6 (right). We observe that GDumb outperforms 8 out of 11 approaches even though it is trained in CI-CL manner.

Table 6. (TI-Offline-Disjoint) Performance on C1 (left) and C2 (middle). (TI-Onlione-Disjoint) Performance on D (right).

Method	Parameters	Regularization	Accuracy
No stored samples			
mean-IMM [33]	3.5M	none	32.42
mode-IMM [33]	9.0M	dropout	42.41
SI [5]	3.5M/9.0M	L2/dropout	43.74
HAT [51]	3.5M/9.0M	L2	44.19
EWC [6]	613K	none	45.13
LwF [3]	9.0M	L2	48.11
EBLL [52]	9.0M	L2	48.17
MAS [32]	3.5M/9.0M	none	48.98
PackNet [53]	613K/3.5M	L2/dropout	55.96
k=4500			
GEM [7]	613K/3.5M	none/dropout	44.23
GDumb	834K	cutmix	45.50
iCARL [4]	613K/3.5M	dropout	48.55
k=9000			
GEM [7]	613K/3.5M	none/dropout	45.27
iCARL [4]	613K/3.5M	dropout	49.94
GDumb	834K	cutmix	**57.27**

(C2)

Method	MNIST
(k)	(4400)
GEM [7]	98.42 ± 0.10
EWC [6]	98.64 ± 0.22
SI [5]	99.09 ± 0.15
Online EWC [29]	99.12 ± 0.11
MAS [32]	99.22 ± 0.21
DGR [28]	99.50 ± 0.03
LwF [3]	99.60 ± 0.03
DGR+Distil [28]	99.61 ± 0.02
RtF	99.66 ± 0.03
GDumb	**99.77 ± 0.03**

(C1)

Method	CIFAR100
(k)	(1105)
RWalk [8]	40.9 ± 3.97
EWC [6]	42.4 ± 3.02
Base	42.9 ± 2.07
MAS [32]	44.2 ± 2.39
SI [5]	47.1 ± 4.41
iCARL [4]	50.1
S-GEM [36]	56.2
PNN [26]	59.2 ± 0.85
GEM [7]	61.2 ± 0.78
A-GEM [36]	63.1 ± 1.24
TinyER [34]	68.5 ± 0.65
GDumb	60.3 ± 0.85

(D)

Class Incremental Online CL with Joint Tasks (Form. E): We now measure impact of imbalanced data stream with blurry task boundaries [37]. Results are presented in Table 7 (left). We outperform competing models by over 10% and 16%, overwhelming surpassing complicated methods attuned to this benchmark. This demonstrates that GDumb works well even when almost all simplifying assumptions are removed.

Table 7. (CI-Online-Joint) Performance on E (left). Note, this is particularly challenging as the tasks here are non-disjoint (blurry task boundary) with class-imbalance. On the (right), we benchmark resource consumption in terms of training time and memory usage. Memory cost is provided in terms of the total parameters P, the size of the minibatch B, the total size of the network hidden state H (assuming all methods use the same architecture), the size of the episodic memory M per task. GDumb, at the very least, is 7.5× times faster than the existing efficient CL formulations.

Method	MNIST	CIFAR10
Reservoir [43]	69.12	-
GSS-Clust [37]	-	25.0
FSS-Clust [37]	-	26.0
GSS-IQP [37]	76.49	29.6
GSS-Greedy [37]	77.96	29.6
GDumb (Ours)	**88.93**	**45.8**

(E)

Method	Train time	Memory (Train)	Memory (Test)
Base	105s	P + B*H	P + B*H
EWC	250s	4*P + B*H	P + B*H
PNN	409s	2*P*T + B*H*T	2*P*T + B*H*T
GEM	5238s	P*T + (B+M)*H	P + B*H
A-GEM	449s	2*P + (B+M)*H	P + B*H
GDumb	**60s**	**P + M*H**	**P + B*H**

(Resources)

4.2   Resources Needed

It is important that our approach is in the ballpark of online continual learning constraints of memory and compute usage to achieve its performance. We benchmark our resource consumption against the efficient CL algorithms in Table 7 (right) benchmarked with a V100 GPU on formulation E [36]. We observe that we require only 60s on a slower GTX 1070 GPU (and 350s on a 4790 i7 CPU), performing several times efficient than various recently proposed algorithm. Note that sampling time is negligible, while testing time is not included in the above.

4.3   Potential Future Extensions

Active Sampling: Given an importance value $v_t \in \mathrm{R}^+$ (by active learner) along with sample (x_t, y_t) at time t, we can extend our sampler by having the objective of storing most important samples (maximizing $\sum_{i=1}^{|\mathcal{D}_c|} v_i$) for any given class c in its storage of size k. This will allow an algorithm to reject less important samples. Of course, it is not clear how to learn to quantify *importance* of a sample.

Dynamic Probabilistic Masking: It is possible to extend masking in GDumb beyond CI-CL and TI-CL to dynamic task hierarchies across video/scene types useful in recently proposed settings [40]. Since GDumb applies a mask (given a context) only at inference, we can dynamically adapt to the context. Similarly, we can extend GDumb beyond deterministic oracles ($m_i \in \{0, 1\}$) to probabilistic one ($m_i \in [0, 1]$). This delivers a lot of flexibility to handle diverse extensions like cost-sensitive classification, class-imbalance among others.

5   Conclusion

In this work, we provided a general view of a continual image classification problem. We then proposed a simple and general approach with minimal restrictions and empirically showed that it outperforms almost all the complicated state-of-the-art approaches *in their own formulations* for which they were specifically designed. We hope that our approach serves the purpose of a strong baseline to benchmark the effectiveness of any newly proposed CL algorithm. Our solution also raises various concerns to be investigated: (1) Even though there are plenty of research articles focused on specific scenarios relating CL problem, are we really progressing in the right direction? (2) Which formulation to focus on? and (3) Do we need different experimental formulations, more complex than the current ones, so that the effectiveness of recent CL models, if they are, is pronounced?

Acknowledgements. AP would like to thank Aditya Bharti, Shyamgopal Karthik, Saujas Vaduguru, and Aurobindo Munagala for helpful discussions. PHS and PD thank EPSRC/MURI grant EP/N019474/1, and Facebook (DeepFakes grant) for their support. This project was supported by the Royal Academy of Engineering under the Research Chair and Senior Research Fellowships scheme. PHS and PD also acknowledge FiveAI UK.

References

1. McCloskey, M., Cohen, N.J.: Catastrophic interference in connectionist networks: The sequential learning problem. In: Psychology of Learning and Motivation (1989)
2. Goodfellow, I.J., Mirza, M., Xiao, D., Courville, A., Bengio, Y.: An empirical investigation of catastrophic forgetting in gradient-based neural networks. arXiv preprint arXiv:1312.6211 (2013)
3. Li, Z., Hoiem, D.: Learning without forgetting. TPAMI 40(12), 2935–2947 (2017)
4. Rebuffi, S.A., Kolesnikov, A., Sperl, G., Lampert, C.H.: icarl: incremental classifier and representation learning. In: CVPR (2017)
5. Zenke, F., Poole, B., Ganguli, S.: Continual learning through synaptic intelligence. ICML **70**, 3987 (2017)
6. Kirkpatrick, J., et al.: Overcoming catastrophic forgetting in neural networks. PNAS **114**(13), 3521–3526 (2017)
7. Lopez-Paz, D., Ranzato, M.: Gradient episodic memory for continual learning. In: NeurIP (2017)
8. Chaudhry, A., Dokania, P.K., Ajanthan, T., Torr, P.H.: Riemannian walk for incremental learning: understanding forgetting and intransigence. In: ECCV (2018)
9. De Lange, M., et al.: Continual learning: a comparative study on how to defy forgetting in classification tasks. arXiv preprint arXiv:1909.08383 (2019)
10. Scheirer, W., Rocha, A., Sapkota, A., Boult, T.: Towards open set recognition. TPAMI **35**(7), 1757–1772 (2012)
11. Aljundi, R., Caccia, L., Belilovsky, E., Caccia, M., Charlin, L., Tuytelaars, T.: Online continual learning with maximally interfered retrieval. In: NeurIPS (2019)
12. Jin, X., Du, J., Ren, X.: Gradient based memory editing for task-free continual learning (2020)
13. Dhar, P., Vikram Singh, R., Peng, K.C., Wu, Z., Chellappa, R.: Learning without memorizing. In: CVPR (2019)
14. Zhang, J., et al.: Class-incremental learning via deep model consolidation. In: WACV (2020)
15. Yu, L., et al.: Semantic drift compensation for class-incremental learning. In: CVPR (2020)
16. Wu, Y., et al.: Large scale incremental learning. In: CVPR (2019)
17. Hou, S., Pan, X., Loy, C.C., Wang, Z., Lin, D.: Learning a unified classifier incrementally via rebalancing. In: CVPR (2019)
18. Castro, F.M., Marín-Jiménez, M.J., Guil, N., Schmid, C., Alahari, K.: End-to-end incremental learning. In: ECCV (2018)
19. Belouadah, E., Popescu, A.: Il2m: class incremental learning with dual memory. In: ICCV (2019)
20. Zhao, B., Xiao, X., Gan, G., Zhang, B., Xia, S.T.: Maintaining discrimination and fairness in class incremental learning. In: CVPR (2020)
21. Douillard, A., Cord, M., Ollion, C., Robert, T., Valle, E.: Small-task incremental learning. ECCV (2020)
22. Liu, Y., Su, Y., Liu, A.A., Schiele, B., Sun, Q.: Mnemonics training: multi-class incremental learning without forgetting. In: CVPR (2020)
23. Rajasegaran, J., Hayat, M., Khan, S., Khan, F.S., Shao, L.: Random path selection for incremental learning. In: NeurIPS (2019)
24. Rajasegaran, J., Khan, S., Hayat, M., Khan, F.S., Shah, M.: itaml: an incremental task-agnostic meta-learning approach. In: CVPR (2020)

25. Abati, D., Tomczak, J., Blankevoort, T., Calderara, S., Cucchiara, R., Bejnordi, B.E.: Conditional channel gated networks for task-aware continual learning. In: CVPR (2020)
26. Rusu, A.A., et al.: Progressive neural networks. arXiv preprint arXiv:1606.04671 (2016)
27. Yoon, J., Lee, J., Yang, E., Hwang, S.J.: Lifelong learning with dynamically expandable network. In: ICLR (2018)
28. Shin, H., Lee, J.K., Kim, J., Kim, J.: Continual learning with deep generative replay. In: NeurIPS (2017)
29. Schwarz, J., et al.: Progress & compress: a scalable framework for continual learning. ICML (2018)
30. Yoon, J., Kim, S., Yang, E., Hwang, S.J.: Scalable and order-robust continual learning with additive parameter decomposition. In: ICLR (2020)
31. Nguyen, C.V., Li, Y., Bui, T.D., Turner, R.E.: Variational continual learning. In: ICLR (2018)
32. Aljundi, R., Babiloni, F., Elhoseiny, M., Rohrbach, M., Tuytelaars, T.: Memory aware synapses: learning what (not) to forget. In: ECCV (2018)
33. Lee, S.W., Kim, J.H., Jun, J., Ha, J.W., Zhang, B.T.: Overcoming catastrophic forgetting by incremental moment matching. In: NeurIPS (2017)
34. Chaudhry, A., et al.: Continual learning with tiny episodic memories. ICML-W (2019)
35. Chaudhry, A., Gordo, A., Lopez-Paz, D., Dokania, P.K., Torr, P.: Using hindsight to anchor past knowledge in continual learning (2020)
36. Chaudhry, A., Ranzato, M., Rohrbach, M., Elhoseiny, M.: Efficient lifelong learning with a-gem. In: ICLR (2019)
37. Aljundi, R., Lin, M., Goujaud, B., Bengio, Y.: Gradient based sample selection for online continual learning. In: NeurIPS (2019)
38. Tulving, E.: Episodic memory: from mind to brain. Ann. Rev. Psychol. **53**(1), 1–25 (2002)
39. Norman, K.A., O'Reilly, R.C.: Modeling hippocampal and neocortical contributions to recognition memory: a complementary-learning-systems approach. Psychol. Rev. **110**(4), 611 (2003)
40. Ren, M., Iuzzolino, M.L., Mozer, M.C., Zemel, R.S.: Wandering within a world: online contextualized few-shot learning. arXiv preprint arXiv:2007.04546 (2020)
41. Ji, X., Henriques, J., Tuytelaars, T., Vedaldi, A.: Automatic recall machines: internal replay, continual learning and the brain. arXiv preprint arXiv:2006.12323 (2020)
42. Hsu, Y.C., Liu, Y.C., Kira, Z.: Re-evaluating continual learning scenarios: a categorization and case for strong baselines. In: NeurIPS-W (2018)
43. Riemer, M., et al.: Learning to learn without forgetting by maximizing transfer and minimizing interference. In: ICLR (2019)
44. Rolnick, D., Ahuja, A., Schwarz, J., Lillicrap, T.P., Wayne, G.: Experience replay for continual learning. In: NeurIPS (2019)
45. Loshchilov, I., Hutter, F.: Sgdr: stochastic gradient descent with warm restarts. In: ICLR (2017)
46. Yun, S., Han, D., Oh, S.J., Chun, S., Choe, J., Yoo, Y.: Cutmix: regularization strategy to train strong classifiers with localizable features. In: ICCV (2019)
47. Yin, H., et al.: Dreaming to distill: data-free knowledge transfer via deepinversion. In: CVPR (2020)
48. Zeno, C., Golan, I., Hoffer, E., Soudry, D.: Task agnostic continual learning using online variational bayes. arXiv preprint arXiv:1803.10123 (2018)

49. Hocquet, G., Bichler, O., Querlioz, D.: Ova-inn: continual learning with invertible neural networks. IJCNN (2020)
50. van de Ven, G.M., Tolias, A.S.: Generative replay with feedback connections as a general strategy for continual learning. arXiv preprint arXiv:1809.10635 (2018)
51. Serra, J., Suris, D., Miron, M., Karatzoglou, A.: Overcoming catastrophic forgetting with hard attention to the task. ICML (2018)
52. Rannen, A., Aljundi, R., Blaschko, M.B., Tuytelaars, T.: Encoder based lifelong learning. In: CVPR (2017)
53. Mallya, A., Lazebnik, S.: Packnet: adding multiple tasks to a single network by iterative pruning. In: CVPR (2018)
54. Huang, G., Liu, Z., Van Der Maaten, L., Weinberger, K.Q.: Densely connected convolutional networks. In: CVPR (2017)

Learning Lane Graph Representations
for Motion Forecasting

Ming Liang[1(✉)], Bin Yang[1,2], Rui Hu[1], Yun Chen[1], Renjie Liao[1,2], Song Feng[1], and Raquel Urtasun[1,2]

[1] Uber ATG, Pittsburgh, USA
{ming.liang,byang10,rui.hu,yun.chen,rjliao,songf,urtasun}@uber.com
[2] University of Toronto, Toronto, Canada

Abstract. We propose a motion forecasting model that exploits a novel structured map representation as well as actor-map interactions. Instead of encoding vectorized maps as raster images, we construct a lane graph from raw map data to explicitly preserve the map structure. To capture the complex topology and long range dependencies of the lane graph, we propose LaneGCN which extends graph convolutions with multiple adjacency matrices and along-lane dilation. To capture the complex interactions between actors and maps, we exploit a fusion network consisting of four types of interactions, actor-to-lane, lane-to-lane, lane-to-actor and actor-to-actor. Powered by LaneGCN and actor-map interactions, our model is able to predict accurate and realistic multi-modal trajectories. Our approach significantly outperforms the state-of-the-art on the large scale Argoverse motion forecasting benchmark.

Keywords: HD map · Motion forecasting · Autonomous driving

1 Introduction

Autonomous driving has the potential to revolutionize transportation. Self-driving vehicles (SDVs) have to accurately predict the future motions of other traffic participants in order to safely operate. High Definition maps (HD-maps) provide extremely useful geometric and semantic information for motion forecasting, as the behaviors of actors largely depend on the map topology. For example, a vehicle is unlikely to take a left turn when there is not a left turn lane nearby. Effectively exploiting HD maps is essential for motion forecasting models to produce plausible and accurate trajectories.

First attempts exploit HD maps as heuristics [42]. Actors are first associated with lanes and all candidate motion paths are then generated based on map topology. In this way, the prediction results are constrained by the map. However,

Electronic supplementary material The online version of this chapter (https://doi.org/10.1007/978-3-030-58536-5_32) contains supplementary material, which is available to authorized users.

© Springer Nature Switzerland AG 2020
A. Vedaldi et al. (Eds.): ECCV 2020, LNCS 12347, pp. 541–556, 2020.
https://doi.org/10.1007/978-3-030-58536-5_32

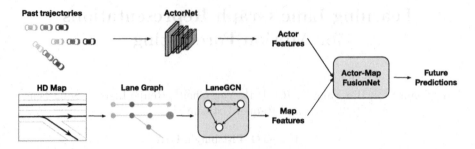

Fig. 1. Our approach: We construct a lane graph from raw map data and use LaneGCN to extract map features. In parallel, ActorNet extracts actor features from observed past trajectories. We then use FusionNet to model the Interactions between actors themselves and the map, and predict the future trajectories.

this approach can not capture rare and non-compliant behaviours, which while not very likely, might be safety critical.

Recent works [3,5–7,14,23,29,38] use machine learning to learn semantic representations from maps. To enable HD maps to be processed by neural networks the map data is rasterized to create image-like raster inputs. Map topology is implicitly encoded as lines, masks or colours, which are then processed by a 2D Convolutional Neural Network (CNN). These learned map features were shown to provide useful context information for motion forecasting. However, these approach has two disadvantages. First, the rasterization process inevitably results in information loss. Second, maps have a graph structure with complex topology which 2D convolution may be very inefficient to capture. For example, a lane of interest may extend for a long range in the lane direction. To capture this information, the receptive field has to be very large, covering not only the intended area, but also large areas outside the lane. Furthermore, lane pairs in the same or opposite directions have completely different semantic meanings and dependencies, although the lanes in both pairs are spatially close to each other.

In this paper we made three main contributions: (1) Instead of using rasterization, we construct a lane graph from vectorized map data, thus avoiding information loss. We then propose the Lane Graph Convolutional Network (LaneGCN), which effectively captures the complex topology and long range dependencies of the lane graph. (2) Based on LaneGCN, our motion forecasting model captures all possible actor-map interactions. In particular, we represent both actors and lanes as nodes in the graph and use a 1D CNN and LaneGCN to extract the features for the actor and lane nodes respectively, and then exploit spatial attention and another LaneGCN to model four types of interactions: *actor-to-lane*, *lane-to-lane*, *lane-to-actor* and *actor-to-actor*. We refer the reader to Fig. 1 for an illustration of our approach. (3) We conduct experiments on the large-scale Argoverse motion forecasting benchmark [9], and show significant improvements over the state-of-the-art.

2   Related Work

In this section, we review work on map representations, learning map representations for autonomy tasks, and graph convolutional networks.

Map Representations: HD maps capture both the lane geometry as well as their connectivity. [21] proposes to parameterize the lane boundaries as a set of polylines, and exploit a Recurrent Neural Network (RNN) to extract them from sensor data. [28] further extends the polyline representation to a more structured para3ization. Instead of modelling the geometry of each lane, [22] proposes to parameterize the unknown lane graph as a Directed Acyclic Graphical model (DAG), which is more robust and able to handle more complex topology like branching. In addition to modelling the geometry, [32,33] encode different lane types in a graphical model to better exploit their appearance features. [11] parameterizes the road layout using an undirected graph, showcasing outstanding performance in large-scale city scale road topology.

Learning Map Representations for Autonomy: Rasterization based map representations have been extensively used. [10,12,14] rasterize map elements (roads, crosswalks) as layers and encode the lane direction with different colors. [3,8] encode roadmap, traffic lights and speed limits in rasterized bird's eye view images. [23] encodes the history of static entities, dynamic entities and semantic map information in a top-down spatial grid. HDNet [38] exploits the road mask as input feature to improve object detection performance. Rasterized maps have been fused with LiDAR point clouds to perform joint perception and prediction [4,27,29] as well as end-to-end motion planning [35,40,41]. While raster map representations are popular, an alternative is to use vectorized map features. [9] uses the distance along the centerlines and offset from the centerlines as input to their nearest neighbours regression and LSTM [20] models. [1,34] use 1D CNN and LSTM to encode lane features. In contrast, our model constructs a lane graph from vectorized map data, and extracts multi-scale topology features using the proposed LaneGCN. In concurrent work VectorNet[16], two graph networks are used to extract actor/lane features and model global interactions, respectively. There are two major differences between VectorNet and LaneGCN. First, VectorNet uses vanilla graph networks with undirected full connections, while we build a sparsely connected lane graph following the map topology and propose task specific multi-type and dilated graph operators. Second, VectorNet uses polyline-level nodes for interaction, while our LaneGCN uses polyline segments as map nodes to capture higher resolution. Note that in our approach nodes in different polylines can interact with each other through dilated connections.

Graph Convolutional Networks: Graph Convolutional Networks (GCNs) [13,15,19,26,30,36] have been shown to be effective for graph representation learning. They generalize the 2D convolution on grids to arbitrary graphs via the so called graph convolution. Different from 2D convolution, which operates on neighbors in a local grid, graph convolution operates on the neighboring nodes defined by the graph structure, typically described in the form of an adjacency matrix. We draw inspiration from GCNs and propose LaneGCN, which is a

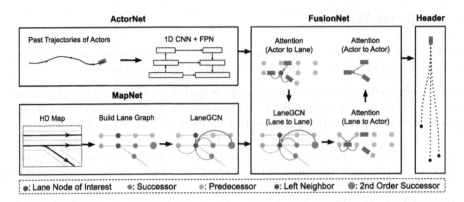

Fig. 2. Overall architecture: Our model is composed of four modules. (1) ActorNet receives the past actor trajectories as input, and uses 1D convolution to extract actor node features. (2) MapNet constructs a lane graph from HD maps, and uses a LaneGCN to exact lane node features. (3) FusionNet is a stack of 4 interaction blocks. The actor to lane block fuses real-time traffic information from actor nodes to lane nodes. The lane to lane block propagates information over the lane graph and updates lane features. The lane to actor block fuses updated map information from lane nodes to actor nodes. The actor to actor block performs interactions among actors. We use another LaneGCN for the lane to lane block, and spatial attention layers for the other blocks. (4) The prediction header uses after-fusion actor features to produce multi-modal trajectories.

specialized version designed for lane graphs. In our model, we introduce multiple adjacency matrices and multi-scale dilated convolutions, which are effective in capturing the complex topology and long-range dependencies of the lane graph.

3   Lane Graph Representations for Motion Forecasting

In this section, we propose a novel motion forecasting model that learns structured map representations and fuses the information of traffic actors and HD maps taking into account their interactions. In the following, we explain the four modules that compose our model, *i.e.,* how to compute actor features with **ActorNet**, how to represent the map via **MapNet**, how to fuse the information from both actors and the map with **FusionNet**, and finally how to predict the final motion forecasting trajectories through the **Prediction Header**. We refer the reader to Fig. 2 for an illustration of the overall architecture.

3.1   ActorNet: Extracting Traffic Participant Representations

We assume actor data is composed of the observed past trajectories of all actors in the scene. Each trajectory is represented as a sequence of displacements $\{\Delta\mathbf{p}_{-(T-1)}, \dots, \Delta\mathbf{p}_{-1}, \Delta\mathbf{p}_0\}$, where $\Delta\mathbf{p}_t$ is the 2D displacement from time step $t-1$ to t, and T is the trajectory size. All coordinates are defined in the Bird's Eye View (BEV), as this is the space of interest for traffic agents. For trajectories with sizes smaller than T, we pad them with zeros. We add a binary $1 \times T$ mask

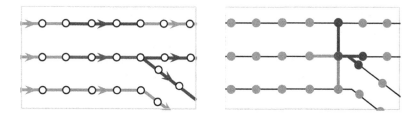

Fig. 3. Lane graph construction from vectorized map data. **Left**: The lane centerline of interest, its predecessor, successor, left and right neighbor are denoted with red, orange, blue, purple, and green lines, respectively. Each centerline is given as a sequence of BEV points (hollow circles). Right: Derived lane graph with an example lane node. The lane node of interest, its predecessor, successor, left and right neighbor are denoted with red, orange, blue, purple and green circles respectively. See Sect. 3.2 for more information. (Color figure online)

to indicate if the element at each step is padded or not and concatenate it with the trajectory tensor, resulting in an input tensor of size $3 \times T$.

While both CNNs and RNNs can be used for temporal data, here we use an 1D CNN to process the trajectory input for its effectiveness in extracting multi-scale features and efficiency in parallel computing. The output of ActorNet is a temporal feature map, whose element at $t = 0$ is used as the actor feature. The network has 3 groups/scales of 1D convolutions. Each group consists of 2 residual blocks [18], with the stride of the first block as 2. We then use a Feature Pyramid Network (FPN) [31] to fuse the multi-scale features, and apply another residual block to obtain the output tensor. For all layers, the convolution kernel size is 3 and the number of output channels is 128. Layer normalization [2] and the Rectified Linear Unit (ReLU) [17] are used after each convolution.

3.2   MapNet: Extracting Structured Map Representation

We use a novel deep model, called MapNet, to learn structured map representations from vectorized map data. This contrasts previous approaches, which encode the map as a raster image and apply 2D convolutions to extract features. MapNet consists of two steps: (1) building a lane graph from vectorized map data; (2) applying our novel LaneGCN to the lane graph to output the map features.

Map Data: In this paper, we adopt a simple form of vectorized map data as our representation of HD maps. Specifically, the map data is represented as a set of lanes and their connectivity. Each lane contains a centerline, *i.e.*, a sequence of 2D BEV points, which are arranged following the lane direction (see Fig. 3, top). For any two lanes which are directly reachable, 4 types of connections are given: *predecessor*, *successor*, *left neighbour* and *right neighbour*. Given a lane A, its predecessor and successor are the lanes which can directly travel to A and from A respectively. Left and right neighbours refer to the lanes which can be directly reached without violating traffic rules. This simple map format provides essential geometric and semantic information for motion forecasting, as vehicles generally plan their routes by reference to lane centerlines and their connectivity.

Lane Graph Construction: Instead of encoding maps as raster images, we derive a lane graph from the map data as the input. In designing the lane graph, we expect its nodes to have a fine resolution. Given any actor location, we query the lane graph and find its nearest nodes to retrieve accurate map information. From this point of view, it is not an optimal choice to directly use the lane centerlines as the nodes.

We refer the reader to Fig. 3 for an example of the lane graph construction. We first define a lane node as the straight line segment formed by any two consecutive points (grey circles in Fig. 3) of the centerline. The location of a lane node is the averaged coordinates of its two end points. Following the connections between lane centerlines, we also derive 4 connectivity types for the lane nodes, *i.e., predecessor, successor, left neighbour* and *right neighbour.* For any lane node A, its predecessor and successor are defined as the neighbouring lane nodes that can travel to A or from A respectively. Note that one can reach the first lane node of a lane l_A from the last lane node of lane l_B if l_B is the predecessor of l_A. Left and right neighbours are defined as the spatially closest lane node measured by ℓ_2 distance on the left and on the right neighbouring lane respectively. We denote the lane nodes with $V \in \mathbb{R}^{N \times 2}$, where N is the number of lane nodes and the ith row of V is the BEV coordinates of the ith node. We represent the connectivity with 4 adjacency matrices $\{A_i\}_{i \in \{\text{pre,suc,left,right}\}}$, with $A_i \in \mathbb{R}^{N \times N}$. We denote $A_{i,jk}$, as the element in the jth row and kth column of A_i. Then $A_{i,jk} = 1$ if node k is an i-type neighbor of node j.

LaneConv Operator: A natural operator to handle lane graphs is the graph convolution [36]. The most widely used graph convolution operator [26] is defined as $Y = LXW$, where $X \in \mathbb{R}^{N \times F}$ is the node feature, $W \in \mathbb{R}^{F \times O}$ is the weight matrix, and $Y \in \mathbb{R}^{N \times O}$ is the output. The graph Laplacian matrix $L \in \mathbb{R}^{N \times N}$ takes the form $L = D^{-1/2}(I + A)D^{-1/2}$, where I, A and D are the identity, adjacency and degree matrices respectively. I and A account for self connection and connections between different nodes. All connections share the same weight W, and the degree matrix D is used to normalize the output. However, this vanilla graph convolution is inefficient in our case due to the following reasons. First, it is not clear what kind of node feature will preserve the information in the lane graphs. Second, a single graph Laplacian can not capture the connection type, *i.e.,* losing the directional information carried by the connection type. Third, it is not straightforward to handle long range dependencies , *e.g.,* akin dilated convolution, within this form of graph convolution. Motivated by these challenges, we introduce our novel specially designed operator for lane graphs, called *LaneConv.*

Node Feature: We first define the input feature of the lane nodes. Each lane node corresponds to a straight line segment of a centerline. To encode all the lane node information, we need to take into account both the shape (size and orientation) and the location (the coordinates of the center) of the corresponding line segment. We parameterize the node feature as follows,

$$\mathbf{x}_i = \text{MLP}_{\text{shape}} \left(\mathbf{v}_i^{\text{end}} - \mathbf{v}_i^{\text{start}} \right) + \text{MLP}_{\text{loc}} \left(\mathbf{v}_i \right), \tag{1}$$

where MLP indicates a multi-layer perceptron and the two subscripts refer to shape and location, respectively. $\mathbf{v}_i$ is the location of the ith lane node, *i.e.*, the center between two end points, $\mathbf{v}_i^{\text{start}}$ and $\mathbf{v}_i^{\text{end}}$ are the BEV coordinates of the node i's starting and ending points, and $\mathbf{x}_i$ is the ith row of the node feature matrix X, denoting the input feature of the ith lane node.

LaneConv: The node feature above only captures the local information of a line segment. To aggregate the topology information of the lane graph at a larger scale, we design the following LaneConv operator

$$Y = XW_0 + \sum_{i \in \{\text{pre,suc,left,right}\}} A_i X W_i, \tag{2}$$

where A_i and W_i are the adjacency and the weight matrices corresponding to the ith connection type respectively. Since we order the lane nodes from the start to the end of the lane, A_{suc} and A_{pre} are matrices obtained by shifting the identity matrix one step towards upper right (non-zero superdiagonal) and lower left (non-zero subdiagonal). A_{suc} and A_{pre} can propagate information from the forward and backward neighbours whereas A_{left} and A_{right} allow information to flow from the cross-lane neighbours. It is not hard to see that our LaneConv builds on top of the general graph convolution and encodes more geometric (*e.g.*, connection type/direction) information. As shown in our experiments this improves over the vanilla graph convolution.

Dilated LaneConv: Since motion forecasting models usually predict the future trajectories of actors with a time horizon of several seconds, actors with high speed could have moved a long distance. Therefore, the model needs to capture the long range dependency along the lane direction for accurate prediction. In regular grid graphs, a dilated convolution operator [39] can effectively capture the long range dependency by enlarging the receptive field. Inspired by this operator, we propose the *dilated LaneConv* operator to achieve a similar goal for irregular graphs.

In particular, the k-dilation LaneConv operator is defined as follows,

$$Y = XW_0 + A_{\text{pre}}^k X W_{\text{pre},k} + A_{\text{suc}}^k X W_{\text{suc},k}, \tag{3}$$

where A_{pre}^k is the kth matrix power of A_{pre}. This allows us to directly propagate information along the lane for k steps, with k a hyperparameter. Since A_{pre}^k is highly sparse, one can efficiently compute it using sparse matrix multiplication. Note that the dilated LaneConv is only used for predecessor and successor, as the long range dependency is mostly along the lane direction.

LaneGCN: Based on the dilated LaneConv, we further propose a multi-scale LaneConv operator and use it to build our LaneGCN. Combining Eq. (2) and (3) with multiple dilations, we get a multi-scale LaneConv operator with C dilation sizes as follows

$$Y = XW_0 + \sum_{i \in \{\text{left,right}\}} A_i X W_i + \sum_{c=1}^{C} \left(A_{\text{pre}}^{k_c} X W_{\text{pre},k_c} + A_{\text{suc}}^{k_c} X W_{\text{suc},k_c} \right), \tag{4}$$

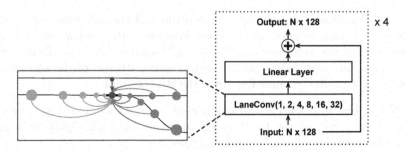

Fig. 4. LaneGCN architecture. Our LaneGCN is a stack of 4 multi-scale LaneConv residual blocks, each of which consists of a LaneConv(1,2,4,8,16,32) and a linear layer with a residual connection [18]. All layers have 128 feature channels.

where k_c is the cth dilation size. We denote LaneConv($k_1, \cdots, k_C$) this multi-scale layer. The architecture of LaneGCN is shown in Fig. 4. The network is composed of 4 LaneConv residual [18] blocks, which are the stack of a LaneConv(1, 2, 4, 8, 16, 32) and a linear layer, as well as a shortcut. All layers have 128 feature channels. Layer normalization [2] and ReLU [17] are used after each LaneConv and linear layer.

3.3 FusionNet

In this section we propose a network to fuse the information of the actor and lane nodes given by ActorNet and MapNet, respectively. The behaviour of an actor strongly depends on its context, *i.e.*, other actors and the map. Although the interactions between actors has been explored by previous work, the interactions between the actors and the map, and map conditioned interactions between actors have received much less attention. In our model, we use spatial attention and LaneGCN to capture a complete set of actor-map interactions (see Fig. 2).

We build a stack of four fusion modules to capture all information flows between actors and lane nodes, *i.e.*, actors to lanes (A2L), lanes to lanes (L2L), lanes to actors (L2A) and actors to actors (A2A). Intuitively, A2L introduces real-time traffic information to lane nodes, such as blockage or usage of the lanes. L2L updates lane node features by propagating the traffic information over the lane graph. L2A fuses updated map features with real-time traffic information back to the actors. A2A handles the interactions between actors and produces the output actor features, which are then used by the prediction header for motion forecasting.

We implement L2L using another LaneGCN, which has the same architecture as the one used in our MapNet (see Sect. 3.2). In the following we describe the other three modules in detail. We exploit a spatial attention layer [37] for A2L, L2A and A2A. The attention layer applies to each of the three modules in the same way. Taking A2L as an example, given an actor node i, we aggregate the features from its context lane nodes j as follows

$$\mathbf{y}_i = \mathbf{x}_i W_0 + \sum_j \phi(\text{concat}(\mathbf{x}_i, \Delta_{i,j}, \mathbf{x}_j)W_1)W_2, \tag{5}$$

with $\mathbf{x}_i$ the feature of the ith node, W a weight matrix, ϕ the composition of layer normalization and ReLU, and $\Delta_{ij} = \text{MLP}(\mathbf{v}_j - \mathbf{v}_i)$, where $\mathbf{v}$ denotes the node location. The context nodes are defined to be the lane nodes whose ℓ_2 distance from the actor node i is smaller than a threshold. The thresholds for A2L, L2A and A2A are set to 7 m, 6 m, and 100 m respectively. Each of A2L, L2A and A2A has two residual blocks, which consist of a stack of the proposed attention layer and a linear layer, as well as a residual connection. All layers have 128 output feature channels.

3.4   Prediction Header

Taking the after-fusion actor features as input, a multi-modal prediction header outputs the final motion forecasting. For each actor, it predicts K possible future trajectories and their confidence scores. The header has two branches, a regression branch to predict the trajectory of each mode and a classification branch to predict the confidence score of each mode. For the mth actor, we apply a residual block and a linear layer in the regression branch to regress the K sequences of BEV coordinates:

$$O_{m,\text{reg}} = \{(\mathbf{p}_{m,1}^k, \mathbf{p}_{m,2}^k, ..., \mathbf{p}_{m,T}^k)\}_{k \in [0, K-1]} \tag{6}$$

where $\mathbf{p}_{m,i}^k$ is the predicted mth actor's BEV coordinates of the kth mode at the ith time step. For the classification branch, we apply an MLP to $\mathbf{p}_{m,T}^k - \mathbf{p}_{m,0}$ to get K distance embeddings. We then concatenate each distance embedding with the actor feature, apply a residual block and a linear layer to output K confidence scores, $O_{m,\text{cls}} = (c_{m,0}, c_{m,1}, ..., c_{m,K-1})$.

3.5   Learning

As all the modules are differentiable, we can train the model in an end-to-end way. We use the sum of classification and regression losses to train the model

$$L = L_{\text{cls}} + \alpha L_{\text{reg}}, \tag{7}$$

where $\alpha = 1.0$. Given K predicted trajectories of an actor, we find a positive trajectory $\hat{k}$ that has the minimum final displacement error, $i.e.$, the Euclidean distance between the predicted and ground truth locations at the final time step. For classification, we use the max-margin loss:

$$L_{\text{cls}} = \frac{1}{M(K-1)} \sum_{m=1}^{M} \sum_{k \neq \hat{k}} \max(0, c_{m,k} + \epsilon - c_{m,\hat{k}}) \tag{8}$$

where ϵ is the margin and M is the total number of actors. For regression, we apply the smooth $\ell 1$ loss on all predicted time steps:

$$L_{\text{reg}} = \frac{1}{MT} \sum_{m=1}^{M} \sum_{t=1}^{T} \text{reg}(\mathbf{p}_{m,t}^{\hat{k}} - \mathbf{p}_{m,t}^*) \tag{9}$$

where $\mathbf{p}_t^*$ is the ground truth BEV coordinates at time step t, $\text{reg}(\mathbf{x}) = \sum_i d(x_i)$, x_i is the ith element of $\mathbf{x}$, and $d(x_i)$ is the smooth $\ell 1$ loss defined as

$$d(x_i) = \begin{cases} 0.5x_i^2 & \text{if } \|x_i\| < 1 \\ \|x_i\| - 0.5 & \text{otherwise,} \end{cases} \tag{10}$$

where $\|x_i\|$ denotes the ℓ_1 norm of x_i.

4   Experimental Evaluation

We evaluate our model on the large scale Argoverse [9] motion forecasting benchmark, which is publicly available and provides vectorized map data. We first compare our model with the state-of-the-art and show significant improvements in all metrics. We then conduct ablation studies on the architecture and LaneConv operators, and show the advantage of our model design choices. Finally, we show qualitative results and discuss future directions.

4.1   Experimental Settings

Dataset: Argoverse [9] is a motion forecasting benchmark with over 30K scenarios collected in Pittsburgh and Miami. Each scenario is a sequence of frames sampled at 10 HZ. Each sequence has an interesting object called "agent", and the task is to predict the future locations of agents in a 3 s future horizon. The sequences are split into training, validation and test sets, which have 205942, 39472 and 78143 sequences respectively. These splits have no geographical overlap. For the training and validation sets, each sequence lasts for 5 s. The first 2 s are used as input data and the other 3 s are used as ground truth for models to predict. For the test set, only the first 2 s are provided. Each frame is given as the centroid coordinates of all objects in the scene. The actor data is a trajectory of 20 time steps. The map data is a set of lane centerlines and their connectivity. We use both actor and map data in the way described in Sects. 3.1 and 3.2, without any other preprocessing step. We did not use the other map data such as the rasterized drivable area map and ground height map provided with the benchmark.

Metrics: We employ two extensively used motion forecasting metrics, *Average Displacement Error* (ADE) is defined as the ℓ_2 distance between the predicted and ground truth locations, averaged over all steps. *Final Displacement Error* (FDE) is defined as the ℓ_2 distance between the predicted and ground truth locations at the last step in the predicted horizon. As motion forecasting is by nature multi-modal, Argoverse uses the minimum ADE (minADE) and minimum FDE (minFDE) of the top K predictions as the metrics. When K=1, minADE

Table 1. Results on Argoverse motion forecasting benchmark (test set)

Model	K = 1			K = 6		
	minADE	minFDE	MR	minADE	**minFDE**	MR
Argoverse Baseline [9]	2.96	6.81	0.81	2.34	5.44	0.69
Argoverse Baseline (NN) [9]	3.45	7.88	0.87	1.71	3.29	0.54
Holmes (*7th*) [24]	2.91	6.54	0.82	1.38	2.66	0.42
cxx (*3rd*) [1]	1.91	4.31	0.66	0.99	1.71	0.19
uulm-mrm (*2nd*) [12,14]	1.90	4.19	0.63	0.94	1.55	0.22
Jean (*1st*) [1,34]	1.86	4.18	0.63	0.93	1.49	0.19
Our model	**1.71**	**3.78**	**0.59**	**0.87**	**1.36**	**0.16**

and minFDE are equal to the deterministic ADE and FDE. Argoverse benchmark allows up to 6 predictions, and the online server ranks the entries with minFDE with K = 6. We use minADE and minFDE for K = 1 and K = 6 as the main metrics. When comparing our model with top entries on the leaderboard, we also show *Miss Rate* (MR), which is the ratio of predictions (the best mode) whose final location is more than 2.0 m away from the ground truth.

Implementation Details: We use all actors and lanes whose distance from the agent is smaller than 100 m as the input. The coordinate system in our model is the BEV centered at the agent location at $t = 0$. We use the orientation from the agent location at $t = -1$ to the agent location at $t = 0$ as the positive x axis. We train the model on 4 TITAN-X GPUs using a batch size of 128 with the Adam [25] optimizer with an initial learning rate of 1×10^{-3}, which is decayed to 1×10^{-4} at 32 epochs. The training process finishes at 36 epochs and takes about 11.5 h. All our results are based on the same model, whose architecture and hyper-parameters are described in Sect. 3.

4.2   Results

Comparison with the State-of-the-Art: We compare our model with four top entries and two official baselines on the Argoverse motion forecasting leaderboard. We submit our result at the time of ECCV submission (2020/03/15). The metrics are minADE, minFDE and MR for K = 1 and K = 6, and the leaderboard is ranked by minFDE for K = 6. As shown in Table 1, our model significantly outperforms all other models in all metrics. Among the compared methods, uulm-mrm encodes the input data using a rasterization approach [12,14]. They represent actor states, lanes and the drivable area with a synthesized image, which is then processed by a 2D CNN. In this approach, map topology and actor-map interactions are both implicitly learned by 2D convolution. In contrast, our model explicitly learns structured map features and performs actor-map fusion. Jean and cxx encode actors and lanes with 1D CNN and/or LSTM, and use attention [37] to fuse the features. In their models, lanes are encoded independently so the global map topology is not captured. Moreover, there is no actor

Table 2. Ablation study results of modules

Backbone		FusionNet				K = 1		K = 6	
ActorNet	MapNet	L2A	A2L	L2L	A2A	minADE	minFDE	minADE	minFDE
✓						1.90	4.38	0.91	1.66
✓					✓	1.58	3.61	0.79	1.29
✓	✓	✓				1.55	3.52	0.76	1.23
✓	✓	✓	✓	✓		1.39	3.05	0.72	1.10
✓	✓	✓	✓	✓	✓	**1.35**	**2.97**	**0.71**	**1.08**

to lane and lane to lane fusion. In contrast, our model learns the lane features using the LaneConv, which captures the multi-scale topology of the lane graph.

Importance of Each Module: In Table 2, we show the results of using Actor-Net as the baseline and progressively adding more modules. Three observations can be drawn from the results. First, all modules improve the performance of the model, demonstrating the effectiveness of both LaneGCN and our overall architecture. Second, the information flow from actors to maps brings useful traffic information which benefits the motion forecasting performance, as the incorporation of A2L and L2L significantly outperforms L2A only. Third, A2L, L2L and L2A also facilitates the interaction between actors, which can be seen from the smaller gain of adding A2A to this combination (from 4th row to 5th row) compared to adding A2A to ActorNet alone (from 1st row to 2nd row). Intuitively, the information of different actors is propagated over the lane graph and leads to effective map conditioned interactions.

Lane Graph Operators: In Table 3, we show the results of the ablation study on lane graph operators. The baseline model uses the combination of A2L, L2L and L2A. We start from the vanilla graph convolution (GraphConv), and evaluate the effect of adding each component of the LaneConv block (see Fig. 4), including the residual block, multi-type connections and dilation. The last row is the LaneConv used in our model (fourth row of Table 2). All these components significantly improve the performance. The residual block only adds about 7% parameters, but effectively facilitates the training. Both multi-type connections and dilation significantly boost the performance, demonstrating the clear advantage of LaneConv over vanilla graph convolution.

Qualitative Results: In Fig. 5, we compare qualitatively our model to other methods on 4 hard cases. The results of other models are adapted from the slides of Argoverse motion forecasting competition [1]. As the examples are from the test set and we have no access to the labels, in our results we did not show the ground truth trajectory. The first row shows a case where the baselines miss the mode. While the other methods fail to capture the right turn prediction, our model produces a mode which nicely follows the right turn centerline. The second row shows a case where the agent is waiting to perform an unprotected left turn for the first 2 s. Due to the lack of actor motion history, maps are important

Table 3. Ablation study results of lane graph operators

Component				K=1		K=6	
GraphConv	Residual	Multi-Type	Dilate	minADE	minFDE	minADE	minFDE
✓				1.72	3.93	0.82	1.41
✓		✓		1.59	3.59	0.77	1.24
✓			✓	1.46	3.29	0.74	1.16
✓	✓			1.53	3.48	0.79	1.33
✓	✓	✓		1.48	3.33	0.74	1.19
✓	✓		✓	1.41	3.12	0.73	1.14
✓	✓	✓		**1.39**	**3.05**	**0.72**	**1.10**

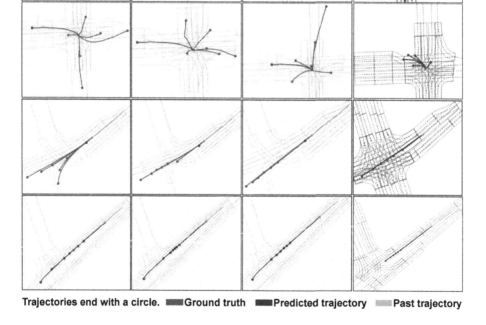

Trajectories end with a circle. ▬Ground truth ▬Predicted trajectory ▬Past trajectory

Fig. 5. Qualitative results on hard cases. From top to bottom, these hard cases involve missing the right turn mode, lacking history information, extreme deceleration and acceleration, respectively. See the text for more information.

for the model to produce reasonable trajectories. The other models produce divergent trajectories, some of which are non-traffic-rule compliant. In contrast, our model produces reasonable trajectories following the lane topology. The third row shows a case of a car decelerating and coming to a stop at the intersection. Our model produces a mode with more deceleration then the baselines and all the modes reasonably follow the lane. The fourth row shows a case of extreme acceleration. None of the models captures this case well, possibly because there is not enough information to make this prediction.

Overall, these results suggest that LaneGCN effectively learns structured map representations, which are used by the model to predict realistic trajectories. One potential way to improve our model is to incorporate more map information into the lane graph. Currently our model uses the centerlines and their connectivity. Other map information, such as traffic lights and traffic signs, provides useful information for motion forecasting, which is well illustrated by the second and third cases in Fig. 5. To account for new map data, our model can be easily extended by introducing new nodes and connections. We will explore this direction in future work.

5   Conclusion

In this paper, we propose a novel motion forecasting model to learn lane graph representations and perform a complete set of actor-map interactions. Instead of using a rasterized map as input, we construct a lane graph from vectorized map data and propose the LaneGCN to extract map topology features. We use spatial attention and the LaneGCN to fuse the information of both actors and lanes. We conduct experiments on the large scale Argoverse motion forecasting benchmark. Our model significantly outperforms the state-of-the-art. In the future we plan to explore the incorporation of other map data.

References

1. ArgoAI challenge. NeurIPS Workshop on Machine Learning for Autonomous Driving. https://slideslive.com/38923162/argoai-challenge (2019)
2. Ba, J.L., Kiros, J.R., Hinton, G.E.: Layer normalization. arXiv:1607.06450 (2016)
3. Bansal, M., Krizhevsky, A., Ogale, A.: Chauffeurnet: Learning to drive by imitating the best and synthesizing the worst. arXiv:1812.03079 (2018)
4. Casas, S., Gulino, C., Liao, R., Urtasun, R.: Spatially-aware graph neural networks for relational behavior forecasting from sensor data. In: ICRA (2020)
5. Casas, S., Gulino, C., Suo, S., Luo, K., Liao, R., Urtasun, R.: Implicit latent variable model for scene-consistent motion forecasting. In: Proceedings of the European Conference on Computer Vision (ECCV) (2020)
6. Casas, S., Gulino, C., Suo, S., Urtasun, R.: The importance of prior knowledge in precise multimodal prediction. In: IROS (2020)
7. Casas, S., Luo, W., Urtasun, R.: Intentnet: Learning to predict intention from raw sensor data. In: Conference on Robot Learning, pp. 947–956 (2018)

8. Chai, Y., Sapp, B., Bansal, M., Anguelov, D.: Multipath: Multiple probabilistic anchor trajectory hypotheses for behavior prediction. arXiv:abs/1910.05449 (2019)
9. Chang, M.F., et al.: Argoverse: 3D tracking and forecasting with rich maps. In: Proceedings of the IEEE Conference on Computer Vision and Pattern Recognition, pp. 8748–8757 (2019)
10. Chou, F.C., et al.: Predicting motion of vulnerable road users using high-definition maps and efficient convnets. arXiv:abs/1906.08469 (2019)
11. Chu, H., et al.: Neural turtle graphics for modeling city road layouts. In: ICCV (2019)
12. Cui, H., et al.: Multimodal trajectory predictions for autonomous driving using deep convolutional networks. In: 2019 International Conference on Robotics and Automation (ICRA), pp. 2090–2096 (2018)
13. Defferrard, M., Bresson, X., Vandergheynst, P.: Convolutional neural networks on graphs with fast localized spectral filtering. In: Advances in Neural Information Processing Systems, pp. 3844–3852 (2016)
14. Djuric, N., et al.: Motion prediction of traffic actors for autonomous driving using deep convolutional networks. arXiv:1808.05819 (2018)
15. Duvenaud, D.K., et al.: Convolutional networks on graphs for learning molecular fingerprints. In: Advances in Neural Information Processing Systems, pp. 2224–2232 (2015)
16. Gao, J., et al.: Vectornet: Encoding HD maps and agent dynamics from vectorized representation. In: Proceedings of the IEEE/CVF Conference on Computer Vision and Pattern Recognition, pp. 11525–11533 (2020)
17. Glorot, X., Bordes, A., Bengio, Y.: Deep sparse rectifier neural networks. In: Proceedings of the Fourteenth International Conference on Artificial Intelligence and Statistics, pp. 315–323 (2011)
18. He, K., Zhang, X., Ren, S., Sun, J.: Deep residual learning for image recognition. In: Proceedings of the IEEE Conference on Computer Vision and Pattern Recognition, pp. 770–778 (2016)
19. Henaff, M., Bruna, J., LeCun, Y.: Deep convolutional networks on graph-structured data. arXiv:1506.05163 (2015)
20. Hochreiter, S., Schmidhuber, J.: Long short-term memory. Neural Comput. 9(8), 1735–1780 (1997)
21. Homayounfar, N., Ma, W.C., Lakshmikanth, S.K., Urtasun, R.: Hierarchical recurrent attention networks for structured online maps. In: 2018 IEEE/CVF Conference on Computer Vision and Pattern Recognition, pp. 3417–3426 (2018)
22. Homayounfar, N., Ma, W.C., Liang, J., Wu, X., Fan, J., Urtasun, R.: Dagmapper: Learning to map by discovering lane topology. In: ICCV (2019)
23. Hong, J., Sapp, B., Philbin, J.: Rules of the road: Predicting driving behavior with a convolutional model of semantic interactions. In: Proceedings of the IEEE Conference on Computer Vision and Pattern Recognition, pp. 8454–8462 (2019)
24. Huang, X., et al.: Diversity-aware vehicle motion prediction via latent semantic sampling. arXiv:1911.12736 (2019)
25. Kingma, D.P., Ba, J.: Adam: A method for stochastic optimization. arXiv:1412.6980 (2014)
26. Kipf, T.N., Welling, M.: Semi-supervised classification with graph convolutional networks. arXiv:1609.02907 (2016)
27. Li, L., et al.: End-to-end contextual perception and prediction with interaction transformer. In: IROS (2020)
28. Liang, J., Homayounfar, N., Ma, W.C., Wang, S., Urtasun, R.: Convolutional recurrent network for road boundary extraction. In: CVPR (2019)

29. Liang, M., et al.: PnPNet: End-to-end perception and prediction with tracking in the loop. In: Proceedings of the IEEE Conference on Computer Vision and Pattern Recognition, pp. 11553–11562 (2020)
30. Liao, R., Zhao, Z., Urtasun, R., Zemel, R.S.: LanczosNet: Multi-scale deep graph convolutional networks. arXiv:1901.01484 (2019)
31. Lin, T.Y., Dollár, P., Girshick, R.B., He, K., Hariharan, B., Belongie, S.J.: Feature pyramid networks for object detection. In: 2017 IEEE Conference on Computer Vision and Pattern Recognition (CVPR), pp. 936–944 (2016)
32. Máttyus, G., Wang, S., Fidler, S., Urtasun, R.: Enhancing road maps by parsing aerial images around the world. In: 2015 IEEE International Conference on Computer Vision (ICCV), pp. 1689–1697 (2015)
33. Máttyus, G., Wang, S., Fidler, S., Urtasun, R.: HD maps: Fine-grained road segmentation by parsing ground and aerial images. In: 2016 IEEE Conference on Computer Vision and Pattern Recognition (CVPR), pp. 3611–3619 (2016)
34. Mercat, J., Gilles, T., Zoghby, N.E., Sandou, G., Beauvois, D., Gil, G.P.: Multi-head attention for multi-modal joint vehicle motion forecasting. arXiv:1910.03650 (2019)
35. Sadat, A., Casas, S., Ren, M., Wu, X., Dhawan, P., Urtasun, R.: Perceive, predict, and plan: Safe motion planning through interpretable semantic representations. In: Proceedings of the European Conference on Computer Vision (ECCV) (2020)
36. Shuman, D.I., Narang, S.K., Frossard, P., Ortega, A., Vandergheynst, P.: The emerging field of signal processing on graphs: Extending high-dimensional data analysis to networks and other irregular domains. IEEE Signal Process. Mag. **30**(3), 83–98 (2013)
37. Vaswani, A., et al.: Attention is all you need. In: Advances in Neural Information Processing Systems, pp. 5998–6008 (2017)
38. Yang, B., Liang, M., Urtasun, R.: HDNET: Exploiting HD maps for 3D object detection. In: Conference on Robot Learning, pp. 146–155 (2018)
39. Yu, F., Koltun, V.: Multi-scale context aggregation by dilated convolutions. arXiv:1511.07122 (2015)
40. Zeng, W., et al.: End-to-end interpretable neural motion planner. In: Proceedings of the IEEE Conference on Computer Vision and Pattern Recognition (2019)
41. Zeng, W., Wang, S., Liao, R., Chen, Y., Yang, B., Urtasun, R.: DSDNet: Deep structured self-driving network. In: ECCV (2020)
42. Ziegler, J., et al.: Making bertha drive—an autonomous journey on a historic route. IEEE Intell. Transp. Syst. Mag. **6**(2), 8–20 (2014)

What Matters in Unsupervised Optical Flow

Rico Jonschkowski[1,2(✉)], Austin Stone[1,2], Jonathan T. Barron[2],
Ariel Gordon[1,2], Kurt Konolige[1,2], and Anelia Angelova[1,2]

[1] Robotics at Google, Mountain View, USA
[2] Google AI, Mountain View, USA
{rjon,austinstone,barron,gariel,konolige,anelia}@google.com

Abstract. We systematically compare and analyze a set of key components in unsupervised optical flow to identify which photometric loss, occlusion handling, and smoothness regularization is most effective. Alongside this investigation we construct a number of novel improvements to unsupervised flow models, such as cost volume normalization, stopping the gradient at the occlusion mask, encouraging smoothness before upsampling the flow field, and continual self-supervision with image resizing. By combining the results of our investigation with our improved model components, we are able to present a new unsupervised flow technique that significantly outperforms the previous unsupervised state-of-the-art and performs on par with supervised FlowNet2 on the KITTI 2015 dataset, while also being significantly simpler than related approaches.

1 Introduction

Optical flow is a key representation in computer vision that describes the pixel-level correspondence between two images. Since optical flow is useful for estimating motion, disparity, and semantic correspondence, improvements in optical flow directly benefit downstream tasks such as visual odometry, stereo depth estimation, and object tracking. The performance of optical flow techniques has recently seen dramatic improvements, due to the widespread adoption of deep learning. Because ground-truth labels for dense optical flow are difficult to obtain for real image pairs, supervised optical flow techniques are primarily trained using synthetic data [5]. Although models trained on synthetic data often generalize well to real images, there is an inherent mismatch between these two data sources that those approaches may struggle to overcome [17,28].

Though non-synthetic data for training *supervised* optical flow techniques is scarce, the data required to train an *unsupervised* model is abundant: all

Electronic supplementary material The online version of this chapter (https://doi.org/10.1007/978-3-030-58536-5_33) contains supplementary material, which is available to authorized users.

A. Vedaldi et al. (Eds.): ECCV 2020, LNCS 12347, pp. 557–572, 2020.
https://doi.org/10.1007/978-3-030-58536-5_33

that training requires is unlabeled video, of which there are countless hours freely available on the internet. If an unsupervised approach could leverage this abundant and diverse real data, it would produce an optical flow model that does not suffer from any mismatch between its training data and its test data, and could presumably produce higher-quality results. The core assumption shared by unsupervised optical flow techniques is that an object's appearance does not change as it moves, which allows these models to be trained using unlabeled video as follows: The model is used to estimate a flow field between two images, that flow field is used to warp one image to match the other, and then the model weights are updated so as to minimize the difference between those two images – and to accommodate some form of regularization.

Although all unsupervised optical flow methods share this basic idea, their details vary greatly. In this work we systematically compare, improve, and integrate key components to further our understanding and provide a unified framework for unsupervised optical flow. Our contributions are:

1. We systematically compare key components of unsupervised optical flow, such as photometric losses, occlusion estimation techniques, self-supervision, and smoothness constraints, and we analyze the effect of other choices, such as pretraining, image resolution, data augmentation, and batch size.
2. We propose four improvements to these key components: cost volume normalization, gradient stopping for occlusion estimation, applying smoothness at the native flow resolution, and image resizing for self-supervision.
3. We integrate the best performing improved components in a unified framework for unsupervised optical flow (UFlow for short) that sets a new state of the art – even compared to substantially more complex methods that estimate flow from multiple frames or co-train flow with monocular or stereo depth estimation. To facilitate future research, our source code is available at https://github.com/google-research/google-research/tree/master/uflow.

2   Related Work

The motion between an object and a viewer causes apparent movement of brightness patterns in the image [7]. Optical flow techniques attempt to invert this relationship to recover a motion estimate [16]. Classical methods infer optical flow for a pair of images by minimizing a loss function that measures photometric consistency and smoothness [3,10,24]. Recent approaches reframe optical flow estimation as a learning problem in which a CNN-based model regresses from a pair of images to a flow field [5,11]. Some models incorporate ideas from earlier methods, such as cost volumes and coarse-to-fine warping [20,25,32]. These supervised approaches require representative training data with accurate optical flow labels. Though such data can be generated for rigid objects with known geometry [6,19], recovering this ground truth flow for arbitrary scenes is laborious, and requires approaches as unusual as manually painting scenes with textured fluorescent paint and imaging it under ultraviolet light [1]. Since such approaches scale poorly, supervised methods have mainly relied on synthetic data for training, and often for evaluation [2,4,5]. Synthesizing "good" training data (such that learned models generalize to real images) is itself a hard research problem, requiring careful consideration of scene content, camera motion, lens distortion, and sensor degradation [17].

Unsupervised approaches circumvent the need for labels by optimizing photometric consistency with some regularization [22,34], similar to the classical optimization-based methods mentioned above. Where traditional methods solve an optimization problem for each image pair, unsupervised learning jointly optimizes an objective across all pairs in a dataset and learns a function that regresses a flow field from images. This approach has two advantages: 1) inference is fast because optimization is only performed during training, and 2) by jointly optimizing across the whole train set, information is shared across image pairs which can potentially improve performance. This unsupervised approach was extended to use edge-aware smoothness [30], a bi-directional Census loss [18], different forms of occlusion estimation [12,18,30], self-supervision [14,15], and estimation from multiple frames [12,15]. Other extensions introduced geometric reasoning through epipolar constraints [35] or by co-training optical flow with depth and ego-motion models from monocular [21,33,36] or stereo input [29].

These works have pushed the state of the art and generated a range of ideas for unsupervised optical flow. But since each of them evaluates a different *combination* of ideas, it is unclear how *individual* ideas compare to each other and which ideas combine well together. For example, the methods OAFlow [30] and DDFlow [14] use different photometric losses and different ways to mask occlusions, and OAFlow uses an edge-aware smoothness loss while DDFlow regularizes learning through self-supervision. DDFlow performs better than OAFlow, but does this mean that every component of DDFlow is better than every component of OAFlow? The ablation studies often presented in these papers show that each novel contribution of each work does indeed improve the performance of each individual model, but they do not provide a guarantee that each such contribution will always improve performance when added to any *other* model. Our work addresses this problem by systematically comparing and combining photometric losses (L1, Charbonnier [24], Census [14,18,35,36], and structural similarity [21,29,33]), different methods for occlusion estimation [3,30], first order and second order edge-aware smoothness [27], and self-supervision [14]. Our work also improves cost volume computation, occlusion estimation, smoothness, and self-supervision and integrates all components into an state of the art framework for unsupervised optical, while being simpler than many proposed methods to form a solid base for future work.

3 Preliminaries on Unsupervised Optical Flow

The task of estimating optical flow can be defined as follows: Given two color images $I^{(1)}, I^{(2)} \in \mathbb{R}^{H \times W \times 3}$, we want to estimate the flow field $V^{(1)} \in \mathbb{R}^{H \times W \times 2}$, which for each pixel in $I^{(1)}$ denotes the relative position of its corresponding pixel in $I^{(2)}$. Note that optical flow is an asymmetric representation of pixel motion: $V^{(1)}$ provides a flow vector for each pixel in $I^{(1)}$, but to find a mapping from image 2 back to image 1, one would need to estimate $V^{(2)}$.

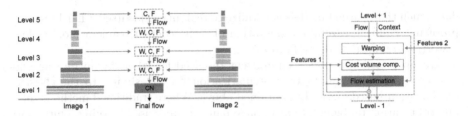

Fig. 1. Model overview. *Left:* Feature pyramids feed into a top-down flow estimation. *Right:* A zoomed in view on a "W, C, F" (warping, cost volume, flow estimation) block

In the context of unsupervised learning, we want to find a function $V^{(1)} = f_\theta(I^{(1)}, I^{(2)})$ with parameters θ learned from a set of image sequences $D = \{(I^{(1)}, I^{(2)}, \ldots, I^{(N)})\}$. Because we lack ground truth flow, we must define a proxy objective $\mathcal{L}(D, \theta)$, such as photometric consistency between $I^{(1)}$ and $I^{(2)}$ after it has been warped according to some estimated $V^{(1)}$. To enforce photometric consistency only for pixels that can be reconstructed from the other image, we must also estimate an occlusion mask $O^{(1)} \in \mathbb{R}^{H \times W}$, for example based on the estimated forward and backward flow fields $O^{(1)} = g(V^{(1)}, V^{(2)})$. $\mathcal{L}(\cdot)$ might also include other terms for, e.g. for smoothness or self-supervision. If $\mathcal{L}(\cdot)$ is differentiable with respect to θ, the parameters that minimize this loss $\theta^* = \arg\min(\mathcal{L}(D, \theta))$ can be recovered using gradient-based optimization.

4   Key Components of Unsupervised Optical Flow

This section compares and improves key components of unsupervised optical flow. We will first discuss a model $f_\theta(\cdot)$, which we base on PWC-Net [25], and improve through *cost-volume normalization*. Then we go through different components of the objective function $\mathcal{L}(\cdot)$: occlusion-aware photometric consistency, smoothness, and self-supervision. Here, we propose improvements to each component: *stopping the gradient at the occlusion mask, computing smoothness at the native flow resolution,* and *image resizing for self-supervision*. We end this section by discussing data augmentation and optimization.

As shown in Fig. 1, our model feeds images $I^{(1)}$ and $I^{(2)}$ into a shared CNN that generates a feature pyramid, where features are used as input for warping (W), cost volume computation (C), and flow estimation (F). At each level ℓ, the estimated flow $V^{(1,\ell+1)}$ from the level above is upscaled, passed down to the lower level as $\hat{V}^{(1,\ell+1)}$, and then used to warp $F^{(2,\ell)}$, the features of image 2. The warped features $w(F^{(2,\ell)}, \hat{V}^{(1,\ell)})$ together with $F^{(1,\ell)}$ are used to compute a cost volume. The cost volume considers feature correlations for all pixels and all 81 combinations of shifting $w(F^{(2,\ell)}, \hat{V}^{(1,\ell)})$ up to 4 pixels up/down and left/right. This results in a cost volume $C^\ell \in \mathbb{R}^{\frac{W}{2^\ell} \times \frac{H}{2^\ell} \times 81}$ that describes how closely each pixel in $F^{(1,\ell)}$ resembles the 81 pixels around its location in $F^{(2,\ell)}$. The cost volume, the features from image 1, the higher level flow and the *context* – the output of the second to last layer of the flow estimation network – are fed into a CNN that estimates a flow $V^{(1,\ell)}$. After a number of flow estimation levels, there is a final stage of flow refinement at level two in which the flow and context are fed into a *context network* (CN), which is a stack of dilated convolutions.

Model Shrinking, Level Dropout and Cost Volume Normalization: PWC-Net was designed for supervised learning of optical flow [25]. To deal with increased memory requirements for unsupervised learning due to bi-directional losses, occlusion estimation, and self-supervision, we remove level six, use 32 channels in all levels, and add residual connections to all flow estimation modules (the "+" in the bottom right of Fig. 1). Additionally, we dropout residual flow estimation at all levels to further regularize learning, i.e. we randomly pass the resized and rescaled flow estimate from the level above directly to the level below.

Another difference when using this model for unsupervised rather than supervised learning is that unsupervised losses are typically only imposed on the final output (presumably because photometric consistency and other objectives work better at higher resolutions). But without supervised losses on intermediate flow predictions, the model has difficulty learning flow estimation at higher levels. We found that this is caused by very low values in the estimated cost volumes as a result of vanishing feature activations at higher levels.

We address this problem by *cost volume normalization*. Let us denote features for image i at level ℓ as $F^{(i,\ell)} \in \mathbb{R}^{\frac{H}{2^\ell} \times \frac{W}{2^\ell} \times d}$. The cost volume between images 1 and 2 for all image locations (x, y) and all considered image shifts (u, v) is the inner product of the normalized features of the two images:

$$C^{(\ell)}_{x,y,u,v} = \sum_d \left(\frac{F^{(1,\ell)}_{x,y,d} - \mu^{(1,\ell)}}{\sigma^{(1,\ell)}} \right) \left(\frac{F^{(2,\ell)}_{x+u,y+v,d} - \mu^{(2,\ell)}}{\sigma^{(2,\ell)}} \right) . \tag{1}$$

Where $\mu^{(i,\ell)}$ and $\sigma^{(i,\ell)}$ are the sample mean and standard deviation of $F^{(i,\ell)}$ over its spatial and feature dimensions. We found that cost volume normalization improves convergence and final performance in unsupervised optical flow. These findings are consistent with prior work that used a similar form of normalization to improve geometric matching [23].

Unsupervised Learning Objectives: Defining a learning objective $\mathcal{L}(\cdot)$ that specifies the task of learning optical flow without having access to labels is the core problem of unsupervised optical flow. Similar to related work [14,15,18,30], we train our model by estimating optical flow and applying the respective losses in both directions. In this work we consider a learning objective that consists of three terms: occlusion-aware photometric consistency, edge-aware smoothness, and self-supervision, which we will now discuss in detail.

Photometric Consistency: The photometric consistency term encourages the estimated flow to align image patches with a similar appearance by penalizing photometric dissimilarity. The metric for measuring appearance similarity is critical for any unsupervised optical flow technique. Related approaches use three different objectives here (sometimes in combination), (i) the generalized Charbonnier loss [12,18,22,30,34,35], (ii) the structural similarity index (SSIM) loss [21,29,33], and (iii) the Census loss [18,35,36]. We compare all three losses in this paper. The generalized Charbonnier loss [24] is $\mathcal{L}_C = \frac{1}{n} \sum \left((I^{(1)} - w(I^{(2)})^2 + \epsilon^2 \right)^\alpha$. Our experiments use $\epsilon = 0.001$ and $\alpha = 0.5$ and also compare to using a modified L1 loss $\mathcal{L}_{L1} = \sum |I^{(1)} - w(I^{(2)}) + \epsilon'|$ with $\epsilon' = 10^{-6}$. For the SSIM [31] loss, we use an occlusion-aware implementation from recent work [9]. For the Census loss, we use a soft Hamming distance on Census-transformed image patches [18]. Based on the empirical results discussed below, we use the Census loss unless otherwise stated. All photometric losses are computed using an occlusion-masked average over all pixels [30].

Occlusion Estimation: By definition, occluded regions do not have a correspondence in the other image, so they should be discounted when computing the photometric loss. Related approaches estimate occlusions by (i) checking for consistent forward and backward flow [30], (ii) using the range map of the backward flow [3], and (iii) learning a model for occlusion estimation [12]. We are considering and comparing the first two variants here and improve the second variant through gradient stopping. In addition to taking into account occlusions, we also mask "invalid" pixels whose flow vectors point outside of the frame of the image [30]. The forward-backward consistency check defines occlusions a pixels for which the flow and the back-projected backward flow disagree by more than a threshold, such that the occlusion mask is defined as $O^{(1)} = \mathbb{1}_{|V^{(1)} - w(V^{(2)})|^2 < \alpha_1 (|V^{(1)}|^2 - |w(V^{(2)})|^2) + \alpha_2}$, where $\alpha_1 = 0.01$ and $\alpha_2 = 0.5$ [26]. An alternative approach computes a "range map" $R^{(i)} \in \mathbb{R}^{H \times W}$ – a soft histogram of how many pixels in the other image map onto a given pixel, which is constructed by having each flow vector distribute a total weight of 1 to the four pixels around its end point according to a bilinear kernel [30]. Pixels that none of the reverse flow vectors point to are assumed to have no correspondence in the other image, and are therefore occluded. As proposed by Wang *et al.* [30], we compute an occlusion mask $O^{(i)} \in \mathbb{R}^{W \times H}$ by thresholding the range map at 1. Based on the empirical results below, we use range-map based occlusion estimation by default, but use the forward-backward consistency check on KITTI, where it significantly improves performance.

Gradient Stopping at Occlusion Masks: Although prior work does not mention this issue [30], we found that propagating the gradient of the photometric loss into the occlusion estimation consistently degraded performance or caused divergence when the occlusion estimation was differentiable, as is the case for range-map based occlusion. This behavior is to be expected because when computing the occlusion-weighted average over photometric dissimilarity, there should be a gradient towards masking pixels with high photometric error. We address this problem by stopping the gradient at the occlusion mask, which eliminates divergence and improves performance.

Smoothness: Different forms of smoothness are commonly used to regularize optical flow in traditional methods [3,10,24] as well as most recent unsupervised approaches [12,18,21,22,29,30,33–36]. In this work, we consider edge-aware first and second order smoothness [27], where flows are encouraged to align their boundaries with visual edges in the image $I^{(1)}$. Formally, we define kth order smoothness as:

$$\mathcal{L}_{smooth(k)} = \frac{1}{n} \sum \exp\left(-\frac{\lambda}{3} \sum_c \left|\frac{\partial I_c^{(1,\ell)}}{\partial x}\right|\right) \left|\frac{\partial^k V^{(1,\ell)}}{\partial x^k}\right| + \exp\left(-\frac{\lambda}{3} \sum_c \left|\frac{\partial I_c^{(1,\ell)}}{\partial y}\right|\right) \left|\frac{\partial^k V^{(1,\ell)}}{\partial y^k}\right| . \quad (2)$$

Where λ modulates edge weighting based on $I_c^{(1,\ell)}$ for color channel $c \in [0,2]$. By default, we use first order smoothness on Flying Chairs and Sintel and second order smoothness on KITTI, which we ablate in different experiments.

Smoothness at Flow Resolution: A question that we have not seen addressed is at which level ℓ, smoothness should be applied. Since we follow the commonly used method of estimating optical flow at $\ell = 2$, i.e. at a quarter of the input resolution, followed by upsampling through bilinear interpolation, our model produces piece-wise linear flow fields. As a result, only every fourth pixel can possibly have a non-zero second order derivative, which might not be aligned with the corresponding image edge and thereby reduce the effectiveness of edge-aware smoothness. To address this, we apply smoothness at level $\ell = 2$ where flow is generated and downsample the image instead of upsampling the flow. This of course does not affect evaluation, which is done at the original image resolution.

Self-supervision: The idea of self-supervision in unsupervised optical flow is to generate optical flow labels by applying the learned model on a pair of images, then modify the images to make flow estimation more difficult and train the model to recover the originally estimated flow [14,15]. Since we see the main utility of this technique in learning flow estimation for pixels that go out of the image boundary – where cost-volume computation is not informative and photo-metric losses do not apply – we build on and improve ideas about self-supervised image crops [14]. For our self-supervised objective, we apply our model on the full images, crop the images by removing 64 pixels from each edge, apply the model again, and use the cropped estimated flow from the full images as supervision for flow estimation from the cropped images. We define the self-supervision objectives as an occlusion-weighted Charbonnier loss, that takes into account only pixels that have low forward-backward consistency in the "student" flow from cropped image and high forward-backward consistency in the "teacher" flow from the original images, similar to DDFlow [14].

Continual Self-supervision and Image Resizing: Unlike related work, we do not first train and then freeze a teacher model to supervise a separate student model but rather have a single model that supervises itself, which simplifies the approach, reduces the required memory, and allows the self-supervision signal to improve continually. To stabilize learning, we stop gradients of the self-supervision objectives to be propagated into the "teacher" flow. Additionally, we resize the image crops to match the original resolution before feeding them into the model (and we rescale the self-generated flow labels accordingly) to make the self-supervision examples more representative of the problem of extrapolating flow beyond the image boundary in the original size.

Optimization: To train our model $f_\theta(\cdot)$ we minimize a weighted sum of losses:

$$\mathcal{L}(D, \theta) = w_{photo} \cdot \mathcal{L}_{photo} + w_{smooth} \cdot \mathcal{L}_{smooth} + w_{self} \cdot \mathcal{L}_{self}, \qquad (3)$$

where $\mathcal{L}_{photo}$ is our photometric loss, $\mathcal{L}_{smooth}$ is smoothness regularization, and $\mathcal{L}_{self}$ is the self-supervision Charbonnier loss. We set w_{photo} to 1 for experiments using the Census loss and to 2 when we compare to the SSIM, Charbonnier, or L1 losses. We set w_{self} to 2 when using first order, and to 4 for second order smoothness and use an edge-weight of $\lambda = 150$. We use $w_{self} = 0$ during the first half of training, linearly increase it 0.3 during the next 10% of gradient steps and keep it constant afterwards.

RGB image values are scaled to $[-1, 1]$, and augmentated by randomly swapping the color channels and randomly shifting the hue. Sintel images are additionally randomly flipped up/down and left/right. All models are trained using with Adam [13] ($\beta_1 = 0.9$, $\beta_2 = 0.999$, $\epsilon = 10^{-8}$) with a learning rate of 10^{-4} for m steps, followed by another $\frac{1}{5}m$ steps during which the learning rate is exponentially decayed to 10^{-8}. All ablations use $m = 50K$ with batch size 32, but the final model was trained using $m = 1M$ with batch size 1, which produced slightly better performance as described below. Either way, the training takes about three days. Experiments on Sintel and KITTI start from a model that was first trained on Flying Chairs.

5   Experiments

We evaluate our model on the standard optical flow benchmark datasets: Flying Chairs [5], Sintel [4], and KITTI 2012/2015 [6,19]. We divide Flying Chairs and Sintel according to its standard train/test split. For KITTI, we train on the multi-view extension on the KITTI 2015 dataset, and we do not train on any data from KITTI 2012 because it does not have moving objects.

Related work is inconsistent in their use of train/test splits. For Sintel, it is common to train on the training set, report the benchmark performance on the test set, and evaluate ablations on the training set only (because test set labels are not public), which does not test generalization very well. Others "download the Sintel movie and extract ~10,000 images" [15] including the test set images, which is intended to demonstrate the ability of unsupervised methods to train on raw video data, but unfortunately also includes the benchmark test images in the training set. For KITTI, other works train on the raw KITTI dataset with and without excluding the evaluation set, or most commonly train on frames 1–8 and 13–20 of the multi-view extension of KITTI 2012/2015 datasets and evaluate on frames 10/11. But this split can mask overfitting to the trained sequences – either in the ablation results or also in the benchmark results, when the multiview-extensions of both the train and the test set are used. We therefore adopt the training regimen of Zhong *et al.* [35] and train two models for each dataset, one on the training set and one on test set (or for KITTI on their multiview extension) and evaluate these models appropriately.

Following the conventions of the KITTI benchmark, we report endpoint error ("EPE") and error rates ("ER"), where a prediction is considered erroneous if its EPE is > 3 pixels and if the distance between the predicted point and the true end point is > 5% of the length of the true flow vector. We compute these metrics for all pixels ("occ" in the KITTI benchmark, which we call "all" in this paper). We use the common practice of pretraining on the train split of the Flying Chairs dataset before training on Sintel/KITTI. We evaluate on all images in the native resolution, but have the model perform inference on a resolution that is divisible by 32, output at a four times smaller resolution, and then resize the output to the original resolution for evaluation. On KITTI, we observe that performance improves when using a square input resolution instead of a resolution in the original aspect ratio – perhaps because KITTI is dominated by horizontal motion. Accordingly, we use the following resolutions in our experiments: Flying Chairs: 384×512, Sintel: 448×1024, KITTI: 640×640.

6 Results

We evaluate our model in an extensive comparison and ablation study, from which we identify the best combination of components, tested in the "full" setting, which is often different from the components that work best individually in our "minimal" setting (more details below). We then compare our resulting model to the best published methods on unsupervised optical flow, and show that it outperforms all methods on all benchmarks.

Ablations and Comparisons of Key Components. To determine which aspects of unsupervised optical flow are most important, we perform an extensive series of ablation studies. We find that a) occlusion-masking, self-supervision, and smoothness are all important, b) level dropout and cost volume normalization improve performance, c) the Census loss outperforms other photometric losses, d) range-map based occlusion estimation requires gradient stopping to work, c) edge-aware smoothness and smoothness level matters significantly, d) self supervision helps especially for KITTI, and is improved by our changes, e) losses might be the current performance bottleneck, f) changing the resolution can substantially improve results, g) data augmentation and pretraining are helpful.

In each ablation study we train one model per domain (on Flying Chairs, KITTI-test, and Sintel-test), and evaluate those on the corresponding validation split from the same domain, taking into account occluded and non-occluded pixels "(all)". To estimate the noise in our results, we trained models with six different random seeds for each domain and computed their standard deviations per metric: Flying Chairs: 0.0162, Sintel Clean: 0.0248, Sintel Final: 0.0131, KITTI-2015: 0.0704, 0.0718%. We now describe the findings of each study.

Core Components: Table 1 shows how performance varies as each core component of our model (occlusion masking, smoothness, and self-supervision) is removed. We see that every component contributes to the overall performance. Since the utility of different components depends on what other components are used, all following experiments compare to the "minimal" (first row) and "full" (last row) versions of our method. Qualitative results for rows 9, 4, 3, and 1 are shown in Fig. 2 (from left to right). Note how the flow error ΔV increases with each removal of a core component.

Table 1. Core components: OM: occlusion masking, SM: smoothness, SS: self-supervision; "div.": divergence

OM	SM	SS	Chairs test	Sintel *train* Clean	Final	KITTI-15 *train* all	noc	ER%
–	–	–	3.58	4.20	6.80	13.07	2.47	21.21
–	–	✓	2.99	3.34	5.18	11.36	2.30	18.61
–	✓	–	2.84	3.37	5.19	11.37	2.17	19.31
–	✓	✓	2.74	3.12	4.56	3.28	2.08	9.97
✓	–	–	3.28	3.78	5.85	div.	div.	div.
✓	–	✓	2.91	3.26	4.72	3.02	2.11	9.89
✓	✓	–	2.63	3.20	4.63	4.15	2.05	13.15
✓	✓	✓	**2.55**	**3.00**	**4.18**	**2.94**	**1.98**	**9.65**

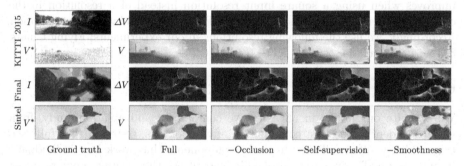

Ground truth   Full   −Occlusion   −Self-supervision   −Smoothness

Fig. 2. Qualitative ablation results of our model on random images not seen during training. Flow quality deteriorates as we progressively ablate core components

Model Improvements: Table 2 shows that level dropout (LD) and cost volume normalization (CVN) improve performance in the full setting (but not generally in the minimal setting). CVN appears to be more important for Chairs and Sintel while LD helps most for KITTI.

Table 2. Model improvements. CVN: cost volume normalization, LD: level dropout

	CVN	LD	Chairs test	Sintel *train* Clean	Final	KITTI-15 *train* all	noc	ER%
Minimal	–	–	5.01	4.52	6.67	13.30	2.72	21.69
	–	✓	5.29	4.40	**6.59**	12.75	2.49	21.30
	✓	–	4.86	**4.19**	6.69	13.294	2.59	21.54
	✓	✓	**3.58**	4.20	6.80	13.07	**2.47**	**21.21**
Full	–	–	3.78	3.41	4.70	39.09	30.19	98.77
	–	✓	3.21	3.45	4.61	2.96	**1.96**	9.77
	✓	–	**2.54**	3.07	4.31	3.16	2.04	10.35
	✓	✓	2.55	**3.00**	**4.18**	**2.94**	1.98	**9.65**

Photometric Losses: Table 3 compares commonly used photometric losses and shows that it is important to test every component with the full method, rather than looking at isolated performance. By itself, the commonly-used Charbonnier loss works better, but in the full setting, it underperforms the simpler L1 loss. For KITTI, Census works best in both settings. But for Sintel (in particular Sintel

Table 3. Photometric losses. Best results of L1 and Charbonnier underlined

	Method	Chairs test	Sintel *train* Clean	Final	KITTI-15 *train* all	noc	ER%
Minimal	L1	<u>4.27</u>	5.51	7.74	17.02	6.11	32.96
	Charbonnier	4.31	<u>5.50</u>	<u>7.64</u>	<u>16.94</u>	<u>6.09</u>	<u>32.84</u>
	SSIM	**3.51**	**4.01**	**5.41**	11.99	2.46	21.72
	Census	3.54	4.23	6.98	**11.66**	**2.37**	**21.15**
Full	L1	<u>2.83</u>	<u>4.23</u>	<u>5.75</u>	<u>5.53</u>	<u>3.17</u>	<u>18.65</u>
	Charbonnier	2.86	4.24	5.81	5.56	3.21	18.82
	SSIM	**2.54**	3.08	4.52	3.29	2.04	10.41
	Census	2.61	**3.00**	**4.20**	**3.08**	**2.01**	**10.01**

Final), the SSIM loss significantly outperforms Census in the minimal setting (5.41 vs. 6.98) but does not perform as well when used with all components in the full setting.

Occlusion Estimation: Table 4 compares different approaches to occlusion estimation (forward-backward consistency and range maps). We see that range-map based occlusion consistently diverges unless we stop the gradient of the photometric loss. But when gradients are stopped, this method works well, especially for Flying Chairs and Sintel Clean. Forward-backward consistency works best for KITTI, especially if not applied from the beginning.

Smoothness: Prior work suggests that photometric and smoothness losses taken together work better at higher resolutions [8]. But our analysis of the smoothness loss alone shows an advantage of applying this loss at the resolution of flow estimation, rather than at the image resolution, in particular for Flying Chairs and Sintel (Table 5). Our results also show that first order smoothness works better on Chairs and Sintel while second order smoothness works better on KITTI (Table 6). We see that context is important because in the minimal setting, the best second order smoothness weight for KITTI is 8, but in the full setting, it is 2. Comparing different edge-weights λ (Eq. 2) in Table 7, we see that nonzero edge-weights improve performance, particularly in the full setting. To our surprise, the simple strategy of only optimizing the Census loss and second order smoothness without edge-awareness, occlusion, or self-supervision (first row) produces performance on KITTI that improves on previous the state of the art.

Table 4. Occlusion estimation. RM: range-map based occllusion, FB: forward-backward consistency check

	Method	Chairs test	Sintel train Clean	Final	KITTI-15 train all	noc	ER%
Minimal	None	3.51	4.15	6.69	12.89	2.41	21.17
	RM (w/o grad stop)	div.	div.	div.	div.	div.	div.
	RM (w/ grad stop)	**3.27**	3.78	5.86	10.65	2.29	18.76
	FB (from step 1)	3.57	**3.71**	**4.83**	**8.99**	2.16	**17.71**
	FB (after 20% steps)	3.49	3.76	4.92	9.75	**2.13**	18.38
Full	None	2.73	3.84	5.13	3.28	2.10	10.07
	RM (w/o grad stop)	div.	div.	div.	div.	div.	div.
	RM (w/ grad stop)	**2.58**	**3.01**	4.25	3.10	2.04	9.86
	FB (from step 1)	3.28	3.49	4.45	2.96	1.99	9.65
	FB (after 20% steps)	3.14	3.12	**4.13**	**2.88**	**1.95**	**9.54**

Table 5. Level for smoothness loss

	Smoothn level	Chairs test	Sintel train Clean	Final	KITTI-15 train all	noc	ER%
Minimal	0	3.05	4.10	5.22	12.16	2.32	20.33
	1	2.94	3.65	**5.07**	11.94	2.24	19.98
	2	**2.85**	**3.33**	5.21	**11.43**	**2.23**	**19.38**
Full	0	2.87	3.65	4.63	2.95	2.02	9.87
	1	2.74	3.13	4.29	2.96	**1.99**	9.78
	2	**2.58**	**3.00**	**4.24**	**2.93**	**1.99**	**9.63**

Table 6. Comparison of weights for first/second order smoothness

	w_{smooth} 1st	2nd	Chairs test	Sintel train Clean	Final	KITTI-15 train all	occ	ER%
Minimal	0	0	4.55	4.16	6.84	div.	div.	div.
	0	2	3.13	3.77	6.32	11.37	2.17	19.33
	0	8	4.02	3.50	6.08	**7.27**	**2.11**	**14.70**
	4	0	**2.85**	**3.35**	**5.05**	7.23	2.30	18.58
	16	0	4.37	4.78	6.03	9.58	4.09	22.82
Full	0	0	2.92	3.27	4.77	2.92	2.07	9.75
	0	2	2.79	div.	div.	**2.93**	1.99	**9.61**
	0	8	2.75	3.33	4.77	2.94	**1.91**	9.85
	4	0	**2.60**	**3.00**	**4.17**	5.39	2.03	16.58
	16	0	3.68	4.22	5.30	8.71	4.01	21.52

Table 7. Smoothness edge-weights

	λ	Chairs test	Sintel train Clean	Final	KITTI-15 train all	noc	ER%
Minimal	0	4.93	6.00	6.65	**4.15**	2.36	12.50
	10	4.33	5.32	6.12	4.22	**2.17**	**12.28**
	150	**2.83**	**3.36**	**5.12**	11.41	2.21	19.37
Full	0	4.87	5.78	6.40	3.86	2.84	11.81
	10	3.75	4.62	5.34	3.14	2.11	10.27
	150	**2.56**	**3.02**	**4.20**	**2.87**	**1.95**	**9.59**

Table 8. Self-supervision ablation

	Self-supervision	Chairs test	Sintel train Clean	Final	KITTI-15 train all	noc	ER%
Minimal	None	3.48	4.10	6.62	13.05	2.48	21.23
	No resize	3.16	3.53	5.67	12.87	2.35	20.22
	Frozen teacher	3.10	3.36	5.24	**8.11**	2.38	**13.90**
	Default	**2.99**	**3.34**	**5.18**	11.36	2.30	18.61
Full	None	2.67	3.18	4.60	4.10	2.02	12.95
	No resize	**2.51**	3.14	4.48	3.53	2.02	11.13
	Frozen teacher	2.66	3.04	4.24	2.99	1.99	9.70
	Default	2.61	**2.99**	**4.23**	**2.86**	**1.95**	**9.57**

Table 9. Losses on Sintel for zero flow, ground truth flow, and predicted flow

	Flow	L1 noc	all	SSIM noc	all	Census noc	all	SM all	Census + SM noc	all
Clean	Zero	.146	.161	.927	.946	3.160	3.193	0	3.160	3.193
	GT	.031	.052	.191	.241	2.041	2.122	.032	2.073	2.154
	UFlow	.031	.042	.203	.247	2.06	2.130	.024	2.085	2.154
Final	Zero	.126	.142	.731	.751	3.037	3.075	0	3.037	3.075
	GT	.034	.055	.185	.233	2.086	2.154	.063	2.149	2.217
	UFlow	.032	.037	.167	.226	2.044	2.091	.045	2.089	2.136

Self-Supervision: In Table 8 we ablate the use of self-supervision and our proposed changes, and confirm that self-supervision on image crops is instrumental in achieving good results on KITTI, where errors are dominated by fast motion near the image edges. We also see that self-supervision is most effective when the image crop is resized as proposed by our method. Freezing the teacher network, as done in other works, seems to be important only when not using the other regularizing components. With these components in place, sharing the same model for both student and teacher appears to be beneficial.

Loss Comparison to Ground Truth: Photometric loss functions used in unsupervised optical flow rely on the brightness consistency assumption: that pixel intensities in the camera image are invariant to motion in the world. But photometric consistency is an imperfect indicator of flow quality (e.g. in regions of shadows and specularity). To analyze this issue, we compute photometric and smoothness losses not only for the flow field produced by our model, but also for a flow field filled with zeros and for the ground truth flow. Table 9 shows that our model is able to achieve comparable or better photometric consistency (and overall loss) than the ground truth flow. This trend is more pronounced on Sintel Final, which we believe violates the consistency assumption more than Sintel Clean. This result suggests that the loss functions currently used may be a limiting factor in unsupervised methods.

Resolution: Table 10 shows, perhaps surprisingly, that estimating flow at a different resolution and aspect ratio can substantially improve performance on KITTI-15 (2.93 vs. 3.80), presumably because the motion field in this dataset is dominated by horizontal motion. We have not observed this effect in other datasets.

Table 10. Resolution

Resolution	KITTI-15 *train*		
	all	noc	ER%
Min. 384 × 1280	13.25	2.79	21.38
640 × 640	12.91	2.42	21.17
Full 384 × 1280	3.80	2.13	10.88
640 × 640	**2.93**	**1.96**	**9.61**

Data Augmentation: Table 11 evaluates the importance of color augmentation (color channel swapping and hue randomization) for all domains, as well as image flipping for Sintel. The results show that both augmentation techniques improve performance, in particular image flipping for Sintel (which is a much smaller dataset than Chairs or KITTI).

Table 11. Data augmentation. F: image flipping up/down and left/right (not used for KITTI), C: color augmentation

	F	C	Chairs *test*	Sintel *train* Clean	Final	KITTI-15 *train* all	noc	ER%
Minimal	–	–	**3.47**	4.39	**6.56**	13.27	2.56	22.13
	–	✓	3.56	4.38	6.58	**13.07**	**2.47**	**21.21**
	✓	–	3.49	4.23	6.73	–	–	–
	✓	✓	3.58	**4.20**	6.80	–	–	–
Full	–	–	2.53	3.84	5.14	3.06	2.03	9.82
	–	✓	2.61	3.78	5.23	**2.94**	**1.98**	**9.65**
	✓	–	2.57	3.02	4.22	–	–	–
	✓	✓	**2.55**	**3.00**	**4.18**	–	–	–

Pretraining: Pretraining is a common strategy in supervised [5,25] and unsupervised [14,35] optical flow. The results in Table 12 confirm that pretraining on Chairs improves performance on Sintel and KITTI.

Table 12. Pretraining on Chairs

Pretraining on Chairs	Sintel *train* Clean	Final	KITTI-15 *train* all	noc	ER%
Min. –	4.41	7.53	**12.93**	**2.44**	21.24
✓	4.20	6.80	13.07	2.47	**21.21**
Full –	3.38	4.81	3.08	2.04	10.00
✓	**3.00**	**4.18**	**2.94**	**1.98**	**9.65**

Gradient Steps and Batch Size: All experiments up to this point have trained the model 60K steps at a batch size of 32. Table 13 shows a comparison to another training regime that trains longer with smaller batches, which consistently improves performance. We use this regime for our comparison to other published methods.

Table 13. Gradient steps (S) and batch size (B)

	S	B	Chairs test	Sintel *train* Clean	Final	KITTI-15 *train* all	noc	ER%
test	60K	32	{3.16}	3.04	4.23	2.92	**1.96**	9.71
	1.2M	1	{2.82}	**3.01**	**4.09**	**2.84**	**1.96**	**9.39**
train	60K	32	2.57	{2.47}	{3.92}	{2.74}	{1.87}	{9.04}
	1.2M	1	**2.55**	{2.50}	{3.39}	{2.71}	{1.88}	{9.05}

Comparison to State of the Art: We show qualitative results in Fig. 3 and quantitatively evaluate our model trained on KITTI and Sintel data in the corresponding benchmarks in Table 14, where we compare against state-of-the-art techniques for unsupervised and supervised optical flow. Results not reported by prior work are indicated with "–".

Table 14. Our model (yellow) compared to state of the art. Supervised models in gray fine-tune on their evaluation domain, which is often not possible in practice. Braces indicate models whose training set includes its evaluation set, and so are not comparable: "()" trained on the labeled evaluation set, "{}" trained on the unlabeled evaluation set, and "[]" trained on data related to the evaluation set (e.g. < 5 frames away in KITTI, or having the same content in Sintel). The best unsupervised and supervised (without finetuning) results are in bold. Methods that use additional modalities are denoted with MDM: mono depth/motion, SDM: stereo depth/motion, MF: multi-frame flow

Method	Sintel Clean [4] EPE train	test	Sintel Final [4] EPE train	test	KITTI 2012 [6] EPE train	test	KITTI 2015 [19] EPE train	EPE (noc) train	ER in % train	test
(A) FlowNet2-ft [11]	(1.45)	4.16	(2.01)	5.74	(1.28)	1.8	(2.30)	–	(8.61)	11.48
(B) PWC-Net-ft [25]	(1.70)	3.86	(2.21)	5.13	(1.45)	1.7	(2.16)	–	(9.80)	9.60
(C) SelFlow-ft [15]	(1.68)	[3.74]	(1.77)	{4.26}	(0.76)	1.5	(1.18)	–	–	8.42
(D) VCN-ft [32]	(1.66)	2.81	(2.24)	4.40	–	–	(1.16)	–	(4.10)	6.30
(E) FlowNet2 [11]	2.02	3.96	3.14	6.02	4.09	–	9.84	–	28.20	–
(F) PWC-Net [25]	2.55	–	3.93	–	4.14	–	10.35	–	33.67	–
(G) VCN [32]	2.21	–	3.62	–	–	–	**8.36**	–	**25.10**	–
(H) Back2Basics [34]	–	–	–	–	11.30	9.9	–	–	–	–
(I) DSTFlow [22]	{6.16}	10.41	{7.38}	11.28	[10.43]	12.4	[16.79]	[6.96]	[36.00]	[39.00]
(J) OAFlow [30]	{4.03}	7.95	{5.95}	9.15	[3.55]	[4.2]	[8.88]	–	–	[31.20]
(K) UnFlow [18]	–	–	7.91	10.21	3.29	–	8.10	–	23.27	–
(L) GeoNet [33] (MDM)	–	–	–	–	–	–	10.81	8.05	–	–
(M) DF-Net [36] (MDM)	–	–	–	–	3.54	4.4	{8.98}	–	{26.01}	{25.70}
(N) CCFlow [21] (MDM)	–	–	–	–	–	–	5.66	–	20.93	25.27
(O) MFOccFlow [12] (MF)	{3.89}	7.23	{5.52}	8.81	–	–	[6.59]	[3.22]	–	22.94
(P) UnOS [29] (SDM)	–	–	–	–	1.64	**1.8**	5.58	–	–	18.00
(Q) EPIFlow [35]	3.94	7.00	5.08	8.51	2.61	3.4	5.56	2.56	–	16.95
(R) DDFlow [14]	{2.92}	6.18	{3.98}	7.40	[2.35]	3.0	[5.72]	[2.73]	–	14.29
(S) SelFlow [15] (MF)	[2.88]	[6.56]	{3.87}	{6.57}	[1.69]	2.2	[4.84]	[2.40]	–	14.19
(T) UFlow-test	**3.01**	–	**4.09**	–	**1.58**	–	**2.84**	**1.96**	**9.39**	–
(U) UFlow-train	{2.50}	**5.21**	{3.39}	**6.50**	1.68	1.9	{2.71}	{1.88}	{9.05}	**11.13**

Among unsupervised approaches (H-U), our model sets a new state of the art for Sintel Clean (5.21 vs. 6.18), Sintel Final (6.50 vs. 7.40), and KITTI-15 (11.13% vs. 14.19%) – where, for a lack of comparability, we had to disregard

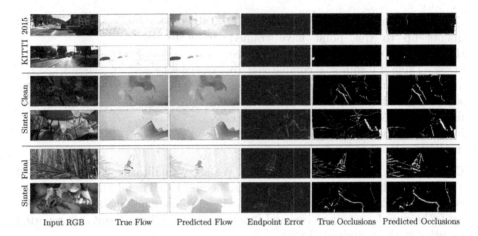

Input RGB   True Flow   Predicted Flow   Endpoint Error   True Occlusions   Predicted Occlusions

Fig. 3. Results for our model on random examples not seen during training taken from KITTI 2015 and Sintel Final. These qualitative results show the model's ability to estimate fast motions, relatively fine details, and substantial occlusions

results in braces that came from (partially) training on the test set. UFlow is only outperformed (1.8 vs. 1.9) on KITTI-12, which does not include moving objects, by a stereo-depth and motion based approach (P).

The top-performing supervised models finetuned on data from the evaluation domain (models A-D) do outperform our unsupervised model, as one may expect. But on KITTI-15, our model *performs on par with the supervised FlowNet2*. Of course, fine-tuning on the domain is only possible because the KITTI training data also contains ground-truth flow, which we ignore but which supervised techniques require. This sort of supervision is hard to obtain (KITTI being virtually the only non-synthetic dataset with this information), which demonstrates the value of unsupervised flow techniques such as ours. Without access to the ground truth labels of the test domain, our unsupervised method compares more favorably to its supervised counterparts, significantly outperforming them on KITTI. Our final experiment analyses cross-domain generalization in more detail.

Table 15 evaluates out-of-domain generalization by training and evaluating models across three datasets. While performance is best when training and test data are from the same domain, our model shows good generalization. It consistently outperforms DDFlow and it outperforms the supervised PWC-Net in all but one generalization task (training on Chairs and testing on Sintel Clean).

Table 15. Generalization across datasets. Performance when training on one dataset and testing on different one (gray if same)

	Method	Chairs test	Sintel *train* Clean	Sintel *train* Final	KITTI-15 *train* all	KITTI-15 *train* noc	KITTI-15 *train* ER%
Train on Chairs	PWC-Net [25]	2.00	**3.33**	4.59	13.20	–	41.79
	DDFlow [14]	2.97	4.83	4.85	17.26	–	–
	UFlow-test	{2.82}	4.36	5.12	15.68	7.96	32.69
	UFlow-train	2.55	3.43	**4.17**	**11.27**	**5.66**	30.31
Train on Sintel	PWC-Net [25]	3.69	(1.86)	(2.31)	10.52	–	30.49
	DDFlow [14]	3.46	{2.92}	{3.98}	12.69	–	–
	UFlow-test	3.39	3.01	4.09	**7.67**	**3.77**	**17.41**
	UFlow-train	**3.25**	{2.50}	{3.39}	9.40	4.53	20.02
Train on KITTI	DDFlow [14]	6.35	6.20	7.08	[5.72]	–	–
	UFlow-test	5.25	6.34	7.01	2.84	1.96	9.39
	UFlow-train	**5.05**	**5.58**	**6.31**	{2.71}	{1.88}	{9.05}

7   Conclusion

We have presented a study into what matters in unsupervised optical flow that systematically analyzes, compares, and improves a set of key components. This study results in a range of novel observations about these components and their interactions, from which we integrate the best components and improvements into a unified framework for unsupervised optical flow. Our resulting UFlow model substantially outperforms the state of the art among unsupervised methods and performs on par with the supervised FlowNet2 on the challenging KITTI 2015 benchmark, despite not using any labels. In addition to its strong performance, our method is also significantly simpler than many related approaches, which we hope will make it useful as a starting point for further research into unsupervised optical flow. Our code is available at https://github.com/google-research/google-research/tree/master/uflow.

References

1. Baker, S., Scharstein, D., Lewis, J.P., Roth, S., Black, M.J., Szeliski, R.: A database and evaluation methodology for optical flow. IJCV **92**, 1–31 (2011)
2. Barron, J.L., Fleet, D.J., Beauchemin, S.S.: Performance of optical flow techniques. IJCV **12**, 43–77 (1994)
3. Brox, T., Bruhn, A., Papenberg, N., Weickert, J.: High accuracy optical flow estimation based on a theory for warping. In: Pajdla, T., Matas, J. (eds.) ECCV 2004. LNCS, vol. 3024, pp. 25–36. Springer, Heidelberg (2004). https://doi.org/10.1007/978-3-540-24673-2_3
4. Butler, D.J., Wulff, J., Stanley, G.B., Black, M.J.: A naturalistic open source movie for optical flow evaluation. In: Fitzgibbon, A., Lazebnik, S., Perona, P., Sato, Y., Schmid, C. (eds.) ECCV 2012. LNCS, vol. 7577, pp. 611–625. Springer, Heidelberg (2012). https://doi.org/10.1007/978-3-642-33783-3_44
5. Dosovitskiy, A., et al.: FlowNet: learning optical flow with convolutional networks. In: ICCV (2015)
6. Geiger, A., Lenz, P., Urtasun, R.: Are we ready for autonomous driving? In: CVPR. The KITTI Vision Benchmark Suite (2012)
7. Gibson, J.J.: The Perception of the Visual World. Houghton Mifflin, Boston (1950)
8. Godard, C., Mac Aodha, O., Firman, M., Brostow, G.J.: Digging into self-supervised monocular depth estimation. In: ICCV (2019)
9. Gordon, A., Li, H., Jonschkowski, R., Angelova, A.: Depth from videos in the wild: unsupervised monocular depth learning from unknown cameras. In: ICCV (2019)
10. Horn, B.K.P., Schunck, B.G.: Determining optical flow. Artif. Intell. (1981)
11. Ilg, E., Mayer, N., Saikia, T., Keuper, M., Dosovitskiy, A., Brox, T.: Flownet 2.0: evolution of optical flow estimation with deep networks. In: CVPR (2017)
12. Janai, J., Güney, F., Ranjan, A., Black, M.J., Geiger, A.: Unsupervised learning of multi-frame optical flow with occlusions. In: ECCV (2018)
13. Kingma, D.P., Ba, J.: Adam: a method for stochastic optimization. In: ICLR (2015)
14. Liu, P., King, I., Lyu, M.R., Xu, J.: DDFlow: learning optical flow with unlabeled data distillation. In: AAAI (2019)
15. Liu, P., Lyu, M.R., King, I., Xu, J.: Selflow: self-supervised learning of optical flow. In: CVPR (2019)

16. Lucas, B.D., Kanade, T.: An iterative image registration technique with an application to stereo vision. In: DARPA Image Understanding Workshop (1981)
17. Mayer, N., et al.: What makes good synthetic training data for learning disparity and optical flow estimation? IJCV **126**, 942–960 (2018)
18. Meister, S., Hur, J., Roth, S.: Unflow: unsupervised learning of optical flow with a bidirectional census loss. In: AAAI (2018)
19. Menze, M., Heipke, C., Geiger, A.: Joint 3d estimation of vehicles and scene flow. In: ISPRS Workshop on Image Sequence Analysis (2015)
20. Ranjan, A., Black, M.J.: Optical flow estimation using a spatial pyramid network. In: CVPR (2017)
21. Ranjan, A., et al.: Competitive collaboration: joint unsupervised learning of depth, camera motion, optical flow and motion segmentation. In: CVPR (2019)
22. Ren, Z., Yan, J., Ni, B., Liu, B., Yang, X., Zha, H.: Unsupervised deep learning for optical flow estimation. AAAI (2017)
23. Rocco, I., Arandjelovic, R., Sivic, J.: Convolutional neural network architecture for geometric matching. In: CVPR (2017)
24. Sun, D., Roth, S., Black, M.J.: Secrets of optical flow estimation and their principles. In: CVPR (2010)
25. Sun, D., Yang, X., Liu, M.-Y., Kautz, J.: PWC-Net: CNNs for optical flow using pyramid, warping, and cost volume. In: CVPR (2018)
26. Sundaram, N., Brox, T., Keutzer, K.: Dense point trajectories by GPU-accelerated large displacement optical flow. In: Daniilidis, K., Maragos, P., Paragios, N. (eds.) ECCV 2010. LNCS, vol. 6311, pp. 438–451. Springer, Heidelberg (2010). https://doi.org/10.1007/978-3-642-15549-9_32
27. Tomasi, C., Manduchi, R.: Bilateral filtering for gray and color images. In: ICCV (1998)
28. Torralba, A., Efros, A.A.: Unbiased look at dataset bias. In: CVPR (2011)
29. Wang, Y., Wang, P., Yang, Z., Luo, C., Yang, Y., Xu, W.: UnOS: unified unsupervised optical-flow and stereo-depth estimation by watching videos. In: CVPR (2019)
30. Wang, Y., Yang, Y., Yang, Z., Zhao, L., Wang, P., Xu, W.: Occlusion aware unsupervised learning of optical flow. In: CVPR (2018)
31. Wang, Z., Bovik, A.C., Sheikh, H.R., Simoncelli, E.P.: Image quality assessment: from error visibility to structural similarity. IEEE Trans. Image Process. **13**, 600–612 (2004)
32. Yang, G., Ramanan, D.: Volumetric correspondence networks for optical flow. In: NeurIPS (2019)
33. Yin, Z., Shi, J.: GeoNet: unsupervised learning of dense depth, optical flow and camera pose. In: CVPR (2018)
34. Yu, J.J., Harley, A.W., Derpanis, K.G.: Back to basics: unsupervised learning of optical flow via brightness constancy and motion smoothness. In: Hua, G., Jégou, H. (eds.) ECCV 2016. LNCS, vol. 9915, pp. 3–10. Springer, Cham (2016). https://doi.org/10.1007/978-3-319-49409-8_1
35. Zhong, Y., Ji, P., Wang, J., Dai, Y., Li, H.: Unsupervised deep epipolar flow for stationary or dynamic scenes. In: CVPR (2019)
36. Zou, Y., Luo, Z., Huang, J.-B.: DF-Net: unsupervised joint learning of depth and flow using cross-task consistency. In: ECCV (2018)

Synthesis and Completion of Facades from Satellite Imagery

Xiaowei Zhang$^{(\boxtimes)}$, Christopher May, and Daniel Aliaga

Purdue University, West Lafayette, USA
{zhan2597,may5,aliaga}@purdue.edu

Abstract. Automatic satellite-based reconstruction enables large and widespread creation of urban areas. However, satellite imagery is often noisy and incomplete, and is not suitable for reconstructing detailed building facades. We present a machine learning-based inverse procedural modeling method to automatically create synthetic facades from satellite imagery. Our key observation is that building facades exhibit regular, grid-like structures. Hence, we can overcome the low-resolution, noisy, and partial building data obtained from satellite imagery by synthesizing the underlying facade layout. Our method infers regular facade details from satellite-based image-fragments of a building, and applies them to occluded or under-sampled parts of the building, resulting in plausible, crisp facades. Using urban areas from six cities, we compare our approach to several state-of-the-art image completion/in-filling methods and our approach consistently creates better facade images.

Keywords: Image synthesis and completion · Inverse procedural modeling · Satellite imagery

1 Introduction

Urban inverse procedural modeling is beneficial for many simulation, training, and entertainment applications. Using satellite data enables large scale, potentially global reconstructions. However, satellite data is challenging to work with due to limitations in resolution, noise, complex camera models, partial coverage, and occlusions. These aspects hinder high quality urban reconstruction.

Our key observation is that buildings in dense urban areas typically exhibit a regular, grid-like facade structure. We exploit this observation via a machine learning-based inverse procedural modeling approach to determine procedural parameters for a number of facade grammars in the presence of incomplete data. The grammars are then applied to the faces of reconstructed 3D building models during a facade completion phase. This methodology significantly improves the

Electronic supplementary material The online version of this chapter (https://doi.org/10.1007/978-3-030-58536-5_34) contains supplementary material, which is available to authorized users.

© Springer Nature Switzerland AG 2020
A. Vedaldi et al. (Eds.): ECCV 2020, LNCS 12347, pp. 573–588, 2020.
https://doi.org/10.1007/978-3-030-58536-5_34

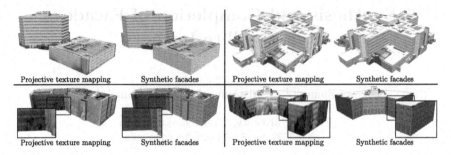

| Projective texture mapping | Synthetic facades | Projective texture mapping | Synthetic facades |
| Projective texture mapping | Synthetic facades | Projective texture mapping | Synthetic facades |

Fig. 1. *Examples of facade synthesis and completion.* Our method automatically creates procedural facades from satellite-based images despite noise, occlusions, and incomplete coverage.

resilience to occluded/noisy images and produces more accurate facade layouts as compared to alternative segmentation-based methods. Since satellite images have a very limited off-nadir view (e.g., at most 20 to 40°), and building surface coverage is limited (e.g., the orbital path of the satellite is not able to capture all building sides), often only fragments of a building are seen. Furthermore, facades that are observed may only be seen at very oblique angles, resulting in low resolution and stretched facade images. Nonetheless, a procedural approach has the ability to recreate the observed portion as well as create a plausible synthesized facade reconstruction of the occluded/not-sampled fragments. The result is plausible, complete building facades.

Our approach takes as input 3D building models obtained from point-clouds (e.g., [19]), as well as satellite image fragments projected onto the faces of the building models. The image fragments are used together with trained deep networks to find a representative sample of a facade with minimal noise, and infer its style and procedural parameters. The parameters are then used to complete the rest of the facade, and potentially other non-observed facades of a building. In the end, our approach produces complete facade layouts applied to building models. Figure 1 shows example results of our approach. Since we have a procedural output (instead of an image), we can zoom-in to any part of the facade and still have a crisp result, as observed in the close-up views.

Our results yield improvements over other methods applied to the same data. Over our six test areas, each spanning 1–$2\,\mathrm{km}^2$, our method is consistently better than the prior work we compare to quantitatively and qualitatively, and the average accuracy of several performance metrics is 85.4% despite significant occlusions, noise, and strong blurriness. Further, our deep networks are trained on a new dataset of rectified satellite facade views with ground truth segmentation that we also offer as a contribution. As far as we know, our work is the first pipeline to handle façade reconstruction based on satellite imagery despite the occlusions and resolution limitations of such imagery.

Our main contributions include: (1) A machine learning based pipeline addressing occlusion and regularity for satellite facade patterns. (2) A facade completion technique to generate plausible facade layouts based on the predicted grammars and building geometry. (3) A satellite facade dataset with ground truth window and door segmentation.

2   Related Work

Related work can be divided into building-envelope reconstruction, facade reconstruction, and forward/inverse procedural modeling. Musialski et al. [18] provides a review of urban reconstruction. Despite having the highest-resolution commercially available satellite imagery (i.e., WorldView3), the main structure of a building occupies on average 90×90 pixels on the ground plane and on average the best observation of a facade is 20 pixels tall. Aside from the relatively low resolution of satellite imagery, there are several other aspects that differentiate satellite-based multi-view stereo reconstruction from ground/aerial multi-view stereo reconstruction [21,22]. First, satellites use scan-line sensors producing images with a different projection model than standard frame cameras. Usually a rational polynomial coefficient (RPC) model is used. Such RPCs are hard to calibrate, require iterative processes, need many ground control points, and performing 3D to 2D as well as 2D to 3D mapping is difficult [34]. Second, the image quality can vary a lot due to a number of factors, including the viewing angles of satellite sensors are greatly limited by the orbit (i.e., not very off-nadir), images of an area might be days/weeks/months apart yielding different illumination and potentially physical changes, and radiometric quality is lower despite attempts of atmospheric corrections (see Fig. 2). While our work does not address the problem of 3D building reconstruction, building geometry is reconstructed automatically from a SOTA multi-view stereo point cloud obtained from satellite images, similar to and by extending [13,32]. It's important to note that the above limitations affect the quality of the reconstructed models, which are used by our facade synthesis method. Thus we cannot expect to have perfect building geometry with which to produce synthetic facade layouts.

Fig. 2. *Satellite image and facade closeups.* Example satellite image and views of some typical facades.

Almost all facade reconstruction methods use ground or aerial imagery, typically rectified and rectangular. Many approaches have been followed (e.g., using dynamic programming [3], using lattices [23], using matrix approximations [29], and inferring grammars from pre-labelled segments [7,12,15]). However, these methods do not perform well for our very under-sampled facades. For example, see our comparisons in the results section.

More recently, deep learning based facade parsing has obtained excellent results for ground-level imagery. For example, Liu et al. [14] and Fathalla et al. [6] perform facade segmentation but assume high-resolution frontal views. Nishida et al. [20] further assumes hand-specified building silhouettes and their facade stage depends on having clear boundaries between floors and between columns. Further, none of these account for the significant occlusions in satellite-based facades. Kelly et al. [10] could automatically and realistically decorate buildings by synthesizing geometric details/textures. However, their work requires style references (e.g., façade and roof textures, window layouts) and such references from satellite would be very low-resolution and heavily occluded. Kozinski et al. [11] (and partially Mathias et al. [16]) include provisions for occlusions but depend on many assumed structural priors for numerous object classes and SIFT feature vectors. On average the facades we encounter are only 20×90 pixels in size (often significantly worse) and thus make it prohibitive to determine such detailed structure. Image-to-image translation, such as Isola et al. [9] and Zhu et al. [35], has been proposed but does not support all of regularity, occlusions, and satellite data. From the semantic segmentation point of view, facade parsing could also be considered as a segmentation task. Many papers (e.g. DeepLabv3+ [2], EncNet [31], etc.) have shown great success with segmentation, but none of them use satellite facade data. Thus we trained those neural networks from scratch using our created satellite facade dataset (see Results section) and observe that these state-of-the-art segmentation neural networks also suffer from the low-quality of satellite facade data and cannot generate crisp facades.

Filling-in missing pixels of an image, often referred as image in-painting or completion, is an important task in computer vision. Deep learning and GAN-based approaches (e.g., DeepFill [30], PICNet [33]) have achieved promising results in this task. However, image in-painting is ill-suited for resolving shadows and occlusions in satellite facade images. First, detection of these areas is a very challenging problem, especially for satellite data. Second, even assuming these areas could be detected automatically, image in-painting approaches cannot infer correctly due to the low quality of satellite facade data. We also show in the Results section comparisons to these approaches.

Inverse procedural modeling (IPM) attempts to determine the procedure (e.g., rules and/or parameter values) yielding a desired geometric output. IPM has been used to stochastically derive a procedural model [24,26], infer Manhattan-world buildings from aerial imagery [28], or arbitrary buildings from polygonal data [1,4,5]. However, none of these methods have been used to infer building facade layouts from satellite data.

3   Facade Synthesis

While there might be 1–20 satellite images observing portions of buildings, there is usually not a high quality satellite observation of every facade on a building due to shadows, foliage/occlusions, and limited resolution. Thus simply applying satellite images to building faces via projective texture mapping is inadequate. Further, such texture mapping depends on very accurate image-to-image registration, geometric modeling, and complete coverage of all building facades. Our approach attempts to overcome these issues by synthesizing procedural facades using a selected subset of the available satellite imagery, and then applying these facades across the entire building. This approach has the following advantages:

- *Crisp Results.* The produced facade details will be crisp and visible at any resolution.
- *Exploits Best Observations.* Without relying on accurate RPCs and image registration, we choose the best, potentially fragmented, observations of each building and use it to obtain facade details.
- *Completes Missing Fragments.* Even if a facade/fragment is missing, we can fill-in the facade with details from a partial observation (or in worst case with details from neighboring facades).

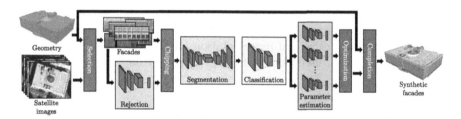

Fig. 3. *Pipeline.* The pipeline of our multi-stage approach for facade completion and synthesis.

We provide an overview of the proposed procedural facade approach in Fig. 3 and in the following we describe the pipeline starting with our selection method, followed by our deep-learning based facade style classification and parameter estimation, and finally our facade and building completion.

3.1   Selection

In a first stage, we choose the satellite image that has low grazing angle and does not have much dark pixels as the best view of the facade, and the resulting image is used as input to the rest of the pipeline. In many cases, even the best observation of a facade is not useful due to noise, shadows, trees, and occlusions.

Thus we employ a deep-learning based rejection model to prevent further processing of any such facades. Rejected facades will not undergo classification or parameter estimation, but can still receive synthetic facade layouts as part of the completion phase (Sect. 3.3).

Our rejection network is based on a pre-trained ResNet [8] model, in which we modify the last fully connected layer to two classes: one for "good" facades to be accepted and the other for "bad" facades to be rejected. We used 120 examples of "good" facades from our facade data set and 120

Fig. 4. *Accept or reject.* The first row shows facades that our rejection model will accept. The second row shows facades that will be rejected.

examples of "bad" facades, resulting in 1920 training images in total after applying data augmentation such as flip, rotation, random crop and intensity variations. The model performs with 92% accuracy when tested on 200 test images. Figure 4 shows some examples of accepted and rejected facades.

3.2 Classification and Parameter Estimation

In a second stage, our approach estimates the style and parameters of an equivalent procedural facade representation. Our method extracts a "chip" from the selected facade image because i) satellite-based images often suffer from occlusions and thus assuming a full facade view would be prohibitive, and ii) otherwise the parameter space would be unnecessarily large as the number of floors/windows may vary significantly yet the spacing between floors and windows is regular. The procedural representation for the entire facade is obtained from the chip and then used during the next stage to complete each facade.

Chip Extraction. To choose the best chip to extract, we divide the original facade image into a set of N tiles each of size 6×6 m. Each chip is formed by selecting a tile as the center and then varying the chip size to 6, 12, or 18 m and varying the aspect ratio (e.g., 1:1, 1:2, or 2:1). In total, $9N$ different candidate chips are produced for each facade. Please see Fig. 5 for a visual depiction. We evaluate each chip by

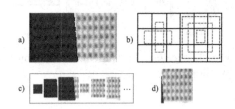

Fig. 5. *Chip extraction.* a) Original facade. b) Division of a) into tiles and demonstration of how chips are formed. c) Apply b) to a). d) The best chip.

passing it through our rejection network and evaluating its rejection score. The chip with the lowest rejection score is considered to be the cleanest chip found for the facade, and is selected to represent this facade further in the pipeline.

Segmentation. During segmentation, we only label each pixel as belonging to window/door or non-window/non-door since other facade classes are usually not visible in satellite imagery. During development, we experimented with several state-of-the-art deep-network based semantic segmentation models (e.g., DeepLabv3+ [2], EncNet [31], and Pix2Pix [9]). Please see Segmentation models in the Results section for quantitative and qualitative comparisons among these architectures. We found that the architecture of Pix2Pix [9] performs among the best ones, and in particular we specify the generator architecture to consist of ResNet blocks, the discriminator architecture to be 34×34 PatchGAN, and the input image size to be 96×96. We train the segmentation network from scratch using our own manually created satellite facade dataset. Specifically, we train with 120 facade images (960 after applying the aforementioned data augmentation) along with ground truth from our dataset.

After segmentation, we have binary segmented chip facades with two labels: one representing windows and doors (black), and one representing the building wall (white). Using a binary representation eases the burden for deep-network based recognition and parameter estimation. In addition, we apply some image processing techniques to further refine the segmented image. First we perform a small amount of dilation (e.g. rectangular dilation with a kernel size of 3 pixels) to reduce some of the noisy black window/door pixels. Next, since some facades are not perfectly rectified (due to errors in image registration and/or geometry), we perform a global image rotation computed automatically to force rows of windows/doors to be horizontal. Further, each window/door is replaced by a filled-in version of its rectangular bounding box. The end result is a binary image with rectangular windows and doors representing the facade, and serves as the input to our recognition and estimation networks.

Grammar Classification and Estimation. We represent a synthetic facade by one of six possible grammars each with a number of parameters, defined in a systematic fashion. While a single grammar with many parameters might be able to express more facades we found its generality to result in overall lower quality given the low-resolution nature of our facade imagery. For our grammar classification, a facade may contain doors and windows, or only windows. Further, the windows can be arranged as a grid of disjoint windows, as columns of vertically abutting windows, or as rows of horizontally abutting windows (see Fig. 2 and Fig. 6). Since window shapes are hard to differentiate with satellite data, we treat all windows as rectangles.

Which grammar a facade belongs to, along with the parameters for said grammar, is determined with a set of deep networks based on ResNet [8]. There is a classification network, which determines the grammar, followed by six parameter estimation networks, for determining the parameters specific to each grammar. The classification network is a ResNet [8] with modification of the last fully-connected layer to the number of grammars. The final output layer of this network yields confidence values for each of the aforementioned grammars. After classifying a facade via this network, the segmented facade chip is then sent through the parameter estimation network that corresponds to the highest confidence value in the classification output.

To robustly find the procedural parameters for the classified grammar, we use a separate deep network for each individual grammar, all of which are also based on ResNet [8]. They differ only in the last fully-connected layer, where we modify the number of parameters to match that of the grammar. We also use mean squared error as the loss function for our estimation networks. The predicted parameters (e.g., window rows, columns, relative size, etc.) altogether yield a synthetic facade that is similar to the input image.

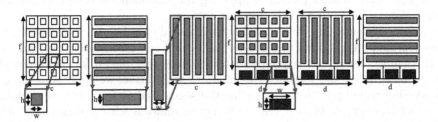

Fig. 6. *Grammars.* Our grammars of (1–3) three styles of only windows and (4–6) three styles with doors at the base. "f" stands for the number of floors. "c" is the number of column boundaries. "d" is number of doors. "h" is the relative height and "w" is the relative width. Please see the close-ups for additional parameters in the different grammars.

To train the estimation networks by systematically iterating over possible facade parameter configurations, we synthesized 200,000, 20,000, 20,000, 400,000, 50,000, and 50,000 facades from grammars 1 to 6 in Fig. 6, respectively, based on the different number of parameters for each. We also perform data augmentation accounting for noise and errors in the segmentation (i.e., up to 10% noise such as perturbation of boundaries in windows/doors) and randomly remove up to 10% of windows/doors. To train the classification network, we collected 108,000 images in total from the aforementioned training images, distributed evenly among all six grammars.

Optimization. After recognition and parameter estimation, we perform a coarse-to-fine refinement for each chip. Segmentation suffers from noise, shadows, trees, and occlusions. Fortunately, our parameter estimation network is able to recover a procedural facade that fills-in occluded content though there might be an overall translation or scale error. Thus, we define an objective function, using F-score [25], as:

$$F = \frac{2 \cdot precision \cdot recall}{precision + recall} \qquad P^* = \underset{P}{\mathrm{argmax}} F, \qquad (1)$$

In the above, P stands for the grammar parameters in Fig. 6, P^* is the optimal parameter set, *Accuracy* is the percentage of pixels labelled accurately,

Precision and *Recall* are computed by considering the label windows/doors as positive and the label wall as negative. *Precision* is the number of true positives divided by the sum of true and false positives (e.g., how correct is the windows and doors labelling in our results). *Recall* is the number of true positives divided by the sum of true positives and false negatives where for false negatives we use the number of incorrectly labeled wall pixels (e.g., how many windows and doors pixels our result can correctly label). Overall, F is essentially the harmonic mean of *Precision* and *Recall*.

Our optimizer tries to maximize this function using Monte Carlo stochastic optimization (e.g. altering P such as the number of floors, windows and window size) so as to create a synthetic facade that improves the F-score with respect to the segmentation result. Please see Optimization in Results section for details and comparisons.

3.3 Completion

In a third and final stage, our method applies the estimated procedural parameters to all facades and generates windows and doors with the estimated sizes and spacing. Although the prior step determined parameters for rectangular chips, the actual facades on the buildings are not limited to rectangles but instead may have irregular shapes. To this end, we logically divide a building facade into a set of horizontally-adjacent rectangular sections. Since doors only appear at the bottom of a facade, we partition each rectangular section, that touches ground level, into two subsections: a door subsection extending from the bottom of the facade up to the door height, and a window subsection covering the remainder. Doors are placed horizontally-centered in the door subsections and sized according to the estimated parameters. The window subsections are then further subdivided into window cells, also sized and spaced according to the estimated parameters, with one window placed into each cell. The tallest window subsections determine vertical window placement such that building floors are level across all sections.

Since each chip's parameters are estimated independently, neighboring facades will in general have different door/window sizes and spacing, and potentially different grammars. To remedy this issue, we first group facades together based on similar heights. All facades within each group are then forced to use the grammar of the highest scoring facade in the group, scored according to the grammar classification confidence value from the previous stage, with parameter values averaged over matching grammars in the group.

The resulting facades have windows and doors, which are colored according to the average window/door color as determined by the segmentation. Similarly, the facade wall is colored according to the average non-window color.

4 Results

Our method is implemented using OpenCV, OpenGL, and PyTorch, and it runs on an Intel i7 workstation with NVIDIA GTX 1080 cards. We have applied our method to six test areas in the United States captured by WorldView3 satellite images: a portion of (A1) Jacksonville, Florida $(2.0\,\mathrm{km}^2)$, (A2) UC San Diego, California $(1\,\mathrm{km}^2)$, (A3) San Fernando, California $(1\,\mathrm{km}^2)$, (A4) Omaha, Nebraska $(2.2\,\mathrm{km}^2)$, (A5) San Diego, California $(1.2\,\mathrm{km}^2)$ and (A6) USC, California $(2\,\mathrm{km}^2)$. Collectively, the areas have a few hundred buildings and medium to tall buildings and have from 20 to a few hundred windows/doors each. Our method runs automatically yielding facades for 14 buildings per minute. The training time for our classification network is about 12 h, and the training time for our estimation networks from grammars 1) to 6) is about 20 h, 3 h, 3 h, 36 h, 8 h, and 8 h, respectively.

Dataset. In order to train our neural network models, evaluate our method, and compare with other methods, we present a dataset of real satellite facades, which includes about 400 rectified images of facades from the aforementioned six areas, which have been manually annotated with two different labels: one for windows/doors and the other for the walls. Because of the low-quality of these facades, even humans can't precisely do the segmentation. Thus, mis-segmentation and misalignment always exist. Further, we carefully refine the annotations for 61 facade images and use those facades as a test data set for evaluating models/methods.

Pipeline Steps. We show example pipeline steps in Fig. 7 which includes chip extraction results, segmentation results, image processing results and our final facade completion results. Additional example facades are in supplemental figures. Our paper video also shows the pipeline and example results.

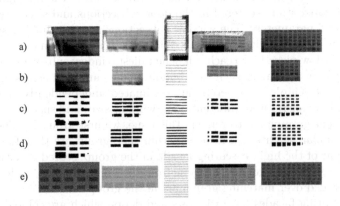

Fig. 7. *Pipeline steps.* a) Selected facade images. b) Facade chips. c) Results of using our segmentation model b). d) Images after applying dilation, rotation and replacement of windows/doors with filled-in rectangular bounding boxes and then being fed to our neural networks. e) Synthesized facades.

Table 1. *Segmentation quantitative comparison.* Pixel accuracy, precision, recall and F-score metrics evaluated on 61 facades for models from b) to g). Those terms are defined in optimization section.

Model	Accuracy	Precision	Recall	F-score
b)	0.843665	0.756	0.747	0.742
c)	0.8482	0.795	0.712	0.742
d)	0.866343	0.836	0.741	0.771
e)	0.846425	0.802	0.696	0.732
f)	0.849911	0.776	0.725	0.740
g)	0.870966	0.864	0.709	0.766

Segmentation Models. We test satellite facade segmentation on three state-of-the-art neural network architectures: Pix2Pix [9], Deep Labv3+ [2] and Enc-Net [31]. We train these architectures from scratch using our data set and also customize the hyper-parameters to fit our segmentation problem. For Pix2Pix we also try different generator and discriminator architectures which could support different sizes of input images. See supplemental table and supplement Fig. 2 for specific configurations and qualitative comparisons. Please see Table 1 for quantitative comparisons. Based on this comparison, we perceive Pix2Pix_96 to work best and it is the segmentation model we use in our approach.

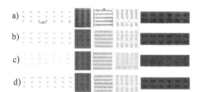

Fig. 8. *Optimization qualitative results.* a) Original facades. b) Manually created ground truth. c) Our results without optimization. d) Our results with optimization.

Table 2. *Optimization quantitative comparison.* Pixel accuracy, precision, recall, F-score and blob accuracy evaluated on 61 facades for models c) and d) in Fig. 8.

Method	Accuracy	Precision	Recall	F-score	Blob
c)	0.725	0.556	0.673	0.597	0.810
d)	0.880	0.818	0.834	0.815	0.923

Optimization. We evaluate 61 facade images using both our method without optimization and our method with optimization. Thus we show that we improve pixel accuracy, precision, recall, F-score and blob accuracy by perturbing grammar parameters. The blob accuracy is the window count accuracy defined as:

$$Blob = 1 - \frac{|Our\_Window\_Count - Ground\_Truth\_Window\_Count|}{Ground\_Truth\_Window\_Count}, \quad (2)$$

Please see Fig. 8 and Table 2 for qualitative and quantitative comparisons. In summary, with optimization our metrics improve from 0.69 to 0.85, an improvement of 16% on average.

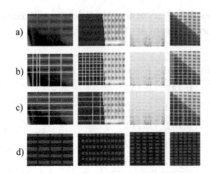

Table 3. *Facade quantitative comparison.* We evaluate Mean Absolute Error (MAE) and Mean Relative Error (MRE) of the number of floors and the number of windows per floor on 61 facades for c) and d) in Fig. 9.

Method	MAE		MRE	
	#floors	#windows	#floors	#windows
c)	0.770	0.770	15.8%	12.1%
d)	0.246	0.164	4.2%	3.9%

Fig. 9. *Facade subdivision comparison.* We provide a) satellite-based facades to b) an image-based approach, c) Nishida et al. [20], and d) Ours.

Comparisons. We compare our approach to several state-of-the-art methods. First, in Fig. 9 we show a visual comparison between the facade subdivision of b) an image-gradient-based approach (e.g., [17]), c) Nishida et al. [20] (retrained using the same training set as our approach), and d) our method. We highlight that Nishida et al. [20] (and also Teboul et al. [27]) essentially make use during their processing pipeline of an image-gradient based method similar to [17] (thus we include the image-gradient comparison). We also include facade quantitative comparisons in Table 3.

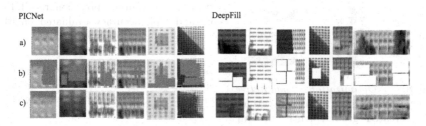

Fig. 10. *Image in-painting.* a) Original facades. b) Rectangular areas to be filled-in. c) Results after inpainting.

Second, we test two state-of-the-art neural network architectures for image inpainting/completion: DeepFill [30] and PICNet [33]. With DeepFill determining which part to "fill" is an unaddressed challenge and thus for this comparison we manually select occluded, shadowed and/or tree-covered areas. In PICNet, we use the random rectangular mask generation method they provide (e.g., select a sufficient number of rectangles within the image to most likely performed all necessary in-filling). Please see Fig. 10 for visual results. While the methods are able to place content in the occluded areas, there are still significant artifacts which will hinder subsequent facade process.

To evaluate the facade processing ability directly using the segmentation model and image in-painting model, we evaluate performance using our 61 test images qualitatively and quantitatively. To be specific, for the segmentation model, we choose the aforementioned Pix2Pix_96 and apply it to the facade images directly. Then, we dilate each window/door to occupy a rectangular bounding box. For the image in-painting model, we choose DeepFill [30] and complete the facade images with manually selected masks. Then we apply the segmentation model to the completed facade images and we also use a version of the windows/doors dilated to rectangles. The quantitative metrics include

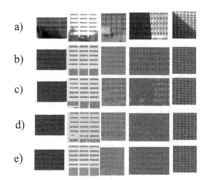

Fig. 11. *Facade comparisons.* Comparison to SOTA methods on facade parsing. a) Input satellite facades. b) Manually created ground truth. c) The results of applying Pix2Pix_96 to a). d) The results of applying Pix2Pix_96 to image completed by DeepFill [30]. e) Ours.

pixel accuracy, precision, recall, and blob accuracy. In Fig. 11 and Table 4, we show details of comparing our method to the segmentation model and the image in-painting model.

Table 4. *Quantitative comparison.* Pixel accuracy, precision, recall, F-score and blob accuracy evaluated for models from c) to e) in Fig. 11. We evaluated c) and e) on 61 facades in the left table. However the right table shows applying d) to 22 facades (22 out of 61 facades are occluded and suitable for image in-painting.) and we manually set the mask as best as possible.

Method	Accuracy	Precision	Recall	F-score	Blob
c)	0.835	0.695	0.868	0.758	0.891
e)	0.880	0.818	0.834	0.815	0.923

Method	Accuracy	Precision	Recall	F-score	Blob
c)	0.802	0.705	0.797	0.728	0.840
d)	0.806	0.803	0.612	0.677	0.875
e)	0.843	0.768	0.828	0.783	0.918

Examples. Finally, we show in Fig. 12 many close-ups of reconstructed buildings as well as an overall view of one area (A1). Views of our additional areas (A2) and more buildings are in supplemental figures.

Fig. 12. *Examples.* We show a view of a reconstructed area A1 within Google Earth and close-ups of our buildings.

5   Conclusions and Future Work

We have presented a method to automatically synthesize crisp and regular building facades from satellite imagery. Facades are classified into one of several procedural grammars, and the corresponding parameters are estimated using trained neural networks. The resulting grammars are applied to building models, resulting in complete, plausible facades that are free of the noise, occlusions, and partial coverage that is inherent in satellite data. Our comparisons to other approaches shows the improvement of our method. However, our approach has some limitations. First, for facades whose styles are outside our defined grammars, we could give our best guess. Second, for facades with logos, we didn't show those areas.

Our approach has several avenues of future work. First, we would like to incorporate more general grammar sets to capture finer details. Second, we would also like to incorporate a more sophisticated wall/window color treatment. Finally, we are also interested in estimated and procedural facade textures to give the resulting buildings more details.

Acknowledgements. This research was supported in part by the Intelligence Advanced Research Projects Activity (IARPA) via Department of Interior/ Interior Business Center (DOI/IBC) contract number D17PC00280. Additional support came from National Science Foundation grants #10001387 and #1835739.

References

1. Bokeloh, M., Wand, M., Seidel, H.P.: A connection between partial symmetry and inverse procedural modeling. ACM Trans. Graph. **29** (2010)
2. Chen, L.-C., Zhu, Y., Papandreou, G., Schroff, F., Adam, H.: Encoder-decoder with atrous separable convolution for semantic image segmentation. In: Ferrari, V., Hebert, M., Sminchisescu, C., Weiss, Y. (eds.) ECCV 2018. LNCS, vol. 11211, pp. 833–851. Springer, Cham (2018). https://doi.org/10.1007/978-3-030-01234-2_49

3. Cohen, A., Schwing, A.G., Pollefeys, M.: Efficient structured parsing of facades using dynamic programming. In: IEEE Computer Vision and Pattern Recognition, pp. 3206–3213 (2014)
4. Demir, I., Aliaga, D.G., Benes, B.: Procedural editing of 3D building point clouds. In: 2015 IEEE International Conference on Computer Vision (ICCV), pp. 2147–2155, December 2015. https://doi.org/10.1109/ICCV.2015.248
5. Demir, I., Aliaga, D.G., Benes, B.: Coupled segmentation and similarity detection for architectural models. ACM Trans. Graph. **34**(4), 1–11 (2015)
6. Fathalla, R., Vogiatzis, G.: A deep learning pipeline for semantic facade segmentation. In: Proceedings of the British Machine Vision Conference 2016, BMVC 2017, September 2017. c 2017. The copyright of this document resides with its authors. It may be distributed unchanged freely in print or electronic forms. http://publications.aston.ac.uk/id/eprint/31805/
7. Gadde, R., Marlet, R., Paragios, N.: Learning grammars for architecture-specific facade parsing. Int. J. Comput. Vis. **117**(3), 290–316 (2016). https://doi.org/10.1007/s11263-016-0887-4
8. He, K., Zhang, X., Ren, S., Sun, J.: Deep residual learning for image recognition. CoRR abs/1512.03385 (2015). http://arxiv.org/abs/1512.03385
9. Isola, P., Zhu, J.Y., Zhou, T., Efros, A.A.: Image-to-image translation with conditional adversarial networks. In: IEEE Computer Vision and Pattern Recognition, pp. 1125–1134 (2017)
10. Kelly, T., Guerrero, P., Steed, A., Wonka, P., Mitra, N.J.: FrankenGAN: guided detail synthesis for building mass-models using style-synchonized GANs. ACM Trans. Graph. **37**(6) (2018). https://doi.org/10.1145/3272127.3275065
11. Kozinski, M., Gadde, R., Zagoruyko, S., Obozinski, G., Marlet, R.: A MRF shape prior for facade parsing with occlusions. In: IEEE Computer Vision and Pattern Recognition, pp. 2820–2828 (2015)
12. Koziński, M., Obozinski, G., Marlet, R.: Beyond procedural facade parsing: bidirectional alignment via linear programming. In: Cremers, D., Reid, I., Saito, H., Yang, M.-H. (eds.) ACCV 2014. LNCS, vol. 9006, pp. 79–94. Springer, Cham (2015). https://doi.org/10.1007/978-3-319-16817-3_6
13. Leotta, M.J., et al.: Urban semantic 3D reconstruction from multiview satellite imagery. In: Proceedings of the IEEE/CVF Conference on Computer Vision and Pattern Recognition (CVPR) Workshops, June 2019
14. Liu, H., Zhang, J., Zhu, J., Hoi, S.C.H.: DeepFacade: a deep learning approach to facade parsing. In: International Joint Conference on Artificial Intelligence, pp. 2301–2307 (2017)
15. Martinovic, A., Van Gool, L.: Bayesian grammar learning for inverse procedural modeling. In: IEEE Computer Vision and Pattern Recognition, pp. 201–208 (2013)
16. Mathias, M., Martinović, A., Van Gool, L.: ATLAS: a three-layered approach to facade parsing. Int. J. Comput. Vis. **118**(1), 22–48 (2016). https://doi.org/10.1007/s11263-015-0868-z
17. Müller, P., Zeng, G., Wonka, P., Van Gool, L.: Image-based procedural modeling of facades. ACM Trans. Graph. **26**(3), 85–es (2007). https://doi.org/10.1145/1276377.1276484
18. Musialski, P., Wonka, P., Aliaga, D.G., Wimmer, M., Van Gool, L., Purgathofer, W.: A survey of urban reconstruction. Comput. Graph. Forum **32**, 146–177 (2013)
19. Nguatem, W., Mayer, H.: Modeling urban scenes from pointclouds. In: IEEE International Conference on Computer Vision, pp. 3837–3846 (2017)
20. Nishida, G., Bousseau, A., Aliaga, D.G.: Procedural modeling of a building from a single image. Comput. Graph. Forum **37**, 415–429 (2018)

21. Ozcanli, O.C., Dong, Y., Mundy, J.L., Webb, H., Hammoud, R., Tom, V.: A comparison of stereo and multiview 3-D reconstruction using cross-sensor satellite imagery. In: IEEE Computer Vision and Pattern Recognition Workshops, pp. 17–25 (2015)

22. Qin, R.: Automated 3D recovery from very high resolution multi-view satellite images. In: ASPRS (IGTF) Annual Conference, p. 10 (2017)

23. Riemenschneider, H., et al.: Irregular lattices for complex shape grammar facade parsing. In: IEEE Computer Vision and Pattern Recognition, pp. 1640–1647 (2012)

24. Ritchie, D., Mildenhall, B., Goodman, N.D., Hanrahan, P.: Controlling procedural modeling programs with stochastically-ordered sequential Monte Carlo. ACM Trans. Graph. **34**(4), 1–11 (2015)

25. Sasaki, Y.: The truth of the f-measure. Teach Tutor Mater, January 2007

26. Talton, J.O., Lou, Y., Lesser, S., Duke, J., Měch, R., Koltun, V.: Metropolis procedural modeling. ACM Trans. Graph. **30**(2), 1–14 (2011)

27. Teboul, O., Kokkinos, I., Simon, L., Koutsourakis, P., Paragios, N.: Shape grammar parsing via reinforcement learning. In: IEEE Computer Vision and Pattern Recognition, pp. 2273–2280 (2011)

28. Vanegas, C.A., Aliaga, D.G., Beneš, B.: Building reconstruction using manhattan-world grammars. In: IEEE Computer Vision and Pattern Recognition (2010)

29. Yang, C., Han, T., Quan, L., Tai, C.L.: Parsing façade with rank-one approximation. In: IEEE Computer Vision and Pattern Recognition, pp. 1720–1727 (2012)

30. Yu, J., Lin, Z., Yang, J., Shen, X., Lu, X., Huang, T.S.: Generative image inpainting with contextual attention. CoRR abs/1801.07892 (2018). http://arxiv.org/abs/1801.07892

31. Zhang, H., et al.: Context encoding for semantic segmentation. In: The IEEE Conference on Computer Vision and Pattern Recognition (CVPR), June 2018

32. Zhang, X., May, C., Nishida, G., Aliaga, D.: Progressive regularization of satellite-based 3D buildings for interactive rendering. In: Symposium on Interactive 3D Graphics and Games, I3D 2020. Association for Computing Machinery, New York (2020)

33. Zheng, C., Cham, T.J., Cai, J.: Pluralistic image completion. In: Proceedings of the IEEE Conference on Computer Vision and Pattern Recognition, pp. 1438–1447 (2019)

34. Zheng, E., Wang, K., Dunn, E., Frahm, J.M.: Minimal solvers for 3D geometry from satellite imagery. In: IEEE International Conference on Computer Vision, pp. 738–746 (2015)

35. Zhu, J.Y., Park, T., Isola, P., Efros, A.A.: Unpaired image-to-image translation using cycle-consistent adversarial networks. In: IEEE International Conference on Computer Vision, pp. 2223–2232 (2017)

Mapillary Planet-Scale Depth Dataset

Manuel López Antequera[1]($\boxtimes$) (iD), Pau Gargallo[1] (iD), Markus Hofinger[2] (iD),
Samuel Rota Bulò[1] (iD), Yubin Kuang[1] (iD), and Peter Kontschieder[1] (iD)

[1] Facebook, Menlo Park, USA
mlop@fb.com

[2] Institute of Computer Graphics and Vision, Graz University of Technology,
Graz, Austria

Abstract. Learning-based methods produce remarkable results on single image depth tasks when trained on well-established benchmarks, however, there is a large gap from these benchmarks to real-world performance that is usually obscured by the common practice of fine-tuning on the target dataset. We introduce a new depth dataset that is an order of magnitude larger than previous datasets, but more importantly, contains an unprecedented gamut of locations, camera models and scene types while offering metric depth (not just up-to-scale). Additionally, we investigate the problem of training single image depth networks using images captured with many different cameras, validating an existing approach and proposing a simpler alternative. With our contributions we achieve excellent results on challenging benchmarks before fine-tuning, and set the state of the art on the popular KITTI dataset after fine-tuning.

The dataset is available at mapillary.com/dataset/depth.

1 Introduction

The availability of large-scale training datasets has significantly contributed to the rise of deep learning based approaches in computer vision. Starting with ImageNet [6] for image classification, also the quality of object detection [3,8] or semantic-, instance- and panoptic segmentation algorithms [5,18,22,31] has been greatly improved within a few years only. Yet, metric-accurate, large-scale, natural image datasets are still to come for the task of monocular depth estimation, most likely because they cannot be collected with commodity hardware in a straightforward way. Research in monocular depth estimation therefore predominantly use smaller, less varied or up-to-scale datasets for training [11,17,30]. Unsupervised methods are achieving remarkable results [12,13], but their performance still lags behind that of supervised methods.

For validation of single image depth methods, an important benchmark is the Make3D [24] dataset, comprising laser scans coupled with RGB images. Although dated, it is still a reference benchmark in the field. Recently, modern hardware has been used in a similar fashion to produce very high-quality datasets to be used as benchmarks for single-image depth methods such as DIODE [29] and iBims-1 [15]. Please refer to Table 1 for an overview of depth datasets.

© Springer Nature Switzerland AG 2020
A. Vedaldi et al. (Eds.): ECCV 2020, LNCS 12347, pp. 589–604, 2020.
https://doi.org/10.1007/978-3-030-58536-5_35

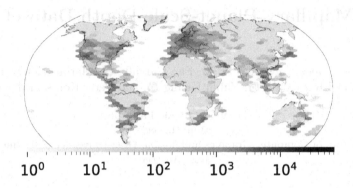

Fig. 1. Global distribution of the Mapillary Planet-Scale Depth (MPSD) dataset

Fig. 2. Sample images and depth values from the proposed MPSD dataset

In this paper we introduce a novel, virtually arbitrarily scalable dataset – MPSD – providing training data for monocular depth estimation. Our dataset is solely derived from Mapillary's publicly available image database[1].

The dataset is generated by running monocular Structure-from-Motion and multi-view stereo we obtain dense depth for eligible images. Our dataset, containing images from all over the world, is larger, more complex and diverse than any previously published depth dataset. It currently comprises $\approx 750,000$ images, extracted from over $50,000$ individual 3D reconstructions captured by a broad range of camera types with different focal lengths (Fig. 4) and distortion characteristics, in a broad set of environments (Fig. 1) and weather conditions, seasons, times of day, viewpoint and with real noise and motion patterns (Figs. 2 and 3).

Training with such a dataset is not straightforward, as it is necessary to account for the heterogeneous cameras used to capture the images. This is a problem often overlooked and only recently studied by Fácil et al. [9], where they demonstrate the advantage of explicitly accounting for the camera intrinsics during training. We successfully use their proposed CAM-Convs to train using our dataset, and also suggest an alternative, simpler technique to deal with the problem of multi-camera training.

---

[1] Currently holding $\approx 10^9$ images and corresponding GPS positions.

Table 1. Overview of depth datasets The proposed Mapillary Planet-Scale Depth (MPSD) dataset is large and diverse enough to effectively transfer onto several target datasets without fine-tuning. Refer to Sect. 4 for more details.

Dataset	n. Images	Source	Extent	Metric
Make3D [24]	534	Lidar	Palo Alto	yes
iBims-1 [15]	100	Lidar	Various scenes	yes
DIODE [29]	26 k	Lidar	25 Scenes	yes
KITTI [11]	94 k	Lidar	Karlsruhe	yes
WSVD [30]	1.5 M	Stereo	7 k videos	no
Cityscapes [5]	25 k	Stereo	50 Cities	yes
MegaDepth [17]	130 k	SfM	200 Scenes	no
MPSD	750 k	SfM	50 k Scenes(Fig. 1)	yes

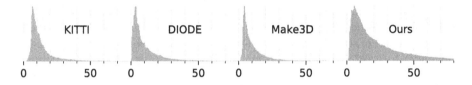

Fig. 3. Volume-normalized depth (m) distributions on several datasets

At the core of our work, we discuss the challenges to be tackled during the generation of MPSD from *real-world* data and how to take advantage of it in modern deep-learning based algorithms. We particularly address how to:

– Generate a *metric-accurate*-depth dataset from images captured in sub-optimal conditions for structure-from-motion such as low framerate, non-orbital trajectories, and under-constrained camera parameters.
– Effectively train deep neural networks for monocular depth estimation with data from many heterogeneous camera sources.

We conducted exhaustive ablations for the task of monocular depth estimation, proving the superior quality of our dataset against reference benchmarks like KITTI [11], MegaDepth [17], Cityscapes [5], DIODE [29] or Make3D [24]. With our approach and dataset, we achieve new state-of-the-art results for monocular depth prediction on the well-known KITTI benchmark.

2   Dataset

The Mapillary Planet-Scale Depth (MPSD) dataset contains 750,000 images and associated depth maps. It is based on imagery collected from Mapillary[2],

---

[2] Mapillary is a street-level imagery platform hosting images collected by members of their community.

on which we perform monocular structure-from-motion (SfM) to obtain relative camera poses. A multi-view stereo algorithm is then used to compute dense depth. Absolute (metric) depth is recovered from the GPS metadata that is available alongside the images. Although similar approaches to produce depth have been used on phototourism-style [17,27] datasets sourced from photography websites such as Flickr, Mapillary imagery poses new challenges when used in this manner.

Mapillary collects street-level imagery that is uploaded by individual users and organizations. It is a very heterogeneous source, presenting imagery captured in a wide range of conditions and locations with a vast number of cameras, both consumer-grade and professional. This type of imagery is of great interest as the community progresses towards algorithms that are expected to perform beyond small benchmarks. However, recovering depth from Mapillary images cannot be performed by using out-of-the-box SfM pipelines, as it presents some challenges not present in phototourism datasets: All images in Mapillary are uploaded as-is from thousands of different uncalibrated cameras, requiring self-calibration from the SfM pipeline itself. However, most sequences are not valid to perform self-calibration because they are captured using forward-facing cameras and under forward / zooming motion, underconstraining the camera parameters [28]. Moreover, capture is usually performed at a low framerate, increasing the baseline between consecutive frames, which makes the correspondence problem non-trivial.

Due to these sub-optimal conditions for perfoming structure-from-motion, we found no turnkey solution (including the MegaDepth [17] pipeline) that could extract valid depth at scale from Mapillary sequences. Our process is described in the following sections as three stages: 2.1) Global model-wise camera calibration, 2.2) Image search and 2.3) Reconstruction and multi-view stereo.

2.1   Global Model-Wise Camera Calibration

Camera calibration is required for metric 3D reconstruction. Cameras are usually calibrated using a physical printed calibration pattern imaged under a variety of poses with respect to the camera. In the case of Mapillary imagery, this is not possible as there are thousands of independent users and camera calibration is not enforced on them by the Mapillary app used to record images. However, it is also possible to automatically obtain good calibration parameters with monocular SfM for a camera or set of cameras if there are enough images capturing a scene with a layout such that the camera parameters are constrained. This method for automatically calibrating cameras is common in the 'phototourism' scenario where a large number of images capture the same object, generally following orbital-like trajectories. We attempt to obtain camera parameters from Mapillary images in a similar fashion, however, the coverage offered by Mapillary is not optimal for this. Imagery is most often recorded from forward-facing cameras mounted on vehicles driven on roads. Motion is thus mostly linear and without rotation. Naively downloading imagery and attempting to perform reconstructions did not yield stable camera parameters. Instead, we sample sequences that

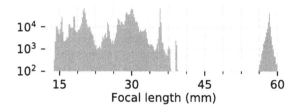

Fig. 4. Distribution of focal lengths (mm) in MPSD

are: 1. Dense enough (less than 5 m and $30°$[3] between consecutive frames) in order to get enough point correspondences. 2. Have enough rotation (cumulative turn of 70+°) in order to better constrain the focal length.

For each sequence, we compute the optimal calibration by running SfM reconstructions iteratively. We first run an incremental SfM algorithm with a default set of intrinsic parameters—focal length equiv. to 30 mm and no radial distortion. The focal length and the first two distortion parameters of the Brown model are optimized during bundle adjustment. The result is a 3D reconstruction and an updated set of camera parameters. These parameters are used as initialization for a new SfM reconstruction, which in turn yields updated camera parameters. We iterate this process several times until the camera parameters stabilize. It is necessary to run the reconstruction process multiple times because improved initial camera parameters can improve the matching step which leads to better tracks and more constraints for the camera parameters.

Since the images at Mapillary are gathered by thousands of users with different devices, we can't obtain camera parameters for all of them as it isn't always the case that we can find adequate imagery on which to perform the aforementioned calibration process. We simplify the problem by assuming that all cameras reporting the same make, model, resolution and focal length will share the calibration parameters. In other words, we ignore differences due to manufacturing tolerances, temperature and so on. This simplification undoubtedly introduces some errors, but it is fundamental to make use of the imagery available in Mapillary, as there is rarely enough coverage from a single user to perform calibration on each user's camera independently.

To ensure that we have not calibrated the camera using an outlier (that is, a device whose calibration parameters deviate substantially from the modal parameters for that camera make and model), we run the calibration process for 10 different sequences for each camera make and model, and visualize the resulting camera parameters. We then manually confirm that calibrations from different sequences yield similar results and select one of them as the valid calibration for all images taken with that make and model, resolution and focal length.[4] Through this process we obtained calibrations for 250 camera models.

---

[3] We obtain an initial estimate of the turning angle as the angle between consecutive segments on the GPS track of the sequence.

[4] Many action cameras and phones are able to capture under different 'modes' with different optics. We use the combination of reported focal length and resolution to disambiguate these modes.

2.2   Image Search

After calibrating a large set of camera models, we then mine Mapillary for images taken using these cameras and use them to perform SfM reconstructions and multi-view stereo to obtain depth. We start by selecting sampling weights w_r from 6 regions: North America: 20%, South America: 15%, Europe: 20%, Asia: 20%, Oceania: 15%, Africa: 10%. Each region is then partitioned into countries, assigning to each country a weight $w_c = w_r \frac{\#\text{ims country}}{\#\text{ims region}}$.

Each country is then partitioned in a regular grid of 156 by 156 km cells and the budget of images for that country is evenly distributed on this grid. We sample images randomly within each cell that: 1. Have been taken with one of the 250 cameras in our calibrated camera set. 2. Have at least 20 neighboring images in a radius of 10 meters (measured by the images' GPS tags).

Each image and its neighbors are then used to perform a 3D reconstruction and multi-view stereo as described in the following section.

Further checks (described below) are performed after reconstructing in order to accept or reject this group of images into the dataset. If not accepted, another image from the cell is sampled. Since many cells are empty (rural areas or nature), we exhaust those and oversample more densely covered cells to satisfy the number of images allocated for that country. This is done to ensure that all of the images in underrepresented areas are used, obtaining as much diversity as possible.

2.3   Reconstruction and Multi-view Stereo

We perform structure from motion to obtain reconstructions of each of the candidate groups of images downloaded from Mapillary as described above. The reconstructions are performed using the OpenSfM [1] library with its default configuration settings. We chose OpenSfM due to prior familiarity, but other software [21,25] could be used to obtain similar results with appropriate settings.

We use the semantic segmentations available for each image in Mapillary to mask out regions that can negatively affect the reconstruction (pedestrians, vehicles, ego-vehicle and sky). After reconstructing, we obtain a set of sparse correspondences as well as relative camera poses.

The reconstruction is aligned to the GPS data associated with each image during bundle adjustment by adding a cost proportional to the squared distance between the GPS position and the reconstructed camera position. This fixes the scale ambiguity and yields metric distances, however, GPS measurements have noise that can affect this scale. In order to reduce the effect of the GPS measurement noise on the metric accuracy of the data, we filter out reconstructions that span a small region: After reconstructing[5], we check that the furthest two images are at least 20 meters apart, otherwise we discard the reconstruction.

In the experimental section of this paper we confirm that training on MPSD does indeed produce networks able to recover metric depth from single images.

---

[5] This check is performed only after reconstructing since it must be performed on images that can be registered to the reconstruction (some of the images in the neighborhood might have failed to reconstruct).

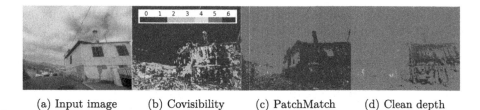

(a) Input image   (b) Covisibility   (c) PatchMatch   (d) Clean depth

Fig. 5. Initial depth as obtained by multi-view stereo using PatchMatch (c) may contain spurious values. We clean the estimated depth by checking for consistency across several neighboring images (b). Only the depth values that are consistent with at least 3 neighbors are kept (d)

Using the relative poses obtained from SfM, we run a Patch-Match based multi-view stereo algorithm [26] to obtain dense depth estimates. This is a simple winner-takes-all stereo algorithm. Different depth and normal values are tested for each pixel and the one that gives the best normalized cross correlation score with the neighboring views is kept. The result is a dense but noisy depth map.

Most of the noise in the depth maps is removed in a post-processing step that checks the consistency between the depth maps of neighboring images. Depth values that are not consistent with at least 2 neighboring views are removed. This reduces the number of pixels for which a depth value is produced. We do not add any smoothness term to produce smoother depth maps nor do we try to inpaint the missing depth values.

The result is thus a set of 'clean' depth values that might be sparse, but that is reliable as shown in Fig. 5. Finally, we discard depth maps in which less than 5% of the pixels have a depth value.

Opposite to what is done in the SfM step, during depth map estimation we do not mask out dynamic objects. The rationale is that we do want to have depth values on dynamic objects for training. While some dynamic objects are moving during capture and can possibly lead to wrong depth values, there are also many static-during-capture dynamic objects for which depth estimation will work. Additionally, when objects move in different directions than the camera, their motion does not satisfy the epipolar constraint and are easily rejected by the MVS algorithm. A notable exception are objects moving along the same road as the camera for which MVS produces scaled depth values. Manual analysis of our dataset finds very few examples of this, and training on MPSD produces networks that can predict depth on moving objects such as cars, a fact that we experimentally determine in Sect. 4.

3   Training with Multiple Cameras

The relationship between real world dimensions and pixels on the image plane for undistorted images as defined by the pinhole model is simple: The depth z

of an object is expressed in terms of its size in pixels the image plane y', its real size in meters y and the focal length of the camera f in pixels: $z = f\frac{y}{y'}$.

Since we are dealing with the prediction of per-pixel depth values, we can simplify this expression for a single pixel ($y' = 1$): $z = fy$. Although single image depth is usually described as an ill-posed problem, it is solvable if y and f are known. It can be decomposed into two smaller problems:1. Finding the focal length of the camera f, 2. Recalling the real dimensions depicted by pixel y'.

Most single-image depth methods deal with a single camera, simplifying the task. Learned models will implicitly memorize the value of the focal length f. This might be sufficient for applications using a single camera, as long as training data gathered with the same device is available. However, methods trained with a single camera will not generalize well to images captured by other devices.

Naively training on a dataset containing multiple cameras negatively impacts performance [9]. We hypothesize that this is because the network must accurately predict the focal length, a difficult task to perform, even when directly supervised to do so [14,20]. Focal length normalization alleviates this problem: The network is trained to predict $y = z/f$, a magnitude that only depends on the real world size of the area represented by the pixel of interest. This is quite effective and it has an intuitive explanation: the real-world size of objects is highly coupled with semantic segmentation, a task that convolutional neural networks excel at. To obtain a metric depth value during deployment, the predictions are multiplied by the focal length.

CAM-Convs. [9] explicitly encode camera intrinsics by concatenating a map of the viewing directions in polar coordinates to each skip-connection in a u-net architecture. CAM-Convs are more general than just applying focal length normalization (although the authors found that it is beneficial to use both techniques in combination as it accelerates convergence), as it can also model different sensor sizes and aspect ratios explicitly. Images from different sensor sizes and resolutions can be resized and even *squashed* if necessary to fit the aspect ratio of the batch, as the network is explicitly informed about this through the appended features. It also adequately models the location of the principal point, enabling training on non-central crops.

In this work we experimentally validate CAM-Convs as a viable option to train single-image depth networks in datasets containing images taken from multiple cameras, while proposing a simpler alternative.

Camera Normalization. We suggest an alternative approach: resizing the images to approximate them being taken by a canonical camera with square pixels, focal length f_c and no radial distortion. With camera normalization, the relationship between the pixel size and the real size for any given object depends only on the depth, simplifying the task of depth prediction. For example, if the canonical focal length is $f_c = 700$ px, an object of height $y = 2$ m will have depth that is inversely proportional to its size in the image $z = 700 * 2/y'$ pixels. By resizing the input images to always have the same focal length, the network only needs to learn to regress the real-world sizes of objects, as the focal length prediction isn't required anymore.

Instead of *informing* the network about the viewing angles as is done when using CAM-Convs, these angles are intrinsically learned by the model, as every input pixel always corresponds to the same viewing angles during training. The network does not need to produce different responses for similarly-looking patches as is the case when using CAM-Convs. Moreover, by resizing images to a fixed focal length, the range of pixel sizes at which objects are represented is reduced. For example, if the focal length is variable, the smallest objects will look even smaller on images with small focal lengths, and the larger objects would look even larger on images with large focal lengths.

Unlike CAM-Convs (that require a u-net like architecture), this approach is independent of the architecture, as it depends only on scaling the input images by a factor of f_c/f. In other ways, our technique is less flexible than CAM-Convs: Images are cropped or padded to a common size to form batches during training, and non-central crops can't be used. However, we didn't find these drawbacks to be of practical relevance on our experiments.

4   Experiments

Architecture. We use a single architecture for all of our experiments, except in those cases where we compare against the implementations offered by other authors. We do so in order to offer a fair comparison and to focus on the differences in datasets and techniques for handling training with several cameras. The network is an encoder-decoder with skip connections (u-net) architecture, with a dilated resnet-50 pre-trained on ImageNet as the encoder. The dilation rates are 1,1,2 and 4 for each of the four residual modules, producing a feature map 16 times smaller than the input image. We use in-place activated batch normalization [2] to reduce the memory footprint during training allowing for large batches. Input size is always 1216×352 pixels, regardless of the scaling and cropping strategy.

After the encoder we append a DeeplabV3 [4] head to aggregate contextual information. It is formed by a set of dilated convolutions with different dilation rates (12, 24 and 36) as well as a global pooling of the features whose outputs are concatenated, batch normalized and convolved together to form an output feature map of the same dimensionality as the input.

This feature map is then upsampled through bilinear interpolation in three stages, each stage doubling the resolution. Features from the matching level in the encoder are concatenated to the upsampled features before being fed to a 'skip module' consisting of a convolution and activation. When using CAM-Convs, the viewing angles and normalized camera coordinates (8 channels) are resized to the corresponding shape and concatenated to the upsampled features and used as input to the skip modules. The final output of the u-net is thus at half of the resolution of the input image. We upsample once more to fit the size of the ground truth before computing the loss or evaluating.

Datasets. We train on MegaDepth as a baseline and compare with our proposed MPSD dataset. When training our own models[6], we only use the subset of MegaDepth ($\approx$ 100k out of $\approx$ 130k images) that contains euclidean depth, as the rest of the dataset only contains *ordinal* (foreground/background) depth relationships).

During training, we use the KITTI validation set to track performance and perform early stopping. We then evaluate in a range of datasets (without any fine-tuning): Make3D [24], DIODE [29](outdoor), Cityscapes [5], MegaDepth [17] and KITTI [11]. We follow the usual practice of filtering out depth values that are unreliably large, removing values larger than 80 meters for all datasets except MegaDepth (no metric depth) and Make3D (70 m).

Scaling and Cropping Strategies. We experiment with both CAM-Convs and our proposed camera normalization. Since CAM-Convs allow training using non-central crops, we evaluate both central and random crops when training using CAM-Convs. As a baseline, we also train our architecture on MegaDepth without any explicit handling of the focal length on the input images or the architecture: The network is fed with undistorted images that are simply resized so that their width is 1216 and then center-cropped to a common size of (1216, 352).

Training Details We train the network to predict the logarithm of the focal length-normalized depth and minimize the loss proposed in [7]:

$$L(z, z^*, f) = \frac{1}{n} \sum_i d_i^2 - \frac{\lambda}{n^2} \left(\sum_i d_i \right)^2 \tag{1}$$

where $d_i = \log(z) - \log(z^*)$. The loss is only evaluated on those pixels with known depth, where n is the number of valid depth points in the image. When training on MegaDepth, we use the fully scale-invariant version with $\lambda = 1$. We note that using a scale-invariant component in the loss leads to faster convergence, even if the training data is metric depth, thus, when training on MPSD, we set $\lambda = 0.5$.

We use stochastic gradient descent with an initial learning rate of 0.015, Nesterov momentum of 0.9 and weight decay of 10^{-4} for a maximum of 200 k steps, stopping early if the performance on the KITTI validation set decreases. The learning rate is decayed on every step following $\mathrm{lr}_i = 0.015(1 - i/200,000)^{0.2}$. The batch size is set to 64, distributed over 8 V-100 GPUs.

MPSD vs. MegaDepth. Training using our dataset (rows 6–9 in Table 2) yields better performance across the board when compared to MegaDepth, with the exception of evaluating on MegaDepth itself (row 5). Although MPSD does not exclusively contain driving scenarios, the type of imagery available in our dataset is mostly street-level imagery similar to KITTI and Cityscapes, which could be an explanation for the large gap. However, networks trained on our dataset also generalize well to Make3D and DIODE which are not datasets captured in

---

[6] We also compare with the model offered by the authors, trained on their full dataset.

Table 2. Results obtained by training on MegaDepth or MPSD. MD-Ordinal is the model trained by the authors of MegaDepth using their full dataset (including ordinal data, supervised with their ordinal loss). All other entries share the architecture from Sect. 4 and are trained with euclidean depth. "mini MPSD" is the MPSD dataset reduced to the size of the euclidean subset of MegaDepth. Scaling strategies are *Naive*: Resize and center crop to a fixed size, *CC*: CAM-Convs, *FF*: camera normalization. Crop strategies are (R)andom and (C)enter. We only report RMSE on methods trained with MPSD, as the scale is arbitrary when trained on MegaDepth. The best result is highlighted in bold. Entries fine-tuned on the target data are marked with *. In those cases, the second-best is also highlighted with bold text.

#	Training set	Scale	Crop	KITTI SILog	rmse	MegaDepth SILog	Cityscapes SILog	rmse	DIODE (outdoor) SILog	rmse	Make3D SILog	rmse
1	MD-Ordinal	-	-	30.1	-	10.8	35.19	-	47.52	-	38.2	-
2	MegaDepth	Naive	C	25.61	-	11.86	65.11	-	42.91	-	59.89	-
3	MegaDepth	CC	R	26.92	-	10.67	62.92	-	50.3	-	54.24	-
4	MegaDepth	CC	C	23.79	-	11.51	60.08	-	47.28	-	55.9	-
5	MegaDepth	FF	C	26.79	-	**9.96***	36.73	-	48.28	-	41.64	-
6	mini MPSD	FF	C	14.89	4.87	17.85	22.61	9.05	44.43	8.44	29.55	5.99
7	MPSD	FF	C	**12.77**	4.21	14.68	19.77	**7.91**	42.2	7.78	**27.49**	**5.54**
8	MPSD	CC	C	13.33	**4.13**	21.5	34.83	12.77	43.04	8.05	54.66	59.45
9	MPSD	FF+C	C	12.8	4.39	**14.04**	**19.52**	8.13	**41.69**	**7.75**	28.07	5.67
10	MPSD+KITTI	FF	C	**9.23***	**3.04***	32.23	27.11	8.58	45.55	10.69	37.56	6.49

driving scenarios. The size and variety in MPSD allows networks to generalize much better to all of the datasets we tried on.

It's Not Only Size that Matters. We carry out an experiment to ascertain if the size of the MPSD dataset is crucial for the performance gains obtained when using it as a training set instead of MegaDepth. To do so, we factor out the size difference between MPSD and MegaDepth. We randomly sample our dataset to reduce it to the same size as the MegaDepth euclidean depth subset (around 100,000 images) and train on it. The results can be found on the sixth entry in Table 2 'mini MPSD': Performance on all the validation sets is only slightly worse than using the full dataset, while still much better than networks train on MegaDepth. We hypothesize that this is because the domain of MegaDepth is quite limited: although it is not a small dataset, the images in it display a small set of monuments and landmarks (the images were reconstructed into 200 distinct reconstructions). In contrast, the images from our dataset have been gathered from more than 50,000 independent reconstructions all over the globe.

Metric Accuracy of MPSD. Although we obtain state of the art results when training on MPSD for the scale-dependant RMSE metric, we perform a simple experiment to determine if there is scale bias in MPSD: Using a network trained exclusively on MPSD, we produce depth predictions on several metric

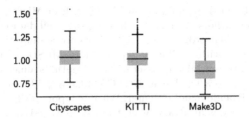

Fig. 6. Scale factors to align predictions of a MPSD-trained model (Entry #7 in Table 2) to the ground truth depth. Although there is some fixed scale that clearly improves results on Make3D, it is not the case for KITTI or Cityscapes, indicating that the depth in MPSD is indeed metric.

depth datasets and compute a scale factor as a least squares solution to align each predicted depth frame with the ground truth depth. The resulting scale factors are aggregated on Fig. 6, with average scales of 1.03, 1.01, and 0.89 for Cityscapes, KITTI and Make3D. The fact that we find a consistent underestimation of the depth on only one of the datasets implies that there is no scale bias in our dataset (networks trained on MPSD underestimate depth more often on Make3D due to the domain gap between MPSD and Make3D).

Scaling and Cropping Strategies. Table 2 collects all of our experimental results for single image depth. We report the scale-invariant SILog [7] score in all cases. Since the MegaDepth dataset does not provide metric depth, we only report the root-mean-square error (in meters) for networks trained on MPSD.

We compare naively resizing the images versus using either CAM-Convs or camera normalization to train on MegaDepth. Our experiments confirm the observations of Fácil et al. [9]: Accounting for the camera intrinsics explicitly greatly benefits training using datasets collected from more than one camera[7].

When comparing the two methods for dealing with multiple cameras, we found no clear winner. Both produce similar results, with the CAM-Convs producing the best results for some datasets and camera normalization in others. We also combined both methods (row #9, Table 2), resulting in the best performance for some of the evaluations. In this case, the CAM-Convs degenerate into a scaled version of Coord-Convs [19], a constant mapping of the viewing directions.

When using CAM-Convs, there is reason to believe that random crops may be used more effectively, as the concatenated viewing directions convey information about the cropped region. We experimented with using random crops during training (see rows #2 and #3 in Table 2) and found no conclusive results. We suspect that this is due to the wide aspect ratio for our crop size, as randomly cropping using this adspect ratio means in practice that the top and bottom of images will be sampled more often (usually containing regions with no ground truth depth like parts of the ego-vehicle or the sky) (Fig. 7).

---

[7] Refer to [9] for a thorough evaluation about the need of accounting for the focal length when training on datasets with multiple cameras.

Fig. 7. MPSD includes valid depth points on dynamic objects such that depth networks trained on MPSD are not *"dynamic-blind"*. These examples are produced by a network trained exclusively on MPSD and evaluated on KITTI.

Table 3. RMSE(m) on the KITTI validation set, separated in to static (traffic signs etc.) and dynamic regions (pedestrians, cars, bicycles etc.) regions

Training set(s)	Static	Dynamic
MegaDepth	93.04	117.98
MPSD	**4.16**	**5.16**
MegaDepth+KITTI	3.74	4.29
MPSD+KITTI	**3.12**	**3.52**

Dynamic Objects. In Sect. 2.3 we described how we include depth from static-during-capture dynamic objects in MPSD. To demonstrate that the included depth is valid (that is, that a network trained on MPSD can recover depth on dynamic objects), we have devised a simple experiment. We trained two versions of our architecture, one on MegaDepth, and one on MPSD. Each version is then fine-tuned on KITTI, leading to four different architectures. We then compare all of them on the KITTI validation set.

For each image, we first run a state-of-the-art segmentation network [23] to separate each pixel into dynamic or static. We then run the depth prediction network on the image and align the depth prediction to the ground truth[8]. Finally, we calculate the RMSE for dynamic and static regions and gather the results in Table 3. The small gap between the dynamic and static regions when training on MPSD indicates the presence of quality annotations on the dynamic regions (Table 4).

Competing in the KITTI Public Benchmark We have reported results after training on our diverse large scale dataset and evaluating on other smaller scale datasets that are well-known in the community without performing any fine-tuning. However, the best-known benchmark to date is still the KITTI depth dataset. The authors offer a test server to ensure fair comparison on a test set with held-out ground truth. Entries in this benchmark use KITTI and (optionally) other data to train their networks.

---

[8] This is so that the MegaDepth-trained network can be included in this comparison.

Table 4. KITTI leaderboard at the time of submission. Simple supervised pre-training on MPSD can outperform methods trained with new techniques such as ordinal regression [10] and planar guidance [16].

Rank	Method	SILog	sqErrorRel	absErrorRel	iRMSE
1	MPSD	**11.12**	**2.07** %	8.99 %	**11.56**
2	GSM (Anon.)	11.23	2.13 %	8.88 %	12.65
3	GSM (Anon.)	11.56	2.25 %	8.99 %	12.44
4	LCI (Anon.)	11.63	2.20 %	9.07 %	12.42
5	BTS [16]	11.67	2.21 %	9.04 %	12.23
6	AcED (Anon.)	11.70	2.45 %	9.54 %	12.51
7	DORN [10]	11.77	2.23 %	**8.78** %	12.98

To compete in the benchmark, we fine-tuned the network trained on MPSD and our proposed camera normalization scaling method (entry #7 on Table 2) on the KITTI training set for 3 epochs, resulting in entry #10 on Table 2. We evaluated the held-out set using this network and submitted our predictions to the official benchmark server, obtaining a SILog score of 11.12, surpassing all other entries at the time of submission. However, it is worth noting that, after fine-tuning, the network performs worse on all the other benchmarks than a network trained only on MPSD. This is evidenced by comparing entries #7 and #10 from Table 2: We believe that future research should focus on the cross-dataset scenario.

5   Conclusion

We have presented the generation procedure of Mapillary Planet-Scale Depth (MPSD), a depth dataset automatically generated from geo-tagged RGB images. The dataset has an unprecedented scale, geographical span, variety in appearance, and range of capturing devices. Additionally, we have addressed the difficulties that arise when using a dataset taken by a large heterogeneous set of cameras when training single-image depth estimation, comparing the existing CAM-Convs with a proposed alternative.

MPSD is larger and more varied than any other publicly available depth dataset, obtaining state-of-the-art results on several benchmarks in the cross-dataset scenario, where no fine-tuning is allowed. We also achieves a new state of the art result on the KITTI single-image depth benchmark by using MPSD to pre-train a depth network that is then fine-tuned on the benchmark.

References

1. OpenSfM. https://github.com/mapillary/OpenSfM

2. Bulo, S.R., Porzi, L., Kontschieder, P.: In-place activated BatchNorm for memory-optimized training of DNNs. In: Proceedings of the IEEE Computer Society Conference on Computer Vision and Pattern Recognition, pp. 5639–5647 (2018). https://doi.org/10.1109/CVPR.2018.00591

3. Caesar, H., et al.: nuScenes: a multimodal dataset for autonomous driving. CoRR arXiv:abs/1903.11027 (2019)

4. Chen, L.C., Papandreou, G., Schroff, F., Adam, H.: Rethinking Atrous Convolution for Semantic Image Segmentation (2017). http://arxiv.org/abs/1706.05587

5. Cordts, M., et al.: The cityscapes dataset for semantic urban scene understanding. In: Proceedings of the IEEE Conference on Computer Vision and Pattern Recognition (CVPR) (2016)

6. Deng, J., et al.: ImageNet: a large-scale hierarchical image database. In: Computer Vision and Pattern Recognition (CVPR) (2009)

7. Eigen, D., Puhrsch, C., Fergus, R.: Depth Map Prediction from a Single Image using a Multi-Scale Deep Network (2014). http://arxiv.org/abs/1406.2283

8. Everingham, M., Van Gool, L., Williams, C.K.I., Winn, J., Zisserman, A.: The Pascal visual object classes (VOC) challenge. Int. J. Comput. Vis. (IJCV) **88**(2), 303–338 (2010)

9. Facil, J.M., et al.: CAM-Convs: Camera-Aware Multi-Scale Convolutions for Single-View Depth (2019). http://arxiv.org/abs/1904.02028

10. Fu, H., Gong, M., Wang, C., Batmanghelich, K., Tao, D.: Deep Ordinal Regression Network for Monocular Depth Estimation (2018). https://doi.org/10.1109/CVPR.2018.00214. http://arxiv.org/abs/1806.02446

11. Geiger, A., Lenz, P., Urtasun, R.: Are we ready for autonomous driving? The KITTI vision benchmark suite. In: 2012 IEEE Conference on Computer Vision and Pattern Recognition, pp. 3354–3361. IEEE (2012). https://doi.org/10.1109/CVPR.2012.6248074. http://ieeexplore.ieee.org/document/6248074/

12. Godard, C., Mac Aodha, O., Brostow, G.: Digging Into Self-Supervised Monocular Depth Estimation (2018). http://arxiv.org/abs/1806.01260

13. Gordon, A., Li, H., Jonschkowski, R., Angelova, A.: Depth from Videos in the Wild: Unsupervised Monocular Depth Learning from Unknown Cameras (2019). http://arxiv.org/abs/1904.04998

14. Hold-Geoffroy, Y., et al.: A perceptual measure for deep single image camera calibration. In: The IEEE Conference on Computer Vision and Pattern Recognition (CVPR) (2017). arxiv:1712.01259. http://arxiv.org/abs/1712.01259

15. Koch, T., Liebel, L., Fraundorfer, F., Körner, M.: Evaluation of CNN-based single-image depth estimation methods. In: Leal-Taixé, L., Roth, S. (eds.) ECCV 2018. LNCS, vol. 11131, pp. 331–348. Springer, Cham (2018). https://doi.org/10.1007/978-3-030-11015-4_25

16. Lee, J.H., Han, M.K., Ko, D.W., Suh, I.H.: From Big to Small: Multi-Scale Local Planar Guidance for Monocular Depth Estimation (2019). http://arxiv.org/abs/1907.10326

17. Li, Z., Snavely, N.: MegaDepth: Learning Single-View Depth Prediction from Internet Photos (2018). https://doi.org/10.1109/CVPR.2018.00218. http://arxiv.org/abs/1804.00607

18. Lin, T., et al.: Microsoft COCO: Common objects in context. CoRR arXiv:abs/1405.0312 (2014)

19. Liu, R., et al.: An Intriguing Failing of Convolutional Neural Networks and the CoordConv Solution (NeurIPS), 1–26 (2018). http://arxiv.org/abs/1807.03247

20. López-Antequera, M., et al.: Deep single image camera calibration with radial distortion. In: Computer Vision and Pattern Recognition (CVPR) (2019)

21. Moulon, P., et al.: Openmvg. https://github.com/openMVG/openMVG
22. Neuhold, G., Ollmann, T., Rota Bulò, S., Kontschieder, P.: The Mapillary vistas dataset for semantic understanding of street scenes. In: International Conference on Computer Vision (ICCV) (2017)
23. Porzi, L., Bulò, S.R., Colovic, A., Kontschieder, P.: Seamless scene segmentation. In: CVPR (2019)
24. Saxena, A., Sun, M., Ng, A.Y.: Make3D: learning 3D scene structure from a single still image. IEEE Trans. Pattern Anal. Mach. Intell. **31**(5), 824–840 (2009). https://doi.org/10.1109/TPAMI.2008.132
25. Schonberger, J.L., Frahm, J.M.: Structure-from-motion revisited. In: IEEE Conference on Computer Vision and Pattern Recognition (CVPR), pp. 4104–4113 (2016). https://doi.org/10.1109/CVPR.2016.445. http://ieeexplore.ieee.org/document/7780814/
26. Shen, S.: Accurate multiple view 3D reconstruction using patch-based stereo for large-scale scenes. IEEE Trans. Image Process. **22**(5), 1901–1914 (2013). https://doi.org/10.1109/TIP.2013.2237921
27. Snavely, N., Seitz, S.M., Szeliski, R.: Photo tourism: exploring photo collections in 3D. In: SIGGRAPH Conference Proceedings, pp. 835–846. ACM Press, New York, NY, USA (2006)
28. Sturm, P.: Critical motion sequences for monocular self-calibration and uncalibrated euclidean reconstruction. In: Proceedings of IEEE Computer Society Conference on Computer Vision and Pattern Recognition (CVPR) (1997). https://doi.org/10.1109/CVPR.1997.609467
29. Vasiljevic, I., et al.: DIODE: A Dense Indoor and Outdoor DEpth Dataset (2019). http://arxiv.org/abs/1908.00463
30. Wang, C., Lucey, S., Perazzi, F., Wang, O.: Web stereo video supervision for depth prediction from dynamic scenes. CoRR arXiv:abs/1904.11112 (2019). http://arxiv.org/abs/1904.11112
31. Yu, F., et al.: BDD100K: A diverse driving video database with scalable annotation tooling. CoRR arXiv:abs/1805.04687 (2018)

V2VNet: Vehicle-to-Vehicle Communication for Joint Perception and Prediction

Tsun-Hsuan Wang[1], Sivabalan Manivasagam[1,2]([✉]), Ming Liang[1], Bin Yang[1,2], Wenyuan Zeng[1,2], and Raquel Urtasun[1,2]

[1] UberATG, Pittsburgh, USA
[2] University of Toronto, Toronto, Canada
{tsunhsuan.wang,manivasagam,ming.liang,byang,wenyuan,urtasun}@uber.com

Abstract. In this paper, we explore the use of vehicle-to-vehicle (V2V) communication to improve the perception and motion forecasting performance of self-driving vehicles. By intelligently aggregating the information received from multiple nearby vehicles, we can observe the same scene from different viewpoints. This allows us to see through occlusions and detect actors at long range, where the observations are very sparse or non-existent. We also show that our approach of sending compressed deep feature map activations achieves high accuracy while satisfying communication bandwidth requirements.

Keywords: Autonomous driving · Object detection · Motion forecast

1 Introduction

While a world densely populated with self-driving vehicles (SDVs) might seem futuristic, these vehicles will one day soon be the norm. They will provide safer, cheaper and less congested transportation solutions for everyone, everywhere. A core component of self-driving vehicles is their ability to perceive the world. From sensor data, the SDV needs to reason about the scene in 3D, identify the other agents, and forecast how their futures might play out. These tasks are commonly referred to as perception and motion forecasting. Both strong perception and motion forecasting are critical for the SDV to plan and maneuver through traffic to get from one point to another safely.

The reliability of perception and motion forecasting algorithms has significantly improved in the past few years due to the development of neural network architectures that can reason in 3D and intelligently fuse multi-sensor data (e.g., images, LiDAR, maps) [28,29]. Motion forecasting algorithm performance has been further improved by building good multimodal distributions [4,6,12,19] that capture diverse actor behaviour and by modelling actor

Electronic supplementary material The online version of this chapter (https:// doi.org/10.1007/978-3-030-58536-5_36) contains supplementary material, which is available to authorized users.

© Springer Nature Switzerland AG 2020
A. Vedaldi et al. (Eds.): ECCV 2020, LNCS 12347, pp. 605–621, 2020.
https://doi.org/10.1007/978-3-030-58536-5_36

Fig. 1. *Left:* Safety critical scenario of a pedestrian coming out of occlusion. V2V communication can be leveraged to use the fact that multiple self-driving vehicles see the scene from different viewpoints, and thus see through occluders. *Right:* Example *V2VSim* Scene. Virtual scene with occluded actor (blue) and SDVs (red and green), Rendered LiDAR from each SDV in the scene. (Color figure online)

interactions [3,25,36,37]. Recently, [5,31] propose approaches that perform joint perception and motion forecasting, dubbed *perception and prediction* (P&P), further increasing the accuracy while being computationally more efficient than classical two-step pipelines.

Despite these advances, challenges remain. For example, objects that are heavily occluded or far away result in sparse observations and pose a challenge for modern computer vision systems. Failing to detect and predict the intention of these hard-to-see actors might have catastrophic consequences in safety critical situations when there are only a few miliseconds to react: imagine the SDV driving along a road and a child chasing after a soccer ball runs into the street from behind a parked car (Fig. 1, left). This situation is difficult for both SDVs and human drivers to correctly perceive and adjust for. The crux of the problem is that the SDV and the human can only see the scene from a single viewpoint.

However, SDVs could have super-human capabilities if we equip them with the ability to transmit information and utilize the information received from nearby vehicles to better perceive the world. Then the SDV could see behind the occlusion and detect the child earlier, allowing for a safer avoidance maneuver.

In this paper, we consider the *vehicle-to-vehicle* (V2V) communication setting, where each vehicle can broadcast and receive information to/from nearby vehicles (within a 70 m radius). Note that this broadcast range is realistic based on existing communication protocols [21]. We show that to achieve the best compromise of having strong perception and motion forecasting performance while also satisfying existing hardware transmission bandwidth capabilities, we should send compressed intermediate representations of the P&P neural network. Thus, we derive a novel P&P model, called *V2VNet*, which utilizes a spatially aware graph neural network (GNN) to aggregate the information received from all the nearby SDVs, allowing us to intelligently combine information from different points in time and viewpoints in the scene.

To evaluate our approach, we require a dataset where multiple self-driving vehicles are in the same local traffic scene. Unfortunately, no such dataset exists. Therefore, our second contribution is a new dataset, dubbed *V2V-Sim* (see Fig. 1, right) that mimics the setting where there are multiple SDVs driving in the area. Towards this goal, we use a high-fidelity LiDAR simulator [33], which uses a large catalog of static 3D scenes and dynamic objects built from real-world data, to

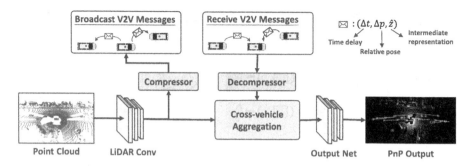

Fig. 2. Overview of V2VNet.

simulate realistic LiDAR point clouds for a given traffic scene. With this simulator, we can recreate traffic scenarios recorded from the real-world and simulate them as if a percentage of the vehicles are SDVs in the network. We show that V2VNet and other V2V methods significantly boosts performance relative to the single vehicle system, and that our compressed intermediate representations reduce bandwidth requirements without sacrificing performance. We hope this work brings attention to the potential benefits of the V2V setting for bringing safer autonomous vehicles on the road. To enable this, we plan to release this new dataset and make a challenge with a leaderboard and evaluation server.

2   Related Work

Joint Perception and Prediction: Detection and motion forecasting play a crucial role in any autonomous driving system. [3–5,25,30,31] unified 3D detection and motion forecasting for self-driving, gaining two key benefits: (1) Sharing computation of both tasks achieves efficient memory usage and fast inference time. (2) Jointly reasoning about detection and motion forecasting improves accuracy and robustness. We build upon these existing P&P models by incorporating V2V communication to share information from different SDVs, enhancing detection and motion forecasting.

Vehicle-to-Vehicle Perception: For the perception task, prior work has utilized messages encoding three types of data: raw sensor data, output detections, or metadata messages that contain vehicle information such as location, heading and speed. [34,38] associate the received V2V messages with outputs of local sensors. [8] aggregate LiDAR point clouds from other vehicles, followed by a deep network for detection. [35,44] process sensor measurements via a deep network and then generate perception outputs for cross-vehicle data sharing. In contrast, we leverage the power of deep networks by transmitting a compressed intermediate representation. Furthermore, while previous works demonstrate results on a limited number of simple and unrealistic scenarios, we showcase the effectiveness of our model on a diverse large-scale self-driving V2V dataset.

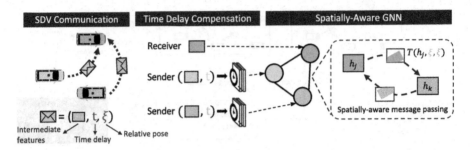

Fig. 3. After SDVs communicate messages, each receiver SDV compensates for time-delay of the received messages, and a GNN aggregates the spatial messages to compute the final intermediate representation.

Aggregation of Multiple Beliefs: In V2V setting, the receiver vehicle should collect and aggregate information from an arbitrary number of sender vehicles for downstream inference. A straightforward approach is to perform permutation invariant operations such as pooling [10,40] over features from different vehicles. However, this strategy ignores cross-vehicle relations (spatial locations, headings, times) and fails to jointly reason about features from the sender and receiver. On the other hand, recent work on graph neural networks (GNNs) has shown success on processing graph-structured data [15,18,26,46]. MPNN [17] abstract commonalities of GNNs with a message passing framework. GGNN [27] introduce a gating mechanism for node update in the propagation step. Graph-neural networks have also be effective in self-driving: [3,25] propose a spatially-aware GNN and an interaction transformer to model the interactions between actors in self-driving scenes. [41] uses GNNs to estimate value functions of map nodes and share vehicle information for coordinated route planning. We believe GNNs are tailored for V2V communication, as each vehicle can be a node in the graph. V2VNet leverages GNNs to aggregate and combine messages from other vehicles.

Active Perception: In V2V perception, the receiving vehicle should aggregate information from different viewpoints such that its field of view is maximized, trusting more the view that can see better. Our work is related to a long line of work in active perception, which focuses on deciding what action the agent should take to better perceive the environment. Active perception has been effective in localization and mapping [13,22], vision-based navigation [14], serving as a learning signal [20,48], and various other robotics applications [9]. In this work, rather than actively steering SDVs to obtain better viewpoint and sending information to the others, we consider a more realistic scenario where multiple SDVs have their own routes but are currently in the same geographical area, allowing the SDVs to see better by sharing perception messages.

3   Perceiving the World by Leveraging Multiple Vehicles

In this paper, we design a novel perception and motion forecasting model that enables the self-driving vehicle to leverage the fact that several SDVs may be present in the same geographic area. Following the success of joint perception and prediction algorithms [3,5,30,31], which we call P&P, we design our approach as a joint architecture to perform both tasks, which is enhanced to incorporate information received from other vehicles. Specifically, we would like to devise our P&P model to do the following: given sensor data the SDV should (1) process this data, (2) broadcast it, (3) incorporate information received from other nearby SDVs, and then (4) generate final estimates of where all traffic participants are in the 3D space and their predicted future trajectories.

Two key questions arise in the V2V setting: (i) what information should each vehicle broadcast to retain all the important information while minimizing the transmission bandwidth required? (ii) how should each vehicle incorporate the information received from other vehicles to increase the accuracy of its perception and motion forecasting outputs? In this section we address these two questions.

3.1   Which Information Should Be Transmitted

An SDV can choose to broadcast three types of information: (i) the raw sensor data, (ii) the intermediate representations of its P&P system, or (iii) the output detections and motion forecast trajectories. While all three message types are valuable for improving performance, we would like to minimize the message sizes while maximizing P&P accuracy gains. Note that small message sizes are critical because we want to leverage cheap, low-bandwidth, decentralized communication devices. While sending raw measurements minimizes information loss, they require more bandwidth. Furthermore, the receiving vehicle would need to process all additional sensor data received, which might prevent it from meeting the real-time inference requirements. On the other hand, transmitting the outputs of the P&P system is very good in terms of bandwidth, as only a few numbers need to be broadcasted. However, we may lose valuable scene context and uncertainty information that could be very important to better fuse the information.

In this paper, we argue that sending intermediate representations of the P&P network achieves the best of both worlds. First, each vehicle processes its own sensor data and computes its intermediate feature representation. This is compressed and broadcasted to nearby SDVs. Then, each SDV's intermediate representation is updated using the received messages from other SDVs. This is further processed through additional network layers to produce the final perception and motion forecasting outputs. This approach has two advantages: (1) Intermediate representations in deep networks can be easily compressed [11,43], while retaining important information for downstream tasks. (2) It has low computation overhead, as the sensor data from other vehicles has already been pre-processed.

In the following, we first showcase how to compute the intermediate representations and how to compress them. We then show how each vehicle should incorporate the received information to increase the accuracy of its P&P outputs.

Algorithm 1. Cross-vehicle Aggregation

1: **input:** representation $\hat{z}_i$, relative pose Δp_i, and time delay $\Delta t_{i \to k}$ for each SDV i
2: **for** each vehicle i **do**
3:     $h_i^{(0)} = CNN(\hat{z}_i, \Delta t_{i \to k}) \parallel 0$ ▷ Compensate time delay, init. node state
4: **end for**
5: **for** l iterations **do**   ▷ Message passing
6:     **for** each vehicle i **do**   ▷ Processed in parallel
7:         $m_{i \to k}^{(l)} = CNN(T(h_i^{(l)}, \xi_{i \to k}), h_k^{(l)}) \cdot M_{i \to k}$ ▷ Spatially transform message
8:         $h_i^{(l+1)} = ConvGRU(h_i^{(l)}, \phi_M([\forall_{j \in N(i)}, m_{j \to i}^{(l)}]))$ ▷ Node state update
9:     **end for**
10: **end for**
11: $z_i^{(L)} = MLP(h_i^{(L)})$ ▷ Output updated intermediate representation

---

3.2   Leveraging Multiple Vehicles

V2VNet has three main stages: (1) a convolutional network block that processes raw sensor data and creates a compressible intermediate representation, (2) a cross-vehicle aggregation stage, which aggregates information received from multiple vehicles with the vehicle's internal state (computed from its own sensor data) to compute an updated intermediate representation, (3) an output network that computes the final P&P outputs. We now describe these steps in more details. We refer the reader to Fig. 2 for our V2VNet architecture.

LiDAR Convolution Block: Following the architecture from [45], we extract features from LiDAR data and transform them into bird's-eye-view (BEV). Specifically, we voxelize the past five LiDAR point cloud sweeps into $15.6 \, \text{cm}^3$ voxels, apply several convolutional layers, and output feature maps of shape $H \times W \times C$, where $H \times W$ denotes the scene range in BEV, and C is the number of feature channels. We use 3 layers of 3×3 convolution filters (with strides of 2, 1, 2) to produce a 4x downsampled spatial feature map. This is the intermediate representation that we then compress and broadcast to other nearby SDVs.

Compression: We now describe how each vehicle compresses its intermediate representations prior to transmission. We adapt Ballé *et al.*'s variational image compression algorithm [2] to compress our intermediate representations; a convolutional network learns to compress our representations with the help of a learned hyperprior. The latent representation is then quantized and encoded losslessly with very few bits via entropy encoding. Note that our compression module is differentiable and therefore trainable, allowing our approach to learn how to preserve the feature map information while minimizing bandwidth.

Cross-vehicle Aggregation: After the SDV computes its intermediate representation and transmits its compressed bitstream, it decodes the representation received from other vehicles. Specifically, we apply entropy decoding to the bit stream and apply a decoder CNN to extract the decompressed feature map. We then aggregate the received information from other vehicles to produce an

Table 1. Detection Average Precision (AP) at IoU = {0.5, 0.7}, prediction with ℓ_2 error at recall 0.9 at different timestamps, and Trajectory Collision Rate (TCR).

Method	AP@IoU ↑		ℓ_2 Error (m) ↓			TCR ↓
	0.5	0.7	1.0 s	2.0 s	3.0 s	$\tau = 0.01$
No Fusion	77.3	68.5	0.43	0.67	0.98	2.84
Output Fusion	90.8	86.3	**0.29**	**0.50**	0.80	3.00
LiDAR Fusion	92.2	88.5	**0.29**	**0.50**	0.79	2.31
V2VNet	**93.1**	**89.9**	**0.29**	**0.50**	**0.78**	**2.25**

updated intermediate representation. Our aggregation module has to handle the fact that different SDVs are located at different spatial locations and see the actors at different timestamps due to the rolling shutter of the LiDAR sensor and the different triggering per vehicle of the sensors. This is important as the intermediate feature representations are spatially aware.

Towards this goal, each vehicle uses a fully-connected graph neural network (GNN) [39] as the aggregation module, where each node in the GNN is the state representation of an SDV in the scene, including itself (see Fig. 3). Each SDV maintains its own local graph based on which SDVs are within range (i.e., 70 m). GNNs are a natural choice as they handle dynamic graph topologies, which arise in the V2V setting. GNNs are deep-learning models tailored to graph-structured data: each node maintains a state representation, and for a fix number of iterations, messages are sent between nodes and the node states are updated based on the aggregated received information using a neural network. Note that the GNN messages are different from the messages transmitted/received by the SDVs: the GNN computation is done locally by the SDV. We design our GNN to temporally warp and spatially transform the received messages to the receiver's coordinate system. We now describe the aggregation process that the receiving vehicle performs. We refer the reader to Algorithm 1 for pseudocode.

We first compensate for the time delay between the vehicles to create an initial state for each node in the graph. Specifically, for each node, we apply a convolutional neural network (CNN) that takes as input the received intermediate representation $\hat{z}_i$, the relative 6DoF pose Δp_i between the receiving and transmitting SDVs and the time delay $\Delta t_{i \to k}$ with respect to the receiving vehicle sensor time. Note that for the node representing the receiving car, $\hat{z}$ is directly its intermediate representation. The time delay is computed as the time difference between the sweep start times of each vehicle, based on universal GPS time. We then take the time-delay-compensated representation and concatenate with zeros to augment the capacity of the node state to aggregate the information received from other vehicles after propagation (line 3 in Algorithm 1).

Next we perform GNN message passing. The key insight is that because the other SDVs are in the same local area, the node representations will have over-lapping fields of view. If we intelligently transform the representations and share information between nodes where the fields-of-view overlap, we can enhance the

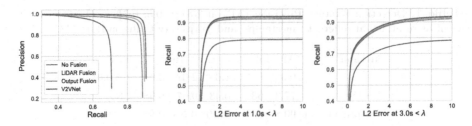

Fig. 4. Left: Detection Precision-Recall (PR) Curve at IoU $= 0.7$. Center/Right: Recall as a function of L_2 Error Prediction at 1.0 s and 3.0 s.

SDV's understanding of the scene and produce better output P&P. Figure 3 visually depicts our spatial aggregation module. We first apply a relative spatial transformation $\xi_{i \to k}$ to warp the intermediate state of the i-th node to send a GNN message to the k-th node. We then perform joint reasoning on the spatially-aligned feature maps of both nodes using a CNN. The final modified message is computed as in Algorithm 1 line 7, where T applies the spatial transformation and resampling of the feature state via bilinear-interpolation, and $M_{i \to k}$ masks out non-overlapping areas between the fields of view. Note that with this design, our messages maintain the spatial awareness.

We next aggregate at each node the received messages via a mask-aware permutation-invariant function ϕ_M and update the node state with a convolutional gated recurrent unit (ConvGRU) (Algorithm 1 line 8), where $j \in N(i)$ are the neighboring nodes in the network for node i and ϕ_M is the mean operator. The mask-aware accumulation operator ensures only overlapping fields-of-view are considered. In addition, the gating mechanism in the node update enables information selection for the accumulated received messages based on the current belief of the receiving SDV. After the final iteration, a multilayer perceptron outputs the updated intermediate representation (Algorithm 1 Line 11). We repeat this message propagation scheme for a fix number of iterations.

Output Network: After performing message passing, we apply a set of four Inception-like [42] convolutional blocks to capture multi-scale context efficiently, which is important for prediction. Finally, we take the feature map and exploit two network branches to output detection and motion forecasting estimates respectively. The detection output is (x, y, w, h, θ), denoting the position, size and orientation of each object. The output of the motion forecast branch is parameterized as (x_t, y_t), which denotes the object's location at future time step t. We forecast the motion of the actors for the next 3 s at 0.5 s intervals. Please see supplementary for additional architecture and implementation details.

3.3   Learning

We first pretrain the LiDAR backbone and output headers, bypassing the cross-vehicle aggregation stage. Our loss function is cross-entropy on the vehicle classification output and smooth ℓ_1 on the bounding box parameters. We apply hard-

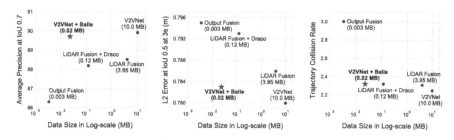

Fig. 5. Compression: Detection (AP at IoU 0.7), Prediction (ℓ_2 error at recall 0.9 at 3.0 s), and Trajectory Collision Rate ($\tau = 0.01$) performance on models with compression module.

negative mining to improve performance. We then finetune jointly the LiDAR backbone, cross-vehicle aggregation, and output header modules on our novel V2V dataset (see Sect. 4) with synchronized inputs (no time delay) using the same loss function. We do not use the temporal warping function at this stage. During training, for every example in the mini-batch, we randomly sample the number of connected vehicles uniformly on $[0, min(c, 6)]$, where c is the number of candidate vehicles available. This is to make sure V2VNet can handle arbitrary graph connectivity while also making sure the fraction of vehicles on the V2V network remains within the GPU memory constraints. Finally, the temporal warping function is trained to compensate for time delay with asynchronous inputs, where all other parts of the network are fixed. We uniformly sample time delay between 0.0 s and 0.1 s (time of one 10 Hz LiDAR sweep). We then train the compression module with the main network (backbone, aggregation, output header) fixed. We use a rate-distortion objective, which aims to maximize the bit rate in transmission while minimizing the distortion between uncompressed and decompressed data. We define the rate objective as the entropy of the transmitted code, and the distortion objective as the reconstruction loss (between the decompressed and uncompressed feature maps).

4 V2V-Sim: A Dataset for V2V Communication

No realistic dataset for V2V communication exists in the literature. Some approaches simulate the V2V setting by using different frames from KITTI [16] to emulate multiple vehicles [8,32,44]. However, this is unrealistic since sensor measurements are at different timestamps, so moving objects may be at completely different locations (e.g., a 1 s. time difference can cause 20 m change in position). Other approaches utilize a platoon strategy for data collection [7,23,35,47], where each vehicle follows behind the previous one closely. While more realistic than using KITTI, this data collection is biased: the perspectives of different vehicles are highly correlated with each other, and the data does not provide the richness of different V2V scenarios. For example, we will never see SDVs coming in the opposite direction, or SDVs turning from other lanes at intersections.

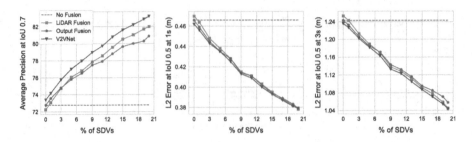

Fig. 6. Density of SDV: AP at IoU = 0.7 and ℓ_2 error at $\{1\,\text{s}, 3\,\text{s}\}$ at highest recall rate at IoU = 0.5 wrt % of SDVs in the scene.

To address these deficiencies, we use a high-fidelity LiDAR simulator, LiDAR-sim [33], to generate our large-scale V2V communication dataset, which we call *V2V-Sim*. LiDARsim is a simulation system that uses a large catalog of 3D static scenes and dynamic objects that are built from real-world data collections to simulate new scenarios. Given a scenario (i.e., scene, vehicle assets and their trajectories), LiDARsim applies raycasting followed by a deep neural network to generate a realistic LIDAR point cloud for each frame in the scenario.

We leverage traffic scenarios captured in the real world ATG4D dataset [45] to generate our simulations. We recreate the snippets in LiDARsim's virtual world using the ground-truth 3D tracks provided in ATG4D. By using the same scenario layouts and agent trajectories recorded from the real world, we can replicate realistic traffic. In particular, at each timestep, we place the actor 3D-assets into the virtual scene according to the real-world labels and generate the simulated LiDAR point cloud seen from the different candidate vehicles (see Fig. 1, right). We define the candidate vehicles to be non-parked vehicles that are within the 70-m broadcast range of the vehicle that recorded the real-world snippet. We generate 5500 25 s snippets collected from multiple cities. We subsample the frames in the snippets to produce our final 46,796/4,404 frames for train/test splits for the V2V-Sim dataset. V2V-Sim has on average 10 candidate vehicles that could be in the V2V network per sample, with a maximum of 63 and a variance of 7, demonstrating the traffic diversity. The fraction of vehicles that are candidates increases linearly w.r.t broadcast range.

5   Experimental Evaluation

In this section we showcase the performance of our approach compared to other transmission and aggregation strategies as well as single vehicle P&P.

Metrics: We evaluate both detection and motion forecasting around the ego-vehicle with a range of: $x \in [-100, 100]\,\text{m}$, $y \in [-40, 40]\,\text{m}$. We include completely occluded objects (0 LiDAR points hit the object), making the task much more challenging and realistic than standard benchmarks. For object detection, we compute Average Precision (AP) and Precision-Recall (PR) Curve at

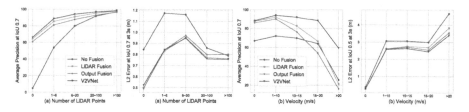

Fig. 7. Performance on objects with (fist two columns) different number of LiDAR point observation (last two columns) different velocities.

Intersection-over-Union (IoU) threshold of 0.7. For motion forecasting, we compute absolute ℓ_2 displacement error of the object center's location at future timestamps (3 s prediction horizon with 0.5 s interval) on true positive detections. We set the IoU threshold to 0.5 and recall to 0.9 (we pick the highest recall if 0.9 cannot be reached) to obtain the true positives. These values were chosen such that we retrieve most objects, which is critical for safety in self-driving. Note that most self-driving systems adopt this high recall as operating point. We also compute Trajectory Collision Rate (TCR), defined as the collision rate between the predicted trajectories of detected objects, where collision occurs when two cars overlap with each other more than a specific IoU (i.e., collision threshold τ). This metric evaluates whether the predictions are consistent with each other. We exclude the other SDVs during evaluation, as those can be trivially predicted.

Baselines: We evaluate the single vehicle setting, dubbed *No Fusion*, which consists of LiDAR backbone network and output headers only, without V2V communication. We also introduce two baselines for V2V communication: *LiDAR Fusion* and *Output Fusion*. *LiDAR Fusion* warps all received LiDAR sweeps from other vehicles to the coordinate frame of the receiver via the relative transformation between vehicles (which is known, as all SDVs are assume to be localized) and performs direct aggregation. We use the state-of-the-art LiDAR compression algorithm Draco [1] to compress *LiDAR Fusion* messages. For *Output Fusion*, each vehicle sends post-processed outputs, i.e., bounding boxes with confidence scores, and predicted future trajectories after non-maximum suppression (NMS). At the receiver end, all bounding boxes and future trajectories are first transformed to the ego-vehicle coordinate system and then aggregated across vehicles. NMS is then applied again to produce the final results.

Experimental Details: For all analysis we set the maximum number of SDVs per scene to be 7 (except for an ablation study measuring how the number of SDVs affect V2V performance in Fig. 6). All models are trained with Adam [24].

Comparison to Existing Approaches: As shown in Table 1, V2V-based models significantly outperform *No Fusion* on detection (~20% at IoU 0.7)

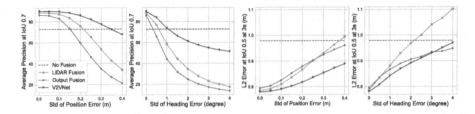

Fig. 8. Robustness on noisy vehicles' relative pose estimates.

and prediction ($\sim$0.2 m ℓ_2 error reduction at 3 s.). *LiDAR Fusion* and *V2VNet* also show strong reduction (20% at 0.01 collision threshold) in TCR. These results demonstrate that all types of V2V communication provide substantial performance gains. Among all V2V approaches, *V2VNet* is either on-par with *LiDAR Fusion* (which has no information loss) or achieves the best performance. V2VNet's slight performance gain over *LiDAR Fusion* may come from using the GNN in the cross-vehicle aggregation stage to reason about different vehicles' feature maps more intelligently than naive aggregation. *Output Fusion*'s drop in performance for TCR is due to the large number of false positives relative to other V2V methods (see detection PR curve Fig. 4, left, at recall >0.6). Figure 4 shows the percentage of objects with an ℓ_2 error at 3 s smaller than a constant. This metric shows similar trends consistent with Table 1.

Compression: Figure 5 shows the tradeoff between transmission bandwidth and accuracy for different V2V methods with and without compression. Draco [1] achieves 33x compression for *LiDAR Fusion*, while our compressed interme-diate representations achieved a 417x compression rate. Note that compression marginally affects the performance. This shows that the intermediate P&P rep-resentations are much easier to compress than LiDAR. Given the message size for one timestamp with a sensor capture rate of 10 Hz, we compute the transmis-sion delay based on V2V communication protocol [21]. At the broadcast range 120 m, the data rate is roughly 25 Mbps. This means sending *V2VNet* messages may induce roughly 9 ms delay, which is very low.

SDV Density: We now investigate how V2V performance changes as a function of % of SDVs in the scene. To make this setting like the real world, for a given 25 s snippet, we choose a fraction of candidate vehicles in the scene to be SDVs for the whole snippet. As shown in Fig. 6, V2V performance increases linearly with the % of SDVs in both detection and prediction.

Number of LiDAR Points, Velocity: As shown in Fig. 7(a) V2V methods boost the performance on completely- and mostly-occluded objects (0 and 1–6 LiDAR points) by over 60% in AP. This is an extremely exciting result, since the main challenges of perception and motion forecasting are objects with very

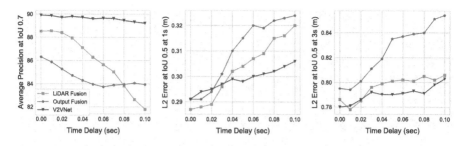

Fig. 9. Effect of time delay in data exchange.

sparse observations. Figure 7(b) shows performance on objects with different velocities. While other V2V methods drop in detection performance as object velocity increases, *V2VNet* has consistent performance gains over *No Fusion* on fast moving objects. *Output Fusion* and *LiDAR Fusion* may have deteriorated due to the rolling shutter of the moving SDV and the motion blur of moving agents during the temporal sweep of the LiDAR sensor. These effects are more severe in the V2V setting, where SDVs may be moving in opposite directions at high speeds while recording moving actors. Although not explicitly tackling such issue, *V2VNet* performs contextual and iterative reasoning on information from different vehicles, which may indirectly handle rolling shutter inconsistencies.

Imperfect Localization: We simulate inaccurate pose estimates by introducing different levels of Gaussian ($\sigma = 0.4$ m) and von Mises ($\sigma = 4°$; $\frac{1}{\kappa} = 4.873 \times 10^{-3}$) noise to position and heading of the transmitting SDVs. As shown in Fig. 8, on both noise types, *V2VNet* outperforms *LiDAR Fusion* and *Output Fusion* in P&P performance. The only exception is *Output Fusion* ℓ_2 error with heading noise larger than $3°$. We hypothesize that *Output Fusion's* performance is better at this setting due to its low-recall (fewer true positives) relative to V2VNet (0.62 vs. 0.73 at $4°$ noise). Fewer true positives can cause lower ℓ_2 error relative to higher recall methods. Degradation from heading noise is more severe than position noise, as subtle rotation in the ego-view will cause substantial misalignment for far-off objects; a vehicle bounding box (5 m $\times$ 2m) rotated by $1°$ with respect to a pivot 70 m away generates an IoU of 0.39 with the original.

Asynchronous Propagation: We simulate random time delay by delaying the messages of other vehicles at random from $\mathcal{U}(0, t)$, where $t = 0.1$. We apply a piece-wise linear velocity model (computed via finite differences) in *Output Fusion* to compensate for time delay. We do not make adjustments for *LiDAR Fusion* as it is non-trivial. As shown in Fig. 9, *V2VNet* demonstrates robustness across different time delays. *Output Fusion* does not perform well at high time delays as the piece-wise linear model used is sensitive to velocity estimates.

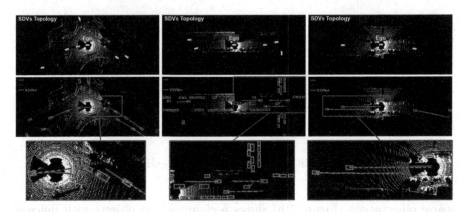

Fig. 10. V2V-Net Qualitative Examples. Left: Occluded car detected; Middle: Perception range increased; Right: Fast car detected.

Mixed Fleet: We also investigate the case that the SDV may receive different types of perception messages (i.e., sensor data, intermediate representation and P&P outputs). We analyze the setting where every SDV (other than the receiving vehicle) has 1/3 chance to broadcast each measurement type. We then perform *Sensor Fusion, V2VNet, Output Fusion* for the relevant set of messages to generate the final output. The result is in between the three V2V approaches: 88.6 AP at IoU = 0.7 for detection, 0.79 m error at 3.0 s prediction, and 2.63 TCR.

Qualitative Results: As shown in Fig. 10, V2VNet can see further and handle occlusion. For example, in Fig. 10 far right, we perceive and motion forecast a high-speed vehicle in our right lane, which can give the downstream planning system more information to better plan a safe maneuver for a lane change. *V2V-Net* also detects many more vehicles in the scene that were originally not detected by *No Fusion* (Fig. 10, middle).

6   Conclusion

In this paper, we have proposed a V2V approach for perception and prediction that transmits compressed intermediate representations of the P&P neural network, achieving the best compromise between accuracy improvements and bandwidth requirements. To demonstrate the effectiveness of our approach we have created a novel *V2V-Sim* dataset that realistically simulates the world when SDVs will be ubiquitous. We hope that our findings will inspire future work in V2V perception and motion forecasting strategies for safer self-driving cars.

Acknowledgments. We gratefully acknowledge James Tu for valuable contributions in the final paper.

References

1. Draco 3d data compression (2019). https://github.com/google/draco
2. Ballé, J., Minnen, D., Singh, S., Hwang, S.J., Johnston, N.: Variational image compression with a scale hyperprior. In: International Conference on Learning Representations (2018)
3. Casas, S., Gulino, C., Liao, R., Urtasun, R.: Spatially-aware graph neural networks for relational behavior forecasting from sensor data. arXiv (2019)
4. Casas, S., Gulino, C., Suo, S., Luo, K., Liao, R., Urtasun, R.: Implicit latent variable model for scene-consistent motion forecasting. In: ECCV (2020)
5. Casas, S., Luo, W., Urtasun, R.: Intentnet: learning to predict intention from raw sensor data. In: Conference on Robot Learning (2018)
6. Chai, Y., Sapp, B., Bansal, M., Anguelov, D.: Multipath: multiple probabilistic anchor trajectory hypotheses for behavior prediction. arXiv (2019)
7. Chen, Q., et al.: DSRC and radar object matching for cooperative driver assistance systems. In: 2015 IEEE Intelligent Vehicles Symposium (IV) (2015)
8. Chen, Q., Tang, S., Yang, Q., Fu, S.: Cooper: cooperative perception for connected autonomous vehicles based on 3D point clouds. arXiv (2019)
9. Chen, S., Li, Y., Kwok, N.M.: Active vision in robotic systems: a survey of recent developments. Int. J. Robot. Res. **30**(11), 1343–1377 (2011)
10. Chen, X., Ma, H., Wan, J., Li, B., Xia, T.: Multi-view 3D object detection network for autonomous driving. In: CVPR (2017)
11. Choi, H., Bajic, I.V.: High efficiency compression for object detection. In: 2018 IEEE International Conference on Acoustics, Speech and Signal Processing (ICASSP) (2018)
12. Cui, H., et al.: Deep kinematic models for physically realistic prediction of vehicle trajectories. arXiv (2019)
13. Davison, A.J., Murray, D.W.: Simultaneous localization and map-building using active vision. PAMI (2002)
14. Davison, A.J.: Mobile robot navigation using active vision. Advances in Scientific Philosophy Essays in Honour of (1999)
15. Duvenaud, D.K., et al.: Convolutional networks on graphs for learning molecular fingerprints. In: NIPS (2015)
16. Geiger, A., Lenz, P., Urtasun, R.: Are we ready for autonomous driving? The kitti vision benchmark suite. In: CVPR (2012)
17. Gilmer, J., Schoenholz, S.S., Riley, P.F., Vinyals, O., Dahl, G.E.: Neural message passing for quantum chemistry. In: Proceedings of the 34th International Conference on Machine Learning-Volume 70 (2017)
18. Hamilton, W., Ying, Z., Leskovec, J.: Inductive representation learning on large graphs. In: NIPS (2017)
19. Jain, A., Casas, S., Liao, R., Xiong, Y., Feng, S., Segal, S., Urtasun, R.: Discrete residual flow for probabilistic pedestrian behavior prediction. arXiv (2019)
20. Jayaraman, D., Grauman, K.: Look-ahead before you leap: end-to-end active recognition by forecasting the effect of motion. In: Leibe, B., Matas, J., Sebe, N., Welling, M. (eds.) ECCV 2016. LNCS, vol. 9909, pp. 489–505. Springer, Cham (2016). https://doi.org/10.1007/978-3-319-46454-1_30
21. Kenney, J.B.: Dedicated short-range communications (DSRC) standards in the united states. Proc. IEEE **99**(7), 1162–1182 (2011)
22. Kim, A., Eustice, R.M.: Active visual slam for robotic area coverage: theory and experiment. Int. J. Robot. Res. **34**(4–5), 457–475 (2015)

23. Kim, S.W., et al.: Multivehicle cooperative driving using cooperative perception: design and experimental validation. IEEE Trans. Intell. Transp. Syst. **16**(2), 663–680 (2014)

24. Kingma, D.P., Ba, J.: Adam: a method for stochastic optimization. In: 3rd International Conference on Learning Representations, ICLR 2015, Conference Track Proceedings, San Diego, CA, USA, 7–9 May 2015 (2015)

25. Li, L., Yang, B., Liang, M., Zeng, W., Ren, M., Segal, S., Urtasun, R.: End-to-end contextual perception and prediction with interaction transformer. In: IROS (2020)

26. Li, R., Tapaswi, M., Liao, R., Jia, J., Urtasun, R., Fidler, S.: Situation recognition with graph neural networks. In: ICCV (2017)

27. Li, Y., Tarlow, D., Brockschmidt, M., Zemel, R.S.: Gated graph sequence neural networks. In: 4th International Conference on Learning Representations, ICLR 2016, Conference Track Proceedings, San Juan, Puerto Rico, 2–4 May 2016 (2016)

28. Liang, M., Yang, B., Chen, Y., Hu, R., Urtasun, R.: Multi-task multi-sensor fusion for 3D object detection. In: CVPR (2019)

29. Liang, M., Yang, B., Wang, S., Urtasun, R.: Deep continuous fusion for multi-sensor 3D object detection. In: ECCV (2018)

30. Liang, M., Yang, B., Zeng, W., Chen, Y., Hu, R., Casas, S., Urtasun, R.: PnpNet: learning temporal instance representations for joint perception and motion forecasting. In: CVPR (2020)

31. Luo, W., Yang, B., Urtasun, R.: Fast and furious: real time end-to-end 3D detection, tracking and motion forecasting with a single convolutional net. In: CVPR (2018)

32. Maalej, Y., Sorour, S., Abdel-Rahim, A., Guizani, M.: Vanets meet autonomous vehicles: a multimodal 3D environment learning approach. In: GLOBECOM 2017–2017 IEEE Global Communications Conference (2017)

33. Manivasagam, S., et al.: Lidarsim: realistic lidar simulation by leveraging the real world. In: CVPR (2020)

34. Rauch, A., Klanner, F., Rasshofer, R., Dietmayer, K.: Car2x-based perception in a high-level fusion architecture for cooperative perception systems. In: 2012 IEEE Intelligent Vehicles Symposium (2012)

35. Rawashdeh, Z.Y., Wang, Z.: Collaborative automated driving: a machine learning-based method to enhance the accuracy of shared information. In: 2018 21st International Conference on Intelligent Transportation Systems (ITSC) (2018)

36. Rhinehart, N., Kitani, K.M., Vernaza, P.: R2p2: a reparameterized pushforward policy for diverse, precise generative path forecasting. In: ECCV (2018)

37. Rhinehart, N., McAllister, R., Kitani, K., Levine, S.: Precog: prediction conditioned on goals in visual multi-agent settings. arXiv (2019)

38. Rockl, M., Strang, T., Kranz, M.: V2V communications in automotive multi-sensor multi-target tracking. In: 2008 IEEE 68th Vehicular Technology Conference (2008)

39. Schlichtkrull, M., Kipf, T.N., Bloem, P., Van Den Berg, R., Titov, I., Welling, M.: Modeling relational data with graph convolutional networks. In: European Semantic Web Conference (2018)

40. Su, H., Maji, S., Kalogerakis, E., Learned-Miller, E.: Multi-view convolutional neural networks for 3D shape recognition. In: ICCV (2015)

41. Sykora, Q., Ren, M., Urtasun, R.: Multi-agent routing value iteration network. In: ICML 2020 (2020)

42. Szegedy, C., et al.: Going deeper with convolutions. In: CVPR (2015)

43. Wei, X., Barsan, I.A., Wang, S., Martinez, J., Urtasun, R.: Learning to localize through compressed binary maps. In: CVPR (2019)

44. Xiao, Z., Mo, Z., Jiang, K., Yang, D.: Multimedia fusion at semantic level in vehicle cooperative perception. In: 2018 IEEE International Conference on Multimedia & Expo Workshops (ICMEW) (2018)

45. Yang, B., Luo, W., Urtasun, R.: Pixor: real-time 3D object detection from point clouds. In: CVPR (2018)

46. Yu, B., Yin, H., Zhu, Z.: Spatio-temporal graph convolutional networks: a deep learning framework for traffic forecasting. arXiv (2017)

47. Yuan, T., et al.: Object matching for inter-vehicle communication systems-an IMM-based track association approach with sequential multiple hypothesis test. IEEE Trans. Intell. Transp. Syst. **18**(12), 3501–3512 (2017)

48. Yun, S., Choi, J., Yoo, Y., Yun, K., Young Choi, J.: Action-decision networks for visual tracking with deep reinforcement learning. In: CVPR (2017)

Training Interpretable Convolutional Neural Networks by Differentiating Class-Specific Filters

Haoyu Liang[1], Zhihao Ouyang[2,4], Yuyuan Zeng[2,3], Hang Su[1(✉)], Zihao He[5], Shu-Tao Xia[2,3], Jun Zhu[1(✉)], and Bo Zhang[1]

[1] Department of Computer Science and Technology, BNRist Center, Institute for AI, THBI Laboratory, Tsinghua University, Beijing 100084, China
{lianghy18,suhangss,dcszj,dcszb}@mails.tsinghua.edu.cn
[2] Tsinghua SIGS, Shenzhen 518055, China
{oyzh18,zengyy19}@mails.tsinghua.edu.cn, xiast@sz.tsinghua.edu.cn
[3] Peng Cheng Laboratory, University of Southern California, Los Angele, USA
[4] ByteDance AI Lab, University of Southern California, Los Angele, USA
[5] Department of CS, University of Southern California, Los Angele, USA
zihaoh@usc.edu

Abstract. Convolutional neural networks (CNNs) have been successfully used in a range of tasks. However, CNNs are often viewed as "blackbox" and lack of interpretability. One main reason is due to the *filter-class entanglement* – an intricate many-to-many correspondence between filters and classes. Most existing works attempt post-hoc interpretation on a pretrained model, while neglecting to reduce the entanglement underlying the model. In contrast, we focus on alleviating filter-class entanglement during training. Inspired by cellular differentiation, we propose a novel strategy to train interpretable CNNs by encouraging *class-specific filters*, among which each filter responds to only one (or few) class. Concretely, we design a learnable sparse Class-Specific Gate (CSG) structure to assign each filter with one (or few) class in a flexible way. The gate allows a filter's activation to pass only when the input samples come from the specific class. Extensive experiments demonstrate the fabulous performance of our method in generating a sparse and highly class-related representation of the input, which leads to stronger interpretability. Moreover, comparing with the standard training strategy, our model displays benefits in applications like object localization and adversarial sample detection. Code link: https://github.com/hyliang96/CSGCNN.

Keywords: Class-specific filters · Interpretability · Disentangled representation · Filter-class entanglement · Gate

H. Liang and Z. Ouyang—contributed equally.

Electronic supplementary material The online version of this chapter (https://doi.org/10.1007/978-3-030-58536-5_37) contains supplementary material, which is available to authorized users.

A. Vedaldi et al. (Eds.): ECCV 2020, LNCS 12347, pp. 622–638, 2020.
https://doi.org/10.1007/978-3-030-58536-5_37

1   Introduction

Convolutional Neural Networks (CNNs) demonstrate extraordinary performance in various visual tasks [13,16,17,23]. However, the strong expressive power of CNNs is still far from being interpretable, which significantly limits their applications that require humans' trust or interaction, e.g., self-driving cars and medical image analysis [3,8].

In this paper, we argue that filter-class entanglement is one of the most critical reasons that hamper the interpretability of CNNs. The intricate *many-to-many correspondence* relationship between filters and classes is so-called *filter-class entanglement* as shown on the left of Fig. 1. As a matter of fact, previous studies have shown that filters in CNNs generally extract features of a mixture of various semantic concepts, including the classes of objects, parts, scenes, textures, materials, and colors [2,45]. Therefore alleviating the entanglement is crucial for humans from interpreting the concepts of a filter [45], which has been shown as an essential role in the visualization and analysis of networks [30] in human-machine collaborative systems [41,44]. To alleviate the entanglement, this paper aims to learn *class-specific filter* which responds to only one (or few) class.

Usually, it is non-trivial to deal with the entanglement, as many existing works show. (1) Most interpretability-related research simply focuses on post-hoc interpretation of filters [2,36], which manages to interpret the main semantic concepts captured by a filter. However, post-hoc interpretation fails to alleviate the filter-class entanglement prevalent in pre-trained models. (2) Many VAEs' variants [6,9,18,20,24] and InfoGAN [10] try to learn disentangled data representation with better interpretability in an unsupervised way. However, they are challenged by [25], which proves that it's impossible unsupervised to learn disentangled features without proper inductive bias.

Despite the challenges above, it's reasonable and feasible to learning class-specific filters in *high convolutional layers* in image classification tasks. (1) It has been demonstrated that high-layer convolutional filters extract high-level semantic features which might relate to certain classes to some extent [40]; (2) the redundant overlap between the features extracted by different filters makes it possible to learn specialized filters [31]; (3) specialized filters demonstrate higher interpretability [45] and better performance [31] in computer vision tasks; (4) [19,28,38] successfully learn class-specific filters in high convolution layers, though, under an inflexible predefined filter-class correspondence.

Therefore, we propose to learn class-specific filters in the last convolutional layer during training, which is inspired by cellular differentiation [35]. Through differentiation, the stem cells evolve to functional cells with specialized instincts, so as to support sophisticated functions of the multi-cellular organism effectively. For example, neural stem cells will differentiate into different categories like neurons, astrocytes, and oligodendrocytes through particular simulations from transmitting amplifying cells. Similarly, we expect the filters in CNN to "differentiate" to disparate groups that have specialized responsibilities for specific tasks. For specifically, we encourage the CNN to build a one-filter to one-class correspondence (differentiation) during training.

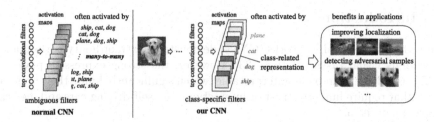

Fig. 1. The motivation of learning class-specific filters. In a normal CNN, each filter responds to multiple classes, since it extracts a mixture of features from many classes [45], which is a symptom of filter-class entanglement. In contrast, we enforce each filter to respond to one (or few) class, namely to be class-specific. It brings better interpretability and class-related feature representation. Such features not only facilitate understanding the inner logic of CNNs, but also benefits applications like object localization and adversarial sample detection.

Specifically, we propose a novel training method to learn class-specific filters. Different from existing works on class-specific filters that predefine filer-class correspondence, our model learns a *flexible* correspondence that assigns only a necessary portion of filters to a class and allows classes to share filters. Specifically, we design a learnable Class-Specific Gate (CSG) structure after the last convolutional filters, which assigns filters to classes and limits each filter's activation to pass only when its specific class(es) is input. In our training process, we periodically insert CSG into the CNN and jointly minimize the classification cross-entropy and the sparsity of CSG, so as to keep the model's performance on classification meanwhile encourage class-specific filters. Experimental results in Sect. 4 demonstrate that our training method makes data representation sparse and highly correlated with the labeled class, which not only illustrates the alleviation of filter-class entanglement but also enhances the interpretability from many aspects like filter orthogonality and filter redundancy. Besides, in Sect. 5 our method shows benefits in applications including improving objects localization and adversarial sample detection.

Contributions. The contributions of this work can be summarized as: (1) we propose a novel training strategy for CNNs to learn a flexible class-filter correspondence where each filter extract features mainly from only one or few classes; (2) we propose to evaluate filter-class correspondence with the mutual information between filter activation and prediction on classes, and moreover, we design a metric based on it to evaluate the overall filter-class entanglement in a network layer; (3) we quantitively demonstrate the benefits of the class-specific filter in alleviating filter redundancy, enhancing interpretability and applications like object localization and adversarial sample detection.

2   Related Works

Existing works related to our work include post-hoc filter interpretation, learning disentangled representation.

Post-hoc Interpretation for Filters is widely studied, which aims to interpret the patterns captured by filters in pre-trained CNNs. Plenty of works visualize the pattern of a neuron as an image, which is the gradient [27,34,40] or accumulated gradient [29,30] of a certain score about the activation of the neuron. Some works determine the main visual patterns extracted by a convolutional filter by treating it as a pattern detector [2] or appending an auxiliary detection module [14]. Some other works transfer the representation in CNN into an explanatory graph [42,43] or a decision tree [1,46], which aims to figure out the visual patterns of filters and the relationship between co-activated patterns. Post-hoc filter interpretation helps to understand the main patterns of a filter but makes no change to the existing filter-class entanglement of the pre-trained models, while our work aims to train interpretable models.

Learning Disentangled Representation refers to learning data representation that encodes different semantic information into different dimensions. As a principle, it's proved impossible to learn disentangled representation without inductive bias [25]. Unsupervised methods such as variants of VAEs [21] and InfoGAN [10] rely on regularization. VAEs [21] are modified into many variants [6,9,18,20,24], while their disentangling performance is sensitive to hyperparameters and random seeds. Some other unsupervised methods rely on special network architectures including interpretable CNNs [45] and CapsNet [33]. As for supervised methods, [37] propose to disentangle with interaction with the environment; [4] apply weak supervision from grouping information, while our work applies weak supervision from classification labels.

Class-specific Filters has been applied in image and video classification task. The existing works focus on improve accuracy, including label consistent neuron [19] and filter bank [28,38]. However, those works predefine an unlearnable correspondence between filters and classes where principally each filter responds to only one class and all classes occupy the same number of filters. In contrast, this paper focuses on the interpretability of class-specific filters, and we propose a more flexible correspondence where similar classes can share filters and a class can occupy a learnable number of filters. Therefore, our learnable correspondence helps reveal inter-class similarity and intra-class variability.

3   Method

Learning disentangled filters in CNNs alleviates filter-class entanglement and meanwhile narrows the gap between human concept and CNN's representations. In this section, we first present an ideal case of class-specific filters, which is a direction for our disentanglement training, and then we elaborate on our method about how to induce filter differentiation towards it in training an interpretable network.

3.1   Ideally Class-Specific Filters

This subsection introduces an ideal case for the target of our filter disentanglement training. As shown in Fig. 2, each filter mainly responds to (i.e. relates to)

normal filters   ideally class-specific filters   relaxed class-specific filters   filters activated by a 'dog'

shared by many classes   ship   dog   cat   plane   ship │ dog   cat │ plane
                                                                        ship & plane   dog & cat

Fig. 2. The intuition of learning class-specific filters. In a standard CNN, a filter extracts a mixture of features from many classes [45], which is a symptom of filter-class entanglement. In contrast, an ideally class-specific filter extracts features mostly from only one class and a relaxed filter can be shared by few classes, For flexibility, we actually apply the relaxed class-specific filter which is allowed to be shared by few classes. Its activation to other classes is weak and has little effect on prediction.

only one class. We call such filters *ideally class-specific* and call disentangling filter towards it in training as *class-specific differentiation* of filters.

To give a rigorous definition of "ideally class-specific" for a convolutional layers, we use a matrix $G \in [0,1]^{C \times K}$ to measure the relevance between filters and classes, where K is the number of filters, C is the number of classes. Each element $G_c^k \in [0,1]$ represents the relevance between the k-th filter and c-th class, (a larger G_c^k indicates a closer correlation). As shown in Fig. 3, the k-th filter extracts features mainly in the c-th class iif $G_c^k = 1$. Denote a sample in dataset D as $(x, y) \in D$ where x is an image and $y \in \{1, 2, ..., C\}$ is the label. Given (x, y) as an input, we can index a row $G_y \in [0,1]^K$ from the matrix G, which can be used as a gate multiplied to the activation maps to shut down those irrelevant channels. Let $\tilde{y}$ be the probability vector predicted by the STD path, and $\tilde{y}^G$ be the probability vector predicted by the LSG path where the gate G_y is multiplied on the activation maps from the penultimate layer. Thus, we call convolutional filters as *ideally class-specific* filters, if there exists a G (all columns G_k are *one-hot*) that raises little difference between the classification performance of $\tilde{y}^G$ and $\tilde{y}$.

3.2   Problem Formulation

In order to train a CNN towards differentiating filters to class-specific meanwhile keep classification accuracy, we introduce a Class-Specific Gate (CSG) path in addition to the standard (STD) path of forward propagation. In the CSG path, channels are selectively blocked with the learnable gates. This path's classification performance is regarded as a regularization for filter differentiation training.

To derive the formulation of training a CNN with class-specific filters, we start from an original problem that learns ideally class-specific filters and then relax the problem for the convenience of a practical solution.

The Original Problem is to train a CNN with ideally class-specific filters. See Fig. 3 for the network structure. The network with parameters θ forward propagates in two paths: (1) the standard (STD) path predicting $\tilde{y}_\theta$, and (2) the CSG path with gate matrix G predicting $\tilde{y}_\theta^G$ where activations of the penultimate layer are multiplied by learnable gates G_y for inputs with label y.

In order to find the gate matrix G that precisely describes the relevance between filters and classes, we search in the binary space for a G that yields the best classification performance through the CSG path, i.e., to solve the

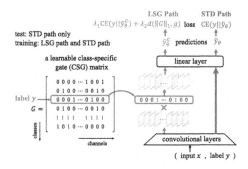

Fig. 3. In Class-Specific Gate (CSG) training, we *alternately* train a CNN through the CSG path and the standard (STD) path. In the CSG path, activations after the penultimate layer (i.e. the last convolution) pass through learnable gates indexed by the label. In training, network parameters and the gates are optimized to minimize the cross-entropy joint with a sparsity regularization for the CSG matrix. In testing, we just run the STD path.

optimization problem $\Phi_0(\theta) = \min\limits_{G} \text{CE}(y||\tilde{y}_\theta^G)$[1] s.t. $\forall k \in \{1, 2, ..., K\}, G_k$ is one-hot. Φ_0 evaluates the performance of the CNN with differentiated filters. Therefore, it is natural to add Φ_0 into training loss as a regularization that forces filters to be class-specific. Thus, we get the following formulation of the *original problem* to train a CNN towards ideally class-specific filters as

$$\min\limits_{\theta} L_0(\theta) = \text{CE}(y||\tilde{y}_\theta) + \lambda_1 \Phi_0(\theta). \tag{1}$$

where the $CE(y||\tilde{y}_\theta)$ ensures the accuracy and $\lambda_1 \Phi_0(\theta)$ encourage sparsity of G.

However, the original problem is difficult to solve in practice. On the one hand, the assumption that each filter is complete one-filter/one-class assumption hardly holds, since it is usual for several classes to share one high-level feature in CNNs; on the other hand, binary vectors in a non-continuous space are difficult to optimize with gradient descent.

Relaxation. To overcome the two difficulties in the original problem above, we relax relax the one-hot vector G^k to a sparse continuous vector $G^k \in [0,1]^C$ where at least one element equals to 1, i.e., $\left\|G^k\right\|_\infty = 1$. To encourage the sparsity of G, we introduce a regularization $d(\left\|G\right\|_1, g)$ that encourages the L1 *vector norm* $\left\|G\right\|_1$ not to exceed the upper bound g when $\left\|G\right\|_1 \geq g$, and has no effect when $\left\|G\right\|_1 < g$. A general form for d is $d(a, b) = \psi(\text{ReLU}(a - b))$, where ψ can be any norm, including L1, L2 and smooth-L1 norm. Besides, we should set $g \geq K$ because $\left\|G\right\|_1 \geq K$ which is ensured by $\left\|G^k\right\|_\infty = 1$. Using the aforementioned relaxation, Φ_0 is reformulated as

$$\Phi(\theta) = \min\limits_{G}\{\text{CE}(y||\tilde{y}_\theta^G) + \mu d(\left\|G\right\|_1, g)\} \quad s.t. \ G \in V_G, \tag{2}$$

---

[1] $\text{CE}(y||\tilde{y}_\theta^G) = -\frac{1}{|D|}\sum_{(x,y)\in D} \log((\tilde{y}_\theta^G)_y)$, where $\tilde{y}_\theta^G$ is a predicted probability vector.

Algorithm 1. CSG Training

1: **for** e in epochs **do**
2:     **for** n in batches **do**
3:         **if** $e \% period \leq epoch\_num\_for\_CSG$ **then**
4:             $\tilde{y}_\theta^G \leftarrow$ prediction through the CSG path with G
5:             $\mathcal{L} \leftarrow \lambda_1 \mathrm{CE}(y || \tilde{y}_\theta^G) + \lambda_2 d(||G||_1, g)$
6:             $G \leftarrow G - \epsilon \frac{\partial \mathcal{L}}{\partial G}$ ▷ update G using the gradient decent
7:             $G^k \leftarrow G^k / ||G^k||_\infty$ ▷ normalize each column of G
8:             $G \leftarrow \mathrm{clip}(G, 0, 1)$
9:         **else**
10:            $\tilde{y}_\theta \leftarrow$ prediction through the STD path
11:            $\mathcal{L} \leftarrow \mathrm{CE}(y || \tilde{y}_\theta)$
12:        **end if**
13:        $\theta \leftarrow \theta - \epsilon \frac{\partial \mathcal{L}}{\partial \theta}$
14:     **end for**
15: **end for**

---

where the set $V_G = \{ G \in [0,1]^{C \times K} : ||G^k||_\infty = 1 \}$ and μ is a coefficient to balance classification and sparsity. Φ can be regarded as a loss function for *filter-class entanglement*, i.e., a CNN with higher class-specificity has a lower Φ.

Replacing Φ_0 in Eq. (1) with Φ, we get $\min_\theta \mathrm{CE}(y || \tilde{y}_\theta) + \lambda_1 \Phi(\theta)$ as an intermediate problem. It is mathematically equivalent if we move $\min_G$ within Φ to the leftmost and replace $\lambda_1 \mu$ with λ_2. Thus, combining Eqs. (1) and (2), we formulate a *relaxed problem* as

$$\min_{\theta, G} L(\theta, G) = \mathrm{CE}(y || \tilde{y}_\theta) + \lambda_1 \mathrm{CE}(y || \tilde{y}_\theta^G) + \lambda_2 d(||G||_1, g) \quad s.t. \ G \in V_G. \quad (3)$$

The relaxed problem is easier to solve by jointly optimizing θ and G with gradient, compared to either the discrete optimization in the original problem or the nested optimization in the intermediate problem. Solving the relaxed problem, we can obtain a CNN for classification with class-specific filters, where G precisely describes the correlation between filters and classes.

3.3 Optimization

To solve the optimization problem formulated in Eq. (3) we apply an approximate projected gradient descent (PGD): when G is updated with gradient, G^k will be normalized by $||G^k||_\infty$ to ensure $||G^k||_\infty = 1$, and then clipped into the range $[0, 1]$.

However, it is probably difficult for the normal training scheme due to poor convergence. In the normal scheme, we predict through both CSG and STD paths to directly calculate $L(\theta, G)$ and update θ and G with gradients of it. Due to that most channels are blocked in the CSG path, the gradient through the CSG path will be much weaker than that of STD path, which hinders converging to class-specific filters.

Table 1. Metrics of the STD CNN (baseline) and the CSG CNN (Ours).

Dataset	Model	C	K	Training	Accuracy	MIS	L1-density	L1-interval[a]
CIFAR-10	ResNet20	10	64	CSG	**0.9192**	**0.1603**	0.1000	$[0.1, 0.1]$
				STD	0.9169	0.1119	-	-
ImageNet	ResNet18	1000	512	CSG	0.6784	**0.6259**	0.0016	$[0.001, 0.1]$
				STD	**0.6976**	0.5514	-	-
PASCAL VOC 2010	ResNet152	6	2048	CSG	**0.8506**	**0.1998**	0.1996	$[0.1667, 0.2]$
				STD	0.8429	0.1427	-	-

[a] They are $[\frac{1}{C}, \frac{g}{CK}]$ – the theoretical convergence interval for L1-density of CSG CNNs, where g is the upper bound for $\|G\|_1$ in Eq. (2). See Appendix A for the derivation. When the dataset has numerous classes like ImageNet, the L1-density can drop much lower than $\frac{g}{CK}$ due to the projection in our approximate PGD.

To address this issue, we propose an *alternate training scheme* that the STD/CSG path works alternately in different epochs. As shown in Algorithm 1, in the epoch for CSG path, we update G, θ with the gradient of $\lambda_1 \text{CE}(y\|\tilde{y}_\theta^G) + \lambda_2 d(\|G\|_1, g)$, and in the STD path we update θ with the gradient of $\text{CE}(y\|\tilde{y}_\theta)$ In this scheme, the classification performance fluctuates periodically at the beginning but the converged performance is slightly better than the normal scheme in our test. Meanwhile, the filters gradually differentiate into class-specific filters.

4 Experiment

In this section, we conduct five experiments. We first delve into CSG training from three aspects, so as to respectively study the effectiveness of CSG training, the class-specificity of filters and the correlation among class-specific filters. Especially, to measure filter-class correspondences we apply the mutual information (MI) between each filter's activation and the prediction on each class. In the following parts, we denote our training method Class-Specific Gate as *CSG*, the standard training as *STD*, and CNNs trained with them as *CSG CNNs* and *STD CNNs*, respectively.

Training. We use CSG/STD to train ResNet-18/20/152s [17] for classification task on CIFAR-10 [22]/ImageNet [11]/PASCAL VOC 2010 [12] respectively. We select six animals from PASCAL VOC and preprocess it to be a classification dataset. The ResNet-18/20s are trained from scratch and the ResNet-152s are finetuned from ImaageNet. See Appendix B for detailed training settings.

4.1 Effectiveness of CSG Training

First of all, we conduct experiments to verify the effectiveness of our CSG training in learning a sparse gate matrix and achieve high class-specificity of filters.

Quantitative Evaluation Metrics. To evaluate the effectiveness of CSG, we calculate 3 metrics: L1-density, mutual information score, and classification accuracy. (1) Accuracy measures the classification performance. (2) To measure the correspondence between filters and classes, we propose the mutual information (MI) matrix $M \in \mathbb{R}^{K \times C}$ where $M_{kc} = \mathrm{MI}(a_k \| \mathbf{1}_{y=c})$ is the MI between a_k – the activation of filter-k and class-c. To calculate the MI, we sample (x, y) across the dataset, a_k (the globally avg-pooled activation map of filter k over all the sampled x) is a continuous variable, and $\mathbf{1}_{y=c}$ is a categorical variable. The estimation method [32] for the MI between them is implemented in the API 'sklearn.feature_selection.mutual_info_classif'. Base on this, we propose a mutual information score $MIS = \mathrm{mean}_k \max_c M_{kc}$ as an overall metric of class-specificity of all filters. Higher MIS indicates higher class-specificity, aka, lower filter-class entanglement. (3) The L1-density $= \frac{\|G\|_1}{KC}$ is the L1-norm of CSG normalized by the number of elements, which measures the sparsity of CSG.

Table 1 shows that CSG CNNs are comparable to or even slightly outperforms STD CNNs in test accuracy, while the CSG CNNs have MIS much higher than STD CNNs and the L1-density of G is limited in its theoretical convergence interval. These metrics quantitatively demonstrate CSG's effectiveness on learning a sparse gate matrix and class-specific filters without sacrifice on classification accuracy.

Visualizing the Gate/MI Matrices. To demonstrate that the relevance between the learned filters and classes is exactly described by the gate matrix, we visualize gate matrix and MI matrices in Fig. 4. (a) demonstrates that CSG training yields a sparse CSG matrix where each filter is only related to one or few classes. (b1,b2) shows that CSG CNN has sparser MI matrices and larger MIS compared to STD CNN. (c) shows the strongest elements in the two matrices almost overlaps, which indicates that CSG effectively

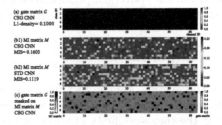

Fig. 4. (a) visualizes the CSG matrix of CSG CNN to verify its sparsity. (b1, b2) compare the MI matrices of CSG/STD CNN and their MIS. (c) is got by overlapping (a) and (b1).

learns filters following the guidance of the CSG matrix. These together verify that CSG training effectively learns a sparse gate matrix and filters focusing on the one or few classes described by the gates.

4.2 Study on Class-Specificity

In this subsection, we study the property of class-specific filters and the mechanism of filter differentiation through experiments on ResNet20 trained in Sect. 4.1.

Indispensability for Related Class. It is already shown by the MI matrix that filters are class-specific, while we further reveal that the filters related to a class are also *indispensable* in recognizing the class. We remove the filters highly

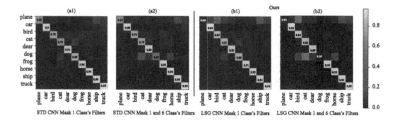

Fig. 5. Classification confusion matrix for STD/CSG models when masking filters highly related to the first class (a1, b1), and the first and sixth classes (a2, b2).

related to certain class(es) referencing the gate matrix (i.e., $G_c^k > 0.5$), and then visualize the classification confusion matrices. For STD CNN, we simply remove 10% filters that have the largest average activation to certain classes across the dataset. As shown in Fig. 5, when filters highly related to "plane" are removed, the CSG CNN fails to recognize the first class "plane"; nevertheless, the STD CNN still manages to recognize "plane". Analogously, when removed the filters highly related to the first and sixth classes, we observe a similar phenomenon. This demonstrates that the filters specific to a class are indispensable in recognizing this class. Such a phenomenon is because the filters beyond the group mainly respond to other classes and hence can't substitute for those filters.

Mechanism of Class-Specific Differentiation. In this part we reveal that the mechanism of filter differentiation by studying the directional similarity between features and gates, and by the way, explain misclassification with the CSG. We use cosine to measure the directional similarity S_y^c between $a_y \in \mathbb{R}^K$ the mean feature (average-pooled feature from class-specific filters) for class y, and G_c the c-th row of the CSG matrix (see Appendix C for details). We calculate S_y^c over all true positive (TP) and false negative (FN) images and

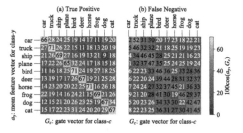

Fig. 6. The cosine similarity between mean feature vectors for a class (y-axis) and a row in the CSG matrix (x-axis), calculated on all TP/FN samples. We reorder classes in CIFAR10 for better visualization.

obtain similarity matrices $S_{TP}, S_{FN} \in [0, 1]^{C \times C}$ respectively, as shown in Fig. 6.

From the figure, we observe two phenomena and provide the following analysis. (1) TP similarity matrix is diagonally dominant. This reveals mechanism of learning class-specifics: CSG forces filters to yield feature vectors whose direction approaches that of the gate vector for its related class. (2) FN similarity matrix is far from diagonally dominant and two classes with many shared features, such as car & truck and ship & plane, have high similarity in the FN similarity matrix. CSG enlightens us that hard samples with feature across classes tend to

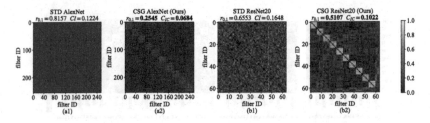

Fig. 7. The correlation matrix of filters (cosine between filters' weights) in AlexNet and ResNet20 trained with STD/CSG. $r_{0.1}$: the ratio of elements≥ 0.1, measures filter redundancy. C_{IC}: the inter-class filter correlation, measures inter-class filter similarity.

be misclassified. Thus, the mechanism of misclassification in the CSG CNN is probably that the features across classes are extracted by the shared filters. To some extent, it proves differentiation is beneficial for accuracy.

4.3   Correlation Between Filters

We further designed several experiments to explore what happen to filters (why they are class-specified). Our analysis shows inter-class filters are approximately orthogonal and less redundant, and the class-specific filters yield highly class-related representation.

Fixing the Gates. To make it convenient to study inter-class filters, we group filters in a tidier way – each class monopolizes m_1 filters and m_2 extra filters are shared by all classes. This setting can be regarded as tightening the constraint on the gate matrix to $G \in \{0,1\}^{C \times K}$, according to Sect. 3.3 (see Appendix E for illustration). The corresponding CSG matrix is fixed during training. We train an AlexNet [23] ($m_1 = 25, m_2 = 6$) and a ResNet20 ($m_1 = 6, m_2 = 4$) by STD and by CSG in this way. Models trained with this setting naturally inherits all features of previous CSG models and has tidier filter groups.

Filter Orthogonality Analysis. To study the orthogonality between filters, we evaluate filter correlation with the cosine of filters' weights. The correlations between all filters are visualized as correlations matrices $\mathcal{C}$ in Fig. 7. In subfigures (a1, b1) for STD models, the filters are randomly correlated with each other. In contrast, in subfigures (a2, b2) for CSG models, the matrices are approximately block-diagonal, which means the correlation between the filters is limited to several class-specific filter groups. This indicates that filters for the same class are highly correlated (non-orthogonal) due to the co-occurrence of features extracted by them, while filters for different classes are almost uncorrelated (orthogonal) for the lack of co-occurrence. See Appendix D for a detailed explanation for orthogonality.

Filter Redundancy. To verify that the redundancy of filters are reduced by CSG, we research the inter-class filter correlation. Let the filter-i, j are correlated if their correlation $C_{i,j} \geq s$ (s is a threshold). We calculate r_s the ratio of such elements in a correlation matrix and plot the results in Fig. 8. It shows that CSG significantly reduces the redundant correlation ($C_{i,j} < s = 0.1$) between most filter pairs from different groups. We explain the reduction of filter redundancy as a natural consequence of encouraging inter-class filters to be orthogonal.

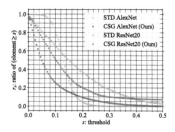

Fig. 8. The ratio of elements larger than a varying threshold in correlation matrices.

For a set of filter groups orthogonal to each other, a filter in any group can not be a linear representation with the filers from other groups. This directly avoids redundant filter across groups. Besides, experiments in [31] also verify the opinion that filter orthogonality reduces filter redundancy.

Highly Class-Related Representation. Based on filter correlation, we further find that filters trained with CSG yield highly class-related representation, namely the representation of an image tends to exactly correspond to its labeled class rather than other classes. Because the implied class is mainly decided on which filters are most activated, meanwhile those filters are less activated by other classes and less correlated to the filters for other classes.

To verify this reasoning, we analyze the correlation between the filters highly activated by each class. First, we pick out m_1 filters that have the strongest activation to class c, denoted as group A_c. We define inter-class filter correlation as the average correlation between filters in different classes' group A_c: $C_{IC} = \frac{1}{C(C-1)m^2} \sum_c \sum_{c \neq c'} \sum_{k \in A_c} \sum_{k' \in A_{c'}} C_{k,k'}$. The results C_{IC} in Fig. 7 show that inter-class filter correlation in CSG is about half of that in STD, both on AlexNet and ResNet20. This demonstrates that different classes tend to activate uncorrelated filters in CSG CNNs, aka, CSG CNNs yield highly class-related representations where the representations for different classes have less overlap.

5 Application

Using the class-specificity of filters, we can improve filters' interpretability on object localization. Moreover, the highly class-related representation makes it easier to distinguish abnormal behavior of adversarial samples.

5.1 Localization

In this subsection, we conduct experiments to demonstrate that our class-specific filters can localize a class better, for CSG training is demonstrated to encourage each filter in the penultimate layer to focus on fewer classes.

Table 2. The performance of localization with resized activation maps in the CSG/STD CNN. For almost all classes, CSG CNN significantly outperforms STD CNN both on Avg-IoU and AP20/AP30.

Localization	Metric	Training	Bird	Cat	Dog	Cow	Horse	Sheep	Total
GradMap for one filter	Avg-IoU	CSG	**0.2765**	**0.2876**	**0.3313**	**0.3261**	**0.3159**	**0.2857**	**0.3035**
		STD	0.2056	0.2568	0.2786	0.2921	0.2779	0.2698	0.2606
	AP20	CSG	**0.6624**	**0.8006**	**0.9170**	**0.8828**	**0.8204**	**0.8089**	**0.8165**
		STD	0.4759	0.7081	0.7556	0.8069	0.7621	0.7764	0.7029
ActivMap for one filter	Avg-IoU	CSG	**0.3270**	**0.4666**	**0.4005**	**0.4228**	**0.3358**	**0.4344**	**0.4290**
		STD	0.2865	0.3815	0.3443	0.3674	0.3020	0.3653	0.3602
	AP30	CSG	**0.5876**	**0.8609**	**0.7502**	**0.8075**	**0.5811**	**0.8606**	**0.8254**
		STD	0.4751	0.7003	0.6365	0.7216	0.5154	0.6972	0.6759
CAMs for all filters	Avg-IoU	CSG	**0.3489**	**0.4027**	**0.3640**	**0.3972**	**0.3524**	**0.3562**	**0.3694**
		STD	0.3458	0.3677	0.3492	0.3516	0.3170	0.3470	0.3483
	AP30	CSG	0.6399	**0.8382**	**0.7197**	**0.7517**	**0.7136**	**0.7073**	**0.7294**
		STD	**0.6495**	0.7832	0.7085	0.6621	0.5825	0.6504	0.6853

Localization method. Gradient maps [34] and activation maps (resized to input size) is a widely used method to determine the area of objects or visual concepts, which not only works in localization task without bounding box labels [2], but also take an important role in network visualization and understanding the function of filters [47]. We study CSG CNNs' performs on localizing object classes with three localization techniques based on filters, including gradient-based saliency map (GradMap) and activation map (ActivMap) for a single filter and classification activation map (CAM) [2] for all filters. See Appendix G for the localization techniques.

Quantitative evaluation. We train ResNet152s to do classification on preprocessed PASCAL VOC and use Avg-IoU (average intersection over union) and AP20/AP30 (average precision 20%/30%) to evaluate their localization. Higher metrics indicate better localization. See Appendix H for a detailed definition of the metrics. The results for localization with one or all filters are shown in Table 2. For localization with one filter, most classes are localized better with CSG CNN. That's because CSG encourages filters to be activated by the labeled class rather than many other classes, which alleviates other classes' interference on GradMaps and ActivMaps for each filters. Furthermore, as a weighted sum of better one-filter activation maps, CSG also outperform STD on CAMs.

Discussion. It is widely recognized [2] that localization reveals what semantics or classes a filter focuses on. Compared with a vanilla filter, a class-specific filter responds more intensively to the region of relevant semantics and less to the region of irrelevant semantics like background. Thus localization is improved. For STD training, confusion on the penultimate layer is caused by all convolutional layers, however, in our CSG training they are jointly trained with back-propagation to disentangle the penultimate layer.

Visualiziation. Besides the quantitative evaluation above, in Fig. 9, we also visualize some images and their CAMs from the STD/CSG CNN on ImageNet [11]. We observe that the CAMs of STD CNN often activate extra area beyond the labeled class. However, CSG training successfully helps the CNN find a more precise area of the labeled objects. Such a phenomenon vividly demonstrates that CSG training improves the performance in localization.

Fig. 9. Visualizing the localization in STD CNN and CSG CNN with CAM [47].

5.2   Adversarial Sample Detection

This subsection shows that the highly class-related representation of our CSG training can promote adversarial samples detection. It is studied [39] that adversarial samples can be detected based on the abnormal representation across layers: in low layers of a network, the representation of an adversarial sample is similar to the original class, while on the high layers it is similar to the target class. Such mismatch is easier to be detected with the class-related representation from CSG, where the implied class are more exposed.

To verify this judgment, we train a random forest [5] with the features of normal samples and adversarial samples extracted by global average pooling after each convolution layers of ResNet20 trained in Sect. 4.1. We generate adversarial samples for random targeted classes by commonly used white-box attacks such as FGSM [15], PGD [26], and CW [7]. See Appendix I for detailed attack settings. We repeat

Table 3. The mean error rate(%) for random forests on adversarial samples detection with the features of CNN.

Num. of training samples		500	1000	2000
FGSM	STD	4.76	4.60	4.15
	CSG (Ours)	**4.50**	**3.94**	**3.84**
PGD	STD	5.03	6.20	4.08
	CSG (Ours)	**4.52**	**3.95**	**3.25**
CW	STD	7.33	6.76	6.46
	CSG (Ours)	**7.03**	**6.18**	**5.87**

each experiment five times and report the mean error rates in Table 3. The experimental results demonstrate that the class-related representation can better distinguish the abnormal behavior of adversarial samples and hence improve the robustness of the model. Further experiments in Appendix J show that CSG training can improve robustness in defending adversarial attack.

6   Conclusion

In this work, we propose a simply yet effective structure – Class-Specific Gate (CSG) to induce filter differentiation in CNNs. With reasonable assumptions about the behaviors of filters, we derive regularization terms to constrain the form of CSG. As a result, the sparsity of the gate matrix encourages class-specific filters, and therefore yields sparse and highly class-related representations, which endows model with better interpretability and robustness. We believe CSG is a promising technique to differentiate filters in CNNs. Referring to CSG's successful utility and feasibility in the classification problem, as one of our future works, we expect that CSG also has the potential to interpret other tasks like detection, segmentation, etc, and networks more than CNNs.

Acknowledgement. This work was supported by the National Key R&D Program of China (2017YFA0700904), NSFC Projects (61620106010, U19B2034, U1811461, U19A2081, 61673241, 61771273), Beijing NSF Project (L172037), PCL Future Greater-Bay Area Network Facilities for Large-scale Experiments and Applications (LZC0019), Beijing Academy of Artificial Intelligence (BAAI), Tsinghua-Huawei Joint Research Program, a grant from Tsinghua Institute for Guo Qiang, Tiangong Institute for Intelligent Computing, the JP Morgan Faculty Research Program, Microsoft Research Asia, Rejoice Sport Tech. co., LTD and the NVIDIA NVAIL Program with GPU/DGX Acceleration.

References

1. Bai, J., Li, Y., Li, J., Jiang, Y., Xia, S.: Rectified decision trees: Towards interpretability, compression and empirical soundness. arXiv preprint arXiv:1903.05965 (2019)
2. Bau, D., Zhou, B., Khosla, A., Oliva, A., Torralba, A.: Network dissection: quantifying interpretability of deep visual representations. In: Proceedings of the IEEE Conference on Computer Vision and Pattern Recognition, pp. 6541–6549 (2017)
3. Bojarski, M., et al.: Explaining how a deep neural network trained with end-to-end learning steers a car. arXiv preprint arXiv:1704.07911 (2017)
4. Bouchacourt, D., Tomioka, R., Nowozin, S.: Multi-level variational autoencoder: learning disentangled representations from grouped observations. In: Thirty-Second AAAI Conference on Artificial Intelligence (2018)
5. Breiman, L.: Random forests. Mach. Learn. **45**(1), 5–32 (2001)
6. Burgess, C.P., et al.: Understanding disentangling in β-vae. arXiv preprint arXiv:1804.03599 (2018)
7. Carlini, N., Wagner, D.: Towards evaluating the robustness of neural networks. In: S&P (2017)
8. Caruana, R., et al.: Intelligible models for healthcare: predicting pneumonia risk and hospital 30-day readmission. In: Proceedings of the 21th ACM SIGKDD International Conference on Knowledge Discovery and Data Mining, pp. 1721–1730. ACM (2015)
9. Chen, T.Q., Li, X., Grosse, R.B., Duvenaud, D.K.: Isolating sources of disentanglement in variational autoencoders. In: Advances in Neural Information Processing Systems, pp. 2610–2620 (2018)

10. Chen, X., et al.: Infogan: Interpretable representation learning by information maximizing generative adversarial nets. In: Advances in Neural Information Processing Systems, pp. 2172–2180 (2016)
11. Deng, J., et al.: ImageNet: a large-scale hierarchical image database. In: Proceedings of the IEEE Conference on Computer Vision and Pattern Recognition (2009)
12. Everingham, M., Van Gool, L., Williams, C.K.I., Winn, J., Zisserman, A.: The PASCAL Visual Object Classes Challenge 2010 (VOC2010) Results. http://www.pascal-network.org/challenges/VOC/voc2010/workshop/index.html
13. Girshick, R.: Fast R-CNN. In: Proceedings of the IEEE International Conference on Computer Vision, pp. 1440–1448 (2015)
14. Gonzalez-Garcia, A., Modolo, D., Ferrari, V.: Do semantic parts emerge in convolutional neural networks? Int. J. Comput. Vis. **126**(5), 476–494 (2018)
15. Goodfellow, I.J., Shlens, J., Szegedy, C.: Explaining and harnessing adversarial examples. In: International Conference on Learning Representations (2014)
16. He, K., Gkioxari, G., Dollár, P., Girshick, R.: Mask R-CNN. In: Proceedings of the IEEE International Conference on Computer Vision, pp. 2961–2969 (2017)
17. He, K., Zhang, X., Ren, S., Sun, J.: Deep residual learning for image recognition. In: Proceedings of the IEEE Conference on Computer Vision and Pattern Recognition, pp. 770–778 (2016)
18. Higgins, I., Matthey, L., Pal, A., Burgess, C., Glorot, X., Botvinick, M., Mohamed, S., Lerchner, A.: beta-vae: learning basic visual concepts with a constrained variational framework. Int. Conf. Learn. Represent. **2**(5), 6 (2017)
19. Jiang, Z., Wang, Y., Davis, L., Andrews, W., Rozgic, V.: Learning discriminative features via label consistent neural network. In: 2017 IEEE Winter Conference on Applications of Computer Vision, pp. 207–216. IEEE (2017)
20. Kim, H., Mnih, A.: Disentangling by factorising. arXiv preprint arXiv:1802.05983 (2018)
21. Kingma, D.P., Welling, M.: Stochastic gradient VB and the variational autoencoder. In: International Conference on Learning Representations (2014)
22. Krizhevsky, A., et al.: Learning multiple layers of features from tiny images. Technical report TR-2009 (2009)
23. Krizhevsky, A., Sutskever, I., Hinton, G.E.: Imagenet classification with deep convolutional neural networks. In: Advances in Neural Information Processing Systems, pp. 1097–1105 (2012)
24. Kumar, A., Sattigeri, P., Balakrishnan, A.: Variational inference of disentangled latent concepts from unlabeled observations. arXiv preprint arXiv:1711.00848 (2017)
25. Locatello, F., et al.: Challenging common assumptions in the unsupervised learning of disentangled representations. arXiv preprint arXiv:1811.12359 (2018)
26. Madry, A., Makelov, A., Schmidt, L., Tsipras, D., Vladu, A.: Towards deep learning models resistant to adversarial attacks. arXiv preprint arXiv:1706.06083 (2017)
27. Mahendran, A., Vedaldi, A.: Understanding deep image representations by inverting them. In: Proceedings of the IEEE Conference on Computer Vision and Pattern Recognition, pp. 5188–5196 (2015)
28. Martinez, B., Modolo, D., Xiong, Y., Tighe, J.: Action recognition with spatial-temporal discriminative filter banks. In: Proceedings of the IEEE International Conference on Computer Vision, pp. 5482–5491 (2019)
29. Mordvintsev, A., Olah, C., Tyka, M.: Inceptionism: going deeper into neural networks (2015). https://research.googleblog.com/2015/06/inceptionism-going-deeper-into-neural.html

30. Olah, C., et al.: The building blocks of interpretability. Distill (2018). https://doi.org/10.23915/distill.00010. https://distill.pub/2018/building-blocks
31. Prakash, A., Storer, J., Florencio, D., Zhang, C.: RePr: improved training of convolutional filters. In: Proceedings of the IEEE Conference on Computer Vision and Pattern Recognition, pp. 10666–10675 (2019)
32. Ross, B.C.: Mutual information between discrete and continuous data sets. PloS one **9**(2), e87357 (2014)
33. Sabour, S., Frosst, N., Hinton, G.E.: Dynamic routing between capsules. In: Advances in Neural Information Processing Systems, pp. 3856–3866 (2017)
34. Simonyan, K., Vedaldi, A., Zisserman, A.: Deep inside convolutional networks: visualising image classification models and saliency maps. arXiv preprint arXiv:1312.6034 (2013)
35. Smith, A.G., et al.: Inhibition of pluripotential embryonic stem cell differentiation by purified polypeptides. Nature **336**(6200), 688–690 (1988)
36. Szegedy, C., et al.: Intriguing properties of neural networks. arXiv preprint arXiv:1312.6199 (2013)
37. Thomas, V., et al.: Disentangling the independently controllable factors of variation by interacting with the world. arXiv preprint arXiv:1802.09484 (2018)
38. Wang, Y., Morariu, V.I., Davis, L.S.: Learning a discriminative filter bank within a cnn for fine-grained recognition. In: Proceedings of the IEEE Conference on Computer Vision and Pattern Recognition, pp. 4148–4157 (2018)
39. Wang, Y., Su, H., Zhang, B., Hu, X.: Interpret neural networks by identifying critical data routing paths. In: Proceedings of the IEEE Conference on Computer Vision and Pattern Recognition, pp. 8906–8914 (2018)
40. Zeiler, M.D., Fergus, R.: Visualizing and understanding convolutional networks. In: Fleet, D., Pajdla, T., Schiele, B., Tuytelaars, T. (eds.) ECCV 2014. LNCS, vol. 8689, pp. 818–833. Springer, Cham (2014). https://doi.org/10.1007/978-3-319-10590-1_53
41. Zhang, Q., Cao, R., Wu, Y.N., Zhu, S.C.: Mining object parts from CNNS via active question-answering. In: Proceedings of the IEEE Conference on Computer Vision and Pattern Recognition, pp. 346–355 (2017)
42. Zhang, Q., Cao, R., Shi, F., Wu, Y.N., Zhu, S.C.: Interpreting CNN knowledge via an explanatory graph. In: Thirty-Second AAAI Conference on Artificial Intelligence (2018)
43. Zhang, Q., Cao, R., Wu, Y.N., Zhu, S.C.: Growing interpretable part graphs on convnets via multi-shot learning. In: Thirty-First AAAI Conference on Artificial Intelligence (2017)
44. Zhang, Q., et al.: Interactively transferring cnn patterns for part localization. arXiv preprint arXiv:1708.01783 (2017)
45. Zhang, Q., Wu, Y.N., Zhu, S.C.: Interpretable convolutional neural networks. In: Proceedings of the IEEE Conference on Computer Vision and Pattern Recognition, pp. 8827–8836 (2018)
46. Zhang, Q., Yang, Y., Ma, H., Wu, Y.N.: Interpreting CNNS via decision trees. In: Proceedings of the IEEE Conference on Computer Vision and Pattern Recognition, pp. 6261–6270 (2019)
47. Zhou, B., Khosla, A., Lapedriza, A., Oliva, A., Torralba, A.: Learning deep features for discriminative localization. In: Proceedings of the IEEE Conference on Computer Vision and Pattern Recognition, pp. 2921–2929 (2016)

EagleEye: Fast Sub-net Evaluation for Efficient Neural Network Pruning

Bailin Li[1], Bowen Wu[2], Jiang Su[1(✉)], and Guangrun Wang[2]

[1] Dark Matter AI Inc., Guangzhou, China
bl-zorro@163.com, sujiang@dm-ai.cn
[2] Sun Yat-sen University, Guangzhou, China
{wubw,wanggrun}@mail2.sysu.edu.cn

Abstract. Finding out the computational redundant part of a trained Deep Neural Network (DNN) is the key question that pruning algorithms target on. Many algorithms try to predict model performance of the pruned sub-nets by introducing various evaluation methods. But they are either inaccurate or very complicated for general application. In this work, we present a pruning method called EagleEye, in which a simple yet efficient evaluation component based on adaptive batch normalization is applied to unveil a strong correlation between different pruned DNN structures and their final settled accuracy. This strong correlation allows us to fast spot the pruned candidates with highest potential accuracy without actually fine-tuning them. This module is also general to plug-in and improve some existing pruning algorithms. EagleEye achieves better pruning performance than all of the studied pruning algorithms in our experiments. Concretely, to prune MobileNet V1 and ResNet-50, EagleEye outperforms all compared methods by up to 3.8%. Even in the more challenging experiments of pruning the compact model of MobileNet V1, Eagle-Eye achieves the highest accuracy of 70.9% with an overall 50% operations (FLOPs) pruned. All accuracy results are Top-1 ImageNet classification accuracy. Source code and models are accessible to open-source community (https://github.com/anonymous47823493/EagleEye).

Keywords: Model compression · Neural network pruning

1 Introduction

Deep Neural Network (DNN) pruning aims to reduce computational redundancy from a full model with an allowed accuracy range. Pruned models usually result in a smaller energy or hardware resource budget and, therefore, are especially meaningful to the deployment to power-efficient front-end systems. However, how to trim off the parts of a network that make little contribution to the model accuracy is no trivial question.

Electronic supplementary material The online version of this chapter (https://doi.org/10.1007/978-3-030-58536-5_38) contains supplementary material, which is available to authorized users.

© Springer Nature Switzerland AG 2020
A. Vedaldi et al. (Eds.): ECCV 2020, LNCS 12347, pp. 639–654, 2020.
https://doi.org/10.1007/978-3-030-58536-5_38

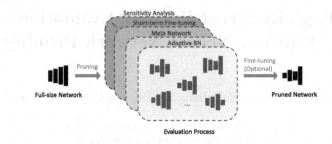

Fig. 1. A generalized pipeline for pruning tasks. The evaluation process unveils the potential of different pruning strategies and picks the one that most likely to deliver high accuracy after convergence.

DNN pruning can be considered as a searching problem. The searching space consists of all legitimate pruned networks, which are referred as sub-nets or pruning candidates. In such space, how to obtain the sub-net with highest accuracy with reasonably small searching efforts is the core of a pruning task.

Particularly, an evaluation process can be commonly found in existing pruning pipelines. Such process aims to unveil the potential of sub-nets so that best pruning candidate can be selected to deliver the final pruning strategy. A visual illustration of this generalization is shown in Fig. 1. More details about the existing evaluation methods will be discussed throughout this work. An advantage of using an evaluation module is fast decision-making because training all subnets, in a large searching space, to convergence for comparison can be very time-consuming and hence impractical.

However, we found that the evaluation methods in existing works are suboptimal. Concretely, they are either inaccurate or complicated.

By saying "inaccurate" , it means the winner sub-nets from the evaluation process do not necessarily deliver high accuracy when they converge [7,13,19]. This will be quantitatively proved in Sect. 4.1 as a correlation problem measured by several commonly used correlation coefficients. To our knowledge, we are the first to introduce correlation-based analysis for sub-net selection in pruning task. Moreover, we demonstrate that the reason such evaluation is inaccurate is the use of sub-optimal statistical values for Batch Normalization (BN) layers [10]. In this work, we use a so-called "adaptive BN" technique to fix the issue and effectively reach a higher correlation for our proposed evaluation process.

By saying "complicated", it points to the fact that the evaluation process in some works rely on tricky or computationally intensive components such as a reinforcement learning agent [7], auxiliary network training [22], knowledge distillation [8], and so on. These methods require careful hyper-parameter tuning or extra training efforts on the auxiliary models. These requirements make it potentially difficult to repeat the results and these pruning methods can be time-consuming due to their high algorithmic complexity.

Above-mentioned issues in current works motivate us to propose a better pruning algorithm that equips with a faster and more accurate evaluation process, which eventually helps to provide the state-of-the-art pruning performance. The main novelty of the proposed EagleEye pruning algorithm is described as below:

- We point out the reason that a so-called "vanilla" evaluation step (explained in Sect. 3.1) widely found in many existing pruning methods leads to poor pruning results. To quantitatively demonstrate the issue, we are the first to introduce a correlation analysis to the domain of pruning algorithm.
- We adopt the technique of adaptive batch normalization for pruning purposes in this work to address the issue in the "vanilla" evaluation step. It is one of the modules in our proposed pruning algorithm called EagleEye. Our proposed algorithm can effectively estimate the converged accuracy for any pruned model in the time of only a few iterations of inference. It is also general enough to plug-in and improve some existing methods for performance improvement.
- Our experiments show that although EagleEye is simple, it achieves the state-of-the-art pruning performance in comparisons with many more complex approaches. In the ResNet-50 experiments, EagleEye delivers 1.3% to 3.8% higher accuracy than compared algorithms. Even in the challenging task of pruning the compact model of MobileNet V1, EagleEye achieves the highest accuracy of 70.9% with an overall 50% operations (FLOPs) pruned. The results here are ImageNet top-1 classification accuracy.

2   Related Work

Pruning was mainly handled by hand-crafted heuristics in early time [13]. So a pruned candidate network is obtained by human expertise and evaluated by training it to the converged accuracy, which can be very time consuming considering the large number of plausible sub-nets. In later chapters, we will show that the pruning candidate selection is problematic and selected pruned networks cannot necessarily deliver the highest accuracy after fine-tuning. Greedy strategy were introduced to save manual efforts [26] in more recent time. But it is easy for such strategy to fall into the local optimal caused by the greedy nature. For example, NetAdapt [26] supposes the layer l_t with the least accuracy drop, noted as d_t, is greedily pruned at step t. However, there may exist a better pruning strategy where $d'_t > d_t$, but $d'_t + d'_{t+1} < d_t + d_{t+1}$. Our method searches the pruning ratios for all layers together in one single step and therefore avoids this issue.

Some other works induce sparsity to weights in training phase for pruning purposes. For example, [25] introduces group-LASSO to introduce sparsity of the kernels and [21] regularizes the parameter in batch normalization layer. [23] ranks the importance of filters based on Taylor expansion and trimmed off the low-ranked ones. The selection standards proposed in these methods are orthogonal to our proposed algorithm. More recently, versatile techniques were proposed to achieve automated and efficient pruning strategies such as reinforcement learning [7], generative adversarial learning mechanism [17] and so on. But the introduced hyper-parameters add difficulty to repeat the experiments and the trail-and-error to get the auxiliary models work well can be time consuming.

The technique of adjusting BN was used to serve for non-pruning purposes in existing works. [14] adapts the BN statistics for target domain in domain adaptation tasks. The common point with our work is that we both notice the batch normalization requires an adjustment to adapt models in a new setting where either model or domain changes. But this useful technique has not been particularly used for model pruning purposes.

3 Methodology

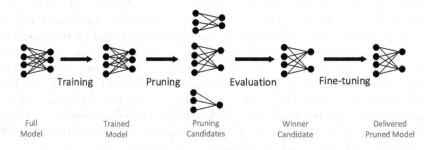

Fig. 2. A typical pipeline for neural network training and pruning

A typical neural network training and pruning pipeline is generalized and visualized in Fig. 2. Pruning is normally applied to a trained full-size network for redundancy removal purposes. An fine-tuning process is then followed up to gain accuracy back from losing parameters in the trimmed filters. In this work, we focus on structured filter pruning approaches, which can be generally formulated as

$$(r_1, r_2, ..., r_L)^* = \argmin_{r_1, r_2, ..., r_L} \mathcal{L}(\mathcal{A}(r_1, r_2, ..., r_L; w)), \quad s.t. \; \mathcal{C} < constraints,$$

(1)

where $\mathcal{L}$ is the loss function and $\mathcal{A}$ is the neural network model. r_l is the pruning ratio applied to the lth layer. Given some constraints $\mathcal{C}$ such as targeted amount of parameters, operations, or execution latency, a combination of pruning ratios $(r_1, r_2, ..., r_L)$, which is referred as pruning strategy, is applied to the full-size model. All possible combinations of the pruning ratios form a searching space. To obtain a compact model with the highest accuracy, one should search through the search space by applying different pruning strategies to the model, fine-tuning each of the pruned model to converged and pick the best one. We consider the pruning task as finding the optimal pruning strategy, denoted as $(r_1, r_2, ..., r_L)^*$, that results in the highest converged accuracy of the pruned model.

Apart from handcraft designing, different searching methods have been applied in previous work to find the optimal pruning strategy, such as greedy algorithm [26,28], RL [7], and evlolutionary algorithm [20]. All of the these methods are guided by the evaluation results of the pruning strategies.

3.1   Motivation

In many published approaches [7,13,19] in this domain, pruning candidates directly compare with each other in terms of evaluation accuracy. The subnets with higher evaluation accuracy are selected and expected to also deliver high accuracy after fine-tuning. However, such intention can not be necessarily achieved as we notice the sub-nets perform poorly if directly used to do inference. The inference results normally fall into a very low-range accuracy, which is illustrated in Fig. 3 left. An early attempt is to randomly generate pruning rates for MobileNet V1 and apply $\mathcal{L}$1-norm based pruning [13] for 50 times. The dark red bars form the histogram of accuracy collected from directly doing inference with the pruned candidates in the same way that [7,13,19] do before fine-tuning. Because our pruning rates are randomly generated in this early attempt, so the accuracy is very low and only for observation. The gray bars in Fig. 4 shows the situation after fine-tuning these 50 pruned networks. We notice a huge difference in accuracy distribution between these two results. Therefore, there are two questions came up to our mind given above observation. The first question is why removal to filters, especially considered as "unimportant" filters, can cause such noticeable accuracy degradation although the pruning rates are random? The natural question to ask next is how strongly the low-range accuracy is positively correlated to the final converged accuracy. These two questions triggered our investigation into this commonly used evaluation process, which is called vanilla evaluation in this work.

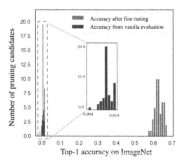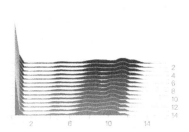

Fig. 3. Left: Histogram for accuracy collected from directly pruning MobileNet V1 and fine-tuning 15 epoches. Right: Evolution of the weight distribution of a pruned MobileNetV1 [9] during fine-tuning on ImageNet [3]. Where X axis presents the magnitude of the $\mathcal{L}$1-norm of kernel, Y axis presents the quantity, Z axis presents the fine-tuning epochs.

Some initial investigations are done to tentatively address the above two questions. Figure 3 right shows that it might not be the weights that mess up the accuracy at the evaluation stage as only a gentle shift in weight distribution is observed during fine-tuning, but the delivered inference accuracy is very different. On the other side, Fig. 4 left shows that the low-range accuracy indeed presents poor correlation with the fine-tuned accuracy, which means that it can be misleading to use evaluated accuracy to guide the pruning candidates selection.

Interestingly, we found that it is the batch normalization layer that largely affects the evaluation. Without fine-tuning, pruning candidates have parameters that are a subset of those in the full-size model. So the layer-wise feature map data are also affected by the changed model dimensions. However, vanilla evaluation still uses Batch Normalization (BN) inherited from the full-size model. The outdated statistical values of BN layers eventually drag down the evaluation accuracy to a surprisingly low range and, more importantly, break the correlation between evaluation accuracy and the final converged accuracy of the pruning candidates in the strategy searching space. A brief training, also called fine-tuning, all pruning candidates and then compare them is a more accurate way to carry out the evaluation [15,20]. However, it is very time-consuming to do the training-based evaluation for even single-epoch fine-tuning due to the large scale of the searching space. We give quantitative analysis later in this section to demonstrate this point.

Firstly, to quantitatively demonstrate the idea of vanilla evaluation and the problems that come with it, we symbolize the original BN [10] as below:

$$y = \gamma \frac{x - \mu}{\sqrt{\sigma^2 + \epsilon}} + \beta, \tag{2}$$

Where β and γ are trainable scale and bias terms. ϵ is a term with small value to avoid zero division. For a mini-batch with size N, the statistical values of μ and σ^2 are calculated as below:

$$\mu_{\mathcal{B}} = E[x_{\mathcal{B}}] = \frac{1}{N} \sum_{i=1}^{N} x_i, \quad \sigma_{\mathcal{B}}^2 = Var[x_{\mathcal{B}}] = \frac{1}{N-1} \sum_{i=1}^{N} (x_i - \mu_{\mathcal{B}})^2. \tag{3}$$

During training, μ and σ^2 are calculated with the moving mean and variance:

$$\mu_t = m\mu_{t-1} + (1-m)\mu_{\mathcal{B}}, \quad \sigma_t^2 = m\sigma_{t-1}^2 + (1-m)\sigma_{\mathcal{B}}^2, \tag{4}$$

where m is the momentum coefficient and subscript t refers to the number of training iterations. In a typical training pipeline, if the total number of training iteration is T, μ_T and σ_T^2 are used in testing phase. These two items are called global BN statistics, where "global" refers to the full-size model.

3.2   Adaptive Batch Normalization

As briefly mentioned before, vanilla evaluation used in [7,13,19] apply global BN statistics to pruned networks to fast evaluate their accuracy potential, which we think leads to the low-range accuracy results and unfair candidate selection. If the global BN statistics are out-dated to the sub-nets, we should re-calculate μ_T and σ_T^2 with adaptive values by conducting a few iterations of inference on part of the training set, which essentially adapts the BN statistical values to the pruned network connections. Concretely, we freeze all the network parameters while resetting the moving average statistics. Then, we update the moving statistics by a few iterations of forward-propagation, using Eq. 4, but without backward propagation. We note the adaptive BN statistics as $\hat{\mu}_T$ and $\hat{\sigma}_T^2$.

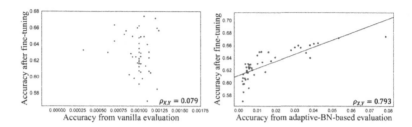

Fig. 4. Correlation between fine-tuning accuracy and inference accuracy gained from vanilla evaluation (left), adaptive-BN-based evaluation (right) based on MobileNet V1 experiments on ImageNet Top-1 classification results.

Figure 4 right illustrates that applying adaptive BN delivers evaluation accuracy that has a stronger correlation, compared to the vanilla evaluation Fig. 4 left.

As another evidence, we compare the distance of BN statistical values between "true" statistics. We consider μ and σ^2 sampled from the validation data as the "true" statistics, noted as μ_{val} and σ_{val}^2 , because they are the real statistical values in the testing phase. Specially, we are not obtaining insights from the validation data, which we think is unfair, but simply showing that our evaluation results are closer to the ground truth compared to the vanilla method. Concretely, we expect $\hat{\mu}_T$ and $\hat{\sigma}_T^2$ to be as close as possible to the "true" BN statistics values,μ_{val} and σ_{val}^2, so they could deliver close computational results. So we visualize the distance of BN statistical values gained from different evaluation methods (see Fig. 5). Each pixel in the heatmaps represents a distance for a type of BN statistics, either μ_{val} or σ_{val}^2, between post-evaluation results and the "true" statistics sampled via one filter in MobileNet V1 [9]. The visual observation shows that adaptive BN provides closer statistical values to the "true" values while global BN is way further. A possible explanation is that the global BN statistics are out-dated and not adapted to the pruned network connections. So they mess up the inference accuracy during evaluation for the pruned networks.

Noticeably, fine-tuning also relieves such problem of mismatched BN statistics because the training process itself re-calculates the BN statistical values in the forward pass and hence fixes the mismatch. However, BN statistics are not trainable values but sampling parameters only calculated in inference time. Our adaptive BN targets on this issue by conducting re-sampling in exactly the inference step, which achieves the same goal but with way less computational cost compared to fine-tuning. This is the main reason that we claim the application

Fig. 5. Visualization of distances of BN statistics in terms of the moving mean and variance. Each pixel refers to the distance of one BN statistics of a channel in MobileNetV1. (a) $\|\mu_T - \mu_{val}\|_2$, distance of moving mean between global BN and the "true" values. (b) distance of moving mean between adaptive-BN and the "true" values $\|\hat{\mu}_T - \mu_{val}\|_2$. (c) $\|\sigma_T^2 - \sigma_{val}^2\|_2$, distance of moving variance between global BN and the "true" values. (d) distance of moving variance between adaptive-BN and the "true" values $\|\sigma_T^2 - \sigma_{val}^2\|_2$

of adaptive BN in pruning evaluation is more efficient than the fine-tuning-based solution.

3.3   Correlation Measurement

As mentioned before, a "good" evaluation process in the pruning pipeline should present a strong positive correlation between the evaluated pruning candidates and their corresponding converged accuracy. Here, we compare two different evaluation methods, adaptive-BN-based and vanilla evaluation, and study their correlation with the fine-tuned accuracy. So we symbolize a vector of accuracy for all pruning candidates in the searching space (Fig. 6) separately using the above two evaluation methods as X_1 and X_2 correspondingly while fine-tuned accuracy is noted as Y. We firstly use Pearson Correlation Coefficient [24] (PCC) $\rho_{X,Y}$, which is used to measure the linear correlation between two variables X and Y, to measure the correlation between $\rho_{X_1,Y}$ and $\rho_{X_2,Y}$.

Since we particularly care about high-accuracy sub-nets in the ordered accuracy vectors, Spearman Correlation Coefficient (SCC) [2] $\phi_{X,Y}$ and Kendall rank Correlation Coefficient (KRCC) [11] $\tau_{X,Y}$ are adopted to measure the monotonic correlation. We compare the correlation between (X_1, Y) and (X_2, Y) in above three metrics with different pruning rates. All cases present a stronger correlation for the adaptive-BN-based evaluation than the vanilla strategy. See richer details about quantitative analysis in Sect. 4.1.

3.4   EagleEye Pruning Algorithm

Based on the discussion about the accurate evaluation process in pruning, we now present the overall workflow of EagleEye in Fig. 6. Our pruning pipeline contains three parts, pruning strategy generation, filter pruning, and adaptive-BN-based evaluation.

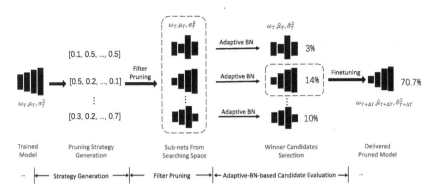

Fig. 6. Workflow of the EagleEye Pruning Algorithm

Strategy generation outputs pruning strategies in the form of layer-wise pruning rate vectors like $(r_1, r_2, ..., r_L)$ for a L-layer model. The generation process follows pre-defined constraints such as inference latency, a global reduction of operations (FLOPs) or parameters and so on. Concretely, it randomly samples L real numbers from a given range $[0, R]$ to form a pruning strategy, where r_l denotes the pruning ratio for the lth layer. R is the largest pruning ratio applied to a layer. This is essentially a Monte Carlo sampling process with a uniform distribution for all legitimate layer-wise pruning rates, i.e. removed number of filters over the number of total filters. Noticeably, other strategy generation methods can be used here, such as the evolutionary algorithm, reinforcement learning etc., we found that a simple random sampling is good enough for the entire pipeline to quickly yield pruning candidates with state-of-the-art accuracy. A possible reason for this can be that the adjustment to the BN statistics leads to a much more accurate prediction to the sub-nets' potential, so the efforts of generating candidates are allowed to be massively simplified. The low computation cost of this simple component also adds the advantage of fast speed to the entire algorithm.

Filter pruning process prunes the full-size trained model according to the generated pruning strategy from the previous module. Similar to a normal filter pruning method, the filters are firstly ranked according to their $\mathcal{L}1$-norm and the r_l of the least important filters are trimmed off permanently. The sampled pruning candidates from the searching space are ready to be delivered to the next evaluation stage after this process.

The adaptive-BN-based candidate evaluation module provides a BN statistics adaptation and fast evaluation to the pruned candidates handed over from the previous module. Given a pruned network, it freezes all learnable parameters and traverses through a small amount of data in the training set to calculate the adaptive BN statistics $\hat{\mu}$ and $\hat{\sigma}^2$. In practice, we sampled 1/30 of the total training set for 100 iterations in our ImageNet experiments, which takes only 10-ish seconds in a single Nvidia 2080 Ti GPU. Next, this module evaluates the performance of the candidate networks on a small part of training set data, called "sub-validation set", and picks the top ones in the accuracy ranking as winner candidates. The correlation analysis presented in Sect. 4.1 guarantees the effectiveness of this process. After a fine-tuning process, the winner candidates are finally delivered as outputs.

4   Experiments

4.1   Quantitative Analysis of Correlation

We use three commonly used correlation coefficient(ρ,σ and τ) to quantitatively measure the relation between X_1, X_2 and Y, which are defined in Sect. 3.3.

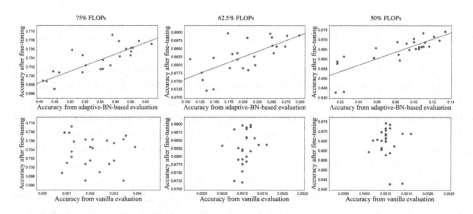

Fig. 7. Vanilla vs. adaptive-BN evaluation: Correlation between evaluation and fine-tuning accuracy with different pruning ratios (MobileNet V1 [9] on ImageNet [3] classification Top-1 results)

Firstly, as mentioned in Sect. 3.1 the poor correlation, presented by Fig. 4 sub-figure, is basically 10 times smaller than adaptive-BN-based results shown in Fig. 4 right sub-figure. This matches with the visual observation that the adaptive-BN-based samples are more trendy while the vanilla strategy tends to give randomly distributed samples on the figure. This means the vanilla evaluation hardly present accurate prediction to the pruned networks about their fine-tuned accuracy.

Based on the above initial exploration, we extend the quantitative study to a larger scale applying three correlation coefficients to different pruning ratios as shown in Table 1. Firstly, the adaptive-BN-based evaluation delivers stronger correlation measured in all three coefficients compared to the vanilla evaluation. In average, ρ is 0.67 higher, ϕ is 0.79 higher and τ is 0.46 higher. Noticeably, the correlation high in ϕ and τ means that the winner pruning candidates selected from the adaptive-based evaluation module are more likely to rank high in the fine-tuned accuracy ranking as ϕ emphasizes the monotonic correlation.

Table 1. Correlation analysis quantified by Pearson Correlation Coefficient $\rho_{X,Y}$, Spearman Correlation Coefficient $\phi_{X,Y}$, and Kendall rank Correlation Coefficient $\tau_{X,Y}$.

FLOPs constraints	$\rho_{X_1,Y}$	$\rho_{X_2,Y}$	$\phi_{X_1,Y}$	$\phi_{X_2,Y}$	$\tau_{X_1,Y}$	$\tau_{X_2,Y}$
Not fixed	0.793	0.079	0.850	0.025	0.679	0.063
75% FLOPs	0.819	−0.038	0.829	−0.030	0.656	−0.003
62.5% FLOPs	0.683	0.250	0.644	0.395	0.458	0.267
50% FLOPs	0.813	0.105	0.803	0.127	0.639	0.122

Especially, the third to fifth rows of Table 1 shows the correlation metrics with different pruning rates (for instance, 75% FLOPs also means 25% pruning rate to operations). The corresponding results are also visualized in Fig. 7. The second row in Table 1 means the pruning rate follows a layer-wise Monte Carlo sampling with a uniform distribution among the legitimate pruning rate options. All the above tables and figures prove that the adaptive-BN-based evaluation shows stronger correlation, and hence a more robust prediction, between the evaluated and fine-tuned accuracy for the pruning candidates.

4.2 Generality of the Adaptive-BN-Based Evaluation Method

The proposed adaptive-BN-based evaluation method is general enough to plug-in and improves some existing methods. As an example, we apply it to AMC [7], which is an automatic method based on Reinforcement Learning mechanism.

AMC [7] trains an RL-agent to decide the pruning ratio for each layer. At each training step, the agent tries applying different pruning ratios (pruning strategy) to the full-size model as an action. Then it directly evaluates the accuracy without fine-tuning, which is noted as vanilla evaluation in our paper, and takes this validation accuracy as the reward. As the RL-agent is trained with the reward based on the vanilla evaluation, which is proved to have a poor correlation to the converged accuracy of pruned networks. So we replace the vanilla evaluation process with our proposed adaptive-BN-based evaluation. Concretely, after pruning out filters at each step, we freeze all learnable parameters and do inference on the training set to fix the BN statistics and evaluate the accuracy of the model on the sub-validation set. We feed this accuracy as a reward to train the RL-agent in place of the accuracy of vanilla evaluation. The experiment about MobileNetV1 [9] on ImageNet [3] classification accuracy is improved from 70.5% (reported in AMC [7]) to 70.7%. It shows that the RL-agent can find a better pruning strategy with the help of our adaptive-BN-based evaluation module.

Another example is the "short-term fine-tune" block in [26], which also can be handily replaced by our adaptiveBN-based module for a faster pruning strategy selection. On the other side, our pipeline can also be upgraded by existing methods such as the evolutionary algorithm used in [20] to improve the basic Monte Carlo sampling strategy. The above experiments and discussion demonstrate the generality of our adaptive-BN-based evaluation module, but can not be analyzed in more detail due to the limited length of this paper.

4.3 Efficiency of Our Proposed Method

Our proposed pruning evaluation based on adaptive BN turn the prediction of sub-net accuracy into a very fast and reliable process, so EagleEye is much less time-consuming to complete the entire pruning pipeline than other heavy evaluation based algorithms. In this part, we compare the execution cost for various state-of-the-art algorithms to demonstrate the efficiency of our method.

Table 2. Comparison of computation costs of various pruning methods in the task where all pruning methods are executed to find the best pruning strategy from 1000 potential strategies (candidates).

Method	Evaluation method	Candidate selection	GPU hours
ThiNet [22]	Finetuning	1000×10 finetune epochs	∼ 8000
NetAdapt [26]	Finetuning	10^4 training iterations	864
Filter pruning [13]	Vanilla	1000×25 finetune epochs	∼ 20000
AMC [26]	Vanilla	Training an RL agent	-
Meta-Pruning [20]	PruningNet	Training an auxiliary network	-
EagleEye	adaptive-BN	<1000×100 inference iterations	25

Table 2 compares the computational costs of picking the best pruning strategy among 1000 potential pruning candidates. As ThiNet [22] and Filter Pruning [13] require manually assigning layer-wise pruning ratio, The final GPU hours are the estimation of completing the pruning pipeline for 1000 random strategies. In practice, the real computation cost highly depends on the expert's heuristic practice of trial-and-error. The computation time for AMC [7] and Meta-pruning can be long because training either an RL network or an auxiliary network itself is time-consuming and tricky. Among all compared methods, EagleEye is the most efficient method as each evaluation takes no more than 100 iterations, which takes 10 to 20 s in a single Nvidia 2080 Ti GPU. So the total candidate selection is simply an evaluation comparison process, which also can be done in negligible time.

4.4 Effectiveness of Our Proposed Method

To demonstrate the effectiveness of EagleEye, we compare it with several state-of-the-art pruning methods on MobileNetV1 and ResNet-50 [4] models tested on the small dataset of CIFAR-10 [12] and the large dataset of ImageNet.

ResNet. Table 3 left shows EagleEye outperforms all compared methods in terms of Top-1 accuracy on CIFAR-10 dataset. To further prove the robustness of our method, we compare the top-1 accuracy of ResNet-50 on ImageNet

Table 3. Pruning results of ResNet-56 (left) and MobileNetV1 (right) on CIFAR-10

Method	FLOPs	Top1-Acc
ResNet-56	125.49M	93.26%
FP [13]	90.90M	93.06%
RFP [1]	90.70M	93.12%
NISP [29]	81.00M	93.01%
GAL [18]	78.30M	92.98%
HRank [15]	88.72M	93.52%
EagleEye	62.23M	**94.66%**

Method	FLOPs	Top1-Acc
0.75 × MobileNetV1		88.07%
FP(our-implement) [13]	26.5M	91.58 %
EagleEye		**91.89%**
0.5 × MobileNetV1		87.51%
FP(our-implement) [13]	12.1M	90.4%
EagleEye		**91.44%**
0.25 × MobileNetV1		84.59%
FP(our-implement) [13]	3.3M	85.81%
EagleEye		**88.01%**

under different FLOPs constraints. For each FLOPs constraint (3G, 2G, and 1G), 1000 pruning strategies are generated. Then the adaptive-BN-based evaluation method is applied to each candidate. We just fine-tune the top-2 candidates and return the best as delivered pruned model. It is shown that EagleEye achieves the best results among the compared approaches listed in Table 4.

ThiNet [22] prunes the channels uniformly for each layer other than finding an optimal pruning strategy, which hurts the performance significantly. Meta-Pruning [20] trains an auxiliary network called "PruningNet" to predict the weights of the pruned model. But the adopted vanilla evaluation may mislead the searching of the pruning strategies. As shown in Table 4, our proposed algorithm outperform all compared methods given different pruned network targets.

MobileNet. We conduct experiments of the compact model of MobileNetV1 and compare the pruning results with Filter Pruning [13] and the directly-scaled models. Please refer to y material for more details about FP implementation and training methods to get the accuracy for the directly-scaled models. Table 3 right shows that EagleEye gets the best results in all cases.

Pruning MobileNetV1 for ImageNet is more challenging as it is already a very compact model. We compare the top-1 ImageNet classification accuracy under the same FLOPs constraint (about 280M FLOPs) and the results are shown in Table 5. 1500 pruning strategies are generated with this FLOPs constraint. Then adaptive-BN-based evaluation is applied to each candidate. After fine-tuning the top-2 candidates, the pruning candidate that returns the highest accuracy is selected as the final output.

AMC [7] trains their pruning strategy decision agent based on the pruned model without fine-tuning, which may lead to a problematic selection on the candidates. NetAdapt [26] searches for the pruning strategy based on a greedy algorithm, which may drop into a local optimum as analysed in Sect. 2. It is shown that EagleEye achieves the best performance among all studied methods again in this task (see Table 5).

Table 4. Comparisions of ResNet-50 and other pruning methods on ImageNet

FLOPs after pruning	Method	FLOPs	Top1-Acc	Top5-Acc
3G	ThiNet-70 [20]	2.9G	75.8%	90.67%
	AutoSlim [28]	3.0G	76.0%	-
	Meta-Pruning [20]	3.0G	76.2%	-
	EagleEye	3.0G	**77.1%**	**93.37%**
2G	0.75 × ResNet-50 [4]	2.3G	74.8%	-
	Thinet-50 [22]	2.1G	74.7%	90.02%
	AutoSlim [28]	2.0G	75.6%	-
	CP [8]	2.0G	73.3%	90.8%
	FPGM [6]	2.31G	75.59%	92.63%
	SFP [5]	2.32G	74.61%	92.06%
	GBN [27]	1.79G	75.18%	92.41%
	GDP [16]	2.24G	72.61%	91.05%
	DCP [30]	1.77G	74.95%	92.32%
	Meta-Pruning [20]	2.0G	75.4%	-
	EagleEye	2.0G	**76.4%**	**92.89%**
1G	0.5 × ResNet-50 [4]	1.1G	72.0%	-
	ThiNet-30 [22]	1.2G	72.1%	88.30%
	AutoSlim [28]	1.0G	74.0%	-
	Meta-Pruning [20]	1.0G	73.4%	-
	EagleEye	1.0G	**74.2%**	**91.77%**

Table 5. Comparisions of MobileNetV1 and other pruning methods on ImageNet

Method	FLOPs	Top1-Acc	Top5-Acc
0.75 × MobileNetV1 [9]	325M	68.4%	-
AMC [7]	285M	70.5%	-
NetAdapt [26]	284M	69.1%	-
Meta-Pruning [20]	281M	70.6%	-
EagleEye	284M	**70.9%**	**89.62%**

5   Discussion and Conclusions

We presented EagleEye pruning algorithm, in which a fast and accurate evaluation process based on adaptive batch normalization is proposed. Our experiments show the efficiency and effectiveness of our proposed method by delivering higher accuracy than the studied methods in the pruning experiments on ImageNet dataset. An interesting work is to further explore the generality of the adaptive-BN-based module by integrating it into many other existing methods and observe the potential improvement. Another experiment that is worth a try is to replace the random generation of pruning strategy with more advanced methods such as evolutionary algorithms and so on.

Acknowledgements. Jiang Su is the corresponding author of this work. This work was supported in part by the National Natural Science Foundation of China (NSFC) under Grant No.U1811463.

References

1. Ayinde, B.O., Zurada, J.M.: Building efficient convnets using redundant feature pruning. ArXiv abs/1802.07653 (2018)
2. Cohen, T.S., Geiger, M., Köhler, J., Welling, M.: Spherical CNNs. In: ICLR (2018)
3. Deng, J., Dong, W., Socher, R., Li, L.J., Li, K., Fei-Fei, L.: Imagenet: a large-scale hierarchical image database. In: 2009 IEEE Conference on Computer Vision and Pattern Recognition, pp. 248–255. IEEE (2009)
4. He, K., Zhang, X., Ren, S., Sun, J.: Deep residual learning for image recognition. In: Proceedings of the IEEE Conference on Computer Vision and Pattern Recognition, pp. 770–778 (2016)
5. He, Y., Kang, G., Dong, X., Fu, Y., Yang, Y.: Soft filter pruning for accelerating deep convolutional neural networks. arXiv preprint arXiv:1808.06866 (2018)
6. He, Y., Liu, P., Wang, Z., Yang, Y.: Pruning filter via geometric median for deep convolutional neural networks acceleration. arXiv preprint arXiv:1811.00250 (2018)
7. He, Y., Lin, J., Liu, Z., Wang, H., Li, L.-J., Han, S.: AMC: AutoML for model compression and acceleration on mobile devices. In: Ferrari, V., Hebert, M., Sminchisescu, C., Weiss, Y. (eds.) ECCV 2018. LNCS, vol. 11211, pp. 815–832. Springer, Cham (2018). https://doi.org/10.1007/978-3-030-01234-2_48
8. He, Y., Zhang, X., Sun, J.: Channel pruning for accelerating very deep neural networks. In: Proceedings of the IEEE International Conference on Computer Vision, pp. 1389–1397 (2017)
9. Howard, A.G., et al.: Mobilenets: efficient convolutional neural networks for mobile vision applications. arXiv preprint arXiv:1704.04861 (2017)
10. Ioffe, S., Szegedy, C.: Batch normalization: accelerating deep network training by reducing internal covariate shift. arXiv preprint arXiv:1502.03167 (2015)
11. Kendall, M.G.: A new measure of rank correlation. Biometrika **30**, 81–93 (1938)
12. Krizhevsky, A.: Learning multiple layers of features from tiny images (2009)
13. Li, H., Kadav, A., Durdanovic, I., Samet, H., Graf, H.P.: Pruning filters for efficient convnets. arXiv preprint arXiv:1608.08710 (2016)
14. Li, Y., Wang, N., Shi, J., Liu, J., Hou, X.: Revisiting batch normalization for practical domain adaptation. arXiv preprint arXiv:1603.04779 (2016)
15. Lin, M., et al.: Hrank: filter pruning using high-rank feature map. ArXiv abs/2002.10179 (2020)
16. Lin, S., Ji, R., Li, Y., Wu, Y., Huang, F., Zhang, B.: Accelerating convolutional networks via global & dynamic filter pruning. In: IJCAI, pp. 2425–2432 (2018)
17. Lin, S., et al.: Towards optimal structured CNN pruning via generative adversarial learning. In: Proceedings of the IEEE Conference on Computer Vision and Pattern Recognition, pp. 2790–2799 (2019)
18. Lin, S., et al.: Towards optimal structured CNN pruning via generative adversarial learning. In: 2019 IEEE/CVF Conference on Computer Vision and Pattern Recognition (CVPR), pp. 2785–2794 (2019)

19. Liu, N., Ma, X., Xu, Z., Wang, Y., Tang, J., Ye, J.: AutoCompress: an automatic DNN structured pruning framework for ultra-high compression rates (2020)
20. Liu, Z., et al.: Metapruning: meta learning for automatic neural network channel pruning. In: 2019 IEEE/CVF International Conference on Computer Vision (ICCV), pp. 3295–3304 (2019)
21. Liu, Z., Li, J., Shen, Z., Huang, G., Yan, S., Zhang, C.: Learning efficient convolutional networks through network slimming. In: Proceedings of the IEEE International Conference on Computer Vision, pp. 2736–2744 (2017)
22. Luo, J.H., Wu, J., Lin, W.: Thinet: a filter level pruning method for deep neural network compression. In: Proceedings of the IEEE International Conference on Computer Vision, pp. 5058–5066 (2017)
23. Molchanov, P., Mallya, A., Tyree, S., Frosio, I., Kautz, J.: Importance estimation for neural network pruning. In: Proceedings of the IEEE Conference on Computer Vision and Pattern Recognition (2019)
24. Soper, H., Young, A., Cave, B., Lee, A., Pearson, K.: On the distribution of the correlation coefficient in small samples. Appendix II to the papers of "student" and RA fisher. Biometrika **11**(4), 328–413 (1917)
25. Wen, W., Wu, C., Wang, Y., Chen, Y., Li, H.: Learning structured sparsity in deep neural networks. In: Advances in Neural Information Processing Systems, pp. 2074–2082 (2016)
26. Yang, T.-J., et al.: NetAdapt: platform-aware neural network adaptation for mobile applications. In: Ferrari, V., Hebert, M., Sminchisescu, C., Weiss, Y. (eds.) ECCV 2018. LNCS, vol. 11214, pp. 289–304. Springer, Cham (2018). https://doi.org/10.1007/978-3-030-01249-6_18
27. You, Z., Yan, K., Ye, J., Ma, M., Wang, P.: Gate decorator: global filter pruning method for accelerating deep convolutional neural networks. In: Advances in Neural Information Processing Systems (NeurIPS) (2019)
28. Yu, J., Huang, T.: Network slimming by slimmable networks: towards one-shot architecture search for channel numbers. arXiv preprint arXiv:1903.11728 (2019)
29. Yu, R., et al.: Nisp: pruning networks using neuron importance score propagation. In: 2018 IEEE/CVF Conference on Computer Vision and Pattern Recognition, pp. 9194–9203 (2017)
30. Zhuang, Z., et al.: Discrimination-aware channel pruning for deep neural networks. In: Advances in Neural Information Processing Systems, pp. 875–886 (2018)

Intrinsic Point Cloud Interpolation via Dual Latent Space Navigation

Marie-Julie Rakotosaona$^{(\boxtimes)}$ and Maks Ovsjanikov

LIX, Ecole Polytechnique, IP Paris, Palaiseau, France
{mrakotos,maks}@lix.polytechnique.fr

Abstract. We present a learning-based method for interpolating and manipulating 3D shapes represented as point clouds, that is explicitly designed to preserve intrinsic shape properties. Our approach is based on constructing a dual encoding space that enables shape synthesis and, at the same time, provides links to the intrinsic shape information, which is typically not available on point cloud data. Our method works in a single pass and avoids expensive optimization, employed by existing techniques. Furthermore, the strong regularization provided by our dual latent space approach also helps to improve shape recovery in challenging settings from noisy point clouds across different datasets. Extensive experiments show that our method results in more realistic and smoother interpolations compared to baselines. Both the code and our pre-trained network can be found online: https://github.com/ mrakotosaon/intrinsic_interpolations.

Keywords: 3D point clouds · 3D reconstruction · Deep learning · Applications · Methodology · Theory

1 Introduction

A core problem in 3D computer vision is to manipulate and analyze shapes represented as point clouds. Compared to other representations such as triangle meshes or dense voxel grids, point clouds are distinguished by their generality, simplicity and flexibility. For these reasons, and especially with the introduction of PointNet and its variants [37,38,42], point clouds have gained popularity in machine learning applications, including point-based *generative models*.

Unfortunately the flexibility of the point cloud representation also comes at a cost, as it does not encode any *topological* or intrinsic metric information of the underlying object. Thus, methods trained on point cloud data can, by their nature, be insensitive to distortion that might appear on generated shapes. This problem is particularly prominent in 3D shape interpolation, where a common

Electronic supplementary material The online version of this chapter (https:// doi.org/10.1007/978-3-030-58536-5_39) contains supplementary material, which is available to authorized users.

© Springer Nature Switzerland AG 2020
A. Vedaldi et al. (Eds.): ECCV 2020, LNCS 12347, pp. 655–672, 2020.
https://doi.org/10.1007/978-3-030-58536-5_39

Fig. 1. Intrinsic point cloud interpolation between points from an incomplete scan with holes (left, reconstructed in first blue column) and points from a noisy mesh (right, reconstructed in last blue column). Our method both reconstructs the shape better and produces a more natural interpolation than a PointNet-based auto-encoder. (Color figure online)

approach is to generate intermediate shapes by interpolating the learned latent vectors. In this case, even if the end-shapes are realistic, the intermediate ones can have severe distortions that are very difficult to detect and correct using only point-based information. More generally, several works have observed that point cloud-based generative models can fail to capture the space of natural shapes [27,33], making it difficult to navigate them while maintaining realism.

In this paper, we introduce a novel architecture aimed specifically at injecting intrinsic information into a generative point-based network. Our method works by learning consistent mappings across the latent space obtained by a point cloud auto-encoder and a feature encoding that captures the *intrinsic* shape structure. We show that these two components can be optimized using shapes represented as triangle meshes during training. The resulting linked latent space combines the strengths of a generative latent model and of intrinsic surface information. Finally, we use the learned networks at test time on raw 3D point clouds that are neither in correspondence with the training shapes, nor contain any connectivity information.

Our approach is general and not only enables smooth interpolations, while avoiding expensive iterative optimization, but also, as we show bellow, leads to more accurate shape reconstruction from noisy point clouds across different datasets. We demonstrate on a wide range of experiments that our approach can significantly improve upon recent baselines in terms of the accuracy of shape recovery as well as realism and smoothness of shape interpolation.

2   Related Work

Shape interpolation, also known as morphing in certain contexts, is a vast and well-researched area of computer vision and computer graphics (see [32] for a survey of the early approaches). Below we review only most relevant works and focus on structure-preserving mesh interpolation, and on recent learning-based methods that operate on point clouds.

Classical methods for 3D shape interpolation have primarily focused on designing well-founded geometric metrics, and associated optimization methods that enable smooth structure-preserving interpolations. Early works in this direction include variants of as-rigid-as-possible interpolation and modeling [2,28,50] and various *representations* of shape deformation that facilitate specific transformation types, e.g. [15,26,34,45,46] among many others.

A somewhat more principled framework is provided by the notion of *shape spaces* [29,36] in which interpolation can be phrased as computing a shortest path (geodesic). In the case of surface meshes, this approach was studied in detail in [30] and then extended in numerous follow-up works, including [16,23–25,48] among others. These approaches enjoy a rich theoretical foundation, but are typically restricted to shapes having a fixed connectivity and can lead to difficult optimization problems at test time.

We also note a recent set of methods based on the formalism of *optimal transport* [6,9,43] which have also been used for shape interpolation. These approaches treat the input shapes as probability measures that are interpolated via efficient optimization techniques.

Somewhat more closely related to ours are data-driven and feature-based interpolation methods. These include interpolation based on hand-crafted features [18,27] or on exploring various *local* shape spaces obtained by analyzing a shape collection [19,39,51]. Such techniques work well if the input shapes are sufficiently similar, but require triangle meshes and dense point-wise correspondences, or a single template that is fitted to all input data to build a statistical model, e.g. [7,8,22].

Most closely related to ours are recent generative models that operate directly on unorganized point clouds [1,33,35]. These methods are often inspired by the seminal work of PointNet and its variants [37,38] and are typically based on autoencoder architectures that allow shape exploration by manipulation in the latent space. Despite significant progress in this area, however, the structure of learned latent spaces is typically not easy to control or analyze. For example, it is well-known (see e.g. [27]) that commonly-used *linear interpolation* in latent space can give rise to unrealistic shapes that are difficult to detect and rectify.

Common approaches to address these issues include extensive data augmentation [21], adversarial losses that penalize unrealistic instances [5,33] or explicit modeling of the metric in the latent space. The latter can be done by computing the Jacobian of the decoder from the latent to the embedding space [12,41] or using feature-based metrics at test time [17,31]. Unfortunately, as we show below, such techniques either lead to difficult optimization problems at test time, or can still result in significant shape distortion.

Contribution. In this paper, we propose to address the challenges mentioned above by building a *dual latent space* that combines a learned point-based autoencoder with another parallel encoding that captures the intrinsic shape metric given by the lengths of edges of triangle meshes, required only during training. This second encoding exploits the insights of mesh-based interpolation techniques [24,30,40] that highlight the importance of *interpolating the intrinsic surface information* rather than the point coordinates. We combine these two encodings by constructing dense networks that "translate" between the two latent spaces, and enable smooth and accurate interpolation without relying on correspondences or solving expensive optimization problems at test time.

3   Motivation and Background

Our main goal is to design a method capable of *efficiently and accurately* interpolating shapes represented as point clouds. This problem is challenging for several key reasons. First, most existing theoretically well-founded axiomatic 3D shape interpolation methods [23–25,30] assume the input shapes to be represented as triangle meshes with fixed connectivity in 1-1 correspondence, and furthermore typically require extensive optimization at test time. On the other hand, learning-based approaches typically embed the shapes in a compact latent space, and interpolate them by linearly interpolating the corresponding latent vectors [1,49]. Although this approach is efficient, the metric in the latent space is typically not well-understood and therefore *linear interpolation* in this space may result in unrealistic and heavily distorted shapes. Classical methods such as Variational Auto-Encoders (VAEs) help introduce regularity into the latent space, and enable more accurate generative models, but offer little control on the distances and thus interpolation in the latent space. To address this challenge, several recent approaches have proposed ways to endow the latent space with a metric and help recover geodesic distances [12,17,31]. However, these methods again typically involve expensive computations such as the Jacobian of the decoder network, and *optimization at test time*.

Within this context, our main goal is to combine the formalism and shape metrics proposed by geometric methods [24,30] with the accuracy and flexibility of data-driven techniques while maintaining efficiency and scalability.

Shape Interpolation Energy. We first recall the intrinsic shape interpolation energy introduced in [30]. Suppose we are given a pair of shapes M, N represented as triangle meshes with fixed connectivity, so that $M = (\mathcal{V}_M, \mathcal{E})$, and $N = (\mathcal{V}_N, \mathcal{E})$, where $\mathcal{V}, \mathcal{E}$ represent the coordinates of the points and the fixed set of edges respectively. An interpolating sequence is defined by a one parameter family $S_t = (\mathcal{V}_t, \mathcal{E})$, such that $\mathcal{V}_0 = \mathcal{V}_M$, and $\mathcal{V}_1 = \mathcal{V}_N$. Denoting by $v_i(t)$ the trajectory of vertex i in S_t, the basic time-continuous intrinsic interpolation energy of S_t is defined as:

$$E_{\text{cont}}(S_t) = \int_{t=0}^{1} \sum_{(i,j)\in\mathcal{E}} \left(\frac{\partial \|v_i(t) - v_j(t)\|_2}{\partial t} \right)^2 dt. \tag{1}$$

This energy measures the integral of the change of all the edge lengths in the interpolation sequence. It can be discretized in time by sampling the interval

$[0 \ldots 1]$ with samples t_k, where $k = 1 \ldots n_k$. When the time samples are uniform, resulting in a discrete set of shapes $\{S_k\}$, this leads to the discrete energy:

$$E_{\text{disc}}(\{S_k\}) = \sum_{k=2}^{n_k} \sum_{ij \in \mathcal{E}} \left(\|v_i(t_k) - v_j(t_k)\|_2 - \|v_i(t_{k-1}) - v_j(t_{k-1})\|_2 \right)^2. \qquad (2)$$

This discrete energy simply measures the sum of the squared differences between lengths of edges across consecutive shapes in the sequence. The authors of [30] argue that computing a shape sequence between M and N that minimizes such a distortion energy results in an accurate interpolation of the two shapes (more precisely in [30] use squared edge lengths and employ an additional weak regularization, which we omit for simplicity and as we have found it to be unnecessary in our case). Note that both the continuous and discrete versions of the energy promote *as-isometric-as-possible* shape interpolations. Specifically they aim to minimize the *intrinsic distortion* by promoting intermediate meshes whose edge lengths interpolate as well as possible the edge lengths of M, N, without requiring the two input shapes to be isometric themselves.

Despite the simplicity and elegance of the intrinsic interpolation energy, minimizing it directly is challenging as it leads to large non-convex optimization problems over vertex coordinates. Indeed, additional regularization is typically required to achieve realistic interpolation across large motions [24,30]. Perhaps even more importantly, the assumption of input shapes having a fixed triangle mesh and being in 1-1 correspondence is very restrictive in practice.

Latent Space Optimization. In the context of data-driven techniques the standard way to manipulate shapes is through operations in the *latent space*. This is done by first training an auto-encoder (AE) architecture and then using the learned latent space for shape manipulation. Specifically, an encoder is trained to associate a *latent vector* l_S to each 3D shape S in a training set via $l_S = \text{enc}(S)$, while the decoder is trained so that $\text{dec}(l_S) \approx S$. Given two shapes M, N, the interpolation is performed by first computing their latent vectors, l_M, l_N and then constructing an interpolating sequence via $S_t = \text{dec}(t l_N + (1-t) l_M)$ [1,49].

Unfortunately, basic *linear interpolation* in the latent space can produce significant artefacts in the resulting reconstructed shapes (see, e.g., Fig. 2). More broadly, the metric (distance) structure of the latent space is not easy to control, as the encoder-decoder architecture is typically trained only to be able to *reconstruct* the shapes, and does not capture any information about distances in the latent space.

3.1 Metric Interpolation in a Learned Space

To overcome this limitation, perhaps the simplest approach is to use a learned latent space, but to compute an interpolating sequence while minimizing the intrinsic distortion energy of the decoded shapes explicitly. Namely, after training an auto-encoder, given the source and target shapes with latent vectors l_M, l_N, one can construct a set of samples l_k in the latent space and *at test time* optimize:

$$\min_{l_1, l_2, \ldots, l_k} E_{\text{disc}}(\{S_k\}), \text{s.t. } S_i = \text{dec}(l_i), i = 1 \ldots k,$$
$$S_0 = \text{dec}(l_M), S_{k+1} = \text{dec}(l_N). \tag{3}$$

This operation employs the fact that a decoder can be trained to always produce shapes that are in 1-1 correspondence, thus making it possible to compare the decoded shapes $\{S_k\}$.

To solve this problem, the samples l_k can be initialized through linear interpolation of l_M, l_N, and Eq. (3) can optimized via gradient descent using the pretrained decoder network. This is more efficient than directly optimizing Eq. (2) through the coordinates of the vertices, as the latent space typically has a much smaller dimensionality. Intuitively, this procedure adjusts the latent vectors to correct the distortion induced by using the Euclidean metric in the latent space. In addition, the use of a pre-trained decoder acts as the regularization (required by purely geometric methods) to produce realistic shapes.

Despite leading to significant improvement compared to the basic linear interpolation in the latent space, this approach has two key limitations 1) it requires potentially expensive optimization at test time, and 2) its accuracy is limited by the initial linear interpolation in the latent space. The latter issue is particularly prominent since the latent space is not related to the intrinsic distortion energy and therefore linear interpolation can be a suboptimal initialization for the problem in Eq. (3).

Intuition. Our main intuition is that in the absence of any constraints, the intrinsic distortion energy E_{disc} is minimized by the family of shapes that linearly interpolates the edge lengths between the source and the target. This, however is not guaranteed to lead to actual 3D shapes, both because integrability conditions must hold to ensure that edges can be assembled into a consistent mesh [47] and because interpolated shapes might not be realistic from the point of view of the training data. Therefore, we build *two* auto-encoder networks that capture, respectively, point coordinates and lengths of edges of underlying meshes (available at training) so that Euclidean distances in the latent space depend linearly on distances between lists of ordered edges. We then build two "translation" or mapping networks across the two latent spaces. Finally, after training these networks, at test time, we linearly interpolate in the edge length latent space, but recover each shape by mapping onto the shape space and reconstructing using the shape decoder. As we show below, this results in smooth and realistic shape interpolation without relying on correspondences or optimization at test time.

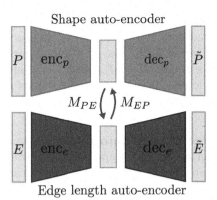

Shape auto-encoder

Edge length auto-encoder

Fig. 2. Linear interpolation in the latent space of the shape AE produces artefacts, as the interpolation is close to linear interpolation of the coordinates.

Fig. 3. Our overall architecture. We build two auto-encoders that capture the shape and edge length structure respectively, as well as two mapping networks M_{PE} and M_{EP} that "translate" across the two latent spaces.

4 Method

4.1 Overview

Figure 3 gives an overview of our network. As mentioned above, it consists of three main building blocks and training steps: a shape auto-encoder, an auto-encoder of the edge lengths of the underlying mesh, and two "translation" networks that enable communication between the two latent spaces. These networks are used at test time to endow given point clouds with intrinsic information which is then used, in particular, for more accurate point cloud interpolation. We assume that the training data is given in the form of triangle meshes with fixed connectivity, while the input at test time consists of unorganized point clouds. In the following section we describe our architecture and the associated losses, while the implementation and experimental details are given in Sect. 5.

4.2 Architecture

Shape Auto-Encoder. Our first building block (Fig. 3 top) consists of a shape auto-encoder, based on the PointNet architecture [37]. We denote the encoder and decoder networks as enc_p and dec_p respectively (we provide the exact implementation details and compare to a VAE in the supplementary). To train this network we use the basic L_2 reconstruction loss, since we assume that the input shapes are in 1-1 correspondence. This leads to the following training loss:

$$L_{rec}(P) = \frac{1}{n} \sum_{i=1}^{n} \| P_i - \tilde{P}_i \|^2, \text{ where } \tilde{P} = \mathrm{dec}_p\left(\mathrm{enc}_p(P)\right). \tag{4}$$

Here P is a training shape, the summation is done over all points in the point cloud, and P_i represents the 3D coordinates of point i.

Importantly, our point-based encoder enc_p inherits the permutation invariance of PointNet [37], which is crucial in real applications. Specifically, this allows us to encode arbitrary point clouds at test time even if they have significantly different sampling and are not in correspondence with the training data.

Edge Length Auto-Encoder. As observed in previous works and as we confirm below, the shape AE can capture the structure of individual shapes, but often fails to reflect the overall structure of *shape space*, which is particularly evident during shape interpolation. We address this by constructing a separate auto-encoder whose latent space captures the *intrinsic* shape information, and by learning mappings across the two latent spaces.

For this, we first build an auto-encoder (enc_e, dec_e) with dense layers that aims to reconstruct a list of edge lengths. Note that since we assume 1-1 correspondence at training time, the list of lengths of edges can be given in canonical (e.g., lexicographic with respect to vertex ids) order. We therefore build an auto-encoder that encodes this list into a compact vector and decodes it back from the latent representation. Our training loss for this part consists of two components: an L_2 error on the predicted edge lengths and an additional term that promotes linearity in the learned latent space:

$$L_e(E_A) = \|\text{dec}_e(\text{enc}_e(E_A)) - E_A\|^2 \tag{5}$$

$$L_{lin}(E_A, E_B) = \left\| \frac{\text{dec}_e(\text{enc}_e(E_A)) + \text{dec}_e(\text{enc}_e(E_B))}{2} \right.$$
$$\left. -\text{dec}_e\left(\frac{\text{enc}_e(E_A) + \text{enc}_e(E_B)}{2}\right) \right\|^2 . \tag{6}$$

Here E_A, E_B are the lists of edge lengths corresponding to the triangle meshes A, B given during training. Our motivation for the loss L_{lin} is to explicitly encourage linear structure, which promotes smoothness of interpolated edge lengths and thus, as we show below, minimizes intrinsic distortion.

Mapping Networks. Given two pretrained auto-encoders described above, we train two dense mapping networks that translate elements between the two latent spaces. We use M_{PE} and M_{EP} to denote the networks that translate an element from the shape (resp. edge) latent space to the edge (resp. shape) latent space.

To define the losses we use to train these two networks, for a training mesh A we let $l_A = enc_p(A)$ denote the latent vector associated with A by the shape encoder. Recall that when training the shape AE we compare A with $dec_p(l_A)$. To train our mapping networks M_{PE} and M_{EP} we instead compare A with $dec_p(M_{EP}(M_{PE}(l_A)))$. In other words, rather than decoding directly from l_A we first map it to the edge length latent space (via M_{PE}). We then map the result back to the shape latent space (via M_{EP}) and finally decode the 3D shape. We denote the shape reconstructed this way by $\tilde{A} = \text{dec}_p(M_{EP}(M_{PE}(\text{enc}_p(A))))$. We compare $\tilde{A}$ to the original shape A, which leads to the following loss:

$$L_{map1}(A) = d^{\text{rot}}(\tilde{A}, A). \tag{7}$$

Here d^{rot} is a *rotation invariant* shape distance comparing the original and reconstructed shape. We use it since the list of edge lengths can only encode a shape up to rigid motion [20]. Specifically, we first compute the optimal rigid transformation between the input shape A and the predicted point cloud $\tilde{A}$ using Kabsh algorithm [4]. We then compute the mean square error between the coordinates after alignment. As shown in [27] this loss is differentiable using the derivative of the Singular Value Decomposition.

Our second loss compares the edge lengths of the reconstructed shape $\tilde{A}$ to the edge lengths of A. For this we use the standard L_2 norm squared:

$$L_{map2}(A) = \|E_A - E_{\tilde{A}}\|_2^2, \tag{8}$$

where E_A denotes the list of edge lengths of shape A.

Our last loss considers a similar difference but starting in the edge length latent space, rather than the shape one. Specifically, given a shape A with the list of edge lengths E_A, we first encode it to the edge length latent space via $\text{enc}_e(E_A)$. We then translate the resulting latent vector to the shape latent space (via M_{EP}) and back to the edge length latent space (via M_{PE}), and finally decode the result using dec_e. This leads to the following loss:

$$L_{map3}(A) = \|\text{dec}_e(M_{PE}(M_{EP}(\text{enc}_e(E_A)))) - E_A\|_2^2, \tag{9}$$

Our overall loss is then simply a weighted sum of three terms $\alpha L_{map1} + \beta L_{map2} + \gamma L_{map3}$ for shapes given at training where γ is non-zero. We evaluate other possible losses in the supplementary materials.

Network Training. To summarize, we train our overall network architecture described in Fig. 3 in three separate steps. First we train the shape-based auto-encoder using the loss given in Eq. (4). Then we train the edge length auto-encoder using the sum of the losses in Eq. (5) and Eq. (6). Finally we train the dense networks M_{EP} and M_{PE} using the sum of the three losses in Eq. (7), Eq. (8), Eq. (9). We also experimented with training the different components jointly but have observed that the problem is both more difficult and the relative properties of the computed latent spaces become less pronounced when trained together, leading to less realistic reconstructions (Sec. 4.1 of the supplementary).

4.3   Navigating the Restricted Latent Space

After training the networks as described above, we use them at test time for shape reconstruction and interpolation. We stress that at test time we do not use the edge encoder and decoder networks enc_e, dec_e, as they require canonical edge ordering. Instead we use the permutation invariant shape auto-encoder and the mapping networks M_{PE}, M_{EP} to better preserve intrinsic shape properties.

Interpolation. Given two possibly noisy unorganized point clouds P_A and P_B we first compute their associated edge-based latent codes: $m_A = M_{PE}(\text{enc}_p(P_A))$ and $m_B = M_{PE}(\text{enc}_p(P_B))$. Here we use the permutation-invariance of our encoder enc_p allowing to encode unordered point sets. We then linearly interpolate between m_A and m_B but use the *shape decoder* dec_p for reconstruction. Thus, we compute a family of intermediate point clouds as follows:

$$P_\alpha = \text{dec}_p\left(M_{EP}\left((1-\alpha)m_A + \alpha m_B\right)\right), \ \alpha \in [0\ldots 1] \tag{10}$$

In other words, we interpolate the latent codes in the edge-based latent space, but perform the reconstruction via the shape decoder dec_p. This allows us to make sure that the reconstructed shapes are both realistic and their intrinsic metric is interpolated smoothly. Note that unlike the purely geometric methods, such as [30], our approach does not rely on the given mesh structure at test time. Instead, we employ the learned edge-based latent space as a proxy for recovering the intrinsic shape structure, which as we show below, is sufficient to obtain accurate and smooth interpolations.

Since the edge length auto-encoder is fully rotation invariant, it is necessary to align the output shapes at test time. We can do so easily by using the same optimal rigid transformation as used to compute Eq. (9).

Reconstruction. Given a point cloud P_A we also use our trained architecture for shape recovery via $S = \text{dec}_p(M_{EP}(M_{PE}(\text{enc}_p(P_A))))$. Here we use the fact that the edge-length latent space helps to regularize the shape space avoiding noisy or distorted output.

4.4   Interpretation

Our approach can be interpreted both in terms of capturing the structure of individual 3D shapes and of the entire *shape space*. For the former, our shape and edge-length auto-encoders help to capture, respectively, the *extrinsic* and *intrinsic* information of the underlying surface. Jointly, they enable more accurate shape recovery and comparison. In this context, our approach is related to methods for reconstructing a shape from its intrinsic metric. This problem, while possible theoretically [20], is computationally challenging and error prone in practice [10,13,14,47]. By using a *learned* latent space our reconstruction is both efficient and leads to realistic results.

In terms of the shape space, the latent vectors of the shape auto-encoder provide a way to parametrize the space of realistic 3D shapes while the edge-length latent space helps to impose a *distance structure* on that space. This is similar to the standard approach in Riemannian geometry [11] where the manifold structure of a space and the *metric* on it are encoded separately. We highlight this interpretation in the supplementary, and leave its complete exploration as exciting future work.

4.5   Unsupervised Training

Our method can be adapted to the unsupervised context where the 1-1 correspondences are not provided during training. Contrary to our main pipeline, we cannot compute the edge lengths directly from the training data. However, we can encourage the model to produce a consistent mesh as described in [21]. We initialize the weights by pre-training on a selected mesh using the reconstruction loss L_{rec} described in (4) and train the model using Chamfer distance and regularization losses to keep the triangulation consistent. Finally, we can train the edge-length auto-encoder by using the output of the shape auto-encoder as training data. We describe this process in detail in the supplementary materials.

5   Results

Datasets. We train our networks on two different datasets: humans and animals. For humans, we use the dataset proposed in [27]. The dataset contains 17440 shapes subsampled to 1k points from DFAUST [8] and SURREAL [44]. The test set contains 10 sub-collections (character + action sequence, each consisting of 80 shapes) that are isolated from the training set of DFAUST and 2000 shapes from SURREAL dataset. During training the area of each shape is normalized to a common value. For animals we sample 12000 shapes from the SMAL dataset [52]. We sample an equal number of shapes from the 5 categories (big cats, horses, cows, hippos, dogs) to build a training set of 10000 shapes and a testset of 2000 shapes. We simplify the shapes from SMAL to 2002 points per mesh. The animal dataset provides challenging shape pairs that are far from being isometric, some of which we highlight in the supplementary video.

5.1   Shape Interpolation

We evaluate our method on our core application of shape interpolation and compare against six different recent baselines. Namely, we compare to three data-driven methods, by performing linear interpolations in the latent spaces of auto-encoders using PointNet [37] and PointNet++ [38] architectures as well

Table 1. We report the mean squared variance of the edge length (EL), per surface area and total shape volume over the interpolations of 100 shape pairs. Our method achieves lowest variance across all intrinsic features among direct inference methods. Note that GD coord. Leads to interpolation with low distortion, as it optimizes the coordinates directly but produces unrealistic shapes (see Fig. 4).

	Direct inference				Optimization based		
	Ours	PointNet	3D-Coded	PointNet++	GD L2	GD EL	GD Coord
EL	**0.231**	0.3510	0.6130	0.2993	0.3631	0.2985	**0.0345**
Area (10^{-4})	**1.261**	1.773	3.137	1.586	1.838	1.714	**0.248**
Volume (10^{-4})	**0.342**	1.613	1.243	335.2	1.483	1.703	**0.152**

as the pre-trained auto-encoder proposed in the state-of-the-art non-rigid shape matching method 3D-CODED [21].

We also compare to three optimization-based geometric methods, by building on the ideas from [12,30,41]. We produce our first two baselines by initializing a linear path in latent space of our shape auto-encoder and optimizing each sample via 1000 steps of gradient descent. We use GD EL to denote the method that optimizes E_{disc} as described in Eq. (3), and G2 L2 to denote the method that minimizes the L^2 variance over the interpolated shape coordinates as described in [41]. Finally we compare to a method simplified from [30] (GD Coord.), in which we first initialize a path by linearly interpolating the coordinates of source and target shapes. Similarly to GD EL, we minimize the discrete interpolation energy E_{disc} using gradient descent on the point coordinates directly.

Remark that GD Coord., GD L2 and GD EL methods all rely on gradient descent to compute each interpolation *at test time*. In other words, these approaches all require to solve a highly non-trivial optimization problem during interpolation, leading to additional computational cost and parameters (learning rate, number of iterations). In contrast, our method outputs a smooth interpolation in a single pass.

To evaluate the interpolations we sample 50 shapes from the DFAUST testset using farthest point sampling. We then test on 100 random pairs from those 50 shapes. We use our pipeline trained with $\alpha = 30$, $\beta = 1200$ and $\gamma = 800$ in the mapping networks loss described in Sect. 4.2. We provide an ablation study on the choice of losses in supplementary materials.

Table 1 shows quantitative comparisons. Given an interpolation path (S_n) obtained by each method, we compute the mean squared variance of various shape features f on the path. We consider three features: lengths of edges, area of faces and overall volume enclosed by the shape (computed from the mesh embedding). For each of these, we compute the sum of the squared differences across all instances in the interpolating sequence:

$$Var_f(S_n) = \frac{1}{n-1} \sum_{i=2}^{n} \|f(S_i) - f(S_{i-1})\|^2. \tag{11}$$

Intuitively, we expect a good interpolation method to result in smooth interpolations which would have low variance across all of the intrinsic shape properties. When comparing with PointNet++ as it inputs normalized bounding boxes, we normalize the total area of each output. The large volume variance of this baseline is primarily due to bad reconstruction quality of the input shapes.

As shown in Table 1 our method produces the best results among the direct data-driven methods and the best results over all the baselines except from GD Coord. This latter method is not data-driven and optimizes edge lengths directly on the coordinates without any constraints. As such, it produces shapes with low distortion but that are not realistic (see Fig. 4). Furthermore, similarly to [30] it requires the input shapes to be represented as meshes in 1-1 correspondence.

In all qualitative figures, we visualize the minimum ratio between the linear interpolation of the ground truth edge lengths and the edge lengths of the

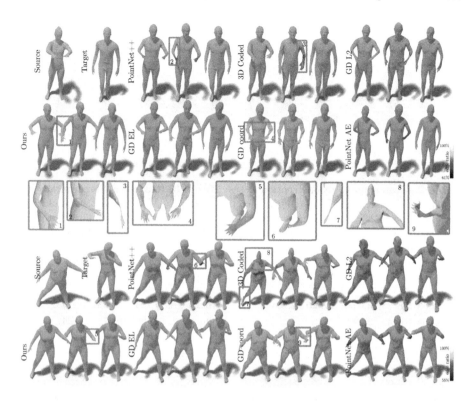

Fig. 4. Qualitative comparison of interpolation on DFAUST testset. We display the edge ratio between the linear interpolation of the target and source edges and the produced interpolation.

produced shapes. We color-code this ratio to highlight areas of highest intrinsic distortion (shown in red). In Fig. 4 we illustrate the interpolated shapes between the input source and target, shown in grey. We observe that PointNet AE and PointNet++ methods tend to produce results that are closer to linear interpolation of the coordinates. As highlighted above, we notice that while GD Coord. has low variance in the interpolated intrinsic features, the reconstructed shapes do not look natural. Overall, our method presents less distortions and more smooth interpolations compared to all baselines. We present more comparisons and evaluations in a video and in the supplementary.

We further evaluate our model on the SMAL dataset. To build the interpolation pairs from the test set, we sample 10 shapes per category by farthest points sampling. We then choose 100 random pairs from that dataset. In Fig. 5 we show results of interpolating between two horses. We observe that linear interpolation in the shape latent space leads to shape distortions such as shorter legs (middle) and wrong shape size estimation (top left). The Shape AE (resp. Ours) produces a edge variance of 2.068 (resp. 1.548). Similarly to above, our method

Fig. 5. Interpolation of two horses from the SMAL dataset.

Table 2. Mean squared reconstruction losses on the humans testset. Edge length reconstruction loss (EL), Point cloud coordinates reconstruction loss (PC) and per triangle area difference.

	EL (10^{-5})	PC (10^{-4})	Area (10^{-8})
PointNet AE	3.023	**2.120**	2.454
Edge Length AE	3.127	–	–
Ours	1.641	2.572	1.562

Table 3. Reconstruction accuracy on the SCAPE dataset. Chamfer distance (CD), mean squared total volume difference and total area difference.

	CD (10^{-3})	volume (10^{-5})	Area
Shape AE	4.703	30.851	0.1382
Ours	4.135	9.47	0.047

shows improvement at interpolating intrinsic information. We provide detailed numerical evaluation of interpolations on SMAL in supplementary materials.

Interpolation in the Unsupervised Case. The unsupervised Shape AE (resp. Ours) produces a edge variance of 0.599 (resp. 0.394). While we observe better results in the supervised setting, our method nevertheless produces quantitative and qualitative improvement over the linear interpolation in latent space. We provide further numerical and qualitative results in the supplementary materials.

5.2   Shape Reconstruction

We also evaluate the accuracy of our model for shape reconstruction on the DFAUST/SURREAL testset. In Table 2, we compare the reconstruction accuracy to the base models. We measure intrinsic features: edge length and per triangle area reconstruction loss, and extrinsic L^2 coordinates reconstruction loss. Our method reconstructs the input shape intrinsic features better that the PointNet AE while producing comparable extrinsic reconstruction loss.

We further evaluate the generalization capacity of our network by evaluating on the SCAPE [3] dataset. For testing we sample 1000 random points from the surface of each mesh. Table 3 shows an improvement in the reconstruction for our method. We observe even higher relative performance when comparing the total volume and total area of the reconstructed shapes which give a sense of the perceived quality of the shapes. Shape distortions are often related to shrunk or disproportional body parts. We show qualitative results on reconstruction in the supplementary materials. Overall, our method produces more precise and natural reconstructions. Finally, as shown in Fig. 1, our method is robust to high levels of noise (left), holes, and missing parts (right). We provide further reconstruction examples in the supplementary materials.

6   Conclusion, Limitations and Future Work

We presented a method for interpolating unorganized point clouds. Key to our approach is a dual latent space that both captures the extrinsic and intrinsic shape information, given by edge lengths provided during training. We demonstrate that our approach leads to significant improvement over existing methods, both in terms of interpolation smoothness and quality of the generated results. In the future, we plan to extend our method to incorporate other features such as semantic classes or segmentations. It would also be interesting to explore our dual encoding space in other applications on images or graphs.

Acknowledgements. Parts of this work were supported by the KAUST CRG-2017-3426 Award and the ERC Starting Grant No. 758800 (EXPROTEA).

References

1. Achlioptas, P., Diamanti, O., Mitliagkas, I., Guibas, L.: Learning representations and generative models for 3D point clouds. In: Dy, J., Krause, A. (eds.) Proceedings of the 35th International Conference on Machine Learning. Proceedings of Machine Learning Research, vol. 80, pp. 40–49. Stockholmsmässan, Stockholm Sweden, 10–15 July 2018
2. Alexa, M., Cohen-Or, D., Levin, D.: As-rigid-as-possible shape interpolation. In: Proceedings of the 27th Annual Conference on Computer Graphics and Interactive Techniques, pp. 157–164. ACM Press/Addison-Wesley Publishing Co. (2000)
3. Anguelov, D., Srinivasan, P., Koller, D., Thrun, S., Rodgers, J., Davis, J.: Scape: shape completion and animation of people. In: ACM Transactions on Graphics (TOG), vol. 24, pp. 408–416. ACM (2005)
4. Arun, K.S., Huang, T.S., Blostein, S.D.: Least-squares fitting of two 3-D point sets. IEEE Trans. Pattern Anal. Mach. Intell. **1**(5), 698–700 (1987)
5. Ben-Hamu, H., Maron, H., Kezurer, I., Avineri, G., Lipman, Y.: Multi-chart generative surface modeling. In: SIGGRAPH Asia 2018 Technical Papers, p. 215. ACM (2018)
6. Benamou, J.D., Brenier, Y.: A computational fluid mechanics solution to the Monge-Kantorovich mass transfer problem. Numer. Math. **84**(3), 375–393 (2000)
7. Bogo, F., Romero, J., Loper, M., Black, M.J.: FAUST: dataset and evaluation for 3D mesh registration. In: Proceedings of the IEEE Conference on Computer Vision and Pattern Recognition, pp. 3794–3801 (2014)
8. Bogo, F., Romero, J., Pons-Moll, G., Black, M.J.: Dynamic FAUST: registering human bodies in motion. In: Proceedings of the IEEE Conference on Computer Vision and Pattern Recognition, pp. 6233–6242, July 2017

9. Bonneel, N., Rabin, J., Peyré, G., Pfister, H.: Sliced and radon Wasserstein Barycenters of measures. J. Math. Imaging Vis. **51**(1), 22–45 (2015)
10. Boscaini, D., Eynard, D., Kourounis, D., Bronstein, M.M.: Shape-from-operator: recovering shapes from intrinsic operators. In: Computer Graphics Forum, vol. 34, pp. 265–274. Wiley Online Library (2015)
11. Carmo, M.P.D.: Riemannian geometry. Birkhäuser (1992)
12. Chen, N., Klushyn, A., Kurle, R., Jiang, X., Bayer, J., van der Smagt, P.: Metrics for deep generative models. arXiv preprint arXiv:1711.01204 (2017)
13. Chern, A., Knöppel, F., Pinkall, U., Schröder, P.: Shape from metric. ACM Trans. Graph. (TOG) **37**(4), 63 (2018)
14. Corman, E., Solomon, J., Ben-Chen, M., Guibas, L., Ovsjanikov, M.: Functional characterization of intrinsic and extrinsic geometry. ACM Trans. Graph. (TOG) **36**(2), 1–17 (2017)
15. Crane, K., Pinkall, U., Schröder, P.: Spin transformations of discrete surfaces. ACM Trans. Graph. (TOG) **30**(4), 104 (2011)
16. Freifeld, O., Black, M.J.: Lie bodies: a manifold representation of 3D human shape. In: Fitzgibbon, A., Lazebnik, S., Perona, P., Sato, Y., Schmid, C. (eds.) ECCV 2012. LNCS, vol. 7572, pp. 1–14. Springer, Heidelberg (2012). https://doi.org/10.1007/978-3-642-33718-5_1
17. Frenzel, M.F., Teleaga, B., Ushio, A.: Latent space cartography: generalised metric-inspired measures and measure-based transformations for generative models. arXiv preprint arXiv:1902.02113 (2019)
18. Gao, L., Chen, S.Y., Lai, Y.K., Xia, S.: Data-driven shape interpolation and morphing editing. Comput. Graph. Forum **36**(8), 19–31 (2017)
19. Gao, L., Lai, Y.K., Huang, Q.X., Hu, S.M.: A data-driven approach to realistic shape morphing. Comput. Graph. Forum **32**(2pt4), 449–457 (2013)
20. Gluck, H.: Almost all simply connected closed surfaces are rigid. In: Glaser, L.C., Rushing, T.B. (eds.) Geometric Topology. LNM, vol. 438, pp. 225–239. Springer, Heidelberg (1975). https://doi.org/10.1007/BFb0066118
21. Groueix, T., Fisher, M., Kim, V.G., Russell, B.C., Aubry, M.: 3D-coded: 3D correspondences by deep deformation. In: Proceedings of the European Conference on Computer Vision (ECCV), pp. 230–246 (2018)
22. Hasler, N., Stoll, C., Sunkel, M., Rosenhahn, B., Seidel, H.P.: A statistical model of human pose and body shape. Comput. Graph. Forum **28**(2), 337–346 (2009)
23. Heeren, B., Rumpf, M., Schröder, P., Wardetzky, M., Wirth, B.: Exploring the geometry of the space of shells. In: Computer Graphics Forum, vol. 33, pp. 247–256. Wiley Online Library (2014)
24. Heeren, B., Rumpf, M., Schröder, P., Wardetzky, M., Wirth, B.: Splines in the space of shells. Comput. Graph. Forum **35**(5), 111–120 (2016)
25. Heeren, B., Rumpf, M., Wardetzky, M., Wirth, B.: Time-discrete geodesics in the space of shells. Comput. Graph. Forum **31**(5), 1755–1764 (2012)
26. Huang, J., et al.: Subspace gradient domain mesh deformation. ACM Trans. Graph. (TOG) **25**(3), 1126–1134 (2006)
27. Huang, R., Rakotosaona, M.J., Achlioptas, P., Guibas, L., Ovsjanikov, M.: OperatorNet: recovering 3D shapes from difference operators. arXiv preprint arXiv:1904.10754 (2019)
28. Igarashi, T., Moscovich, T., Hughes, J.F.: As-rigid-as-possible shape manipulation. ACM Trans. Graph. (TOG) **24**(3), 1134–1141 (2005)
29. Kendall, D.G.: Shape manifolds, procrustean metrics, and complex projective spaces. Bull. Lond. Math. Soc. **16**(2), 81–121 (1984)

30. Kilian, M., Mitra, N.J., Pottmann, H.: Geometric modeling in shape space. ACM Trans. Graph. (TOG) **26**(3), 64 (2007)
31. Laine, S.: Feature-based metrics for exploring the latent space of generative models (2018)
32. Lazarus, F., Verroust, A.: Three-dimensional metamorphosis: a survey. Vis. Comput. **14**(8), 373–389 (1998)
33. Li, C.L., Zaheer, M., Zhang, Y., Poczos, B., Salakhutdinov, R.: Point cloud GAN. arXiv preprint arXiv:1810.05795 (2018)
34. Lipman, Y., Cohen-Or, D., Gal, R., Levin, D.: Volume and shape preservation via moving frame manipulation. ACM Trans. Graph. (TOG) **26**(1), 5 (2007)
35. Liu, X., Han, Z., Wen, X., Liu, Y.S., Zwicker, M.: L2G auto-encoder: understanding point clouds by local-to-global reconstruction with hierarchical self-attention. In: Proceedings of the 27th ACM International Conference on Multimedia, pp. 989–997. ACM (2019)
36. Michor, P.W., Mumford, D.B.: Riemannian geometries on spaces of plane curves. J. Eur. Math. Soc. (2006)
37. Qi, C.R., Su, H., Mo, K., Guibas, L.J.: PointNet: deep learning on point sets for 3D classification and segmentation. In: Proceedings of the CVPR, pp. 652–660 (2017)
38. Qi, C.R., Yi, L., Su, H., Guibas, L.J.: PointNet++: deep hierarchical feature learning on point sets in a metric space. In: Advances in Neural Information Processing Systems, pp. 5099–5108 (2017)
39. von Radziewsky, P., Eisemann, E., Seidel, H.P., Hildebrandt, K.: Optimized subspaces for deformation-based modeling and shape interpolation. Comput. Graph. **58**, 128–138 (2016)
40. Sassen, J., Heeren, B., Hildebrandt, K., Rumpf, M.: Solving variational problems using nonlinear rotation-invariant coordinates. In: Bommes, D., Huang, H. (eds.) Symposium on Geometry Processing 2019- Posters. The Eurographics Association (2019)
41. Shao, H., Kumar, A., Thomas Fletcher, P.: The Riemannian geometry of deep generative models. In: Proceedings of the IEEE Conference on Computer Vision and Pattern Recognition Workshops, pp. 315–323 (2018)
42. Shen, Y., Feng, C., Yang, Y., Tian, D.: Mining point cloud local structures by kernel correlation and graph pooling. In: Proceedings of the IEEE Conference on Computer Vision and Pattern Recognition, pp. 4548–4557 (2018)
43. Solomon, J., et al.: Convolutional Wasserstein distances: efficient optimal transportation on geometric domains. ACM Trans. Graph. (TOG) **34**(4), 66 (2015)
44. Varol, G., et al.: Learning from synthetic humans. In: CVPR (2017)
45. Vaxman, A., Müller, C., Weber, O.: Conformal mesh deformations with Möbius transformations. ACM Trans. Graph. (TOG) **34**(4), 55 (2015)
46. Von Funck, W., Theisel, H., Seidel, H.P.: Vector field based shape deformations. ACM Trans. Graph. (TOG) **25**(3), 1118–1125 (2006)
47. Wang, Y., Liu, B., Tong, Y.: Linear surface reconstruction from discrete fundamental forms on triangle meshes. In: Computer Graphics Forum, vol. 31, pp. 2277–2287. Wiley Online Library (2012)
48. Wirth, B., Bar, L., Rumpf, M., Sapiro, G.: A continuum mechanical approach to geodesics in shape space. Int. J. Comput. Vis. **93**(3), 293–318 (2011)
49. Wu, J., Zhang, C., Xue, T., Freeman, B., Tenenbaum, J.: Learning a probabilistic latent space of object shapes via 3D generative-adversarial modeling. In: Advances in Neural Information Processing Systems, pp. 82–90 (2016)
50. Xu, D., Zhang, H., Wang, Q., Bao, H.: Poisson shape interpolation. Graph. Models **68**(3), 268–281 (2006)

51. Zhang, Z., Li, G., Lu, H., Ouyang, Y., Yin, M., Xian, C.: Fast as-isometric-as-possible shape interpolation. Comput. Graph. **46**, 244–256 (2015)
52. Zuffi, S., Kanazawa, A., Jacobs, D., Black, M.J.: 3D menagerie: modeling the 3D shape and pose of animals. In: IEEE Conference on Computer Vision and Pattern Recognition (CVPR), July 2017

Cross-Domain Cascaded Deep Translation

Oren Katzir[1]([✉]) [iD], Dani Lischinski[2][iD], and Daniel Cohen-Or[1][iD]

[1] Tel-Aviv University, Tel Aviv-Yafo, Israel
orenkatzir@mail.tau.ac.il
[2] Hebrew University of Jerusalem, Jerusalem, Israel

Abstract. In recent years we have witnessed tremendous progress in unpaired image-to-image translation, propelled by the emergence of DNNs and adversarial training strategies. However, most existing methods focus on transfer of *style* and *appearance*, rather than on *shape* translation. The latter task is challenging, due to its intricate non-local nature, which calls for additional supervision. We mitigate this by descending the deep layers of a pre-trained network, where the deep features contain more semantics, and applying the translation between these deep features. Our translation is performed in a cascaded, deep-to-shallow, fashion, along the deep feature hierarchy: we first translate between the deepest layers that encode the higher-level semantic content of the image, proceeding to translate the shallower layers, conditioned on the deeper ones. We further demonstrate the effectiveness of using pre-trained deep features in the context of unconditioned image generation. Our code and trained models will be made publicly available.

Keywords: Unpaired image-to-image translation · Image generation

1 Introduction

In recent years, neural networks have significantly advanced generative image modeling. With the emergence of Generative Adversarial Networks (GANs) [9], image-to-image translation methods have dramatically progressed, revolutionizing applications such as inpainting [41], super resolution [34], domain adaptation [11], and more. In particular, there have been intriguing advances in the setting of unpaired image-to-image translation through the use of cycle-consistency [39,43], as well as other approaches [3,15,20,25]. However, most existing methods acknowledge the difficulty in translating *shapes* from one domain to another, as this might entail drastic geometric deformations. Consider, for example, translating between elephants and giraffes, where one would expect the neck of an elephant to be extended, while the elephant's head should shrink. The challenge is compounded by the fact that, even within the same

Electronic supplementary material The online version of this chapter (https://doi.org/10.1007/978-3-030-58536-5_40) contains supplementary material, which is available to authorized users.

© Springer Nature Switzerland AG 2020
A. Vedaldi et al. (Eds.): ECCV 2020, LNCS 12347, pp. 673–689, 2020.
https://doi.org/10.1007/978-3-030-58536-5_40

domain, images might exhibit extreme variations in object shape and pose, partial occlusions, and contain multiple instances of the object of interest. One might even argue that this translation task is ill-posed to begin with, and at the very least, requires high-level semantics to be accounted for.

conv_3_1   conv_4_1   conv_5_1   conv_5_1   conv_4_1   conv_3_1

Fig. 1. Given an image from domain A (zebras), we extract its deep features using a network pre-trained for classification, specifically VGG-19 pre-trained on ImageNet, and translate them into deep features of domain B (giraffes). We first translate high level semantics (conv_5_1) of the zebra to those of a giraffe, as shown by the inner pair of images. Then, we use a cascade of deep-to-shallow translators, one for each deep feature layer, to translate shallower layers, i.e. conv_4_1 and then conv_3_1. The images were obtained from the deep features by feature inversion networks.

Nonetheless, several image-to-image translation methods address shape deformation, aided by supervision in the form of a foreground mask [21,28]. In contrast, GANimorph [8] and the recently proposed TransGaGa [35] show remarkable translation without requiring additional supervision for several datasets. However, these techniques excel in controlled setting only, where the images are controlled, and the foreground separation is rather simple.

In this paper, we address the problem of unpaired image-to-image translation, without requiring foreground masks, between two different domains, where the objects of interest share some semantic similarity (e.g., four-legged mammals), whose shapes and appearances may, nevertheless, be drastically different. Our key idea is to accomplish the translation task by learning to translate between deep feature maps. Rather than learning to extract the relevant high-level semantic information for the specific pair of domains at hand, we leverage deep features extracted by a network pre-trained for image classification, thereby benefiting from its large-scale fully supervised training.

Our work is motivated by the well-known observation that neurons in the deeper layers of pre-trained classification networks represent larger receptive fields in image space, and encode high-level semantic content [42]. In other words, local activation patterns in the deeper layers may encode very different shapes in size and structure. Furthermore, Aberman et al. [1] show that semantically similar regions from different domains, e.g., dog and cat, have similar activations. That is, the encoding of a cat's eye resembles that of a dog's eye more than that of its tail. These properties are attractive, since they suggest that it might be possible to learn a *semantically consistent* translation between activation patterns produced by images from different domains, and that the resulting (reconstructed) image would be able to change drastically, hopefully bypassing the common difficulties in image-to-image translation methods.

More specifically, we learn to translate between several layers of deep feature maps, extracted from two domains by a pre-trained classification network, namely VGG-19 [30]. The translation is carried out one layer at a time in a deep-to-shallow (coarse-to-fine) *cascaded* manner. For each layer, we adversarially train a dedicated translator that acts in the feature space of that layer. The deepest layer translator effectively learns to translate between semantically similar global structures, such as body shape or head position, as demonstrated by the middle pair of images in Fig. 1. The translator of each shallower layer is conditioned on the translation result of the previous layer, and learns to add fine scale and appearance details, such as texture. At every layer, in order to visualize the generated deep features, we use a network pre-trained for inverting the deep features of VGG-19, following the method of Dosovitskiy and Brox [5]. The images shown in Fig. 1 were generated in this manner.

Our conceptual novelty may be regarded as applying transfer learning between classification and image translation, as we learn to translate high-level semantics, encoded by the deep features extracted by a pre-trained classification network. This is in contrast to existing methods [8,35], which learn to translate the images directly. We compare our method with several state-of-the-art image translation methods. To demonstrate the effectiveness of our approach, we present results for several pairs of domains that share some high-level semantics, yet exhibit drastically different shapes and appearances. These domains are extremely challenging, as images might contain multiple instances of the subject, with cropping and occlusion, and exhibiting a variety of poses. Nevertheless, our translations are semantically consistent, typically preserving the number of instances, and reproducing their poses, partial occlusion or cropping, as shown in Fig. 5. We further demonstrate the power of our transfer learning approach by leveraging the same deep feature spaces to train an unconditioned image generation model.

2 Related Work

Several works [17,39,43] have presented remarkable unpaired image-to-image translations, using a framework commonly referred to as CycleGAN. The key idea is that the ill-posed conditional generative process can be regularized by a cycle-consistency constraint, which forces the translation to perform a bijective mapping. The cycle constraint has become a popular regularization technique for unpaired image-to-image translation. For example, the UNIT framework [24] assumes a shared latent space between the domains and enforces the cycle constraint in the shared latent space. Several works were developed to extend the one-to-one mapping to many-to-many mapping [2,15,20,25]. These methods decompose the encoding space to shared latent space, representing the domain invariant content space, and domain specific style space. Therefore, many translations can be achieved from a single content code by changing the style code of the input image.

Many translation methods share the inability to translate high-level semantics, including different shape geometry. This type of translation is usually necessary in the case of transfiguration, where one aims to transform a specific type of objects without changing the background. Lee et al. [20] and Mejjati et al. [27] learn an attention map and apply translation only on the the foreground object. However, both methods only improve translations that do not deform shapes. Gokaslan et al. [8] succeed in preforming several shape-deforming translations by several modifications to the CycleGAN framework, including using dilated convolutions in the discriminator. However, they do not demonstrate strong shape deformations, such as zebras to elephants or giraffes, as we show in Sect. 4.

Some works [21,28] assume some kind of segmentation is given, and use this segmentation to guide shape deformation translation. However, such segmentation is hard to achieve. In a recent work, Wu et al. [35] disentangle the input images to geometry and appearance spaces, relying on high intra-consistency, and learn to translate each of the two domains separately. However, the variation of geometry and appearance of in-the-wild images is too large to be disentangled successfully[1].

Contrary to the above works, our work leverages a pre-trained network and the translation is applied directly on deep feature maps, thus being guided by high-level semantics. Several image-to-image methods, such as [4,16,38], also incorporate such pre-trained networks, though usually, only as perceptual loss, constraining the translated image to remain semantically close to the input image. Differently, Sungatullina et al. [31] incorporate pre-trained VGG features into the discriminator architecture, to assist in the discrimination phase. Wu et al. [36] use VGG-19 as a fixed encoder in the translation, where only the decoder is learned. Upchurch et al. [33] present the only method, to our knowledge, that actually translates deep features between two domains. However, the translation is not learned, but defined by simply interpolating between the deep features, which restricts the scope of method to highly aligned domains. In another context, Yin et al. [40] train an autoencoder to embed point clouds, and perform translation in the learned embedding. In contrast, we utilize semantics to preform the translation in the much more difficult scenario of images.

Our work shares some similarities with Huang et al. [14], who suggest using a generative adversarial model [9] in a coarse-to-fine manner with respect to a pre-trained encoder. The generation process begins from the deepest features and then recursively synthesizes shallower layers conditioned on the deeper layer, until generating the final image. This method was only applied on small encoders and low resolution images and was not explored for very deep and semantic encoding neural networks such as VGG-19 [30].

Deep image analogies [22] transfer visual attributes between semantically similar images, by feed-forwarding them through a pre-trained network. Their work does not train a generative model; nonetheless, they create new deep features by fusing content features from one image with style features of another. Similarly, Aberman et al. [1] synthesize hybrid images from two aligned images by selecting the dominant deep feature activations.

---

[1] Unfortunately, at the time of this submission the authors of [35] were unable to release their code or train their network on the datasets presented in our paper.

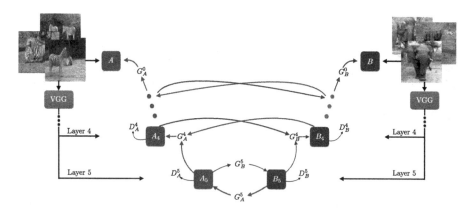

Fig. 2. Translation architecture. We translate between domains A and B starting from the deepest feature maps A_5 and B_5, which encode the highest level semantic content of the images. Translation proceeds from deeper to shallower feature maps until reaching the image itself. The feature maps are extracted by feed forwarding every image through the pre-trained VGG-19 network and sampling five of its layers. The translation of each layer is learned individually, conditioned on the translation result of the next deeper layer (except the deepest layer, whose translation is unconditional).

3 Method

Our general setting is similar to that of previous unpaired image-to-image translation methods. Given images from two domains, A and B, our goal is to learn to translate between them. However, unlike other image-to-image translation methods, we perform the translation on the deep features extracted by a pre-trained classification network, specifically VGG-19 [30].

The translation is carried out in a deep-to-shallow (coarse-to-fine) manner, using a cascade of pairs of translators, one pair per layer. The entire architecture used to train the translators is shown schematically in Fig. 2, while Fig. 3 illustrates the test-time translation (inference) process. Once the deepest feature map has been translated, we translate the next (shallower and less semantic feature map), conditioned on the translated deeper layer. In this manner, the translation of the shallower map preserves the general structure of the translated deeper one, but adds finer details, which are not encoded in the deeper feature maps. We repeat this procedure until the original image level is reached. Below we describe the training and the inference processes in more detail.

Pre-processing: We extract high-level semantic features from input images from both domains, A and B, by feed-forwarding the images through the pre-trained VGG-19 [30] network. Next, we sample five of the resulting deep feature maps, specifically `conv_i_1` ($i = 1, 2, 3, 4, 5$), where each map has progressively coarser spatial resolution, but a larger number channels. We denote the i-th sampled feature map for image $a \in A$ as a_i. Since propagation through the pre-trained

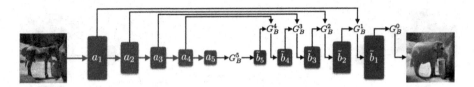

Fig. 3. Translation at test time. The input (left) is fed forward through VGG-19, yielding a set of deep feature maps. Then, we translate each feature map, starting from a_5. The final result is obtained from the shallowest translated map by feature inversion.

VGG-19 network may yield features in any range, while the range of the synthesized features is usually known, we first normalize each channel, of every layer i, by calculating its mean and standard deviation across the domain and clamp the normalized feature values to the range of $[-1, 1]$. While the clamping is a potentially harmful irreversible operation, we did not observe any adverse effect on the results. We use A_i (B_i) to denote set of all normalized deep feature maps of level i, extracted from images in domain A (B).

Inference: We perform the translation in a coarse-to-fine fashion. Thus, the translator from domain A to B, actually consists of a sequence of translators $\{G_B^5, G_B^4, \ldots, G_B^1\}$, where each translator is responsible for translating the i-th feature map layer a_i, from A_i to B_i conditioned on the previously translated deeper layer $\tilde{b}_{i+1}$ (except for the deepest layer translator G_B^5, which is unconditioned). Finally, G_B^0 uses feature inversion to convert $\tilde{b}_1$ to obtain the translated image. The translators G_A^i from domain B_i to A_i are defined symmetrically. The entire inference pipeline is shown in Fig. 3.

Feature Inversion: In all the results we show, e.g., Fig. 1, we visualize the output of the various translators by pre-training a deep feature inversion network (per domain), for each layer $i = 1, \ldots, 5$, following [5]. The network aims to reconstruct the original image given the feature map of a specific layer, regularized by adversarial loss so that the reconstructed image would lie in the manifold of natural images. For more details we refer the reader to [5]. The specific settings used in our implementation are elaborated in the supplementary materials.

Deepest Layer Translation: We begin by translating the deepest feature maps, encoding the highest-level semantic features, i.e., A_5 and B_5, hence, our problem is reduced to translating high-dimensional tensors. Our solution builds upon the commonly used CycleGAN framework [43]. Specifically, we use the three losses proposed in [43]. First, in order to generate deep features in the appropriate domain, we utilize an adversarial domain loss $\mathcal{L}_{adv}$. We simultaneously train two translators G_A^5, G_B^5 which try to fool domain-specific discriminators, D_A^5, D_B^5 (for domains A_5, B_5, respectively). However, differently from [43] and other image translation methods [15,28], we have found LSGAN [26] not to be well-suited for our task, leading to mode collapse or convergence failures. Instead, we found

WGAN-GP [10] more effective, thus, the adversarial loss for translation from X to Y is defined as

$$\mathcal{L}_{adv}\left(G_Y, D_Y, X, Y\right) = \mathop{\mathbb{E}}_{x\sim\mathbb{P}_X}\left[D_Y\left(G_Y\left(x\right)\right)\right] - \mathop{\mathbb{E}}_{y\sim\mathbb{P}_Y}\left[D_Y\left(y\right)\right]$$
$$+\lambda_{gp}\mathop{\mathbb{E}}_{\hat{y}\sim\mathbb{P}_{\hat{Y}}}\left[\left(\left\|\nabla D_Y\left(\hat{y}\right)\right\| - 1\right)^2\right], \tag{1}$$

where $G_Y : X \rightarrow Y$ is the translator, D_Y is the target domain discriminator, $\lambda_{gp} = 10$ in all our experiments, and $\mathbb{P}_{\hat{Y}}$ is defined by uniformly sampling along straight lines between $\tilde{y} \sim G\left(\mathbb{P}_X\right)$ and $y \sim \mathbb{P}_Y$. For more details we refer the reader to [10].

Second, for regularizing the translation to a one-to-one mapping, we add the cycle consistency loss,

$$\mathcal{L}_{cyc}(G_X, G_Y, X, Y) = \mathop{\mathbb{E}}_{x\sim\mathbb{P}_X}\left\|G_X\left(G_Y\left(x\right)\right) - x\right\| + \mathop{\mathbb{E}}_{y\sim\mathbb{P}_Y}\left\|G_Y\left(G_X\left(y\right)\right) - x\right\|, \tag{2}$$

where $\|\cdot\|$ stands for the L_1 norm.

Finally, as in [43], we have found it helpful to use an identity loss, which guides the networks to preserve common high level features,

$$\mathcal{L}_{idty}(G_X, G_Y, X, Y) = \mathop{\mathbb{E}}_{x\sim\mathbb{P}_X}\left\|G_X\left(x\right) - x\right\| + \mathop{\mathbb{E}}_{y\sim\mathbb{P}_Y}\left\|G_Y\left(y\right) - y\right\|. \tag{3}$$

The entire loss combines these components as follows

$$\mathcal{L}^5 = \mathcal{L}_{adv}\left(G_B^5, D_B^5, A_5, B_5\right) + \mathcal{L}_{adv}\left(G_A^5, D_A^5, B_5, A_5\right)$$
$$+ \lambda_{cyc}\mathcal{L}_{cyc}(G_A^5, G_B^5, A_5, B_5) + \lambda_{idty}\mathcal{L}_{idty}(G_A^5, G_B^5, A_5, B_5), \tag{4}$$

where λ_{cyc} and λ_{idty} were set to 100 in all our experiments.

Coarse to Fine Conditional Translation: Consider two successive layers, $a_i \in A_i$ and $a_{i+1} \in A_{i+1}$, where the latter has already been translated, yielding $\tilde{b}_{i+1}$ as the translation outcome (see Fig. 3). We next perform the translation of the layer a_i to yield $\tilde{b}_i$, using the translator G_B^i, conditioned on $\tilde{b}^{i+1}$. Note that G_B^i is effectively a function of all the previously translated layers.

The architecture of G_B^i is schematically shown in Fig. 4. Since shallower layers encode less of the semantic content of the image, it is more difficult to learn how they should be deformed, and thus they are used to transfer "style", while the "content" comes from the already translated deeper layer. Inspired by [15], we add an adaptive instance normalization (AdaIN) [13] component, whose parameters are learned from the current layer. Thus, several layers of G_B^i are normalized according to the AdaIN component. G_A^i, which is designed symmetrically, is learned simultaneously with G_B^i, as shown in Fig. 4(a).

The loss for training these shallower translators is defined similarly to that used for training the deepest translation: it consists of adversarial, cycle consistency, and identity terms. While the adversarial loss is unconditional, similarly to (1), the cyclic loss is now conditioned: $\left\|G_A^i\left(G_B^i\left(a_i, \tilde{b}_{i+1}\right), a_{i+1}\right) - a_i\right\| +$

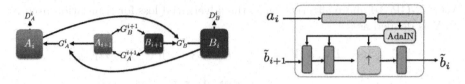

Fig. 4. Translation of layer i is conditioned on the previously translated layer $i + 1$. The two translators G_A^i and G_B^i are trained simultaneously (see left figure), while the $i + 1, \ldots, 5$ translators are fixed. On the right we show the schematic architecture of G_B^i which has two inputs: $a_i \in A_i$ and $\tilde{b}_{i+1}$. a_i is fed-forward through several layers to yield AdaIN parameters which control the generation of $\tilde{b}_i$. Since $\tilde{b}_i$ has twice the spatial size of $\tilde{b}_{i+1}$, we add an upsampling layer marked by $\uparrow$.

$\left\| G_B^i \left(G_A^i \left(b_i, \tilde{a}_{i+1} \right), b_{i+1} \right) - b_i \right\|$, and the same conditioning is used for the identity loss: $\left\| G_A^i \left(a_i, a_{i+1} \right) - a_i \right\| + \left\| G_B^i \left(b_i, b_{i+1} \right) - b_i \right\|$.

We train the pairs of translators one layer at a time, starting from G_A^5 and G_B^5. More details regarding the implementation and the training of the translators are included in the supplementary materials.

4   Experiments

We evaluate our approach on several publicly available datasets: (1) Cat $\leftrightarrow$ Dog faces [20], which contains 871 cat images and 1364 dog images and does not require high shape deformation; (2) Kaggle Cat $\leftrightarrow$ Dog [6] dataset with over $12,500$ images in each domain, where images may contain part of the subject or several instances; (3) MSCOCO dataset [23], specifically, zebra $\leftrightarrow$ elephant and zebra $\leftrightarrow$ giraffe (overall there are 1917 zebras, 2547 giraffes and 2144 elephants). These are extremely challenging datasets, and it should be noted that no previous method has used MSCOCO, without supervision in the form of segmentation.

Our deepest translators, i.e., G_A^5, G_B^5, consist of encoder-decoder structure with several strided convolutional layers followed by symmetric transpose convolutional layers. We use group normalization [37] and ReLU activation function (except the last layer, which is tanh). The conditional generators, consist of learned AdaIN layer, achieved by several strided convolutional layers followed by fully connected layers. The content generator has also several convolutional layers and one single transpose convolutional layer which doubles the spatial resolution (Fig. 4(right)). In practice we only train G^5, G^4, G^3, and apply feature inversion directly on the output of the latter, with negligible degradation. For the exact layer specifics we refer the reader to the supplementary materials, and to our (soon to be published) code. We train each layer for 400 epochs with a fixed learning rate of 0.0001 using the Adam optimizer [18]. On a single RTX 2080, training the entire ensemble of networks (all translators, from the deepest layer to the shallowest layer, and the final feature inversion network), takes around 2.5 days.

MSCOCO
↓ Zebra
↓ Elephant

MSCOCO
↑ Zebra
↑ Elephant

MSCOCO
↓ Giraffe
↓ Zebra

MSCOCO
↑ Giraffe
↑ Zebra

Kaggle
↓ Dog
↓ Cat

Kaggle
↑ Dog
↑ Cat

Fig. 5. Examples of challenging translation results, featuring significant shape deformations.

Several translation examples are presented in Fig. 5. Our translation is able to achieve high shape deformation. Note that our translations are semantically consistent, in the sense that they preserve the pose of the object of interest, and the number of instances is mostly preserved. Furthermore, partial occlusions of such objects, or their cropping by the image boundaries are correctly reproduced. See for example, the translations of the pairs of animals in columns 5–6. More results are provided in the supplementary materials.

4.1 Ablation Study

Below, we analyze the impact of the main elements of our method.

Loss Components. First, we ablate each of our loss components. Figure 6 visualizes the translation of the 5th (deepest) layer with and without cycle, identity and adversarial losses. The best result is obtained by using all of the losses, which balance each other.

Translation Depth. In Fig. 7 we compare between translation results using different VGG-19 layers. Evidently, shallower layers introduce more rigid spatial constraints, restricting the ability of shapes to be changed by the translation. The shallowest layer can hardly change the shape of the input image, which may explain the failure of traditional image translation methods. In Table 1, we use the common FID score [32] to show that the cascaded translation achieves better translation compared to individual layer translation. Additional results are shown in the supplementary materials.

Original | &idty cyc | adv | &adv cyc | &adv idty | adv cyc&idty

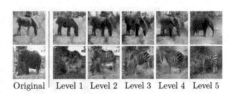

Original | Level 1 | Level 2 | Level 3 | Level 4 | Level 5

Fig. 6. Translation of the 5th (deepest) layer with different loss combinations. Using all three components yields the best result.

Fig. 7. Translation of different VGG layers, separately. Low level semantics translation fails to deform the geometry of the object.

Table 1. FID score comparison of different layer translation. Each translation was trained independently. We compare the FID scores on three datasets, measured both directions per dataset. The two directions appear side-by-side, →/←, at each cell

→/←	Layer 5	Layer 4	Layer 3	Cascaded (ours)
Cat ↔ Dog	126.93/127.53	181.90/164.42	178.13/91.71	**67.58/46.02**
Zebra ↔ Giraffe	167.62/184.37	103.41/53.36	112.43 /68.62	**67.41/39.38**
Zebra ↔ Elephant	101.26/76	105.58/57.34	166.32/113.28	**68.45/47.86**

Original   Alexnet   VGG Fine-tuned   VGG   Original   Alexnet   VGG Fine-tuned   VGG

Fig. 8. Translation with different pre-trained networks. All the networks were pre-trained on ImageNet. VGG was further fine-tuned to classify between zebras and giraffes. Evidently, using this fine-tuned version does not improve the translation results. In addition, translation between AlexNet features fails to produce reasonable results.

Type of Pre-trained Network. While our method is conceptually agnostic to the type of feature extraction network, we rely on the assumption that the extracted features represent high-level semantics. Therefore, we chose the VGG-19 deep features, which are commonly used for image generation tasks [1,5,7,22].

Table 2. FID score comparison. We compare our FID scores against five approaches on three datasets, measured for both translation directions per dataset. The two directions appear side-by-side, →/←, at each cell

→/←	CycleGAN	MUNIT	DRIT	GANimorph	Ours
Cat ↔ Dog	125.75/94.27	159.57/108.51	153.94/139.17	139.17/134.14	**67.58/46.02**
Zebra ↔ Giraffe	**55.65**/58.93	238.06/60.78	59.75/54.06	98.25/120.05	67.41/**39.38**
Zebra ↔ Elephant	86.55/68.44	109.56/80.1	78.01/56.39	99.98/89.74	**68.45/47.86**

Nonetheless, we experimented with a fine-tuned version of VGG-19, as well as a different network architecture, as shown in Fig. 8. We first fine-tuned VGG-19 to classify between zebras and giraffes and trained our translation networks using the resulting features. As can be seen, the translation results are inferior to the results achieved by the standard VGG-19. This may be attributed to VGG-19 fixating on the unique differences between the zebra and giraffe images, unrelated to the translation, such as background. For more about the extracted features, we refer the reader to the supplementary materials. In addition, in Fig. 8, we examine a different network, AlexNet, also pretrained on ImageNet. We observe that the deepest image translation is not able to generate valid shapes of zebras or giraffes. AlexNet uses a stride of 4 in its first convolutional layer. Thus, the resulting features have less spatial encoding, especially at the deeper layers, which may explain the difficulty to invert and translate these features.

4.2 Comparison to Other Methods

We compare our result with several leading image-to-image translation methods, i.e., CycleGAN [43], MUNIT [15], DRIT [20] and GANimorph [8].

Quantitative Comparison: In order to perform a quantitative comparison, we use the FID score [32], as reported in Table 2. Our method achieves the best FID score on five out of the six cross-domain translations for which this score was measured.

Qualitative Comparison: In Fig. 9 we show several challenging translation examples. While traditional image translation methods struggle to preform translations with such drastic shape deformation, our method is able to do so thanks to its use of the pre-trained VGG-19 network.

The success of our method can also be explained and visualized by examining the translated deep features. We feed forward every image, original and translated, through the entire VGG network, extracting the last fully-connected layer (before the classification layer). We project this vector (of size 4096) to 2D, using t-SNE, as shown in Fig. 10 It may be seen that the distribution of the translated vectors (in cyan) is closest to that of the target domain (in red) when using our method.

Fig. 9. Comparison to other image-to-image translation methods. The unpaired translations, from left to right, are zebra ↔ giraffe, elephant ↔ zebra, and Kaggle dog ↔ cat, where every translation has four examples, two in each direction. While previous translation methods struggle to deform the geometry of the source images, our method is able to preform drastic geometric deformation, while preserving the poses of the subjects and the overall composition of the image.

Limitations. Our method achieves translations with significant shape deformation in many previously unattainable scenarios; yet, a few limitations remain. First, the background of the object is not preserved, as the background is encoded in the deep features along with the semantic parts. Also, in some cases the translated deep features may be missing small instances or parts of the object. This may be attributed to the fact that VGG-19 is generally not invertible and was trained to classify a finite set of classes. In addition, since we translate deep features, small errors in the deep translation may be amplified to large errors in the image, while for image-to-image translation method that operate on the image directly, small translation errors would typically be more local. Please note that, similarly to CycleGAN and GANimorph, our translation is deterministic.

4.3 Unconditional Generation via Deep Feature Synthesis

The expressive power of deep features can also be leveraged by unconditional generative models that synthesize the deep features, rather then generating the images directly. Specifically, we demonstrate that such generative models are able to compete with state-of-the-art synthesis networks, especially with respect to higher-lever semantics. We train a variational auto encoder (VAE) [19] to

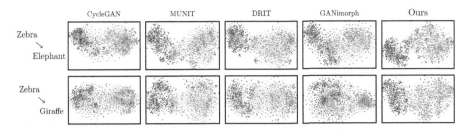

Fig. 10. Comparison of the deepest latent spaces (5th layer), projected using t-SNE. The latent space of the source domain is in blue, and the target domain is in red. The distribution of the translation results (in cyan) is most similar to that of the target domain when using our method. (Color figure online)

generate the conv_5_1 feature maps of zebras (using feature maps extracted by VGG-19 pretrained on ImageNet, as our training data). We then synthesize shallower layers in a cascaded fashion, using a process similar to the one described in Sect. 3 (for more details please refer to the supplementary materials). We refer to the resulting generative model as DEEP-VAE. As shown in Fig. 11, the images generated by DEEP-VAE are not blurry, a phenomenon ordinary VAEs are notoriously known for. We compare our DEEP-VAE to DFC-VAE [12], which uses a perceptual loss for reconstruction, and to VQ-VAE-2 [29], a state-of-the-art VAE synthesis module, which learns a multi-categorical distribution over a learned dictionary elements. As shown in Fig. 11, DFC-VAE fails to produce clear and sharp images, while many of the VQ-VAE-2 results do not contain the main semantic attribute (zebras, giraffes or elephants) at all. This is also evident in the FID scores, shown in Table 3. In order to further demonstrate the generative power of deep features, we also show several examples of latent interpolation in Fig. 12. More results are reported in the supplementary materials.

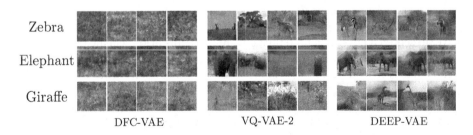

Fig. 11. Synthesis quality comparison. While DCF-VAE is trained using a perceptual loss, it is unable to produce realistic results. VQ-VAE-2 is able to generate higher quality images, however these images rarely contain the main semantic content of the training dataset, i.e. zebras, giraffes and elephants. Our method produces good quality images with the structure of the animal evident in almost all of the generated images.

Table 3. FID score comparison for VAE synthesis

Dataset	DFC-VAE	VQ-VAE-2	DEEP-VAE
Zebra	324	154.41	**57.66**
Elephant	347.93	267.32	**80.2**
Giraffe	346.47	254	**108.02**

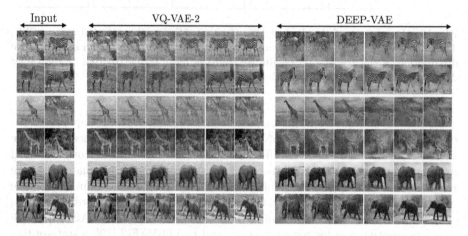

Fig. 12. Latent interpolation between deep features. Two input images are encoded by a trained VAE. Uniform interpolation is preformed between the two encodings, and the decoded result is shown for both VQ-VAE-2 and DEEP-VAE. While DEEP-VAE has a very simple architecture it competes with the state of the art VQ-VAE-2 w.r.t reconstruction and yield interpolation results without ghosting artifacts.

5   Conclusions

Translating between image domains that differ not only in their appearance, but also exhibit significant geometric deformations, is a highly challenging task. We have presented a novel unpaired image-to-image translation scheme that operates directly on pre-trained deep features, where local activation patterns provide a rich semantic encoding of large image regions. Thus, translating between such patterns is capable of generating significant, yet semantically consistent, shape deformations. In a sense, this solution may be thought of as transfer learning, since we make use of features that were trained for a classification task for an unpaired translation task. We have also demonstrated the potential of such transfer learning in the context of unconditional image generation. In the future, we would like to continue exploring the applications of powerful pre-trained deep features for other challenging tasks, possibly in different domains, such as videos, sketches or 3D shapes.

References

1. Aberman, K., Liao, J., Shi, M., Lischinski, D., Chen, B., Cohen-Or, D.: Neural best-buddies: sparse cross-domain correspondence. ACM Trans. Graph. (TOG) **37**(4), 69 (2018)
2. Almahairi, A., Rajeswar, S., Sordoni, A., Bachman, P., Courville, A.: Augmented CycleGAN: Learning many-to-many mappings from unpaired data. arXiv preprint arXiv:1802.10151 (2018)
3. Benaim, S., Wolf, L.: One-sided unsupervised domain mapping. In: Advances in Neural Information Processing Systems, pp. 752–762 (2017)
4. Di, X., Sindagi, V.A., Patel, V.M.: GP-GAN: gender preserving GAN for synthesizing faces from landmarks. In: 2018 24th International Conference on Pattern Recognition (ICPR), pp. 1079–1084. IEEE (2018)
5. Dosovitskiy, A., Brox, T.: Generating images with perceptual similarity metrics based on deep networks. In: Advances in Neural Information Processing Systems, pp. 658–666 (2016)
6. Elson, J., Douceur, J.J., Howell, J., Saul, J.: Asirra: a CAPTCHA that exploits interest-aligned manual image categorization. In: ACM Conference on Computer and Communications Security (2007)
7. Gatys, L.A., Ecker, A.S., Bethge, M.: Image style transfer using convolutional neural networks. In: Proceedings of the IEEE conference on computer vision and pattern recognition, pp. 2414–2423 (2016)
8. Gokaslan, A., Ramanujan, V., Ritchie, D., Kim, K.I., Tompkin, J.: Improving shape deformation in unsupervised image-to-image translation. In: Ferrari, V., Hebert, M., Sminchisescu, C., Weiss, Y. (eds.) ECCV 2018, Part XII. LNCS, vol. 11216, pp. 662–678. Springer, Cham (2018). https://doi.org/10.1007/978-3-030-01258-8_40
9. Goodfellow, I., et al.: Generative adversarial nets. In: Advances in Neural Information Processing systems, pp. 2672–2680 (2014)
10. Gulrajani, I., Ahmed, F., Arjovsky, M., Dumoulin, V., Courville, A.C.: Improved training of Wasserstein GANs. In: Advances in Neural Information Processing Systems, pp. 5767–5777 (2017)
11. Hoffman, J., et al.: Cycada: Cycle-consistent adversarial domain adaptation. arXiv preprint arXiv:1711.03213 (2017)
12. Hou, X., Shen, L., Sun, K., Qiu, G.: Deep feature consistent variational autoencoder. In: 2017 IEEE Winter Conference on Applications of Computer Vision (WACV), pp. 1133–1141. IEEE (2017)
13. Huang, X., Belongie, S.: Arbitrary style transfer in real-time with adaptive instance normalization. In: Proceedings of the IEEE International Conference on Computer Vision, pp. 1501–1510 (2017)
14. Huang, X., Li, Y., Poursaeed, O., Hopcroft, J., Belongie, S.: Stacked generative adversarial networks. In: Proceedings of the IEEE Conference on Computer Vision and Pattern Recognition, pp. 5077–5086 (2017)
15. Huang, X., Liu, M.-Y., Belongie, S., Kautz, J.: Multimodal unsupervised image-to-image translation. In: Ferrari, V., Hebert, M., Sminchisescu, C., Weiss, Y. (eds.) ECCV 2018, Part III. LNCS, vol. 11207, pp. 179–196. Springer, Cham (2018). https://doi.org/10.1007/978-3-030-01219-9_11
16. Ignatov, A., Kobyshev, N., Timofte, R., Vanhoey, K., Van Gool, L.: WESPE: weakly supervised photo enhancer for digital cameras. In: Proceedings of the IEEE Conference on Computer Vision and Pattern Recognition Workshops, pp. 691–700 (2018)

17. Kim, T., Cha, M., Kim, H., Lee, J.K., Kim, J.: Learning to discover cross-domain relations with generative adversarial networks. In: Proceedings of the 34th International Conference on Machine Learning-Volume 70, pp. 1857–1865. JMLR. org (2017)
18. Kingma, D.P., Ba, J.: Adam: A method for stochastic optimization. CoRR abs/1412.6980 (2014)
19. Kingma, D.P., Welling, M.: Auto-encoding variational bayes. arXiv preprint arXiv:1312.6114 (2013)
20. Lee, H.-Y., Tseng, H.-Y., Huang, J.-B., Singh, M., Yang, M.-H.: Diverse image-to-image translation via disentangled representations. In: Ferrari, V., Hebert, M., Sminchisescu, C., Weiss, Y. (eds.) ECCV 2018, Part I. LNCS, vol. 11205, pp. 36–52. Springer, Cham (2018). https://doi.org/10.1007/978-3-030-01246-5_3
21. Liang, X., Zhang, H., Lin, L., Xing, E.: Generative semantic manipulation with mask-contrasting GAN. In: Ferrari, V., Hebert, M., Sminchisescu, C., Weiss, Y. (eds.) ECCV 2018, Part XIII. LNCS, vol. 11217, pp. 574–590. Springer, Cham (2018). https://doi.org/10.1007/978-3-030-01261-8_34
22. Liao, J., Yao, Y., Yuan, L., Hua, G., Kang, S.B.: Visual attribute transfer through deep image analogy. arXiv preprint arXiv:1705.01088 (2017)
23. Lin, T.-Y., et al.: Microsoft COCO: common objects in context. In: Fleet, D., Pajdla, T., Schiele, B., Tuytelaars, T. (eds.) ECCV 2014, Part V. LNCS, vol. 8693, pp. 740–755. Springer, Cham (2014). https://doi.org/10.1007/978-3-319-10602-1_48
24. Liu, M.Y., Breuel, T., Kautz, J.: Unsupervised image-to-image translation networks. In: Advances in Neural Information Processing Systems, pp. 700–708 (2017)
25. Ma, L., Jia, X., Georgoulis, S., Tuytelaars, T., Van Gool, L.: Exemplar guided unsupervised image-to-image translation. arXiv preprint arXiv:1805.11145 (2018)
26. Mao, X., Li, Q., Xie, H., Lau, R.Y., Wang, Z., Paul Smolley, S.: Least squares generative adversarial networks. In: Proceedings of the IEEE International Conference on Computer Vision, pp. 2794–2802 (2017)
27. Mejjati, Y.A., Richardt, C., Tompkin, J., Cosker, D., Kim, K.I.: Unsupervised attention-guided image-to-image translation. In: Advances in Neural Information Processing Systems, pp. 3693–3703 (2018)
28. Mo, S., Cho, M., Shin, J.: InstaGAN: Instance-aware image-to-image translation. arXiv preprint arXiv:1812.10889 (2018)
29. Razavi, A., van den Oord, A., Vinyals, O.: Generating diverse high-fidelity images with VQ-VAE-2. In: Advances in Neural Information Processing Systems, pp. 14837–14847 (2019)
30. Simonyan, K., Zisserman, A.: Very deep convolutional networks for large-scale image recognition. arXiv preprint arXiv:1409.1556 (2014)
31. Sungatullina, D., Zakharov, E., Ulyanov, D., Lempitsky, V.: Image manipulation with perceptual discriminators. In: Ferrari, V., Hebert, M., Sminchisescu, C., Weiss, Y. (eds.) ECCV 2018, Part VI. LNCS, vol. 11210, pp. 587–602. Springer, Cham (2018). https://doi.org/10.1007/978-3-030-01231-1_36
32. Szegedy, C., et al.: Going deeper with convolutions. In: Proceedings of the IEEE Conference on Computer Vision and Pattern Recognition, pp. 1–9 (2015)
33. Upchurch, P., et al.: Deep feature interpolation for image content changes. In: Proceedings of the IEEE Conference on Computer Vision and Pattern Recognition, pp. 7064–7073 (2017)
34. Wang, X., et al.: ESRGAN: enhanced super-resolution generative adversarial networks. In: Leal-Taixé, L., Roth, S. (eds.) ECCV 2018, Part V. LNCS, vol. 11133, pp. 63–79. Springer, Cham (2019). https://doi.org/10.1007/978-3-030-11021-5_5

35. Wu, W., Cao, K., Li, C., Qian, C., Loy, C.C.: TransGaGa: Geometry-aware unsupervised image-to-image translation. arXiv preprint arXiv:1904.09571 (2019)
36. Wu, X., Shao, J., Gao, L., Shen, H.T.: Unpaired image-to-image translation from shared deep space. In: 2018 25th IEEE International Conference on Image Processing (ICIP), pp. 2127–2131. IEEE (2018)
37. Wu, Y., He, K.: Group normalization. In: Ferrari, V., Hebert, M., Sminchisescu, C., Weiss, Y. (eds.) ECCV 2018, Part XIII. LNCS, vol. 11217, pp. 3–19. Springer, Cham (2018). https://doi.org/10.1007/978-3-030-01261-8_1
38. Xu, J., et al.: Unpaired sentiment-to-sentiment translation: A cycled reinforcement learning approach. arXiv preprint arXiv:1805.05181 (2018)
39. Yi, Z., Zhang, H., Tan, P., Gong, M.: DualGAN: unsupervised dual learning for image-to-image translation. In: Proceedings of the IEEE ICCV, pp. 2849–2857 (2017)
40. Yin, K., Chen, Z., Huang, H., Cohen-Or, D., Zhang, H.: LoGAN: Unpaired shape transform in latent overcomplete space. arXiv preprint arXiv:1903.10170 (2019)
41. Yu, J., Lin, Z., Yang, J., Shen, X., Lu, X., Huang, T.S.: Free-form image inpainting with gated convolution. arXiv preprint arXiv:1806.03589 (2018)
42. Zeiler, M.D., Fergus, R.: Visualizing and understanding convolutional networks. In: Fleet, D., Pajdla, T., Schiele, B., Tuytelaars, T. (eds.) ECCV 2014, Part I. LNCS, vol. 8689, pp. 818–833. Springer, Cham (2014). https://doi.org/10.1007/978-3-319-10590-1_53
43. Zhu, J.Y., Park, T., Isola, P., Efros, A.A.: Unpaired image-to-image translation using cycle-consistent adversarial networks. In: Proceedings of the IEEE ICCV, pp. 2223–2232 (2017)

"Look Ma, No Landmarks!" – Unsupervised, Model-Based Dense Face Alignment

Tatsuro Koizumi[1,2]([⊠]) and William A. P. Smith[2]

[1] Canon Inc., Tokyo, Japan
[2] University of York, York, UK
{tk856,william.smith}@york.ac.uk

Abstract. In this paper, we show how to train an image-to-image network to predict dense correspondence between a face image and a 3D morphable model using only the model for supervision. We show that both geometric parameters (shape, pose and camera intrinsics) and photometric parameters (texture and lighting) can be inferred directly from the correspondence map using linear least squares and our novel inverse spherical harmonic lighting model. The least squares residuals provide an unsupervised training signal that allows us to avoid artefacts common in the literature such as shrinking and conservative underfitting. Our approach uses a network that is 10× smaller than parameter regression networks, significantly reduces sensitivity to image alignment and allows known camera calibration or multi-image constraints to be incorporated during inference. We achieve results competitive with state-of-the-art but without any auxiliary supervision used by previous methods.

Keywords: 3D morphable model · Dense correspondence · Face alignment · Landmark · Unsupervised · Self-supervised

1 Introduction

CNN-based face image analysis with a 3D morphable model (3DMM) [7] has recently shown great promise for both 3D face reconstruction from a single image [22,24,31] and dense face alignment [3,8,35,38,39] (i.e. predicting dense correspondence from image pixels to model). These methods are supervised, limiting their application only to labelled images and not providing a general method that can be extended to new object classes.

One line of CNN-based 3D face reconstruction work offers the promise of overcoming this reliance using model-based autoencoders for self-supervision [6,10, 13,28,29]. Here, a 3DMM and a differentiable renderer are used as a model-based

Electronic supplementary material The online version of this chapter (https://doi.org/10.1007/978-3-030-58536-5_41) contains supplementary material, which is available to authorized users.

A. Vedaldi et al. (Eds.): ECCV 2020, LNCS 12347, pp. 690–706, 2020.
https://doi.org/10.1007/978-3-030-58536-5_41

Input   Correspondence   Geometry   Reconstruction   Albedo

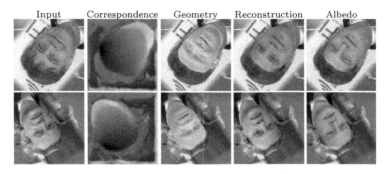

Fig. 1. From a single input image our network learns to predict dense correspondence. From this, we can infer least squares optimal 3DMM geometry and albedo giving high quality reconstructions with 2D transformation invariance.

decoder such that a trainable encoder (a CNN) can learn to regress semantically meaningful model parameters. In principle, model-based autoencoders can be trained in a self-supervised fashion. In practice, most rely on auxiliary supervision in the form of landmarks [6,29,32], paired identity images [10] or ground truth 3D geometry [13]. The Model-based Face Autoencoder (MoFA) of Tewari et al. [29] did demonstrate a completely unsupervised variant but the estimated face is prone to shrinking into the inner face region and requires careful pre-alignment of training images and initialisation of camera parameter predictions such that the initial 3DMM models approximately align with the face images. This makes the approach unable to learn invariance to 2D transformations.

In this paper, we propose a completely unsupervised strategy for learning to fit a 3DMM to a single image. The main difference to previous work is that, instead of image-to-3DMM parameter regression with a contractive CNN, we propose to estimate a dense image-model correspondence map with an image-to-image CNN architecture (Fig. 1). There are significant benefits in doing so:

1. All 3DMM parameters can be estimated from a correspondence map (Sect. 2). Therefore, using a CNN to predict both geometric and photometric parameters, as done in all previous work [6,10,13,28,29], is redundant.
2. The estimated parameters are least squares optimal with respect to the input image and estimated correspondence map. Optimality for a given image is not guaranteed for a parameter regression CNN whose training objective seeks optimality only in aggregate over the whole training set.
3. Image-to-image CNNs are well suited to estimating correspondence maps with invariance to 2D transformations. Intuitively, it is enough for the correspondence CNN to learn "part detectors" with robustness to 2D rotation (convolution layers are already translation invariant). On the other hand, contractive CNNs are ill-suited to directly regressing geometric parameters with 2D transformation invariance [15]. This is because spatial information is lost in contractive layers and fully connected layers must exhaustively represent both features and their locations to reason about geometric parameters.

4. Image-to-image CNNs are much smaller than parameter regression networks due to the lack of fully connected layers. Concretely, we require ~10× fewer parameters than previous CNN based approaches (e.g. 13.4M parameters for our U-Net versus 138M parameters in VGG-face used by [10,29]).
5. Every pixel in the input image can contribute to the losses during training. Previous model-based methods learn only from the parts of the image covered by the geometry of the current 3DMM estimate. In our approach, there is no longer a shortcut for the network to reduce reconstruction loss by shrinking the model to avoid difficult pixels.
6. We defer estimation of actual face geometry. Correspondence is an intermediate representation from which we infer geometry. At test time, if we have access to calibration information or have multiple images from the same camera (e.g. a video), we can exploit these constraints when we finally compute shape from the estimated correspondence map(s). Parameter regression networks cannot do this – they commit to an explanation of the shape and camera parameters for a single image with no way to inject calibration information or constraints post hoc.

Alternatively, our approach can be viewed as a means to learn dense face alignment using model fitting as a form of self-supervision. Correspondence is, in itself, a useful representation. Once trained, the 3DMM can be discarded and the correspondence estimation network used for tasks such as landmarking or semantic segmentation without requiring ground truth labels for supervision.

Our specific novel contributions are as follows. We interpolate a 3DMM to pixel space (Sect. 2.1) then show how to estimate both camera and shape parameters from a correspondence map using linear least squares (Sects. 2.2 and 3.2). We propose an inverse spherical harmonic lighting model enabling simultaneous least squares inverse rendering for both albedo and lighting parameters (Sects. 2.3 and 3.3). Finally, we combine the two least squares solutions with a robust residual loss, a reconstruction loss and priors to enable unsupervised training of our dense alignment network (Sect. 3.4). We make an implementation available[1].

1.1   Related Work

Deep Integration of 3DMMs. The power of deep learning and CNNs has been applied to the task of face model fitting in the last 2 years. Tran et al. [31] use the results of [19] train a CNN discriminatively to regress the same parameters for any single image of the same person. Richardson et al. [22] use synthetic renderings as training data. Both these methods are supervised. MoFA [29] essentially merges analysis-by-synthesis and CNN-based regression in an autoencoder architecture in which the encoder learns the inverse problem, supervised by an appearance error provided by a fixed decoder that implements the forward process. Kim et al. [13] take a similar approach but train on synthetic data that is progressively updated to make it match the distribution of real images. Rather

---

[1] https://github.com/kzmttr/UMDFA.

than require that the appearance of the reconstructed model matches that of the input image, Genova et al. [10] use a face encoder to measure similarity in an identity space. Hence, they do not estimate pose or establish correspondence to the input image, but instead ensure discriminative texture and shape are reconstructed. This can be seen as a self-supervised variant of Tran et al. [31]. A number of extension to MoFA have since been considered. Tewari et al. [28] learn a corrective space to augment the model reconstruction with additional details. Both Tran and Liu [17,32] and Tewari et al. [27] learn the model itself. In contrast to all of these approaches, we do not regress 3DMM parameters. Instead, we regress an intermediate pixel-wise representation of geometry from which geometric and photometric parameters can be directly inferred in a least squares optimal sense. Importantly, all pixels contribute to this solution, not only those covered by a rendering of the model.

Image-to-Image Methods. Going beyond model fitting, a number of methods make pixel-wise predictions. SFSNet [26] infers lighting and normal and albedo maps from single face images. Their training is bootstrapped using synthetic faces sampled from a model. Sela et al. [25] use an image-to-image network to predict facial depth and correspondence to a canonical model. The network is trained entirely supervised using synthetic data and model fitting requires an offline nonrigid registration to the estimated correspondences. Guler et al. [3] and Yu et al. [35] predict dense correspondence maps using an image-to-image network and supervision provided by landmark-based 3DMM fits. Feng et al. [8] predict a UV map from a 3D face to 2D image coordinates. Zhu et al. [38,39] propose the projected normalised coordinate code (PNCC) as a representation for dense correspondence. Crispell and Bazik [5] augment PNCC with a predicted 3D offset. All of these approaches are supervised. Several [5,25,35] fit a model to estimated depth or correspondence, but this is done as an offline, nonlinear optimisation. In contrast, we show how to fit a 3DMM in-network. This means that we can use the residuals as a supervisory signal for the image-to-image network, negating the need for any direct supervision.

2   3DMM Parameters from Image-Model Correspondence

We begin by asking: What can be estimated given dense image-model correspondence alone? Specifically, since we wish to incorporate the estimation process into a network, we are interested in what can be estimated efficiently and in a differentiable manner. Linear least squares satisfies both of these requirements and we use it to estimate optimal geometric and, subsequently, photometric parameters. This necessitates interpolating our 3DMM to pixel space which we explain first.

2.1   Interpolating a 3DMM to UV and Pixel Space

We represent a 3D face based on a 3DMM:

$$\mathbf{v}_j(\boldsymbol{\alpha}) = \sum_{i=1}^{N_s+N_e} \alpha_i \mathbf{s}_j^i + \bar{\mathbf{s}}_j, \quad \mathbf{r}_j(\boldsymbol{\beta}) = \sum_{i=1}^{N_r} \beta_i \mathbf{a}_j^i + \bar{\mathbf{a}}_j \tag{1}$$

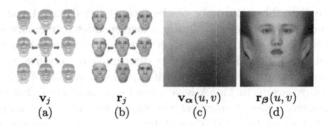

$$\mathbf{v}_j \qquad \mathbf{r}_j \qquad \mathbf{v}_\alpha(u,v) \qquad \mathbf{r}_\beta(u,v)$$
$$\text{(a)} \qquad \text{(b)} \qquad \text{(c)} \qquad \text{(d)}$$

Fig. 2. A 3D morphable model of geometry (a) and albedo (b) can be interpolated to a UV space (c, d) via an embedding. We refer to this as a UV-3DMM.

where $\mathbf{v}_j$ is the 3D position and $\mathbf{r}_j$ is the RGB albedo (or reflectance) of the jth vertex respectively. $\mathbf{s}_j^i$ is ith linear basis of the vertex position and $\bar{\mathbf{s}}_j$ is its mean. In the same manner, $\mathbf{a}_j^i$ is ith linear basis of the vertex albedo and $\bar{\mathbf{a}}_j$ is its mean. α_i and β_i are the ith coefficient of the linear combination with $\boldsymbol{\alpha} = [\alpha_1, \dots, \alpha_{N_s+N_e}]^T$ the stacked shape parameters and $\boldsymbol{\beta} = [\beta_1, \dots, \beta_{N_r}]^T$ the stacked albedo parameters. N_s, N_e and N_r are the number of dimensions for neutral shape, expression and albedo respectively.

UV Interpolation of the 3DMM. We compute a UV embedding for our 3DMM (in practice by flattening the mean shape – see supplementary material for details) such that every vertex is assigned a fixed 2D UV coordinate. Via barycentric interpolation we can compute a linear shape and texture model for any position, $(u, v) \in [-1, 1] \times [-1, 1]$, in UV space. Accordingly, we write $\mathbf{s}^i(u, v)$, $\bar{\mathbf{s}}(u, v)$, $\mathbf{a}^i(u, v)$ and $\bar{\mathbf{a}}(u, v)$ for the interpolated ith shape basis, shape mean, ith albedo basis and albedo mean at arbitrary location in UV space (u, v). Note that (u, v) is continuous and the barycentric interpolation amounts to taking linear combinations of basis and mean values at the original vertex positions.

The 3D position of the model interpolated at UV coordinate (u, v) is:

$$\mathbf{v}_\alpha(u, v) = \mathbf{S}_{u,v} \boldsymbol{\alpha} + \bar{\mathbf{s}}(u, v), \tag{2}$$

where $\mathbf{S}_{u,v} = [\mathbf{s}^1(u, v), \dots, \mathbf{s}^{N_s+N_e}(u, v)]$ are the stacked shape bases for the model interpolated at UV position (u, v). Similarly, we can write the model albedo interpolated at UV position (u, v):

$$\mathbf{r}_\beta(u, v) = \mathbf{A}_{u,v} \boldsymbol{\beta} + \bar{\mathbf{a}}(u, v), \tag{3}$$

where again $\mathbf{A}_{u,v} = [\mathbf{a}^1(u, v), \dots, \mathbf{a}^{N_r}(u, v)]$ are the stacked albedo bases for the model interpolated at UV position (u, v).

We refer to $\mathbf{v}_\alpha(u, v)$ and $\mathbf{r}_\beta(u, v)$ as a *UV-3DMM* (see Fig. 2).

UV Correspondence Map. Now, suppose that we are given a correspondence map between a face image, $\mathbf{i}(x, y)$, and the UV space of our 3DMM, i.e. we are given two maps: $u(x, y)$ and $v(x, y)$ defined for each pixel $(x, y) \in \{1, \dots, W\} \times \{1, \dots, H\}$ in the face image. Each pixel provides a correspondence between image and model. We can now interpolate our 3DMM at each

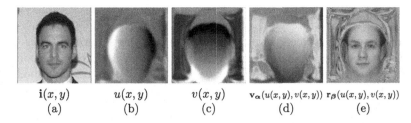

i(x, y)   u(x, y)   v(x, y)   v$_\alpha$(u(x, y), v(x, y)) r$_\beta$(u(x, y), v(x, y))
(a)   (b)   (c)   (d)   (e)

Fig. 3. Using estimated correspondences (b, c) from an image (a) to the UV space of the 3DMM, we can define a pixel-3DMM of geometry (d) and albedo (e) in pixel space as a function of 3DMM parameters.

pixel, via the correspondence map, giving a *pixel-3DMM*: $\mathbf{v}_\alpha(u(x, y), v(x, y))$ and $\mathbf{r}_\beta(u(x, y), v(x, y))$ (see Fig. 3). Details of how the interpolation is efficiently implemented in-network is described in supplementary material.

2.2 Least Squares Shape-from-correspondence

Assume that camera calibration information, i.e. the intrinsic matrix $\mathbf{K} \in \mathbb{R}^{3 \times 3}$ and the extrinsic rotation $\mathbf{R} \in \mathbb{R}^{3 \times 3}$ and translation $\mathbf{t} \in \mathbb{R}^3$, were known. Then, the perspective projection of the 3D position at model UV coordinate (u, v) to pixel position (x, y) is given (up to a scaling) by:

$$\lambda \begin{bmatrix} x \\ y \\ 1 \end{bmatrix} = \mathbf{project}_\alpha(u, v) = \mathbf{K} \begin{bmatrix} \mathbf{R} \ \mathbf{t} \end{bmatrix} \begin{bmatrix} \mathbf{v}_\alpha(u, v) \\ 1 \end{bmatrix}, \tag{4}$$

where λ is an arbitrary scale. Using the Direct Linear Transform [11] we can write (4) as a linear system by taking the cross product between the left and right hand sides and setting equal to the zero vector.

Then, the shape parameters, $\boldsymbol{\alpha}$, minimising the reprojection error can be found by solving the following linear least squares problem:

$$\min_{\alpha} \sum_{x=1}^{W} \sum_{y=1}^{H} \left\| \begin{bmatrix} x \\ y \\ 1 \end{bmatrix}_{\times} \mathbf{project}_\alpha(u(x, y), v(x, y)) \right\|^2, \text{ where } \begin{bmatrix} \mathbf{x} \end{bmatrix}_{\times} = \begin{bmatrix} 0 & -x_3 & x_2 \\ x_3 & 0 & -x_1 \\ -x_2 & x_1 & 0 \end{bmatrix}. \tag{5}$$

Note that the residuals of the least squares solution indicate how well the model can explain a shape consistent with the correspondence map and therefore provide a measure of the plausibility of the correspondence map. In practice, $\boldsymbol{\alpha}$ can also be statistically regularised.

During unsupervised training, we of course do not have access to camera calibration information. We later show how to rewrite (5) such that both optimal shape and camera parameters can be found algebraically using linear least squares by additionally estimating a depth map.

2.3   Least Squares Inverse Rendering

Having computed geometry from correspondence, the surface normals of the shape can be computed. Together with the original image and the correspondence from image to model, this is sufficient to reason about lighting and albedo. We now show how to simultaneously solve for lighting and albedo coefficients using linear least squares.

Spherical Harmonic Lighting. The spherical harmonic (SH) lighting model [20] efficiently describes how a diffuse object appears under arbitrarily complex environment illumination. At a surface point with normal direction $\mathbf{n}$ and RGB albedo $\mathbf{r}$, the RGB colour intensity, $\mathbf{i}$, is given by:

$$\mathbf{i} = \mathbf{r} \odot \mathbf{B}(\mathbf{n})\mathbf{L}, \qquad (6)$$

where $\odot$ denotes element-wise multiplication, $\mathbf{B}(\mathbf{n}) \in \mathbb{R}^{3 \times N_L}$ contains the SH basis vectors which depend only on $\mathbf{n}$

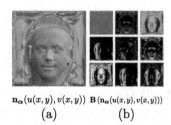

$\mathbf{n}_\alpha(u(x,y),v(x,y))$ $\mathbf{B}(\mathbf{n}_\alpha(u(x,y),v(x,y)))$
(a)   (b)

Fig. 4. From shape parameters α we calculate per-vertex surface normals and interpolate via $u(x,y)$ and $v(x,y)$ to a pixel space normal map (a). From this we define an SH basis in pixel space (b).

and $\mathbf{L} \in \mathbb{R}^{N_L \times 3}$ contains the colour lighting coefficients. For an order 2 approximation, $N_L = 9$ and so there are 27 unknown lighting parameters. This expression is bilinear in diffuse albedo and the spherical harmonic lighting coefficients. This means there is no closed form solution for both optimal albedo and lighting simultaneously. Aldrian and Smith [2] use alternating linear least squares but this requires multiple iterations and is only optimal with respect to the parameters solved for last.

An Inverse Lighting Model. In contrast to the conventional model, we use spherical harmonics to represent *inverse lighting*. That is, a quantity that (when multiplied by the image intensity) removes the effect of shading, giving the diffuse albedo. In other words, we use the spherical harmonic basis functions to represent the reciprocal of diffuse shading:

$$\mathbf{i} \odot \mathbf{B}(\mathbf{n})\mathbf{L} = \mathbf{r}. \qquad (7)$$

This seemingly subtle difference brings a significant practical advantage: it is linear in both lighting and albedo simultaneously so we can solve for both in a single linear least squares formulation. Importantly, we show empirically in supplementary material that this inverse model can explain conventional SH lighting with very low error.

Inverse Rendering with a Correspondence Map. As in the previous section, suppose that we have an estimated correspondence map from a face image to the model. From the geometry estimated by least squares shape-from-correspondence, we can estimate per-vertex surface normals. Then, from the 3DMM UV map we can interpolate a surface normal, $\mathbf{n}_\alpha(u,v)$, at any position in UV space or, given the estimated image-model correspondence maps we can

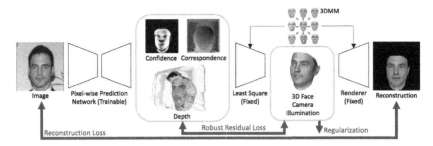

Fig. 5. Overview of proposed architecture. In addition to correspondence our network also predicts a confidence map (for robustness) and depth map (enabling uncalibrated reconstruction). The least squares layer solves first for geometric and then photometric parameters.

interpolate a pixel space normal map $\mathbf{n}_\alpha(u(x,y), v(x,y))$ (see Fig. 4(a)). Given the input face image, $\mathbf{i}(x,y)$, we can now write a linear least squares problem for lighting and albedo parameters:

$$\min_{\mathbf{L},\beta} \sum_{x=1}^{W} \sum_{y=1}^{H} \left\| \mathbf{i}(x,y) \odot \mathbf{B}(\mathbf{n}_\alpha(u(x,y), v(x,y)))\mathbf{L} - \mathbf{r}_\beta(u(x,y), v(x,y)) \right\|^2. \quad (8)$$

3 Self-supervised Learning of Dense Correspondence

We now show how an image-to-image network for dense face alignment can be trained using self-supervision (see Fig. 5). The idea is that the network predicts a correspondence map from which we implement the fitting process described in Sect. 2 as differentiable layers. We use a U-Net [23] as the pixel-wise prediction network though any image-to-image architecture would suffice. The network learns from losses measuring the quality of the fit to the correspondence map as well as an appearance loss computed via differentiable rendering. Some modifications are required to incorporate the least squares solutions into the network which we describe in the following sections.

The various loss functions from which the network learns are combined using weights. We distinguish between those that must be manually chosen (i.e. hyperparameters of our method), denoted by η, and those that are learnt as part of the training, denoted by ω.

3.1 Per-Pixel Confidence

In general, not all of the image will contain face parts. In addition, the face may be occluded by non-face objects such as glasses or unmodelled features such as beards. We do not wish these pixels to contribute to the least squares solutions. Therefore, our network also predicts a scalar confidence map $w(x,y) \in [0,1]$ indicating whether pixel (x,y) is believed to belong to the face. As with correspondence, this is learnt unsupervised without ever providing the network with ground truth face segmentations.

3.2 Uncalibrated Shape-from-correspondence

The least squares solution for geometry in (5) assumed known camera calibration. While this may be available (and can be exploited) at test time, it is not available during unsupervised training. We propose an algebraic solution that allows us to estimate both shape and camera parameters but which requires the network to also estimate a depth map, $z(x, y)$. Again, depth map prediction is learnt unsupervised without any ground truth depth during training. We compute the shape residuals in 3D space by back projection using inverse camera parameters and the estimated depth:

$$\varepsilon_{\text{geo}}(x, y) = \left\| z(x, y)\mathbf{P}[x, y, 1]^T + \mathbf{q} - \mathbf{v}_\alpha(u, v) \right\|_2, \tag{9}$$

where the inverse camera parameters, $\mathbf{P} \in \mathbb{R}^{3\times3}$ and $\mathbf{q} \in \mathbb{R}^3$, are related to standard parameters via $\lambda\mathbf{KR} = \mathbf{P}^{-1}$ and $\lambda\mathbf{Kt} = -\mathbf{P}^{-1}\mathbf{q}$ with λ representing the scale ambiguity. These residuals are linear in the unknown shape parameters and inverse camera parameters.

We can now write the linear least squares system that we solve in-network to compute optimal shape and camera parameters:

$$\boldsymbol{\alpha}^*, \mathbf{P}^*, \mathbf{q}^* = \underset{\boldsymbol{\alpha}, \mathbf{P}, \mathbf{q}}{\arg\min}\, E_{\text{geo}}(\boldsymbol{\alpha}, \mathbf{P}, \mathbf{q}) + R_{\text{geo}}(\boldsymbol{\alpha}, \mathbf{P}, \mathbf{q}), \tag{10}$$

where $E_{\text{geo}} = \sum_{x,y} w(x, y)\varepsilon_{\text{geo}}(x, y)^2$ is the sum of squared residuals from (9), weighted by the estimated per-pixel confidences and $R_{\text{geo}} = \boldsymbol{\alpha}^T \text{diag}(\boldsymbol{\omega}_{\text{geo}})\boldsymbol{\alpha}$ regularises the solution with the statistical prior, weighting each dimension with a learnable weight.

Since (10) is quadratic, optimal $\boldsymbol{\alpha}$, $\mathbf{P}$, and $\mathbf{q}$ can be obtained using the pseudoinverse matrix. Since the pseudoinverse is differentiable, during training loss gradients can be backpropagated through the least squares solution and into the image-to-image network.

3.3 In-Network Least Squares Inverse Rendering

With the optimal shape parameters $\boldsymbol{\alpha}^*$ estimated by geometric least squares, we can compute a per-pixel normal map and write the residuals of fitting our inverse lighting model:

$$\varepsilon_{\text{photo}}(x, y) = \left\| \mathbf{i}(x, y) \odot \mathbf{B}(\mathbf{n}_{\alpha^*}(u(x, y), v(x, y)))\mathbf{L} - \mathbf{r}_\beta(u(x, y), v(x, y)) \right\|_2. \tag{11}$$

We write a linear least squares system, this time for albedo and lighting:

$$\boldsymbol{\beta}^*, \mathbf{L}^* = \underset{\boldsymbol{\beta}, \mathbf{L}}{\arg\min}\, E_{\text{photo}}(\boldsymbol{\beta}, \mathbf{L}) + R_{\text{photo}}(\boldsymbol{\beta}, \mathbf{L}). \tag{12}$$

Once again, $E_{\text{photo}} = \sum_{x,y} w(x, y)\varepsilon_{\text{photo}}(x, y)^2$ is the weighted sum of squared residuals and $R_{\text{photo}} = \boldsymbol{\beta}^T \text{diag}(\boldsymbol{\omega}_{\text{photo}})\boldsymbol{\beta} + \eta_L \|\mathbf{L}\|_{\text{Fro}}^2$ regularises both albedo and lighting parameters. As for geometry, (12) is quadratic and so optimal $\boldsymbol{\beta}$ and $\mathbf{L}$ can be found via the differentiable pseudoinverse.

3.4 Losses

We train our network with four losses (described below):

$$E_{\text{total}} = \eta_{\text{res}} E_{\text{res}} + \eta_{\text{rec}} E_{\text{rec}} + \eta_{\text{stat}} E_{\text{stat}} + \eta_{\text{int}} E_{\text{int}} \tag{13}$$

with $\eta_{\text{rec}} = 1.0$, $\eta_{\text{res}} = 3.0$, $\eta_{\text{stat}} = 1.0$, and $\eta_{\text{int}} = 1.0$.

Least Squares Residuals Loss. The least squares layer in our network solves for optimal shape, albedo, camera and lighting parameters by minimising the geometric (9) and photometric (11) residuals. The network can learn from these residuals since they indicate how consistent the 3DMM fit is with the estimated correspondence map (and depth/confidence maps) and the image. Whereas the least squares layer required a closed form solution and therefore uses linear least squares, the loss used for network training is not so constrained. For this reason, we use a robust loss on the residuals:

$$E_{\text{res}} = \sum_{x,y} \min\left(\varepsilon(x,y), 1\right), \quad \text{where } \varepsilon(x,y) = \eta_{\text{geo}} \varepsilon_{\text{geo}}(x,y) + \eta_{\text{photo}} \varepsilon_{\text{photo}}(x,y),$$
$$\tag{14}$$

and $\eta_{\text{geo}} = 20$ and $\eta_{\text{photo}} = 5$. This loss has an important effect: it encourages the model to expand so that more pixels in the input image can be explained by the model in both geometry and colour. For example, suppose that the pixel-wise network detects an ear with high confidence and estimates good correspondence to the ear region in the model. If the ear of the least squares 3DMM fit is not close to the detected ear pixels, this incurs a residual loss, encouraging the model to expand towards the ear. However, we must make the loss robust since every pixel in the image contributes to it, even background (we do not use the confidence map here). The clamping suppresses the effect from outlier pixels such as occlusion and background.

Reconstruction Loss Based on Differentiable Rendering. We also compute a conventional reconstruction loss using differentiable rendering to compare the fitted model to the image. Without this, the clamped residual loss does not penalise growing the face to fit to background. We render the 3DMM geometry given by the geometry least squares solution. Our differentiable renderer calculates a projection of each vertex as a 2D point on the image as well as its visibility and RGB albedo. We divide the per-vertex RGB albedo by our inverse lighting model to obtain RGB pixel intensities and measure the discrepancy to the sampled intensities:

$$E_{\text{rec}} = \frac{1}{\sum_{j=1}^{N_v} w_j} \sum_{j=1}^{N_v} w_j \left\| \mathbf{i}(x_j, y_j) - \mathbf{r}_j(\boldsymbol{\beta}^*) \oslash \{\mathbf{B}(\mathbf{n}_{\alpha^*}(u(x,y), v(x,y)))\mathbf{L}^*\} \right\|_2,$$
$$\tag{15}$$

where N_v is the number of the vertices and $w_j = 1$ if a vertex is visible, zero otherwise (computed using self occlusion testing and depth testing against a z-buffer). We use differentiable bilinear sampling and $\mathbf{i}(x_j, y_j)$ represents bilinear sampling of the input image at the non-integer pixel position (x_j, y_j) given by projection of vertex $\mathbf{v}_j(\boldsymbol{\alpha}^*)$ using the estimated camera parameters.

Statistical Regularisation Loss. This loss encourages the network to keep the estimated face plausible in terms of the shape and albedo parameters. It is the weighted squared average of the estimated 3DMM coefficients:

$$E_{\text{stat}} = \sum_{i=1}^{N_s + N_e} \omega_r^i (\alpha_i^*)^2 + \sum_{i=1}^{N_r} \omega_s^i (\beta_i^*)^2. \tag{16}$$

Since the 3DMM bases are normalised by their standard deviation, the statistical average of α_i^2 and β_i^2 should be kept to be 1 during training. We do this by controlling the loss weight ω_r^i and ω_s^i (see supplementary material).

Camera Intrinsics Regularisation Loss. Finally, we employ regularisation on the estimated camera intrinsic parameters. This penalises the difference between vertical and horizontal focal length as well as the shear:

$$E_{\text{int}} = \eta_{\text{asp}} \frac{(k_{11} - k_{22})^2}{k_{11}^2 + k_{22}^2} + \eta_{\text{sh}} \frac{k_{12}^2}{k_{11}^2 + k_{22}^2}, \tag{17}$$

where the k_{ij} are the elements of the intrinsic camera parameter matrix $\mathbf{K}$. The first term represents the difference of vertical and horizontal focal length and the second term represents the sheer component. We normalise the loss by the horizontal and vertical focal length to avoid reducing the scale of focal length. We set $\eta_{\text{asp}} = 1.0$ and $\eta_{\text{sh}} = 1.0$.

4   Training

Initialisation. Supervision of our network relies on the difference of appearance between the input image and the estimated face, initial estimation must be enough close to the optimal parameters to obtain meaningful gradient from the loss function. We initialise the network (see supplementary material for details) such that for all inputs it predicts a planar depth map, a correspondence map given by the mean face centred in the image and a binary confidence map given by the rasterisation mask of the centred mean face.

Training Data. We train on ~200k images from pre-aligned CelebA dataset [16]. We augment with random 2D similarity transformations (scale factor: [0.77, 1.3], translation: [−75, 75] pixels horizontal/vertical, rotation [−180°, 180°]). The background region is filled by random images from ImageNet [14] with blended boundary. Finally, we crop the image by 224 × 224 pixels.

Optimisation. We use the Adadelta optimizer [36] with learning rate 0.01, batch size 3, 300k iterations. Network weights and biases are initialised by He initialisation [12]. Training takes approximately 120 h on Nvidia GTX 1080Ti.

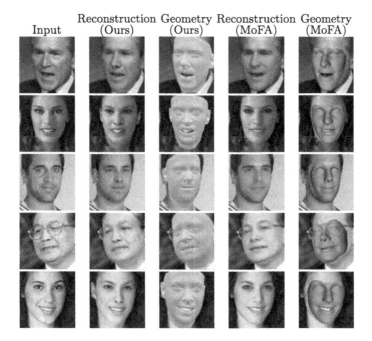

Input | Reconstruction (Ours) | Geometry (Ours) | Reconstruction (MoFA) | Geometry (MoFA)

Fig. 6. Result of MoFA [29] and ours from images in MoFA-test dataset.

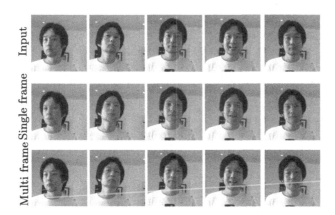

Fig. 7. Result of multiframe aggregation.

5   Experiments

Qualitative Evaluation. We qualitatively evaluate our method based on test images from CelebA dataset (Fig. 6). Our method successfully predicts 3D face including ears under arbitrary 2D similarity transformation. We compare our method with MoFA [29] which can only reconstruct the centre region of a face whereas our method can reconstruct a full head face. Our method also has bet-

Table 1. Quantitative evaluation on NoW dataset [24].

	Median	Mean	Std	Supervision
Tran [31]	1.83	2.33	2.05	Fully supervised
PRNet [8]	1.51	1.99	1.90	Fully supervised
RingNet [24]	1.23	1.55	1.32	Landmarks, ID
Ours	1.52	1.89	1.57	None

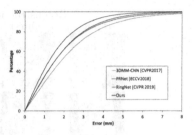

Fig. 8. Cumulative error for the NoW dataset [24].

Table 2. Quantitative evaluation on Stirling/ESRC 3D Face Database [1,9]

	Error (HQ)	Error (LQ)	Error (full)
MTCNN-CNN6-eos [9]	2.70 ± 0.98	2.78 ± 0.95	2.75 ± 0.93
MTCNN-CNN6-3DDFA [9]	2.04 ± 0.67	2.19 ± 0.70	2.14 ± 0.69
SCU-BRL [30]	2.65 ± 0.67	2.87 ± 0.81	2.81 ± 0.80
Ours (w/o E_{int})	2.65 ± 0.98	2.60 ± 0.83	2.62 ± 0.88
Ours	2.39 ± 0.81	2.55 ± 0.82	2.49 ± 0.82

Table 3. Quantitative evaluation on AFLW [18] and AFLW2000-3D [39] dataset. The accuracy is evaluated by the normalized mean error.

Method	AFLW dataset			AFLW2000-3D dataset		
	Mean [0–30]	Mean [0–90]	Std [0–90]	Mean [0–30]	Mean [0–90]	Std [0–90]
LBF [21]	7.17	17.72	10.64	6.17	16.19	9.87
ESR [4]	5.58	12.07	7.33	4.38	11.72	8.04
CFSS [37]	4.68	12.51	9.49	3.44	13.02	10.08
MDM [33]	5.14	13.40	9.72	4.64	13.07	10.07
SDM [34]	4.67	9.19	6.10	3.56	9.37	7.23
3DDFA [39]	4.11	5.60	0.99	2.84	3.79	1.08
Ours (direct)	5.51	16.00	10.74	4.98	16.63	10.98
Ours (fitted)	5.87	18.63	13.20	4.74	18.55	13.38

ter fidelity of reconstruction due to the optimality of the least squares. We also test multiframe aggregation of the pixel-wise prediction (Fig. 7). By optimising multiframe geometry and reflectance to the intermediate output in a single optimisation, superior quality of output can be obtained. See supplementary material for additional qualitative results and comparisons.

Quantitative Evaluation. We quantitatively evaluate our method based on landmarks (Table 3). We follow the evaluation protocol proposed in Zhu et al. [39] and compare our result with supervised facial landmark detection methods. We evaluate landmarks obtained from both direct correspondence and fitted model.

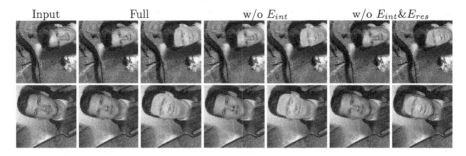

Input   Full   w/o E_{int} w/o E_{int}&E_{res}

Fig. 9. Ablation study to show the contribution of intrinsic parameter regularisation E_{int} and robust residual loss E_{res}. We show input, then for each condition we show overlaid reconstruction followed by overlaid geometry.

Our network shows comparable result to some supervised methods. We quantitatively evaluate our method on the NoW dataset [24] (Table 1, Fig. 8) and Stirling/ESRC 3D Face Database (Table 2) in which the error of reconstructed neutral face shape is calculated. Our method does not outperform other methods that use richer supervision though it is comparable to some supervised methods.

Ablation Study. We investigate the contribution of each loss function qualitatively (Fig. 9) and quantitatively (Table 2). The right column in Fig. 9 shows the result trained by only the reconstruction loss and the statistical regularisation. This is a clear example of shrinking problem, and the robust residual loss significantly improves the problem. From Fig. 9 and Table 2, it is also clear that the intrinsic parameter regularisation enables the reconstruction of plausible and precise shape.

6   Conclusion

We have presented the first method that combines trainable pixel-wise face alignment with differentiable linear least squares to reconstruct a 3D face model. To the best of our knowledge, this is the first method that enables full ear-to-ear face reconstruction under arbitrary in-plane transformation based on unsupervised training. Our approach has further potential of boosting the performance of conventional supervised face alignment methods by harnessing abundant unlabelled images as well as application to other domains in which annotated images are scarce. In future work, our method can be further improved by incorporating an occlusion model, specular reflection, and perceptual metric to alleviate the vulnerability of photometric error based optimisation. It would also be interesting to make the 3DMM learnable [32] or to estimate a corrective function [28] within our framework allowing reconstruction outside the space of the model.

Acknowledgements. W. Smith is supported by a Royal Academy of Engineering/The Leverhulme Trust Senior Research Fellowship.

References

1. Psychological image collection at stirling (PICS). http://pics.stir.ac.uk/
2. Aldrian, O., Smith, W.A.: Inverse rendering of faces with a 3D morphable model. IEEE Trans. Pattern Anal. Mach. Intell. **35**(5), 1080–1093 (2013)
3. Alp Guler, R., Trigeorgis, G., Antonakos, E., Snape, P., Zafeiriou, S., Kokkinos, I.: DenseReg: fully convolutional dense shape regression in-the-wild. In: Proceedings of the IEEE Conference on Computer Vision and Pattern Recognition, pp. 6799–6808 (2017)
4. Cao, X., Wei, Y., Wen, F., Sun, J.: Face alignment by explicit shape regression. Int. J. Comput. Vis. **107**(2), 177–190 (2014). https://doi.org/10.1007/s11263-013-0667-3
5. Crispell, D., Bazik, M.: Pix2face: direct 3D face model estimation. In: Proceedings of the IEEE International Conference on Computer Vision, pp. 2512–2518 (2017)
6. Deng, Y., Yang, J., Xu, S., Chen, D., Jia, Y., Tong, X.: Accurate 3D face reconstruction with weakly-supervised learning: from single image to image set. In: IEEE Computer Vision and Pattern Recognition Workshops (2019)
7. Egger, B., et al.: 3D morphable face models-past, present and future. arXiv preprint arXiv:1909.01815 (2019)
8. Feng, Y., Wu, F., Shao, X., Wang, Y., Zhou, X.: Joint 3D face reconstruction and dense alignment with position map regression network. In: Proceedings of the European Conference on Computer Vision (ECCV), pp. 534–551 (2018)
9. Feng, Z.H., et al.: Evaluation of dense 3D reconstruction from 2D face images in the wild. In: 2018 13th IEEE International Conference on Automatic Face & Gesture Recognition (FG 2018), pp. 780–786. IEEE (2018)
10. Genova, K., Cole, F., Maschinot, A., Sarna, A., Vlasic, D., Freeman, W.T.: Unsupervised training for 3D morphable model regression. In: Proceedings of the IEEE Conference on Computer Vision and Pattern Recognition, pp. 8377–8386 (2018)
11. Hartley, R., Zisserman, A.: Multiple View Geometry in Computer Vision. Cambridge University Press, Cambridge (2003)
12. He, K., Zhang, X., Ren, S., Sun, J.: Delving deep into rectifiers: surpassing human-level performance on imagenet classification. In: Proceedings of the IEEE International Conference on Computer Vision, pp. 1026–1034 (2015)
13. Kim, H., Zollöfer, M., Tewari, A., Thies, J., Richardt, C., Christian, T.: Inverse-FaceNet: deep single-shot inverse face rendering from a single image. In: Proceedings of Computer Vision and Pattern Recognition (CVPR 2018) (2018)
14. Krizhevsky, A., Sutskever, I., Hinton, G.E.: ImageNet classification with deep convolutional neural networks. In: Pereira, F., Burges, C.J.C., Bottou, L., Weinberger, K.Q. (eds.) Advances in Neural Information Processing Systems, vol. 25, pp. 1097–1105. Curran Associates, Inc. (2012). http://papers.nips.cc/paper/4824-imagenet-classification-with-deep-convolutional-neural-networks.pdf
15. Liu, R., et al.: An intriguing failing of convolutional neural networks and the Coord-Conv solution. In: Advances in Neural Information Processing Systems, pp. 9605–9616 (2018)
16. Liu, Z., Luo, P., Wang, X., Tang, X.: Deep learning face attributes in the wild. In: Proceedings of International Conference on Computer Vision (ICCV) (2015)
17. Tran, L., Liu, X.: Nonlinear 3D face morphable model. In: IEEE Computer Vision and Pattern Recognition (CVPR), Salt Lake City, UT (2018)

18. Koestinger, M., Wohlhart, P., Roth, P.M., Bischof, H.: Annotated facial landmarks in the wild: a large-scale, real-world database for facial landmark localization. In: Proceedings of the First IEEE International Workshop on Benchmarking Facial Image Analysis Technologies (2011)
19. Piotraschke, M., Blanz, V.: Automated 3D face reconstruction from multiple images using quality measures. In: Proceedings of the CVPR, pp. 3418–3427 (2016)
20. Ramamoorthi, R., Hanrahan, P.: An efficient representation for irradiance environment maps. In: Proceedings of the SIGGRAPH, pp. 497–500 (2001)
21. Ren, S., Cao, X., Wei, Y., Sun, J.: Face alignment at 3000 FPS via regressing local binary features. In: Proceedings of the IEEE Conference on Computer Vision and Pattern Recognition, pp. 1685–1692 (2014)
22. Richardson, E., Sela, M., Kimmel, R.: 3D face reconstruction by learning from synthetic data. In: Proceedings of the 3DV, pp. 460–469 (2016)
23. Ronneberger, O., Fischer, P., Brox, T.: U-Net: convolutional networks for biomedical image segmentation. In: Navab, N., Hornegger, J., Wells, W.M., Frangi, A.F. (eds.) MICCAI 2015. LNCS, vol. 9351, pp. 234–241. Springer, Cham (2015). https://doi.org/10.1007/978-3-319-24574-4_28
24. Sanyal, S., Bolkart, T., Feng, H., Black, M.: Learning to regress 3D face shape and expression from an image without 3D supervision. In: Proceedings of the IEEE Conference on Computer Vision and Pattern Recognition (CVPR), June 2019
25. Sela, M., Richardson, E., Kimmel, R.: Unrestricted facial geometry reconstruction using image-to-image translation. In: Proceedings of the IEEE International Conference on Computer Vision, pp. 1576–1585 (2017)
26. Sengupta, S., Kanazawa, A., Castillo, C.D., Jacobs, D.W.: SfSNet: learning shape, reflectance and illuminance of faces 'in the wild'. In: Proceedings of the ECCV (2018)
27. Tewari, A., et al.: FML: face model learning from videos. arXiv preprint arXiv:1812.07603 (2018)
28. Tewari, A., et al.: Self-supervised multi-level face model learning for monocular reconstruction at over 250 Hz. In: The IEEE Conference on Computer Vision and Pattern Recognition (CVPR) (2018)
29. Tewari, A., et al.: MoFA: model-based deep convolutional face autoencoder for unsupervised monocular reconstruction. In: The IEEE International Conference on Computer Vision (ICCV) (2017)
30. Tian, W., Liu, F., Zhao, Q.: Landmark-based 3D face reconstruction from an arbitrary number of unconstrained images. In: 2018 13th IEEE International Conference on Automatic Face & Gesture Recognition (FG 2018), pp. 774–779. IEEE (2018)
31. Tran, A.T., Hassner, T., Masi, I., Medioni, G.: Regressing robust and discriminative 3D morphable models with a very deep neural network. In: Proceedings of the CVPR, pp. 5163–5172 (2017)
32. Tran, L., Liu, X.: On learning 3D face morphable model from in-the-wild images. IEEE Trans. Pattern Anal. Mach. Intell. (2019, to appear)
33. Trigeorgis, G., Snape, P., Nicolaou, M.A., Antonakos, E., Zafeiriou, S.: Mnemonic descent method: a recurrent process applied for end-to-end face alignment. In: Proceedings of the IEEE Conference on Computer Vision and Pattern Recognition, pp. 4177–4187 (2016)
34. Yan, J., Lei, Z., Yi, D., Li, S.Z.: Learn to combine multiple hypotheses for accurate face alignment. In: 2013 IEEE International Conference on Computer Vision Workshops, pp. 392–396 (2013)

35. Yu, R., Saito, S., Li, H., Ceylan, D., Li, H.: Learning dense facial correspondences in unconstrained images. In: Proceedings of the IEEE International Conference on Computer Vision, pp. 4723–4732 (2017)
36. Zeiler, M.D.: ADADELTA: an adaptive learning rate method. CoRR abs/1212.5701 (2012)
37. Zhu, S., Li, C., Change Loy, C., Tang, X.: Face alignment by coarse-to-fine shape searching. In: Proceedings of the IEEE Conference on Computer Vision and Pattern Recognition, pp. 4998–5006 (2015)
38. Zhu, X., Lei, Z., Liu, X., Shi, H., Li, S.Z.: Face alignment across large poses: a 3D solution. In: Proceedings of the IEEE Conference on Computer Vision and Pattern Recognition, pp. 146–155 (2016)
39. Zhu, X., Liu, X., Lei, Z., Li, S.Z.: Face alignment in full pose range: a 3D total solution. IEEE Trans. Pattern Anal. Mach. Intell. 41(1), 78–92 (2017)

Online Invariance Selection for Local Feature Descriptors

Rémi Pautrat[1], Viktor Larsson[1]([✉]), Martin R. Oswald[1], and Marc Pollefeys[1,2]

[1] Department of Computer Science, ETH Zurich, Zurich, Switzerland
{remi.pautrat,vlarsson}@inf.ethz.ch
[2] Microsoft MR & AI, Zurich, Switzerland

Abstract. To be invariant, or not to be invariant: that is the question formulated in this work about local descriptors. A limitation of current feature descriptors is the trade-off between generalization and discriminative power: more invariance means less informative descriptors. We propose to overcome this limitation with a disentanglement of invariance in local descriptors and with an online selection of the most appropriate invariance given the context. Our framework (https://github.com/rpautrat/LISRD) consists in a joint learning of multiple local descriptors with different levels of invariance and of meta descriptors encoding the regional variations of an image. The similarity of these meta descriptors across images is used to select the right invariance when matching the local descriptors. Our approach, named Local Invariance Selection at Runtime for Descriptors (LISRD), enables descriptors to adapt to adverse changes in images, while remaining discriminative when invariance is not required. We demonstrate that our method can boost the performance of current descriptors and outperforms state-of-the-art descriptors in several matching tasks, when evaluated on challenging datasets with day-night illumination as well as viewpoint changes.

Keywords: Local descriptors · Invariance · Visual localization

1 Introduction

Sparse features detection and description is at the root of many computer vision tasks: Structure-from-Motion (SfM), Simultaneous Localization and Mapping (SLAM), image retrieval, tracking, etc. They offer a compact representation in terms of memory storage and allow for efficient image matching, and are thus well suited for large-scale applications [14,35,36]. These features should however be able to cope with real world conditions such as day-night changes [44], seasonal variations [34] and matching across large baselines [40].

Electronic supplementary material The online version of this chapter (https://doi.org/10.1007/978-3-030-58536-5_42) contains supplementary material, which is available to authorized users.

© Springer Nature Switzerland AG 2020
A. Vedaldi et al. (Eds.): ECCV 2020, LNCS 12347, pp. 707–724, 2020.
https://doi.org/10.1007/978-3-030-58536-5_42

Fig. 1. Importance of invariance among descriptors. SIFT descriptors (left) perform well on rotated images (top), but are outperformed by Upright SIFT descriptors (middle) when no rotation is present (bottom). We propose a method (right) that automatically selects the proper invariance during matching time.

To be able to do matching in extreme scenarios, the successive feature detectors and descriptors have become more and more invariant [23]. The Harris corner detector [12] was already invariant to rotations, but not to scale. The SIFT detector and descriptor [20] was one of the first to achieve invariance with respect to scale, rotation and uniform light changes. More recently, learned descriptors have been able to encode invariance without handcrafting it. On the one hand, patch-based descriptors can become invariant to transforms when estimating the shape of the patch [10,25,29,43]. On the other hand, recent dense descriptors leverage the power of large convolutional neural networks (CNN) to become more general and invariant. Most of them are trained on images with many variations in the training set, either obtained through data augmentation [8], with large databases of challenging images [9,42] or with style transfer [31]. They can also directly encode the invariance in the network itself [19]. The general trend in descriptor learning is thus to capture as much invariance as possible.

While feature detectors should generally be invariant to be repeatable under different scenarios [44], the same is not necessarily true for descriptors [41]. There is a direct trade-off for descriptors between generalization and discriminative power. More invariance allows a better generalization, but produces descriptors that are less informative. Figure 1 shows that the rotation variant descriptor Upright SIFT performs better than its invariant counterpart SIFT when only small rotations are present in the data. We argue that the best level of invariance depends on the situation. As a consequence, this questions the recent trend of jointly learning detector and descriptor: they may have to be dissociated if one does not want the descriptor to be as invariant as the detector.

In this work we focus on learning descriptors only and propose to select at runtime the right invariance given the context. Instead of learning a single generic descriptor, we compute several descriptors with different levels of invariance. We then propose a method to automatically select the most suitable invariance during matching. We achieve this by leveraging the local descriptors to learn meta descriptors that can encode global information about the variations present in the image. At matching time, the local descriptors distances are weighted by the similarity of these meta descriptors to produce a single descriptor distance.

Matches based on this distance can then be filtered using standard heuristics such as ratio test or mutual nearest neighbor.

Overall, our method, named Local Invariance Selection at Runtime for Descriptors (LISRD - pronounced as lizard), brings flexibility and interpretability into the feature description. When some image variations are known to be limited for a given application, one may directly use the most discriminative descriptor among all our learned local descriptors. However, it is usually hard to make such an assumption about the inter-image variations, and LISRD can instead automatically select the best invariance independently for each local region. Hence we are able to distinguish between different levels of variations within the same image (e.g. if half of the image is in the shadow but not the other half) and we show that this can improve the matching capabilities in comparison to using a single descriptor. The meta descriptors formulation is also not restricted to our proposed learned local descriptors, but can be easily generalized to most keypoint detectors and descriptors, as shown in Fig. 1 where it is applied to SIFT and Upright SIFT. Furthermore, the meta description only adds a small overhead to the current pipelines of keypoint detection and description in terms of runtime and memory consumption, which makes it suitable for real time applications. In summary, this work makes the following **contributions**:

- We show how to learn several local descriptors with multiple variance properties through a single network, in a similar spirit as in multi-task learning.
- We propose a light-weight meta descriptor approach to automatically select the best invariance of the local descriptors given the context.
- Our concept of meta descriptor and general approach of invariance selection can be easily transferred to most feature point detectors and descriptors, which we demonstrate for learned as well as traditional handcrafted descriptors.

2   Related Work

Learned Local Feature Descriptors. The recent progress in deep learning has enabled learned local descriptors to outperform the classical baselines by a large margin [8,9,21,31]. Following the classical approach, early works run a CNN on a small image region around the point of interest to get a patch descriptor [24,29,38]. The patch is not restricted to square areas, but can encode spatial transforms, such as affine [25] and polar [10] ones. The network is often optimized with a triplet loss using heuristics to extract positive and negative patches [3,11,22,39], or by directly maximizing the average precision (AP) [13]. Working on sparse features also gives the possibility to leverage both the visual context of the image and the spatial relationships between the keypoint locations [21]. More recently, descriptors extracted densely by CNN architectures from full images have shown both fast inference time and high performance on matching and retrieval tasks, and can jointly detect a heatmap of keypoints. Some works detect keypoints and describe them in parallel, such as Super-Point [8] and R2D2 [31], with for the latter an additional reliability map keeping track of the most informative locations in the image. Another approach is to use the features of the network as dense descriptors to subsequently detect keypoints, based on those features [9,28,42]. DELF [28] selects the keypoints using a learned attention, D2-Net [9] retrieves the maximum responses of the descriptor feature map across all channels, while UR2KID [42] clusters the channels in different groups and extracts keypoints based on their L2 responses. Even though jointly estimating the keypoints and descriptors allows a faster prediction and yields descriptors that are more correlated to the keypoints, the consequence is that detector and descriptor will share the same invariance. Therefore, we choose to focus exclusively on descriptor learning in this work.

Invariance in Feature Descriptors. Selecting an online invariance for binary descriptors is the core idea of BOLD [2], where a subset of the binary tests is chosen at runtime for each image patch to maximize the invariance to small affine transformations. Similarly, the general trend of most recent learned methods is to obtain descriptors as invariant as possible to any image variations. LIFT [43] mimics SIFT to achieve rotation invariance by estimating the keypoints, their orientation and finally their descriptor. Invariance to specific geometric changes can be achieved through group convolutions [7] by clustering the different geometrical transformations into specific groups [19]. However, the usual strategy is to incorporate as much diversity in the training data as possible. Illumination invariance can for example be obtained by training on images with multiple lighting conditions [15]. Photometric and homographic data augmentations also increase robustness to illumination and viewpoint changes [8]. Similarly, R2D2 [31] improves the robustness to day-night changes by synthesizing night images with style transfer and also to viewpoint changes by leveraging flow between close-by images [30]. Methods like D2-Net [9] and UR2KID [42] leverage a large database of images with multiple conditions and non planar viewpoint changes thanks to SfM data [17]. In this work, we adopt a mixture of the previously mentioned methods, namely the same synthesized night images as in [31], homographic augmentation, and training on datasets with multiple illumination changes [27].

Multi-task Learning in Description and Matching Tasks. Using a single network to achieve multiple and related tasks in feature description and matching is not new. Jointly learning the detector and descriptor [8,9,31] is already multi-task learning that makes the descriptors more discriminative at the predicted keypoint locations. HF-Net [32] unifies the detection of feature points, local and also global descriptors for image retrieval using multi-task distillation with a teacher network. Methods such as SuperGlue [33] and ContextDesc [21] can leverage both visual and geometric context in their descriptors in order to get a more consistent matching between images. UR2KID [42] bypasses the need of keypoint supervision during training and directly optimizes the descriptors jointly for local matching and image retrieval. In our approach, multiple descriptors are also learned in parallel, but instead of differing in their scope, they differ in their level of invariance. Furthermore, unlike previous hierarchical global-to-local approaches, our method relies on local descriptors first and leverages global information only to refine the local matching.

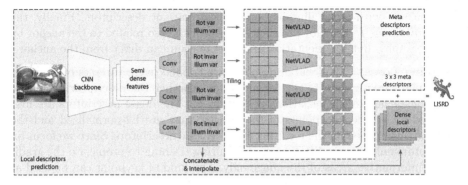

Fig. 2. Overview of our network architecture. Our network computes four local dense descriptors with diverse invariances and aggregates them through a NetVLAD layer [1] to obtain a regional description of the variations of the image.

3 Learning the Best Invariance for Local Descriptors

Our approach to select the most relevant variance for local feature descriptors consists in two steps. First, we design a network to learn several dense descriptors, each with a different type of invariance (see Sect. 3.1). Second, we propose a strategy in Sect. 3.2 to determine the best invariance to use when matching the local descriptors. Figure 2 provides an overview of the full architecture.

3.1 Disentangling Invariance for Local Descriptors

Many properties of an image have an influence on descriptors, but disentangling all of them would be intractable. We focus here on two factors known to have a large impact on descriptors performance: rotation and illumination. Our framework can however be generalized to other kinds of variations, for instance scaling. Since each of the two factors can either be variant or invariant, there are four possible combinations of variance with respect to illumination and rotation. We show in the following that the variant versions of descriptors are more discriminative since they are more specialized, while the invariant ones are trading the discriminative power for better generalization capabilities.

Network Architecture. Our network is inspired by SuperPoint [8], with slight modifications. It takes RGB images as input, computes semi-dense features with a shared backbone of convolutions and is then divided into 4 heads predicting a semi-dense descriptor each, one per combination of variance, as shown in Fig. 2. Since most computations are redundant between the 4 local descriptors, the shared backbone reduces the number of weights in the network and offers an inference time competitive with the current learned descriptors.

Dataset Preparation. The training dataset is composed of triplets of images. The first one, the *anchor image* I^A, is taken from a large database of real images. The *variant image* I^V is a warped version of the anchor by a homography without rotation and with equal illumination to train variant descriptors. Finally, the *invariant image* I^I used for invariant descriptors is also related to the anchor by a homography, but its orientation and illumination can differ from the anchor.

Training Losses. The local descriptors are trained using variants of the margin triplet ranking loss [5,24], depending on whether the descriptor should be invariant or not to the variations present in I^I. The dense descriptors are first sampled on selected keypoints of the images, they are L2-normalized and the losses are computed on the resulting set of feature descriptors. Since we focus on descriptors only, we use SIFT keypoints during training to propagate the gradient in informative areas of the image only. Any kind of keypoint can be used at inference time nonetheless, as demonstrated in Sect. 4.5.

Formally, given two images I^a and I^b related by a homography $\mathcal{H}$ and n keypoints $\mathbf{x}^a_{1..n}$ in image I^a, we warp each point to image I^b using the homography: $\mathbf{x}^b_{1..n} = \mathcal{H}(\mathbf{x}^a_{1..n})$. This yields a set of n correspondences between the two images, where we can extract the descriptors from each dense descriptor map: $\mathbf{d}^a_{1..n}$ and $\mathbf{d}^b_{1..n}$. Let us define a generic triplet loss $L_T(I^a, I^b, \text{dist})$ between I^a and I^b, given a descriptor distance $\text{dist}(\mathbf{x}^a, \mathbf{x}^b)$. The triplet loss first enforces a correct correspondence $(\mathbf{x}^a_i, \mathbf{x}^b_i)$ to be close in descriptor space through a positive distance

$$p_i = \text{dist}(\mathbf{x}^a_i, \mathbf{x}^b_i). \tag{1}$$

Additionally, the triplet loss increases the negative distance n_i between $\mathbf{x}^a_i$ and the closest point in I^b which is at least at a distance T from the correct match $\mathbf{x}^b_i$. This distance is computed symmetrically across the two images and the minimum is kept:

$$n_i = \min(\text{dist}(\mathbf{x}^a_i, \mathbf{x}^b_{n_b(i)}), \text{dist}(\mathbf{x}^b_i, \mathbf{x}^a_{n_a(i)})), \tag{2}$$

with $n_b(i) = \arg\min_{j \in [1,n]}(\text{dist}(\mathbf{x}^a_i, \mathbf{x}^b_j))$ s.t. $||\mathbf{x}^a_i - \mathbf{x}^b_j||_2 > T$, and similarly for $n_a(i)$. Given a margin M, the triplet margin loss is then defined as

$$L_T(I^a, I^b, \text{dist}) = \frac{1}{n} \sum_{i=1}^{n} \max(M + (p_i)^2 - (n_i)^2, 0). \tag{3}$$

In our case, the loss L_I for invariant descriptors is an instance of this generic triplet loss between the anchor image I^A and the invariant image I^I, for the L2 descriptor distance:

$$L_I = L_T(I^A, I^I, ||\mathbf{d}^A - \mathbf{d}^I||_2). \tag{4}$$

The loss L_V for variant descriptors is based on the full triplet of images: I^A, I^I and I^V. It enforces variant descriptors to be different between the anchor and the invariant image, while preserving similarity between the anchor and the variant image. Its positive loss is the distance in descriptor space of positive matches between I^A and I^V, and similarly for the negative distance between I^A and I^I:

$$L_V = \frac{1}{n} \sum_{i=1}^{n} \max(fM + ||\mathbf{d}^A_i - \mathbf{d}^V_i||_2^2 - ||\mathbf{d}^A_i - \mathbf{d}^I_i||_2^2, 0), \tag{5}$$

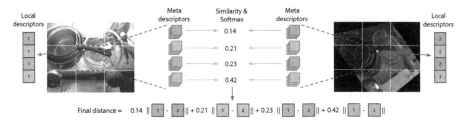

Fig. 3. The LISRD descriptor distance between two points is the sum of the four local descriptors distances, weighted by the similarity of the meta descriptors.

where f is a factor controlling at which point the anchor and the invariant images are different. For rotation changes, $f = \min(1, \frac{\theta_I}{\theta_{max}})$, where θ_I is the absolute angle of rotation between the anchor and the invariant image and θ_{max} is a hyper-parameter representing the threshold beyond which the two images should be considered different. This threshold ensures that only large rotations are penalized by the loss. It is hard to quantify the difference in illumination between two real images, so we set $f = 1$ when the illumination differs between the anchor and invariant image.

When a descriptor d in the set $\mathcal{D}$ of descriptors is supposed to be invariant to all changes (illumination and/or rotation) between I^A and I^I, we use L_I. Otherwise, L_V is used. We define $L_{I/V}(d)$ as the selected loss and the total loss for local descriptors as

$$L_l = \frac{1}{|\mathcal{D}|} \sum_{d \in \mathcal{D}} L_{I/V}(d). \tag{6}$$

3.2 Online Selection of the Best Invariance

Given the local descriptors of the previous section, this section explores how to pick the most relevant invariance when matching images. Since it would be costly to recompute and compare the image variations for every pair of images to be matched, we propose to rely solely on the information contained in the descriptors to perform the selection. A naive approach would be to separately compute the similarity of the different local descriptors and to pick the most similar ones. However, the invariance selection would gain by having more context than the information of a single local descriptor and should be consistent with neighboring descriptors. Therefore, we propose to extract regional descriptors from the local ones and to use them to guide the invariance selection.

The local descriptors are thus gathered in neighboring areas through a NetVLAD layer [1] to get a meta descriptor sharing the same kind of invariance as the subset of local descriptors, but with more context than a single local descriptor. Thus, having similar meta descriptors means sharing the same level of variations. The neighboring areas are created by tiling the image into a $c \times c$ grid and computing a meta descriptor for each tile. Hence, we get four meta descriptors per tile, which are then L2 normalized.

When matching the local descriptors of a tile, the four similarities between the meta descriptors are computed with a scalar product and we can rank the four local descriptors according to these similarities. Instead of making a hard choice by taking only the closest local descriptor, we use a soft assignment. A softmax operation is applied to the four similarities, to get four weights summing to one. These weights are then used to compute the distance between the local descriptors as shown in Fig. 3. More precisely, suppose that we want to compute the distance in descriptor space between point $\mathbf{x}^a$ in image I^a and point $\mathbf{x}^b$ in image I^b. Point $\mathbf{x}^a$ is associated with 4 local descriptors $\mathbf{d}^a_{1..4}$ and 4 meta descriptors $\mathbf{m}^a_{1..4}$ corresponding to the region where $\mathbf{x}^a$ lies, and similarly for $\mathbf{x}^b$. Then the final descriptor distance between $\mathbf{x}^a$ and $\mathbf{x}^b$ is

$$\text{dist}(\mathbf{x}^a, \mathbf{x}^b) = \sum_{i=1}^{4} \frac{\exp\left((\mathbf{m}^a_i)^\intercal \cdot \mathbf{m}^b_i\right)}{\sum_{j=1}^{4} \exp\left((\mathbf{m}^a_j)^\intercal \cdot \mathbf{m}^b_j\right)} \|\mathbf{d}^a_i - \mathbf{d}^b_i\|_2. \tag{7}$$

Thus, the similarity of the meta descriptors acts as a weighting of the local descriptors distances and can put a stronger emphasis on one specific variance when the corresponding meta descriptors have a high similarity. Matching is then performed with this descriptor distance, and can easily be refined with ratio test [20] or mutual nearest neighbor.

Training Loss. The 4 NetVLAD layers are trained with a weak supervision based on another instance of the triplet loss L_T between I^A and I^I with the distance defined above:

$$L_m = L_T(I^A, I^I, \text{dist}) \tag{8}$$

Thanks to this weak supervision, there is no need to explicitly supervise the meta descriptors, which would require knowing the amount of rotation and illumination for every tile in the image. The total loss of the network is finally a combination of the local and meta descriptors, weighted by a factor λ:

$$L = L_l + \lambda L_m. \tag{9}$$

3.3   Training Details

Datasets. To train descriptors with different levels of variance in terms of rotation and illumination, datasets presenting all possible combinations of changes are needed. Control over the amount of changes is also required in order to know which loss between L_I and L_V should be used for each descriptor. We use in total four datasets to accomplish that. Illumination variations are obtained through the multi illumination dataset in the wild [27] and the style transferred night images of the Aachen day dataset [31]. Both offer pairs of images with fixed viewpoint and different illuminations. Images with fixed illumination come from the MS COCO dataset [18] and the day flow images from the Aachen dataset [31]. For all datasets except the latter, the images are augmented with random homographies containing translation, scaling, rotation and perspective distortion, similarly as in [8]. For the day images of Aachen, the flow is used to create the correspondences and we consider that these images contain only small rotations and no major illumination changes. Overall, there is an equal distribution of images with and without illumination changes, and of rotated and non rotated images.

Implementation Details. We describe here the details of our architecture. The backbone network, inspired by the VGG16 [37], is composed of successive 3×3 convolutional layers with channel size 64-64-64-64-128-128-256-256. Each conv layer is followed by ReLU activation and batch normalization. Every two layers, a 2×2 average pooling with stride 2 is applied to reduce the spatial resolution by 2. For an image of size $H \times W \times 3$, the output feature map will have a size of $H/8 \times W/8 \times 256$. The local descriptor heads are all composed of the following operations: 3×3 conv of channel size 256 - ReLU - Batch Norm - 1×1 conv of channel size 128. The final dimension of each local descriptor is thus $H/8 \times W/8 \times 128$, and each concatenated descriptor is 512-dimensional. The semi-dense descriptors can then be bilinearly interpolated to the locations of any keypoint. Note that in order to achieve a better robustness to scale changes, one can also detect the keypoints and describe them at multiple image resolutions and aggregate the results in the original image resolution, similarly as in [9] and [31]. The NetVLAD layers consists in 8 clusters of 128-dimensional descriptors, hence a meta descriptor size of 1024. We used $c \times c = 3 \times 3$ tiles per image.

The network is trained on RGB images resized to 240×320 with the following hyper-parameters: distance threshold $T = 8$, $\theta_{max} = \frac{\pi}{4}$, margin $M = 1$, loss factor $\lambda = 1$. It comprises roughly 3.7M parameters, which are optimized with the Adam solver [16] (learning rate = 0.001 and $\beta = (0.9, 0.999)$). In practice, the local descriptors are pre-trained first and then fine-tuned by an end-to-end training with the meta descriptors. At test time, a single forward pass on a GeForce RTX 2080 Ti with 480×640 images takes 6ms on average.

4   Experimental Results

We present here experiments validating the relevance of our method. Section 4.2 highlights the importance of learning different invariances, validates the proposed approach with an ablation study, and shows that LISRD can be extended to other descriptors such as SIFT and Upright SIFT. LISRD is then compared to the state of the art on a benchmark homography dataset (Sect. 4.3), on a challenging dataset with diverse conditions where the presence or lack of invariance is essential (Sect. 4.4) and on a visual localization task in the real world (Sect. 4.5).

4.1   Metrics

Since we want to compare the performance of the descriptors only, all the following metrics are computed on SIFT keypoints if not stated otherwise. The metrics are computed on pairs of images resized to 480×640 and related by a known homography. Resizing is performed by upscaling/downscaling the images

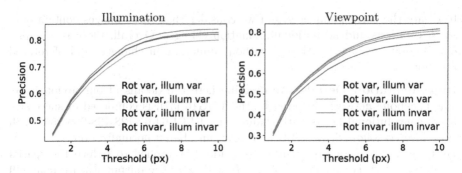

Fig. 4. Precision on HPatches of the 4 local descriptors. Variant ones are better when invariance is not needed (e.g. rotation for the illumination dataset).

to have each edge greater or equal respectively to 480 and 640, and a central crop is applied to get the target resolution. We keep a maximum of 1000 points among the keypoints shared between the two views and matches are obtained after mutual nearest neighbor filtering.

Homography Estimation. We follow the procedure of [8] to compute a homography estimation score. Given a pair of images, RANSAC is used to fit a homography between the clouds of matched keypoints. The score is obtained by warping the four corners of the first image $\hat{c}_{1...4}$ with the predicted homography and comparing their distance to the same points $c_{1...4}$ warped by the ground truth homography. The homography is considered as correct when the average distance is below a threshold ϵ, which is set to 3 pixels in all experiments: HEstimation $= \frac{1}{4} \sum_{i=1}^{4} ||\hat{c}_i - c_i||_2 \le \epsilon$.

Precision. Precision (also known as mean matching accuracy) is the percentage of correct matches over all the predicted matches [9,31]. We use by default a threshold of 3 pixels to consider a match to be correct.

Recall. Recall is the ratio of correctly predicted matches over the total number of ground truth matches, where a ground truth correspondence is the *closest* point within an error threshold of 3 pixels. A predicted match with the *second closest* point but still within the correct threshold is considered as incorrect.

4.2   Method Validation

Impact of the Different Invariances. One can check the validity of our approach by comparing the 4 local descriptors. We use the HPatches dataset [4], which is standard in descriptor evaluation. It is composed of 116 sequences of 5 pairs of images, with either viewpoint changes (given by a known homography) or illumination changes with fixed viewpoint. Figure 4 shows the comparison between the 4 descriptors in terms of precision. On viewpoint changes, the illumination variant descriptors are superior as the lighting is fixed in these images and they are thus more discriminative. Since HPatches contains few rotations,

Table 1. Ablation study on the HPatches dataset.

	HEstimation
Best of the 4	0.778
Greedy	0.774
Hard assignment	0.762
No tiling	0.752
5 × 5 tiles	0.773
Single desc	0.766
LISRD (ours)	**0.784**

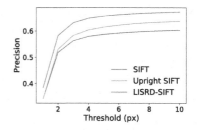

Fig. 5. Variants of SIFT vs. our method fusing them (LISRD-SIFT). Precision is computed on HPatches viewpoint.

there is no significant difference in terms of rotation invariance and being rotation variant brings a small advantage on average. The precision on illumination changes shows that the best performing descriptors are the illumination invariant ones and that being rotation variant helps since the viewpoint is fixed. Thus there is no descriptor outperforming the others in all cases, and our hypothesis that variant descriptors are more discriminative than invariant ones is validated.

Ablation Study. To confirm the benefit of our online selection of invariance and choice of parameters, we compare LISRD on homography estimation on the HPatches dataset with other selection methods of the local descriptors as well as with variants of our approach (Table 1). *Best of the 4* computes the metrics for the 4 local descriptors separately and picks the best score. *Greedy* computes the pairwise distances of all points for each local descriptor and greedily chooses the local descriptor with smallest distance for each pair of points. *Hard assignment* selects the local descriptor that maximizes the meta descriptor similarity, instead of choosing a soft assignment as in our proposed method. *No tiling* and *5 × 5 tiles* are variants of our method with no tiling or with *5 × 5* tiles for the meta descriptors. Finally, *Single desc* is a descriptor trained with exactly the same architecture as ours, but with the 4 local descriptors concatenated and trained with invariance in both illumination and rotation.

On the full HPatches dataset, *Best of the 4* corresponds to the descriptor invariant to both illumination and rotation, as both changes are present. However, our selection method can still leverage the other descriptors: for example an illumination variant descriptor for the viewpoint part. The disparity between LISRD and *Greedy* and *Hard assignment* highlights the added value of the meta descriptors, and shows that a soft assignment can better leverage the 4 descriptors at the same time. Finally, the comparison with *Single desc* confirms our hypothesis that disentangling the types of invariance is beneficial compared to learning a single invariant descriptor with the same number of weights.

Generalization to Other Descriptors. LISRD can be easily generalized to other kinds of descriptors, and not only to our proposed learned local descriptors. We demonstrate this by applying our approach to the duo of local descriptors

SIFT and Upright SIFT – SIFT without rotation invariance, as presented in Fig. 1. Instead of having four local descriptors, there are only two of them, one invariant to rotation and one variant, and similarly for the meta descriptors. Our method is evaluated against SIFT and Upright SIFT on the viewpoint part of HPatches. This dataset contains indeed sequences with no rotation, where Upright SIFT performs better, and other sequences with strong rotations, where SIFT takes over. Figure 5 shows that our method can effectively leverage both SIFT and Upright SIFT and outperforms the two.

4.3 Descriptor Evaluation on HPatches

This section compares the performance of LISRD against state-of-the-art local descriptors on the benchmark dataset HPatches. Since our approach requires global context from full images, we cannot run it on the patch level dataset. We use the full sequences of images instead, similarly as in [8,9,31]. We consider the following baselines: Root SIFT with the default Kornia[1] implementation; HardNet [24] (trained on the PS-dataset [26]), SOSNet [39] (trained on the Liberty dataset of UBC Phototour [6]), SuperPoint (SP) [8], D2-Net [9], R2D2 [31] and GIFT [19] with the authors implementation. Since we want to evaluate the descriptors only, SIFT keypoints are detected in the images and for each method, we extract the local descriptors at these locations. For Root SIFT, HardNet and SOSNet, we sample 32×32 patches at each SIFT keypoint and rotate them according to the SIFT orientation. As we want to evaluate the impact of rotation and illumination invariance only, we use single scale implementations for all methods[2]. Our method could however be made scale invariant using similar multi-scale approaches as in [9,31].

Table 2. Comparison to the state of the art on HPatches. Homography estimation, precision and recall are computed for error thresholds of 3 pixels. The best score is in bold and the second best one is underlined.

		Root SIFT	HardNet	SOSNet	SP	D2-Net	R2D2	GIFT	Ours
HP Illum	HEstimation	0.898	0.884	<u>0.919</u>	0.877	0.818	0.916	**0.923**	0.884
	Precision	0.554	0.574	0.591	0.629	0.650	**0.666**	0.573	<u>0.665</u>
	Recall	0.431	0.483	0.519	0.565	0.564	<u>0.580</u>	0.521	**0.655**
HP View	HEstimation	0.644	0.688	**0.742**	0.651	0.553	0.627	<u>0.715</u>	0.688
	Precision	0.515	0.582	<u>0.598</u>	0.595	0.564	0.550	0.552	**0.626**
	Recall	0.350	0.422	<u>0.448</u>	0.446	0.382	0.371	0.429	**0.495**

---

[1] https://kornia.github.io/.

[2] In the case of GIFT, which is both rotation and scale invariant, we sample images with scale 1 to make it rotation invariant only.

The results are summarized in Table 2. Overall, LISRD ranks among the two best methods in precision and recall. The possibility to leverage rotation variant descriptors on the fixed pairs of the illumination part and to alternatively select the right level of lighting invariance given the amount of illumination changes probably explains our superior performance on the illumination part. Note the comparison with SuperPoint, whose architecture and training procedure are very similar to LISRD, and where our method displays better results in all metrics, thus showing the gain of our approach. The weaker results in homography estimation can be explained by a limitation of our method. Since our meta descriptors have a very coarse spatial resolution (3×3 grid), if one of them fails to pick the right invariance, this will impact all the matches of its region. Thus, the correct matches predicted by LISRD can in that case become very concentrated in a specific part of the image, which makes the homography estimation with RANSAC less accurate. This issue could be avoided with a finer tiling of the meta descriptors, but at the price of a reduced global context.

4.4   Evaluation in Challenging and Cross-Modal Situations

The HPatches dataset offers a fair benchmark, but is limited to only few rotations and medium illumination changes. Our approach is designed to be used in a variety of scenarios and with changing conditions, so that all our local descriptors can be leveraged. In order to evaluate our method on such a versatile task, we designed a new benchmark dataset, based on the day-night image matching (DNIM) dataset [44]. This dataset is composed of sequences of images of a fixed camera taking pictures at regular time intervals and across day and night, with a total of 1722 images. For each sequence, the image with timestamp closest to noon is taken as day reference and the image closest to midnight as night reference. We create two benchmarks, where the images of each sequence are paired with either the day reference or the night one. We then synthetically warp the pairs with the same homography sampling scheme as in [8] with an equal distribution of homographies with and without rotations. We plan to release the homographies used in this benchmark to let other researchers compare with their own methods. Examples of images and matches for this dataset can be found in the supplementary material.

Table 3 and Fig. 6 show the evaluation with the state-of-the-art descriptors, using SuperPoint keypoints. LISRD can adapt its invariance to illumination and rotations to alternatively select the most relevant descriptor and it outperforms the other methods by a large margin both in terms of precision and recall.

Table 3. Evaluation on a use case where invariance selection matters. Homography estimation, precision and recall are computed with SuperPoint keypoints on a dataset with day-night changes and various levels of rotation. Selecting the relevant variant or invariant descriptors boosts the precision and recall of our method compared to the previous state-of-the-art methods.

		Root SIFT	HardNet	SOSNet	SP	D2-Net	R2D2	GIFT	Ours
Day ref	HEstimation	0.121	**0.199**	0.178	0.146	0.094	0.170	0.187	0.198
	Precision	0.188	0.232	0.228	0.195	0.195	0.175	0.152	**0.291**
	Recall	0.112	0.194	0.203	0.178	0.117	0.162	0.133	**0.317**
Night ref	HEstimation	0.141	**0.262**	0.211	0.182	0.145	0.196	0.241	**0.262**
	Precision	0.238	0.366	0.297	0.264	0.259	0.237	0.236	**0.371**
	Recall	0.164	0.323	0.269	0.255	0.182	0.216	0.209	**0.384**

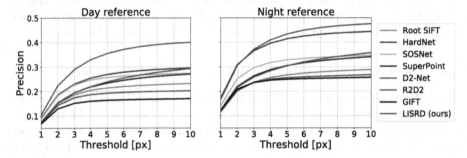

Fig. 6. Precision curves on the DNIM dataset [44] augmented with rotations. LISRD leverages its variant and more discriminative descriptors whenever possible and is thus more accurate than the state-of-the-art descriptors for all pixel error thresholds.

4.5 Application to Localization in Challenging Conditions

A typical application of image matching including adverse conditions such as strong illumination changes and wide baselines is the visual localization task. We evaluate our method on the local feature challenge of CVPR 2019 based on the Aachen Day-Night dataset [34]. The goal is to localize 98 night time query images as accurately as possible, 20 day images per query with known camera pose. As the keypoint quality is essential in this task, we compare our method with other descriptors for various types of keypoints: SIFT, SuperPoint and D2-Net multi-scale (MS). The numbers for the baseline methods are taken from the benchmark on the official website[3]. The results in Table 4 show that our method is not limited to SIFT keypoints and can effectively improve the performance of local descriptors in challenging conditions. Note in particular the improvement over SuperPoint, which shares a similar architecture as ours.

---

[3] https://www.visuallocalization.net/.

Table 4. Visual localization performance on the Aachen Day-Night dataset [34]. We report the percentage of correctly localized queries for various distance and orientation error thresholds for SIFT, SuperPoint and D2-Net multi-scale (MS). Our method shows a good generalization when evaluated on different keypoints (KP) and can improve the original descriptor performance.

Error threshold	SIFT KP		SuperPoint KP		D2-Net KP	
	Up-Root SIFT	Ours	SuperPoint	Ours	D2-Net (MS)	Ours (MS)
$0.5m, 2°$	54.1	**72.4**	73.5	**78.6**	67.3	**73.5**
$1m, 5°$	66.3	**82.7**	79.6	**86.7**	87.8	**88.8**
$5m, 10°$	75.5	**94.9**	88.8	**98.0**	**100.0**	99.0

5 Conclusion

We presented a novel approach to learn local feature descriptors able to adapt to multiple variations in images, while remaining discriminative. We unified the learning of several local descriptors with multiple levels of invariance and of meta descriptors leveraging regional context to guide the local descriptors matching.

While restricted to illumination and rotation invariance, our framework can be generalized to more variations, at the cost of an exponentially growing number of descriptors however. A future direction of work would be to reduce the amount of redundancy between each descriptor by enforcing a stronger disentanglement separating each factors of variation. Since our approach is able to enforce different levels of invariance, one can also add another head to our network to predict invariant keypoints, while keeping discriminative descriptors, thus solving the current issue in joint learning of invariant detectors and descriptors.

Overall, this work is a first step towards disentangled descriptors. Separating the types of invariances paves the way to a full disentanglement of the factors of variations of images and could lead to flexible and interpretable local descriptors.

Acknowledgments. This work has been supported by an ETH Zurich Postdoctoral Fellowship and Innosuisse funding (Grant No. 34475.1 IP-ICT).

References

1. Arandjelović, R., Gronat, P., Torii, A., Pajdla, T., Sivic, J.: NetVLAD: CNN architecture for weakly supervised place recognition. In: Computer Vision and Pattern Recognition (CVPR) (2016)
2. Balntas, V., Tang, L., Mikolajczyk, K.: Bold - binary online learned descriptor for efficient image matching. In: Computer Vision and Pattern Recognition (CVPR) (2015)
3. Balntas, V., Johns, E., Tang, L., Mikolajczyk, K.: PN-Net: conjoined triple deep network for learning local image descriptors. In: Computer Vision and Pattern Recognition (CVPR) (2016)

4. Balntas, V., Lenc, K., Vedaldi, A., Mikolajczyk, K.: Hpatches: a benchmark and evaluation of handcrafted and learned local descriptors. In: Computer Vision and Pattern Recognition (CVPR) (2017)
5. Balntas, V., Riba, E., Ponsa, D., Mikolajczyk, K.: Learning local feature descriptors with triplets and shallow convolutional neural networks. In: British Machine Vision Conference (BMVC) (2016)
6. Brown, M., Hua, G., Winder, S.: Discriminative learning of local image descriptors. Trans. Pattern Anal. Mach. Intell. (PAMI) (2010)
7. Cohen, T., Welling, M.: Group equivariant convolutional networks. In: International Conference on Machine Learning (ICML) (2016)
8. DeTone, D., Malisiewicz, T., Rabinovich, A.: Superpoint: self-supervised interest point detection and description. In: Computer Vision and Pattern Recognition Workshops (CVPRW) (2018)
9. Dusmanu, M., et al.: D2-Net: a trainable CNN for joint detection and description of local features. In: Computer Vision and Pattern Recognition (CVPR) (2019)
10. Ebel, P., Mishchuk, A., Yi, K.M., Fua, P., Trulls, E.: Beyond Cartesian representations for local descriptors. In: International Conference on Computer Vision (ICCV) (2019)
11. Han, X., Leung, T., Jia, Y., Sukthankar, R., Berg, A.: Matchnet: unifying feature and metric learning for patch-based matching. In: Computer Vision and Pattern Recognition (CVPR) (2015)
12. Harris, C., Stephens, M.: A combined corner and edge detector. In: Proceedings of Fourth Alvey Vision Conference (1988)
13. He, K., Lu, Y., Sclaroff, S.: Local descriptors optimized for average precision. In: Computer Vision and Pattern Recognition (CVPR) (2018)
14. Heinly, J., Schönberger, J.L., Dunn, E., Frahm, J.M.: Reconstructing the world* in six days *(as captured by the Yahoo 100 million image dataset). In: Computer Vision and Pattern Recognition (CVPR) (2015)
15. Kaliroff, D., Gilboa, G.: Self-supervised unconstrained illumination invariant representation. In: arXiv (2019)
16. Kingma, D., Ba, J.: Adam: a method for stochastic optimization (2014)
17. Li, Z., Snavely, N.: Megadepth: learning single-view depth prediction from internet photos. In: Computer Vision and Pattern Recognition (CVPR) (2018)
18. Lin, T.-Y., Maire, M., Belongie, S., Hays, J., Perona, P., Ramanan, D., Dollár, P., Zitnick, C.L.: Microsoft COCO: common objects in context. In: Fleet, D., Pajdla, T., Schiele, B., Tuytelaars, T. (eds.) ECCV 2014. LNCS, vol. 8693, pp. 740–755. Springer, Cham (2014). https://doi.org/10.1007/978-3-319-10602-1_48
19. Liu, Y., Shen, Z., Lin, Z., Peng, S., Bao, H., Zhou, X.: GIFT: learning transformation-invariant dense visual descriptors via group CNNs. In: Advances in Neural Information Processing Systems (NeurIPS) (2019)
20. Lowe, D.G.: Distinctive image features from scale-invariant keypoints. Int. J. Comput. Vis. (IJCV) **60** (2004)
21. Luo, Z., et al.: Contextdesc: local descriptor augmentation with cross-modality context. In: Computer Vision and Pattern Recognition (CVPR) (2019)
22. Luo, Z., Shen, T., Zhou, L., Zhu, S., Zhang, R., Yao, Y., Fang, T., Quan, L.: GeoDesc: learning local descriptors by integrating geometry constraints. In: Ferrari, V., Hebert, M., Sminchisescu, C., Weiss, Y. (eds.) ECCV 2018. LNCS, vol. 11213, pp. 170–185. Springer, Cham (2018). https://doi.org/10.1007/978-3-030-01240-3_11
23. Mikolajczyk, K., et al.: A comparison of affine region detectors. Int. J. Comput. Vis. (IJCV) (2005)

24. Mishchuk, A., Mishkin, D., Radenović, F., Matas, J.: Working hard to know your neighbor's margins: local descriptor learning loss. In: Advances in Neural Information Processing Systems (NIPS) (2017)
25. Mishkin, D., Radenović, F., Matas, J.: Repeatability is not enough: learning affine regions via discriminability. In: Ferrari, V., Hebert, M., Sminchisescu, C., Weiss, Y. (eds.) ECCV 2018. LNCS, vol. 11213, pp. 287–304. Springer, Cham (2018). https://doi.org/10.1007/978-3-030-01240-3_18
26. Mitra, R., et al.: A Large Dataset for Improving Patch Matching. arXiv (2018)
27. Murmann, L., Gharbi, M., Aittala, M., Durand, F.: A multi-illumination dataset of indoor object appearance. In: International Conference on Computer Vision (ICCV) (2019)
28. Noh, H., Araujo, A., Sim, J., Weyand, T., Han, B.: Large-scale image retrieval with attentive deep local features. In: International Conference on Computer Vision (ICCV) (2017)
29. Ono, Y., Trulls, E., Fua, P., Yi, K.M.: LF-Net: learning local features from images. In: Advances in Neural Information Processing Systems (NIPS) (2018)
30. Revaud, J., Weinzaepfel, P., Harchaoui, Z., Schmid, C.: EpicFlow: edge-preserving interpolation of correspondences for optical flow. In: Computer Vision and Pattern Recognition (CVPR) (2015)
31. Revaud, J., Weinzaepfel, P., de Souza, C.R., Humenberger, M.: R2D2: repeatable and reliable detector and descriptor. In: Advances in Neural Information Processing Systems (NeurIPS) (2019)
32. Sarlin, P.E., Cadena, C., Siegwart, R., Dymczyk, M.: From coarse to fine: robust hierarchical localization at large scale. In: Computer Vision and Pattern Recognition (CVPR) (2019)
33. Sarlin, P.E., DeTone, D., Malisiewicz, T., Rabinovich, A.: Superglue: learning feature matching with graph neural networks. In: Computer Vision and Pattern Recognition (CVPR) (2020)
34. Sattler, T., et al.: Benchmarking 6DOF outdoor visual localization in changing conditions. In: Computer Vision and Pattern Recognition (CVPR) (2018)
35. Sattler, T., et al.: Are large-scale 3D models really necessary for accurate visual localization? In: Computer Vision and Pattern Recognition (CVPR) (2017)
36. Schönberger, J.L., Pollefeys, M., Geiger, A., Sattler, T.: Semantic visual localization. In: Computer Vision and Pattern Recognition (CVPR) (2017)
37. Simonyan, K., Zisserman, A.: Very deep convolutional networks for large-scale image recognition. In: International Conference on Learning Representations (ICLR) (2014)
38. Tian, Y., Fan, B., Wu, F.: L2-Net: Deep learning of discriminative patch descriptor in Euclidean space. In: Computer Vision and Pattern Recognition (CVPR) (2017)
39. Tian, Y., Yu, X., Fan, B., Wu, F., Heijnen, H., Balntas, V.: SOSNET: second order similarity regularization for local descriptor learning. In: Computer Vision and Pattern Recognition (CVPR) (2019)
40. Tola, E., Lepetit, V., Fua, P.: Daisy: an efficient dense descriptor applied to wide baseline stereo. Trans. Pattern Anal. Mach. Intell. (PAMI) 32 (2010)
41. Wu, C., Li, X., Frahm, J., Pollefeys, M.: 3D model matching with viewpoint-invariant patches (VIP). In: Computer Vision and Pattern Recognition (CVPR) (2008)
42. Yang, T.Y., Nguyen, D.K., Heijnen, H., Balntas, V.: Ur2kid: Unifying retrieval, keypoint detection, and keypoint description without local correspondence supervision. In: arXiv (2020)

43. Yi, K.M., Trulls, E., Lepetit, V., Fua, P.: LIFT: learned invariant feature transform. In: Leibe, B., Matas, J., Sebe, N., Welling, M. (eds.) ECCV 2016. LNCS, vol. 9910, pp. 467–483. Springer, Cham (2016). https://doi.org/10.1007/978-3-319-46466-4_28

44. Zhou, H., Sattler, T., Jacobs, D.W.: Evaluating local features for day-night matching. In: Hua, G., Jégou, H. (eds.) ECCV 2016. LNCS, vol. 9915, pp. 724–736. Springer, Cham (2016). https://doi.org/10.1007/978-3-319-49409-8_60

Rethinking Image Inpainting via a Mutual Encoder-Decoder with Feature Equalizations

Hongyu Liu[1], Bin Jiang[1(✉)], Yibing Song[2(✉)], Wei Huang[1], and Chao Yang[1]

[1] College of Computer Science and Electronic Engineering, Hunan University, Changsha, China
{kumapower,jiangbin,hwei,yangchaoedu}@hnu.edu.cn
[2] Tencent AI Lab, Shenzhen, China
yibingsong.cv@gmail.com

Abstract. Deep encoder-decoder based CNNs have advanced image inpainting methods for hole filling. While existing methods recover structures and textures step-by-step in the hole regions, they typically use two encoder-decoders for separate recovery. The CNN features of each encoder are learned to capture either missing structures or textures without considering them as a whole. The insufficient utilization of these encoder features hampers the performance of recovering both structures and textures. In this paper, we propose a mutual encoder-decoder CNN for joint recovery of both. We use CNN features from the deep and shallow layers of the encoder to represent structures and textures of an input image, respectively. The deep layer features are sent to a structure branch, while the shallow layer features are sent to a texture branch. In each branch, we fill holes in multiple scales of the CNN features. The filled CNN features from both branches are concatenated and then equalized. During feature equalization, we reweigh channel attentions first and propose a bilateral propagation activation function to enable spatial equalization. To this end, the filled CNN features of structure and texture mutually benefit each other to represent image content at all feature levels. We then use the equalized feature to supplement decoder features for output image generation through skip connections. Experiments on benchmark datasets show that the proposed method is effective to recover structures and textures and performs favorably against state-of-the-art approaches.

Keywords: Deep image inpainting · Feature equalizations

This work is done partially when H. Liu is an intern at Tencent AI Lab. The results and code are available at https://github.com/KumapowerLIU/Rethinking-Inpainting-MEDFE.

The original version of this chapter was revised: City and country of the second affiliation was corrected from "Bellevue, USA" to "Shenzhen, China". The correction to this chapter is available at https://doi.org/10.1007/978-3-030-58536-5_47

Electronic supplementary material The online version of this chapter (https://doi.org/10.1007/978-3-030-58536-5_43) contains supplementary material, which is available to authorized users.

1 Introduction

There is a need to recover missing contents in corrupted images for visual aesthetics improvement. Deep neural networks have advanced image inpainting by introducing semantic guidance to fill hole regions. Different from the traditional methods [2,3,7,8] that propagate uncorrupted image contents to the hole regions via patch-based image matching, deep inpainting methods [13,25] utilize CNN features in different levels (i.e., from low-level features to high-level semantics) to produce more meaningful and globally consistent results.

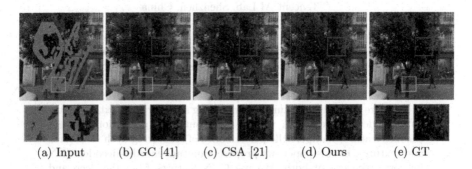

(a) Input   (b) GC [41]   (c) CSA [21]   (d) Ours   (e) GT

Fig. 1. Visual comparison on the Paris StreetView dataset [6]. GT is the ground truth image. The proposed inpainting method is effective to reduce blur and artifacts within and around the hole regions, which are brought by inconsistent structure and texture features.

The encoder-decoder architecture is prevalent in existing deep inpainting methods [13,19,25,38]. However, a direct utilization of the end-to-end training and prediction processes generates limited results. This is due to the challenging factor that the hole region is completely empty. Without sufficient image guidance, an encoder-decoder is not able to reconstruct the whole missing content. An alternative is to use two encoder-decoders to separately learn missing structures and textures in a step-by-step manner. These two-stage methods [21,24,26,27,29,40,41] typically generate an intermediate image with recovered structures in the first stage (i.e., encoder-decoder), and send this image to the second stage for texture generation. Although structures and textures are produced on the output image, their appearances are not consistent. Figure 1 shows an example. The inconsistent structures and textures within hole regions produce blur and artifacts as shown in (b) and (c). Meanwhile, the recovered contents are not coherent to the uncorrupted contents around the hole boundaries (e.g., the leaves). This limitation is because of the independent learning of CNN features representing structures and textures. In practice, the structures and textures correlate with each other to formulate the image contents. Without considering their coherence, existing methods are not able to produce visually pleasing results.

In this work, we propose a mutual encoder-decoder to jointly learn CNN features representing structures and textures. The features from the deep layers of the encoder contain structure semantics while the features from the shallow layers contain texture details. The hole regions of these two features are filled via two separate branches. In the CNN feature space, we use a multi-scale filling block within each branch for hole filling. Each block consists of 3 partial convolution streams with progressively increased kernel sizes. After hole filling in these two features, we propose a feature equalization method to ensure the structure and texture features consistent with each other. Meanwhile, the equalized features are coherent with the features of uncorrupted image content around the hole boundaries. The proposed feature equalization consists of channel reweighing and bilateral propagation. We concatenate two features first and perform channel reweighing via attention exploration [12]. The attentions across two features are set to be consistent after channel equalization. Then, we propose a bilateral propagation activation function to equalize the feature consistency in the whole feature maps. This activation function uses elements on the global feature maps to propagate channel consistency (i.e., feature coherence across the hole boundaries), while using elements within local neighboring regions to maintain channel similarities (i.e., feature consistency within the hole). To this end, we fuse the texture and structure features together to reduce inconsistency in the CNN feature maps. The equalized features then supplement the decoder features in all the feature levels via encoder-decoder skip connections. The feature consistency is then reflected in the reconstructed output image, where the blur and artifacts are effectively removed around the hole regions as shown in Fig. 1(d). Experiments on the benchmark datasets show that the proposed method performs favorably against state-of-the-art approaches.

We summarize the contributions of this work as follows:

- We propose a mutual encoder-decoder network for image inpainting. The CNN features from the shallow layer are learned to represent textures and the features from deep layers represent structures.
- We propose a feature equalization method to make structure and texture features consistent with each other. We first reweigh channels after feature concatenation and propose a bilateral propagation activation function to make the whole feature consistent.
- Extensive experiments on the benchmark datasets show the effectiveness of the proposed inpainting method in removing blur and artifacts caused by inconsistent structure and texture features. The proposed method performs favorably against state-of-the-art inpainting approaches.

2 Related Works

Empirical Image Inpainting. The empirical image inpainting methods [1,3, 18] based on diffusion techniques propagate the neighborhood appearances to the missing regions. However, they only consider surrounding pixels of missing regions, which can only deal with small holes in background inpainting tasks and may fail to generate meaningful structures. In contrast, methods [2,4,5,28,36] based on patch match fill missing regions by transferring similar and relevant patches from the remaining image region to the hole region. Although empirical methods perform well to handle small holes on the background inpainting task, they are not able to generate semantically meaningful content. When the hole region is large, these methods suffer from a lack of semantic guidance.

Deep Image Inpainting. Image inpainting based on deep learning typically involves the generative adversarial network [9] to supplement visual perceptual guidance for hole filling. Pathak et al. [25] first bring adversarial training [9] to inpainting and demonstrate semantic hole-filling. Iizuka et al. [13] propose local and global discriminators, assisted by dilated convolution [39] to improve the inpainting quality. Nazeri et al. [24] propose EdgeConnect that predicts salient edges for inpainting guidance. Song et al. [29] utilize a segmentation prediction network to generate segmentation guidance for detail refinement around the hole region. Xiong et al. [34] present foreground-aware inpainting, which involves three stages, i.e., contour detection, contour completion and image completion, for the disentanglement of structure inference and content hallucination. Ren et al. [26] introduce a structure-aware network, which splits the inpainting task into two parts: structure reconstruction and texture generation. It uses appearance flow to sample features from contextual regions. Yan et al. [37] speculate the relationship between the contextual regions in the encoder layer and the associated hole region in the decoder layer for better predictions. Yu et al. [40] and Song et al. [27] search for a collection of background patches with the highest similarity to the generated contents in the first stage prediction. Liu et al. [20] address this inpainting task via exploiting the partial convolutional layer and mask-update operation. Following the [20], Yu et al. [41] present gate convolution that learns a dynamic mask-updating mechanism and combines with the SN-PatchGAN discriminator to achieve better predictions. Liu et al. [21] propose coherent semantic attention, which considers the feature coherency of hole regions to guarantee the pixel continuity in image level. Wang et al. [32] propose a generative multi-column convolutional neural network (GMCNN) that uses varying receptive fields in branches. Different from existing deep inpainting methods, our method produces CNN features to consistently represent structures and textures to reduce blur and artifacts around the hole region.

3   Proposed Algorithm

Figure 2 shows the pipeline of the proposed method. We use one mutual encoder-decoder to jointly learn structure and texture features and equalize them for consistent representation. The details are presented in the following:

3.1   Mutual Encoder-Decoder

We use an encoder-decoder for end-to-end image generation to fill holes. The structure of this encoder-decoder is a simplified generative network [14], where

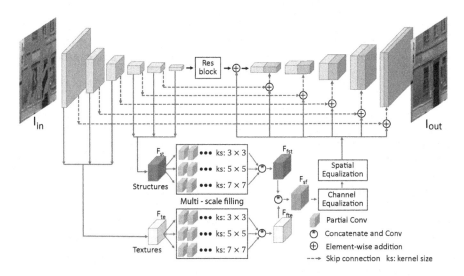

Fig. 2. The overview of the proposed pipeline. We use a mutual encoder-decoder to jointly recover structures and textures during hole filling. The deep layer features of the encoder are reorganized as structure features, while the shallow layer features are reorganized as texture features. We fill holes in multi-scales within the CNN feature space and equalize output features in both channel and spatial domains. The equalized features contain consistent structure and texture features at different CNN feature levels, and supplement the decoder via skip connections for output image generation.

there are 6 convolutional layers in the encoder and 5 convolutional layers in the decoder, respectively. Meanwhile, 4 residual blocks [10] with dilated convolutions are set between the encoders and decoders. The dilated convolutions [13,24] increase the size of the receptive field to perceive encoder features.

In the encoder, we reorganize the CNN features from deep layers as structure features where the semantics reside. Meanwhile, we reorganize the CNN features from shallow layers as texture features to represent image details. We denote the structure features as F_{st} and the texture features as F_{te} as shown in Fig. 2. The reorganization process is to resize and transform the CNN feature maps from different convolutional layers to the same size, and concatenate them accordingly.

After CNN feature reorganization, we design two branches (i.e., the structure branch and the texture branch) to separately perform hole filling on F_{te} and F_{st}. The architectures of these two branches are the same. In each branch, there are 3 parallel streams to fill holes in multiple scales. Each stream consists of 5 partial convolutions [20] with the same kernel size while the kernel size differs among different streams. By using different kernel sizes, we perform multi-scale filling in each branch for the input CNN features. The filled features from 3 streams (i.e., 3 scales) are concatenated and mapped to the same size of the input feature map via a 1×1 convolution. We denote the output of the structure branch as F_{fst}, and the output of the texture branch as F_{fte}. To ensure the hole filling to focus on the textures and structures, we incorporate supervisions on F_{fst} and F_{fte}. We use a 1×1 convolution to separately map F_{fst} and F_{fte} to a color image I_{ost} and a color image I_{ote}, respectively. The pixel-wise L_1 loss can be written as follows:

$$
\begin{aligned}
L_{rst} &= \|I_{ost} - I_{st}\|_1 \\
L_{rte} &= \|I_{ote} - I_{gt}\|_1
\end{aligned}
\tag{1}
$$

where I_{gt} is the ground truth image and I_{st} is the structure image of I_{gt}. We use an edge-preserving smoothing method RTV [35] to generate I_{st} following [26].

The hole regions in F_{te} and F_{st} are filled via structure and texture branches, individually. The feature representations in F_{fte} and F_{fst} are not consistent to reflect the recovered structures and textures. This inconsistency leads to blur and artifacts within and around the hole regions as shown in Fig. 1. To mitigate these effects, we concatenate F_{fte} and F_{fst} first, and make a simple fusion to generate F_{sf} via a 1×1 convolutional layer. The texture and structure representations in F_{sf} are corrected via feature equalization at different CNN feature levels (i.e., across shallow to deep CNN layers).

3.2   Feature Equalizations

We equalize the fused CNN features F_{sf} in both channel and spatial domains. The channel equalization follows the squeeze and excitation operation [12] to ensure that the attentions within each channel of F_{sf} are the same. As the reweighed channels are influenced by both structure and texture representations in F_{sf}, the consistent attentions indicate that these representations are set to be consistent as well. We propagate channel equalization to the spatial domain via the proposed bilateral propagation activation function (BPA).

Formulation. BPA is inspired by the edge-preserving image smoothing [30] to generate response values based on spatial and range distances. It can be written as follows:

$$
y_i^s = \frac{1}{C(x)} \sum_{j \in s} g_{\alpha_s}(\|j - i\|) x_j
\tag{2}
$$

$$
y_i^r = \frac{1}{C(x)} \sum_{j \in v} f(x_i, x_j) x_j
\tag{3}
$$

$$
y_i = q(y_i^s, y_i^r)
\tag{4}
$$

where x_i is the feature channel at position i of input feature x, x_j is a neighboring feature channel around i at position j, y_i^s and y_i^r are the feature channels after spatial and range similarity measurements. We set the normalization factor as $C(x) = N$, where N is the number of positions in x. We use q to denote the concatenation and channel reduction of y_i^s and y_i^r via a 1×1 convolutional layer.

The bilateral propagation utilizes the distances of feature channels from both spatial and range domains. We explore j within a neighboring region s, which is set as the same spatial size as the input feature for global propagation. The spatial contributions from neighboring feature channels are adjusted via a Gaussian function g_{α_s}. When computing y_i^r, we measure the similarities between feature channels x_i and x_j via $f(.)$ within a neighboring region v around i. The size of v is 3×3. To this end, the bilateral propagation considers both global continuity via y_s^i and local consistency via y_r^i.

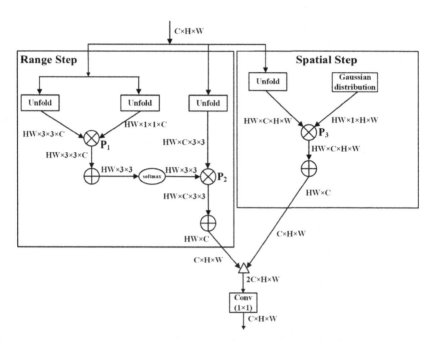

Fig. 3. The pipeline of the bilateral propagation activation function. We denote the broadcast dot product operation as $\otimes$, element-wise addition in the selected channel as $\oplus$, and the concatenation as $\triangle$. For two matrices with different dimensions, broadcast operations first broadcast features in each dimension to match the dimensions of the two matrices.

During the range similarity computation step, we define the pairwise function $f(.)$ as a dot product operation, which can be written as follows:

$$f(x_i, x_j) = (x_i)^T (x_j). \tag{5}$$

The proposed bilateral propagation shares similarity to the non-local block [31] that for each i, $\frac{1}{C(x)} f(x_i, x_j)$ becomes the softmax computation along dimension j. The difference resides on the region design of propagation. The non-local block uses feature channels from all the positions to generate y_i and the similarity is only measured between x_i and x_j. In contrast, BPA considers both feature channel similarity and spatial distance between x_i and x_j during bilateral weight computation. In addition, we use a global region s to compute spatial distance while using a local region v to compute range distance. The advantage of global and local region selections is that we ensure both long-term continuity in the whole spatial region and local consistency around the current feature channel. The boundaries of hole regions are unified with the neighboring image content and the contents within the hole regions are set to be consistent.

Implementations. Figure 3 shows how bilateral propagation operates in the network. The range step corresponds to the computation of y_i^r in Eq. 3 and the spatial step corresponds to y_i^s in Eq. 2. During range computation, the operations until the element-wise multiplication P_1 represent Eq. 5 at all spatial locations. We use the unfold function in PyTorch to reshape the features to vectors (i.e., $HW \times 3 \times 3 \times C$) for obtaining all the neighboring x_j for each x_i, so that we can make efficient element-wise matrix multiplications. Similarly, the operations until P_2 represent the term $\sum_j f(x_i, x_j) \cdot x_j$ in Eq. 3. During spatial computation, the operations until P_3 represent the term $\sum_j g_{\alpha_s}(\|j - i\|)x_j$. As a result, the bilateral propagation operation can be efficiently executed via the element-wise matrix multiplications and additions shown in Fig. 3.

3.3   Loss Functions

We introduce several loss functions to measure structure and texture differences including pixel reconstruction loss, perceptual loss, style loss, and relativistic average LS adversarial loss [16] during training. We also employ a discriminator with local and global operations to ensure local-global contents consistency. And the spectral normalization [23] is applied in both local and global discriminators to achieve stable training.

Pixel Reconstruction Loss. We measure the pixel-wise difference from two aspects. The first one is the loss terms illustrated in Eq. 1 where we add supervisions on the texture and structure branches. The second one measures the similarity between the network output and the ground truth, which can be written as follows:

$$L_{re} = \|I_{out} - I_{gt}\|_1 \tag{6}$$

where I_{out} is the finally predicted image by the network.

Perceptual Loss. To capture the high-level semantics and simulate human perception of images quality, we utilize the perceptual loss [15] L_{perc} defined on the ImageNet-pretrained VGG-16 feature backbone:

$$L_{prec} = \mathbb{E}\Big[\sum_i \frac{1}{N_i}\|\Phi_i(I_{out}) - \Phi_i(I_{gt})\|_1\Big] \tag{7}$$

where Φ_i is the activation map of the i-th layer of the VGG-16 backbone. In our work, Φ_i corresponds to the activation maps from layers ReLu1_1, ReLu2_1, ReLu3_1, ReLu4_1, and ReLu5_1.

Style Loss. The transposed convolutional layers from the decoder will bring artifacts that resemble checkerboard. To mitigate this effect, we introduce the style loss. Given feature maps of size $C_j \times H_j \times W_j$, we compute the style loss as follows:

$$L_{style} = \mathbb{E}_j \left[\| G_j^{\Phi}(I_{out}) - G_j^{\Phi}(I_{gt}) \|_1 \right] \tag{8}$$

where G_j^{Φ} is a $C_j \times C_j$ Gram matrix constructed from the selected activation maps. These activation maps are the same as those used in the perceptual loss.

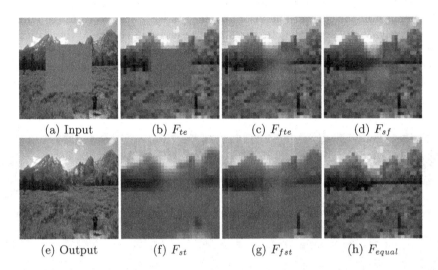

Fig. 4. Visualization of the feature map response. The input and output images are shown in (a) and (e), respectively. We use a 1×1 convolutional layer to map high dimensional feature maps to the color images as shown in (b)–(d) and (f)–(h).

Relativistic Average LS Adversarial Loss. We follow [40] to utilize global and local discriminators for perception enhancement. The relativistic average LS adversarial loss is adopted for our discriminators. For the generator, the adversarial loss is defined as:

$$L_{adv} = -\mathbb{E}_{x_r}[\log(1 - D_{ra}(x_r, x_f))] - \mathbb{E}_{x_f}[\log(D_{ra}(x_f, x_r))] \tag{9}$$

where $D_{ra}(x_r, x_f) = \text{sigmoid}(C(x_r) - \mathbb{E}_{x_f}[C(x_f)])$ and $C(.)$ indicates the local or global discriminator without the last sigmoid function. To this end, real and fake data pairs (x_r, x_f) are sampled from the ground-truth and output images.

Total Losses. The whole objective function of the proposed network can be written as:

$$L_{total} = \lambda_r L_{re} + \lambda_p L_{prec} + \lambda_s L_{style} + \lambda_{adv} L_{adv} + \lambda_{st} L_{rst} + \lambda_{te} L_{rte} \tag{10}$$

where λ_r, λ_p, λ_s, λ_{adv}, λ_{st} and λ_{te} are the tradeoff parameters. In our implementation, we empirically set $\lambda_r = 1$, $\lambda_p = 0.1$, $\lambda_s = 250$, $\lambda_{adv} = 0.2$, $\lambda_{st} = 1$, $\lambda_{te} = 1$.

3.4 Visualizations

We use a structure branch and a texture branch to separately fill holes in CNN feature space. Then, we perform feature equalization to enable consistent feature representations in different feature levels for output image reconstruction. In this section, we visualize the feature maps during different steps to show whether they correspond to our objectives. We use a 1×1 convolutional layer to map CNN feature maps to color images for a clear display.

Figure 4 shows the visualization results. The input image is shown in (a) with a mask in the center. The visualized F_{te} and F_{st} are shown in (b) and (f), respectively. We observe that textures are preserved in (b) while the structures are in (f). By multi-scale hole filling, the hole regions in F_{fte} and F_{fst} are effectively reduced as shown in (c) and (g). After equalization, the hole regions in (h) are effectively filled and the equalized features contribute to the decoders to generate the output image as shown in (e).

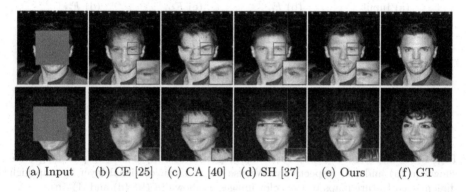

(a) Input   (b) CE [25]   (c) CA [40]   (d) SH [37]   (e) Ours   (f) GT

Fig. 5. Visual evaluations for filling center holes. Our method performs favorably against existing approaches to retain both structures and textures.

4 Experiments

We evaluate our method on three datasets: Paris StreetView [6], Place2 [43] and CelebA [22]. We follow the training, testing, and validation splits of these three datasets. Data augmentation such as flipping is also adopted during training. Our model is optimized by the Adam optimizer [17] with a learning rate of 2×10^{-4} on a single NVIDIA 2080TI GPU. The training process of the CelebA model, Paris StreetView model and Place2 model are stopped after 6 epochs, 30 epochs and 60 epochs, respectively. All the masks and images for training and testing are with the size of 256×256.

We compare our method with six state-of-the-art method: CE [25], CA [40], SH [37], CSA [21], SF [26] and GC [41]. For a fair evaluation on model generalization abilities, we conduct experiments on filling center holes and irregular holes on the input images. The center hole is brought by a mask that covers the image center with a size of 128×128. We obtain irregular masks from PConv [20]. These masks are in different categories according to the ratios of the hole regions versus the entire image size (i.e., below 10%, from 10% to 20%, etc.). For holes in the image center, we compare with CA [40], SH [37] and CE [25] on the CelebA [22] validation set. We choose these three methods because they are more effective to fill holes in the image center than fill irregular holes. When handling irregular holes on the input images, we compare with CSA [21], SF [26] and GC [41] using Paris StreetView [6] and Place2 [43] validation datasets.

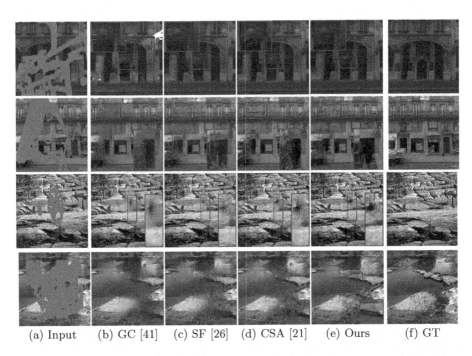

(a) Input   (b) GC [41]   (c) SF [26]   (d) CSA [21]   (e) Ours   (f) GT

Fig. 6. Visual evaluations for filling irregular holes. Our method performs favorably against existing approaches to retain both structures and textures.

4.1 Visual Evaluations

The visual comparison on the results for filling center holes are in Fig. 5 and the results for filling irregular holes are in Fig. 6. We also display ground truth images in (f) to show the actual image content. In Fig. 5, the input images are shown in (a). The results produced by CE and CA contain distorted structures and blurry textures as shown in (b) and (c). Although more visually pleasing contents are generated in (d), the semantics remain unreasonable. By utilizing

Table 1. Numerical evaluations on the CelebA dataset where the inputs are with centering hole regions. ↓ indicates lower is better while ↑ indicates higher is better.

	CE	CA	SH	Ours
FID↓	52.17	37.61	29.72	**25.51**
PSNR↑	8.53	23.65	26.10	**26.32**
SSIM↑	0.137	0.870	0.902	**0.910**

Table 2. Numerical comparisons on the Place2 dataset. ↓ indicates lower is better while ↑ indicates higher is better.

	Mask	GC	SF	CSA	Ours
FID↓	10–20%	19.04	8.78	7.85	**6.91**
	20–30%	28.45	16.38	13.95	**8.06**
	30–40%	40.71	27.54	25.74	**19.36**
	40–50%	60.72	40.93	38.74	**28.79**
PSNR↑	10–20%	27.10	29.50	**31.31**	31.13
	20–30%	25.18	27.22	28.66	**28.87**
	30–40%	22.51	24.37	25.01	**25.34**
	40–50%	20.35	21.90	22.54	**22.81**
SSIM↑	10–20%	0.929	0.926	0.954	**0.957**
	20–30%	0.878	0.885	0.918	**0.923**
	30–40%	0.823	0.802	0.843	**0.854**
	40–50%	0.670	0.678	0.702	**0.719**

consistent structure and texture features, our method is effective to generate results with realistic textures.

Figure 6 shows the comparison for filling irregular holes, which is more challenging than filling centering holes. The results from GC contain noisy patterns shown in (b). The details are missing and the structures are distorted in (c) and (d). These methods are not effective to recover image contents without bringing in obvious artifacts (i.e., the second row around the door regions). In contrast, our method learns to represent structures and textures in a consistent formation. The results shown in (e) indicate the effectiveness of our method to produce visually pleasing contents. The evaluations on filling both centering holes and irregular holes indicate that our method performs favorably against existing hole filling approaches.

4.2   Numerical Evaluations

We conduct numerical evaluations on the Place2 dataset with different mask ratios. Besides, we evaluate numerically on the CelebA dataset with centering holes in the input images. There are 100 validation images from the "valley" scene category chosen for evaluations. In CelebA, we randomly choose 500 images for evaluation. For the evaluation metrics, we follow [26] to use SSIM [33] and PSNR. Moreover, we introduce FID (Frechet Inception Distance) metric [11] as it indicates the perceptual quality of the results. The evaluation results are shown in Tables 1 and 2. Our method outperforms existing methods to fill centering holes. Meanwhile, favorable performance is achieved by our method to fill irregular holes under various hole versus image ratios.

Human Subject Evaluation. We follow [42] to involve over 35 volunteers for evaluating the results on CelebA, Place2 and Paris StreetView datasets. The volunteers are all image experts with image processing background. There are 20 questions for each subject. In each question, the subject needs to select the most realistic result from 4 results generated by different methods without knowing the hole region in advance. We tally the votes and show the statistics in Table 3. Our method performs favorably against existing methods.

Table 3. Human Subject Evaluation results. Each subject selects the most realistic result without knowing hole regions in advance.

	CE	CA	SH	GC	SF	CSA	Ours
Paris StreetView	N/A	N/A	N/A	5.3%	21.0%	29.8%	**43.7%**
Place2	N/A	N/A	N/A	3.0%	25.0%	29.6%	**42.4%**
CelebA	1.2%	2.0%	40.4%	N/A	N/A	N/A	**56.4%**

Table 4. Ablation study on the Paris StreetView dataset. Our performance is improved by using structure and texture branches.

	Ours without textures	Ours without structures	Ours
FID↓	30.37	27.46	**25.10**
PSNR↑	22.80	22.96	**23.38**
SSIM↑	0.818	0.823	**0.833**

Table 5. Ablation study on the Place2 dataset. Non-local aggregation improves our baseline while feature equalization makes further improvement.

	Ours without equalization	Non-local aggregation	Ours
FID↓	29.11	24.07	**21.26**
PSNR↑	23.14	23.64	**24.57**
SSIM↑	0.837	0.848	**0.852**

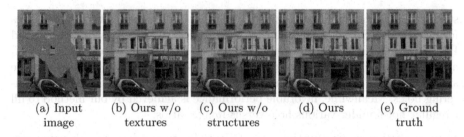

|(a) Input image|(b) Ours w/o textures|(c) Ours w/o structures|(d) Ours|(e) Ground truth|

Fig. 7. Abalation studies on structure and texture branches. A joint utilization of these two branches improves the content quality.

|(a) Input image|(b) Ours w/o equalization|(c) Non-Local aggregation|(d) Ours|(e) Ground truth|

Fig. 8. Ablation studies on feature equalizations. More realistic and visually pleasing contents are generated via feature equalizations.

5   Ablation Study

Structure and Texture Branches. To evaluate the effects of structure and texture branches, we use each of these branches separately for network training. For fair comparisons, we expand the channel number of the texture and structure branch outputs via additional convolutions. So the single branch output contains the same size as that of F_{sf}. As shown in Fig. 7, the output of our method without a texture branch contains rich structure information (i.e., the window in the red and green boxes) while the textures are missing. In comparison, the output of our method without a structure branch does not contain meaningful structure (i.e., the window in the red and green boxes). By utilizing both branches, our method achieves favorable results on both structures and textures. Table 4 shows the similar numerical performance on the Paris StreetView dataset where these two branches improve our method significantly.

Feature Equalizations. We show the contributions of feature equalizations by removing them from the pipeline and showing the performance degradation. Moreover, we show that the bilateral propagation activation function (BPA) is more effective to fill hole regions than the Non-local attentions [31]. As shown in Fig. 8, without using equalization our method generates visually unpleasant contents and visible artifacts. In comparison, the contents generated by [31] are more natural. However, the recovered contents are still blurry and inconsistent because the Non-local block ignores the local coherency and global distance of features. This limitation is effectively solved via our method with feature equalizations. Similar performance has been shown numerically in Table 5 where our method achieves favorable results.

6   Concluding Remarks

We propose a mutual encoder-decoder with feature equalizations to correlate filled structures with textures during image inpainting. The shallow and deep layer features are reorganized as texture and structure features, respectively. In the CNN feature space, we introduce a texture branch and a structure branch to fill holes in multi-scales and fuse the outputs together via feature equalizations. During equalization, we first ensure consistent attentions among individual channels and propagate them to the whole spatial feature map region via the proposed bilateral propagation activation function. The experiments carried out over the benchmark datasets have shown the effectiveness of the proposed method when compared to state-of-the-art approaches on filling both regular and irregular hole regions.

Acknowledgements. This work is partially supported by the National Natural Science Foundation of China under Grant No. 61702176.

References

1. Ballester, C., Bertalmio, M., Caselles, V., Sapiro, G., Verdera, J.: Filling-in by joint interpolation of vector fields and gray levels. TIP **10**, 1200–1211 (2001)
2. Barnes, C., Shechtman, E., Finkelstein, A., Goldman, D.: PatchMatch: a randomized correspondence algorithm for structural image editing. In: SIGGRAPH (2009)
3. Bertalmio, M., Sapiro, G., Caselles, V., Ballester, C.: Image inpainting. In: SIGGRAPH (2000)
4. Criminisi, A., Pérez, P., Toyama, K.: Region filling and object removal by exemplar-based image inpainting. TIP **13**, 1200–1212 (2004)
5. Darabi, S., Shechtman, E., Barnes, C., Goldman, D.B., Sen, P.: Image melding: combining inconsistent images using patch-based synthesis. ACM Trans. Graph. **31**, 18 (2012)
6. Doersch, C., Singh, S., Gupta, A., Sivic, J., Efros, A.A.: What makes Paris look like Paris? Commun. ACM **58**, 103–110 (2015)
7. Efros, A., Freeman, W.: Image quilting for texture synthesis and transfer. In: SIGGRAPH (2001)
8. Efros, A., Freeman, W.: Texture synthesis by nonparametric sampling. In: ICCV (2001)
9. Goodfellow, I., et al.: Generative adversarial nets. In: NIPS (2014)
10. He, K., Zhang, X., Ren, S., Sun, J.: Deep residual learning for image recognition. In: CVPR (2016)
11. Heusel, M., Ramsauer, H., Unterthiner, T., Nessler, B., Hochreiter, S.: GANs trained by a two time-scale update rule converge to a local Nash equilibrium. In: NIPS (2017)
12. Hu, J., Shen, L., Sun, G.: Squeeze-and-excitation networks. In: CVPR (2018)

13. Iizuka, S., Simo-Serra, E., Ishikawa, H.: Globally and locally consistent image completion. In: SIGGRAPH (2017)
14. Isola, P., Zhu, J.Y., Zhou, T., Efros, A.A.: Image-to-image translation with conditional adversarial networks. In: CVPR (2017)
15. Johnson, J., Alahi, A., Fei-Fei, L.: Perceptual losses for real-time style transfer and super-resolution. In: Leibe, B., Matas, J., Sebe, N., Welling, M. (eds.) ECCV 2016. LNCS, vol. 9906, pp. 694–711. Springer, Cham (2016). https://doi.org/10.1007/978-3-319-46475-6_43
16. Jolicoeur-Martineau, A.: The relativistic discriminator: a key element missing from standard GAN. In: ICLR (2018)
17. Kingma, D.P., Ba, J.: Adam: a method for stochastic optimization. arXiv preprint arXiv:1412.6980 (2014)
18. Levin, A., Zomet, A., Weiss, Y.: Learning how to inpaint from global image statistics. In: ICCV (2003)
19. Li, Y., Liu, S., Yang, J., Yang, M.H.: Generative face completion. In: CVPR (2017)
20. Liu, G., Reda, F.A., Shih, K.J., Wang, T.-C., Tao, A., Catanzaro, B.: Image inpainting for irregular holes using partial convolutions. In: Ferrari, V., Hebert, M., Sminchisescu, C., Weiss, Y. (eds.) ECCV 2018. LNCS, vol. 11215, pp. 89–105. Springer, Cham (2018). https://doi.org/10.1007/978-3-030-01252-6_6
21. Liu, H., Jiang, B., Xiao, Y., Yang, C.: Coherent semantic attention for image inpainting. In: ICCV (2019)
22. Liu, Z., LuoPi, n., Wang, X., Tang, X.: Deep learning face attributes in the wild. In: ICCV (2015)
23. Miyato, T., Kataoka, T., Koyama, M., Yoshida, Y.: Spectral normalization for generative adversarial networks. arXiv preprint arXiv:1802.05957 (2018)
24. Nazeri, K., Ng, E., Joseph, T., Qureshi, F.Z., Ebrahimi, M.: EdgeConnect: generative image inpainting with adversarial edge learning. In: ICCV Workshops (2019)
25. Pathak, D., Krahenbuhl, P., Donahue, J., Darrell, T., Efros, A.: Context encoders: feature learning by inpainting. In: CVPR (2016)
26. Ren, Y., Yu, X., Zhang, R., Li, T.H., Liu, S., Li, G.: StructureFlow: image inpainting via structure-aware appearance flow. In: ICCV (2019)
27. Song, Y., Yang, C., Lin, Z., Liu, X., Huang, Q., Li, H., Kuo, C.-C.J.: Contextual-based image inpainting: infer, match, and translate. In: Ferrari, V., Hebert, M., Sminchisescu, C., Weiss, Y. (eds.) ECCV 2018. LNCS, vol. 11206, pp. 3–18. Springer, Cham (2018). https://doi.org/10.1007/978-3-030-01216-8_1
28. Song, Y., Bao, L., He, S., Yang, Q., Yang, M.H.: Stylizing face images via multiple exemplars. CVIU 162, 135–145 (2017)
29. Song, Y., Yang, C., Shen, Y., Wang, P., Huang, Q., Kuo, J.: SPG-Net: segmentation prediction and guidance network for image inpainting. arXiv preprint arXiv:1805.03356 (2018)
30. Tomasi, C., Manduchi, R.: Bilateral filtering for gray and color images. In: CVPR (1998)
31. Wang, X., Girshick, R., Gupta, A., He, K.: Non-local neural networks. In: CVPR (2018)
32. Wang, Y., Tao, X., Qi, X., Shen, X., Jia, J.: Image inpainting via generative multi-column convolutional neural networks. In: NIPS (2018)
33. Wang, Z., Bovik, A.C., Sheikh, H.R., Simoncelli, E.P.: Image quality assessment: from error visibility to structural similarity. TIP 13, 600–612 (2004)
34. Xiong, W., Yu, J., Lin, Z., Yang, J., Lu, X., Barnes, C., Luo, J.: Foreground-aware image inpainting. In: CVPR (2019)

35. Xu, L., Yan, Q., Xia, Y., Jia, J.: Structure extraction from texture via relative total variation. SIGGRAPH **31**, 139 (2012)
36. Xu, Z., Sun, J.: Image inpainting by patch propagation using patch sparsity. TIP **19**, 1153–1165 (2010)
37. Yan, Z., Li, X., Li, M., Zuo, W., Shan, S.: Shift-Net: image inpainting via deep feature rearrangement. In: Ferrari, V., Hebert, M., Sminchisescu, C., Weiss, Y. (eds.) Computer Vision – ECCV 2018. LNCS, vol. 11218, pp. 3–19. Springer, Cham (2018). https://doi.org/10.1007/978-3-030-01264-9_1
38. Yeh, R., Chen, C., Lim, T., Johnson, M.H., Do, M.N.: Semantic image inpainting with perceptual and contextual losses. arXiv preprint arXiv:1607.07539 (2016)
39. Yu, F., Koltun, V.: Multi-scale context aggregation by dilated convolutions. arXiv preprint arXiv:1511.07122 (2015)
40. Yu, J., Lin, Z., Yang, J., Shen, X., Lu, X., Huang, T.S.: Generative image inpainting with contextual attention. In: CVPR (2018)
41. Yu, J., Lin, Z., Yang, J., Shen, X., Lu, X., Huang, T.S.: Free-form image inpainting with gated convolution. In: ICCV (2019)
42. Zeng, Y., Fu, J., Chao, H., Guo, B.: Learning pyramid-context encoder network for high-quality image inpainting. In: CVPR (2019)
43. Zhou, B., Lapedriza, A., Khosla, A., Oliva, A., Torralba, A.: Places: a 10 million image database for scene recognition. PAMI **40**, 1452–1464 (2017)

TextCaps: A Dataset for Image Captioning with Reading Comprehension

Oleksii Sidorov[1]([✉]), Ronghang Hu[1,2], Marcus Rohrbach[1],
and Amanpreet Singh[1]

[1] Facebook AI Research, Menlo Park, USA
acecreamu@gmail.com, {mrf,asg}@fb.com
[2] University of California, Berkeley, USA
ronghang@eecs.berkeley.edu

Abstract. Image descriptions can help visually impaired people to quickly understand the image content. While we made significant progress in automatically describing images and optical character recognition, current approaches are unable to include written text in their descriptions, although text is omnipresent in human environments and frequently critical to understand our surroundings. To study how to comprehend text in the context of an image we collect a novel dataset, TextCaps, with 145k captions for 28k images. Our dataset challenges a model to recognize text, relate it to its visual context, and decide what part of the text to copy or paraphrase, requiring spatial, semantic, and visual reasoning between multiple text tokens and visual entities, such as objects. We study baselines and adapt existing approaches to this new task, which we refer to as *image captioning with reading comprehension*. Our analysis with automatic and human studies shows that our new TextCaps dataset provides many new technical challenges over previous datasets.

1 Introduction

When trying to understand man-made environments, it is not only important to recognize objects but also frequently critical to read associated text and comprehend it in the context to the visual scene. Knowing there is "a red sign" is not sufficient to understand that one is at "Mornington Crescent" Station (see Fig. 1(a)), or knowing that an old artifact is next to a ruler is not enough to know that it is "40 mm wide" (Fig. 1(c)). Reading comprehension in images is crucial for blind people. As the VizWiz datasets [5] suggest, 21% of questions visually-impaired people asked about an image were related to the text in it. Image captioning plays an important role in starting a visual dialog with a blind user allowing them to ask for further information as required. In addition, text out of context (*e.g.* *'5:43p'*) may be of little help, whereas scene description (*e.g.* 'shown on a departure tableau') makes it substantially more meaningful.

Electronic supplementary material The online version of this chapter (https://doi.org/10.1007/978-3-030-58536-5_44) contains supplementary material, which is available to authorized users.

A. Vedaldi et al. (Eds.): ECCV 2020, LNCS 12347, pp. 742–758, 2020.
https://doi.org/10.1007/978-3-030-58536-5_44

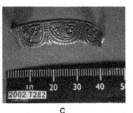

a

Model: a macdonald ' s sign that is on a brick wall

Human: A tile wall with a red circle on it reading Mornington Crescent

b

Model: a sign that has the time of 12 : 37 on it

Human: A kiosk of track 13 of Metra which states that the 5:43 train has moved tracks

c

Model: a ruler that has the number 2003 on it

Human: An old artifact being measured by a ruler that shows it is around 40 millimeters wide

Fig. 1. Existing captioning models cannot read! The *image captioning with reading comprehension* task using data from our TextCaps dataset and BUTD model [4] trained on it. (Color figure online)

In recent years, with the availability of large labelled corpora, progress in image captioning has seen steady increase in performance and quality [4,10,12,13,34] and reading scene text (OCR) has matured [8,16,19,21,31]. However, while OCR only focuses on written text, state-of-the-art image captioning methods focus only on the visual objects when generating captions and fail to recognize and reason about the text in the scene. For example, Fig. 1 shows predictions of a state-of-the-art model [4] on a few images that require reading comprehension. The predictions clearly show an inability of current state-of-the-art image captioning methods to read and comprehend text present in images. Incorporating OCR tokens into a sentence is a challenging task, as unlike conventional vocabulary tokens which depend on the text before them and therefore can be inferred, OCR tokens often can not be predicted from the context and therefore represent independent entities. Predicting a token from vocabulary and selecting an OCR token from the scene are two rather different tasks which have to be seamlessly combined to tackle this task.

Considering the images and reference captions in Fig. 1, we can breakdown what is needed to successfully describe these images: First, detect and extract text/OCR tokens[1] (*'Mornington Crescent'*, *'moved track'*) as well the visual context such as objects in the image (*'red circle'*, *'kiosk'*). Second, generate a grammatically correct sentence which combines words from the vocabulary and OCR tokens. In addition to the challenges in normal captioning, *image captioning with reading comprehension* can include the following technical challenges:

1. Determine the relationships **between different OCR tokens** and between **OCR tokens and the visual context,** to decide if an OCR token should be mentioned in the sentence and which OCR tokens should be joined together (*e.g.* in Fig. 1b: "5:35" denotes the current time and should not be joined with

---

[1] The remainder of the manuscript we refer to the text in an image as "OCR tokens", where one token is typically a word, i.e. a group of characters.

"ON TIME"), based on their (a) *semantics* (Fig. 2b), (b) *spatial* relationship (Fig. 1c), and (c) *visual* appearance and context (Fig. 2d).

2. **Switching multiple times** during caption generation between the words from the model's vocabulary and OCR tokens (Fig. 1b).
3. **Paraphrasing and inference** about the OCR tokens (Fig. 2 bold).
4. Handling of OCR tokens, including ones never seen before (**zero-shot**).

While this list should not suggest a temporal processing order, it explains why today's models lack capabilities to comprehend text in images to generate meaningful descriptions. It is unlikely that the above skills will naturally emerge through supervised deep learning on existing image captioning datasets as they are not focusing on this problem. In contrast, captions in these datasets are collected in a way that implicitly or explicitly avoids mentioning specific instances appearing in the OCR text. To study the novel task of image captioning with reading comprehension, we thus believe it is important to build a dataset containing captions which require reading and reasoning about text in images. We find the COCO Captioning dataset [9] not suitable as only an estimated 2.7% of its captions mention OCR tokens present in the image, and in total there are less than 350 different OCRs (i.e. the OCR vocabulary size), moreover most OCR tokens are common words, such as "stop", "man", which are already present in a standard captioning vocabulary. Meanwhile, in Visual Question Answering, multiple datasets [6,23,30] were recently introduced which focus on text-based visual question answering. This task is harder than OCR recognition and extraction as it requires understanding the OCR extracted text in the context of the question and the image to deduce the correct answer. However, although these datasets focus on text reading, the answers are typically shorter than 5 words (mainly 1 or 2), and, typically, all the words which have to be generated are either entirely from the training vocabulary *or* OCR text, rather than requiring switching between them to build a complete sentence. These differences in task and dataset do not allow training models to generate long sentences. Furthermore and importantly, we require a dataset with human collected reference sentences to validate and test captioning models for *reading comprehension*.

Consequently, in this work, we contribute the following:

- For our novel task *image captioning with reading comprehension*, we collect a new dataset, **TextCaps**, which **contains 142,040 captions** on 28,408 images and requires models to read and reason about text in the image to generate coherent descriptions.
- We analyse our dataset, and find it has **several new technical challenges for captioning**, including the ability to switch multiple times between OCR tokens and vocabulary, zero-shot OCR tokens, as well as paraphrasing and inference about OCR tokens.
- Our evaluation shows that **standard captioning models fail on this new task**, while the state-of-the-art TextVQA [30] model, M4C [17], when trained with our dataset TextCaps, gets encouraging results. Our ablation study shows that it is important to take into account all semantic, visual, and spatial information of OCR tokens to generate high-quality captions.

– We conduct **human evaluations** on model predictions which show that there is a **significant gap between the best model and humans**, indicating an exciting avenue of future image captioning research.

2   Related Work

Image Captioning. The Flickr30k [35] and COCO Captions [9] dataset have both been collected similarly via crowd-sourcing. The COCO Captions dataset is significantly larger than Flickr30k and acts as a base for training the majority of current state-of-the-art image captioning algorithms. It includes 995,684 captions for 164,062 images. The annotators of COCO were asked "Describe all the important parts of the scene" and "Do not describe unimportant details", which resulted in COCO being focused on objects which are more prominent rather than text. SBU Captions [24] is an image captioning dataset which was collected automatically by retrieving one million images and associated user descriptions from Flickr, filtering them based on key words and sentence length. Similarly, Conceptual Captions (CC) dataset [27] is also automatically constructed by crawling images from web pages together with their ALT-text. The collected annotations were extensively filtered and processed, e.g. replacing proper names and titles with object classes (*e.g.* man, city), resulting in 3.3 million image-caption pairs. This simplifies caption generation but at the same time removes fine details such as unique OCR tokens. Apart from conventional paired datasets there are also datasets like NoCaps [1], oriented to a more advanced task of captioning with zero-shot generalization to novel object classes.

While our TextCaps dataset also consists of image-sentence pairs, it focuses on the text in the image, posing additional challenges. Specifically, text can be seen as an additional modality, which models have to read (typically using OCR), comprehend, and include when generating a sentence. Additionally, many OCR tokens do not appear in the training set, but only in the test (zero-shot). In concurrent work, [15] collect captions on VizWiz [5] images but unlike TextCaps there isn't a specific focus on reading comprehension.

Optical Character Recognition (OCR). OCR involves in general two steps, namely (i) detection: finding the location of text, and (ii) extraction: based on the detected text boundaries, extracting the text as characters. OCR can be seen as a subtask for our *image captioning with reading comprehension* task as one needs to know the text present in the image to generate a meaningful description of an image containing text. This makes OCR research an important and relevant topic to our task, which additionally requires to understand the importance of OCR token, their semantic meaning, as well as relationship to visual context and other OCR tokens. Recent OCR models have shown reliability and performance improvements [8,16,19,21,31]. However, in our experiments we observe that OCR is far from a solved problem in real-world scenarios present in our dataset.

Visual Question Answering with Text Reading Ability. Recently, three different text-oriented datasets were presented for the task of Visual Question Answering. TextVQA [30] consists of 28,408 images from selected categories of Open Images v3 dataset, corresponding 45,336 questions, and 10 answers for each question. Scene Text VQA (ST-VQA) dataset [6] has a similar size of 23,038 images and 31,791 questions but only one answer for each question. Both these datasets were annotated via crowd-sourcing. OCR-VQA [23] is a larger dataset (207,572 images) collected semi-automatically using photos of book covers and corresponding metadata. The rule generated questions were paraphrased by human annotators. These three datasets require reading and reasoning about the text in the image while considering the context for answering a question, which is similar in spirit to TextCaps. However, the image, question and answer triplet is not directly suitable for generation of descriptive sentences. We provide additional quantitative comparisons and discussion between our and existing captioning and VQA datasets in Sect. 3.2.

3 ◎ TextCaps Dataset

We collect TextCaps with the goal of studying the novel task of *image captioning with reading comprehension*. Our dataset allows us to test captioning models' reading comprehension ability and we hope it will also enable us to teach image captioning models how "to read", *i.e.*, allow us to design and train image captioning algorithms which are able to process and include information from the text in the image. In this section, we describe the dataset collection and analyze its statistics. The dataset is publicly available at textvqa.org/textcaps.

3.1 Dataset Collection

With the goal of having a diverse set of images, we rely on images from Open Images v3 dataset (CC 2.0 license). Specifically, we use the same subset of images as in the TextVQA dataset [30]; these images have been verified to contain text through an OCR system [8] and human annotators [30]. Using the same images as TextVQA additionally allows multi-task and transfer learning scenarios between OCR-based VQA and image captioning tasks. The images were annotated by human annotators in two stages.[2]

Annotators were asked to describe an image in one sentence which would require reading the text in the image.[3]

---

[2] The full text of the instructions as well as screenshots of the user interface are presented in the Supplemental (Sec. F).

[3] Apart from direct copying, we also allowed indirect use of text, *e.g.* inferring, paraphrasing, summarizing, or reasoning about it (see Fig. 2). This approach creates a fundamental difference from OCR datasets where alteration of text is not acceptable. For captioning, however, the ability to reason about text can be beneficial.

Evaluators were asked to vote yes/no on whether the caption written in the first step satisfies the following requirements: requires reading the text in the image; is true for the given image; consists of one sentence; is grammatically correct; and does not contain subjective language. The majority of 5 votes was used to filter captions of low quality. The quality of the work of evaluators was controlled using gold captions of known good/bad quality.

Five independent captions were collected for each image. An additional 6th caption was collected for the test set only to estimate human performance on the dataset. The annotators did not see previously collected captions for a particular image and did not see the same image twice. In total, we collected 145,329 captions for 28,408 images. We follow the same image splits as TextVQA for training (21,953), validation (3,166), and test (3,289) sets. An estimation performed using ground-truth OCR shows that on average, 39.5% out of all OCR tokens present in the image are covered by the collected human annotations.

Fig. 2. Illustration of TextCaps captions. The bold font highlights instances which do not copy the text directly but require paraphrasing or some inference beyond copying. Underlined font highlights copied text tokens.

3.2 Dataset Analysis

We first discuss several properties of the TextCaps qualitatively and then analyse and compare its statistics to other captioning and OCR-based VQA datasets.

Qualitative Observations. Examples of our collected dataset in Fig. 2 demonstrate that our image captions combine the textual information present in the image with its natural language scene description. We asked the annotators to read and use text in the images but we did not restrict them to directly copy the text. Thus, our dataset also contains captions where OCR tokens are not present directly but were used to infer a description, *e.g.* in Fig. 2a "Rice is winning" instead of "Rice has 18 and Ecu has 17". In a human evaluation of 640 captions we found that about 20% of images have at least one caption (8% of captions) which require more challenging reasoning or paraphrasing rather than just direct copying of visible text. Nevertheless, even the captions which require copying text directly can be complex and may require advanced reasoning as illustrated in multiple examples in Fig. 2. The collected captions are not limited to trivial template "Object X which says Y". We have observed various types of relations between text and other objects in a scene which are impossible to formulate without reading comprehension. For example, in Fig. 2: "A *score board* shows Rice with 18 points vs. ECU with 17 points" (a), "*Box* of Hydroxycut on sale for only 17.88 at a *store*" (b), "Two *light switches* are both in off position" (e).

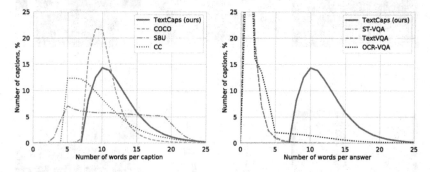

Fig. 3. Distribution of caption/answer lengths in Image Captioning (left) and VQA (right) datasets. VQA answers are significantly shorter than image captions and mostly concentrated within 5 words limit.

Dataset Statistics. To situate TextCaps properly w.r.t. other image captioning datasets, we compare TextCaps with other prominent image captioning datasets, namely COCO [9], SBU [24], and Conceptual Captions [27], as well as reading-oriented VQA datasets TextVQA [30], ST-VQA [6], and OCR-VQA [23]. The average caption length is 12.0 words for SBU, 9.7 words for Conceptual Captions, and 10.5 words for COCO, respectively. The average length for TextCaps is 12.4, slightly larger than the others (see Fig. 3). This can be explained by the fact that captions in TextCaps typically include both scene description as well as the text from it in one sentence, while conventional captioning datasets only cover the scene description. Meanwhile, the average answer length is 1.53 for TextVQA, 1.51 for ST-VQA and 3.31 for OCR-VQA – much smaller than the captions in

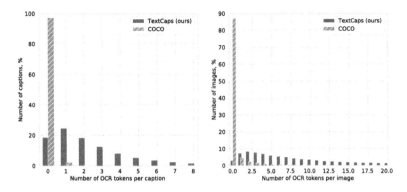

Fig. 4. Distribution of OCR tokens in COCO and TextCaps captions (left) and images (right). In total, COCO contains **2.7%** of captions and **12.7%** of images with at least one OCR token, whereas TextCaps – **81.3%** and **96.9%**.

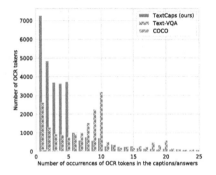
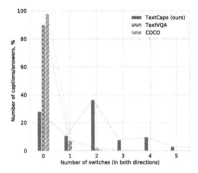

(a) **OCR frequency distribution** shows how many OCR tokens occur once, twice, *etc.* TextCaps has the largest amount of unique and rare (< 5) OCR tokens. Note that TextVQA has 10 answers for each question which are often identical.

(b) **Number of switches between OCR ⇌ Vocab** illustrates the technical complexity of the datasets. An approach which cannot make switches will be sufficient for most of COCO captions and TextVQA but not for TextCaps.

Fig. 5. Analysis of OCR in our dataset vs. others

our dataset. Typical answers like *'yes'*, *'two'*, *'coca cola'* may be sufficient to answer a question but insufficient to describe the image comprehensively.

Figure 4 compares the percentage of captions with a particular number of OCR tokens between COCO and TextCaps datasets.[4] TextCaps has a much larger number of OCR tokens in the captions as well as in the images compared to COCO (note the high percentage at 0). A small part (2.7%) of COCO captions which contain OCR tokens is mostly limited to one token per caption; only 0.38% of captions contain two or more tokens. Whereas in TextCaps, multi-

---

[4] Note that OCR tokens are extracted using Rosetta OCR system [8] which cannot guarantee exhaustive coverage of all text in an image and presents just an estimation.

word reading is much more common (56.8%) which is crucial for capturing real-world information (*e.g.* authors, titles, monuments, *etc.*). Moreover, while COCO Captions contain less than 350 unique OCR tokens, TextCaps contains 39.7k of them.

We also measured the frequency of OCR tokens in the captions. Figure 5a illustrates the number of times a particular OCR token appears in the captions. More than 9000 tokens appear only once in the whole dataset. The curve drops rapidly after 5 occurrences and only a small part of tokens occur more than 10 times. Quantitatively, 75.7% of tokens are presented less then 5 times, and only 12.9% are presented more than 10 times. The distribution specifically demonstrates the large variance in text occurring in natural images which is challenging to model using a fixed word vocabulary. In addition to this long-tailed distribution, we find that an impressive number of 2901 of 6329 unique OCR tokens appearing in the test set captions, have neither appeared in the training nor validation set (i.e. they are "zero-shot") which makes it necessary for models to be able to read new text in images. TextCaps dataset also creates new technical challenges for the models. Figure 5b illustrates that due to the common use of OCR tokens in the captions, models required to switch between OCR and vocabulary words often. The majority of the TextCaps captions require to switch twice or more, whereas most COCO and TextVQA outputs can be generated even without any switches.

4   Benchmark Evaluation

4.1   Baselines

Our baselines aim to illustrate the gap between performance of conventional state-of-the-art image captioning models (BUTD [4], AoANet[18]) in comparison to recent architectures which incorporate reading (M4C [17]).

Bottom-Up Top-Down Attention Model (BUTD). [4] is a widely used image captioning model based on Faster R-CNN [26] object detection features (Bottom-Up) in conjunction with attention-weighted LSTM layers (Top-Down).

Attention on Attention Model (AoANet). [18] is a current SoTA captioning algorithm which uses the attention-on-attention module (AoA) to create a relation between attended vectors in both encoder and decoder.

M4C-Captioner. M4C [17] is a recent model with state-of-the-art performance on the TextVQA task. The model fuses different modalities by embedding them into a common semantic space and processing them with a multimodal transformer. Apart from that, unlike conventional VQA models where a prediction is made via classification, it enables iterative answer decoding with a dynamic pointer network [22,33], allowing the model to generate a multi-word answer, which is not limited to a fixed vocabulary. This feature makes it also suitable for reading-based caption generation. We adapt M4C to our task by removing the question input and directly use its multi-word answer decoder to generate a caption conditioned on the detected objects and OCR tokens in the image (we refer to this model as **M4C-Captioner** and illustrate it in Fig. 6).

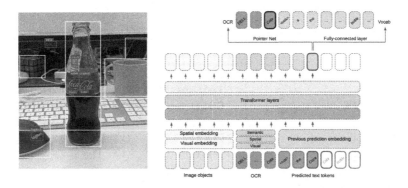

Fig. 6. M4C-Captioner architecture for the image captioning with reading comprehension task.

M4C-Captioner Ablations. In comparison to its full version, we also evaluate a restricted version of this model without access to OCR results (referred to as **M4C-Captioner w/o OCRs**), where we use an empty OCR token list as input to the model. Additionally, we experiment with removing the pointer network (described in details in [17]) from M4C-Captioner, so that the model still has access to OCR features but cannot directly copy OCR tokens, and must use its fixed vocabulary for caption generation (referred to as **M4C-Captioner w/o copying**). As multiple types of features are used for OCR tokens in M4C-Captioner by default (same as in [17]), we further study the impact of each OCR feature type and use only **spatial** information (4-dimensional relative bounding box coordinates $[x_{min}, y_{min}, x_{max}, y_{max}]$ of OCR tokens), **semantic** information (FastText [7] and PHOC [2]), and **visual** (Faster R-CNN [26]) features in different experiments. Additionally, we use ground truth OCR tokens annotated by humans (referred to as **M4C-Captioner w/ GT OCRs**) for training and prediction[5] to study the influence of mistakes of automatic OCR methods.

Human Performance. In addition to our baselines, we provide an estimate of human performance by using the same metrics on the TextCaps test set to benchmark the progress that models still need to make. As discussed in Sect. 4.3, we collected one more caption for each image in the test set. The metrics are then calculated by averaging the results over 6 runs, each time leaving out one caption as a prediction, similar to [14]. On the test set, we use the same approach to evaluate machine-generated captions, so numbers are comparable.

---

[5] This includes a small number of images without GT-OCRs (Supplemental Sec. A).

Table 1. Performance of our baselines on our TextCaps dataset. M4C-Captioner significantly benefits from OCR inputs and achieves the highest CIDEr score, suggesting that it is important to copy text from image on this task. However, there is still a large gap between the current machine performance and human performance, which we hope can be closed by future work.

#	Method	Trained on	TextCaps validation set metrics				
			B-4	M	R	S	C
1	BUTD [4]	COCO	12.4	13.3	33.7	8.7	24.2
2	BUTD [4]	TextCaps	20.1	17.8	42.9	11.7	41.9
3	AoANet [18]	COCO	18.1	17.7	41.4	11.2	32.3
4	AoANet [18]	TextCaps	20.4	18.9	42.9	13.2	42.7
5	M4C-Captioner	COCO	12.3	14.2	34.8	9.2	30.3
6	M4C-Captioner	TextVQA	0.1	4.4	11.3	2.8	16.9
7	M4C-Captioner w/o OCRs	TextCaps	15.9	18.0	39.6	12.1	35.1
8	M4C-Captioner w/o copying	TextCaps	18.2	19.2	41.5	13.1	49.2
9	M4C-Captioner (OCR semantic)	TextCaps	21.4	20.4	44.0	14.1	69.0
10	M4C-Captioner (OCR spatial)	TextCaps	21.7	20.6	44.6	13.7	72.0
11	M4C-Captioner (OCR visual)	TextCaps	22.5	21.3	45.3	14.4	84.0
12	M4C-Captioner (OCR semantic & visual)	TextCaps	23.4	21.5	45.8	14.9	86.0
13	M4C-Captioner	TextCaps	**23.3**	**22.0**	**46.2**	**15.6**	**89.6**
14	M4C-Captioner (w/ GT OCRs)	TextCaps	26.0	23.2	47.8	16.2	104.3
#	Method	Trained on	TextCaps test set metrics					
			B-4	M	R	S	C	H
15	BUTD [4]	TextCaps	14.9	15.2	39.9	8.8	33.8	1.4
16	AoANet [18]	TextCaps	15.9	16.6	40.4	10.5	34.6	1.4
17	M4C-Captioner	TextCaps	**18.9**	**19.8**	**43.2**	**12.8**	**81.0**	**3.0**
18	M4C-Captioner (w/ GT OCRs)	TextCaps	21.3	21.1	45.0	13.5	97.2	3.4
19	Human	–	24.4	26.1	47.0	18.8	125.5	4.7

B-4: BLEU-4; M: METEOR; R: ROUGE_L; S: SPICE; C: CIDEr; H: human evaluation

4.2   Experimental Setup

[6]We follow the default configurations and hyper-parameters for training and evaluation of each baseline. For AoANet we use original implementation and feature extraction technique. For BUTD [4], we use the implementation and hyper-parameters from MMF [28,29]. For M4C-Captioner [17], we follow the same implementation details as used for TextVQA task [17]. We train both models for the same number of iterations on the TextCaps training set. During caption generation, we remove the <unk> token (for unknown words).

Datasets. We first evaluate the models trained using COCO dataset on TextCaps to demonstrate how existing datasets and models lack reading comprehension. Then we train and evaluate each baseline using TextCaps.

---

[6] Code for experiments is available at https://git.io/JJGuG.

Metrics. Apart from automatic captioning metrics including BLEU [25], METEOR [11], ROUGE_L [20], SPICE [3], and CIDEr [32], we also perform human evaluation. We collect 5000 human scores on a Likert scale from 1 to 5 for a random sample of 200 images and compute median score for each caption. Figure 7 shows that ranking of the sentence quality is the same as for automatic metrics. Moreover, all the metrics show very high correlation with human scores but CIDEr and METEOR have the highest. For comparison between different methods, we focus on the CIDEr, which puts more weight on informative n-grams in the captions (such as OCR tokens) and less weight on commonly occurring words with TF-IDF weighting.

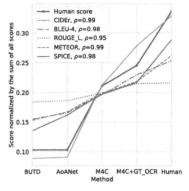

Fig. 7. Human evaluation in comparison to automatic metrics.

Fig. 8. Illustration of positive and negative predictions from different models on TextCaps validation set. For M4C-Captioner, square brackets indicate tokens copied from OCR. While most of the time OCR tokens are very important for correct copying of the text from the images, for common terms such as "pepsi" or "pence", the model sometimes prefer to select them from the vocabulary.

4.3   Results

TextCaps Dataset. It can be observed in results (Table 1) that the BUTD model trained on the COCO captioning dataset (line 1) achieves the lowest CIDEr score, indicating that it fails to describe text in the image. When trained on the TextCaps dataset (line 2), the BUTD model has higher scores as expected, since there is no longer a domain shift between training and evaluation. AoANet (line 3, 4), which is a stronger captioning model, outperforms BUTD but still cannot handle reading comprehension and largely underperforms M4C-Captioner. For the M4C-Captioner model, there is a large gap (especially in CIDEr scores) between training with and without OCR inputs (line 13 vs. 7). Moreover, "M4C-Captioner w/o copying" (line 8) is worse than the full model (line 13) but better than the more restricted "M4C-Captioner w/o OCRs" (line 7). The results indicate that it is important to both encode OCR features **and** be able to directly copy OCR tokens. We also observe (in line 13 vs. 9–12) that it is important for a model to use spatial, visual, and semantic features of OCR tokens together, especially in the complex combinations of OCR tokens where both spatial relation and semantics play an important role in finding a connection between words. However, on the test set, we still notice a large gap between the best machine performance (line 17) and the human performance (line 19) on this task. Also, using ground-truth OCRs (line 18) reduces this gap but still does not close it, suggesting that there is room for future improvement in both better reasoning and better text recognition.

a

M4C-Captioner (trained on COCO):
a red brick building with a sign on it
M4C-Captioner (COCO + TextCaps):
a red sign hangs above a brick building with the
name [frizerie] on it

b

M4C-Captioner (trained on COCO):
a blue sign is posted on the side of a building
M4C-Captioner (COCO + TextCaps):
a blue sign that says [uare] [alive] on it

c

M4C-Captioner (trained on COCO):
a large white board docked in a harbor
M4C-Captioner (COCO + TextCaps):
a boat that says [deffaner] on the side is
docked by the water

Fig. 9. Examples of M4C-Captioner's predictions on COCO data when trained on COCO and TextCaps. It can be observed that *despite of availability of OCR module in both cases*, using TextCaps pushes model to read the text. Square brackets indicate tokens copied from OCR.

Figure 8 shows qualitative examples from different methods. It can be seen that BUTD and M4C-Captioner without OCR inputs rarely mention text in the image except for common brand logos such as "pepsi" that are easy to recognize visually. On the other hand, the full M4C-Captioner approach learns to read text in the image and mention it in its generated captions.[7] Moreover,

---

[7] More predictions from M4C-Captioner are presented in Supplemental (Fig. F.1).

M4C-Captioner learns and recognizes relations between objects and is able to combine multiple OCR tokens into one complex description. For e.g., in Fig. 8(d) the model uses a OCR token to correctly name a player who is blocking another player; in Fig. 8(e) the model attempts to include and combine multiple tokens into a single message ("the *track* is *moved* in *Kenosha*" instead of "the word *moved*, the word *track*, and the word *Kenosha* are on the sign"). In Fig. 8(b) prediction is constructed fully from vocabulary, and even then the model counts similar objects and returns "two pepsi bottles" instead of "pepsi bottle and pepsi bottle". We also observe a large amount of mistakes in model predictions. Many mistakes are due to wrong scene understanding and object identification, which is a common problem in captioning algorithms. We also observe placing OCR tokens in the wrong object or semantic context in the caption (Fig. 8(c, e)), incorrect repetition of an OCR token in a caption (Fig. 8(a, e)), or insufficient use of them (Fig. 8(f)) by the model. Some mistakes (as "number *3*" in Fig. 8(d) are due to the errors of OCR detection algorith m and not the captioning model. This points to many potential directions for future development on this challenging generative task, which requires visual and textual understanding, requiring new model designs, conceptually different from previously existing captioning models.

Transferring to COCO. We further qualitatively show that when integrated with other datasets such as COCO [9], our dataset also enables text-based captioning on other datasets. In this setting, we experiment training M4C-Captioner (Table 1's best) on both TextCaps dataset and COCO dataset together. We balance the number of samples seen by the model from both COCO and TextCaps during training, and apply the trained model on the COCO validation set. COCO Captions mostly focus on visual objects but we show several examples where reading is necessary to describe the scene in Fig. 9. When trained on the union of our dataset and COCO, the M4C-Captioner learns to generate captions containing text present in the images. On the other hand, the same model only describes visual objects without mentioning any text when trained on COCO alone. Quantitative results can be found in Supplemental (Sec. C).

5   Conclusion

Image captioning with reading comprehension is a novel challenging task requiring models to read text in the image, recognize the image content, and comprehend both modalities jointly to generate a succinct image caption. To enable models to learn this ability and study this task in isolation, we collected TextCaps with 142k captions. The captions include a mix of objects and/or visual scene descriptions in relation to OCR tokens copied or rephrased from the images. In most cases, OCR tokens have to be copied and related to the visual scene, but sometimes the OCR tokens have to be understood, and sometimes spatial or visual reasoning between text and objects in the image is required, as shown in our ablation study. Our analysis also points out several challenges of this dataset: Different from other captioning datasets, nearly all our captions require integration of OCR tokens, many are unseen ("zero-shot"). In contrast to TextVQA datasets, TextCaps requires generating long sentences and involves new technical challenges, including many switches between OCR and vocabulary tokens.

We find that current state-of-the-art image captioning models cannot read when trained on existing captioning dataset. However, when adapting the recent M4C VQA model to our task and training it on our TextCaps dataset, we are able to generate impressive captions on both TextCaps and COCO, which involve copying multiple OCR tokens and correctly integrating them in the captions. Our human evaluation confirms the result of the automatic metrics with very high correlation, and also shows that human captions are still significantly better than automatically generated ones, leaving room for many advances in future work, including better semantic understanding between image and text content, missing reasoning capabilities, and reading long text or single characters.

We hope our dataset with challenge server, available at textvqa.org/textcaps, will encourage the community to design better image captioning models for this novel task and address its technical challenges, especially increasing their usefulness for assisting visually disabled people.

Acknowledgments. We would like to thank Guan Pang and Mandy Toh for helping us with OCR ground-truth collection. We would also like to thank Devi Parikh for helpful discussions and insights.

References

1. Agrawal, H., et al.: nocaps: novel object captioning at scale. In: International Conference on Computer Vision (ICCV) (2019)
2. Almazán, J., Gordo, A., Fornés, A., Valveny, E.: Word spotting and recognition with embedded attributes. IEEE Trans. Pattern Anal. Mach. Intell. **36**(12), 2552–2566 (2014)
3. Anderson, P., Fernando, B., Johnson, M., Gould, S.: SPICE: semantic propositional image caption evaluation. In: Leibe, B., Matas, J., Sebe, N., Welling, M. (eds.) ECCV 2016. LNCS, vol. 9909, pp. 382–398. Springer, Cham (2016). https://doi.org/10.1007/978-3-319-46454-1_24
4. Anderson, P., et al.: Bottom-up and top-down attention for image captioning and visual question answering. In: Proceedings of the IEEE Conference on Computer Vision and Pattern Recognition, pp. 6077–6086 (2018)
5. Bigham, J.P., et al.: Vizwiz: nearly real-time answers to visual questions. In: Proceedings of the 23nd Annual ACM Symposium on User Interface Software and Technology, pp. 333–342. ACM (2010)
6. Biten, A.F., et al.: Scene text visual question answering. arXiv preprint arXiv:1905.13648 (2019)
7. Bojanowski, P., Grave, E., Joulin, A., Mikolov, T.: Enriching word vectors with subword information. Trans. Assoc. Comput. Linguist. **5**, 135–146 (2017)
8. Borisyuk, F., Gordo, A., Sivakumar, V.: Rosetta: large scale system for text detection and recognition in images. In: ACM SIGKDD International Conference on Knowledge Discovery & Data Mining, pp. 71–79. ACM (2018)
9. Chen, X., et al.: Microsoft coco captions: data collection and evaluation server. arXiv preprint arXiv:1504.00325 (2015)

10. Chen, Y.C., et al.: Uniter: learning universal image-text representations. arXiv preprint arXiv:1909.11740 (2019)
11. Denkowski, M., Lavie, A.: Meteor universal: language specific translation evaluation for any target language. In: Proceedings of the Ninth Workshop on Statistical Machine Translation, pp. 376–380 (2014)
12. Devlin, J., Chang, M.W., Lee, K., Toutanova, K.: Bert: pre-training of deep bidirectional transformers for language understanding. In: NAACL-HLT (2019)
13. Goyal, P., Mahajan, D.K., Gupta, A., Misra, I.: Scaling and benchmarking self-supervised visual representation learning. In: International Conference on Computer Vision, abs/1905.01235 (2019)
14. Goyal, Y., Khot, T., Summers-Stay, D., Batra, D., Parikh, D.: VQA 2.0 evaluation. https://visualqa.org/evaluation.html
15. Gurari, D., Zhao, Y., Zhang, M., Bhattacharya, N.: Captioning images taken by people who are blind. arXiv preprint arXiv:2002.08565 (2020)
16. He, T., Tian, Z., Huang, W., Shen, C., Qiao, Y., Sun, C.: An end-to-end textspotter with explicit alignment and attention. In: Proceedings of the IEEE Conference on Computer Vision and Pattern Recognition, pp. 5020–5029 (2018)
17. Hu, R., Singh, A., Darrell, T., Rohrbach, M.: Iterative answer prediction with pointer-augmented multimodal transformers for TextVQA. arXiv preprint arXiv:1911.06258 (2019)
18. Huang, L., Wang, W., Chen, J., Wei, X.Y.: Attention on attention for image captioning. In: IEEE International Conference on Computer Vision, pp. 4634–4643 (2019)
19. Li, H., Wang, P., Shen, C.: Towards end-to-end text spotting with convolutional recurrent neural networks. In: Proceedings of the IEEE International Conference on Computer Vision, pp. 5238–5246 (2017)
20. Lin, C.Y.: Rouge: a package for automatic evaluation of summaries. In: Text summarization Branches Out, pp. 74–81 (2004)
21. Liu, X., Liang, D., Yan, S., Chen, D., Qiao, Y., Yan, J.: Fots: fast oriented text spotting with a unified network. In: Proceedings of the IEEE Conference on Computer Vision and Pattern Recognition, pp. 5676–5685 (2018)
22. Lu, J., Yang, J., Batra, D., Parikh, D.: Neural baby talk. In: Proceedings of the IEEE Conference on Computer Vision and Pattern Recognition, pp. 7219–7228 (2018)
23. Mishra, A., Shekhar, S., Singh, A.K., Chakraborty, A.: OCR-VQA: visual question answering by reading text in images. In: ICDAR (2019)
24. Ordonez, V., Kulkarni, G., Berg, T.L.: Im2Text: describing images using 1 million captioned photographs. In: Neural Information Processing Systems (NIPS) (2011)
25. Papineni, K., Roukos, S., Ward, T., Zhu, W.J.: Bleu: a method for automatic evaluation of machine translation. In: Proceedings of the 40th Annual Meeting on Association for Computational Linguistics, pp. 311–318. Association for Computational Linguistics (2002)
26. Ren, S., He, K., Girshick, R., Sun, J.: Faster R-CNN: towards real-time object detection with region proposal networks. In: Advances in Neural Information Processing Systems, pp. 91–99 (2015)
27. Sharma, P., Ding, N., Goodman, S., Soricut, R.: Conceptual captions: a cleaned, hypernymed, image alt-text dataset for automatic image captioning. In: Proceedings of the 56th Annual Meeting of the Association for Computational Linguistics (vol. 1: Long Papers), pp. 2556–2565 (2018)
28. Singh, A., et al.: MMF: a multimodal framework for vision and language research (2020). https://github.com/facebookresearch/mmf

29. Singh, A., et al.: Pythia-a platform for vision & language research. In: SysML Workshop, NeurIPS, vol. 2018 (2018)
30. Singh, A., et al.: Towards VQA models that can read. In: Proceedings of the IEEE Conference on Computer Vision and Pattern Recognition, pp. 8317–8326 (2019)
31. Smith, R.: An overview of the tesseract OCR engine. In: International Conference on Document Analysis and Recognition (ICDAR 2007), vol. 2, pp. 629–633. IEEE (2007)
32. Vedantam, R., Lawrence Zitnick, C., Parikh, D.: Cider: consensus-based image description evaluation. In: Proceedings of the IEEE Conference on Computer Vision and Pattern Recognition, pp. 4566–4575 (2015)
33. Vinyals, O., Fortunato, M., Jaitly, N.: Pointer networks. In: Advances in Neural Information Processing Systems, pp. 2692–2700 (2015)
34. Wang, A., Singh, A., Michael, J., Hill, F., Levy, O., Bowman, S.R.: Glue: a multi-task benchmark and analysis platform for natural language understanding. In: Proceedings of International Conference on Learning Representations (2019)
35. Young, P., Lai, A., Hodosh, M., Hockenmaier, J.: From image descriptions to visual denotations: new similarity metrics for semantic inference over event descriptions. Trans. Assoc. Comput. Linguist. **2**, 67–78 (2014)

It Is Not the Journey
But the Destination: Endpoint
Conditioned Trajectory Prediction

Karttikeya Mangalam[1]($\boxtimes$), Harshayu Girase[1], Shreyas Agarwal[1],
Kuan-Hui Lee[2], Ehsan Adeli[3], Jitendra Malik[1], and Adrien Gaidon[2]

[1] University of California, Berkeley, USA
mangalam@cs.berkeley.edu
[2] Toyota Research Institute, Ann Arbor, USA
[3] Stanford University, Stanford, USA

Abstract. Human trajectory forecasting with multiple socially interacting agents is of critical importance for autonomous navigation in human environments, e.g., for self-driving cars and social robots. In this work, we present Predicted Endpoint Conditioned Network (PECNet) for flexible human trajectory prediction. PECNet infers distant trajectory endpoints to assist in long-range multi-modal trajectory prediction. A novel non-local social pooling layer enables PECNet to infer diverse yet socially compliant trajectories. Additionally, we present a simple "truncation-trick" for improving diversity and multi-modal trajectory prediction performance. We show that PECNet improves state-of-the-art performance on the Stanford Drone trajectory prediction benchmark by ∼20.9% and on the ETH/UCY benchmark by ∼40.8% (Code available at project homepage: https://karttikeya.github.io/publication/htf/).

Keywords: Multimodal trajectory prediction · Social pooling

1 Introduction

Predicting the movement of dynamic objects is a central problem for autonomous agents, be it humans, social robots [1], or self-driving cars [2]. Anticipation by prediction is indeed required for smooth and safe path planning in a changing environment. One of the most frequently encountered dynamic objects are humans. Hence, predicting human motion is of paramount importance for navigation, planning, human-robot interaction, and other critical robotic tasks. However, predicting human motion is nuanced, because humans are not inanimate entities evolving under Newtonian laws [3]. Rather, humans have the will to exert causal forces to change their motion and constantly adjust their paths as they

Electronic supplementary material The online version of this chapter (https://doi.org/10.1007/978-3-030-58536-5_45) contains supplementary material, which is available to authorized users.

© Springer Nature Switzerland AG 2020
A. Vedaldi et al. (Eds.): ECCV 2020, LNCS 12347, pp. 759–776, 2020.
https://doi.org/10.1007/978-3-030-58536-5_45

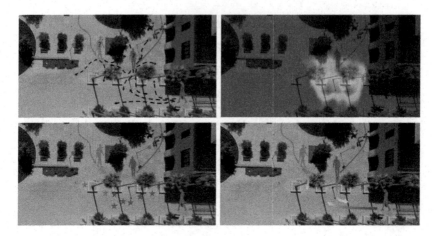

Fig. 1. Imitating the Human Path Planning Process. Our proposed approach to model pedestrian trajectory prediction (top left) breaks down the task in two steps: (a) inferring the local endpoint distribution (top right), and then (b) conditioning on sampled future endpoints (bottom left) for jointly planning socially compliant trajectories for all the agents in the scene (bottom right).

navigate around obstacles to achieve their goals [4]. This complicated planning process is partially internal, and thus makes predicting human trajectories from observations challenging. Hence, a multitude of aspects should be taken into account beyond just past movement history, for instance latent predetermined goals, other moving agents in the scene, and social behavioral patterns.

In this work, we propose to address **human trajectory prediction by modeling intermediate stochastic goals we call endpoints**. We hypothesize that three separate factors interact to shape the trajectory of a pedestrian. First, we posit that pedestrians have some understanding of their long-term desired destination. We extend this hypothesis to sub-trajectories, i.e. the pedestrian has one or multiple intermediate destinations, which we define as potential endpoints of the local trajectory. These sub-goals can be more easily correlated with past observations to predict likely next steps and disentangle potential future trajectories.

Second, the pedestrian plans a trajectory to reach one of these sub-goals, taking into account the present scene elements. Finally, as the agent go about executing a plan, the trajectory gets modified to account for other moving agents, respecting social norms of interaction.

Following the aforementioned intuition, we propose to decompose the trajectory prediction problem into two sub-problems that also motivate our proposed architecture (Fig. 1). First, given the previous trajectories of the humans in the scene, we propose to estimate a latent belief distribution modeling the pedestrians' possible endpoints. Using this estimated latent distribution, we sample plausible endpoints for each pedestrian based on their observed trajectory. A socially-compliant future trajectory is then predicted, conditioned not only on the pedestrian and their immediate neighbors' histories (observed trajectories) but also everybody's estimated endpoints.

In conclusion, our contribution in this work is threefold. **First**, we propose a socially compliant, endpoint conditioned variational auto-encoder that closely imitates the multi-modal human trajectory planning process. **Second**, we propose a novel self-attention based social pooling layer that generalizes previously proposed social pooling mechanisms. **Third**, we show that our model can predict stable and plausible intermediate goals that enable setting a new state-of-the-art on several trajectory prediction benchmarks, improving by **20.9%** on SDD [5] & **40.8%** on ETH [6] & UCY [7].

2   Related Work

There have been many previous studies [8] on how to forecast pedestrians' trajectories and predict their behaviors. Several previous works propose to learn statistical behavioral patterns from the observed motion trajectories for future trajectory prediction [9–18]. Since then, many studies have developed models to account for agent interactions that may affect the trajectory—specifically, through scene and/or social information. Recently, there has been a significant focus on multi-modal trajectory prediction to capture different possible future trajectories given the past. There has also been some research on goal-directed path planning, which consider pedestrians' goals while predicting a path.

2.1   Context-Based Prediction

Many previous studies have imported environment semantics, such as crosswalks, road, or traffic lights, to their proposed trajectory prediction scheme. Kitani et al. [19] encode agent-space interactions by a Markov Decision Process (MDP) to predict potential trajectories for an agent. Ballan et al. [20] leverage a dynamic Bayesian network to construct motion dependencies and patterns from training data and transferred the trained knowledge to testing data. With the great success of the deep neural network, the Recurrent Neural Network (RNN) has become a popular modeling approach for sequence learning. Kim et al. [21] train a RNN combining multiple Long Short-term Memory (LSTM) units to predict the location of nearby cars. These approaches incorporate rich environment cues from the RGB image of the scene for pedestrians' trajectory forecasting.

Behaviour of surrounding dynamic agents is also a crucial cue for contextual trajectory prediction. Human behavior modeling studied from a crowd perspective, *i.e.*, how a pedestrian interacts with other pedestrians, has also been studied widely in human trajectory prediction literature. Traditional approaches use *social forces* [22–25] to capture pedestrians' trajectories towards their goals with attractive forces, while avoiding collisions in the path with repulsive forces. These approaches require hand-crafted rules and features, which are usually complicated and insufficiently robust for complicated high-level behavior modeling. Recently, many studies applied Long Short Term Memory (LSTM [26]) networks to model trajectory prediction with the social cues. Alahi et al. [27] propose a Social LSTM which learns to predict a trajectory with joint interactions. Each pedestrian is modeled by an individual LSTM, and LSTMs are connected with their nearby individual LSTMs to share information from the hidden state.

2.2 Multimodal Trajectory Prediction

In [28,29], the authors raise the importance of accounting for the inherent multimodal nature of human paths *i.e.*, given pedestrians' past history, there are many plausible future paths they can take. This shift of emphasis to plan for multiple future paths has led many recent works to incorporate multi-modality in their trajectory prediction models. Lee et al. [28] propose a conditional variational autoencoder (CVAE), named DESIRE, to generate multiple future trajectories based on agent interactions, scene semantics and expected reward function, within a sampling-based inverse optimal control (IOC) scheme. In [29], Gupta et al. propose a Generative Adversarial Network (GAN) [30] based framework with a novel social pooling mechanism to generate multiple future trajectories in accordance to social norms. In [31], Sadeghian et al. also propose a GAN based framework named SoPhie, which utilizes path history of all the agents in a scene and the scene context information. SoPhie employs a social attention mechanism with physical attention, which helps in learning social information across the agent interactions. However, these socially-aware approaches do not take into account the pedestrians' ultimate goals, which play a key role in shaping their movement in the scene. A few works also approach trajectory prediction via an inverse reinforcement learning (IRL) setup. Zou et al. [32] applies Generative Adversarial Imitation Learning (GAIL) [33] for trajectory prediction, named Social-Aware GAIL (SA-GAIL). With IRL, the authors model the human decision-making process more closely through modeling humans as agents with states (past trajectory history) and actions (future position). SA-GAIL generates socially acceptable trajectories via a learned reward function.

2.3 Conditioned-on-Goal

Goal-conditioned approaches are regarded as inverse planning or *prediction by planning* where the approach learns the final intent or goal of the agent before predicting the full trajectory. In [34], Rehder *et al.* propose a particle filtering based approach for modeling destination conditioned trajectory prediction and use explicit Von-Mises distribution based probabilistic framework for prediction. Later in a follow-up work, [35] Rehder *et al.* further propose a deep learning based destination estimation approach to tackle intention recognition and trajectory prediction simultaneously. The approach uses fully Convolutional Neural Networks (CNN) to construct the path planning towards some potential destinations which are provided by a recurrent Mixture Density Network (RMDN). While both the approaches make an attempt for destination conditioned prediction, a fully probabilistic approach trains poorly due to unstable training and updates. Further, they ignore the presence of other pedestrians in the scene which is key for predicting shorter term motions which are missed by just considering the environment. Rhinehart et al. [36] propose a goal-conditioned multi-agent forecasting approach named PRECOG, which learns a probabilistic forecasting model conditioned on drivers' actions intents such as ahead, stop, etc. However, their approach is designed for vehicle trajectory prediction, and thus conditions on semantic goal states. In our work, we instead propose to utilize destination position for pedestrian trajectory prediction.

In [37], Li *et al.* posit a Conditional Generative Neural System (CGNS), the previous established state-of-the-art result on the ETH/UCY dataset. They propose to use variational divergence minimization with soft-attention to predict feasible multi-modal trajectory distributions. Even more recently, Bhattacharyya *et al.* [38] propose a conditional flow VAE that proposed a general normalizing flow for structured sequence prediction and applies it to the problem of trajectory prediction. Concurrent to our work, Deo *et al.* [39] propose P2TIRL, a Maximum Entropy Reinforcement Learning based trajectory prediction module over a discrete grid. The work [38] shares state-of-the-art with [39] on the Stanford Drone Dataset (SDD) with the TrajNet [40] split. However, these works fail to consider the human aspect of the problem, such as interaction with other agents. We compare our proposed PECNet with all three of the above works on both the SDD & ETH/UCY datasets.

3   Proposed Method

In this work, we aim to tackle the task of human trajectory prediction by reasoning about all the humans in the scene jointly while also respecting social norms. Suppose a pedestrian p^k enters a scene $\mathcal{I}$. Given the previous trajectory of p for past t_p steps, as a sequence of coordinates $\mathcal{T}_p^k := \{\mathbf{u}^k\}_{i=1}^{t_p} = \{(x^k, y^k)\}_{i=1}^{t_p}$, the problem requires predicting the future position of p^k on $\mathcal{I}$ for next t_f steps, $\mathcal{T}_f^k := \{\mathbf{u}^k\}_{i=t_p+1}^{t_p+t_f} = \{(x, y)\}_{i=t_p+1}^{t_p+t_f}$.

As mentioned in Sect. 1, we break the problem into two daisy chained steps. First, we model the sub-goal of p^k, i.e. the last observed trajectory points of p^k say, $\mathcal{G}^k = \mathbf{u}^k|_{t_p+t_f}$ as a representation of the predilection of p^k to go its pre-determined route. This sub-goal, also referred to as the endpoint of the trajectory, the pedestrian's desired end destination for the current sequence. Then in the second step, we jointly consider the past histories $\{\mathcal{T}_p^k\}_{k=1}^{\alpha}$ of all the pedestrians $\{p^k\}_{k=1}^{\alpha}$ present in the scene and their estimated endpoints $\{\mathcal{G}^k\}_{k=1}^{\alpha}$ for predicting socially compliant future trajectories $\mathcal{T}_f^k$. In the rest of this section we describe in detail, our approach to achieve this, using the endpoint estimation VAE for sampling the future endpoints $\mathcal{G}$ and a trajectory prediction module to use the sampled endpoints $\hat{\mathcal{G}}^k$ to predict $\mathcal{T}_f$.

3.1   Endpoint VAE

We propose to model the predilection of the pedestrian as a sub-goal endpoint $\mathcal{G} := \mathbf{u}_{t_f} = (x_{t_f}, y_{t_f})$ which is the last observed trajectory point for pedestrian p^k. First, we infer a distribution on $\mathcal{G}$ based on the previous location history $\mathcal{T}_i$ of p^k using the Endpoint VAE.

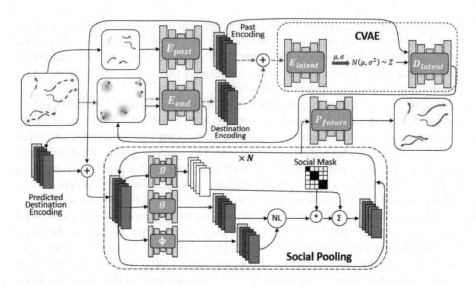

Fig. 2. Architecture of PECNet: PECNet uses past history, $\mathcal{T}_i$ along with ground truth endpoint $\mathcal{G}_c$ to train a VAE for multi-modal endpoint inference. Ground-truth endpoints are denoted by $\star$ whereas $\mathbf{x}$ denote the sampled endpoints $\hat{\mathcal{G}}_c$. The sampled endpoints condition the social-pooling & predictor networks for multi-agent multimodal trajectory forecasting. Red connections denote the parts utilized only during training. Shades of the same color denote spatio-temporal neighbours encoded with the block diagonal social mask in social pooling module. Further Details in Sect. 3.1. (Color figure online)

As illustrated in Fig. 2, we extract the previous history $\mathcal{T}_i^k$ and the ground truth endpoint $\mathcal{G}^k$ for all pedestrian p_k in the scene. We encode the past trajectory $\mathcal{T}_i^k$ of all p_k independently using a past trajectory encoder $\mathbf{E}_{past}$. This yields us $\mathbf{E}_{past}(\mathcal{T}_i)$, a representation of the motion history. Similarly, the future endpoint $\mathcal{G}^k$ is encoded with an Endpoint encoder $\mathbf{E}_{end}$ to produce $\mathbf{E}_{end}(\mathcal{G}^k)$ independently for all k. These representations are concatenated together and passed into the latent encoder $\mathbf{E}_{latent}$ which produces parameter $(\boldsymbol{\mu}, \boldsymbol{\sigma})$ for encoding the latent variable $z = \mathcal{N}(\boldsymbol{\mu}, \boldsymbol{\sigma})$ of the VAE. Finally, we sample possible latent future endpoints from $\mathcal{N}(\boldsymbol{\mu}, \boldsymbol{\sigma})$, concatenate it with $\mathbf{E}_{past}(\mathcal{T}_i)$ for past context and decode using the latent decoder $\mathbf{D}_{latent}$ to yield our guesses for $\hat{\mathcal{G}}^k$. Since the ground truth $\mathcal{G}^k$ belongs to the future, and is unavailable at test time, during evaluation we sample z from $\mathcal{N}(0, \sigma_T \mathbf{I})$, concatenate with $\mathbf{E}_{past}(\mathcal{T}_i)$ (as done in training) and then use the learned $\mathbf{D}_{latent}$ to estimate the future $\hat{\mathcal{G}}^k$. This is illustrated in Fig. 2 where the red connections are only used in the training and not in the evaluation phase.

Truncation Trick: In [41], Brock *et al.* introduce the 'Truncation Trick' as a method of trade-off between the fidelity and variety of samples produced by the generator in BigGAN. In this work, we propose an analogous trick for evaluation phase in multi-modal trajectory forecasting where the variance of the latent endpoint sampling distribution is changed according to the number of samples (K) allowed for multi-modal prediction. In a situation requiring few shot multi-modal prediction, such as under computation constraints, where only a few samples $(K = 1, 2$ or 3) are permissible, we propose to use $\sigma_T = 1$ and truncate the sampling distribution at $\pm c\sqrt{K-1}$. In contrast, in situations where a high number of predictions are to be generated (such as $K = 20$, a standard setting on benchmarks) we propose to use $\sigma_T > 1$ with no truncation. We posit that this procedure allows simple adjustment of prediction diversity in favor of overall performance for different K, thereby providing a simple method of achieving good performance in all settings without requiring any retraining.

3.2   Endpoint Conditioned Trajectory Prediction

Using the sampled estimate of the *endpoints* $\hat{\mathcal{G}}$ from Endpoint VAE, we employ the endpoint encoder $\mathbf{E}_{end}$ once again (within the same forward pass) to obtain encodings for the sampled endpoints $\mathbf{E}_{end}(\hat{\mathcal{G}}^k)$. This is used along with prediction network to *plan* the path $\mathcal{T}_f$ starting to $\mathcal{G}$ thereby predicting the future path.

Note that, another design choice could have been that even during training, the ground truth $\mathbf{E}_{end}(\mathcal{G}^k)$ are used to predict the future $\mathcal{T}_f$. This seems reasonable as well since it provides cleaner, less noisy signals for the downstream social pooling & prediction networks while still training the overall module end to end (because of coupling through $\mathbf{E}_{past}$). However, such a choice will decouple training of the Endpoint VAE (which would then train only with KL Divergence and AWL loss, refer Sect. 3.3) and social pooling network (which would then train only with ATL loss, refer Sect. 3.3) leading to inferior performance empirically.

The sampled endpoints' representations $\mathbf{E}_{end}(\hat{\mathcal{G}}^k)$ are then concatenated with corresponding $\mathbf{E}_{past}(\mathcal{T}_i)$ (as in Sect. 3.1) and passed through N rounds of social pooling using a social pooling mask $\mathbf{M}$ for all the pedestrians in the scene jointly. The social pooling mask $\mathbf{M}$ is $\alpha \times \alpha$ block diagonal matrix denoting the social neighbours for all $\{p_i\}_{i=1}^{\alpha}$ pedestrians in the scene. Mathematically,

$$\mathbf{M}[i,j] = \begin{cases} 0 & \text{if } \min_{1 \leq m,n \leq t_p} \|\mathbf{u}_m^i - \mathbf{u}_n^j\|_2 > t_{dist} \\ 0 & \text{if } \min_{1 \leq m \leq t_p} |\mathcal{F}(\mathbf{u}_0^i) - \mathcal{F}(\mathbf{u}_m^j)| * \min_{1 \leq m \leq t_p} |\mathcal{F}(\mathbf{u}_m^i) - \mathcal{F}(\mathbf{u}_0^j)|) > 0 \quad (1) \\ 1 & \text{otherwise} \end{cases}$$

where $\mathcal{F}(.)$ denoted the actual frame number the trajectory was observed at. Intuitively, $\mathbf{M}$ defines the spatio-temporal neighbours of each pedestrian p_i using proximity threshold t_{dist} for distance in space and ensure temporal overlap. Thus, the matrix $\mathbf{M}$ encodes crucial information regarding social locality of different trajectories which gets utilized in attention based pooling as described below.

Social Pooling: Given the concatenated past history and sampled way-point representations $X_k^{(1)} = (\mathbf{E}_{past}(\mathcal{T}_k^k), \mathbf{E}_{end}(\hat{\mathcal{G}}^k))$ we do N rounds of social pooling where the $(i+1)$th round of pooling recursively updates the representations $X_k^{(i)}$ from the last round according to the non-local attention mechanism [42]:

$$X_k^{(i+1)} = X_k^{(i)} + \frac{1}{\sum\limits_{j=1}^{\alpha} \mathbf{M}_{ij} \cdot e^{\phi(X_k^{(i)})^T \theta(X_j^{(i)})}} \sum_{j=1}^{\alpha} \mathbf{M}_{ij} \cdot e^{\phi(X_k^{(i)})^T \theta(X_j^{(i)})} \mathbf{g}(X_k^{(i)}) \quad (2)$$

where $\{\theta, \phi\}$ are encoders of X_k to map to a learnt latent space where the representation similarity between p_i and p_j trajectories is calculated using the embedded gaussian $\exp(\phi(X_k)^T \theta(X_j))$ for each round of pooling. The social mask, $\mathbf{M}$ is used point-wise to allow pooling only on the spatio-temporal neighbours masking away other pedestrians in the scene. Finally, $\boldsymbol{g}$ is a transformation encoder for X_k used for the weighted sum with all other neighbours. The whole procedure, after being repeated N times yields $X_k^{(N)}$, the pooled prediction features for each pedestrian with information about the past positions and future destinations of all other neighbours in the scene.

Our proposed social pooling is a novel method for extracting relevant information from the neighbours using non-local attention. The proposed social non local pooling (S-NL) method is permutation *invariant* to pedestrian indices as a useful inductive bias for tackling the social pooling task. Further, we argue that this method of learnt social pooling is more robust to social neighbour mis-identification such as say, mis-specified distance (t_{dist}) threshold compared to previously proposed method such as max-pooling [29], sorting based pooling [31] or rigid grid-based pooling [27] since a learning based method can ignore spurious signals in the social mask $\mathbf{M}$.

The pooled features $X_k^{(N)}$ are then passed through the prediction network $\mathbf{P}_{future}$ to yield our estimate of rest of trajectory $\{\mathbf{u}^k\}_{k=t_p+1}^{t_p+t_f}$ which are concatenated with sampled endpoint $\hat{\mathcal{G}}$ yields $\hat{\mathcal{T}}_f$. The complete network is trained end to end with the losses described in the next subsection.

3.3   Loss Functions

For training the entire network end to end we use the loss function,

$$\mathcal{L}_{PECNet} = \lambda_1 \underbrace{D_{KL}(\mathcal{N}(\boldsymbol{\mu}, \boldsymbol{\sigma}) \| \mathcal{N}(0, \mathbf{I}))}_{KL\ Div\ in\ latent\ space} + \lambda_2 \underbrace{\|\hat{\mathcal{G}}_c - \mathcal{G}_c\|_2^2}_{AEL} + \underbrace{\|\hat{\mathcal{T}}_f - \mathcal{T}_f\|^2}_{ATL} \quad (3)$$

where the KL divergence term is used for training the Variational Autoencoder, the Average endpoint Loss (AEL) trains $\mathbf{E}_{end}$, $\mathbf{E}_{past}$, $\mathbf{E}_{latent}$ and $\mathbf{D}_{latent}$ and the Average Trajectory Loss (ATL) trains the entire module together.

4   Experiments

4.1   Datasets

Stanford Drone Dataset: Stanford Drone Dataset [5] is a well established benchmark for human trajectory prediction in bird's eye view. The dataset consists of 20 scenes captured using a drone in top down view around the university

Table 1. Comparison of our method against several recently published multi-modal baselines and previous state-of-the-art method (denoted by *) on the Stanford Drone Dataset [5]. '-S' & '-TT' represents ablations of our method without social pooling & truncation trick. We report results for in pixels for both $K = 5$ & 20 and for several other K in Fig. 5. † denotes concurrent work. Lower is better.

	SoPhie	S-GAN	DESIRE	CF-VAE*	P2TIRL†	SimAug†	O-S-TT	O-TT	Ours	PECNet (ours)
K	20	20	5	20	20	20	20	20	5	20
ADE	16.27	27.23	19.25	12.60	12.58	10.27	10.56	10.23	12.79	**9.96**
FDE	29.38	41.44	34.05	22.30	22.07	19.71	16.72	16.29	25.98	**15.88**

campus containing several moving agents like humans and vehicles. It consists of over 11, 000 unique pedestrians capturing over 185, 000 interactions between agents and over 40, 000 interactions between the agent and scene [5]. We use the standard test train split as used in [29,31,39] and other previous works.

ETH/UCY: Second is the ETH [6] and UCY [7] dataset group, which consists of five different scenes – ETH & HOTEL (from ETH) and UNIV, ZARA1, & ZARA2 (from UCY). All the scenes report the position of pedestrians in world-coordinates and hence the results we report are in metres. The scenes are captured in unconstrained environments with few objects blocking pedestrian paths. Hence, scene constraints from other physical non-animate entities is minimal. For bench-marking, we follow the commonly used leave one set out strategy i.e., training on four scenes and testing on the fifth scene [29,31,37].

4.2   Implementation Details

All the sub-networks used in proposed module are Multi-Layer perceptrons with ReLU non-linearity. Network architecture for each of the sub-networks are mentioned in Fig. 3. The entire network is trained end to end with the $\mathcal{L}_{E\text{-}VAE}$ loss using an ADAM optimizer with a batch size of 512 and learning rate of 3×10^{-4} for all experiments. For the loss coefficient weights, we set $\lambda_1 = \lambda_2 = 1$. We use $N = 3$ rounds of social pooling for Stanford Drone Dataset and $N = 1$ for ETH &

	Network Architecture
$\mathbf{E}_{way}$	$2 \rightarrow 8 \rightarrow 16 \rightarrow 16$
$\mathbf{E}_{past}$	$16 \rightarrow 512 \rightarrow 256 \rightarrow 16$
$\mathbf{E}_{latent}$	$32 \rightarrow 8 \rightarrow 50 \rightarrow 32$
$\mathbf{D}_{latent}$	$32 \rightarrow 1024 \rightarrow 512 \rightarrow 1024 \rightarrow 2$
ϕ, θ	$32 \rightarrow 512 \rightarrow 64 \rightarrow 128$
g	$32 \rightarrow 512 \rightarrow 64 \rightarrow 32$
$\mathbf{P}_{predict}$	$32 \rightarrow 1024 \rightarrow 512 \rightarrow 256 \rightarrow 22$

Fig. 3. Network architecture details for all the sub-networks used in the module.

UCY scenes. Using social masking, we perform the forward pass in mini-batches instead of processing all the pedestrians in the scene in a single forward pass (to aboid memory overflow) constraining all the neighbours of a pedestrian to be in the same mini-batch.

Metrics: For prediction evaluation, we use the Average Displacement Error (ADE) and the Final Displacement Error (FDE) metrics which are commonly used in literature [25,27,29,37]. ADE is the average ℓ_2 distance between the predictions and the ground truth future and FDE is the ℓ_2 distance between the predicted and ground truth at the last observed point. Mathematically,

$$ADE = \frac{\sum_{j=t_i+1}^{t_p+t_f+1} \|\hat{\mathbf{u}}_j - \mathbf{u}_j\|_2}{t_f} \qquad FDE = \|\hat{\mathbf{u}}_{t_p+t_f+1} - \mathbf{u}_{t_p+t_f+1}\|_2 \qquad (4)$$

where $\mathbf{u}_j$, $\hat{\mathbf{u}}_j$ are the ground truth and our estimated position of the pedestrian at future time step j respectively.

Baselines: We compare our PECNet against several published baselines including previous state-of-the-art methods briefly described below.

- Social GAN (S-GAN) [29]: Gupta *et al.* propose a multi-modal human trajectory prediction GAN trained with a variety loss to encourage diversity.
- SoPhie [31]: Sadeghian *et al.* propose a GAN employing attention on social and physical constraints from the scene to produce human-like motion.
- CGNS [37]: Li *et al.* posit a Conditional Generative Neural System (CGNS) that uses conditional latent space learning with variational divergence minimization to learn feasible regions to produce trajectories. They also established the previous state-of-the-art result on the ETH/UCY datasets.
- DESIRE [28]: Lee *et al.* propose an Inverse optimal control based trajectory planning method that uses a refinement structure for predicting trajectories.
- CF-VAE [38]: Recently, a conditional normalizing flow based VAE proposed by Bhattacharyya *et al.* pushes the state-of-the-art on SDD further. Notably, their method also does not also rely on the RGB scene image.
- P2TIRL [39]: A concurrent work by Deo *et al.* proposes a method for trajectory forecasting using a grid based policy learned with maximum entropy inverse reinforcement learning policy. They closely tie with the previous state-of-the-art [38] in ADE/FDE performance.
- SimAug [43]: More recently, a concurrent work by Liang *et al.* proposes to use additional 3D multi-view simulation data adversarially, for novel camera view adaptation. [43] improves upon the P2TIRL as well, with performance close to PECNet's base model. However our best model (with pooling and truncation) still achieves a better ADE/FDE performance.
- Ours-TT: This represents an ablation of our method without using the truncation trick. In other words, we set σ_T to be identically 1 for all K settings. Truncation trick ablations with different K are shown in Fig. 5 & Table 1.
- Ours-S-TT: This represents an ablation of our method without using both the social pooling module and the truncation trick *i.e.* the base PECNet. We set $\sigma_T = 1$ and $N = 0$ for the number of rounds of social pooling and directly transmit the representations to $\mathbf{P}_{future}$, the prediction sub-network.

4.3   Quantitative Results

In this section, we compare and discuss our method's performance against above mentioned baselines on the ADE & FDE metrics.

Stanford Drone Dataset: Table 1 shows the results of our proposed method against the previous baselines & state-of-the-art methods. Our proposed method achieves a superior performance compared to the previous state-of-the-art [38,39] on both ADE & FDE metrics by a significant margin of **20.9%**. Even without using the proposed social pooling module & truncation trick (OUR-S-TT), we achieve a very good performance (10.56 ADE), underlining the importance of future endpoint conditioning in trajectory prediction. As observed by the difference in performance between Ours-S-TT and Our-TT, the social pooling module also plays a crucial role, boosting performance by 0.33 ADE ($\sim$2.1%). Note that, while both P2TIRL [39] & SimAug [43] are concurrent works, we compare with their methods' performance as well in Table 1 for experimental comprehensiveness. All reported results averaged for 100 separate trials.

Table 2. Quantitative results for various previously published methods and state-of-the-art method (denoted by *) on commonly used trajectory prediction datasets. Both ADE and FDE are reported in metres in world coordinates. 'Our-S-TT' represents ablation of our method without social pooling & truncation trick.

	S-GAN		SoPhie		CGNS*		S-LSTM		Ours - S - TT		PECNet (ours)	
	ADE	FDE	ADE	FDE	ADE	FDE	ADE	FDE	ADE	FDE	ADE	FDE
ETH	0.81	1.52	0.70	1.43	0.62	1.40	1.09	2.35	0.58	0.96	**0.54**	**0.87**
HOTEL	0.72	1.61	0.76	1.67	0.70	0.93	0.79	1.76	0.19	0.34	**0.18**	**0.24**
UNIV	0.60	1.26	0.54	1.24	0.48	1.22	0.67	1.40	0.39	0.67	**0.35**	**0.60**
ZARA1	0.34	0.69	0.30	0.63	0.32	0.59	0.47	1.00	0.23	0.39	**0.22**	**0.39**
ZARA2	0.42	0.84	0.38	0.78	0.35	0.71	0.56	1.17	0.24	0.35	**0.17**	**0.30**
AVG	0.58	1.18	0.54	1.15	0.49	0.97	0.72	1.54	0.32	0.54	**0.29**	**0.48**

ETH/UCY: Table 2 shows the results for evaluation of our proposed method on the ETH/UCY scenes. We follow the leave-one-out evaluation protocol with $K = 20$ as in CGNS [37]/Social-GAN [29]. All reported numbers are *without* the truncation trick. In this setting too, we observe that our method outperforms previously proposed methods, including the previous state-of-the-art [37]. We push the state-of-the-art on average by $\sim$**40.8%** with the effect being the most on HOTEL (74.2%) and least on ETH (12.9%). Also, without the social pooling & truncation trick (OUR-S-TT) the performance is still superior to the state-of-the-art by 34.6%, underlining the usefulness of conditioning on the endpoint in PECNet.

Conditioned Way-Point Positions & Oracles: For further evaluation of our model, we condition on future trajectory points other than the last observed point which we refer to as *way-points*. Further, to decouple the errors in inferring the conditioned position from errors in predicting a path to that position, we use a destination (endpoint) oracle. The destination oracle provides ground truth information of the conditioned position to the model, which uses it to predict the rest of the trajectory. All of the models, with and without the destination oracle are trained from scratch for each of the conditioning positions.

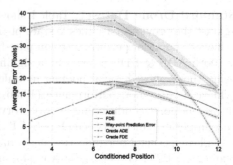

Fig. 4. Conditioned way-point positions & oracles: We evaluate the performance of the proposed method against the choice of future conditioning position on ADE & FDE metrics. Further, we evaluate the performance of a destination oracle version of the model that receives perfect information on conditioned position for predicting rest of the trajectory.

Referring to Fig. 4, we observe several interesting and informative trends that support our earlier hypotheses. (A) As a sanity check, we observe that as we condition on positions further into the future, the FDE for both the Oracle model & the proposed model decrease with a sharp trend after the 7th future position. This is expected since points further into the future provide more information for the final observed point. (B) The ADE error curves for both the oracle and the proposed model have the same decreasing trend albeit with a gentler slope than FDE because the error in predicting the other points (particularly the noisy points in the middle of the trajectory) decreases the gradient. (C) Interestingly, our model's ADE and FDE is not significantly different from that of the Oracle model for points close in the future and the error in the two models are approximately the same until about the 7th future position. This suggests that till around the middle of the future, the conditioned way-points do not hold significant predictive power on the endpoint and hence using our noisy guesses vs. the oracle's ground truth for their position does not make a difference.

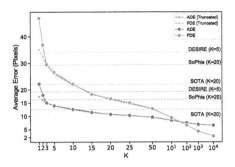

Fig. 5. Performance across K: ADE & FDE performance of our method against number of samples used for evaluation. Several previous baselines are mentioned as well with their number of samples used. Our method significantly outperforms the state-of-the-art reaching their performance with much lesser number of samples & performing much better with same number of samples as theirs ($K = 20$).

Way-Point Prediction Error: The way-point position error is the ℓ_2 distance between the prediction of location of the conditioned position and its ground truth location (in the future). Referring to Fig. 4, we observe an interesting trend in the way-point error as we condition on points further into the future. The way-point prediction error increases at the start which is expected since points further into the future have a higher variance. However, after around the middle (7th point) the error plateaus and then even slightly decreases. This lends support to our hypothesis that pedestrians, having predilection towards their destination, exert their will towards it. Hence, *predicting the last observed way-point allows for lower prediction error than* way-points in the middle! This in a nutshell, confirms the motivation of this work.

Effect of Number of Samples (K): All the previous works use $K = 20$ samples (except DESIRE which uses $K = 5$) to evaluate the multi-modal predictions for metrics ADE & FDE. Referring to Fig. 5, we see the expected decreasing trend in ADE & FDE with time as K increases. Further, we observe that our proposed method achieves the same error as the previous works with much smaller K. Previous state-of-the-art achieves 12.58 [39] ADE using $K = 20$ samples which is matched by PECNet at half the number of samples, $K = 10$. This further lends support to our hypothesis that conditioning on the inferred way-point significantly reduces the modeling complexity for multi-modal trajectory forecasting, providing a better estimate of the ground truth.

Lastly, as K grows large ($K \to \infty$) we observe that the FDE slowly gets closer to 0 with more number of samples, as the ground truth $\mathcal{G}_c$ is eventually found. However, the ADE error is still large (6.49) because of the errors in the rest of the predicted trajectory. This is in accordance with the observed ADE (8.24) for the oracle conditioned on the last observed point (*i.e.* 0 FDE error) in Fig. 4.

Design Choice for VAE: We also evaluate our design choice of using the inferred future way-points $\hat{\mathcal{G}}_c$ for training subsequent modules (social pooling & prediction) instead of using the ground truth $\mathcal{G}_c$. As mentioned in Sect. 3.2, this is also a valid choice for training PECNet end to end. Empirically, we find that such a design achieves 10.87 ADE and 17.03 FDE. This is worse ($\sim$8.8%) than using $\hat{\mathcal{G}}_c$ which motivates our design choice for using $\hat{\mathcal{G}}_c$ (Sect. 3.2).

Truncation Trick: Fig. 5 shows the improvements from the truncation trick for an empirically chosen hyperparameter $c \approx 1.2$. As expected, small values of K gain the most from truncation, with the performance boosting from 22.85 ADE (48.8 FDE) to 17.29 ADE (35.12 FDE) for $K = 1$ ($\sim$24.7%).

4.4 Qualitative Results

In Fig. 6, we present several visualizations of PECNet predictions. As shown, PECNet can produce diverse predictions taking into account the past motion history & inferred endpoints. In Fig. 7, we present animations of several socially compliant predictions. The visualizations show that along with good metric performance PECNet also performs rich multi-modal multi-agent forecasting.

Fig. 6. Visualizing Multimodality: We show visualizations for some multi-modal and diverse predictions produced by PECNet. White represents the past 3.2 s while red & cyan represents predicted & ground truth future respectively over next 4.8 s. Predictions capture a wide-range of plausible trajectory behaviours while discarding improbable ones like, endpoints opposite to pedestrian's direction of motion. (Color figure online)

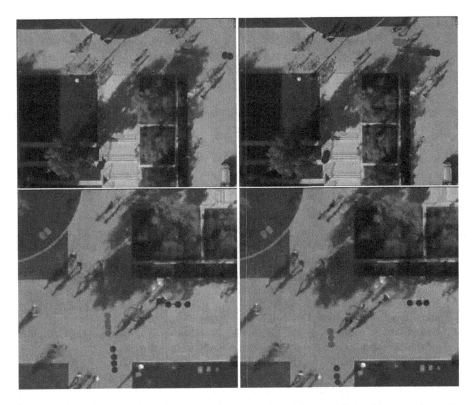

Fig. 7. Social Interaction Animation: Circles show the past 3.2 s & stars show the future 19.2 s (top) & 4.8 s (bottom) for both ground truth (left) & predictions (right). On top, PECNet neatly captures the purple pedestrian's overtake of the red pedestrian predicting a smooth cut-in trajectory while blue's trajectory remains unaffected (a neighbour in social mask **M**). At the bottom, the blue pedestrian avoids collision at the intersection by speeding up the trajectory that was originally linear which PECNet accurately captures (see Supplementary material). Animation best viewed in *Adobe Acrobat Reader*. (Color figure online)

5   Conclusion

In this work we present PECNet, a Pedestrian endpoint conditioned trajectory prediction network. We show that PECNet predicts rich and diverse multi-modal socially compliant trajectories across a variety of scenes. Furthermore, we perform extensive ablations on several design choices such as endpoint conditioning position, number of samples, and choice of training signal to pinpoint the performance gains from PECNet. We also introduce the "truncation trick" [41] for trajectory prediction, a simple method for boosting trajectory prediction accuracy in the few-shots regime. Finally, we benchmark PECNet across multiple datasets including Stanford Drone Dataset [5], ETH [6], and UCY [7], in all of which PECNet achieved the state-of-the-art.

References

1. Bennewitz, M., Burgard, W., Thrun, S.: Learning motion patterns of persons for mobile service robots. In: Proceedings of the 2002 IEEE International Conference on Robotics and Automation (Cat. No. 02CH37292), vol. 4, pp. 3601–3606. IEEE (2002)
2. Thrun, S., Burgard, W., Fox, D.: Probabilistic Robotics. Intelligent Robotics and Autonomous Agents Series. MIT Press, Cambridge (2005)
3. Baker, C.L., Saxe, R., Tenenbaum, J.B.: Action understanding as inverse planning. Cognition 113(3), 329–349 (2009)
4. Ziebart, B.D.,et al.: Planning-based prediction for pedestrians. In: 2009 IEEE/RSJ International Conference on Intelligent Robots and Systems, pp. 3931–3936. IEEE (2009)
5. Robicquet, A., Sadeghian, A., Alahi, A., Savarese, S.: Learning social etiquette: human trajectory understanding in crowded scenes. In: Leibe, B., Matas, J., Sebe, N., Welling, M. (eds.) ECCV 2016. LNCS, vol. 9912, pp. 549–565. Springer, Cham (2016). https://doi.org/10.1007/978-3-319-46484-8_33
6. Pellegrini, S., Ess, A., Van Gool, L.: Improving data association by joint modeling of pedestrian trajectories and groupings. In: Daniilidis, K., Maragos, P., Paragios, N. (eds.) ECCV 2010. LNCS, vol. 6311, pp. 452–465. Springer, Heidelberg (2010). https://doi.org/10.1007/978-3-642-15549-9_33
7. Lerner, A., Chrysanthou, Y., Lischinski, D.: Crowds by example. In: Computer Graphics Forum, vol. 26, pp. 655–664. Wiley Online Library (2007)
8. Rudenko, A., Palmieri, L., Herman, M., Kitani, K.M., Gavrila, D.M., Arras, K.O.: Human motion trajectory prediction: a survey. arXiv e-prints (2019)
9. Kruse, E., Wahl, F.M.: Camera-based observation of obstacle motions to derive statistical data for mobile robot motion planning. In: Proceedings of the 1998 IEEE International Conference on Robotics and Automation (Cat. No. 98CH36146), vol. 1, pp. 662–667. IEEE (1998)
10. Liao, L., Fox, D., Hightower, J., Kautz, H., Schulz, D.: Voronoi tracking: location estimation using sparse and noisy sensor data. In: Proceedings of the 2003 IEEE/RSJ International Conference on Intelligent Robots and Systems (IROS 2003),(Cat. No. 03CH37453), vol. 1, pp. 723–728. IEEE (2003)
11. Bennewitz, M., Burgard, W., Cielniak, G., Thrun, S.: Learning motion patterns of people for compliant robot motion. Int. J. Robot. Res. 24(1), 31–48 (2005)
12. Tay, M.K.C., Laugier, C.: Modelling smooth paths using Gaussian processes. In: Laugier, C., Siegwart, R. (eds.) Field and Service Robotics. STAR, vol. 42, pp. 381–390. Springer, Heidelberg (2008). https://doi.org/10.1007/978-3-540-75404-6_36
13. Käfer, E., Hermes, C., Wöhler, C., Ritter, H., Kummert, F.: Recognition of situation classes at road intersections. In: 2010 IEEE International Conference on Robotics and Automation, pp. 3960–3965. IEEE (2010)
14. Aoude, G., Joseph, J., Roy, N., How, J.: Mobile agent trajectory prediction using Bayesian nonparametric reachability trees. In: Infotech@ Aerospace 2011, p. 1512 (2011)
15. Keller, C.G., Gavrila, D.M.: Will the pedestrian cross? A study on pedestrian path prediction. IEEE Trans. Intell. Transp. Syst. 15(2), 494–506 (2013)
16. Goldhammer, M., Doll, K., Brunsmann, U., Gensler, A., Sick, B.: Pedestrian's trajectory forecast in public traffic with artificial neural networks. In: 2014 22nd International Conference on Pattern Recognition, pp. 4110–4115. IEEE (2014)

17. Xiao, S., Wang, Z., Folkesson, J.: Unsupervised robot learning to predict person motion. In: 2015 IEEE International Conference on Robotics and Automation (ICRA), pp. 691–696. IEEE (2015)
18. Kucner, T.P., Magnusson, M., Schaffernicht, E., Bennetts, V.H., Lilienthal, A.J.: Enabling flow awareness for mobile robots in partially observable environments. IEEE Robot. Autom. Lett. **2**(2), 1093–1100 (2017)
19. Kitani, K.M., Ziebart, B.D., Bagnell, J.A., Hebert, M.: Activity forecasting. In: Fitzgibbon, A., Lazebnik, S., Perona, P., Sato, Y., Schmid, C. (eds.) ECCV 2012. LNCS, vol. 7575, pp. 201–214. Springer, Heidelberg (2012). https://doi.org/10.1007/978-3-642-33765-9_15
20. Ballan, L., Castaldo, F., Alahi, A., Palmieri, F., Savarese, S.: Knowledge transfer for scene-specific motion prediction. In: Leibe, B., Matas, J., Sebe, N., Welling, M. (eds.) ECCV 2016. LNCS, vol. 9905, pp. 697–713. Springer, Cham (2016). https://doi.org/10.1007/978-3-319-46448-0_42
21. Kim, B.D., Kang, C.M., Kim, J., Lee, S.H., Chung, C.C., Choi, J.W.: Probabilistic vehicle trajectory prediction over occupancy grid map via recurrent neural network. In: 2017 IEEE 20th International Conference on Intelligent Transportation Systems (ITSC), pp. 399–404. IEEE (2017)
22. Helbing, D., Molnar, P.: Social force model for pedestrian dynamics. Phys. Rev. E **51**(5), 4282 (1995)
23. Mehran, R., Oyama, A., Shah, M.: Abnormal crowd behavior detection using social force model. In: 2009 IEEE Conference on Computer Vision and Pattern Recognition, pp. 935–942. IEEE (2009)
24. Yamaguchi, K., Berg, A.C., Ortiz, L.E., Berg, T.L.: Who are you with and where are you going? In: CVPR 2011, pp. 1345–1352. IEEE (2011)
25. Alahi, A., Ramanathan, V., Fei-Fei, L.: Socially-aware large-scale crowd forecasting. In: Proceedings of the IEEE Conference on Computer Vision and Pattern Recognition, pp. 2203–2210 (2014)
26. Hochreiter, S., Schmidhuber, J.: Long short-term memory. Neural Comput. **9**(8), 1735–1780 (1997)
27. Alahi, A., Goel, K., Ramanathan, V., Robicquet, A., Fei-Fei, L., Savarese, S.: Social LSTM: human trajectory prediction in crowded spaces. In: Proceedings of the IEEE Conference on Computer Vision and Pattern Recognition, pp. 961–971 (2016)
28. Lee, N., Choi, W., Vernaza, P., Choy, C.B., Torr, P.H.S., Chandraker, M.: Desire: Distant future prediction in dynamic scenes with interacting agents. In: Proceedings of the IEEE Conference on Computer Vision and Pattern Recognition, pp. 336–345 (2017)
29. Gupta, A., Johnson, J., Fei-Fei, L., Savarese, S., Alahi, A.: Social GAN: socially acceptable trajectories with generative adversarial networks. In: Proceedings of the IEEE Conference on Computer Vision and Pattern Recognition, pp. 2255–2264 (2018)
30. Goodfellow, I., et al.: Generative adversarial nets. In: Advances in Neural Information Processing Systems, pp. 2672–2680 (2014)
31. Sadeghian, A., Kosaraju, V., Sadeghian, A., Hirose, N., Rezatofighi, H., Savarese, S.: SoPhie: an attentive GAN for predicting paths compliant to social and physical constraints. In: Proceedings of the IEEE Conference on Computer Vision and Pattern Recognition, pp. 1349–1358 (2019)
32. Zou, H., Su, H., Song, S., Zhu, J.: Understanding human behaviors in crowds by imitating the decision-making process. In: Thirty-Second AAAI Conference on Artificial Intelligence (2018)

33. Ho, J., Ermon, S.: Generative adversarial imitation learning. In: Advances in Neural Information Processing Systems, pp. 4565–4573 (2016)
34. Rehder, E., Kloeden, H.: Goal-directed pedestrian prediction. In: Proceedings of the IEEE International Conference on Computer Vision Workshops, pp. 50–58 (2015)
35. Rehder, E., Wirth, F., Lauer, M., Stiller, C.: Pedestrian prediction by planning using deep neural networks. In: 2018 IEEE International Conference on Robotics and Automation (ICRA), pp. 1–5. IEEE (2018)
36. Rhinehart, N., McAllister, R., Kitani, K., Levine, S.: PRECOG: PREdiction conditioned on goals in visual multi-agent settings. arXiv preprint arXiv:1905.01296 (2019)
37. Li, J., Ma, H., Tomizuka, M.: Conditional generative neural system for probabilistic trajectory prediction. arXiv preprint arXiv:1905.01631 (2019)
38. Bhattacharyya, A., Hanselmann, M., Fritz, M., Schiele, B., Straehle, C.-N.: Conditional flow variational autoencoders for structured sequence prediction. arXiv preprint arXiv:1908.09008 (2019)
39. Deo, N., Trivedi, M.M.: Trajectory forecasts in unknown environments conditioned on grid-based plans. arXiv preprint arXiv:2001.00735 (2020)
40. Sadeghian, A., Kosaraju, V., Gupta, A., Savarese, S., Alahi, A.: TrajNet: towards a benchmark for human trajectory prediction. arXiv preprint (2018)
41. Brock, A., Donahue, J., Simonyan, K.: Large scale GAN training for high fidelity natural image synthesis. arXiv preprint arXiv:1809.11096 (2018)
42. Wang, X., Girshick, R., Gupta, A., He, K.: Non-local neural networks. In: Proceedings of the IEEE Conference on Computer Vision and Pattern Recognition, pp. 7794–7803 (2018)
43. Liang, J., Jiang, L., Hauptmann, A.: SimAug: learning robust representations from 3D simulation for pedestrian trajectory prediction in unseen cameras (2020)

Learning What to Learn for Video Object Segmentation

Goutam Bhat[1(✉)], Felix Järemo Lawin[2], Martin Danelljan[1],
Andreas Robinson[2], Michael Felsberg[2], Luc Van Gool[1], and Radu Timofte[1]

[1] CVL, ETH Zürich, Zürich, Switzerland
goutam.bhat@vision.ee.ethz.ch
[2] CVL, Linköping University, Linköping, Sweden

Abstract. Video object segmentation (VOS) is a highly challenging problem, since the target object is only defined by a first-frame reference mask during inference. The problem of how to capture and utilize this limited information to accurately segment the target remains a fundamental research question. We address this by introducing an end-to-end trainable VOS architecture that integrates a differentiable few-shot learner. Our learner is designed to predict a powerful parametric model of the target by minimizing a segmentation error in the first frame. We further go beyond the standard few-shot learning paradigm by learning what our target model should learn in order to maximize segmentation accuracy. We perform extensive experiments on standard benchmarks. Our approach sets a new state-of-the-art on the large-scale YouTube-VOS 2018 dataset by achieving an overall score of 81.5, corresponding to a 2.6% relative improvement over the previous best result. The code and models are available at https://github.com/visionml/pytracking.

1 Introduction

Semi-supervised Video Object Segmentation (VOS) is the problem of performing pixel-wise classification of a set of target objects in a video sequence. With numerous applications in e.g. autonomous driving [30,31], surveillance [7,9] and video editing [24], it has received significant attention in recent years. VOS is an extremely challenging problem, since the target objects are only defined by a reference segmentation in the first video frame, with no other prior information assumed. The VOS method therefore must utilize this very limited information about the target in order to perform segmentation in the subsequent frames.

While most state-of-the-art VOS methods employ similar image feature extractors and segmentation decoders, a number of approaches [15,25,29,33]

G. Bhat and F. J. Lawin—Both authors contributed equally.

Electronic supplementary material The online version of this chapter (https://doi.org/10.1007/978-3-030-58536-5_46) contains supplementary material, which is available to authorized users.

© Springer Nature Switzerland AG 2020
A. Vedaldi et al. (Eds.): ECCV 2020, LNCS 12347, pp. 777–794, 2020.
https://doi.org/10.1007/978-3-030-58536-5_46

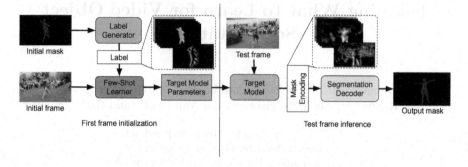

Fig. 1. An overview of our VOS approach. Given the annotated first frame, our few-shot learner optimizes the parameters of a target model which is tasked with predicting an encoding of the target mask (left). In subsequent test frames, the mask encoding output by the target model is utilized by the segmentation decoder to generate the final segmentation (right). Crucially, the label for the few-shot learner is generated by our label generator which is trained end-to-end jointly with the decoder. This allows us to learn what the target model should learn to output to the decoder in order to maximize segmentation accuracy.

have been proposed to utilize the reference frame annotation to perform segmentation. A promising direction is to employ feature matching techniques [14,15,25,33] in order to compare the reference frame regions with new images to segment. Such feature matching layers greatly benefit from their efficiency and differentiability. This allows the design of fully end-to-end trainable architectures, which has been shown to be important for segmentation performance [15,25,33]. However, in order to generalize to novel objects, feature matching techniques rely on a powerful and generic feature embedding, which can be difficult to learn. Instead of only relying on a pre-trained embedding, we investigate an alternative direction, where target-specific network parameters are learned during *inference*, in order to better integrate object appearance information.

We propose a novel VOS method, based on an effective few-shot learner that captures object information in a compact parametric target model. Given a test frame, our target model first predicts an intermediate representation of the target mask, which is then input to a segmentation decoder that generates the final prediction. To achieve a powerful model of the target object from the limited first frame annotation, our few-shot learner is designed to explicitly optimize an error between the target model prediction and a ground truth reference. Owing to the efficiency and differentiability of the few-shot learner, our approach can perform the inference-time learning without compromising the end-to-end training capability. Compared to the embedding based approaches, the inference-time learning in our approach provides greater adaptivity and generalizability to novel objects and scenarios.

We further address the problem of what intermediate mask representation the target model should be trained to predict in order to maximize segmentation accuracy. The standard optimization-based few-shot learning strategy forces the target model to learn to only generate an object mask output. However, directly learning to predict the segmentation mask from a single sample is difficult. More importantly, this approach limits the target-specific information sent to the segmentation decoder to be a single channel mask. To address this important issue, we further propose to *learn what to learn*. That is, our approach learns to generate a multi-channel mask encoding which is used by the few-shot learner as labels to train the target model. This enables our target model to provide richer target-specific information to the segmentation decoder in the test frames. Furthermore, in order to guide the learner to focus on the most crucial aspect of the target, we also learn to predict spatial importance weights for different elements in the few-shot learning objective. Since our optimization-based learner is differentiable, all modules in our architecture can be trained end-to-end by maximizing segmentation accuracy on annotated VOS videos. An overview of our video object segmentation approach is shown in Fig. 1.

Contributions: Our main contributions are listed as follows. (**i**) We propose a novel VOS architecture, based on an optimization-based few-shot learner. (**ii**) Our few-shot learner predicts the target model parameters in an efficient and differentiable manner, enabling end-to-end training. (**iii**) We go beyond standard few-shot learning approaches to further learn what the target model should learn in order to maximize segmentation accuracy. (**iv**) We utilize our learned mask encoding to design a light-weight bounding box initialization module, allowing our approach to generate segmentation masks using only a reference bounding box as input.

We perform comprehensive experiments on the YouTube-VOS [39] and DAVIS [27] benchmarks. Our approach sets a new state-of-the-art on the large-scale YouTube-VOS 2018 dataset, achieving an overall score of 81.5. We further provide detailed ablative analyses, showing the impact of each component in the proposed method.

2   Related Work

In recent years, progress within video object segmentation has surged, leading to rapid performance improvements. Benchmarks such as DAVIS [27] and YouTube-VOS [39] have had a significant impact on this development.

Target Models in VOS: Early works mainly adapted semantic segmentation networks to the VOS task through online fine-tuning [5,13,22,28,38]. However, this strategy easily leads to overfitting to the initial target appearance and impractically long run-times. More recent methods [14,18,24,25,33,35,37] therefore integrate target-specific appearance models into the segmentation architecture. In addition to improved run-times, many of these methods can also benefit from full end-to-end learning, which has been shown to have a crucial impact on performance [15,25,33]. Generally, these works train a target-agnostic segmentation decoder that is conditioned on a target model. The latter integrates information about the target object, deduced from the initial image-mask pair. The target model predicts target-specific information for the test frame, which is then provided to the target-agnostic segmentation decoder to obtain the final prediction. Crucially, in order to achieve end-to-end training of the entire network, the target model needs to be differentiable.

While most VOS methods share similar feature extractors and segmentation decoders, several different strategies for encoding and exploiting the reference frame target information have been proposed. In RGMP [24], a representation of the target is generated by encoding the reference frame. This representation is then concatenated with the current-frame features, before being input to the segmentation decoder. Similarly, OSNM [40] predicts an attention vector from the reference frame and ground-truth target mask, which combined with a spatial guidance map is used to segment the target. The approach in [18] extends RGMP to jointly process multiple targets using an instance specific attention generator. In [15], a light-weight generative model is learned from embedded features corresponding to the initial target labels. The generative model is then used to classify features from the incoming frames. The target models in [14,33] directly store foreground features and classify pixels in the incoming frames through feature matching. The recent STM approach [25] performs feature matching within a space-time memory network. It implements a read operation, which retrieves information from the encoded memory through an attention mechanism. This information is then sent to the segmentation decoder to predict the target mask. The method [37] predicts template correlation filters given the input target mask. Target classification is then performed by applying the correlation filters on the test frame. Lastly, the recent method [29] trains a target model consisting of a two-layer neural network using the Conjugate Gradient method.

Meta-learning for VOS: Since the VOS task itself includes a few-shot learning problem, it can be addressed with techniques developed for meta-learning [3,10,17]. A few recent attempts [1,20] follow this direction. The method [1] learns a classifier using k-means clustering of segmentation features in the training frame. In [20], the final layer of a segmentation network is predicted by closed-form ridge regression [3], using the reference example pair. Meta-learning based techniques have been more commonly adopted in the related field of visual tracking [4,6,8,26]. The method in [26] performs gradient based adaptation to the current target, while [6] learns a target specific feature space online which is combined with a Siamese-based matching network. The recent work [4] proposes an optimization-based meta-learning strategy, where the target model directly generates the output classifications scores. In contrast to these previous approaches, we integrate a differentiable optimization-based few-shot learner to capture target information for the VOS problem. Furthermore, we go beyond standard few-shot and meta-learning techniques by learning what the target model should learn in order to generate accurate segmentations.

3   Method

In this section, we present our method for video object segmentation (VOS). First, we describe our few-shot learning formulation for VOS in Sect. 3.1. In Sect. 3.2 we then describe our approach to learn what the few-shot learner should learn. Section 3.3 details our target module and the internal few-shot learner. Our segmentation architecture is described next in Sect. 3.4. The inference and training procedures are detailed in Sects. 3.5 and 3.6, respectively. Finally, Sect. 3.7 describes how our approach can be easily extended to perform VOS with only a bounding box initialization.

3.1   Video Object Segmentation as Few-Shot Learning

In VOS, the target object is only defined by a reference target mask given in the first frame. No other prior information about the test object is assumed. The VOS method therefore needs to exploit the given first-frame annotation in order to segment the target in each subsequent frame. To address this core problem, we first consider a general class of VOS architectures formulated as $S_\theta(I, T_\tau(I))$, where θ denotes the learnable parameters. The network S_θ takes the current image I along with the output of a target model T_τ. While S_θ itself is target-agnostic, it is conditioned on T_τ, which integrates information about the target object, encoded in its parameters τ. The target model generates a target-aware output that is used by S_θ to predict the final segmentation. The target model parameters τ are obtained from the initial image I_0 and its given mask y_0, which defines the target object itself. We denote this as a function $\tau = A_\theta(I_0, y_0)$. The key challenge in this VOS formulation is in the design of T_τ and A_θ.

We note that the pair (I_0, y_0) in the above formulation constitutes a training sample for learning to segment the given target. However, this training sample is only given during inference. Hence, a few-shot learning problem naturally arises within VOS. We adopt this view to develop our approach. In relation to few-shot learning, A_θ constitutes the internal learning method, which generates the parameters τ of the target model T_τ from a single example pair (I_0, y_0). While there exist a diverse set of few-shot learning methodologies, we aim to find the target model parameters τ that minimizes a supervised learning objective ℓ,

$$\tau = A_\theta(x_0, y_0) = \arg\min_{\tau'} \ell(T_{\tau'}(x_0), y_0). \tag{1}$$

Here, the target model T_τ is learned to output the segmentation of the target object in the initial frame. In general, we operate on a deep representation of the input image $x = F_\theta(I)$, generated by e.g. a ResNet architecture. Given a new frame I during inference, the object is segmented as $S_\theta(I, T_\tau(F_\theta(I)))$. In other words, the target model is applied to the new frame to generate a first segmentation. This output is further refined by S_θ, which can additionally integrate powerful pre-learned knowledge from large VOS datasets.

The main advantage of the optimization-based formulation (1) is that the target model parameters are predicted by directly minimizing the segmentation error in the first frame. This ensures robust segmentation prediction in the coming frames, since consecutive video frames are highly-correlated. For practical purposes, however, the target model prediction (1) also needs to be efficient. Further, to enable end-to-end training of the entire VOS architecture, we wish the learner A_θ to be *differentiable*. While this is challenging in general, different strategies have been proposed in the literature [3,4,17], mostly in the context of meta-learning. We detail the employed approach in Sect. 3.3. In the next section, we first address another fundamental limitation of the formulation in Eq. (1).

3.2 Learning What to Learn

In the approach discussed in the previous section, the target model T_τ learns to predict an initial segmentation mask of the target object from the first frame. This mask is then refined by the network S_θ, which possesses strong learned segmentation priors. However, S_θ is not limited to operate on an approximate target mask in order to perform target-conditional segmentation. In contrast, any information that alleviates the task of the network S_θ to identify and accurately segment the target object is beneficial. Predicting only a single-channel mask thus severely limits the amount of target-specific information that can be passed to the network S_θ. Ideally, the target model should predict multi-channel activations which can provide strong target-aware cues in order to guide the network S_θ to generate accurate segmentations. However, this is not possible in the standard few-shot learning setting (1), since the output of the target model T_τ is directly defined by the available ground-truth mask y_0. In this work, we address this issue by learning what our internal few-shot learner should learn.

Instead of directly employing the first-frame mask y_0 as labels in our few-shot learner, we propose to *learn* these labels. To this end, we introduce a trainable label generator $E_\theta(y)$ that takes the ground-truth mask y as input and predicts the labels for the few-shot learner. The target model is thus predicted as,

$$\tau = A_\theta(x_0, y_0) = \arg\min_{\tau'} \ell\big(T_{\tau'}(x_0), E_\theta(y_0)\big). \tag{2}$$

Unlike in (1), the labels $E_\theta(y_0)$ generated by encoding the ground-truth mask y_0 can be multi-dimensional, allowing the target model T_τ to predict a richer target mask representation in the test frames.

The formulation (2) assigns equal weight to all elements in the few-shot learning loss $\ell\big(T_\tau(x_0), E_\theta(y_0)\big)$. However, this might not be optimal for maximizing the final segmentation accuracy. For instance, it is often beneficial to assign higher weights to target regions in case of small objects, to account for an imbalanced training set. Similarly, it might be beneficial to assign lower weights to ambiguous regions such as object boundaries, and let the segmentation network S_θ handle them. We allow such flexibility in our loss by introducing a weight predictor module $W_\theta(y)$. Similar to E_θ, it takes the ground-truth mask y as input and predicts the importance weight for each element in the loss $\ell\big(T_\tau(x_0), E_\theta(y_0)\big)$. Thus, our weight predictor can guide the few-shot learner to focus on the most crucial aspects of the ground truth label $E_\theta(y)$.

We have not yet fully addressed the question of *how* to learn the label generator E_θ, and the weight predictor W_θ. Ideally, we wish to train all parameters θ in our segmentation architecture in an end-to-end manner on annotated VOS datasets. This requires back-propagating the error measured between the final segmentation output $\tilde{y}_t = S_\theta(I_t, T_\tau(F_\theta(I_t)))$ and the ground truth y_t on a test frame I_t. However, this is feasible only if the internal learner (2) is efficient and differentiable w.r.t. both the underlying features x *and* the parameters of the label generator E_θ and weight predictor W_θ. We address these open questions in the next section, to achieve an efficient and end-to-end trainable VOS architecture.

3.3 Few-Shot Learner

In this section, we detail our target model T_τ and the internal few-shot learner A_θ. The target model $T_\tau : \mathbb{R}^{H \times W \times C} \rightarrow \mathbb{R}^{H \times W \times D}$ is trained to map C-dimensional deep features x to a D-dimensional encoding of the target mask with the same spatial resolution $H \times W$. We require T_τ to be efficient and differentiable. To ensure this, we employ a linear target model $T_\tau(x) = x * \tau$, where $\tau \in \mathbb{R}^{K \times K \times C \times D}$ constitutes the weights of a convolutional layer with kernel size K. While such a target model is simple, it operates on high dimensional deep feature maps. Consequently, it is capable of predicting a rich encoding of the target mask, leading to improved segmentation performance, as shown in our experiments (see Sect. 4). Moreover, while a more complex target module has larger capacity, it is also prone to overfitting and is computationally more costly to learn.

The parameters of the target model are obtained using our internal few-shot learner A_θ by minimizing the squared error between the output of the target model $T_\tau(x)$ and the generated ground-truth labels $E_\theta(y)$, weighted by the element-wise importance weights $W_\theta(y)$,

$$L(\tau) = \frac{1}{2} \sum_{(x_t, y_t) \in \mathcal{D}} \left\| W_\theta(y_t) \cdot \left(T_\tau(x_t) - E_\theta(y_t) \right) \right\|^2 + \frac{\lambda}{2} \|\tau\|^2. \tag{3}$$

Here, $\mathcal{D} = \{(x_t, y_t)\}_{t=0}^{Q-1}$ is a set of feature-mask pairs (x_t, y_t) of size Q. While it usually contains a single ground-truth annotated frame, it is often useful to include additional frames by, for instance, self-annotating new images in the video. The scalar λ is a learned regularization parameter.

As a next step, we design a differentiable and efficient few-shot learner that minimizes (3) as $\tau = A_\theta(\mathcal{D}) = \arg\min_{\tau'} L(\tau')$. We note that (3) is a convex quadratic objective in τ. Thus, it has a well-known closed-form solution, which can be expressed in either primal or dual form. However, both options lead to computations that are intractable when aiming for acceptable frame-rates, requiring extensive matrix multiplications and solutions to linear systems. Moreover, these methods cannot directly utilize the convolutional structure of the problem. Instead we find an approximate solution of (3) by applying steepest descent iterations, previously also used in [4]. Given a current estimate τ^i, it finds the step-length α^i that minimizes the loss in the gradient direction $\alpha^i = \arg\min_\alpha L(\tau^i - \alpha g^i)$. Here, $g^i = \nabla L(\tau^i)$ is the gradient of (3) at τ^i. The optimization iteration can then be expressed as,

$$\tau^{i+1} = \tau^i - \alpha^i g^i, \quad \alpha^i = \frac{\|g^i\|^2}{\sum_t \|W_\theta(y_t) \cdot (x_t * g^i)\|^2 + \lambda \|g^i\|^2},$$

$$g^i = \sum_t x_t *^{\mathrm{T}} \left(W_\theta^2(y_t) \cdot \left(x_t * \tau^i - E_\theta(y_t) \right) \right) + \lambda \tau^i. \quad (4)$$

Here, $*^{\mathrm{T}}$ denotes the transposed convolution operation. A detailed derivation is provided in the supplementary material.

Note that all computations in (4) are easily implemented using standard neural network operations. Since all operations are differentiable, the resulting target model parameters τ^i after i iterations are differentiable w.r.t. all network parameters θ. Our internal few-shot learner is implemented as a network module $A_\theta(\mathcal{D}, \tau^0) = \tau^N$, that performs N iterations of steepest descent (3), starting from a given initialization τ^0. Thanks to the rapid convergence of steepest descent, we only need to perform a handful of iterations during training and inference. Moreover, our optimization-based formulation allows the target model parameters τ to be efficiently updated with new samples by simply adding them to $\mathcal{D}$ and applying a few iterations (4), starting from the current parameters $\tau^0 = \tau$.

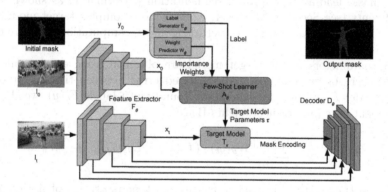

Fig. 2. An overview of our segmentation architecture. It contains a few-shot learner, which generates a parametric target model T_τ from the initial frame annotation. The parameters τ are computed by minimizing the loss (3), using the labels predicted by E_θ. The elements of the loss are weighted using the importance weights predicted by W_θ. In the incoming frames, the target model predicts the mask encoding, which is processed along with image features by our decoder D_θ to produce the final mask.

3.4   Video Object Segmentation Architecture

Our VOS method is implemented as a single end-to-end network, illustrated in Fig. 2. It is composed of a deep feature extractor F_θ, label generator E_θ, loss weight predictor W_θ, target model T_τ, few-shot learner A_θ and the segmentation decoder D_θ. As previously mentioned, θ denotes the network parameters learned during the offline training, while τ are the target model parameters that are *predicted* by the few-shot learner module during inference. The following sections describes the individual modules. More details are provided in the supplement.

Feature Extractor F_θ: We employ a ResNet-50 network as backbone feature extractor F_θ. Features from F_θ are input to both the decoder module D_θ and the target model T_τ. For the latter, we employ the third residual block, which has a spatial stride of $s = 16$. These features are first fed through an additional conv. layer that reduces the dimension to $C = 512$, before given to T_τ.

Few-Shot Label Generator E_θ: Our label generator E_θ predicts the ground truth label for the few-shot learner by encoding the input target mask. The latter is mapped to the resolution of the deep features as $E_\theta : \mathbb{R}^{sH \times sW \times 1} \rightarrow \mathbb{R}^{H \times W \times D}$, where H, W and D are the height, width and dimensionality of the target model features and s is the feature stride. We implement the proposed mask encoder E_θ as a conv-net, decomposed into a generic mask feature extractor for processing the input mask y and a prediction layer for generating the final label.

Weight Predictor W_θ: The weight predictor $W_\theta : \mathbb{R}^{sH \times sW \times 1} \rightarrow \mathbb{R}^{H \times W \times D}$ generates weights for the internal loss (3). It is implemented as a conv-net that takes the target mask y as input, similar to E_θ. In our implementation, W_θ shares the mask feature extractor with E_θ.

Target Model T_τ and Few-Shot Learner A_θ: We implement our target model T_τ as convolutional filter with a kernel size of $K = 3$. The number of output channels D is set to 16. Our few-shot learner A_θ (see Sect. 3.3) predicts the target model parameters τ in the first frame by applying steepest descent iterations (4). On subsequent test frames, we apply the predicted target model $T_\tau(x)$ to obtain target mask encodings, which are then provided to the segmentation decoder.

Segmentation Decoder D_θ: This module takes the output of the target model T_τ along with backbone features as input and predicts the final segmentation mask. Our approach can in principle be combined with any decoder architecture. For simplicity, we employ a decoder network similar to the one used in [29]. We adapt this network to process a multi-channel target mask encoding as input.

3.5   Inference

In this section, we describe our inference procedure. Given a test sequence $\mathcal{V} = \{I_t\}_{t=0}^Q$, along with the first frame annotation y_0, we first create an initial training set $\mathcal{D}_0 = \{(x_0, y_0)\}$ for the few-shot learner, consisting of the single sample pair. Here, $x_0 = F_\theta(I_0)$ is the feature map extracted from the first frame. The few-shot learner then predicts the parameters $\tau_0 = A_\theta(\mathcal{D}_0, \tau^0)$ of the target model by minimizing the internal loss (3). We set the initial estimate of the target model $\tau^0 = 0$ to all zeros. Note that the ground-truth $E_\theta(y_0)$ and importance weights $W_\theta(y_0)$ for the minimization problem (3) are predicted by our network.

The learned model τ_0 is then applied on the subsequent test frame I_1 to obtain a mask encoding $T_{\tau_0}(x_1)$. This encoding is then processed by the decoder module, along with the image features, to generate the mask prediction $\tilde{y}_1 = D_\theta(x_1, T_{\tau_0}(x_1))$. In order to adapt to the changes in the scene, we further update our target model using the information from the processed frames. This is achieved by extending the training set $\mathcal{D}_0$ with the new sample $(x_1, \tilde{y}_1)$, where the predicted mask $\tilde{y}_1$ serves as the pseudo-label for the frame I_1. The extended training set $\mathcal{D}_1$ is then used to obtain new target model parameters $\tau_1 = A_\theta(\mathcal{D}_1, \tau_0)$. Note that instead of predicting τ_1 from scratch, our optimization-based learner allows us to efficiently update the previous target model τ_0. Specifically, we apply additional $N_{\text{update}}^{\text{inf}}$ steepest-descent iterations (4) with the new training set $\mathcal{D}_1$. The updated T_{τ_1} is then applied on the next frame I_2. This process is repeated till the end of the sequence.

Details: Our few-shot learner A_θ employs $N_{\text{init}}^{\text{inf}} = 20$ iterations in the first frame and $N_{\text{update}}^{\text{inf}} = 3$ iterations in each subsequent frame. Our few-shot learner formulation (3) allows an easy integration of a global importance weight for each frame in the training set $\mathcal{D}$. We exploit this flexibility to integrate an exponentially decaying weight η^{-t} to reduce the impact of older frames. We set $\eta = 0.9$ and ensure the weights sum to one. We ensure a maximum $K_{\max} = 32$ samples in the few-shot training set $\mathcal{D}$, by removing the oldest sample. We always keep the first frame since it has the reference target mask y_0. Each frame in the sequence is processed by first cropping a patch that is 5 times larger than the previous estimate of target, while ensuring the maximal size to be equal to the image itself. The cropped region is resized to 832×480 with preserved aspect ratio. If a sequence contains multiple targets, we independently process each in parallel and merge the predicted masks using the soft-aggregation operation [24].

3.6 Training

To train our end-to-end network architecture, we aim to simulate the inference procedure employed by our approach, described in Sect. 3.5. This is achieved by training the network on mini-sequences $\mathcal{V} = \{(I_t, y_t)\}_{t=0}^{Q-1}$ of length Q. These are constructed by sampling frames from annotated VOS sequences. In order to induce robustness to fast appearance changes, we randomly sample frames in temporal order from a larger window of Q' frames. As in inference, we create the initial few-shot training set from the first frame $\mathcal{D}_0 = \{(x_0, y_0)\}$. This is used to learn the initial target model parameters $\tau_0 = A_\theta(\mathcal{D}_0, 0)$ by performing $N_{\text{init}}^{\text{train}}$ steepest descent iterations. In subsequent frames, we use $N_{\text{update}}^{\text{train}}$ iterations to update the model as $\tau_t = A_\theta(\mathcal{D}_t, \tau_{t-1})$. The final prediction $\tilde{y}_t = D_\theta(x_t, T_{\tau_{t-1}}(x_t))$ in each frame is added to the few-shot train set $\mathcal{D}_t = \mathcal{D}_{t-1} \cup \{(x_t, \tilde{y}_t)\}$. All network parameters θ are trained by minimizing the per-sequence loss,

$$\mathcal{L}_{\text{seq}}(\theta; \mathcal{V}) = \frac{1}{Q-1} \sum_{t=1}^{Q-1} \mathcal{L}\Big(D_\theta\big(F_\theta(I_t), T_{\tau_{t-1}}(F_\theta(I_t))\big), y_t\Big). \qquad (5)$$

Here, $\mathcal{L}(\tilde{y}, y)$ is the employed loss between the prediction $\tilde{y}$ and ground-truth y. We compute the gradient of the final loss (5) by averaging over multiple mini-sequences in each batch. The target model parameters τ_{t-1} in (5) are predicted by our few-shot learner A_θ, and therefore depend on the network parameters of the label generator E_θ, weight predictor W_θ, and feature extractor F_θ. These modules can therefore be trained end-to-end by minimizing the loss (5).

Details: Our network is trained using the YouTube-VOS [39] and DAVIS [27] datasets. We use mini-sequences of length $Q = 4$ frames, generated from video segments of length $Q' = 100$. We employ random flipping, rotation, and scaling for data augmentation. We then sample a random 832×480 crop from each frame. The number of steepest-descent iterations in the few-shot learner A_θ is set to $N_{\text{init}}^{train} = 5$ for the first frame and $N_{\text{update}}^{\text{train}} = 2$ in subsequent frames. We use the Lovasz [2] segmentation loss in (5). We initialize our backbone ResNet-50 with the Mask R-CNN [11] weights from [23] (see the supplementary for analysis). All other modules are initialized using [12]. Our network is trained using ADAM [16]. We first train our network for 70k iterations with the backbone weights fixed. The complete network, including the backbone feature extractor, is then trained for an additional 80k iterations. For evaluation on DAVIS, we further fine-tune the network using only the DAVIS training split. The entire training takes 48 hours on 4 Nvidia V100 GPUs. Further details are provided in the supplementary.

3.7 Bounding Box Initialization

In many practical applications, it is too costly to generate an accurate reference-frame annotation to perform VOS. We therefore follow the recent trend [34,36] of using weaker supervision by only assuming the target bounding box in the first frame. By exploiting our learned mask encoding, we show that our architecture can accommodate this setting with only a minimal addition. Analogously to the label generator E_θ, we introduce a bounding box encoder $B_\theta(b_0, x_0)$. It takes a mask-representation b_0 of the initial box along with backbone features x_0 as input and predicts a target mask encoding in the same D-dimensional output space of E_θ and T_τ. This allows us to exploit our existing decoder network in order to predict the target mask in the initial frame as $\tilde{y}_0 = D_\theta(x_0, B_\theta(b_0, x_0))$. VOS is then performed using the same procedure as described in Sect. 3.5, by simply replacing the ground-truth mask y_0 with the predicted mask $\tilde{y}_0$. Our box encoder B_θ consists of two residual blocks followed by a conv-layer and is easily trained by freezing the other parameters in the network. Thus, we only need to sample single frames during training and minimize the segmentation loss $\mathcal{L}(D_\theta(x_0, B_\theta(b_0, x_0)), y_0)$. As a result, we gain the ability to perform VOS with box-initialization without losing performance in the standard VOS setting.

Details: We train the box encoder on images from MSCOCO[19] and YouTube-VOS for $50,000$ iterations, while freezing the pre-trained components of the network. During inference we reduce the impact of the first frame annotation by setting $\eta = 0.8$ and remove it from the memory after $K_{\max}$ frames. For best performance, we only update the target model every fifth frame with $N_{\text{update}}^{\text{inf}} = 5$.

4   Experiments

We evaluate our approach on the two standard VOS benchmarks: YouTube-VOS and DAVIS 2017. Detailed results are provided in the supplementary material. Our approach operates at 14 FPS on single object sequences.

4.1   Ablative Analysis

Here, we analyze the impact of the key components in the proposed VOS architecture. Our analysis is performed on a validation set consisting of 300 sequences randomly sampled from the YouTube-VOS 2019 training set. For simplicity, we do not train the backbone ResNet-50 weights in the networks of this comparison. The networks are evaluated using the mean Jaccard $\mathcal{J}$ index (IoU). Results are shown in Table 1. Qualitative examples are visualized in Fig. 3.

Baseline: Our baseline constitutes a version where the target model is trained to directly predict an initial mask, which is subsequently refined by the decoder D_θ. That is, the ground-truth employed by the few-shot learner is set to the reference mask. Further, we do not back-propagate through the learning of the target model during offline training and instead only train the decoder module D_θ. Thus, we do not perform end-to-end training through the learner.

End-to-End Training: Here, we exploit the differentiablity of our few-shot learner to train the underlying features used by the target model in an end-to-end manner. That is, we train the conv. layer which processes the backbone features (see Sect. 3.4). Learning specialized features for the target model provides a substantial improvement of +3.0 in $\mathcal{J}$ score. This clearly demonstrates the importance of end-to-end learning capability provided by our few-shot learner.

Label Generator E_θ: Instead of training the target model to predict an initial segmentation mask, we here employ the proposed label generator E_θ to learn what the target model should learn. This allows training the target model to output richer representation of the target mask, leading to an improvement of +1.4 in $\mathcal{J}$ score over the version which does not employ the label generator.

Weight Predictor W_θ: In this version, we additionally include the proposed weight predictor W_θ to obtain the importance weights for the internal loss (3). Using the importance weights leads to an additional improvement of +0.9 in $\mathcal{J}$ score. This shows that our weight predictor learns to predict what the internal learner should focus on, in order to generate a robust target model.

Table 1. Ablative analysis on a validation set of 300 videos sampled from the YouTube-VOS 2019 training set. We analyze the impact of **end-to-end training**, the **label generator** and the **weight predictor** by incrementally adding them one at a time.

	Baseline	+End-to-end	+Label Generator E_θ	+Weight Predictor W_θ
$\mathcal{J}$ Score (%)	74.5	77.5	78.9	79.8

4.2 State-of-the-Art Comparison

In this section, we compare our method, denoted **LWL**, with state-of-the-art. Since many approaches employ additional segmentation datasets during training, we always indicate whether additional data is used. We report results for the standard version of our approach which employs additional data (as described in Sect. 3.6), and a second version that is only trained on the train split of the specific dataset. For the latter version, we initialize the backbone ResNet-50 with ImageNet pre-training instead of the MaskRCNN backbone weights.

Table 2. State-of-the-art comparison on the large-scale YouTube-VOS 2018 validation set. Our approach LWL outperforms all previous methods, both when using with additional training data and when training only on YouTube-VOS 2018 train split.

	Additional Training Data					Only YouTube-VOS training					
	LWL	STM [25]	SiamRCNN [34]	PreMVOS [21]	OnAVOS [32]	OSVOS [5]	**LWL**	STM [25]	FRTM [29]	AGAME [15]	AGSSVOS [18]	S2S [38]
$\mathcal{G}$ (overall)	**81.5**	79.4	73.2	66.9	55.2	58.8	**80.2**	68.2	72.1	66.1	71.3	64.4
$\mathcal{J}_{seen}$	**80.4**	79.7	73.5	71.4	60.1	59.8	**78.3**	–	72.3	67.8	71.3	71.0
$\mathcal{J}_{unseen}$	**76.4**	72.8	66.2	56.5	46.1	54.2	**75.6**	–	65.9	61.2	65.5	55.5
$\mathcal{F}_{seen}$	**84.9**	84.2	–	–	62.7	60.5	**82.3**	–	76.2	69.5	75.2	70.0
$\mathcal{F}_{unseen}$	**84.4**	80.9	–	–	51.4	60.7	**84.4**	–	74.1	66.2	73.1	61.2

YouTube-VOS [39]: We evaluate our approach on the YouTube-VOS 2018 validation set, containing 474 sequences and 91 object classes. Out of these, 26 classes are *unseen* in the training dataset. The benchmark reports Jaccard $\mathcal{J}$ and boundary $\mathcal{F}$ scores for *seen* and *unseen* categories. Methods are ranked by the overall $\mathcal{G}$-score, obtained as the average of all four scores.

Among previous approaches, STM [25] obtains the highest overall $\mathcal{G}$-score of 79.4 (see Table 2). Our approach LWL significantly outperforms STM with a relative improvement of over 2.6%, achieving an overall $\mathcal{G}$-score of 81.5. Without the use of additional training data, the performance of STM is notably reduced to an overall $\mathcal{G}$-score of 68.2. FRTM [29] and AGSS-VOS [18] achieve stronger performance of 72.1 and 71.3 respectively, when employing only YouTube-VOS data for training. Our approach outperforms all previous methods by a margin of over 8.1% in this setting. Remarkably, this version even outperforms all previous methods trained with additional data, achieving a $\mathcal{G}$-score of 80.2. This clearly demonstrates the strength of our few-shot learner. Furthermore, our approach achieves an improvement of 9.7% and 10.3% on the $\mathcal{J}_{unseen}$ and $\mathcal{F}_{unseen}$

Fig. 3. Qualitative results of our VOS method. Our approach provides accurate segmentations in very challenging scenarios, including occlusions (row 1 and 3), distractor objects (row 1, and 2), and appearance changes (row 1, 2 and 3). Row 4 shows an example failure case, due to severe occlusions and very similar objects.

scores respectively, over FRTM. This demonstrates the superior generalization capability of our approach to classes that are unseen during training.

DAVIS 2017 [27]: The DAVIS 2017 validation set contains 30 videos. In addition to our standard training setting (see Sect. 3.6), we provide results of our approach when using only the DAVIS 2017 training set. Methods are evaluated in Table 3 in terms of mean Jaccard $\mathcal{J}$ and boundary $\mathcal{F}$ scores, along with the overall score $\mathcal{J}\&\mathcal{F}$. Our approach achieves similar performance to STM, with a marginal 0.2 lower overall score, when using additional training data. However, when employing only DAVIS 2017 training data, the performance of STM is significantly reduced. In contrast, our approach outperforms all previous methods in this setting, with an improvement of 5.5% over the second best method FRTM and 31.3% over STM in terms of $\mathcal{J}\&\mathcal{F}$.

Table 3. State-of-the-art comparison on the DAVIS 2017 validation dataset. Our approach LWL is on par with the best performing method STM, while significantly outperforming all previous methods with only the DAVIS 2017 training data.

	Additional Training Data								Only DAVIS 2017 training				
	LWL	STM [25]	SiamRCNN [34]	PreMVOS [21]	FRTM [29]	AGAME [15]	FEELVOS [33]	AGSSVOS [18]	LWL	STM [25]	FRTM [29]	AGAME [15]	AGSSVOS [18]
$\mathcal{J}\&\mathcal{F}$	81.6	**81.8**	74.8	77.8	76.7	70.0	71.5	67.4	**74.3**	43.0	68.8	63.2	66.6
$\mathcal{J}$	79.1	**79.2**	69.3	73.9	73.8	67.2	69.1	64.9	**72.2**	38.1	66.4	–	63.4
$\mathcal{F}$	84.1	**84.3**	80.2	81.7	79.6	72.7	74.0	69.9	**76.3**	47.9	71.2	–	69.8

Bounding Box Initialization: Finally, we evaluate our approach on VOS with bounding box initialization on YouTube-VOS 2018 and DAVIS 2017 validation sets. Results are reported in Table 4. We compare with the recent Siam-RCNN [34] and Siam-Mask [36]. Our approach LWL achieves a relative improvement of over 3% in terms of $\mathcal{G}$-score over the previous best method Siam-RCNN

Table 4. State-of-the-art comparison with box-initialization on YouTube-VOS 2018 and DAVIS 2017 validation sets. LWL outperforms existing methods on both datasets.

Method	YouTube-VOS 2018					DAVIS 2017		
	$\mathcal{G}$	$\mathcal{J}_{seen}$	$\mathcal{J}_{unseen}$	$\mathcal{F}_{seen}$	$\mathcal{F}_{unseen}$	$\mathcal{J}\&\mathcal{F}$	$\mathcal{J}$	$\mathcal{F}$
LWL	**70.4**	**73.0**	**63.0**	**75.9**	**70.0**	**70.8**	**68.2**	73.5
Siam-RCNN [34]	68.3	69.9	61.4	–	–	70.6	66.1	**75.0**
Siam-Mask [36]	52.8	62.2	45.1	58.2	47.7	56.4	54.3	58.5

on YouTube-VOS. Similarly, on DAVIS 2017, LWL outperforms Siam-RCNN with a $\mathcal{J}\&\mathcal{F}$-score of 70.8. Our approach remarkably outperforms several recent methods employing mask initialization in Table 2 and Table 3, demonstrating that it can readily generalize to the box-initialization setting.

5   Conclusions

We present a novel VOS approach that integrates an optimization-based few-shot learner. The learner predicts a compact target model by minimizing a few-shot objective in the first frame. Given a test frame, the target model outputs a mask encoding which is used by a decoder to predict the target mask. Our learner is differentiable, ensuring an end-to-end trainable VOS architecture We go beyond standard few-shot learning by also learning what the target model should learn in order to maximize segmentation accuracy. This is achieved by designing modules that predict the label and importance weights in the few-shot objective.

Acknowledgments. This work was partly supported by the ETH Zürich Fund (OK), a Huawei Technologies Oy (Finland) project, an Amazon AWS grant, Nvidia, ELLIIT Excellence Center, the Wallenberg AI, Autonomous Systems and Software Program (WASP) and the SSF project Symbicloud.

References

1. Behl, H.S., Najafi, M., Arnab, A., Torr, P.H.S.: Meta learning deep visual words for fast video object segmentation. In: NeurIPS 2019 Workshop on Machine Learning for Autonomous Driving (2018)
2. Berman, M., Rannen Triki, A., Blaschko, M.B.: The lovász-softmax loss: a tractable surrogate for the optimization of the intersection-over-union measure in neural networks. In: Proceedings of the IEEE Conference on Computer Vision and Pattern Recognition, pp. 4413–4421 (2018)
3. Bertinetto, L., Henriques, J.F., Torr, P., Vedaldi, A.: Meta-learning with differentiable closed-form solvers. In: International Conference on Learning Representations (2019)
4. Bhat, G., Danelljan, M., Van Gool, L., Timofte, R.: Learning discriminative model prediction for tracking. In: Proceedings of the IEEE International Conference on Computer Vision, pp. 6182–6191 (2019)

5. Caelles, S., Maninis, K.K., Pont-Tuset, J., Leal-Taixé, L., Cremers, D., Van Gool, L.: One-shot video object segmentation. In: 2017 IEEE Conference on Computer Vision and Pattern Recognition (CVPR), pp. 5320–5329. IEEE (2017)

6. Choi, J., Kwon, J., Lee, K.M.: Deep meta learning for real-time target-aware visual tracking. In: Proceedings of the IEEE International Conference on Computer Vision, pp. 911–920 (2019)

7. Cohen, I., Medioni, G.: Detecting and tracking moving objects for video surveillance. In: Proceedings. 1999 IEEE Computer Society Conference on Computer Vision and Pattern Recognition (Cat. No PR00149), vol. 2, pp. 319–325. IEEE (1999)

8. Danelljan, M., Van Gool, L., Timofte, R.: Probabilistic regression for visual tracking. In: Proceedings of the IEEE Conference on Computer Vision and Pattern Recognition (2020)

9. Erdélyi, A., Barát, T., Valet, P., Winkler, T., Rinner, B.: Adaptive cartooning for privacy protection in camera networks. In: 2014 11th IEEE International Conference on Advanced Video and Signal Based Surveillance (AVSS), pp. 44–49. IEEE (2014)

10. Finn, C., Abbeel, P., Levine, S.: Model-agnostic meta-learning for fast adaptation of deep networks. In: Proceedings of the 34th International Conference on Machine Learning-Volume 70, pp. 1126–1135. JMLR. org (2017)

11. He, K., Gkioxari, G., Dollár, P., Girshick, R.B.: Mask r-cnn. In: 2017 IEEE International Conference on Computer Vision (ICCV), pp. 2980–2988 (2017)

12. He, K., Zhang, X., Ren, S., Sun, J.: Delving deep into rectifiers: surpassing human-level performance on imagenet classification. In: ICCV (2015)

13. Hu, P., Wang, G., Kong, X., Kuen, J., Tan, Y.P.: Motion-guided cascaded refinement network for video object segmentation. In: Proceedings of the IEEE Conference on Computer Vision and Pattern Recognition, pp. 1400–1409 (2018)

14. Hu, Y.-T., Huang, J.-B., Schwing, A.G.: VideoMatch: matching based video object segmentation. In: Ferrari, V., Hebert, M., Sminchisescu, C., Weiss, Y. (eds.) ECCV 2018. LNCS, vol. 11212, pp. 56–73. Springer, Cham (2018). https://doi.org/10.1007/978-3-030-01237-3_4

15. Johnander, J., Danelljan, M., Brissman, E., Khan, F.S., Felsberg, M.: A generative appearance model for end-to-end video object segmentation. In: IEEE Conference on Computer Vision and Pattern Recognition (CVPR) (2019)

16. Kingma, D., Ba, J.: Adam: a method for stochastic optimization. In: International Conference on Learning Representations, December 2014

17. Lee, K., Maji, S., Ravichandran, A., Soatto, S.: Meta-learning with differentiable convex optimization. In: CVPR (2019)

18. Lin, H., Qi, X., Jia, J.: Agss-vos: attention guided single-shot video object segmentation. In: Proceedings of the IEEE International Conference on Computer Vision, pp. 3949–3957 (2019)

19. Lin, T.-Y., et al.: Microsoft COCO: common objects in context. In: Fleet, D., Pajdla, T., Schiele, B., Tuytelaars, T. (eds.) ECCV 2014. LNCS, vol. 8693, pp. 740–755. Springer, Cham (2014). https://doi.org/10.1007/978-3-319-10602-1_48

20. Liu, Y., Liu, L., Zhang, H., Rezatofighi, H., Reid, I.: Meta learning with differentiable closed-form solver for fast video object segmentation. arXiv preprint arXiv:1909.13046 (2019)

21. Luiten, J., Voigtlaender, P., Leibe, B.: PReMVOS: proposal-generation, refinement and merging for video object segmentation. In: Jawahar, C.V., Li, H., Mori, G., Schindler, K. (eds.) ACCV 2018. LNCS, vol. 11364, pp. 565–580. Springer, Cham (2019). https://doi.org/10.1007/978-3-030-20870-7_35

22. Maninis, K.K., et al.: Video object segmentation without temporal information. IEEE Trans. Pattern Anal. Mach. Intell. (TPAMI) **41**(6), 1515–1530 (2018)

23. Massa, F., Girshick, R.: maskrcnn-benchmark: Fast, modular reference implementation of Instance Segmentation and Object Detection algorithms in PyTorch. https://github.com/facebookresearch/maskrcnn-benchmark (2018). Accessed 04 Sep 2019

24. Oh, S.W., Lee, J.Y., Sunkavalli, K., Kim, S.J.: Fast video object segmentation by reference-guided mask propagation. In: 2018 IEEE/CVF Conference on Computer Vision and Pattern Recognition, pp. 7376–7385. IEEE (2018)

25. Oh, S.W., Lee, J.Y., Xu, N., Kim, S.J.: Video object segmentation using space-time memory networks. In: Proceedings of the IEEE International Conference on Computer Vision (2019)

26. Park, E., Berg, A.C.: Meta-tracker: fast and robust online adaptation for visual object trackers. In: Proceedings of the European Conference on Computer Vision (ECCV), pp. 569–585 (2018)

27. Perazzi, F., Pont-Tuset, J., McWilliams, B., Van Gool, L., Gross, M., Sorkine-Hornung, A.: A benchmark dataset and evaluation methodology for video object segmentation. In: Computer Vision and Pattern Recognition (2016)

28. Perazzi, F., Khoreva, A., Benenson, R., Schiele, B., Sorkine-Hornung, A.: Learning video object segmentation from static images. In: Proceedings of the IEEE Conference on Computer Vision and Pattern Recognition, pp. 2663–2672 (2017)

29. Robinson, A., Lawin, F.J., Danelljan, M., Khan, F.S., Felsberg, M.: Learning fast and robust target models for video object segmentation (2020)

30. Ros, G., Ramos, S., Granados, M., Bakhtiary, A., Vazquez, D., Lopez, A.M.: Vision-based offline-online perception paradigm for autonomous driving. In: 2015 IEEE Winter Conference on Applications of Computer Vision, pp. 231–238. IEEE (2015)

31. Saleh, K., Hossny, M., Nahavandi, S.: Kangaroo vehicle collision detection using deep semantic segmentation convolutional neural network. In: 2016 International Conference on Digital Image Computing: Techniques and Applications (DICTA), pp. 1–7. IEEE (2016)

32. Voigtlaender, P., Leibe, B.: Online adaptation of convolutional neural networks for video object segmentation. In: BMVC (2017)

33. Voigtlaender, P., Leibe, B.: Feelvos: fast end-to-end embedding learning for video object segmentation. In: IEEE Conference on Computer Vision and Pattern Recognition (CVPR) (2019)

34. Voigtlaender, P., Luiten, J., Torr, P.H., Leibe, B.: Siam r-cnn: visual tracking by re-detection. In: Proceedings of the IEEE Conference on Computer Vision and Pattern Recognition (CVPR) (2020)

35. Vondrick, C., Shrivastava, A., Fathi, A., Guadarrama, S., Murphy, K.: Tracking emerges by colorizing videos. In: Ferrari, V., Hebert, M., Sminchisescu, C., Weiss, Y. (eds.) ECCV 2018. LNCS, vol. 11217, pp. 402–419. Springer, Cham (2018). https://doi.org/10.1007/978-3-030-01261-8_24

36. Wang, Q., Zhang, L., Bertinetto, L., Hu, W., Torr, P.H.: Fast online object tracking and segmentation: a unifying approach. In: Proceedings of the IEEE Conference on Computer Vision and Pattern Recognition, pp. 1328–1338 (2019)

37. Wang, Z., Xu, J., Liu, L., Zhu, F., Shao, L.: Ranet: ranking attention network for fast video object segmentation. In: Proceedings of the IEEE International Conference on Computer Vision, pp. 3978–3987 (2019)

38. Xu, N., et al.: YouTube-VOS: sequence-to-sequence video object segmentation. In: Ferrari, V., Hebert, M., Sminchisescu, C., Weiss, Y. (eds.) ECCV 2018. LNCS, vol. 11209, pp. 603–619. Springer, Cham (2018). https://doi.org/10.1007/978-3-030-01228-1_36
39. Xu, N., et al.: Youtube-vos: A large-scale video object segmentation benchmark. arXiv preprint arXiv:1809.03327 (2018)
40. Yang, L., Wang, Y., Xiong, X., Yang, J., Katsaggelos, A.K.: Efficient video object segmentation via network modulation. Algorithms **29**, 15 (2018)

Correction to: Rethinking Image Inpainting via a Mutual Encoder-Decoder with Feature Equalizations

Hongyu Liu, Bin Jiang, Yibing Song, Wei Huang, and Chao Yang

Correction to:
Chapter "Rethinking Image Inpainting via a Mutual
Encoder-Decoder with Feature Equalizations"
in: A. Vedaldi et al. (Eds.): *Computer Vision – ECCV 2020*,
LNCS 12347, https://doi.org/10.1007/978-3-030-58536-5_43

In the originally published version of chapter 43, the second affiliation stated a wrong city and country. This has been corrected.

The updated version of this chapter can be found at
https://doi.org/10.1007/978-3-030-58536-5_43

Author Index

Printed in the United States
By Bookmasters